The Emergent Method

A Modern Science Approach to the Phenomenology and Ethics of Emergentism

(Or... *The Emergentist's Guide to the Galaxy*)

by
Michael E. Kean

First Published in Australia by Michael E. Kean

Printed in USA.

Cover design is a free stock photo under Creative Commons CC0 license from Pixabay. See https://pixabay.com/en/above-adventure-aerial-air-amazing-736879/.

National Library of Australia Cataloguing-in-Publication entry:

Creator: Kean, Michael E., author.

Title: The Emergent Method : A Modern Science Approach to the Phenomenology and Ethics of Emergentism / Michael E. Kean.

ISBN: 978-0-9945868-2-7 (Hardback US Trade)

ISBN: 978-0-9945868-1-0 (Paperback A4)

ISBN: 978-0-9945868-0-3 (eBook)

Notes: Includes bibliographical references and index (Hardback and Paperback)

Subjects: Emergence (Philosophy)

Phenomenology

Existentialism

Dewey Number: 142.7

Also available:

ISBN: 978-0-9945868-4-1 (Kindle)

ISBN: 978-0-9945868-3-4 (PDF)

This edition:

ISBN: 978-0-9945868-5-8 (Paperback Royal)

For Lorraine,
Philip, Jeremy, Stephen and Andrew

TABLE OF CONTENTS

Contents

Contents

LIST OF FIGURES

Contents

LIST OF TABLES

Preface

The Emergent Method is part of a philosophical framework for helping us realise our potential. It provides a naturalistic and personal methodology for finding and following our highest purposes.

The following pages touch on many areas of human endeavour but the focus is on discussing, building, and explaining the framework of concepts that contribute to an understanding of the Emergent Method. We will explore a number of seemingly disparate subjects, with a view to integrating them in such a way that you can reap benefits in your everyday life and thereby enrich your life. As we explore these subjects, you will discover that the concept of emergence links them all together. Emergence is the key to a new worldview that is slowly replacing the excessive reductionism of the past.

Areas addressed include:

- Observations of reality[1] and reflections on the nature of existence[2] and being
- Human behaviour in terms of perception, instincts, consciousness and concept formation
- Societal structures, sustainability, and reflections on our emerging global systems
- Ethics and values
- Setting and achieving personal goals

None of these subjects is new in itself and many authors have explored these subjects in much detail before. Recently several authors have pulled some of these subjects together. Two very different authors in this vein would be Richard Dawkins, using a Darwinian approach, and Eckhart Tolle, using a quasi-Buddhist approach. In fact, the synthesis of seemingly disparate ideas, as opposed to the specialist analysis of ideas, seems to be a fast-growing trend in modern thinking. Let's face it - a large amount of information that cannot be integrated into a simple whole is just information overload and useless to our limited brains. What we want to know is how we can quickly combine the information out there to enhance our wellbeing.

In discussing such information, the following chapters become a voyage of discovery towards a better understanding of our world and our standing within the world. The chapters suggest ways in which we can all take part in a much-needed and urgent realignment and redevelopment of our world.

I am not writing for a narrow academic audience – I am writing to a more generalised audience. My aim is to raise awareness of emergence, promote its discussion and enhance its application in all arenas – academic or personal. My aim is to write an easy-to-understand, highly practical book about emergentism within the popular science/philosophy genres.

* * *

Recently I was walking through Lane Cove National Park with some friends when we noticed several signs warning of poisonous fox bait laid down in the area. It sparked a discussion with my eight-year-old fellow-walker. We talked about how early settlers introduced foxes into the Australian bush and how devastating foxes were to the local fauna. We also discussed that if the authorities were to allow the fox to be ultimately successful in its easy conquest of Australian native species, then the logical result would be a steep near-term increase in the fox population. The extinction of Australian fauna would follow, and then the demise of the Australian fox population itself. Through artificially introduced fauna, a local diverse ecology that took millions of years to develop could simply disappear. So we went on to discuss the importance of ecological near-balance within each environment, whether Europe's or Australia's. The fox fitted within its home environment in Europe, where the fauna had coevolved with the fox, but not in Australia. Likewise, Australian creatures were adapted to the Australian environment and not to the European. It was not a matter of which environment or species was better, just a matter of local ecological diversity and richness, which is closely linked to our wellbeing.

Similarly, our bodies and genes are biologically linked to the environments in which we live, and just like the fox, we need biological balance with all else in the changing environment if we are to survive. Are we like the Australian fox, grossly out of balance with our environment today? Are our naturally selected genes more closely linked to a long-lost past environment, radically different to the one we are building for ourselves today? Have we enjoyed a steep near-term increase in population just to see all our gains lost within the next century and our environment likewise reduced to poverty? Has our recently gained consciousness as a species made us lose touch with our instincts and in turn caused us to fall seriously out of balance with our natural environment? Does our consciousness ultimately require near-balance with our planetary environment if we are to survive on this planet in the long term?

It is obvious we need to start thinking about how our human consciousness may have caused serious imbalance with our planetary environment, at least temporarily, and how we might bring it back into evolutionary balance. Are our societies growing in resilience or increasingly fragile? We all have an interest in promoting our societies' diversity and strength. We can do this through the worldview and values by which we choose to live. The Emergent Method offers a new way so that, with appropriately emerging values to guide us, we can build a nurturing environment that promotes our wellbeing.

The last line of environmentalist Tim Flannery's book, *Here on Earth,* says,

"But I am certain of one thing – if we do not strive to love one another, and to love our planet as much as we love ourselves, then no further human progress is possible here on Earth."[3]

It is on this thought that we must now ruminate and act.

---

[1] See Glossary: Reality and Existence

[2] Ibid.

[3] Flannery, Tim, *Here on Earth* (The Text Publishing Company, Melbourne, Australia, 2010), 280

Introduction: What is Emergentism?

The philosophy of emergentism has come to prominence in the last decade or so, but what is it? Many readers will usually look for a quick explanation through Wikipedia's article on the topic. For this reason, I will use Wikipedia's article[1] to structure my introductory remarks. The exercise will help contrast my thoughts on the topic with the very influential thoughts of Wikipedia's authors and the authors of the *Emergent Properties* article found in the Stanford Encyclopedia of Philosophy[2], to which they often refer. Please refer to these articles online if you find my comments a little disjointed. Many specialist terms and their uses in this book are addressed in the Glossary.

What is emergentism? I define emergentism as ***a philosophical theory or approach that emphasises how order arises in everything.*** The Wikipedia article begins with a statement that emergentism is a belief in emergence. I disagree. Emergentism is a philosophical theory and emergence is a scientific theory like every other tentative fact of science.

Emergentism does not particularly involve a philosophy of mind, although it does involve ubiquitous explanations of reality that necessarily incorporate the area of consciousness. Emergentism may be contrasted with scientific reductionism, as both articles suggest, although it is not anti-reductionist like atheism is anti-theistic.

Both Wikipedia's article and the Stanford article note that a property of a system is said to be emergent if it is a genuinely new outcome of a system's interactions with its internal and external environment. However, by its title, the Stanford article perhaps unwittingly reduces the concept of emergence to properties of systems, which is misleading. This is partly because the system cannot be a point of reference if it too changes through emergence. For instance, many emergent properties are the remnant outcomes of scaffold systems that have disappeared long ago. A simple example is the process that led to the formation of a bird nest. The original processes are not reachable through a reductionist analysis of the modern systemic outcomes (birds and nests). The classically unreachable nature of original systems that explain modern ones is exemplified by the problem of trying to unscramble eggs. Nevertheless, we can use process models (or 'recipes') that mimic the original process to reconstruct the modern system after the fact. Emergence is thus a bigger concept than the properties of current systems treated as points of reference.

Emergence in combination with Heisenberg's Uncertainty Principle[3] (discussed in Part 1) assumes unavoidable and unstoppable interaction. Further, the independence of any system is perhaps the most erroneous concept we hold as humans. This error in common sense is perhaps brought about by the way we casually do science '*ceteris paribus*', or 'with all other things held constant'. A process-modelling approach to emergent phenomena recognises that phenomena can never be independent from the dynamic background '*omnia mutantur*', or because 'all things change'. Thus modern emergentism, as I would define it and to which I would ascribe, rejects the classical framework of causality by recognising that order always arises out of relative disorder rather than an orderly regress of causes. The way order is seen to arise out of relative disorder is always through unstoppable movement and unavoidable interpenetration.

Modern emergentism is thus an application of unfolding modern science and mathematics (Heisenberg's Uncertainty Principle, Einstein's General Relativity[4], Darwin's Evolution, Gödel's Incompleteness Theorems, Mandelbrot's Fractals[5], etc.) to the study of how order arises, whether it is in such things as galaxies, biology, moral codes, or logic. It is not the mere taxonomy of observed properties or qualia (qualities, or first-person experiences of properties).

Merleau-Ponty presented a typically existential view on the topic of qualia back in 1945,[6] which I suspect would still satisfy most emergentists today. He says qualities are elements for consciousness rather than of consciousness, that they have a situated sense, rather than being just mute impressions, and that they are never pure or complete in themselves outside of their environmental contexts.[7] We will discuss his phenomenological view in detail in Pillar 2.

I think most would agree that emergentism is a processual view of phenomena. The theory typically assumes that deep explanations of phenomena cannot be reached without a processual approach. However, while emergentists would hold an essential processual view of the universe, as I do, this view does not preclude a certain materiality. Matter is essential to the matter-spacetime dance in our universe described by General Relativity and essential to our human mind/body experience too.

Nevertheless, a simplistic example of a logical emergence is provided by the following two statements, which are a particular form of the Epimenides paradox:[8]

- The following sentence is false.
- The preceding sentence is true.

Both sentences seem logical in their own right, but when we put the two component sentences together into a combined system of logic, suddenly they become something else – something paradoxical. Due to their interdependence, the logic of the sum is different to the sum of the logical parts.

As an emergentist, I would agree that dualism is a conceptual error. That is, the mind/body simply expresses the monistic relationship between space and matter described by General Relativity at a higher level of complexity. Everything interpenetrates. It would be wrong to suggest that the mind emerges from the brain: minds and brains emerge together just as all arrangements of space and matter must emerge together. I would further assert that mind/brain emergence is simply a sophisticated instance of relativistic space/matter emergence.

The compartmentalisation of independent properties, including of time or of mind, hides a host of other conceptual errors that I hope to address in the coming pages, partly through a discussion of fractals, chaos theory, fuzzy logic[9], an environmentally-grounded phenomenology of perception, and how consciousness emerges from sensorimotor perception.

In the interpenetrating yet indeterminate monism that is our world, I would suggest an approach of tracing independent causes-and-effects is crippling, and epiphenomena[10], which again incorporate an assumption of dualistic independence, e.g. of consciousness to its environment, is erroneous. The following chapters intend to shed considerable light on an emergent theory of brain/mind. More deeply, a rejection of classical science's cause-and-effect in preference for a monistic world of uncertainty in which order can emerge from relative and interpenetrating disorder without an infinite regress, which is to say that the evolution of the mind/body is possible, renders the *property dualism* of epiphenomenalism irrelevant.

It makes sense that a serious emergentist will seek to develop a whole-of-nature epistemology that incorporates the mind/body problem, as the following pages seek to do. In doing so, emergentists will not require a special science for each of the vast number of fractal levels, layers, or schemas of nature-mind-body as is suggested by the Wikipedia article; nor will they reject the modern science approach as it develops.

While I see many feedback and feed-forward loops in the perceived layers or strata of emergent systems, the question of causal direction is a problematic one. For instance, most philosophers and biological scientists would probably reject a unidirectional cause-and-effect that is either purely upwardly causal or downwardly causal in a complex ecological system. However, I believe the same rejection should apply to all areas of science and philosophy. All phenomena are emergent rather than the results of classical cause-and-effect. This position also renders the discussion of mental causality superfluous. Complexity is ubiquitous. The concept of cause-and-effect is a remnant of classical science that has not yet passed away in our thinking. We need to move away from it as better processual methods and explanations are articulated.

The Wikipedia article's mention of Alfred Whitehead's Process Philosophy reasonably casts it as a deist or panpsychist philosophy more than a modern emergentist philosophy as briefly described so far. A 'genuinely novel' or strong emergence, as opposed to a 'weak' emergence that is accountable in terms of the preceding history of the cosmos, seems to limit emergence in a way that denies emergence as a fundamental property of a monistic, relativistic, and relatively disordered cosmos. However, despite his philosophical views and controversial criticisms of Einstein's Special and General Relativity, much of the related work of Reginald T. Cahill's Process Physics seems compatible with the weak yet ubiquitous emergentism I support.

Rather than merely hypothesise an explanation of existence, Cahill and his colleagues have developed self-referencing process models (recipes) that seem to successfully mimic various aspects of the emergence of our networked web of quantum existence from an underlying structure.[11] Cahill characterises this structure as informational, but I would characterise it as stochastic and proto-informational to clearly separate it from panpsychist overtones. Cahill's Figure 4 has been adapted for our purposes here in Figure 1.

The diagram may be thought of as a system of emerging truths arising from what Cahill calls a Stochastic Process System (SPS), such as our very early expanding universe seen as a relativistic information system. Other examples of SPSs could include embryonic development, childhood psychological development, and Australia's socio-political development. SPSs also include the process-modelling of such systems - Cahill's process-modelling of the first quantum moments of a universe such as ours being a case in point. The future problems/algorithms offered to quantum computers for their resolution would provide other examples.

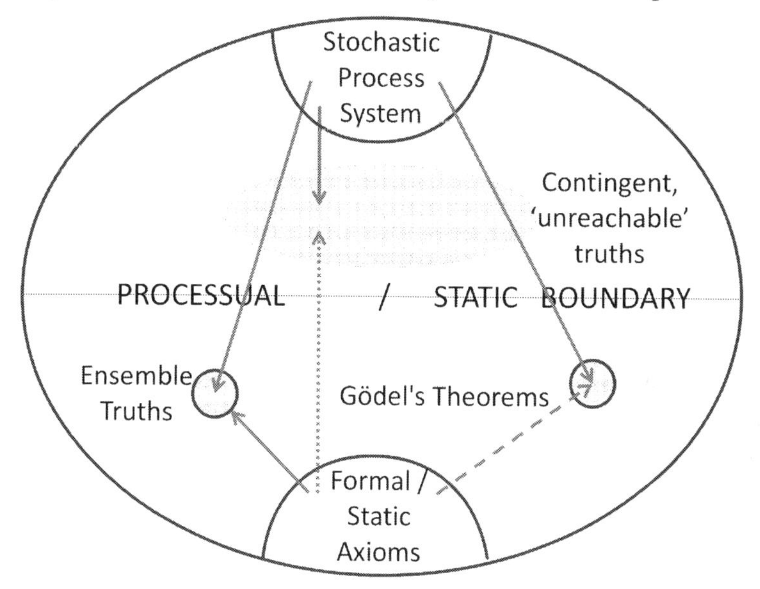

Figure 1: The Processual and Analytical Approaches to Epistemology

The shaded areas in Figure 1 (inside the two circles and in the oval-shaped cloud) represent basic truths or facts arising from the SPS. As a consequence of Gödel's Theorems (discussed in Part 1), these truths are not fully reachable analytically or 'a priori' (that is, 'before experience'). That is, the formal axioms of classical science, mathematics, philosophy, or digital computing represented at the bottom of Figure 1 bar our knowledge of deeper truths. The dashed and dotted lines indicate this limitation of the analytical or static approach. The dashed line indicates that we know the axiomatic formalism hides deeper truths within itself. The dotted line indicates we have no formal idea of the unreachable truths outside the boundaries of the classical/static realm. The classical approach has access to what Cahill calls ensemble truths, i.e. parcels of truths reliant on axiomatic bases, indicated by the solid line to the boundary of the small circle on the left. The two small circles provide examples of classical paradigms or theories.

However, an emergent processual approach (whether natural or artificially modelled) has access to all realms of truth in its becoming because it escapes Gödel limitations. It is experiential or a posteriori, rather than a priori. The three solid lines emanating from the SPS indicate this. Nevertheless, any SPS is limited to the here-and-now by its processing time.[12] The need for preparation and processing time is why the old analogue computers of university engineering laboratories never made it to the home. Still, there are ways we can have quick access to the processual truths limited by time's scarcity. These ways would include the use of non-classical complex numbers, fractals, fuzzy logic, etc. and process-modelling via the algorithms of future quantum computers, although the mimicking of nature's SPSs is not necessarily easy.

Cahill's approach seems worthy of further development. What Cahill's approach seems to make clear is that an emergent process-modelling of existence, in addition to a reductionist analysis of existence, must go hand-in-hand with a Big Science testing, if the test results are to be meaningful and continue to yield an appropriate return on investment. Big Science testing occurs, for example, in the Large Hadron Collider (LHC) at CERN, near Geneva, Switzerland, and at the Advanced Laser Interferometer Gravitational-Wave Observatory (Advanced LIGO) in Louisiana and Washington in the USA. Other areas of scientific endeavour that need a shake-up of classical, reductionist thinking are many, and even areas like artificial intelligence research and quantum computing research may need inclusion in the list.

For instance, are we likely to learn more about koalas from a 15 minute documentary on their behaviours in their natural habitat, recorded in digital TV's standard definition, or do you think we would learn more in 15 minutes from 900 still photos taken from the same documentary, but recorded in ultra-high definition? Even if the still photos were accompanied by the same audio feed in a synchronised manner, I think you will agree that the movie will be the better medium of knowledge transfer in the given timeframe. However, if I extended the window of learning to 4 hours, in which time we could run the movie 16 times, or study each photo for 16 seconds rather than 1 second, you might change your answer. Ideally, if both modes of presentation are made available, we could extract maximum information. This contrast demonstrates the difference between a process-modelling approach to doing science (which, like the movie, captures the process dynamism) and the reductionist approach (which, like the high-resolution stills, captures the static structural detail). We need both approaches, just as the position and the velocity of a particle are both useful pieces of information, but I believe the emergent process-modelling approach currently needs more attention, whether through the developments of quantum computing or other more conventional paths.

The Wikipedia article suggests the emergence of the properties of water is an inferior example of emergence to alternatives. Any presentation of emergentism that characterises, for instance, the property of water under standard conditions, as a worse and older example of emergence than alternatives is problematic. That is, emergence is not a special case in any way; it is fundamental, like entropy or General Relativity. Entropy is just the other side of the emergence coin, and in some cases to discriminate between the two is a matter of anthropomorphic perspective. A thorough investigation of the emergence of water would be just as rewarding as the study of any other physical property. What would arise from such studies would be an incomplete tree of existence growing from the simplest and earliest arrangements of matter-spacetime, comparable with evolution's tree of life proposed on a genetic basis.

While the perception of the property of, say, wetness of water in standard conditions is perhaps marvellous and inexplicable to us as sentient beings, the physical and chemical arrangement embedded in the environment that gives rise to this property is not (in typically reductionist terms). That is, we must be careful to delineate an explanation of human perception from a scientific explanation of perceived properties. There may be an explanatory gap in emergent qualia even with many of the emergent scientific factors in hand and well understood. In this case, the difficulty is with the mind/body/environment experience or belief more than the science. A full discussion of emergentism must take into account the apparent difference between the reportable first-person view and the falsifiable third-person view.

Similarly, the Wikipedia article's division between simple emergence and controversial emergence is not a function of the phenomenon observed, but of the judgments of the observer. In like manner, the confirmation of isolated gravitational waves in deep space should not make us think that local gravitational effects are less important or of a different kind – gravitational waves are all around us, affecting everything we do. Further, when a star explodes and releases heavier elements into its environment, is the process entropic because the star order is broken down or emergent in local oases because heavier elements are released? Your perspective does not impinge on the efficacy of the concepts of entropy or emergence in any way.

In the Game of Life, discussed in the Wikipedia article and in Pillar 3, the wits of one programmer are pitted against another in a shared programmable workspace. The fact that some interactions emerge that are unexpected does not make them exceptional, just as the set play of one football team competing with another over the same field can give rise to unexpected outcomes (such as an 'own goal'). In all these senses, emergence is not something magical. Rather, emergence is the normal outcome of a relativistic universe.

The Wikipedia and Stanford articles present a history of emergentism that spans contributions by John Stuart Mill, C.D. Broad, C. Lloyd Morgan, Samuel Alexander, and Jaegwon Kim. I believe a more rewarding introduction to the rise of emergentism as a philosophy at the beginning of the twentieth century might begin with a study of the overthrow of classical science by Einstein, Gödel, Heisenberg, and many others. An earlier overthrow of classical art by the French impressionists and contemporary post-impressionists might also add to this picture of emergentism's early environmental setting. Such developments accompanied a disavowal of classical science by Edmund Husserl, and his emerging or processual phenomenology within a descriptive psychology, which was soon taken up by philosophers of the existential variety.[13]

This was important because I believe that modern emergentism, unlike many of the contributions mentioned by the Wikipedia and Stanford articles, has made a clean break from the classical constraints of older forms of emergentism, other philosophies, and other sciences, to directly address the substantial challenges of modern science and mathematics.

In hindsight, I would suggest that the modern philosophy of emergentism as it applies to the mind/body problem largely picks up where the existentialist phenomenologists such as Merleau-Ponty left off in the post WWII period. However, an emergent, or at least processual, model of reality itself probably dates back to Heraclitus of Ephesus (540-480 BCE) who suggested that the world does not consist of stable things, but is always in a state of flux. To put this idea another way, what needs a deeper explanation is not movement and change, but the appearance of stability.[14] It is amazing to think that this is still a very controversial concept 2,500 year later. It has very deep implications for how we do science.

Should we assume a stable 'tick-tock' universe that conforms to ideal laws and thereby explain movement and change, or do we have the nub of the issue the wrong way around? I agree with the processual approach of Heraclitus – we often still have the nub of the issue the wrong way around. I also note in passing that modern neuroscientific research, such as presented by Antonio Damasio[15], is closely compatible with emergentism's and Merleau-Ponty's processual approach to an understanding of the mind/body.

Part 1 will discuss and explain emergence, emergentism, and most specifically, the Emergent Method. It will address many of the issues raised here. Part 2 will deal with how we can apply emergentism to a philosophy of mind/body and consider the many ramifications of such a worldview in the human context.

---

[1] See https://en.wikipedia.org/wiki/Emergentism ; accessed 28th January 2016

[2] O'Connor, Timothy and Wong, Hong Yu, "Emergent Properties", The Stanford Encyclopedia of Philosophy (Summer 2015 Edition), Edward N. Zalta (ed.), URL = <http://plato.stanford.edu/archives/sum2015/entries/properties-emergent/>.

[3] See Glossary entry

[4] See Glossary entry

[5] See Glossary entries of the same names

[6] See Merleau-Ponty, Maurice, *Phenomenology of Perception* (Routledge, New York, 2014 (Original edition in French in 1945)), 52, 217, 235, 238, 243, 319, 324, 331, 393

[7] Ibid., See p.5

[8] See Hofstadter, Douglas R., *Gödel, Escher, Bach: An Eternal Golden Braid* (Penguin Books, USA, 1983), 21

[9] See Glossary entries of the same names

[10] See Glossary: *Epiphenomenalism*

[11] See Cahill, Reginald T., *Process Physics* (Flinders University, Adelaide, 2003 (available on the web))

[12] An SPS may also be indeterminate in terms of Turing's Halting Problem. That

is, Turing proved no algorithm exists that always correctly decides whether, for any given program and input, the program comes to an orderly end when run with that input. The essence of Turing's proof was that any such algorithm can be made to contradict itself, and therefore cannot be determinate. That is, feedback and feedforward loops complicate an algorithm's ability to operate deterministically.

[13] See Merleau-Ponty, Maurice, *Phenomenology of Perception* (Routledge, New York, 2014 (Original edition in French in 1945)), Preface, lxxi

[14] See http://www.quotationspage.com/quotes/Heraclitus/ which attributes the statement "All is flux, nothing stays still" to Heraclitus through the third century manuscript, *Diogenes Laertius, Lives of Eminent Philosophers*, Book 9

[15] See Damasio, Antonio, *Descartes' Error* (Vintage Books, London, 2006 (originally 1994)) and Damasio, Antonio, *Self Comes to Mind* (Vintage Books, New York, 2010)

PART 1: Interpenetrating Concepts

In this part, we will discuss various observations regarding reality, to gain an appreciation of how the process of emergence can transform a disorganised universe into one that houses human moral agents. We will consider the significant hurdles faced at this stage of our human development and introduce the idea of a 'virtuous global tradition' gathering pace over millennia. We will then move on to discuss our development in terms of evolving genes and memes. Finally, Part 1 will propose a seven-stage framework to take forward our quest for reaching personal potential through the Emergent Method.

The 'Is' and the 'Ought'

The topic of wellbeing is quickly growing in personal and corporate interest. However, a scientific and philosophical basis to wellbeing has perhaps been slow in coming. This is probably due to fear of the naturalistic fallacy, which is related to David Hume's idea that we cannot deduce *'ought'* from *'is'*, or we cannot define a morality (an *'ought'*) from a study of nature alone (what *'is'*).[1] Hume was in part responding to the religious assertion of his day that God provided the Creation for humanity's moral instruction. Of course, this simple linking of morality and nature is misconceived. Nature does not make explicit any innate 'good' or 'evil'. It blindly advances through an evolutionary trial-and-error just as we usually do. In this sense, Hume was clearly correct.

Perhaps in another sense, nature often seems to get things wonderfully right, at least locally and temporarily. More correctly, many homeostatic systems statistically emerge from the surrounding vastness to maintain a certain dynamic equilibrium that is impressive and not easily overturned. In addition, the adaptive successes of these orderly systems do find value and meaning in our eyes. We often find quite incredible the ability of nature to fashion something new from what already exists.

What we cannot ignore is that the natural environment has also shaped us and holds us - we are part of nature. We are a manifestation or emergent property of nature, just as is a star, rock or sea. Therefore, it makes a lot of sense that we try to understand this shaping and holding process if we are going to find wellbeing and take our place as conscious agents who claim moral values within our planetary home.

The Emergent Method has a strong basis in what 'is' and incorporates an ethical system. It infers many specific approaches to the issue of 'ought', but with less rigidity or sense of fixed purpose and arrogant authority than is typical of some ethical systems. The Emergent Method assists us in answering the very personal question, 'What ought I to do in this wonderfully emergent universe?' but it does so without insisting on an imposed and unchanging dogma. More poignantly and with an even deeper moral sense, the Emergent Method assists us in answering the closely related question, 'What is it that I love and care to do in this wonderfully emergent universe?' That is, the Emergent Method exists to help us find and live the good life.

I believe that if enough people use the Emergent Method, then this can lead to the self-generating wellbeing of our entire species and entire planet. It can assist all of us in answering the same essentially ethical question, 'What do I love and care to do?' at the societal and civilizational levels.

Much of the problem with other ethical systems is that they do not actually start with what 'is' before they suggest the 'ought'. Even worse, what 'is' is seen through the prism of a pre-established 'ought'. Further, unlike nature, or what 'is', ethical systems typically present moral standards that are not explicitly open to trial-and-error revision. A modifiable ethos or ethical system just does not seem ethical. However, once we open up any ethical framework to trial-and-error revision, the supposed fixed gap between 'is' and 'ought' (as well as the dissonance between the conscious and subconscious this often triggers) begins to narrow or move. It seems the dichotomy between 'is' and 'ought' is maintained largely because we of whatever life-stance (atheist, agnostic, believer or other), culture, or civilisation, disallow what really 'is' to slowly modify the preconceived 'ought'.

Without suggesting that nature is somehow right in a fixed sense, disallowing a natural trial-and-error flow in the maintenance of an ethical system also suggests the cloistered cultural narrative that governs much of our behaviour can be unrealistic and detrimental to our wellbeing. More personally, a self that is seen as independent of nature and is prepared to use deceit or violence to achieve its urgent, narrow, self-gratifying ends will eventually bring despair. This leads us to at least one important question: Can a slowly evolving ethical system, subject to change akin to the genetic shuffling that occurs within evolving species, indeed be ethical? I would firmly answer 'yes!' and suggest that this is actually essential to an efficacious ethical system and to our sustainable survival and wellbeing as individuals, as societies and as a planetary ecology.

The concept of self-organising emergence is what links the 'is' and 'ought' together. Alternatively, to put this statement another way, both the 'is' and the 'ought' arise in the fullest sense of reality due to the emergent interrelationships of that reality. We have the Scientific Method to help us keep our claims as to what 'is' realistic, and now we have the Emergent Method to help us keep our psychosocial claims as to the 'ought' realistic. This is an exciting and achievable outcome because we know that in the physiology and workings of our healthy brains, there is no debilitating conflict between the processing of the tentative facts of nature and the often questionable values of morality.

While we have all heard that the left brain deals more with rational issues and the right brain more with intuitive issues, the brain does not deal with factual and moral processes in separate or compartmentalised systems, but rather often uses the same systems to process both kinds of inputs (nominally factual and nominally moral). That is, facts and values, or 'is' and 'ought' assessments, have roughly equal standing within the brain's emergent pattern-matching and intention-assigning systems. To look at the brain's physiology another way, while it is true to say certain neurons in each hemisphere of the brain fire in response to certain inputs, human consciousness arises out of not just the location of neurons. Consciousness also arises out of the sensitivity and frequency of neuronal firings, the number (or cloud or manifold) of neurons involved, and their interconnections with other brain/body regions. Generally the larger the dynamic manifold incorporating more area of the left and right hemispheres, the deeper the sense of self-awareness, time and space.[2] So the 'is' and the 'ought' are naturally brought together in the human brain's physiology and our unfolding flow of consciousness.

Moreover, I would dare to say that if this was not the case, we would be far less aware of facts or values. That is, it is the bringing together of 'is' and 'ought' in harmonic waves in the left and right hemispheres that gives both facts and values their dimensionality in our awareness – exactly like left and right channel give us a sense of stereo sound or slightly staggered inputs to left and right eye give us 3D television. So to speak of left or right brain dominance in the cases of certain functional areas, such as processing mathematical problems or language, perhaps conveys a distorted idea. Without one hemisphere leading and the other following in asynchronous cycles, we have a rather more mono or bland mental imagery. For instance, several researchers have confirmed that with the brain's hemispheres cut off from each other via the normally connecting bridge, the corpus callosum, the right hemisphere is capable of rudimentary language processing, but has little syntactical abilities. With both hemispheres working together, we seamlessly juxtapose the strengths, weaknesses, opportunities and threats of both facts and values within emergent consciousness. That is, within the daily 'is', our minds often find mundane answers to the question of what we 'ought' to do quite easily. However, as the 'is' becomes more of our own making in our unnatural human environment, so we also need to pay more careful attention to the evolution of the shared 'ought'.

The Emergent Dance

When I started researching and writing, I knew I would try to introduce new philosophical arrangements of ideas, and perhaps even a new philosophy. However, it turns out that to solve the problems facing us we need far more – it turns out that we need a new worldview that accounts for reality in ways very different to our past traditions and common sense. This new worldview has many parts to it. I now understand that my task is by no means an easy one, so please be patient with me, as I try to paint a beautiful picture you have never seen before.

Many scientists, especially in the past, believed everything is reducible to one kind of common substance we call energy, radiation, or light. Perhaps the most radical thought I have to make about nature is that this focus on a single essence is wrong. The essence of reality is rather the conservation of dynamic energy/information, or the balance between energy/information in and energy/information out. That is, I would suggest that a precarious, imperfect, and multifaceted balance is the essence of reality, rather than the two parts of anything that may contribute to that balance. This simple observation has mind-blowing connotations in terms of a suitable worldview. Nevertheless, while a worldview with a dynamic balance or equipoise rather than a material object as its central platform is a little strange to Western ears it is quite familiar to Eastern traditions.

"We are stardust", as the old Woodstock song by Joni Mitchell declares. The idea is that we are made from the same matter that was once contained in stars, much like the ones we see in the night sky. However, when we dreamily look into that beautiful night sky, we do not just see the stars – we also see the gaps in between them. The apparent void makes up the majority of the universe's vista. Without the space between stars, all the stars would be clumped together in an expressionless and meaningless single mass. The night sky is full of wonder and beauty not just because of the matter it contains, but also because of the space that differentiates and complements matter.

Likewise, the continually unfolding arrangements of atoms in our stardust bodies house vastly more space between atoms than those atoms take up themselves. So here again, it is the lively dance of matter in the apparent void that makes life a possible, amazing fact. It is the organising chaos of the dance between matter and space (in the sense of mathematical Chaos Theory[3]), simultaneously at both the inner and outer levels, that fills life with wonder and meaning, and if meaning, then purpose too. If purpose, then moral agency as well in the mind of the one that is able to reflect on that apparent meaning.

What is the nature of the 'is' (we ask this question first, before asking of the 'ought')? What is the nature of this organising chaos or dance between space and matter? First is ***movement***. Nothing stays still; stillness is impossible – at best, we can speak of a relative stillness. Movement is an intention that merely lacks direction or arrangement. That is, all movement is proto-intentional, and thus proto-meaningful and proto-valuable.

Nature's dance is also ***vast***; billions of stars dance across the heavens and something in the order of 1,000 trillion, trillion molecules make up the workings of just one human body. The dance is also ***complex***[4] – all levels seamlessly interconnect and interrelate. That is, everything seems to interrelate and ***interpenetrate***.

Interpenetrating Concepts

How each living human body can function as a highly coordinated whole is truly mind-boggling, I am sure you will agree. The mind/body acts and moves in unison with clear proto-intention even before we become aware of it. I would face a very complex logistical exercise indeed, if I tried to manage a distribution warehouse operation with the same proficiency as is effortlessly achieved by the healthy mind/body in processing its daily inputs and outputs.

If we scratch below the surface of nature's workings, we notice a few more things. Nature always interacts locally, that is, not necessarily geographically, but systemically. Another way to say this is that nature always acts within the local boundaries of its systems, such as its weather patterns. That is, Nature is ***locally systemic*** within a relatively non-systemic backdrop. If nature as a whole is not intelligent, but rather proto-intelligent, then it does seem at least locally logistical in its processes.

Further, in those systems, we notice that nature always finds a way forward, without hesitation. It always finds a similarity or difference at a slightly soft or open boundary that it can seem to exploit with proto-intention, just like my mind/body. That is, its lack of perfect consistency or symmetry, or its chaos, provides it with systemic potential. We could say that nature seems to thrive on the strengths and weaknesses in its own systems. It waxes and wanes, and then advances, because it is ***asymmetrical*** and ***asynchronous***. In some localities at least, growing order seems balanced on a hair trigger, ready to take off.

Nature often repeats the same local patterns of its partly symmetrical dance (e.g. in the growth of a crystal or an embryo) over and over again, such that we notice the resultant pattern on the macroscopic or grand scale. That is, nature is often ***algorithmic***. A good example would be the expanding ripple caused by a single stone thrown into a calm pond. Less tidy or more asymmetrical examples of all five aspects – vast, complex, locally systemic, asymmetric and algorithmic - would be the patterns of stars in the night sky orbiting their galactic cores, or Earth's weather patterns.

This last example introduces another important aspect of nature's dance - the Butterfly Effect (that is, nature is ***chaotic***). This phenomenon captures the idea that a small change in one place in a complex web of systems can have large effects across that web. Further, by making just a small change to initial conditions, the systemic outcome can be quite different. We had an example of the Butterfly Effect in June 2011, with the temporary closure of Australian airports caused by very fine but aircraft-engine-crippling ash particles carried undispersed across the Pacific and Antarctica from Chile's erupting Cordon Caulle volcano. In this case, aircraft engines in Australia were made temporarily local to the volcano in Chile by the expanding and then contracting manifold of interactions that connected them.

Nevertheless even vastness, complexity, asymmetry, local algorithmic interactivity and chaos do not fully explain nature's dance. So far, all these things can be largely explained by Newton's laws. However, at the microscopic level of particle physics, for example as studied in the LHC, another fundamentally important aspect of Nature's dance is evident – its quantum mechanics and quantum fields. Quantum Theory is the basis of all our modern electronic devices, yet even its experts do not easily explain it to the rest of us. Quantum Theory predicts that a photon, under certain conditions, can occupy an indeterminate position in space that a wavelike probability curve can best describe. Further, two photons can be subject to a quantum entanglement that causes them to act in a kind of systemic unison even though they may be far apart in geographical space. That is, at many levels of Nature's everyday dance, classical science's cause-and-effect explanations are insufficient. At our current level of understanding, or at the local level of which we're currently aware, nature is ***unpredictable***.

This breakdown of classical explanations at the microscopic level suggests that under certain conditions, a particular certain end is not fully determined by the beginning and intervening circumstances. That is, given a certain beginning and certain intervening circumstances, the quantum-mechanical end can be ***uncertain***. This has important ramifications in terms of our understanding of time itself. Time is not universally constant as classical science once suggested, but is local, variable and can be quantum-mechanically indeterminate or entangled as well. Further, since our universe had a quantum-mechanical beginning in the particle physics of the Big Bang, Quantum Theory is vitally important to our understanding of how the entire universe, proceeding from the Big Bang, has danced with time.

Figure 2 is a representation of the results of a recent experiment re-establishing one of the most basic effects of quantum mechanics.

It shows that photons passed through a screen set up to enable quantum interference can behave like waves instead of particles on their exit. That is, instead of a typical disorderly scatter diagram, like the pattern formed by a can of paint spilt from a ladder onto a cement floor, the flow of photons exiting the multi-slit buckyball screen group themselves into wavelike peaks and troughs of likelihood.

The wavelike peaks are indicated by the mauve in Figure 2 and the wavelike cancellations or troughs are indicated by the gaps between the mauve areas in the figure. These waves subsequently dissolve upon disturbance of the calm quantum interference conditions. That is, multiple individual photons' momentums can somehow link in a wavelike pattern under some conditions. For this to occur, each photon seems to be unpredictably synchronised with all others. It is unpredictable in the sense that we cannot predict the position and path taken by each photon, as we can do for each droplet of paint splashed across the cement floor.

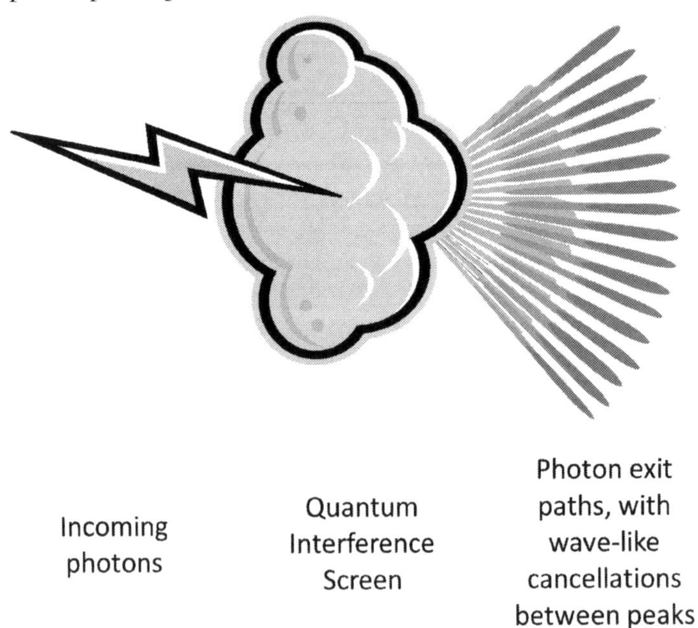

Figure 2: Austrian Buckyball Experiment[5]

It is as if each photon recognises the presence of its neighbours going through a similar experience and this is enough to form the link between them. This is like water molecules rippling in a calm pond or like carbon ions forming a lattice between themselves while they are brought together at a certain temperature, pressure and density to form a diamond. However, experimenters have established that if they control the incoming photons such that only a single photon passes through the interference screen, it will still follow an unpredictable wavelike path instead of a classical one, as if all the other photons were still present. This does not happen to water molecules or ferrous ions – spreading lattices will not emerge if insufficient molecules or ions are present. The action of water molecules clearly indicates that the ripple across the pond correlates with the presence of the water ions at the local level. That is, the rippling property of water can be tracked through a detailed analysis of the interacting water molecules that are relatively still (not involved in laminar or turbulent flow). The astonishingly counter-intuitive behaviour of a single photon travelling through the interference screen suggests a factor in the environment separate from the photon is in play. To some scientists, most particularly physicist David Deutsch, it suggests a many-worlds view of the universe in which *shadow* photons present in almost identical parallel universes interfere with the *present* photon just enough to produce and explain the quantum effect.[6]

An evolving de Broglie-Bohm Theory, perhaps arising from Feynman's "alternative histories" formulation of quantum theory, suggests instead that a web of residual quantum-mechanical interactions in space maintains the wavelike lattice. The rest of the universe (all its sub-atomic particles) including the screen, coupled to the present photon, establishes this web. By this theory, multiverse interference is not required or denied. By this theory, the photon does behave something like the water molecule because it is not really alone. That is, the photon, as one kind of subatomic particle, swims in a calm sea of other subatomic particles all around it, something like a single water particle jiggles in its calm sea at the molecular level. Removing the other photons only deactivates a negligible portion of the subatomic sea. Nevertheless, the majority of scientists remain open-minded as to the best explanation of the quantum world's probabilistic effects, perhaps awaiting a better explanation in the future.

However, these explanations of quantum uncertainty are worth further consideration. Both explanations introduce a noisy and holistic 'hidden variable' into analyses of quantum behaviours that comes from outside an analysis of the reductionist quantum world itself. That is, these explanations suggest we cannot fully understand the microscopic without understanding the macroscopic, whether our universe or multiverse. Incidentally, this was true of the water molecules too. For instance, if you recall your first year of high school physics, put a barrier in the ripple tank and the ripple pattern on top of the water will be predictably modified by the presence of the barrier. Likewise, both photon behaviour explanations suggest that whilst there is microscopic indeterminism, it is at least partly resolved by the cosmic or multiversal. That is, just as Einstein's Theory of General Relativity suggested, reductionist building blocks or particles (having mass and thus considered 'real' by many) help explain local space's macroscopic behaviours, and holistic cosmic interactions, arrangements or waves in space help explain matter's local and microscopic behaviours. It seems reasonable to say more generally that this universe and any universes that may be beyond it help explain the locally-indeterminate path of the lone photon: The photon is not independent and does not act alone. Even though it is massless, that is, the photon is virtual space-stuff and not tangible matter-stuff, it is nevertheless caught in nature's web.

Going back to the Butterfly Effect, what if we replace the single flap of a butterfly's wings with a galactic quantum-mechanical event? What if this event results in a single photon, perhaps emanating from a cosmic ray, landing in a place on Earth where it is able to cause a beneficial genetic mutation quite different to what classical physics might predict? Further, what if that mutation helped cause the speciation of *Homo sapiens* some six million years ago (when our evolutionary branch separated from other primates)? After the event, wouldn't we be right, looking back, in thinking that the universe seemed able to lift itself to a higher level of self-organisation where, by comparison, everything beforehand looked less sophisticated and less meaningful? Wouldn't we be right in thinking that the universe seemed able to form something from relatively nothing? Wouldn't we be right to say that nature, because it is probabilistic rather than classical, ***is able to bootstrap itself***, something like the bootstrap program coded in the BIOS of a PC is initiated when the PC is switched on? It is this bootstrap program in the PC that enables the operating system to be loaded into memory and then a myriad of applications to be selected by the PC user.

Similarly, in the scenario just described, the photon enabled the new *Homo sapiens* species to come into being and then live its life.

Perhaps to understand the astounding ability of the universe to bootstrap itself via SPSs, we need to understand a little of the relationship between quantum mechanical and classical physics. Quantum mechanics is not restricted to the microscopic scale – it operates at the macroscopic scale as well. It is just that its more unusual wavelike effects tend to cancel out at the macroscopic particle-like scale before we notice them. Thus, what is usually left at the macroscopic scale is something that can be described almost as well using Newton's laws. Rather tellingly, Newton's equations are not stated at the quantum or microscopic level. They are stated at a 'universal' and non-probabilistic level, which is the domain of their efficacy.

However, at the microscopic level, there is always a remnant quantum effect or noise unexplained by Newton. Further, according to the Butterfly Effect, there is always a chance of such effects acting as non-trivial triggers in a complex and vast system, a little like the single point of turbulent flow in a water pipe that can quickly spread and overtake the prior laminar flow. Here then we begin to get a glimpse of the power of Nature, our universe and our future to become serendipitously or destructively unpredictable. It seems science now teaches us that apparent miracles and their opposites can occur within partially bounded systems under certain constrained conditions. Conveniently, we might call such apparent miracles of greater orderliness 'negentropy'. The interpenetrating asynchronicity of space with respect to matter is necessary if we are to enjoy the dance's serendipity, at the cost of its destructivity (which we might often label 'entropy').

From our human perspective, it seems we may know and control those things that are more structured, deterministic and certain, but so far cannot as easily model, understand or control those things that are more unstructured, probabilistic or uncertain. To some extent and for some time at least, we have to be content to go along for the ride. This two-part essence of reality, that is, the more deterministic and the probabilistic, perhaps best captured by the quantum-mechanical wave-particle model of light and matter, is the core nature of our dance in the commonly-shared setting of space and time.

Interpenetrating Concepts

Niels Bohr's famous atomic model illustrates the continuity of order within chaos from the galactic to the atomic. The similarity of organisation of solar systems and galaxies with their objects in approximately circular, elliptical or spherical and cloud-like orbits, to the inner atomic worlds arranged in similar manner, provide some evidence. That is, nature seems essentially 'orbital', 'cyclical' and 'cellular' in all of its systemic structures, whether living or non-living.

A large part of the subtle wonder of Nature's dance is that the more Newtonian macroscopic whole, in its emerging self-organising dance, seems often able to harness or incorporate, albeit unwittingly, the quantum mechanical and microscopic chaos in fulfilment of orderly ends. Nature seems to display a blind proto-intelligence as we see it reach out and link local particles to not-so-local waves. We tend to quickly assign cause then meaning and purpose, to these macroscopically observed patterns of growing orderliness. In fact, there is no harnessing going on or even necessary. It is just the so-called law of large numbers at work. As we add many individual probabilities together within a large and roughly homogenous population, statistical characteristics emerge. The law of large numbers is thus a description of emergence itself. This so-called law doesn't have an individual cause; it has an origin arising from spatial commonalities within the local population or group. It describes how structure with semantic value and a limited syntax is bootstrapped from a relative lack of structure.

For instance, all 25 year old drivers have certain driving characteristics regardless of their individual uniqueness or relationships with each other. That is, each member of the group, like each photon, has relationships with the universe that are similar. Even though they each display local or 'particulate' uniqueness, the drivers are caught in nature's web or 'conforming wave' in a characteristic fashion. These emergent characteristics or demographics are quite often very predictable within limits of confidence and error across multiple large and stable samples of populations. These characteristics, whilst dependent on a host population, can become somewhat independent of any particular sample population. This is because we do not derive the characteristics from the sample or individual driver alone, but rather, from the common interrelationships or interactions each member has with its environment (the world at large). This idea of interactions again brings us back to the idea of common waves or interactions that link particles together. Another example of this order-in-the-midst-of-chaos would include the recognisable similarities of every unique member of each animal species on our planet, no matter how we arrange or select sample groups.

A beautiful wave of more deterministic order, or the similarity through common arrangement, seems to compete with the probabilistic disorder, or the particulate uniqueness, in our universe. We often perceive a local 'rule' of order-over-disorder, not restricted to life forms. Every member of each type of galaxy, cloud or geological formation displays the same characteristic. For example, we see the characteristic of order-over-disorder in the signature shape of mountain ranges of differing ages or in the crater-pocked landscapes of our moon compared to similar landscapes of other atmosphere-free satellites of other planets.

If this observable similarity resulting from manifest order were not the case, we would have insurmountable difficulty in forming logical abstractions based on repeated observations, and thus conceptualising our universe. For instance, how would we identify and name stratocumulus clouds if they always had varying basic characteristics of shape, altitude, etc.? How would we recognise diamonds if their lattice arrangements of carbon ions changed markedly in the short term at standard temperature and pressure? Our knowledge depends on the universe's macroscopic order and our own conceptualisations that have a kind of orderly synchronisation with it. More deeply, our orderly knowledge necessarily coexists and coevolves with a partially orderly universe. Our awareness of our own sharply conscious rationality and individually across the entire population of *Homo sapiens* seems to be the supreme testimony to the reality, in a relative sense, of the universe's local order in the midst of its vast disorder.

How can this organising chaos of the dance in the unfolding present be? How can a seemingly probabilistic universe at the detailed, local level also demonstrate striking but perhaps temporal order at the higher, widespread level? How does it become roughly homogenous within partially open or unbounded groupings, networks and lattice structures? What is this common nature of the dance that elegantly binds galaxies and black holes, planets and stars, molecular lattices, expanding ripples in a pond, electrons and nuclei, or subatomic particles, together, but in so doing enables their components individualistic, unique, and therefore partly disordered or chaotic expression?

The basic answer, we will soon explore, is Einstein's aforementioned notion of Relativity, or the continuous tension between matter and space that explains gravity (and all the other classical Newtonian 'forces' that are now more correctly visualised as 'interactions' or 'balances'). As already noted, the process of real-time algorithmic feedback[7] within populations of subatomic particles also encapsulates a large part of the answer. This feedback could be from new order to prior disorder, from spacetime to matter, or from an individual's genetic improvements to his or her descendants within the old population. The feedback could also be from new or improved concepts to old concepts (or mores), within a single mind or within a society.

Another important part of the answer is the underlying similarity or conformity of all subatomic interactions or waves. It is the dynamic arrangements of subatomic particles in the waves of space that makes them unique, not so much the subatomic particles or waves themselves. Without intending any kind of transcendental connotation, the universe, by continually moving towards greater local energy effectiveness and efficiency, seems to be able to store knowledge of itself within its relativistic arrangements. More profoundly, there seems to be a deep similarity between how we gather and store knowledge and how the universe gathers and maintains its palpable order. That is, a relativistic recognition and exploitation of similarity and difference seems to be at the core of establishing order, whether blind or self-aware.

The often negentropic nature of the emergent dance of nature is in the changing system (or conforming wave, manifold, or arrangement) rather than its material components. A simple example: Which added brick in a wall is the one that provides you with the system's shelter? Change the environment, without changing a single brick, and you are likely to change your answer – because you and the brick are part of the wider dynamic web of interactions. The answer to the question thus requires a statistical and perhaps normative approach. Anthropomorphically, we could say the wall acts as a kind of living cell that must deal with its probabilistic environmental contingencies, for example by experimenting with its arrangement of bricks, to promote its effectiveness to its users.

Interestingly, exactly the opposite is true of the destructive, entropic effect. Its action is in between the bricks, as it were, right in the nitty-gritty of each particulate bond or boundary inside the wall's systemic arrangement. For instance, entropic radioactive decay is in the local workings of each weak nuclear interaction. To persist under this scenario, the wall must ensure any cancerous cement or loose brick is efficiently replaced. Relatively speaking, this is not a probabilistic exercise, but more a deterministic and positivist one.

This idea of action at every level (microscopic and macroscopic) also suggests that nature does not calculate as humans do, it just blindly responds to its self-contained stimuli arising from its multifaceted relationships and interconnections, local and cosmic. The feedback / feed-forward mechanisms that reinforce the dance include the Higgs[8], strong nuclear, weak nuclear, electromagnetic, and gravitational interactions of physics' Standard Model[9], and its collection of subatomic particles, as well as natural selection at the biological level, perception at the unconscious level, and conscious imaginings at the level of the human mind.

Notice how each level of wave-particle coagulation seems inconceivable at any of the comparatively disordered lower levels. For instance, if we were quarks and gluons floating around in the subatomic plasma overwhelmed by the strong nuclear force at a time very close to our universe's beginning, we would probably find it unbelievable that our nuclear sophistication arose from an indeterminate soup of quantum fluctuations.[10] Just as unbelievable would be the likelihood of much more complex electromagnetic interactions eventually resulting in a vast world of highly-information-incorporating molecules, arising in our futures. Electromagnetism and the weak nuclear interaction emerging from mere plasma would perhaps seem as unlikely as the macroscopic effects of gravity emerging from coagulations of diverse molecules in space, life emerging from mere molecules-under-the-effects-of-gravity, or perception and self-aware consciousness arising from mere biology.

This thought perhaps perfectly captures the essence of what is meant by the idea of emergence. At each new level, the emergent properties are the result of a revolutionary but positive mutation from a prior configuration of local matter-spacetime. That is, everything, including space, matter and time, dynamically emerges from a prior relative disorder. Emergence is the ability of the universe to continually bootstrap itself, forming diversified islands, systems or levels of organisation that defy the prior levels of surrounding order. Emergence does not happen to existence, it happens in existence, within its layers or between its layers of relative order.

Emergence is a very large and rather controversial topic in terms of its definition and reach. It incorporates all the concepts of irrepressible movement, vastness, complexity, local systemicity, asymmetry, algorithms, General Relativity, quantum theory and statistical mechanics. It also has big implications in terms of the beginnings of the universe and causality itself. Complex arrangements of matter bring about complex measured properties, which often feedback to affect those arrangements. I hope to demonstrate that the phenomenon of emergence, the natural complement of entropy in an expanding universe, is also the basis of the development of human personality and consciousness.

The eventual arrangement of hydrogen and oxygen into linked molecules of H_2O in our early universe brought about the statistical property or behaviour of what we perceive as wetness in water. Now wetness on a smooth inclined surface can cause someone like me to lose his footing while out hiking, and fall to the ground. From this simple example, we can see that not only are the arrangements and behaviours of matter emergent, but so are the causes of matter. In a quantum-mechanical world natural causes are not simple, linear or classically determinate like in a game of billiards. They operate at many levels of space-matter's arrangements simultaneously, are complex, nonlinear and partly indeterminate. Causes are also subjective or anthropomorphic ideas: What I consider causal, you may not, and vice versa. In complex systems such as in organisms, emergent causes are part of the blindly self-organising or dynamically bootstrapping mechanism as well. They help the organism learn how to set up new boundaries to better control and diversify its viability risk.

I think that like Darwin, we do not have to go back to an original cause of life to see that life evolves, or, in this case, to go back to original causes to see that causes are emergent. If new causes emerge in complex systems then new risks and Darwinian struggles also emerge. This is something the insurance industry learned through the Global Financial Crisis, with the rise of more complex financial products interacting in the relative disorder of global financial markets. Today financial markets are more aware of the risks these new interactions imply, and so have applied new algorithms or discount factors, or otherwise acted to mitigate these new risks. This ability of organisms to mould or incorporate new causes and their risky effects is part of the reason why all complex systems seem to take on an intelligence and personality of their own.

Nevertheless for many, emergence and all the other salient features of our universe mentioned above don't get us away from the Ancient Greek idea that there had to be a First Cause that enabled the Big Bang and all else to happen. It seems there had to be God to plug the infinite regress of causes that a linear emergence might otherwise suggest. A basic property of the universe along with all the other properties listed above may indeed be that it has a Maker. If true, this would be the universe's most basic property. Thus, for a long time, I have thought that it is up to the individual to choose sides without any conclusive evidence either way. However, please consider the following.

The first problem, as has already been suggested, is that all properties statistically emerge from a prior vastness that is relatively disordered. If a god is a massless property of existence, then it too emerged from a prior vastness. Nevertheless, the most basic reason why we tend to think there had to be a First Uncaused Cause is because, we not only think that our universe is intrinsically organised or law-abiding, but that this also is the way any beginning must be. In this sense, Christian and other monotheist philosophers have assumed that perfect organisation and symmetry was in the beginning, bundled up in God. Unlike Heraclitus of Ephesus (540-480 BCE) who suggested that what needs explaining is not movement and change, but the appearance of stability, they have assumed the opposite. They have assumed stability and law require no explanation (just discovery), but movement and disorder do need explaining. They have assumed God caused or permitted the first disunity or asymmetry, perhaps via the Big Bang. That is, the Big Bang represented an initial perturbation of perfect order (or ideal Forms, an Ancient Greek idea), to which God responded with His creative acts here on Earth.

Others in the past had a view of the universe as something like a wind-up toy, which began its winding down with the Big Bang. However, we now know this view is too simplistic, partly because it doesn't incorporate the indeterminism of quantum mechanics. Clearly, the monolithic order of the Big Bang was extreme and as this unity of order broke down during entropic expansion, it aided in the local emergence of oases of synergistic organisation. In this sense, emergence is the coagulating and organising diversification of a local monolithic order that is relatively indistinguishable from disorder. However, by the old law of the conservation of energy, that beginning monolithic energy of the Big Bang had to come from something. It could not have been a literal case of something for nothing.

If we look at the question of the Uncaused Cause from a completely different perspective, we arrive at a very different understanding. The problem here has been with the assumption of an original and uni-dependent symmetry or organisation. If we assume that disorganisation and asymmetry is a natural background state of nature or reality and symmetry or organisation is just another local point transitioned within this continually emergent reality, then we get a very different picture of the universe's so-called beginning. In this sense, 'something' is a transient state of disequilibrium or asymmetry, but conversely, 'nothing' is a fleeting state of equilibrium or symmetry. Further, not in disagreement with Einstein's General Relativity, something and nothing will depend on each other for their temporal or fleeting occurrences.

According to this relativistic, interpenetrating and emergent view of reality, we don't need a beginning symmetry, organisation, or single thing to understand it. In this sense, to point to one particular disorganised beginning rather than another is meaningless – any transient and meaningless beginning will do: Beginnings arising from advancing and alternating states of perceived equilibrium and disequilibrium are illusive. Disorderly and multi-dependent mutations could have come and gone for eons before anything interesting happened against the backdrop of whatever level of energy and disorder existed in the local region or in total. More eons could have passed before any of those interesting phenomena gained any emergent continuity. That is, all causes emerge from antecedent effects or states, including chaotic or probabilistic ones. In other words, disorderly effects came first, and all orderly causes emerged later from them. In this sense, disorganisation is the necessary precursor of causation. This is what the Stochastic Process System of Figure 1 was also intended to suggest. That is, a random and uncertain process is the necessary precursor of all systemic 'truths'.

Further, causation is a relative term that requires our anthropomorphic definition and our assessments of an arrangement's meaningfulness. With an emergent view of the universe, we do not have to see reality in terms of orderly causes and their straight-jacketed effects anymore, thus circumventing the whole problem of the first cause creating the original disturbance. If meaningless asymmetry and multi-dependent disturbance is a normal state of reality then we only need to explain abnormal local symmetry and temporal organisation within vast cycles of statistical disorganisation.

This idea of emergent local order and thus meaning is crucial to understanding your own likes and dislikes, or what you love and care to do. That is, the concept of emergence is central to defining, pursuing, and achieving goals that are important to you.

In summary, everything wonderfully emerges from relatively meaningless and tentative beginnings. Evidence from the universe would seem to teach us that nothing is created in the biblical sense; everything is naturally or artificially moulded or mutated from what already is, but with many surprising outcomes in terms of properties.

Before I summarise the most important scientific discoveries of the last 150 years, I think it might be useful to consider the most important objections to modern science raised by the Western classical world.

Aquinas

I am inserting this section partly for the sake of Christians and others familiar with Aquinas' so-called evidences for God, but more importantly to explain the central role of Physics' Uncertainty Principle in our modern, quantum understanding of reality. This understanding of reality is revolutionary when we compare it to Ancient Greek or Medieval thought.

St Thomas Aquinas (1225-1274), a theologian and Dominican priest, sought to prove that 'God is' empirically. Many Christians today still hold to these arguments, at least in part. William Lane Craig, a modern theologian and Christian apologist, is one such person. His debate with Dr Lawrence Krauss on the topic of 'Is there evidence of God?' can be found at http://www.youtube.com/watch?v=TqANWuXQ3Z0 (last accessed May 2014). It is remarkable to note that much of Craig's argument still relies on the medieval arguments of Aquinas. I will consider Craig's arguments that are beyond those of Aquinas as well, although only briefly. For your convenience, to summarise the arguments of Aquinas, I will rely on the Wikipedia article at http://en.wikipedia.org/wiki/Quinque_viae (last accessed May 2016). This article commendably supplies the original Latin transcripts and English translations from which the summaries are drawn.

I am setting out to refute Aquinas' so-called five ways of proving God in this section. However, be aware Christian reader, if I fail to effectively refute Aquinas' five assertions, you cannot use the surviving ones to 'come to God' in a basic way. In fact, any surviving proof might be an obstacle to your ongoing faith. I make this warning based on Heb. 11:6, which says, "But without faith it is impossible to please him: for he that cometh to God must believe that he is."[11] This means that 'God is' is the demanded axiom of the Christian religion. Such an axiomatic life-stance starts with a tightly held presumption, and then goes about trying to reconcile, defend or even prove it. This presumption is the measure of Christian righteousness, but in terms of the Scientific Method, it is scientific sin. Science includes many axioms in its theoretical systems, but none of them is demanded. Such demands would be considered so arrogant as to be unscientific. Nevertheless, let's begin.

Argument 1: The unmoved mover (not used by William Lane Craig):

a. Some things are in motion.
b. A thing cannot, in the same respect and in the same way, move itself: it requires a mover.
c. An infinite regress of movers is impossible.
d. Therefore, there is an unmoved mover from whom all motion proceeds.
e. This mover is what we call God.

Reply:

i. The Uncertainty Principle suggests all things (energy) must be in motion. This is true even if there is a hidden variable that explains quantum uncertainty. This means there cannot be a first movement of things that proceeds from an original source. Thus, the observation in 'a' is misstated and the logic in 'b' is not established. Aquinas made a scientific mistake with respect to things he thought he observed in a steady and stationary state.

ii. It is likely that the unstoppable quantum motion resulting from uncertainty, inadequately intimated by Aquinas as impossible in 'c', is the outcome of the interpenetration of infinity and nullity and the boundary narrowly defined as existence that sits between them. In more familiar terms, ceaseless motion is the result of the relativistic wave-particle nature of all things.

iii. Aquinas advocates the idea of an unmoved mover ('d'), or a 'something for nothing', which requires a violation of the law of the conservation of energy/information. In contrast, the theoretical baryogenesis of the very early universe fulfils the Uncertainty Principle and the conservation of energy/information.

iv. Aquinas assigns the impossible 'unmoved mover' to the Judeo-Christian conception of God (in 'e') - a god-of-the-gaps. He could have just as easily assigned this impossibility to any other god or conception. Thus, the logic of 'e' is inconclusive.

v. The so-called proof fails and is thus rejected.

Argument 2: The first cause (also used by William Lane Craig)

a. Some things are caused.
b. Everything that is caused is caused by something else.
c. An infinite regress of causation is impossible.
d. Therefore, there must be an uncaused cause of all that is caused.
e. This cause is what we call God.

Reply:

i. The Uncertainty Principle suggests you cannot know beforehand the exact position and momentum of anything simultaneously, so the idea of classically separable causes and effects is not correct in the quantum or relativistic world. Thus the local observation in 'a' is uncertain or probabilistic and the logic in 'b' not established (anything is caused by everything else, not everything by something else). This means the idea of an absolute first cause in 'd' is not established, that is, not required. We begin with effects and anthropomorphically assign meaningful causes. Aquinas made a scientific mistake with respect to causes because he thought causes preceded effects in a classical fashion.

ii. It is likely that the inability to describe all causes with certainty is the outcome of the interpenetration of infinity and nullity and the boundary narrowly defined as existence that sits between them. In more familiar terms, the failure of the cause-effect model is the result of the relativistic wave-particle nature of all things.

iii. Aquinas advocates the idea of an uncaused cause ('d'), or a 'something for nothing', which requires a violation of the law of the conservation of energy/information.

iv. Aquinas assigns the impossible 'uncaused cause' to the Judeo-Christian conception of God (in 'e') – a god-of-the-gaps. He could have just as easily assigned this impossibility to any other god or conception. Thus, the logic of 'e' is inconclusive.

v. The so-called proof fails and is thus rejected.

The Calvinist philosophical theologian Jonathan Edwards (1705-1758) argued in his book, Freedom of the Will, as follows:

> *"But if once this grand principle of common sense be given up, that what is not necessary in itself must have a Cause; and we begin to maintain, that things which heretofore have not been, may come into existence, and begin to be of themselves, without any cause; all our means of ascending in our arguing from the creature to the Creator, and all our evidence of the Being of God, is cut off at one blow."*[12]

The reply above would disagree with the 'common sense' of Edwards but agree with the thing Edwards fears, which is that things can naturally evolve or emerge without the need of any organised or meaningful Agent. I would also agree that this suggests "all our evidence of the Being of God is cut off at one blow."

Argument 3: The argument of contingency (also used by William Lane Craig)

a. Many things in the universe may either exist or not exist and are all finite. Such things are called contingent beings.
b. It is impossible for everything in the universe to be contingent, for then there would be a time when nothing existed, and so nothing would exist now, since there would be nothing to bring anything into existence, which is clearly false.

c. Therefore, there must be a necessary being whose existence is not contingent on any other being or beings.
d. This being is whom we call God.

Reply:

i. The Uncertainty Principle (and the Theory of General Relativity) suggests the idea of classical determinacy and uni-dependent or linearly-traceable contingency is incorrect, so the premise of the proof is not correct. Thus the observation in 'a' is misstated. Further, 'a' assumes a disconnected universe in which finite things exist independently rather than interdependently, which is not in agreement with observation. For instance, every subatomic particle of our human bodies orbits around the black hole(s) at the centre of our galaxy about 26.7 thousand light years away. Therefore, our bodies cannot be considered ultimately independent of their environment. Likewise, the path of a lone photon through a quantum interference screen cannot be assumed independent of its environment. This also means the total environment is not independent of local behaviours. (We shall consider Kurt Gödel's Incompleteness Theorems, which extends the idea of dependency or incompleteness to the non-material world, later). Therefore, the logic of 'b' is not established: All things real and virtual can be contingent on each other without beginning. Aquinas made a scientific mistake with respect to determinacy and contingency because he thought he observed fully determinate and uni-dependent contingencies.

ii. It is likely that the inability to describe all contingencies with certainty is the outcome of the interpenetration of infinity and nullity and the fuzzy boundary narrowly defined as existence that sits between them. In more familiar terms, local and cosmic indeterminacy or lack of uni-dependent contingency is the result of the relativistic wave-particle nature of all things.

iii. Aquinas advocates the idea of a non-contingent, non-evolved or non-emergent being (in 'c'), or a 'something for nothing', which requires a violation of the law of the conservation of energy/information.

iv. Aquinas assigns the impossible 'non-contingent being' to the Judeo-Christian conception of God (in 'd') - a god-of-the-gaps. He could have just as easily assigned this impossibility to any other god or conception. Thus, the logic of 'e' is inconclusive.

v. The so-called proof fails and is thus rejected.

Argument 4: The argument from degree (not used by William Lane Craig):

a. Varying perfections of varying degrees may be found throughout the universe.
b. These degrees assume the existence of an ultimate standard of perfection.
c. Therefore, perfection must have a pinnacle.
d. This pinnacle is whom we call God.

Reply:

i. The Uncertainty Principle suggests that instabilities and uncertainties cannot be eliminated. Thus, all material perfections are impossible, and all judgments of imperfection anthropomorphic, so the premise of the proof is not correct in the quantum world. Thus the observation and logic in 'a' and 'b' is not established. Aquinas made a scientific mistake with respect to perfection because he thought his subjectively judged standard of perfection was objective. He seemed to naively follow the ancient Greek notion of ideal Forms.

ii. It is likely that the inability to describe anything with perfection is the outcome of the interpenetration of infinity and nullity and the boundary narrowly defined as existence that sits between them. In more familiar terms, local imperfection or incompleteness is the result of the relativistic wave-particle nature of all things.

iii. Aquinas advocates the idea of a pinnacle of perfection (in 'b' and 'c'), which is unsupported by local evidence. For instance, a theoretical perfect circle has an infinite number of tangents. Conversely, local imperfection in a measured circle's tangents implies an imperfect circle in totality as well. Likewise, the Uncertainty Principle suggest imperfection in the pinnacle (or cosmic totality) because of its tested confirmation of imperfection in the local (or subatomic minimum).

iv. Aquinas assigns his notion of dualistic perfection to the Judeo-Christian conception of God (in 'd') - a god-of-the-gaps. He could have just as easily assigned this unprovable notion to any other god or conception. Thus, the logic of 'd' is inconclusive.

v. The so-called proof fails and is thus rejected.

Argument 5: The teleological argument or argument of design (also used by William Lane Craig)

a. All natural bodies in the world act towards ends.
b. These objects are in themselves unintelligent.
c. Acting towards an end is a characteristic of intelligence.
d. Therefore, there exists an intelligent being that guides all natural bodies towards their ends.
e. This being is whom we call God.

An alternative to this argument by design also provided by Wikipedia is:

a. All natural bodies follow laws of conduct.
b. These objects are themselves unintelligent.
c. Laws of conduct are characteristic of intelligence.
d. Therefore, there exists an intelligent being that created the laws for all natural bodies.
e. This being is whom we call God.

Reply:

i. The Uncertainty Principle suggests that indeterminacies cannot be eliminated. Thus, things do not act towards fully deterministic ends. Ends and beginnings are illusory. Objects act towards probabilistic ends at any local level because the macroscopic classical world is an outcome of the microscopic quantum world, and vice versa. We could also address this question in terms of Gödel's Incompleteness Theorems, which would more strongly suggest that no determinism is fully describable in stringent mathematical law, and thus any kind of law. Therefore, 'a' is not supported in both versions.

ii. All things move, and through movement, display a level of systemic proto-intention or proto-intelligence. Some living things have evolved a self-aware intelligence. Intelligence is observed to emerge in objects from systemic arrangements of proto-unintelligent components (e.g. sperm, eggs and food). More deeply, the emergent intelligence of anything is made possible because the cosmos is vastly self-contained and dynamically interrelated. Exploiting dynamic interrelationships is the core of intelligence, and cosmic interrelationship is inescapable. We are bathed in cosmic interrelationships or proto-intelligence and everything contributes to this state of cosmic reality. Our intelligence emerges from this bath in phases and layers, just as local geological strata slowly build the landscapes upon which we live. Therefore, 'b' is misstated in both versions.

iii. Natural bodies manifest or display laws of our own conception, they do not follow them. This was part of the error of the ancient Greek idea of dualist ideal Forms, refuted previously. The basis of proto-intelligence is in Heisenberg's unstoppable movements and Einstein's unavoidable Relativity. Thus, 'c' is misconceived in both versions.

iv. All living things evolve and all non-living things emerge. These are facts. Darwin's explanation of these facts suggests living things evolve through the mechanism of natural or artificial selection.[13] That is, they may have an intentional agent such as a pigeon breeder, as in the case of artificial selection, but this is not necessary in the case of natural selection. Thus, by the Uncertainty Principle and Natural Selection 'd' is not established in both versions. Aquinas made a scientific mistake with respect to determinacy and design because he was unaware of the facts of quantum physics, relativity, evolution and emergence and because he was likely influenced by the false Ancient Greek notion of dualist ideal Forms.

v. Aquinas assigns incorrectly perceived determinacy to the Judeo-Christian conception of God. He could have just as easily assigned this incorrect notion to any other god or conception. Thus, the logic of 'e' in both versions is inconclusive.

vi. The so-called proof fails and is thus rejected.

We might argue that just as we can trace our universe back to the Big Bang, so we can trace the first movement, first cause, first contingency, original perfection and original determinacy back to the same event, the Big Bang. Thus, it seems these five so-called proofs may regress back to the first Planck Time quantum of the local Big

Bang. Yet even here, we must contend with the Uncertainty Principle. Things cannot become still, unaffected, nothing, perfect, or deterministic at the Big Bang. Necessary motion, effects, indeterminacy and uncertainty must be present at the local Big Bang and extend past it, because energy must be conserved. If there was impossible stillness or nothing at the beginning of the first Planck moment of the Big Bang, there would have been nothing at the end of it. By the law of the conservation of energy/information, the Big Bang would not have advanced to the next moment.

Aquinas himself seemed to partly appreciate this point in Argument 3b, but not fully, because he departed from it in Argument 3c. That is, a more consistent argument could have stated that the contingency/energy/information of all things interpenetrates with God's contingency/energy/information. How was the Higgs boson found? Primary because of the past failure to disconfirm the theory of the conservation of energy/information, and thus the tentative scientific belief that energy/information must be conserved. The most fundamental reason why many scientists do not accept the supernatural is because they tentatively rely on the conservation of energy/information. The hunt for the Higgs boson was thus primarily a hunt for missing energy/information. This was the source of the hunt's success.

Faced with the unknown, we can either tentatively rely on what has served us well to date to apply to the problem, or we can give up, and say the answer must be a 'god of the gaps'. The failure to fully appreciate the conservation of energy/information represents a basic error made by Aquinas, more basic that his ignorance of the Uncertainty Principle. Perhaps even more disturbing, Aquinas' error was seeking an answer to the problem of the infinite regress too quickly. He and Jonathan Edwards jumped to a solution through an abandonment of the conservation of energy/information and an appeal to God. Emergent order from relative disorder (vast nothingness) solves the problem of the infinite regress.

Considering their times, they can perhaps be forgiven for this abandonment, although the ancient Greeks did not totally miss appreciating the concept of energy/information conservation at an early stage. Richard Janko reports in his *Empedocles: On Nature* (2004), that Empedocles (490-430 BCE) wrote, "nothing comes to be or perishes". It is sad that when it comes to the moment of the

Big Bang, or any other 'first' moment, many today are still prepared to believe otherwise without any evidence.

There are several issues here. Firstly is the idea that an infinite regress may be addressed through the mechanism of a cycle in which the nominal end is also a nominal beginning, but I will get back to this. More basically, the infinite regress is solved by understanding what it implies. It implies that we have a uni-dependent or linearly-dependent cosmos and that a fundamental interdependency is not allowed. However, once we accept that nature has a fundamental interdependency or balance (/imbalance), and this is its basic nature at all stages, then the idea of universal cycles acting on many interacting, cell-like, and nested levels becomes obvious. Einstein captured this idea of an interdependent universe in his Theory of Relativity, which is the theory of the basic interdependency of matter and space. That is, we can either solve the infinite regress through appreciating Einstein's monistic General Relativity, or by clinging to the old idea of uni-dependency (classical cause and effect) and introducing a forced spiritual dualism.

However, then we must understand the ramifications of the chosen solution to the (false) problem of the infinite regress. In the case of General Relativity, it is through the formation of the interacting cycle, but in the case of duality, it is through the formation of God. Here again, the theist is ready to declare a no-go zone, by claiming a transcendental realm in which supernatural laws apply that are not subject to scientific methods of examination, so we cannot question the formation of God. The alternative is simple emergence of the cycle because of the interdependent and non-linear nature of General Relativity. The good news is that we can see the interpenetration of infinity and nullity (or the asymmetrical and asynchronous wave-particle nature of everything) on a daily basis in the multifaceted emergence of all things. The relativistic wave-particle interactions are ubiquitous and incredibly consistent. This means they do not require a dualism or 'god-of-the-gaps' for matter and space, energy and information, the direct and indirect, the real and the virtual, or the reductionist and holistic, to dance.

In contrast, the dualistic solution to uni-dependence is to leave the erroneous view of nature's uni-dependence in place and solve the problem it presents by introducing the necessary relativism outside of natural reality, in a 'god-of-the-gaps'. This makes one error, uni-dependence, into two errors, uni-dependence and dualistic god-dependence, without addressing the observable and fundamental interdependence of matter and space.[14]

Further, the continuing interactions of infinity and nullity suggest a mutually restrained, or relativistic, relationship between them that forbids any actual infinity (as opposed to a potential infinity, to use Aristotle's framework), and thus suppresses an actual infinite regress of any kind (not just the kind imagined by classically linear cause and effect).

In summary, Aquinas' arguments of original non- or god-movement, classical causation, non-or god- contingency, perfection and determinacy, which more correctly framed are concepts of necessary movement, effects, contingency, imperfection, and indeterminacy, must go back to the Big Bang, but by the law of the conservation of energy/information, must extend past it. There is no reason why a limited cosmic cycle of (serial and/or parallel) universes is not possible if it the simple result of the relativistic tension or balance between infinity and nullity in all possible dimensions, which I would posit as an explanation of the observed wave-particle nature of real and virtual reality. Without tentative acceptance and continuing testing of such an explanation in this realm, I suspect we wander into a detrimental fantasy.

Additional Argument 1: William Lane Craig argues for objective moral values and duties, which he says must come from God.

Reply:

This is not a proof, just an assertion or claim. It is easily refuted by Pinker (2011), who demonstrates how moral values are changing and perhaps improving over time. If God is unchanging, then his morals would be known as unchanged over time, whether or not they were followed.

However, Pinker demonstrates that this was clearly not the case. Morals emerge. Further, they can't become explicit except through intersubjective agents. More generally, confirmations of General Relativity and the Uncertainty Principle prove wrong all uni-dependent and absolute claims to real-world objectivity. The so-called proof is not established and thus, rejected.

Craig needed this so-called proof added to the list because he was aware of the Christian god-of-the-gaps problem, that is, the inconclusive nature of the last step in each so-called proof of Aquinas. Craig assumes the moral code of Christianity is consistent, unique, and uni-dependent. Thus, if he can attach the specific God of Christianity to the proofs of Aquinas, then he can help affirm them as a whole.

However, he has not addressed the issue of the ideal uniqueness of the Christian moral code at all. I am sure this moral code follows an indeterminate probability distribution amongst the world's Christians as well. Uncertainty is ubiquitous, even in the moral sphere.

Additional Argument 2: William Lane Craig argues that the historical facts concerning Jesus Christ proves there is a God.

Reply:

This is not a proof, just a claim. The discontinuity and uncertainty of the evidence from ~2,000 years ago means it is not credible today in terms of proof. The so-called proof is not established and thus, rejected. Craig needed this proof added to the list again because he was aware of the problem with the inconclusive nature of the last step in each so-called proof of Aquinas. Uncertainty is ubiquitous, even in the sphere of historical 'facts'. In summary, our understanding of reality, gained through modern science, has fundamentally changed the way we consider explanations of the ancient, medieval, and classical worlds.

New Science

We still tend to act as if we live in a classical, cause-and-effect world, whether we talk of religious doctrines, what is taught in our schools, the common sense on our streets, or the philosophies and social sciences of many academics. However, there has been a scientific revolution - a paradigm shift…

- Quantum Field Theory and physicist Werner Heisenberg's Uncertainty Principle: You cannot know beforehand both the position and velocity of a particle. Everything we perceive is indeterminate (and never stays still).
- Einstein's Theory of General Relativity: Physicist John Wheeler – "Matter tells space how to curve. Space tells matter how to move."[15] Everything we perceive is relativistic - there are no unidimensional absolutes, not even of time itself. The universe does not behave like a simple machine or billiards table. More importantly, the universe is not 'uni-dependent', like a factory assembly line with inputs at one end and outputs at the other, or like the mistaken idea

of an infinite regress of causes. Our universe is relativistic and thus interdependent at every turn on two variables rather than one – space and matter (or more deeply, excursions towards nullity and excursions towards infinity). This means all behavioural styles are dependent on everything else. Everything is everything, as the Donny Hathaway song title goes. Every cause is dependent on effects and vice versa (although, as we have seen, the idea of classical cause and effect is deeply flawed). The very small is dependent on the very big and vice versa. The direct is dependent on the indirect and vice versa. The real is dependent on the virtual and vice versa. The more determinate local space-coding is dependent on the more indeterminate local matter-arrangement and vice versa. You, your life, your intelligence, and your free-will are dependent on your environment and vice versa. This also calls the idea of local environmental objectivity into question. It seems our environment is partially intersubjective, just as we are (more correctly, we are intersubjective with the environment). This also hints at the notion that the dichotomous framework of the subjective and the objective is also flawed, as we will discuss later.

- Mathematician Kurt Gödel's Incompleteness Theorems: A fully specified mathematical system (including its axioms and proof theorems) cannot demonstrate its own completeness. It cannot be both consistent and syntactically complete. Every deterministic framework of logic we can conceive relies on an inherent indeterminism for its becoming. That is, every immaterial / virtual logic we conceive is also interdependent and incomplete. This is partly because structures of any variety, material or informational/logical, are remnant properties of original bootstrap scaffolding or systems 'seeding' that is now unreachable through classical means. Further, the more we go back to basics, the more we go back to probabilistic, uncertain, and disconnected disorder (proto-matter, proto-space and proto-information). Thus, the only way to better understand current structures and escape Gödel incompleteness is not through endless reductionism, but through the process-modelling approach of emergentism.

- Genetics and Darwin's Theory,[16] which established that you need an SPS of genetic uncertainty or genetic mutation but not an agent for evolution: That is, an uncertain, dynamic and sometimes trigger-edge relativity between a species and its environment is a better explanation of the rise of a species than any absolutism. Likewise, the emergence of all things (including the cosmos itself) depends on quantum indeterminacy and asynchronous relativity (or slight imbalance).
- Science philosopher Karl Popper's rejection of induction (e.g. 'all observed swans are white therefore all swans are white'), and insistence on explanation, trial-and-error testing, falsification, and deduction as the solid basis of the Scientific Method (but perhaps not so much the basis of common sense). Through testing the 'null hypothesis' we don't discover certain knowledge, we reject certain errors in knowledge and tentatively acquiesce to the uncertain remainder. Only the fittest theories survive. The place of induction is thus as an important but uncertain method of hypothesis proposal in the cycles of scientific development.[17]
- Several mathematical concepts that suggest reality cannot be adequately described in terms of the classical law of identity ('A is A'), the related classical principle of non-contradiction ('if A is A, then it is not B'), or the related classical law of the 'excluded middle' (i.e. you can't have a half-truth; 'either A is A, or it is not A'). These mathematical concepts are as follows:

    - Fuzzy Logic:[18] All real-world facts are a matter of degree. Real-world bivalence (for example, something is either true or false) is a myth. The real world is multivalent, interdependent, and fuzzy. This means uni-dependent cause-and-effect (and ceteris paribus) is misleading.
    - Complex Numbers: These consist of real numbers and 'imaginary' numbers that are multiples of the 'square root of minus one'. Reality has indeterminate states that exist and emerge from between the cracks of the classical ones we can reach through following normal laws and reductionist techniques. The

'square root of minus one' is a mathematical technique that enables physicists, engineers, and technicians to access these otherwise hidden or discontinuous states of existence outside of the classically understood world.

- Chaos Theory: Many real-world, non-linear (or non-classical) and dynamic systems, with aperiodic equilibrium states that seem to have unpredictable behaviour, are nevertheless describable by deterministic equations that involve emergent feedback (which in mathematical equations, is represent by the double forward/backward arrow symbol that we could interpret to mean 'emerges with'). However, if the initial conditions are undefined, then so is the system's behaviour. Further, if the initial conditions are changed a little, the system's behaviour can change a lot. Examples of chaotic systems in nature include the distribution of stars in space, clouds in the sky, and leaves on a tree.
- Fractals:[19] These can be real-world things that due to self-similarity (i.e. feedback between dimensions) take on a vast number of partial dimensions of length in between the three we generally recognise (height, depth and breadth). Fractal geometries are geometries that look the same on differing scales of length, or are 'self-similar'. Cauliflowers, human brains, the branching of veins and arteries in human anatomy, fern fronds, and cloud formations are typical examples, but there are innumerably more.

- Such self-similar geometries have dimensionality of between 2 and 3 (i.e. less than 3 because the dimensions have a certain interdependency). Fractals exist because the 3 dimensions are not independent. That is, height, length, and width, along with time, are interdependent, interpenetrating concepts. All animals, including humans, are self-similar in appearance around a left-right axis. Dimensional self-similarity (not just in spacetime, but also in behavioural variables) makes nature's genetic programming and algorithmic implementation of genetic instructions, and our process-modelling of such systems, much simpler. Nevertheless, such systems display a vast diversity due to a myriad of environmentally-introduced variables. An allometric example of a fractal relationship between dimensions would be Kleiber's Law, which scales the metabolic rate to the ¾ power of the mass of an animal across a very wide sample of species. Since the mass of an animal is related to its volume, this suggests the time dimension of an animal is related to its space dimensions.

- A famous theoretical fractal using a simple complex number feedback equation is the Mandelbrot Set (see Figure 3). This set is generated using the simple formula z_{n+1} 'emerges with' $z_n^2 + c$, where c is a complex number and $z_0 = 0$ is used as the starting point for each c. If z_n diverges or tends towards infinity, then its real and imaginary components (plotted on the horizontal and vertical axes, respectively) sit outside the set, or outside the black area of Figure 3, and if they converge or tend towards nullity without settling on it, they sit inside this black area. That is, the set excludes infinities and nullities much like existence in our cosmic cycle of serial and/or parallel universes must do.

- The set, which reflects some level of fractal self-similarity between the real on the horizontal axis and the imaginary on the vertical axis, suggests our relativistic cycle of universes could be similar, that is fractal, in nature. Could we say the interdependent real and imaginary parts of the Mandelbrot Set are analogous to the relativistic real/direct and virtual/indirect parts of the cosmic cycle? Could we also say that reality at the broadest level includes existence within such a set (the black areas of Figure 3) plus the solutions tending towards nullity or infinity ultimately disallowed in existence, which we could perhaps characterise as 'the non-existent limits of reality' (represented by all the other areas of Figure 3)?

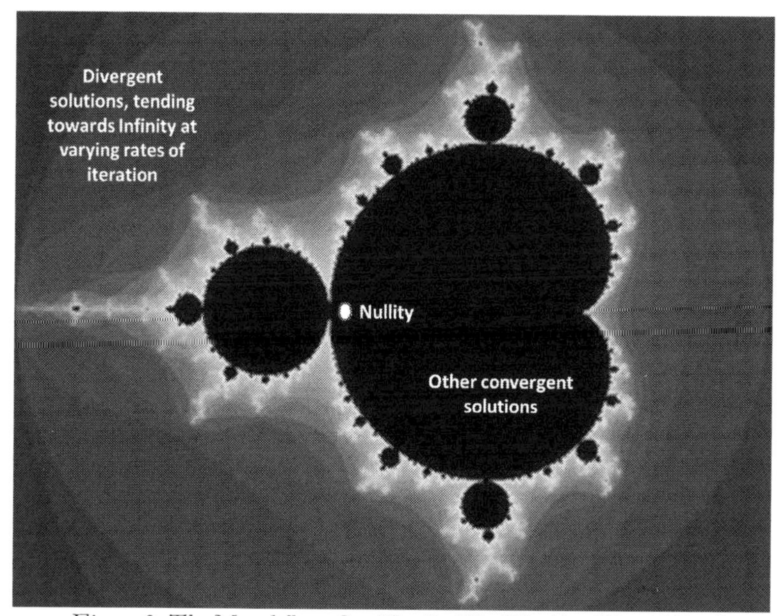

Figure 3: The Mandelbrot Set (Credit: The Worlds of David Darling, accessed 29/01/2015)

In summary, because there is only one reality, every system within the cosmos is also a continuously moving or iterating, mutually dependent, uncertain, incomplete and evolutionary (or emergent) system - which we can only ever know tentatively.

Further, the incomplete local environment we find ourselves in and to which we can apply our imaginations, is pregnant with emergent possibility. The only determinacy in reality is in small strands or oases of indirect 'coding', such as in DNA, rather than in any arrangement of matter. That is, a monolithic determinism does not exist and 'coded determinacy' emerges; it emerges locally, imperfectly and indirectly in spacetime's organising of material things (and vice versa).

Big Morality

It is the relativistic or fuzzy nature of the universe, which in this context is the intimate bonding or meeting of the more certain, determinate or significant spacetime and the more uncertain, indeterminate or insignificant matter, that is at the very heart of the philosophy of emergentism. By the Uncertainty Principle, it is clear that the more certain is not separable from the uncertain in reality at all. It therefore seems reasonable that the universe, and its beyond-momentary existence, is necessarily and sufficiently monistic, with inseparable components we conceptualise as more certain and less certain, or signal and noise. Further, the universe's monistic instability or noise is essential to all motion, borders, apparent causation, intention, growth, development, unique expression, freedom, intelligence, and morality - which we will discuss, but it will take some time.

Whether emergence's complex process is one of subatomic particles organising themselves into molecules, an animal's embryonic development, or how the body's regulatory systems and perceptions interact with emergent consciousness, is not the main point. The salient point is to become more aware and respectful of the proto-intentional, amazingly interactive and emergent nature of the various local processes or systems always at work in nature's various levels of synergy and in each of us. Everything works together, which means even our conscious choices cannot be ours alone. Nevertheless, our choices are unique in the time-and-space node we each occupy. This also means we should question all forms of uni-dependent thinking.

These various processes or systems operating simultaneously at different levels of order highlight the seemingly unknowable relationship between the levels. These seemingly unknowable borders frame what we perceive to be certain at each of the levels. For example, how are we to understand the transition from the universe's quantum-mechanical beginning when space-matter-time was still indeterminate to the sub-nuclear plasma, between the sub-nuclear and the inorganic, between the inorganic and organic life, or between organic life and consciousness, respectively? How are we to understand the almost seamless transition from the non-conscious autonomic systems within the human body to subconsciousness and then self-aware consciousness within the mind?

Pondering how each of these transitions was and is locally accomplished together, over time and concurrently in the present, without reality fragmenting into meaningless disorder or falling into infinity or nullity, is to truly consider the wonder of nature's interpenetrating, relativistic and emergent dance, which is creative only in the weak sense. The unifying driver of emergence that links all the other transitional levels together is simply Heisenberg's unstoppable movement, with which comes unstoppable proto-intention, awaiting opportunity and direction. As per Figure 3, it seems that in each case of emergence a self-sustaining order or a systematic use of local information slowly arises in oases intimately connected to a vast and less synergistic or more monolithic surrounding.

These nominally supervening transitions and their emergent properties in the self-contained vastness of possibilities within our universe changed and continue to change everything. However, perhaps we need one more transition to complete our picture. This transition is from human consciousness-plus-instincts that are still products of a largely natural environment (and a classical understanding of it), to a state of higher consciousness across all societies that sustainably coevolves with our fuzzy and diverse environment at a cosmic level.

If meaningful causes emerge from relatively meaningless effects, then laws emerge from the observations of arising organisation rather than being determined beforehand. That is, the ability of the universe to bootstrap itself means the so-called Laws of Nature (or our naming of them) also emerge. Further, these laws can collapse again if the local environmental conditions no longer provide the requisite relational constraints. The universe neither follows laws nor makes laws. If anything, it just continually and approximately manifests them. Laws are not conditions of existence; they are approximated properties of the things that exist, just as wetness is a property of water or consciousness is a property of human beings. Nature's laws are the patterns we have detected in local and statistically significant temporal arrangements, nothing more. The laws we name are neither a complete set nor perfect.

In essence, without intending any disrespect, personal moral codes and societal laws, as well as religious commandments, are no different. More specifically, moral values are not independent determinants of personal or organisational behaviour. Rather, moral values are the dependent properties, manifestations or named heuristics of people, organisations and their relational behaviours.

Interpenetrating Concepts

A person's values, like nature's laws, are meaningless without expression. Just as we can incorporate emergent causes within our risk management systems, so we can use our heuristic laws of Nature and incorporate emergent moral codes into our planning for future (moral) contingencies.

This picture of morals enables us to see that morals emerge like everything else. Just as Professor David Christian and Bill Gates have painted so effectively for their secondary school students across the globe the message of Big History,[20] so we can now tell the story of Big Morality. Big History pulls together expertise from numerous fields to answer such questions as, 'how did we get here?', and 'where are we headed?' It offers a framework or reference system to students of history that cannot be gained from considering dated listings of historical incidents alone. The students achieve insights into history by noting how various events roughly supervene on prior events, which in turn help the students gain a personal context.

Naturally, the story of Big History begins somewhere near the Big Bang; likewise, Big Morality. However, Big Morality would ask more personal questions like, 'how did I come to be the way I am?', 'why do I love to do the things I love doing?', 'what should I do?' and 'what should my family and my society do?' It focuses more on a discovery of 'My Story' rather than 'History'. It verges towards questions that are personal, rather than just questions of natural science or social science. If I were to rename this Part 1, it would be, *The Interpenetrating Concepts of Big Morality.*

From a naturalistic perspective, if the essential nature of the universe is monistic and self-organising emergence, and the universe intimately holds us within this emergence, then how can our essential nature and behaviour as human beings be anything different? Emergent self-actualisation should describe our human nature and potential quite well.

Space

It can be incredibly enjoyable to focus on the beauty of one small star shining in the vast array of our Milky Way: Its very insignificance seems to make it more precious to us than its larger neighbours, like a single diamond in a jeweller's carefully selected complementary setting brilliantly reflecting and refracting the light that shines upon it. Both the star and the diamond are precious because their simple symmetry-mixed-with-asymmetry, or setting and resultant liveliness, grabs our attention and makes us feel or know something seemingly profound…

Space seems to grant all the stars their freedom of self-expression amongst their fellows. Perhaps our selection of that one small star or diamond for our special attention makes us feel or know that we too belong and can have impact and purpose in a changing universe that dwarfs us in size but nevertheless intimately includes us in our unique form, time and place.

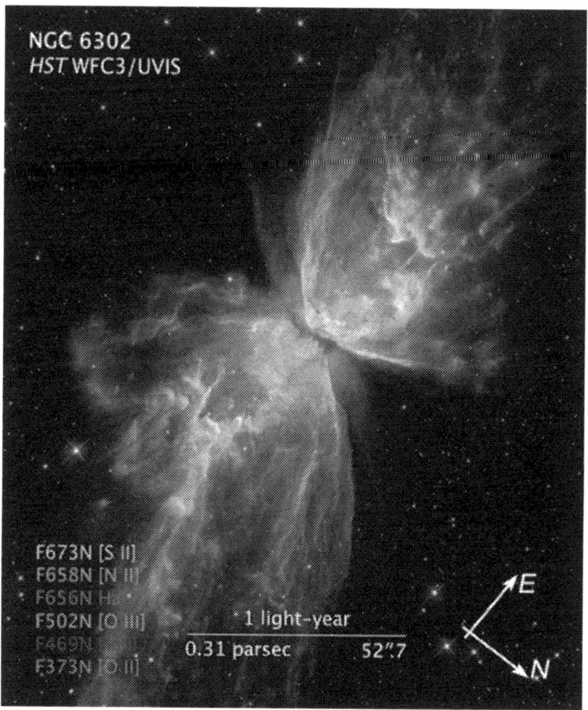

Figure 4: NGC 6302 (Credit: NASA, ESA, and the Hubble SM4 ERO Team)

We can learn from this interplay with the star or diamond, that smallness does not equate with separation and personal insignificance. We seem to confirm that the difference between those things perceived as significant or interconnected and insignificant or separated at any moment is often just the way we consider the setting or space around them. Perhaps this is similar to the quiet voice of a musical instrument that stands out from the silence around it. It is the complex pattern of interacting harmony, relationship, or fit with the local environment, and our reference point, which touches us more deeply than the star's or diamond's mere size.

If we thought a little more deeply, that presumed small star could be a phenomenon like Figure 4's NGC 6302 that stretches across an area of space more than two light years wide. As always, intersubjective labels such as small, insignificant, or separate are simple functions of our limited sensorimotor perceptions and sensory elaboration. Physical size is just a function of matter density. It is conceivable that all the particles of your body or mine could be spread out in an orderly fashion across an area of space as large as NGC 6302 or conversely, that all the particles of NGC 6302 could be squeezed into a plasma the size of your body or mine.

In each case, the original unique order, beauty, and fit that emerged through eons of interactions and disconnections, much as a cake emerges from the irreversible baking of its ingredients, would be lost and dramatically different processes would now describe each case's local interactions. Thus, in an important sense, it is not matter that is unique or beautiful, but rather its emergent arrangements and interrelationships in time and space that grant it this quality. On the other hand, space without matter is just a gurgling nonsense of quantum fluctuations.

I will use the term 'universe' to describe all the outcomes of the Big Bang disequilibrium to its approximate reversal in the future, and the 'cosmos' to include the universe and anything that may exist outside this temporal Big Bang excursion from equilibrium (either serially or in parallel). On this basis, modern scientists generally agree that our universe is most probably finite in size. It seems to be evidently true that this phase of our universe began as a notionally finite phenomenon. That is, we can assume a finite amount of potential energy converted into a finite amount of radiation in the Big Bang 13.8 billion years ago, otherwise the Big Bang would itself be without beginning rather than at a distant instant in time past.

Note: A closed and finite universe is like a two dimensional sheet of paper curled up into a sphere with no edges, except that it applies to the universe's three dimensions. By way of contrast, a flat and infinite universe would be analogous to the same edgeless sheet of paper laid out flat. Scientists now believe that the universe became much more flat after the Big Bang, such that today it is almost perfectly flat, i.e. closer and closer to the nature of an eternally expanding universe rather than one that can end in a gravitational Big Crunch. Nevertheless, helium-4 becomes superconducting at 2.2K and helium-3 at 0.002K. Currently, the temperature of deep space is about 2.7K. Can we be sure of what will happen to the properties of all fleeting subatomic particles around 0.002K? Alternatively, is it possible that the universe one day expands at such a fast rate that as they form, matter and antimatter are separated and partially persist in distinct oases rather than mutually annihilate, just as each likely did in the first moments after the Big Bang? If this happened, could a gravitational Big Crunch soon follow after all?

Another reason why we may assume that the universe is finite is because while infinity exists as a mathematical concept, it is not directly observable in matter's arrangements; singularities within black holes, where spacetime and matter seem to break down, remain finite and black holes themselves remain differing in size. That is, all physical things seem to find finite resolution in reality even though they are affected (constrained and organised) by the indirect mathematical limits of zero, infinity, the square root of minus one, etc. in all their relationships, interactions and iterations. In limit theory, such systemic operations never reach or touch their asymptotes. It seems to me that anything infinite is somehow a shadow or harmonic ripple of existence. Any actual nullity also retires from existence.

The idea of an indirect reality is an interesting one. Is there an infinite indirect reality? Aquinas and the ancient Greeks thought there was, but Einstein would suggest that the indirect world of space's interactions is constrained by matter almost as much as the direct world of matter is constrained by space. That is, even though the indirect world may be unconstrained by entropy or decay, it is constrained by material opportunity. Thus, the infinite seems inexpressible in the components of matter or space. David Deutsch says in his book of the same title, that we are always at the "beginning of infinity".[21] This suggests only a limited unfolding sense of infinity may be expressed in the space-matter realm. It would be an emergent tendency towards infinity, rather than an expressed infinity in existence's becoming.

Perhaps a good example here is the preservation of a gene in DNA across thousands of generations. Genes are indirect and matter-organising lengths of code in the midst of relative disorganisation (noncoding DNA) that are implemented in matter through the production of proteins in life-forms. Even though they don't decay themselves, they do depend on the survival of life-forms for their organisational consistency and longevity. Gene codes are subject to mutation and extinction, but not decay. Exactly the same is true of all of space's coded information, or what I call indirect reality.

However, I would like to suggest that there remains a latent infinity and nullity that is inexpressible in the harmonic substrata of matter-spacetime, something like noncoding DNA can remain latent in the human genome but is ready to spring into use when adaptive mutations happen along. What I am suggesting here is that inexpressible infinity and nullity are necessary limits or shadows of existence, just as we can see the diverging ripples outside the black area, and the black area converging on the white dot, of Figure 3. Interdependent matter/spacetime existence seems to supervene this shadowy framework or field. It exists something like the outline of the Mandelbrot Set in Figure 3, but also temporarily extends past and inside the outline in its fuzzy expression. The tension or relativity between infinity and nullity that precipitates existence as we know it also provides it with its potential. Full reality is thus temporal existence interacting with its informational potential (or proto-information) in the inexpressible field between infinity and nullity.

If the direct was infinite in an absolute sense, then the indirect would not exist, and vice versa. This suggests that actual infinity and the absolute independence of anything is impossible for us to be here, having this discussion. This also suggests we may have wrongly assumed infinity and nullity were independent of each other in the past, when really they never were; they were always caught in a relativistic balance; they always coexisted with a kind of border or mutually constraining and enabling boundary between them, which we call expressible and relativistic existence. That is, infinity and nullity have always been relativistic at their boundary – the boundary we call existence, which at its sharpest we might relate to the black area of the object in Figure 3. If these two lost their relativity, this would enable the actual infinity or the infinite regress of the one or the other. This suggests to me, at the most basic level, even before or outside of the Big Bang, a fully unconstrained infinity or nullity has so far never been seen in reality; only the constrained infinity and nullity captured in the endless relativistic tension or asymmetrical balance between them has been possible. This further suggests that our old concepts of an independent infinity or nullity may be wrong, even in the fullest sense of reality. Besides also always being at the beginning of infinity, the track we ride is always along or near the fuzzy boundary fence between infinity and nullity.

It follows from this line of reasoning that absolute independence is perhaps the most erroneous concept our human minds maintain. (I suspect the reason for this core mind deception is the need to conjure up and maintain the self, but we shall discuss this idea later).

Further, this suggests that even our cosmic cycle (of serial and/or parallel universes) cannot be an expression of a perfect cycle. Something like the Mandelbrot Set, it must be in the form of an incomplete or partly indeterminate cycle that has no meaningful end to its iterations. That is, just as we cannot draw a perfect circle without defining infinite tangents, so our cosmic cycle cannot be specified with infinite precision. This is because, according to the Uncertainty Principle, our cosmos cannot be specified with infinite local precision, which also indicates an impossible cosmic precision (in the same sense that a finite and imperfect set of local tangents can only express an imperfect circle). To put this idea another way, the interaction of infinity and nullity that leads to the fuzzy border of existence between them is not fully expressible in existence. Yet in a similar way to the noncoding DNA that can provide the genetic space in which organisms can evolve, this proposed characteristic of fractal relativity provides the potential field within which existence can emerge and systems can arise with greater degrees of freedom.

The incomplete resolution of the cosmic total may be due to its inexpressible infinity and nullity, its speed-of-light delay, its decaying and indeterminate material contents, but also due to the limited opportunities of its locally determinate cosmic interactions. This limit on locally determinate space was intuitively the case in the first quantum moment of our universe, when there was no difference between the very big and the very small, so why would this change with the expansion of that original finite cell? The relativity of the very big (e.g. the effects of gravity) and the very small (e.g. real subatomic particles) must remain true for the entire expansive universe and its place in the cosmic cycle.

The path each of us takes depends on the entire cosmos across time and space, but the entire cosmos also depends on the path that we take. The resolution of the cosmos depends on each of us, but the resolution of each of us also depends on the cosmic total. Remove any one of us, as an arrangement of energy and information, and the rules of their interactive conservation would be broken, that is, the fabric of reality would be torn apart. This is a reverberating issue we must consider again and again in the coming pages. For better or for worse, we are part of a fuzzy cosmic web from which we can never be divorced. We are nodes or foci much more than we are objects.

An unbroken deterministic cosmic cycle of indirect coding and information (in the components of space) would present us with the oddity that eventually everything, every life, and every experience, would be repeated every cosmic cycle – 'groundhog day' would be real, without any relief. This is because if the cosmic cycle is self-contained and deterministic, then eventually it will repeat itself. That is, once every possible unfolding combination of finite subatomic particles had happened once, the series (the cosmic cycle) would be bound to happen again. This seems to be the implication of Minkowski's Block Universe view, which suggests we can view the cosmic cycle as something like a loaf of bread we could slice up into independent pieces of local spacetime.

By unbroken deterministic cosmic cycle, I mean one that can be described by a deterministic fractal formula, as per Mandelbrot's Set. A relativistic cosmic cycle that is indeterminate in matter and broken up in strands of order and relative disorder caused by the defects matter represents, which is the one proposed here, might complicate the "block" cosmos view, but not necessarily eliminate it. The repeated patterns of subatomic arrangements in spacetime would be recognisable in local strands and perhaps in whole universes, like those recognisable repetitions in the locally magnified iterations of the Mandelbrot Set.

However, I would contend that while a relatively discontinuous and multi-fractal block cosmos (i.e. a shuffling of layered fractals expressed through spacetime codes) makes sense, we would never be able to slice it up into perfectly independent parts of matter-spacetime. This is because at the local level, spacetime, according to relativity and quantum mechanics, is not just lacking universal simultaneity and interdependent with the local matter-spacetime existence border, but if it is like the Mandelbrot Set, then it is also interdependent with the broken up harmonic reflections of infinity-nullity on each side of the set's border. Thus, each spacetime moment only becomes partially resolved with reference to every other expressible configuration of matter-spacetime in the full cosmic cycle (past and future); there remain the potential parts of the set in nested dimensions, infinity and nullity that, like the square root of minus one, cannot be expressed or resolved in our classical 4D of spacetime.

This modified Block Universe view suggests we may picture, at best, shuffled, incomplete, and deterministic strands of 'being in time' at the widest level of our cyclical cosmos. This tight collection of strands of organised existence would be like the genes, including noncoding DNA, of the human genome, which even if fully specified is also incomplete in terms of the organisational potential it houses. Both existence and the human genome are subject to restrictions or opportunities in potential imposed by the physical constraints of, and indeterminacy of, matter's arrangements. We may also picture a probabilistic and emergent 'becoming in time' at any local level of each person or each here-and-now of existence.

We are locked into experiencing reality in the cosmos' intersubjective 'here' and 'now'. I say intersubjective because existence is the borderline subset of its 'more objective' reality and its interpenetrating potential, just as our intersubjective agency is the subset of our more objective reality and our interpenetrating potential.

We are the emergent properties of our cosmos, but could we also claim that the cosmos is an extension of the intersubjective self in a way that is not solipsist, but actually mutual? We, and all existence, cannot really escape this problematic slicing up of the cosmos (in spacetime/velocity or space/position). Every time we do this imperfect slicing and dicing of the actual and potential, we reduce our experience and knowledge to further uncertainty and incompleteness. Nevertheless, this fuzzy slicing at boundaries also enables local emergence in the cosmos, and is thus the source of all systemic 'freedoms'.

The way matter achieves its finite and partial resolution is complex and fascinating at all levels of spacetime-matter coagulation. As we have seen, the resolution of the contradiction between something and nothing/zero or something and everything/infinity can be thought of in terms of a boundary problem, or in terms of dynamic, finite systems blindly seeking equilibrium or homeostasis through emergent self-organisation. That is, even the tiniest something creates a border or boundary between itself and anything else. Something (any system or arrangement) is thus always limited or bounded and thus incomplete in the sense that everything else exists separate to it, and it is only the elusive combination of something and everything else in the cosmic cycle and its shadowy harmonics or potential that represents completeness. The best way to understand reality is perhaps as a relativistic arranging or attempted balancing of real and virtual particles within an inexpressible dynamic framework or field of infinity and nullity, rather than as a finite materialism that advances in discrete steps. Perhaps the delay and infinity-nullity harmonics that fix the relativity of matter and space are the measure of matter's imperfection but space's opportunity, from the local to the cosmic level.

This also hints at the idea that a bounded something (including the cosmic cycle of serial and parallel universes itself) is also incomplete in the sense of Gödel's Incompleteness Theorems. Gödel showed us that all formal mathematical systems (including in those systems their axioms and their theorems) are incomplete, or incapable of proving all truths about all the relations and statements they contain. They are also incapable of demonstrating their own consistency because they rely on a static syntax and external axioms.

Some would say that explaining something by using unexplained axioms is no explanation at all – which would seem to undermine all of our mathematical certainty. However, I'm sure Gödel would have agreed, a good explanation is better than no explanation and a tentative, incomplete knowledge, carefully built on prior incomplete knowledge, is better than no knowledge if it is reliable in all the ways we dare to use it.

The reason why a something cannot demonstrate its own consistency is that to do so would depend not only on itself, but also on everything else (actual and potential) – just like the path in space and time of a lone photon in the Austrian Buckyball Experiment of Figure 2. However, here we run into another problem, because many would also say that an explanation that depends on everything also explains nothing. What a conundrum. It seems we are stuck with our incompleteness and our null hypotheses. All we can do is deal with our challenges ever more proficiently as we apply ourselves to learn more about local arrangements in local oases.

Now if something (material like us or immaterial like the massless photon or a mathematical theorem, it does not matter) is incomplete, then it is also subject to instability, uncertainty, or probability. That is, while a natural or artificial system may always achieve a certain level of order, there will always be a certain level of disorder, indeterminacy, or uncertainty caused by the necessary limitations, boundaries, or partial dependencies of that system.

In classical thermodynamics, we call the universe's march towards disorder entropy or decay. In quantum mechanics, we might describe the prevalence of disorder through the Uncertainty Principle, perhaps best demonstrated by the quantum fluctuations of so-called empty space. In cosmology, a bubbling, disorderly multiverse arising out of the inflationary model of the universe might be posited as a solution to the problems posed by our finite universe's beginning. In living systems, we call inherent disorder mortality and in conscious systems, we might call the struggle against disorder a search for meaning and (moral) value.

In all these systems, at all levels and in all dimensions material or immaterial, the successful local struggle and advance against disorder can be labelled examples of emergence. That is, emergence stumbles upon new levels of order that temporarily stave off the problems at or just beyond the boundary. It does this by diversifying, i.e. by forming and incorporating new and often more complex or carefully differentiating boundaries. These new, persistent boundaries are on average more congruent with the local environment than the previous version of each system. That is, some non-living systems seem to adapt to their environments just as do some living systems. This adaptation is partly explained by the arrangement of our cosmos in which the local is partly determined by the cosmic, rather than being independent of it. A persistent self-referencing system creating nested boundaries through its unstoppable iterations is likely to become cellular, networked (as in a neural network), and adaptive - just like a living system.

Now, let's get back to the topic of space, perhaps the ultimate boundary. If we, as beings conscious at the macroscopic level, could assign any direct attribute to space, then it would be its apparent independence from matter. Space seems inert in every way at our conscious level. To us, space seems a symbol of all that is background, constant, or unchangeable. Space cannot be recognised, understood, or valued until objects fill it. Likewise, silence cannot be appreciated until sounds disrupt it. To our common sense, space just logically is. However, our common sense is often misleading; we perceive an independence or inertness of space that is not real. At the quantum-mechanical and cosmic levels beyond our consciousness, space intimately moves and perhaps even precipitates 'matter defects', as in the quantum foam of the early inflationary universe. The extraordinary expansion of our early universe transformed its spatial quantum vacuum energy into matter that was not annihilated by nearby antimatter.[22]

As matter formed within the primordial fields or interactions, was dispersed, and then brought together under the five emerging interactions (strong nuclear, electromagnetic, Higgs, weak nuclear, and gravity), matter gathered and took up positions within the bounds of the Big Bang's advancing shockwave. At the same time large tracts of space between the clumps of matter opened up. That is, the space that was intimately formed in the expanding universe with matter's quickly changing configurations and bounded systems became apparently less intimately related to that matter over time, and so today, we tend to perceive space and matter as independent. However, if our universe is finite, then clearly they are not – because quite simply everything in existence that is not matter must be space and vice versa. That is, except for solutions in reality that approach inexpressible infinity or nullity, space and matter must fit almost perfectly together hand-in-glove like two pieces of the one existence jigsaw puzzle. Assuming for now no losses to an external multiverse or other dimensions, tangible matter must be close to the complement of space, except for the asymmetry or asynchronicity that divides them (or the relativity that arranges them at up to the speed of light). This is the basis of the theory of Supersymmetry. It would also be true to say that each of us must be very close to the complement of all else in existence.

The point here is that matter, space and time are all essential parts of the same temporary and dynamic fabric that we call our universe. Just as we cannot fully appreciate the pure note of a musical instrument except within a background of relative silence, so material things cannot manifest themselves without immaterial but essential space (or virtual particles) that houses the five forces or interactions (described in the following dot points) we currently recognise, and any others we may not recognise yet.[23]

To put this concept another way, just as matter and energy are interchangeable according to Einstein's famous formula $E = mc^2$, so is matter/energy and space, but not in the static sense the simple formula may suggest. In this sense, there is no such thing as empty space. In fact, scientists tell us that if space were empty then the value of matter's interacting fields and their rates of change would be exactly zero, which would violate physics' Uncertainty Principle.

According to the theory of Supersymmetry, space via its interaction wave-particles houses the complement of real matter and its array of particles, as well as many other of its own virtual particles. In the Standard Model, such wave-particles already include:

- the massless photons that are the quanta of electromagnetic interaction;
- the massive but fleeting and thus virtual W and Z bosons that mediate the weak nuclear interactions of radioactive decay;
- the massless gluons that mediate the strong nuclear interactions;
- the Higgs bosons and the Higgs field or interaction that gives many subatomic particles their mass; and
- the theoretical graviton that mediates gravity.

However, such theories as Supersymmetry and Garrett Lisi's more recent Exceptionally Simple Theory of Everything suggest there may be many more virtual particles that do not necessarily interact with matter directly, but rather pop into and out of existence in an uncertain fashion, again according to the Uncertainty Principle. For example, some scientists call the virtual particles that should correspond with the measurable effects of dark matter WIMPs or Weakly Interacting Massive Particles.

Whether these theories are correct or not in all their detail, we are still left with the idea that if this universe is indeed finite in size, then space, as the idealised difference between real particles, houses a set of complementary but virtual information, captured in its interactions, that perhaps imperfectly describes all matter in an intangible way. That is, immaterial space would seem to house interaction, movement, or behaviour codes for all matter something like DNA houses codes for all living matter. A material's unique refractive index and astronomical spectroscopy provide examples. This would further suggest that when matter reads the space codes, as chemicals read the code in DNA, all sorts of tangible lattices and other forms become possible and arise. However, these nominal space codes would work very differently to DNA codes because

DNA codes can be passively stored for later use, whereas matter's space codes cannot be – they are constantly manifesting themselves in matter's dynamic arrangements.

Matter's space codes are stored in the configurations of matter and matter supplies space with its coding. Matter and space are locked together in a continuously unfolding, asymmetrical, uncertain, and asynchronous feedback loop or system. This monistic loop is a relativistic system consisting of a kind of strange tension between infinity and nullity, the real and virtual, energy and information, the direct and indirect, the arranged and disarranged, or the very big and the very small.

Einstein's theory of General Relativity captures the interpenetrating nature of matter and space. As previously noted, the physicist John Wheeler once summarised Einstein's explanation of the relationship between matter and space in terms of matter telling space how to bend and space telling matter how to move.[24] (At a minimum, the result is a contingent explanation of gravity.) Matter tells space how to 'take position' or 'be' but space tells matter how to 'take velocity' or 'do'. Matter determines the environment, structure, coded information, or place for interacting space, but then space replies by commanding or instructing matter how to move. In this alternative sense, matter is the blind, energetic initiator and space is the enlightened follower, although the initiator-follower distinction is admittedly fuzzy and arbitrary in any cycle.

I use the emotive word enlightened because as a follower rather than initiator, space would have the advantage of experiencing the preceding structural change before providing a processual change, proto-intention and thus proto-intelligence to matter. If consciousness is what it is like to have subjective experience, then space is also proto-consciousness. In this sense, virtual space needs matter to provide its coding platform and real matter needs space to provide it with movement and direction. We might also add that wherever there is a dynamic interpenetration of the coder (or enlightened controller), and the coded (or controlled initiator), there is evidence of a kind of general relativity at work between information and energy. Note: All life systems display this kind of dynamic interpenetration, but your smart phone does not.

It seems easy to become even more anthropomorphic at this point and suggest that matter provides space with the measureable object and spacetime provides matter with drive. This is not so whimsical really, because I would suggest that the unexplained drive or struggle for survival in every living cell owes its existence to the asymmetrical and relativistic tension between matter and space, which seems to seek energy/information equilibrium in the midst of its unrelenting and inexorable movement. Life's struggle seems to be a relativistic phenomenon that simply extends the dynamic (entropic and emergent) relationship between initiating matter and reflective or proto-intentional space. Each of us is caught in the same relativistic web that exists between nullity and infinity.

General Relativity is a fundamental concept because it describes the two essential degrees of freedom in the behaviours of all systems in all conditions, whether living or non-living. The concept is in agreement with Heisenberg's Uncertainty Principle that says we may know beforehand a particle's position or a particle's velocity, but not both simultaneously. In other words, another basic feature of our emergent universe is that the direct provides limiting stability and order to the indirect, but the indirect provides delimiting dynamism and potential (or lack of stability) to the direct.

In one sense, there is a necessary separation of powers between the direct and indirect of existence, which is equivalent to the separation of powers between the delegate and the electorate found in all representational systems of government. In this analogy, the delegate is the dynamic doer, like matter/energy, whereas the electorate holds and endorses the policy or code, like space/information. Without this separation of powers, existence would vanish down an infinite (or null) rabbit hole. With this separation of powers, there is local indeterminacy and deterministic fuzziness that extends to the level of the so-called 'block' cosmos.

Further, there must be a finite delay between space's coded instruction and its run on matter's platform for the proto-intelligent self-reflection to occur, that is, for matter to supply back to space its new position after carrying out space's instruction. If there is no asynchronous asymmetry, then the relationship between the initiator and follower or coded and coder collapses and the emergence of a new reality in the next moment cannot occur. That is, without quantum delay, measured at a minimum in lots of Planck Time, the universe's 'becoming' would not be possible. Without the universe's becoming, blind or self-aware learning would be impossible. Time (or asynchronicity) is a necessary aspect and manifestation of emergence (and entropy). Without Planck delay there is no control of space over matter or embodiment of measured effect by matter reflected back to space. Time is an essential manifestation of the slightly asymmetrical arrangement of the cosmic cycle that reflects its being.

The fundamental asymmetry between nullity and infinity in full reality, and then matter and space in actual existence, gives rise to all reconfiguration and motion, and thus to the time dimension, the speed of light, mass and gravity. I hope that the confirmations of the (first) Higgs Field bring us closer to understanding the underlying relativistic and perhaps purely proto-informational balance between Infinity and Nullity, in a manner not fully divorced from some of the propositions, models and general approaches of Process Physics.[25]

It seems time in our finite cosmos is an emergent property of matter-and-space's relativistic arrangement within the quantum, cosmic boundaries of nullity and infinity, much like temperature emerges as a sufficient number of gas molecules are confined under pressure within a volume. Perhaps we could say that the unbreakable, incomplete connection between the local and cosmic (that likely explains the lone photon's path in the Buckyball Experiment of Figure 2) arose out of the seed of the first universal moment, where the difference between the local and universal (or nullity and infinity) was a maximum of just 10^{-43} seconds.

Figure 5 is a simplified representation of how our cosmic cycle may be structured in detail (just imagine the right hand side eventually joins around the back to the left hand side). It suggests our cosmic cycle has always been part of the relativistic, interdependent, and cyclical tension between infinity and nullity. Perhaps some time before the Big Bang, some kind of proto-matter and anti-proto-matter fleetingly came into being (at the left hand side of the diagram) and then self-annihilated in a series of proto-universes. We can picture our universe (in the middle of the diagram) as something following a bell shape rotated -90 degrees, with the open-end facing to the right (future). At our Big Bang, some form of dynamic mutation happened that enabled baryogenesis. The early inflationary mechanism of our universe is not important to this immediate discussion.

What is important is to see diagrammatically the relativistic nature of the infinity-something-nullity relationship that was perhaps virtual up until sometime before the Big Bang and then became real thereafter. The diagram suggests that nothing is a fleeting state of equilibrium, whereas something is a transient state of disequilibrium, menaced by entropy, but brought to higher order by emergence. Existence is transient because as it approaches infinite proportions it collapses. Nevertheless, shadowy, relativistically-constrained infinity seems to represent something like the potential junk codes from which each universe, through local dynamic mutation, may be formed.

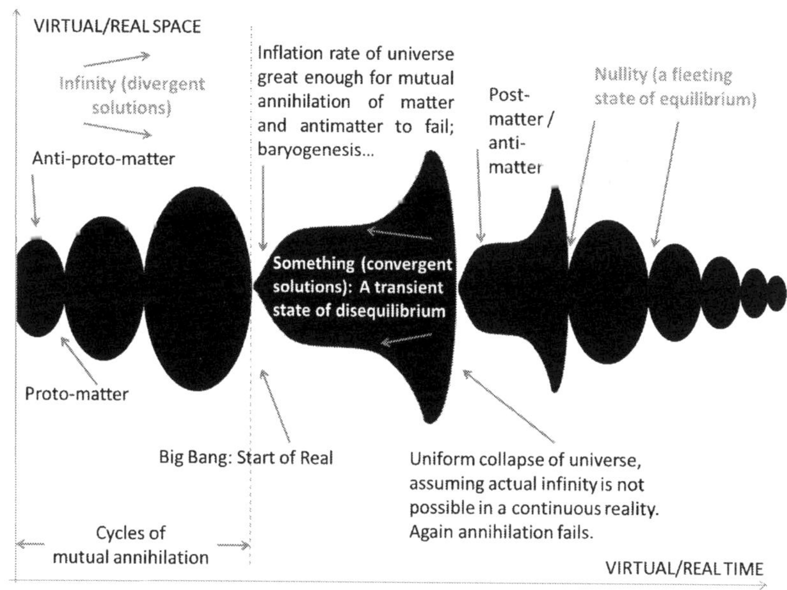

Figure 5: Our Cosmic Cycle: A relativistic struggle between Infinity and Nullity

In the big picture, our discussion so far suggests that one day our universe's transient state of disequilibrium will come to an end and necessarily be followed by the next phase in the cosmic cycle. If the idea of an accelerating expansion of the universe leading to matter and antimatter failing to mutually annihilate is feasible, then the open and curved bell-face would come to an abrupt closure and end. This would happen, in the given scenario, at the moment when the cosmic velocity was sufficient for annihilation to fail across the right-hand-side bell-face, except for minute distortions caused by any lingering clumps of matter. At this point, almost uniform gravitational collapse would occur. This would be the kernel of the next cosmic phase (the next universe) and would perhaps explain the high uniformity of the measured background radiation of our universe in the WMAP, if a similar process preceded our universe.[26] However, if this mechanism is incorrect, then another mechanism must take its place to achieve an equivalent end – an escape from the expression of an actual or uni-dependent infinity.

As the diagram suggests, the next phase in the cosmic cycle could be another phase of something real that is only a slight variation of our current universe. Eventually, mutations could mean a future universe dips again into virtual reality. We could also picture the cosmic cycle through a concept of mutually dependent sources, sinks, and their shadowy field lines, as in fluid dynamics. Alternatively, we might see the cosmic cycle's progression as something like the fractal readout (similar at various time scales) on a cardiograph monitor.

Figure 5 also suggests that in our universe, except for an underlying asymmetry that explains its initial existence, for every push, there is a pull. Nothing can stay still. Movement is amplified and attenuated from the local subatomic level to the cosmic level. Movement is likely then reflected back via background field harmonics to the local subatomic level, as in an echo chamber or pond of water, but never quite eliminated. Action or proto-intention is unavoidable, and it is powerful when organisation emerges. For us, this means influence and power are always available for our exploitation in an organising setting.

Figure 5 also suggests everything is interconnected and we can exploit those connections as we learn more about them. This is at the heart of the Emergent Method. It means we are necessary participants in our universe's local entropy and emergence, alongside every other life, system and subatomic particle (real or virtual) at all levels of organisation or disorganisation.

I now wish to explore, as it were in slow motion, the relativistic relationship between matter and space, as well as matter and time, in Figure 6 and in Figure 7. The diagrams are an attempt to conceptualise how the present local moment emerges in matter at the quantum level of space and time. This emergent becoming within the thickness of each passing moment is pictured as a cellular cycle consisting of four phases (in Figure 6, Systemic Separating, Entropic Degenerating, Negentropic Arranging, then, Systemic Incorporating). By the end of each cycle or orbit, reality is locally changed, yet remains consistent at a rate up to the speed of light.

The left sides in Figure 6 and Figure 7 seem more influenced by Nullity and the right sides by Infinity.

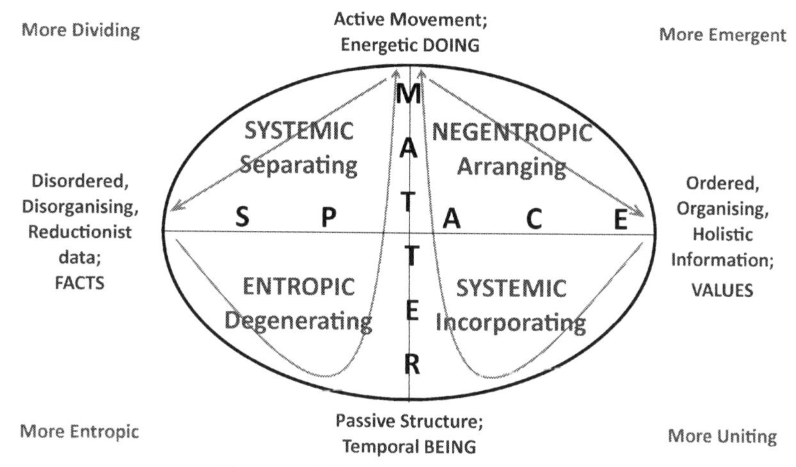

Figure 6: The Becoming, in Planck Length

Both diagrams show the essential attribute of matter as having dynamism, movement, or energy, just as Einstein and Heisenberg would suggest. Thus, matter can vary from having relatively passive structure to active movement, or fleeting 'being' to energetic 'doing'. Matter cannot be absolutely passive or infinitely energetic.

Space is assigned the essential attribute of the mover, framer or codifier of matter. We now know that it gives matter its mass through the Higgs Field and by other means. It can, via its forces or interactions, either push matter apart or pull matter together. Space can therefore oversee disorder or order in its interactions or relationships with matter. This relationship with matter can produce arrangements that vary from relatively disordered/fragmented to highly ordered, or classically reductionist to ideally holistic. Space can neither fully disorder nor fully order matter in any attempted slice of time.

Could it be that the relativistic relationship between matter and space is the manifestation of the inexpressible and proto-informational tension between the Infinity and Nullity at a sub-quantum level? Could it be that the theorised 11 dimensions of Superstring Theory (10 spatial dimensions and 1 time dimension) really only create a 'fit' for this underlying tension in our particular universe, rather than directly address it? Just as the shape of a French horn explains the note it produces, so Superstring Theory hopes to explain our universe, but what explains the French horn itself, that is, what explains the 11-dimension bundle itself? What explains the seven nested, inexpressible dimensions needed by Superstring Theory to explain matter and its interactions in our universe? Could there be infinitely more potential dimensions than 11 in our universe that we can't reach yet? Could the answer ultimately be tied to the inexpressible and proto-informational field that exists between Infinity and Nullity?

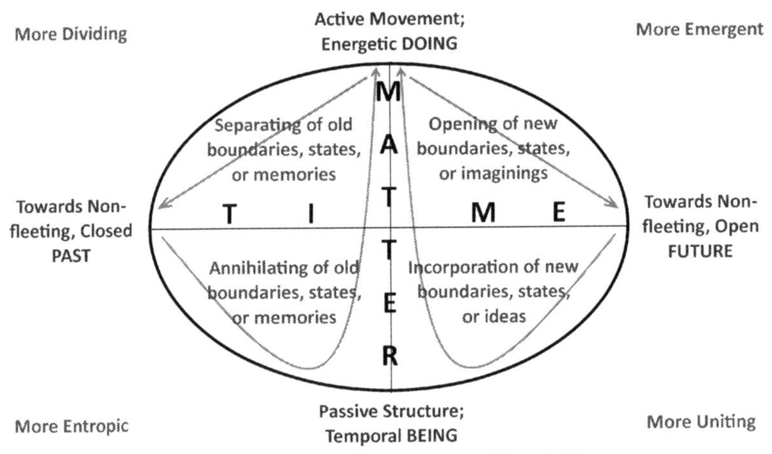

Figure 7: The Becoming, in Planck Time

Time emerges locally from the relativistic, uncertain and dynamic relationship between matter and space. Anthropomorphic time is assigned the attribute of advancing locally from future to past regardless of the observer's current location in time. There is an assigned arrow of time because we assign meaning and value to both negentropy and entropy. We do not tend to see negentropy as an advancement and entropy as a retreat, or both as a relativistic block. We see all as meaningful development or progress rather than partly meaningful and partly meaningless. The observations that we cannot unscramble an egg or un-grow an oak tree back to its seed also promotes this view. Whether this is a correct way to view all change is a little irrelevant; it is certainly convenient for intentional lifeforms. (It also assists the mind in the creation of the illusion of independence and thus the self, which I will discuss later).

Similarly, in a local sense, we could think of space as spreading from ahead to behind. In this 'arrow of spacetime' view, the future or ahead is seen as open, the past or behind as closed, and the local, intersubjective 'now' or 'here' is seen as the observation of a fleeting 'becoming'. It is an interesting mental exercise, isn't it, to think this way? It seems wrong to think of behind as closed, or the past as not closed like any other option in space. What does this say about our generalising, common-sense notions of spacetime and the possibility of a fractal block cosmos? Couldn't we remove the 'arrow of spacetime' view to suggest that just as a point behind is equivalent to a point ahead, so is local past roughly equivalent to local future in a cosmic cycle? Couldn't we also say that time is caught up in many local loops, at many emergent levels, of which the cosmic cycle is just one? That is, couldn't we say everything happens in its own time?

It is interesting to note that whether we talk of time or space (or any of its interactions), the cyclical or dynamic becoming in each of the four quantum quadrants of Figure 6 and Figure 7 are similar. That is, there is a (1) separating and (2) breaking down of old states, with their boundaries, and (3) an emergent opening up and then (4) incorporation of new states into the given environment. We could combine space and time into the one spacetime concept without loss of meaning in the four quadrants of 'becoming'.

In all of this probabilistic activity at the quantum level, there is a tentatively assumed conservation of energy and information across the quanta and perhaps 11 dimensions of local reality.

We will consider the ramifications of this model in terms of human nature later, but I do want to stress that we seek to understand matter, space and time so we can have a reliable framework for understanding life, the body, the mind, perception, consciousness and self. We do this so that we each may answer more knowledgably the very personal and big questions, 'What do I love to do?' and 'What ought I to do?'

For now, we will begin with a concept of a philosophy based on moment-by-moment reality that specifically recognises the undeniable role of space in everything from the subatomic to the universal. Space by definition is beyond or outside the material and tangible yet clearly takes up much more volume than the material or tangible in our universe and is indispensable to its emergent character at all levels. The immaterial and intangible, in the form of space, apparently made up of its virtual particles in the form of either dark matter or dark energy, is a little disputed fact of our universe – and by volume and even mass (even if only fleeting), the overwhelmingly biggest fact of our universe if true. Many scientists believe atoms make up only about 5% of our universe, slightly controversial dark matter about 23%, and little known dark energy the remaining 72%. Space seems to force us to accept that the beyond-physical really does exist. To me, this further supports the ideas that everything emerges from the imperfect balance of reality, something like everything within and outside the Mandelbrot Set emerges from the iterations of its fractal formula.

If physical matter is a direct fact of our universe, (i.e. it exists as tangible things), then space is an indirect fact. That is, space is a fact that tangible things do not directly express except through their movements or behaviours. Space is logically and informationally indispensable to matter and its perceptible styles. The apparent void has an indirect effect, as in gravity, electromagnetic radiation, and the other interactions relating to the inverse square of the distance between two objects in space. Space is an apparently empirical but actually a transient and informational fact. It is not a tangible and direct fact, at least not at the macroscopic level.

Every object subsumes a space within it just as every musical sound subsumes a silence within its frequencies. This leads us to an important question. What is matter? Einstein's equation at low velocities gives us the simple answer – every moving mass is a store of energy ($E = mc^2$). It is a temporal arrangement and filter of space's fleeting 'particles', just as white light split in a prism reveals its inherent frequencies.

After the confirmation of the existence of the Higgs boson in 2013, we now know space and matter are bound together (like silence and music) in a monistic view of existence. Mistaken duality causes us to underestimate the emergent potential of space, where transient something and fleeting nothing (or dynamic energy and organised arrangement, or fact and value at the level of mind/body) collide, sometimes mutually annihilate, but sometimes rearrange themselves and coalesce. Mistaken duality causes us to assume supernatural alternatives and miss the power of relativity and the limited freedom quantum indeterminism provides.

It seems the best way to understand space is to see it as consisting of partially discrete, yet virtual and complementary, subatomic states capable of wave-particle behaviour. At the cutting edge of knowledge in particle physics, matter and space seem more and more alike or less and less distinguishable from each other in their interpenetrating wave-particle union.[27] Treating all of space's virtual particles and matter's real particles as belonging to a unified and fully interpenetrating group is the basis of the Exceptionally Simple Theory of Everything and other ToEs. Some scientists would go as far as to argue that the universe is likely made of only one kind of underlying, gooey, and active stuff called strings. These strings exhibit matter-like reductionist behaviour in some arrangements and space-like or wave-like holistic behaviour in others. I picture open and closed strings as the proto-informational linkages between inexpressible infinity and nullity.

In summary, existence includes matter and its inertia displaced by its quantum-mechanical, self-referencing and iterative dance with space through time. Space, or the classically immaterial, unfolding from the Big Bang's shockwave and the universe's accelerating expansion, frames and intimately supports everything, including us, but seems to have only an indirect dynamic effect on anything. It is not easy to understand the nature of space and its local interactions with matter and antimatter, either scientifically or philosophically, but we do understand more than in the past. We understand enough to embrace the concept of something emerging from relatively or anthropomorphically nothing, even if the reason for anything rather than nothing, or the separation of powers between infinity and nullity so as to eliminate their mutual annihilation, remains a relativistic mystery.

The indirect logical interactions our scientists describe, from strong nuclear to gravitational, do not exist apart from space. Their local sum across the universe, along with the virtual particles shuffling in and out of momentary existence, is expanding space itself. Spacetime is not tangible. It includes the informational component of matter's existence and participates in matter's precipitation and impetus.

As noted earlier, the so-called laws of our universe tend to suggest that such laws exist in an abstract mathematical place separate from existence, but they don't and they can't. Spacetime's logic interacts locally everywhere with matter with algorithmic, asymmetric, self-referencing and self-consistent feedback. There is no system of laws outside of messy reality. Matter upholds the laws themselves, as surely as matter is describable in terms of those laws. That is, the laws of our universe are matter- or platform- dependent and self-contained within this reality. These laws are human abstractions of observed patterns or deterministic codes locally embedded in spacetime by matter. Conversely, we might say that matter is free of law except in the organised yet platform-independent dynamic arrangements it forms in spacetime. This is not some kind of great mystery. It is our anthropomorphism that separates meaningful causes from their relatively meaningless antecedent effects. It is our anthropomorphic tendencies that demands laws be attached to all that we find meaningful. Our own irrepressible movements will always find meaning and value in what we perceive and thus drive the development of all our purposes and laws.

In other words, existence is not something conforming to a centrally implemented blueprint, set of laws, or intelligent design as classical Newtonian physics and ancient Platonists tend to characterise it. Rather, existence emerges locally at every level logically and contemporaneously with the dynamic interactions of expanding space and matter. In these local interactions, we note many recurring patterns or styles across vast populations of subatomic particles, but always associated with subtle and uncertain aberrations driven by strange quantum-mechanical effects, possibly inexpressible extra dimensions, non-physical limits of local information and timing (or asynchronicity) and physical limits of local size, arrangement, cosmological self-containment, etc.

Just as the common gene codes of every cell in embryonic development largely determine the role populations of stem cells will play locally, so the local spacetime system or codes and local systemic environment determines the role of populations of real subatomic particles. Just as genes are expressed with help from the local chemical environment and occasional adaptive mutation, so the local spacetime codes are expressed with a level of local quantum-mechanical uncertainty in matter. However, it would give a false impression to suggest that this spacetime coding exists independently from matter; it does not. Neither do genetic codes exist separately from living matter. Matter embeds both coding systems and both emerge from matter, as surely as matter's properties or styles emerge from their codes. That is, the locally deterministic software is inseparable from the locally probabilistic hardware; the two interpenetrate dynamically. This is the core nature of General Relativity.

As an interesting aside, it seems we will never be able to develop intelligent robots until we allow their software and hardware designs to recursively interact at their lowest levels. However, at that point we will no longer be able to claim the agency for what they become, because the agency would revert to the local environment (and its adaptive mechanism of natural selection).

The idea that space is made up of discrete particles further supports the idea that space, much like DNA, can be seen as containing discrete codes that can be blindly extracted by matter's platforms to gain information about the local environment and thus to exploit that environment. Further, matter's platforms affect spacetime and the information it holds or represents just as surely as spacetime brings dynamism to matter's platforms. Information is not just patterns of extracted data we use; as we hold it, it also changes us physically. That is, information and matter's platforms are both emergent within the one feedback loop or arrangement. This concept is essential to understanding life and human consciousness. We have to rid ourselves of the idea of space's independence, or anything else's independence, if we are to understand how human perception works and how the brain/body moves with unified intention within its environment.

Emergence can cause systems to arise whose focus is static arrangements of matter and properties such as shape, shelter and beauty, or it can cause systems to arise whose focus is dynamic arrangements of matter, such as flight in birdlife. Flight is an indirect property enjoyed by a bird because it uses its wings.

Another example of a complex natural system whose focus is interaction rather than matter is consciousness in organisms. Note: Some writers define the brain as the 'standard' physical organ, the mind as the configured brain in terms of the neuronal and cortical arrangements unique to each individual, and consciousness as the subjective awareness an individual experiences as a result of the mind/body's activities. I don't object to this, but have a simpler approach. I simply define the brain as the physical and direct organ however uniquely configured in the individual, and the mind as the non-physical and indirect organ with which we cogitate and experience consciousness. By this definition, it seems true, due to their asymmetry and asynchronicity, that configurations of matter drive minds to evolve in brains just as much as configurations of space drive brains to evolve in minds. That is, the mind/brain is an extension of the relativistic relationship between space and matter.

Here then we hint at another basic feature of our emergent universe: The idea of related systems coming together to form loops of matter or space within other loops of matter or space at various emergent levels. It seems that the cellular networking of organisms, or the neural networking of the brain, have their bases in the cellular and nested node-and-link or web-like nature of matter-spacetime at all levels. When we have this view of ***nested direct and indirect loops or orbits*** within nature, we get much closer to understanding human consciousness and its emergence within nature's beautiful web. This realisation is probably what urged Douglas Hofstadter, professor of cognitive science, to write his ground-breaking book, *"I Am a Strange Loop"*.[28] However, perhaps Hofstadter did not go far enough. Perhaps the whole cosmic cycle must be a fractal Strange Loop or proto-neural network, as is perhaps suggested by the evolving De Broglie-Bohm Theory of quantum interference, for his claim to be true.

Freedom

This discussion of space and the freedoms it grants us seems to present us with a puzzle in terms of human free will. Does the cosmic cycle determine everything we do? Far more deeply, are the realms of science and morality just different games we play with different rules (limited cause-and-effect and freewill respectively) from the same deck of cards, as suggested by cognitive scientist Steven Pinker?[29] Alternatively, can we link together nature, life and morality via the concept of locally emergent systems? These are vital questions in terms of our selection of a naturalistic philosophy or worldview.

As noted earlier, the fuzzy incompleteness of the total is a function of the local, and the indeterminism of the local is a function of the total. That is, the fractal block cosmos does not determine everything we are or do at the local level. This false idea is a hang-over of classical thinking that suggests a uni-dependent rather than relativistic universe. Put another way, where classical thinking might suggest a cosmic cycle pictured as a perfectly deterministic circle, relativity would suggest a cosmic cycle more akin to the fuzzy outline of the Mandelbrot Set as well as everything converging within it and diverging outside it.

Does this mean a level of free choice is possible within the to-and-fro of matter and space? Are our biorhythms and cycles of consciousness that contribute to wilful action more like the circle or more like the Mandelbrot Set? Is free will real, or does it only appear to be real? Statisticians would typically suggest we treat the local environment, including our free will, as subject to varying levels of finite probability. That is, for our local human purposes, nothing is purely determinate or purely indeterminate. This intimates free will is not real to the point of total independence from all around it, but neither does it lack altogether the possibility to express local degrees of freedom; free will is likely to be fuzzy and emergent just like everything else; it is likely to stretch the classical laws of thought (identity, non-contradiction, and the excluded middle).

Rather than suggesting that the 'ought' can be derived from the 'is', or the 'ought' is a wholly separate game from the 'is', I reckon that the 'ought' or morality, much like bird flight, necessarily emerges from within the 'is' or nature. I also suggest that because we live in a self-contained cosmos, where everything within it necessarily bumps up against everything else, assigned meanings and morality in turn also affect the emergence of the 'natural' environment. This is of central importance to an understanding of human consciousness. This also means the natural environment is not as objective as we might sometimes assume.

Is free-will something like a biased and illusory perception of the probabilistic cause-and-effect that really happens at the lowest levels of wave/particle coagulation? Alternatively, is free will actually making an impact at the microscopic level of molecules involved in the brain's processes? Can both options be possible? As Douglas Hofstadter asks in his book, "Who shoves whom around inside the cranium?"[30]

The emergent properties of a relativistic, self-contained, and fractal cosmos means both options can be possible asynchronously and asymmetrically. Emergence makes possible islands or oases of free choice, just as it make possible degrees of freedom (or valued organisation) in any system. That is, a level of mental causation and free will is possible very much because the universe is not like a carefully designed assembly line with inputs at one end and outputs at the other. Put simply, I reject an independently reductionist worldview broken down into misleading clicks of Planck Time, or an independently holistic worldview that is ignorant of its indeterminate

Planck Length basis and trigger. I am in favour of a relativistic worldview in which, subject to delay introduced by the speed of light, the cosmic total constrains the local and vice versa. That is, proto-intention is communally probabilistic rather than autocratically determined. Relativity enables probabilities, which in turn, enable unique properties or freedoms to emerge within the demographics of dynamically organised populations of particles. This means 'free' will is an emergent manifestation of the environmental feedback system that produces and harbours it. This means the question Hofstadter asks about who is boss inside my head is a slightly misleading one, because it suggests a bivalent answer, mind or matter, whereas the multivalent and interpenetrating mind, body and environment are all involved in the answer.

The version of free choice to which I ascribe emerges from orderly systems due to relativistic interaction, that is, because the universe is not uni-dependent. It can do this very much because the universe is in the continuous habit of bootstrapping itself from within its unstable and changing local boundaries. It can do this because there is a delay between action, reaction and counter-action. It can do this because of interpenetration at up to the speed of light. However, we should not see this as something magical or as something for nothing. Rather, we should see fractal feedback and the weak emergence of everything at nominally supervening levels, including the emergence of matter and free will, as normal parts of natural processes within our complex and self-contained cosmos. Our local and emergent intersubjectivity (or indeterminacy and incompleteness) is the oasis from which our free will is able to bootstrap itself. That is, an essential feature of free will is that it is an aspect of the intersubjective, incomplete, or locally-constrained but environmentally-embedded, 'first-person self'. That environment includes the 'third-person selves' of others.

To put the idea of emergent personal freedom another way, each temporal system has normal operating conditions that if maintained, should allow the system to operate and endure. Outside these operating limits, the cybernetic system breaks down. That is, the limits of the system impose a restriction outside those limits but also grant a certain kernel or degree of freedom to the system within those oasis-like or cellular limits. Thus, free will is only free to the extent it remains within the operating limits granted by the systems and environment that produce it. Moreover, emergence enables the freedoms or abilities of a system to increase and diversify. Our species' history and future is very much that of the universe – we started out as very closed systems with few freedoms, but as we expand our claims against the limits at our borders and succeed in so doing, we become more open and close to the ideals of infinity and unlimited freedom. The cybernetic oasis grows through our successful free enterprises.

There's an old football saying in Australia: "The only place you can lose a game is out there on the field, but the only place you can win the game is in your heart!" I think this wonderfully juxtaposes the relationship between the macroscopic social 'third person' level and the microscopic personal 'first person' level. It captures an essence of the Emergent Method quite well.

It seems that genetic evolution may explain the emergence of our animal drives (which are systemic extensions of the local inorganic proto-intentions towards an imperfect balance between infinity and nullity), but not directly our feelings of free will. The will is a concept that arises because through consciousness we can observe our evolved drives and systemic freedoms. Further, our emerging conscious observations affect the ongoing evolution of those animal drives; they can't do otherwise in a self-contained system. The name we give to these naturally or artificially selected and consciously observed human drives or proto-intentions arising from the microscopic world within is will.

As we will discuss in more detail later, the notion of free will arises from a cultural and psychological structure or set of loops in the macroscopic environment engrafted onto the underlying genetic loop of human behaviour. Psychosocial claims made by us fallible moral agents (within psychosocial loops) can reinforce the idea of free will regardless of, or somewhat separately from, or delayed with respect to, underlying naturalistic or genetic veracity. That is, if the environment including others in society acquiesce to our claims, and if our mountains then move because of those claims, then who is to say those fuzzy claims are untrue? An interesting example of a system of ratified claims is the belief and organisation mustered by the Pharaohs to build the Egyptian pyramids. Psychosocial reality within cultural constraints, so necessary for the development of free will, thus emerges from subconscious underpinnings followed by biosocial acquiescence or affirmation.

As intersubjective observations of drives that can affect future actions, it is a necessary feature of our limited free will that it comes, and is demonstrable, after the fact. That is, free will preserves a fuzzy determinism while it emerges over time; it slips between the cracks of psychosocial action and environmental reaction (or the fractal dimensions of multi-dependent causes-and-effect). To put this idea another way, each self-aware claim is transformed into a normative or positive legitimacy by its worldly success. The explicit worldly value of our claims is the real basis and authority of our free-will and moral agency. Thus, free-will requires a transformation from unique, self-aware claims to self-aware success through a kind of environmental acquiescence or affirmation. Free will is shown to be free only when it transcends the claims of the claimant. The environment in all its forms is thus our necessary partner in our emergent agency. Our interrelationship with the environment at various levels and via various emergent loops enables our free-will to convert from the virtual to the real.

We don't have an independent freedom from the world, but rather an interdependent and relativistic freedom with the world. In other words, our freedom is not ours alone; we share it with the entire cosmos in our local here-and-now. We participate in a separation of powers with/from our environment that defines our systems of free will. We offer to the environment free-will traffic, but the environment is the traffic cop that either allows or disallows our potential free will according to its codes or homeostatic limits. In turn, these limits motivate and influence our free will choices and our free will choices shape the emerging limits of the free will system the local environment encodes. Does this make free will not truly ours and not truly free? I will leave you to decide, but my preference is that we each claim our own unique free will, or the free will of our node in the network, as ours to share.

I prefer to think of our 'free' will as 'fractal' will. That is, it exists in, or makes use of, dimensions or layers between and amongst the four commonly named. It forms patterns of self-similarity between us and our environments at various levels (unconscious to conscious) – and this is what makes it efficacious right down to (or more correctly, with) the subatomic level. Put another way, psychosocial free will systems supervene all the other biological control systems, although the philosophical concept of supervenience hides the complex feedback and feed-forward mechanisms and interdependencies involved here. Perhaps one the greatest benefits provided by The Emergent Method is that it once-and-for-all grounds the concept of free will within our relativistic and fractal web of full reality.

However, eventually, if we humans continue to make selfishly unsustainable claims on our environment and ourselves (our systemic abilities and freedoms) then we will perish. Our observed and claimed will is not fully free. I believe our freedoms are real, but they are found only inside limited, temporal and emergent oases within a vast 'comparative wilderness'. Our freedoms are relative, emergent and temporal, even though we may wrongly hold that they are absolute in terms of human ideals and rights.

Fundamentally, nature and space is blind, i.e. advancing through trial-and-error and not all-seeing. We are a lot like nature, only learning of any possible deeper meaning or purpose in hindsight after we act on our beliefs and observe the outcomes. This means that to us, there is no meaning or value until we realise, claim or accept one after the fact.

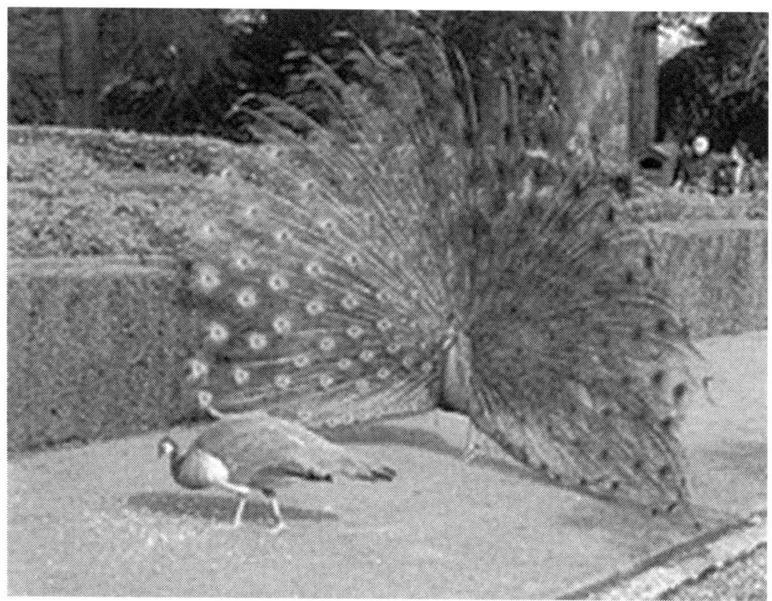

Figure 8: Instinctive two-way communication, meaning, and value

Our eco-psychosocial environments continually suggest value, meaning, and movement to us, which subconsciously we either take on or reject. For example, in Figure 8 the possible female mate of the strutting male peacock finds blind meaning, value and intention in his sexual signalling that calls for an active response.

The beating of the chest and roars of the alpha male also signal to the rest of the troop a clear meaning - who is boss. In each case, instincts have built in the appropriate shortcut signalling, interpreting, and prompting of behaviour. That is, the dynamic animal world is full of coded and self-evolved meanings and value without even seeking or articulating them – because shortcut significations percolating up from irrepressible movements in matter and style in spacetime promote the peacock's or alpha male's proficiency in reality's arrangements.

We humans are largely no different to other animals at the subconscious level. It is just that we can often understand each other's eco-psychosocial claims more consciously and often see the meanings, beliefs, and values in those appeals more reflectively. This is our gift.

Another way to think of self and free will is not as dubious claims mutually legitimised after the fact, but as gifts. Not gifts from an omnipresent, omnipotent and omniscient god, but gifts from evolved consciousness-plus-instincts, whose beginnings is in the emerging real-and-virtual particles of matter and space. In which case the virtual particles of the mind leading to indirect consciousness are not omnipresent - they share their presence with the real particles of the direct brain and body. Nor are the virtual particles omnipotent - their potency grows only in tandem with the emerging arrangements of the brain's real particles. Finally, they are not omniscient either – the virtual particles of the mind are only explicitly learning of themselves through incarnate consciousness. Perhaps we could argue that the cosmic cycle or fractal block cosmos is omnipresent, omnipotent and omniscient, which being self-contained, it surely is in a fuzzy sense, but it lacks an internal modelling of itself. That is, the cosmos must wait on emergent and incarnate consciousness to make it self-aware (and perhaps self-deluded).

When we think of gifts, we think of something freely given, i.e. without expectation of returns. If a gift is given with some attached expectation of reciprocity then it seems like it is no longer a gift, but rather some kind of social contract or game. By this definition, some might agree that there are no gifts in this world - there are no free lunches.

Maybe this view is correct: Consciousness, self-awareness, free will, knowledge, meaning, and morality are not given to us as gifts, but as parts of a social contract or a broader naturalistic contract, we share with all members of our deep ecology – past, present, and future.[31] If self, free will, or moral agency are parts of an unwritten contract rather than a gift or claim-after-the-fact then they carry responsibilities as well as rights. Sometimes we are too quick to implement our game-theory strategies or claim our rights but too slow to bear our responsibilities for outcomes under a contract.

Maybe it's time to think we, as self-aware agents, cannot freely enter into and leave the fraternity with our deep ecology. Maybe we should realise that we, and our intersubjective experiences of free will, are locked into a fraternity that has reciprocating obligations attached to it. Protection of our environment's biodiversity and resilience, and the acceptance of our position or social fabric within this deep ecology, is perhaps just as much our duty as our privilege.

A little less anthropomorphic, our fitness for survival is perhaps equivalent to our interoperable compatibility with Gaia's broad emerging systems, including at the level of the intersubjective self. Gaia is a concept that treats all of Earth as a single organism with many living and non-living components, much like we might treat a bird and its nest, a horse and its rider, or a car and its driver, as a single entity.[32]

What about if we were to treat all these tentative observations of fuzzy limits, spacetime, and matter as a philosophical guide? By fuzzy limits, I mean unexpressed nullity, infinity, nested dimensions, and imaginary numbers in nature's systems. What would these observations suggest about our human dance with existence and its potential? What would this interpenetrating, probabilistic, and quantum-mechanical or asymmetrical, iterative, and emergent philosophy have to say about our consciousness and our moral values? What would this philosophy suggest about an optimal or at least naturalistic and positive moral code that seeks to maximise human wellbeing?

I suggest the answer will include the idea that we desperately need an emergent and more inclusive definition of the oasis of intersubjective self. Just as a mountain may be separately identified, but never independently of its mountain-range environment, so we should view the node of self. This in itself is a radical departure from typical Western thinking.

Axioms

Many philosophies do not start with our limited knowledge of reality. That is, many are sceptical – in that they claim any certainty or knowledge is impossible. Alternatively, if such a certain claim would itself represent a paradox, they at least advocate a behaviour not inconsistent with the claim that reality is always uncertain, unknowable or only ever intersubjectively known. The tendency is to suggest that because the universe's characteristics impart no infinitely precise knowledge, each of us may arbitrarily seek, create, or assign value to knowledge. While there is a hint of veracity here, knowledge and its accrual are not random or things that can be created from nothing. Our accumulated judged truths, even if incomplete, tentative, and emergent, are hard-won-by.

It seems incredibly ironic that our elaborate level of consciousness as a species, which enables us to openly receive lessons from the universe, now enables us to make the choice not to listen, know, judge or tentatively trust in such lessons. We, perhaps like rebellious teenagers, want to rebel against our own evolved nature as rational and volitional, as well as emotional and social, beings. Yet discovering incomplete knowledge and making rational choices based on that statistically-limited knowledge are important ways our species, so much more than any other animal species, survives and flourishes, or finds success and happiness. The controversial philosopher, Ayn Rand, seems to take this view a step further. She proclaimed,

> *"My philosophy [objectivism], in essence, is the concept of man as a heroic being, with his own happiness as the moral purpose of his life, with productive achievement as his noblest activity, and reason as his only absolute".*[33]

I agree with the sentiment but am not so sure we hold any absolutes: Reason and its pursuit, like everything else, is relative and emergent rather than absolute.

Other philosophies assert the dualistic transcendence of existence, at least by some of those that have departed this life, which presents a conundrum to us, the living stuck on this side of the veil across such an asserted but unknowable reality. To be fair, the living may have opportunity to pierce that veil if, indeed, it truly exists but the method is always an intersubjective one via intuition or revelation, unanchored in tentative reality. We will discuss the fundamental problems of all demanded axiomatic life-stances in a later chapter.

Both the sceptics and mystics, to use Ayn Rand's terms, deny the ultimate validity of a tentatively objective reality of limited certainty to us, the living on this side of any veil. Emergentism is a philosophy that embraces the idea of a somewhat objective reality but is also ready to accept that those things perceived today as tangible and direct are not the full extent of reality. Reality is an interpenetration of the real, the virtual, and their inexpressible limits, perhaps best demonstrated by the wave-particle model of light. This concept of the interlocked real and virtual extends to the study and understanding of the mind/body model of humankind, human sensorimotor perception, consciousness, and wellbeing. In this wider sense, reality encompasses the so-called objective and the intersubjective. We will discuss this curious point more fully in Part 2.

A few philosophies do claim to start and rest in a knowable reality – some not very nice ones because they have sometimes led their fallible adherents to despair or suicide, perhaps in the face of a perceived overwhelming guilt, responsibility or contradiction as moral agents (variants of existentialism, surrealism, nihilism, absurdism, etc.) rather than assurance and hope. Nevertheless, if I were to name an antecedent philosophy to the kind of emergentism espoused here, then existentialism, especially as described by Maurice Merleau-Ponty in his Phenomenology of Perception would be one very important contributor. Merleau-Ponty was perhaps unique in not just questioning the claims of classical science and philosophy as they applied to theories of mind at the time (~1945), but also in presenting a viable alternative that I gladly receive and build upon.

A philosophy that does begin with a classical axiom of objective existence and is hopeful of the heroic and rational in humanity is Ayn Rand's objectivism. However, it seems to have little room for the indirect and inexpressible parts of reality so important to emergentism. Later we will discuss why this might be the case.

The following pages will also demonstrate that emergentism, while considering carefully the interpenetration of direct and indirect reality, does not have to be unrealistic or appeal to things outside knowable reality, or be spiritually transcendental or religious. Rather, the study of emergentism suggests that Gaia can only ever advance towards an approximate divinity in nature's vast interpenetrating relationships. We can slowly add value only through natural and artificial emergence. Further, the cosmic cycle is incapable of slicing itself up into deterministic modes of self-reflection, which suggests the cosmic cycle is incapable of panpsychism. That is, temporal deterministic loops of coding must emerge locally in a relativistic cosmos.

Emergentism does not have to be classically realist either. Yet if emergentism as a discussed idea doesn't stand up in scientific reality, for example if it doesn't agree with accepted scientific evidence, then the philosophy or its logic perhaps fails in its current form. However, it is also open to new findings as they arise. The philosophy presented here is at least partially testable against knowable reality. It leads to many testable hypotheses, as any valid theory must. Testability and falsifiability means it is not a con and does not demand that you take it on authority, or demand that you have to wait until you die to confirm its espoused understandings. Testability means this philosophy is useful to the living rather than being a conundrum and can evolve with the experience and knowledge of the living. Dogmatic axiomatic claims such as 'God is', 'God is not', or 'God is Uncertain' do not grant us testability against reality.

However, having said this, testability does rely on reason and on perceiving separateness and differences. A mystical monism rather than materialistic monism might regard our 'perceived-differences' view of the universe as a function of our conscious and calculating minds, always looking for patterns of similarity and difference, rather than a true reflection of reality. Nature's essential feature is its almost complete continuity and connectedness, obviating its need for human-like serial and compartmentalised calculations. Such a mystical view may only see unity and connectedness within here and now, as part of the cosmic cycle. It may not see negative as opposite to positive. Rather it may see all as positive or perhaps neutral and the perception of negative as simply the trap or outcome of the difference-finding view.

While recognising this idea and even having some sympathy for a more holistic view of reality, followers of emergentism also seek to recognise the sufficiency of a monism devoid of innate consciousness within our human experience and limitations. That is, both the reductionist view, otherwise known as upward or normal causality, and the holistic view, otherwise known as downward causality, can lack an emergent dynamism and human efficacy. Local reality, perhaps better described in terms of an emergent causality, always seems to find a way to bootstrap itself from within its differing resources, no matter what worldview or framework we surmise.

Perhaps it might be instructive to consider the axioms of objectivism. As per Gödel's Incompleteness Theorems briefly mentioned earlier, axioms are interesting in terms of understanding the limitations of a humanly devised system, even if it is a system of thought. Wikipedia's Objectivism (Ayn Rand) article reported in 2010 that Rand's philosophy starts with three axioms with respect to existence, identity and consciousness.

An axiom is something that should not be easy to reject based on quasi-objective knowledge in which we separate, compare and distinguish one thing from another, but we can reject, perhaps unfairly, based on a possible or assumed explanation that we cannot currently verify. The basis of such a rejection may lie behind a veil draped across a reality unknowable while we live.

Objectivism's axiom of existence (i.e. existence must be taken as a given in its philosophy) could also be rejected on the basis of an unsuperstitious monism that holds, on an emergent basis, that existence is not a given, but continually emerges from a past that did not contain matter, space or time as we know them today. Objectivism's so-called axiom of irreducible and unexplainable existence can thus appear to stand but actually be faulty if a truer axiom underpins it, such as, 'existence emerges', or 'God alone', immovable and pre-existent as the great "I Am that I Am"[34], or transcendental Consciousness pre-existent yet subject to Buddhism's continual "Dependent Arising"[35].

> *"To be is to be "an entity of a specific nature made of specific attributes." That which has no attributes does not and cannot exist. Hence, the axiom of identity: a thing is what it is. Whereas "existence exists" pertains to existence itself, the law of identity pertains to the nature of an object as being necessarily distinct from other objects … As Rand wrote, "A leaf ... cannot be all red and green at the same time, it cannot freeze and burn at the same time. A is A."*[36]

Again, one-dimensional strings from which matter-space-time are theorised to emerge, or even our dynamic three dimensions of existence in an expanding universe, could defy this 'identity axiom'. The quantum-mechanical effect in which a particle is said to indeterminately occupy space could also contradict this identity axiom. To be generous, at the classical and anthropic level, we could often acquiesce to the claim that existence exists and "A" is "A".

"Rand held that since one is able to perceive something that exists, one's consciousness must exist, "consciousness being the faculty of perceiving that which exists".[37] She saw the attempt by mystics to assert the primacy of consciousness (e.g. as in theism or panpsychism) over the primacy of existence in which all things exist as this or that, as an attempt to escape the responsibility of rational cognition and what she called the absolutism of reality.[38]

Again, I would agree with the sentiment but reject its absolutism because it is difficult to talk of the interdependent fuzzy logic of infinity and nullity from which an uncertain or incomplete existence theoretically emerges in such terms. I also detect a contradiction in the language used here. That is, if we propose that existence exists by axiom, then it is by common sense or intersubjective fiat, but not something ultimately deduced through rational cognition.

Objectivism seems to be an atheistic philosophy not because of any rationally held evidence but because of its axioms, whereas emergentists do not rely on axioms to take tentative life-stances. Again, the life-stance choice is left to you the reader – not because I do not tentatively claim certain life-stances, I do – but because it is your responsibility to choose a life-stance for yourself (- your node in the web of existence).

Nor do emergentists presuppose a supernatural or transcendental existence beyond the veil. Of course, objectivism, fully consistent with its first axiom, would not acknowledge that such a veil exists. Unfortunately, the consistency of an axiomatic assumption or the observations upon which it is based and temporarily confirmed does not ensure its correctness in all circumstances. Western Australian black swans exist in a world otherwise overwhelmed by white ones.

Quite simply, and Gödel's Incompleteness Theorems[39] aside, any system built upon inductive axioms is deniable by those who choose to reject those axioms. Thus in an important way, objectivism's axioms provide a basis to its philosophy that itself can be thought to act as a veil over that which may lie within that axiomatic basis. Emergentists argue that we need to take life-stances that are tentative and processual, and not just static/axiomatic. Further, our life-stances should be interdependent rather than uni-dependent.

The following of emergentism challenges you to consider one, the efficacy of your beliefs in terms of your wellbeing and self-actualisation, and two, the evidence from reality of a certain life-stance. As an emergentist, I see the promotion of your wellbeing, rather than your life-stance as atheist, agnostic, believer or even emergent monist, as the core issue facing you - and all humanity with you.

The emergent understanding of reality also sees all systemic axioms as essentially non-interpenetrating, non-fractal, and non-cyclic views of reality and thus discontinuous rather than self-containing, self-initiating and self-organising – as reality must be if it to escape Gödel-like limitations.

This cosmic cycle can only be dynamically and fractally approaching completeness if it is also self-inclusive and self-absorbed.[40] That is, in the cosmos every level must connect to every other level, at the maximum rate of the speed of light. This suggests the universe's so-called beginning can only ever have been part of a larger cosmic cycle.

Perhaps Gödel has proved to us that if our universe is self-contained (except for its very beginning) and not just some kind of simulated system running within a much grander reality then it cannot be built from various independent modules brought together like the construction of the International Space Station. Rather, our bubble universe must have statistically emerged from the inherent dynamic asymmetry that gave rise to it.

This self-containment seems the only way the universe can continually change state or locally divide and dance but also remain consistent, that is, the only way the finite fabric of reality can remain intact and not get torn to pieces as it disappears down a rabbit hole of infinity or nullity.

Consciousness is Consciousness of Something

Now we shall briefly consider objectivism's third and perhaps most important axiom, in terms of emergentism.

> *"...consciousness is an inherently relational phenomenon. As she puts it, "to be conscious is to be conscious of something," so that an objective reality independent of consciousness must exist for consciousness to be possible, and that there is no possibility of a consciousness that is*

> *conscious of nothing outside itself… "It [consciousness]* ***cannot be aware only of itself — there is no 'itself' until it is aware of something.****" Objectivism holds that the mind cannot create reality, but rather, it is a means of discovering reality."*[41]

Emergentism would largely be in agreement with these concepts. Philosophically, the idea that consciousness is consciousness of something is not a new one.[42] Nevertheless, the axiom that consciousness of nothing outside itself is impossible does express the idea well that we cannot relate to existence without first-person consciousness. To an emergentist, that means our relationship with existence is phenomenological and processual in the first-person sense before it can become more objective and analytical in the third-person sense.

The view that "the mind cannot create reality" probably needs to be understood in context here. I don't think this is to suggest that the mind is epiphenomenal (i.e. it is not nominally causal in emergent reality). Rather, the statement's aim seems to suggest the rejection of the idea of solipsism. Similarly, the idea that consciousness is inherently relational seems a rebuttal of transcendental consciousness and panpsychism.

We cannot be consciously aware of anything until after we sense it. As part of our inherited bodily systems and the values embedded within them, our subconscious sensory elaborations filter the contents of our perceptions before we gain conscious awareness. Our consciousness is something like the visage of our faces, which we cannot see until we look into a mirror. We know our faces are there, but visually at least, we only know it indirectly, whether or not we use mirrors. Likewise, we only know our own consciousness indirectly, reflectively and introspectively, through our more direct observations and identifications of those things that exist. This also hints at the idea that consciousness is an elaborate construct, framework, or balance that does not actually exist at any one point in the brain. That is, consciousness is not uni-dependent, but rather, interdependent. For instance, I cannot understand or perceive and claim myself as subject or an integrated 'I' until I can first perceive myself as object or an individual 'Michael', that is, as one of the many objects around me.

This suggests that the indirect self in self-awareness is actually inseparable from the things that it directly perceives, exactly like awareness of indirect space or silence is inseparable from the perception of direct matter or noise within it, respectively. This further suggests that our early consciousness probably provided us with a more instinctively monistic view of the universe. As a toddler, I may have said to my mother 'Michael is hungry', which shows cognizance of the object Michael. However, when I grew older, I could say to my mother 'I am hungry'. At this point I re-cognized Michael as both object and subject. I perceived 'Michael' as a thing and a reflection - as direct and indirect. I became more indirectly aware. My developed consciousness provided me with a more individuated view of the universe. I now had an intersubjective view of self as an indirect existant (i.e. a matter-object with spatial behaviours and style) and as distinct from other. Self was now a platform upon which I could establish and build capabilities. I now intersubjectively assigned values to things and feelings as they affected me and affected what I wanted in my future. My sense or mind-module of time, that is, now and distant past/future, had matured. This is also when I began to have moral responsibility. With my eco-psychosocial claim to a self and its affirmation within my family and society, the trial-and-error journey towards my adult life and its moral agency had begun.

This drives us to consider whether, like space-matter, the subject-object view of myself is one in which the object/form/role is the true 'me' and the subject/self-aware me is just an empty absence of my form - as empty as classical space itself. Alternatively, is the true 'me' the unique subject/self-aware me, and perhaps like quantum space, the enabler of all that I am? Does my object/form/role arise out of arrangements or filters of my true being, like colours arise out of prismatic filters of white light, or matter arises out of the quantum fluctuations of unstable space?

In terms of our physical and instinctive interconnection with a tentatively objective reality, perhaps the former more objective, third person view is better. But in terms of Gaia's naturalistic contract with us, our moral responsibility under that contract and the fallible moral agency brought about by emergent consciousness, perhaps the latter intersubjective, first person view is more profitable.

Of course, a third consideration is also possible: The true me is a monistic shuffling between nullity and infinity of both the direct and indirect or object and subject. In this view the direct and indirect are bound together in a relativistic uneasiness, with neither fundamentally dominant. A separation of powers that enables fractal adaptation between the direct and indirect arises. In this view, we also see that consciousness and its intersubjectivity have emerged from the more objective and unconscious life-form. Which view of self do you prefer? Do you prefer the view that separates subject from object, or the monistic view that binds subject and object together, not in a superficial blending like paints on a palette, but in a necessary shuffling or relativistic dance between the limits of nullity and infinity? The latter view is preferable because it makes indispensable both sides, direct and indirect, in their dance. It also makes it clear that self is a relativistic extension of the relationship between all direct matter and indirect spacetime.

A feature of this dance is that our intersubjectivity fully circumscribes our moral agency. It seems that intersubjectivity is both the cost and reward of our local moral agency, just as cosmic indeterminism is both the cost and reward of local fuzzy determinism, or matter is the cost and reward of our being. This means that while I could not have changed the missed putt I just made on the green, I can govern choices within my 'intersubjective oasis' to expose the self to new border challenges, or create new neuronal paths in the brain, by spending more time practicing and preparing before the next putt. By this logic, an immature self will have less future opportunity to demonstrate its agency than a mature or experienced one. This also means that we need to encourage and develop characters that choose to think and learn if we are to improve our future prospects as individuals and as a society. Such 'intersubjective learning oases' desperately need to extend and network past the confines of a single person.

It is a small step perhaps to take this discussion of the nature of consciousness and existence to the next level. Are you a separated instance or an integrated part of existence? Religious dualists would argue for the former, theological and materialistic monists, the latter. It is only with an answer to the next question 'Are you an integrated instance of cosmic consciousness?' that spiritual and materialistic monists would part company. Emergentism simply urges us to accept the efficacy and sufficiency of an emergent monistic view, in terms of our wellbeing. This view rejects the idea of a fundamental consciousness or non-emergent intelligence, based on the principle of the conservation of energy and information. It also urges us to come to terms with a relativistic and fractal definition and theory of self with respect to its environment, just as we must come to terms with the complementary space-matter and wave-particle antinomies and all their emergent arrangements and properties. Actually, these things are not antinomies at all, once we free our thinking from uni-dependent constraints and appreciate the innately fuzzy nature of reality.

Our rational minds can only gain a limited knowledge according to the logical principles of concept formation, including the principle of similarity and difference that a largely objective reality and our survival demands. We tend to think that first comes direct sensing or experiencing (or difference detecting with respect to an emergent reference point), then comes trial-and-error identifying (or pattern detecting), then comes believing (or learning / knowing / judging truths / claiming principles) and then comes valuing (or recognition of recurring values). A rational and value-based choosing should logically follow, that is, if we were just rational and volitional – but we're not – we are emotional and social as well. However, the idea is flawed. Our previous beliefs and worldview, along with our genetic predispositions, partially govern our every act and step of sensing and pattern detecting.

Believing (or claiming truths) actually comes before measuring or valuing for the first time (and each time). This is similar to the Scientific Method, wherein ideally the null hypothesis (the inductive belief) is stated before the testing (or measuring and value-assessment) of the hypothesis begins. As you might already guess, the Emergent Method is a way of adapting the Scientific Method, usually applied to the search for scientific certainties, to those personal, first-person hypotheses or beliefs inaccessible to science. The Emergent Method makes us aware of the intersubjective bias in our beliefs as well as in scientific hypotheses. It thus puts emphasis on the importance of the circumspect measuring of the values and concepts we tentatively endorse, and act upon, in order to add value to our lives and the lives of others.

The Emergent Method does this just as the Scientific Method uses testing and peer review to tentatively establish scientific facts and theories, which we then use to develop technologies in order to add value to our lives. However even this assessment of beliefs is subject to confirmation biases, and a host of other cognitive biases that we can collectively group within the mind games of consciousness. Therefore, we need an understanding of human nature and a method that will maintain a healthy scepticism to help safeguard us from such mind games. Just as in science and mathematics, we have no complete or stand-alone proofs, just explanations that usually withstand the test of time and usage.

If sensing or experiencing is a more direct aspect of our instinctive being and believing/valuing more indirect aspects of our conscious self-awareness, then trial-and-error identifying (or pattern-detecting) is the fallible process that binds our more direct outer and indirect inner worlds together. As we will see later, much of this pattern detecting is non-conscious or subconscious, but some of it arises to become a consciously self-aware activity. While we could say that raw experiences are innately objective, the conscious processing and interpretation of our first-person experiences is an intersubjective thing that can only gain a limited objectivity through third-person analysis.

However, wouldn't it be nice if we could also enjoy a kind of personal or first-person objectivity? Within the to-and-fro of the processual cycles and analytical limits of Figure 1, this is an aim of the Emergent Method. This kind of personal objectivity comes from a full appreciation and demystification of the role of the direct and indirect in reality and our intimate interconnection with our deep ecology.

Identifying patterns of similarity and difference through fallible analysis brings order to our otherwise disorganised and chaotic consciousness of the things that exist in our universe. The laws of the universe are not the universe's laws at all, but our narratives or explanations that we use to simplify the universe's complexity so that we might successfully exploit the universe and its/our future. Our successes make our laws seem objective or independently real, but they are not real outside of the intersubjectivity that circumscribes our consciousness and moral agency. We often describe things at a chosen level of complexity *ceteris paribus* (all things being equal), which they are not, ever. The universe knows no such classical perfections within itself. Nevertheless, within the reasonable bounds of certainty and error, we can make some scientific and personal claims sufficient for our needs.

More deeply, locally objective values emerge in living systems before consciousness of values arises. That is, values are a peculiar feature or emergent property of living systems, ultimately because living systems create and defend their borders, unlike non-living systems. Further, intersubjectivity emerges in systems of locally objective facts and values (that is, in living systems).

Finally, our intersubjectivity affects our consciously held values through the strange loops of self. Our paradoxical challenge is quite often to separate the quasi-objective universality of human values from their personal subjectivity. All this makes me think that any god with values must also be an intersubjective and emergent living system selfishly defending its cell-like borders.

Just like the interactions of physics that partly explain our observations of nature's workings, the values that promote our human flourishing seem to exist latently and indirectly within reality, waiting for us to consciously discover and establish them through our perhaps clumsy evolution and emergent consciousness.

Full Reality

Natural science or applied science does not test some things in reality. For instance, outside the realms of applied particle physics and cosmology, natural science does not tend to directly probe space or time because nearly all its formulas work without doing that. Space or the apparent vacuum is just the limiting case or theorised indirect fact of most scenarios. Nevertheless, the limiting case of scientific inquiry can often be the most rewarding or promising area of research. We might think of the wonderful properties of matter near zero degrees Kelvin, near the speed of light, or squeezed to near the Planck length, which is the province of black holes.

However, if we think of the formal sciences (the realms of mathematics and logic), then the indirect concepts of reality (including zero and its inverse, infinity) are explored on a routine basis. In the formal sciences, we casually assume reality to include that which is seemingly beyond existence. Infinity arises in this wider sense of reality most obviously because of indirect zero and the possibility of something divided by zero.

This human ability to ponder an actual infinity that cannot be expressed materially is not to be taken lightly. While the cosmos is constrained by a limiting relativity, our human imagination seems less so constrained. We can conceptualise the infinite and deconstruct relativity. We can break through the relativistic constraint to ponder full reality, that is, expressible relativistic reality and inexpressible reality in its components, harmonics or shadows. This is something no other species can do. It means we can form notionally unlimited (and potentially useful) ideas not available to the rest of Nature. We can also access, manipulate and understand the indirect separately to the direct. We can induce an independence in our imaginations that otherwise does not exist. Our imaginations participate in the quantum processes described in Figure 6 and Figure 7 with a kind of fractal license to practice. This is an amazing feat. Human imagination is an incredible power; its ability to imperfectly split relativistic reality to access full reality is of far greater significance than splitting the atom. Our unlimited imagination sets us apart as potent agents of emergence. More literally, scientists, engineers, and technicians use complex numbers and infinity-nullity to solve real-world problems on a daily basis.

The mathematical revelation of the unconstrained in full reality also helps us understand human consciousness itself. Consciousness does not exist in our individual lives now, in the present moment, without both physical outer-world awareness ('extrospection', or literally looking towards the outer world) and inner-space awareness (introspection or looking towards the inner world). Introspection is that part of consciousness that, like our faces' visage reflected in a mirror, can only be aware of itself after it is aware of something else, i.e. after it extrospects. Extrospection is more or less direct consciousness of something whereas introspection is indirect consciousness of something, or consciousness of the contents of consciousness itself. Introspection is the door to the incomplete, indeterminate, or physically inapplicable states of imagination.

Although even here we have difficulties - if I extrospect without also introspecting, can I really call this an instance of direct consciousness, or is it more correctly an instance of pre-consciousness or subconsciousness or machine-like unconsciousness? It is perhaps only after introspection of the extrospection that we can call the extrospection an instance of conscious awareness. Put simply, consciousness requires us to think recursively, in cycles of relativistic extrospection and introspection. It is rather meaningless to speak of a single instance of extrospection within the mind/brain just as it is rather meaningless to speak of the temperature of a few molecules of gas released into a large container, or to manipulate matter without space, or to measure the direct without the indirect.

The deep philosophical realisation of indirect and emergent consciousness, and thus the proper incorporation of notionally unconstrained human imagination, is as important to the formation of a modern naturalistic philosophy as the inclusion of the indirect concept of zero within mathematics. It is as important as the idea of natural selection to an understanding of the evolution of life, which also has no need of a supernatural component within its indirect, self-referencing and recursive workings.

The understanding of consciousness and the role of imagination within monistic reality solves the mind-brain paradox. That is, philosophers in the past have asked, 'How can intangible mind (or soul) affect tangible brain-matter?' What is the mediating mechanism between the intangible or spiritual and the tangible or natural? Is it miraculous, and so assigned to a god-of-the-gaps?

The line of questioning was wrong-footed in its lack of understanding of relativity and totally blinded by the dualistic perspective. We missed the elephant in the room, which is the ability of human imagination to escape the normal constraints of relativity and thus bamboozle us. As previously noted, imagination is able to think outside of the normally limited and joint expressions of the direct and indirect, and thus falsely separate them as if they are independent. We have sometimes forgotten in the past that if we use our imagination to understand new things, we must also bring back those meanderings to reality to use them in the real world. We must chain our imaginings to the reality of relativity, by which actual infinities are not permitted, to use them in the material realm. This often means rejecting all the engineering solutions that contain imaginary numbers, so that we are left with the real and usable solutions that don't contain imaginary numbers. Likewise, physicists routinely reject unhelpful solutions that contain infinities and retain the interesting solutions that do not.

On the other hand, relativity explains quite simply the fuzzy interaction or interrelationship of the direct and indirect, including that of the body and mind. That is, the mind/body is a simple expression and extension of indirect/direct relativity. This is why our imaginings can be the literal seeds of material outcomes (and vice versa) without any transcendentalism required.

Indirect reality intertwines or entangles consciousness, just as it does existence at every other level (genetic, inorganic, perceptive, etc.) This is crucial to understanding the rational but at times indeterminate continuum that binds our consciousness to full reality. Similarly, the indirect concept of zero completes mathematics and gives mathematics its powerful conceptual methods that enable us to gain knowledge and exploit physical reality so much better than in the ignorant past. We need an understanding and appreciation of relativistic versus full reality.

We also need to know how we emerged and tentatively fit within this fuzzy reality. We need this to establish and extract personal meaning from life and thus success from life. It is not just a matter of how we can fit, but how we inescapably do fit. The universe's direct and indirect reality channels us to fit. It probabilistically causes us to fit, yet still grants us certain freedoms within its imperfections. We somehow belong to this time and space but it somehow belongs to us because without us the fabric of reality would be ripped apart, and perhaps most importantly, because we are the species with fabric-ripping imagination. The practitioner of the Emergent Method recognises this intimate and unavoidable belonging, and realises that the perhaps awkward dynamic fit, which can be a little beyond the human ability to calculate and handle wisely, is nevertheless pregnant with unique potential for you and I, personally.

All our abstractions, concepts and laws regarding reality are approximations and thus subject to correction and evolution, much like our evolved life forms with their innate algorithmic heuristics themselves. We have three components involved in all our concept formations. That which is a more-or-less true reflection of reality as experienced so far within reasonably tight limits of confidence, that which is substantially incomplete, missing, or unrefined, and that which is in error.

For our introspective concept formation (e.g. describing a new kind of feeling) we have a fourth component, not to do with the thing we are measuring, in terms of similarity or difference to things already conceptualised in the past, but the method of measurement – we can't measure introspections directly. We must measure them indirectly or reflectively and recursively with respect to our pre-existing beliefs, concepts and experiences because consciousness is consciousness of something. That is, we cannot have a new feeling about something until first there is a something.

The method of concept formation (direct or indirect) has nothing to do with the validity or otherwise of the other three components of the formed concept itself - confident knowledge, ignorance and error. In fact just as space entangled with matter gives matter its spatial dimensionality, its dynamic beauty and its recurring interactions, so introspection entangled with extrospection gives extrospection its conscious or self-aware dimensionality, its dynamic beauty and its rich, sharable language.

I am not really extrospecting if I do not know I have been extrospecting at some level of consciousness. I am just going through non-volitional, pre-rational, and pre-conscious thought processes. That is, just as space or the void has an indirect impact on material reality (it moves it and lends it a certain dynamic style), likewise our introspective and contemplative awareness on all of our extrospections. We are blind as to the significance of introspective and contemplative awareness until we specifically recognise it and its natural but indirect relationship to physical and direct outer-world awareness.

At a secondary level, we cannot know we are introspecting until we have had experience of previous introspections. This experience emerges from the relative chaos of thoughts around it – it does not suggest an infinite regress. If extrospection is a kind of centripetal force of mind whereby we drink in the surroundings presented to us to form mind/body schemas, then introspection is like running the machinery of thought backwards – it is, as it were, a centrifugal force that projects the same schemas outwards to our mind/body's surroundings. In our imaginings is the seed of the future.

To use another analogy, how does a baby learn to grasp, or learn to walk? In this case if uncontrolled bodily movement is a kind of centripetal force of body whereby the baby witnesses the surroundings presented to it to form a schema, then controlled bodily movement is like running the machinery of body backwards – it is, as it were, a centrifugal force that projects the schema outwards to its body's surroundings. This is another case of abilities emerging from a relatively chaotic background without running into the problem of an infinite regress. That is, the baby learns from its own previous movements. By the Uncertainty Principle movement is a given – it's just a matter of us mastering it in body and in mind. We will discuss the mechanism of the centripetal/centrifugal switch after the original neuronal path is established by our surroundings later.

Perhaps we could also say that introspective awareness binds two objects of extrospective awareness as the inverse square rule describes the binding together of two physical objects in space or an emitter and receiver in the silence. Without such binding, objects have no known relationship to each other or our ongoing stream of consciousness. In this sense, extrospections are like transmitters, making impacts on other extrospections that depend on the strength of the introspections linking them. At a secondary level, some concepts with combined direct and indirect properties become transmitters themselves.

Once we fully appreciate the ramifications of this window into infinite full reality only animals with imagination can know, as opposed to the finite and relativistic reality of which we are a vital part, then I think we can live the grounded, purposeful and fulfilling lives we desire. We can confidently pursue our unfolding destinies within the vagaries of our largely unknown universe.

The direct and indirect do not orbit around each other like equal-mass and twin black holes in space, but rather indirect consciousness orbits around direct consciousness like the lower-mass planets orbit around the dominant mass of the Sun or like space and its interactions frame matter. Asymmetries via shuffling of population attributes are very important in evolution or physics – they enable emergence via beneficial mutation or perturbation. Likewise, asymmetries between extrospections and introspections enable us to learn, just like asymmetries between left and right hemispheres of the brain enable us to make mental images and decisions, as suggested by the neuroscientist Antonio Damasio in his book, Descartes' Error.[43]

Check: Outside of human imagination, does the direct exist and persist anywhere without the indirect? Separating the direct from the indirect is the general form of trying to separate matter from spacetime or the material from the logical. The dualistic worldview would highlight the independence of the direct and indirect. Monism would accept their interdependence and indeed reject their seeming separateness, except in the sense of a partial and fuzzy separation of powers.

By the so-called law of large numbers, nothing, in the sense that Lawrence Krauss (2012) uses this word, will eventually lead to something. By this I mean that vast, dynamic disorganisation or disorder (or the vast statistical nothingness of disordered subatomic particles) will eventually organise itself into a small oasis of something, to more effectively and efficiently deal with the local but transient disequilibrium of energy and information. There is an effective drive towards balance in all areas of the cosmos. This is also true inside the mind/body. We call this innate drive within animals, intention. However, this term could be a little misleading because it suggests it originates in the mind or living organism. Actually, it originates in the interactions of space itself. All irrepressible movement is a form of proto-intention. What is chaotic and meaningless today can arrange itself and emerge as something of intentional value tomorrow.

Is logic powerful if it can find no eventual expression or application in the material? Is the material powerful without logical arrangement? I ask these questions in an attempt to urge you grasp the deep significance and benefit of the dance of interpenetrating and emergent monism.

As our introspective thoughts and concepts change, the way we communicate and interact with ourselves changes, which leads to our goals changing. Likewise as our extrospective thoughts change, the way we communicate and interact with the world around us changes, which leads to the way our goals are fulfilled also changing. It seems that our goals are more consistently fulfilled as we become more in tune with and sure of our unique but interconnected roles in the universe. We need reductionist analysis but we also need the processual or dynamic synthesis of an interpenetrating monism.

Certainly one problem in the way we go about seeking success and happiness is that erroneously, we often tend to focus on direct materiality or indirect now, space or self as if they exist independently and without dynamic relationship. They do not; without intending a transcendental connotation, now, space and self inextricably incorporate us within the single dance of all existence.

Alternatively, we bind both direct and indirect aspects of existence together as if they can be perceived as a simple, direct fact without any inherent and indirect logic: They cannot. If we are to see worthwhile goals fulfilled in our lives, we desperately need to move away from our Euclidean, linear, sequential-processing, flat-world way of thinking and look at our wonderful physical, logical, and impetus-enabling world as it really is. This is a world vast and chaotic yet wholly (if imperfectly) interconnected, massively parallel processing, full of fertile feedback, and its self-contained organisation always iterative and emergent.

This new way of looking and thinking, switching our viewing of the world from the lenses of reductionist analysis or holistic synthesis to all the possible lenses in between, in a kind of dance with full reality, will radically change our extrospections and introspections, so that they become more synergistic, bio-diverse and risk-tolerant. It will free us from axiomatic traps, and allow us to flow more easily within our surrounding dynamic relationships, which will perhaps bring us in tune with our highest goals. The process will be much like what nature often seems to hit upon in the design space[44] or vast possibilities it finds provided for it by our universe's reality within each Planck Length and Planck Time.

Measurements

We could also say that numbers are representations of quantitative measures in nature but also include the issue of the indirect within them. For instance, what is the number seven? It is seven of something (e.g. apples or oranges) in tangible reality with respect to nullity or zero in intangible reality, but with the identification of that something (apple or orange) removed or dropped. All modern arithmetic, like the number seven, is based on the logical and indirect concept of nullity.

The concept of zero oranges is meaningless in an absolute sense; it is only meaningful in a relative sense. Alternatively, zero oranges are equivalent to zero apples, but oranges are never apples. Zero is only a relative, logical, indirect, or informational reality. Zero (and any other absolute we might conceptualise) is meaningless except in relationship to all that truly is. Zero simply marks a relationship or interaction, as does its inverse (infinity) or the square root of 'one less than zero'. However, it was not until arithmetic advanced to the level of incorporating logical and indirect zero as its base or centre that it was able to make huge leaps forward and make large causal impacts in terms of human technology. Mathematical zero is the textbook example of the indirect and inexpressible facilitating impetus and change in the direct.

We might say that the length of the left side of a desk is one metre, plus/minus one millimetre – which might accurately specify the reality that we are able to measure with a ruler, with appropriate confidence and allowance for error. However, within this specification is the notion of a perfect, zero-error or infinitely exact metre, which does not exist in relativistic reality. The idea of a standard of measurement is necessary for practical measurement. It is impossible to adopt a perfect standard, unless that standard itself is ultimately unmeasurable (for instance, by incorporating 'pi', the ratio of the circumference to the diameter of any ideal circle, expressed as a decimal fraction).

So we see here that in our every measurement and thus in our very method of forming concepts with our consciousness, by measuring similarities and differences, we run into the problems of the direct sullied by the indirect, or contingency, imperfection, incompleteness, and uncertainty. We run into the core problem of tangible fact intertwined with the intangible logic of zero and zero's inverse. This is the undeniable nature of reality and our perception of it. Full reality takes the issue a step further because it is partly inexpressible; it is partly potential rather than fully kinetic. It mixes real and possible, or naturally expressible and inexpressible, concepts together. The interconnected becoming of matter in local spacetime does not limit full reality; it gives limited expression to full reality.

Our indirect human logic can exist in conceptual form, but this is never very far ahead of evolved direct form and temporal fact. That is, the indirect world of for example our mathematical logic limits the direct world of for example technological achievement. Conversely, the beauty of the direct often substantially exceeds the logic of that which was (e.g. mathematically) conceived by us - because we can only ever perceive and conceive just a small part of full reality. The direct and indirect, including our human imagination, must always dynamically interpenetrate.

While the measurement of a desk in terms of the metre makes sense, because it was designed by a carpenter with the idea of a metre in his or her mind, does the use of the metre to measure things in nature reflect the same sense of relationship? Does it really make sense from a naturalistic point of view to measure a leaf of a tree in linear metres? The leaf was made chaotically (no two leaves are identical) yet systematically (all the leaves of a tree recognisably belong to that tree, or more specifically, that tree's local fractal formula or algorithmic genetic recipe for leaf making). The leaf was not designed with linear metres as a reference point in mind. As evidenced by its self-similar morphology, we know it blindly and locally emerged as a fractal in less than 4.0 dimensions of spacetime.

This brings us to a far more subtle point. The 3 spatial dimensions we measure universally and linearly do not come into being that way. Like the leaf, and actually with the leaf and the desk, they come into being chaotically, locally, and interdependently. We apply linear, spherical, or some other form of curvilinear coordinates, but this is not the true nature of space or matter. What is the width of a green leaf? Is it a flat dimension, like a dried leaf flattened within a botanist's diary, or does each leaf have its own local, living, nodal, and fractal dimensions and measures of width, height and length?

Wait - aren't we confusing the dimensionality of something in space with the 3 dimensions of space? No. The 3 dimensions are not constant but emerge with the expanding universe as a dynamic and arranged set of subatomic particles housed in a host of differing morphologies and non-living shapes. According to the inflationary theory of early expansion, this has not happened in a classical fashion, as we might expect in a normal explosion. During the early inflationary period, expansion was exponentially quick, then it slowed down due to gravitational and theoretical dark matter effects and now it is speeding up again as those effects become weaker and the effects of theoretical dark energy or the cosmological constant[45] become more apparent.

Further, the 3 dimensions are not unfolding evenly around us. While galaxies are spreading further apart with this quick but gentle expansion, the things within each galaxy are relatively unchanged, because the background effects of the Big Bang and dark matter and dark energy are much weaker than the local interactions within galaxies. Thus Big Bang and dark energy effects that might otherwise tend to rip matter apart, as well as gravity and dark matter effects that might otherwise tend to quietly crunch matter together, actually continually find a local 'near-equilibrium' within matter and its local galactic environment that we might call 'entropy'.

The point here is that the very dimensions of space are changing and this is not just at the galactic scale, it is at the quantum (and sub-quantum) scale as well. Every leaf, as part of the arranged and dynamic set of the universe's subatomic particles, is a fractal manifestation of how the universe has expanded to locally fill its four nominal dimensions of height, breadth, depth and time. Relativity tells us that the leaf is part of the full set of matter that bends space around it. Fractal geometry tells us the 4 dimensions of spacetime are interdependent, and locally manifested in the shape of the leaf. Gravitational waves discovered in 2016 do not just apply at the universal level – that was just the way they were isolated and reliably measured; they apply at the everyday level of your movements and mine.

Over time, our classical measures of space have made us think that space itself, with its 3 dimensions, is linear, uni-dependent, static, only cosmological, non-impactful on matter and impersonal, but in relativistic reality, it is none of these things.

Common-sense and supernatural concepts of reality, or the darker aspects of human imagination unhinged from relativistic reality, have perhaps done us a great disservice, in terms of making us think that spacetime is impersonal, fixed, non-local, and divorced from matter and us. This misconception goes as far as limiting our understanding of the locally self-organising, emergent, and interpenetrating (or all-inclusive) nature of the universe itself. It has limited a perceived meaning of reality we each might extract from this true nature, and thus how each of us fits within the universe's local interconnectedness.

The misconception has also twisted our notion of the self. If we are made of the same stuff as stars and space, then we too must be relativistic, fractal and interpenetrating. We too bend mind/space around and within us as we relate to our environment and our brain's cortical structures. We also move and express a personal style according to the mind/space projected around and within us. We too participate in the formation and manifestation of reality's locally expressed, fuzzy, and fractal dimensionality. The misconception of the impersonal nature of space is no doubt also affecting our concepts of personal values, purpose and self-worth, discussed more fully in the next few chapters. We humans with imagination are part of the unstoppable cosmic engine of change.

Time

We could also speak of our measure of time in hours, just as we have spoken of the measure of length in metres. What is an hour? It is another linear measure of time's passing. Strictly speaking, time doesn't pass; things simply move in local space. As we noted earlier, time emerges in many levels of existence – suggesting that as we move, we are justified in choosing our own time. Our human measure makes us again think that time is linear, uni-dependent, simple, universal, regularly-impactful and impersonal, but again it is none of these things. Time is not the independent variable by which classical science thought we could measure all movements. This is exactly the wrong way around. Time, as a dependent variable of motion (or decay modified by work done), is nonlinear, complex, interdependent, always local, variable, associated with the mass and shape of everything, and very personal.

For instance, a master-clock drives all processes in a PC so that everything can happen in well-governed or well-controlled synchronisation with it. Instructions may be fed to the CPU and processed more or less one-at-a-time. However, a uni-dependent master clock beat, such as an imagined ticking Planck Time, is not the driver of processes in nature. Processes are interconnected in spacetime, yet continually self-referencing and wavelike (as in our bodies' biorhythms) in a way that often makes our digital concept of time measurement seem woefully inappropriate. Control or governance emerges from local proto-intentional space with the process rather than separate from the process.

For example our brain processes, very much like schooling fish, herding cattle, or flocking birds, seem to begin and proceed chaotically, asynchronously, locally, and in a massively parallel fashion (with many players involved). These processes seem to begin with necessary unconscious and subconscious but scattered order or control, and then arrive at a kind of self-referencing, intermediate, and temporal conscious order once they settle on a locally-arising (and then dissipating) peak stability.[46] This temporal stability in emerging consciousness might be compared to the phase-change from liquid to lattice-structured solid spreading through a molten but cooling iron rod, or like the harmonic aligning of left and right channels to a 3D picture as 3D glasses are slowly put in place over the nose by the viewer. The mind process involves literally everything large and small, not just a local, straightjacketed, uni-dependent tick-tock. The mind/brain is an SPS that process-models reality before it enables the analytics of which we are consciously aware.

We might similarly account for time's apparent uncertainty within Planck Lengths. This was a state that affected the entire universe within its first 10^{-43} seconds of existence. Time more truly inherits a wave-particle, big-small nature just like the space and matter with which it emerges. It seems we should begin to conceptualise time as something other than just a sum of fixed, discrete steps of 'Speed of Light divided by Planck Length' (a tick of time equal to 10^{-43} seconds) if we want our concepts and robotic inventions to escape uni-dependency and reflect self-governance.

Cosmologist John Barrow remarked that time is a variable that only emerges when we move beyond the quantum level.[47] This is what the quantum model of particle motion over time most simply suggests (see Figure 9). Time emerges from nonsense like the temperature of a gas mentioned earlier. Time is nonsense in the purely quantum environment, just as temperature is nonsense in a fixed volume with an inadequate number of gas molecules.

What are we to make of the fact that light rays and gravity's theoretical gravitons travelling at approximately 300,000 [km/sec] represent to any observer the boundary between past regions of space and the present regions? It suggests my 'now' is a relative concept that differs to yours, depending on our relative velocities in space (a concept known as relative simultaneity).

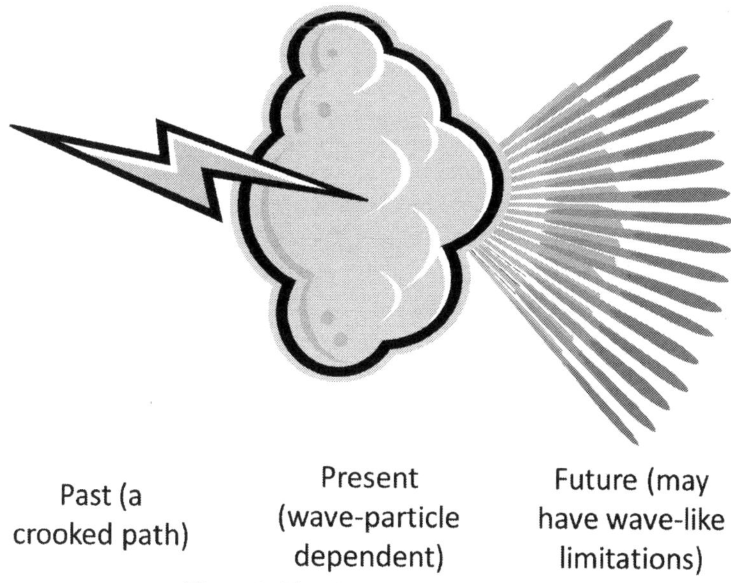

Figure 9: Time's Wave-particle Dependency

What do we make of time passing in indeterminate particles, as suggested by Figure 9? What do we make of time passing at relatively different speeds for different observers in different gravitational environments? What do we make of the mirrored relationship between particles and their antiparticles in terms of their charge, which we can understand in terms of the negative energy modes of the electron field being backward in time? Time (now and thus past/future) is by no means linear or independent from space and matter, or simple or uniform. We should begin to perceive time to have an order and potential that locally, temporally and continually, emerges with space and matter (just as we do).

We might incorrectly conclude then, that all chosen human measures of time and spatial dimension simply involve intersubjective issues of how we measure relationships within reality, but they do not involve issues with respect to our concepts of reality itself, or with respect to how we might fit within that reality. The desk's measured length or age might involve convenient and logical but peripheral measurement issues that don't reflect reality accurately (or that re-organise perceptions of reality to suit human purposes), but how we think about the desk itself, as part of perceived reality, we can assume to not be an issue.

The problem here is that every subatomic particle of the desk indirectly builds space and time/change into the desk – if this was not the case, the desk would not be a desk; it would theoretically be the logical point of singularity within its own black hole. Spacetime makes the desk a desk, and nearly anything else in differing spacetime configurations. Spacetime makes us what we are, but not in a linearly deterministic, perfectly symmetrical or synchronous fashion across the universe. How we think about space and time is important. In a relativistic fashion, space and time enable the stars, the matter inside the desk, the DNA inside the leaf, and the consciousness inside each of us to take on unique arrangements, make unique movements, and follow unique algorithms, whether chosen or not. For living matter, spacetime both compels and enables us to look for local methods or information we may exploit to avoid decay and survive into the next moment.

Our choice is often to experience one thing over another, and often to identify a thing as this or that, but our choice as conscious beings restricted by our times, is always whether to learn and adapt. This simple idea is a key to the success of each of us. We at our best are inclusive but self-initiating, self-organising, self-learning, self-sustaining and self-defining beings, following evolving algorithms of growing complexity, much like the universe of which we are all a part. The exact line of demarcation between our interdependent and emergent causes and our human claims and responsibilities within the naturalistic contract that defines us as a species is fuzzy, just as are the relationships between time and matter. The Emergent Method is about recognising this necessarily fuzzy line of demarcation between our free will and constraints, and allowing these observations to affect us, and guide us towards more sustainable values and wellbeing. Our learning and free will, like the dimensions of a leaf, emerge with time, space, and trajectory because they are interdependent with a fuzzy local determinacy.

The decay or loss of orderliness of all systems, or increasing entropy, is perhaps the logical consequence of the expanding universe.[48] This entropy is perhaps also a reason for the arrow of time (always moving forward) perceived by our animal minds. This could suggest that psychologically perceived time is merely a construct or convenient claim of our human intersubjectivity. However, perhaps all living organisms experience time's propagation in relation to their internal rates of metabolism, which in turn relate to their mass, by Kleiber's Law.[49] That is, it seems there is an objective biological time, even if relativistic, that relates physically to energy translation, mass, entropy, and gravity. If this is true in biological systems, then it is perhaps also true physiologically; the sense of transience could thus be a function of any living system's introsomatic conditions relative to its extrosomatic conditions. This suggests that the dimensions of an organism's volume (allometrically related to its mass) helps to shape the local time dimension perceived by that organism, which perhaps reflects the self-similar or fractal nature of reality in all four dimensions of the organism, including time.

If an animal's time emerges, then it is also mastered just like the skills of grasping, walking, or thinking imaginatively, mentioned earlier. Time happens with us more than to us, but as we experience its chaos all around us, we also learn to rotate the mind machinery of time perceived in the prefrontal cortex the other way. That is, we become proficient managers of our own time with relevant experience.

Entropy begets disorder - it takes things downhill from a local order - it does not let anything stay comfortable. Just as the universe finds a new local equilibrium or homeostasis, entropy messes with it. However, this is a good thing because there can be a formed disequilibrium that in the right circumstances can actually take the universe emergently uphill in the drive towards maximum local stability. There can be a kind of adaptive 'mutation' - one serendipitous change or new property, in the vastness of potential properties, which can spread across the old environment over time.

Compare this mechanism of emergence to the mechanism of natural selection:

1. Repetition (an algorithmic spreading via non-random but chaotic feedback within the orderly system)
2. An occasional mutation or unusual perturbation in the vast web of interactions at its limits or boundaries
3. A vast intermingling of particles in a multitude of conditions wherein some conditions, through the power of the law of large numbers, supply those interacting particles with a measure of continuing homeostasis, equilibrium, or relational 'fit', that we might correlate with Darwin's 'severe struggle for life' in the hostile entropic environment

We need to consider this emergent mechanism when we try to understand the rise or distillation of each of the game-changing interactions, such as spacetime itself, electromagnetism, gravity, life or consciousness. Each of these was a statistically possible move uphill in the vastness of downhill processes over time. We have a law for generally going downhill (entropy) but we don't currently have a widely recognised mechanism for going uphill. If we can name an uphill-going mechanism in the realm of life, natural selection, why can't we name its counterpart in the greater realm of reality, and why not call it natural emergence? Conversely, perhaps we should recognise none of these laws or mechanisms, because entropy, emergence and evolution are all statistically inevitable in the continuous mix of sub-atomic particles that is our vast and expanding universe. Our assigning of a law perhaps exposes our anthropomorphic ethos and intersubjective claim to moral agency.

There is a fundamental link between entropy, self-organising information or negentropy, emergence and our experiences and identifications of space and time. If we don't appreciate that the universe's indirect space and time are ultimately what enable us to metabolise, take shape, and dance, we will not appreciate the personal significance of reality itself. This reality consists of separate yet interpenetrating, self-organising and self-eliminating, direct and indirect, components within the monistic whole. We are part of relativistic reality and are unable to divorce ourselves from it (except in imagination), just as matter, space, and time are unable to be divorced from one another, except perhaps transiently at a theoretical point of singularity within a black hole. As self-aware beings, we are uniquely placed to exploit this wonderfully esoteric knowledge.

Clearly, going uphill depends on going downhill a lot more often (advance by trial-and-error) - that is, until artificial emergence and then artificial selection came along. Like breeders controlling the artificial selection of dog breeds, we, as followers of the Emergent Method, are able to be involved in the processes of our own artificial emergence – of our own self-organising wellbeing. We can slowly learn and through learning, incrementally change ourselves. To some degree of freedom, within some limits and cycles, we can all choose our own unique time and space.

The idea of indirect time requires further clarification: Experienced time exists defiantly only in the present moment (now). The now component of time is emergent and fleeting, with every local motion of matter in space, whereas both past and future are non-fleeting in our intersubjective views. Past and future provide the realistic continuum of time and existence wherein we may store and apply meaning and identity. Without the reference points of past and future, we would simply flow within the process of existence without any possibility of consciousness.

Just as matter hangs in space, we perceive 'now' to hang in the midst of virtual 'past/future'. Past and future provide a logical framework in which we humans can appreciate and indirectly exploit our now. By this logic, we might expand our understanding of existence to include two types of indirect entities – those that are fleeting (or are experiential to consciousness) and those that are non-fleeting and virtual, but provide consciousness with stores of information and meaning (or are identifiable and exploitable reference points). Strangely, it can be shown that modern humans tend to place more value on remembered identifications than on present fleeting experiences.[50]

Should this be the case? Whatever the answer, happiness and success in life seems to be measured by us in the long run not so much in terms of direct material posessions ('having'), but in terms of virtual, non-fleeting identifications and purposes as well as fleeting experiences enjoyed along the way. This observation is crucial to the Emergent Method. Happiness is processual rather than analytical. Life is not truly had; it is only borrowed, as it were, from spacetime. Thus, possessions are never truly had either; they also are only borrowed. Similarly, success and happiness are not had, owned or possessed either; rather, they are experienced, identified and guided by purpose. The good life, just like existence, is a process rather than a place or thing. The interactions of matter and spacetime make reality and make the good life, rather than their idealised components.

Humans may only experience time in the local 'now', but it is not the only way we process it. We process time in two very different ways. We mainly process the near-term present, via the environment experienced by the senses and the body's unconscious responses, by largely subconscious instincts programmed via our genes. Conversely, we mainly process the past/future, with the aid of long-term memory, via the claims and processes of consciousness. We could say that we largely respond to experience in the present moment through subconscious and instinctive/genetic mechanisms whereas our recalled patterns of identifications or decisions associated with past/future, are largely mechanisms of consciousness.

Simply put, human instincts mostly deal with the fleeting aspects of time whereas human consciousness mostly deals with the non-fleeting aspects of time. Crudely, instincts exploit now, but consciousness is needed to exploit past/future. To be successful and happy, we need to marry these two very different aspects of our relationship with time together into the one, harmonious and phenomenological whole.

However, this relationship between time and human nature incorporates two problems with respect to how we process time. Instincts include shortcut biases that often inappropriately choose, deal with, and store experiences and consciousness includes unnecessary or exaggerating mind-games that often incorrectly identify and process recalled or imagined events.

Nevertheless, when we overcome these shortcomings of our human nature, most would agree that indirect experience and indirect identity/purpose/meaning, rather than direct material possessions, are the essential components of a fulfilled and happy life.

Further, when our identifications and rational knowing/valuing in past/future are able to positively coalesce with experience, instinctive choices, and other operations in the present moment, we heighten our feelings of living the good life. That is, human happiness and flourishing is in, or very near, the present moment.

Living too much in the past or too much for the future will not bring us the fulfilment and success we desire.

Perhaps we could take the same concepts and apply them to the concept of space. For instance, humanly experienced space exists exclusively and defiantly only in the 'here'. The perceived 'here' component of space is emergent and fleeting, whereas the 'there' (behind us and in front of us) component is comparatively non-fleeting. It should therefore provide the rational continuum of space wherein we may store and apply information and meaning. Just like 'now' hangs in the midst of past/future, 'here' should hang in the midst of 'behind and ahead' – 'there' should provide a logical framework or reference point in which we humans can appreciate and indirectly exploit the 'here'.

Nevertheless, it is interesting to note that our psychological relationship to here is different to our relationship with now. We tend to see here as ours to own and perhaps defend by way of property rights but we do not tend to equivalently claim now as ours to own. We can also see now as not easily grasped in material things, but we expect here to be indefinitely available in material things through time, even though entropy ensures here is just as fleeting as now at the subatomic level. Further, we tend to see here as a fuzzy conceptual convention and there as a solid reference point, whereas now is seen as more essentially a part of reality's inescapable fabric and the past/future framework as a matter of agreed convention.

Why is this the case, and what should our relationship to here-and-now be, if we are to secure success and happiness? Perhaps in many situations we should dare to own now just as surely as we dare to own here, or perhaps in other situations we should release our individual clench on here just as our clench on now is more communal. That is, perhaps our relationship with here-and-now, just as our relationship with the material-and-immaterial, should be more fluid and cyclical, and less rigid.

Either way, what affects us in terms of past-and-present interdependent causes-and-effects are unique to us, and so an important influence to our sense of personal identity. We fiercely own this identity even though we know we cannot control all the interpenetrating effects that formed it or will take it forward.

Without wanting to suggest a transcendental aspect, it is the universe, as a verging-towards-open system, that seems the owner of our unconscious selves because it houses all those multifaceted causes and effects back to the Big Bang and moving forward into the future. At the same time, our inner world seems to grant us some kind of ownership of our intersubjective selves or self-image, to which our environment often responds. That is, it is perhaps difficult to definitively decide whether we have free will or we don't. However, the uni-dimensional dichotomy of this decision is misplaced. The universe is vastly interconnected, indeterminate, and emergent – and we are an essential part of it. Our moral agency is both free and limited, but it is to be shared if we wish to find wellbeing and live the good life.

Spirit and Soul

The Christian idea of human spirit is also fleeting because it acts or flows in the psychological present moment (and place), whereas the Christian idea of the soul is also a non-fleeting concept wherein we may store meaning (character or personality) that is attributed to some length of past/future.

Note: The Judeo-Christian-Islamic idea of the human spirit is the fleeting life-essence of God's breath, breathed into Adam's body. This combination of spirit and body (wave and particle) enabled Adam to become, or be described as, a living soul. Christian salvation consists not of saving the spirit, but saving the soul that accompanies the combination of spirit and body. To accomplish this, many Christians believe the quickened spirit must inhabit a resurrected and renewed heavenly body – one that still consists of matter but is no longer mortal/imperfect and thus subject to pain, hunger, sadness, and death.

Just as we perceive now to be in the midst of past/future, we can also say that the spirit hangs in the midst of soul. The soul provides a logical framework or reference point in which one might appreciate the spirit. In Freudian terms, we might be able to often relate the spirit to the instinctive, unconscious mind and the soul (particularly its processes) to the conscious mind.

However, by this understanding, the Christian idea of human nature, consisting of spirit and soul apart from the body, would necessarily be an indirect and logical complement of the body, just as spacetime is the indirect and logical complement of matter. You may like to ponder this point for a moment or two, because it would suggest the indirect spirit and soul could not exist without a direct body that bends the indirect around it.

While it is not wrong to speak of fleeting now/here and non-fleeting past/future/there, or fleeting spirit and non-fleeting soul as parts of united continuums, it is wrong to blur their component natures as if blended like paints on an artist's palette. They are not linearly additive, just as we cannot simply separate and add back together the solutions inside and outside the Mandelbrot Set. Rather, all the states of space, time and human nature are continually and probabilistically shuffled together by dynamic matter, and their properties are emerging in relativistic feedback arrangements between matter and its interactions, just as offspring shuffle the gene-codes of their parents in their emergent expression.

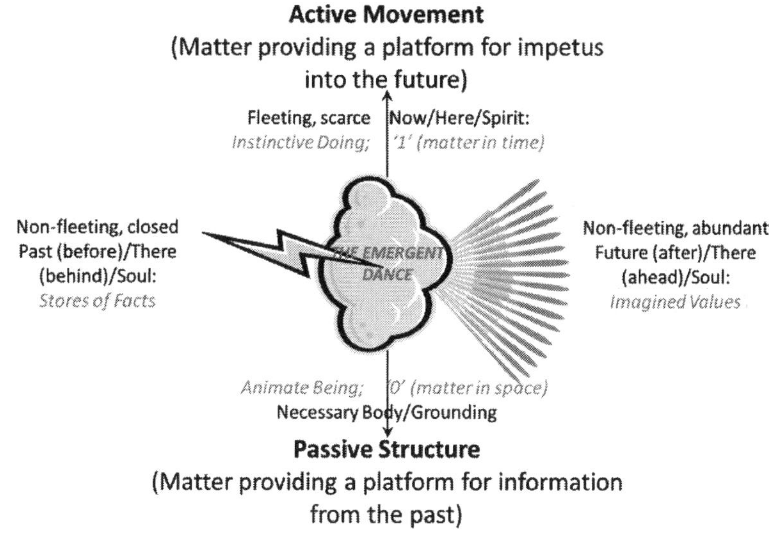

Figure 10: Our Emergent Dance

Figure 10 represents a humanly efficacious view of reality that perhaps presents a development of the relationships between matter, space, and time given in Figure 6: The Becoming, in Planck Length and Figure 7: The Becoming, in Planck Time. It is an existential, self-aware, or intersubjective point of view. As represented in the figure, matter is direct, classically locked into the here-and-now and subject to both entropy and emergence. Captured in our bodies, matter is our animate being.

Conversely, space and time are seen as not locked into the here-and-now, but seem to transcend it through there (behind and ahead) and past/future (before and after). They seem like indirect and virtual enablers of entropy or negentropy. Captured in our bodies, space and time is our instinctive doing and spirit. There (behind or ahead), the past/future (before or after) and the soul (memories or imagination) are not 'being' or 'doing'. They are indirect stores of potential, opportunity, value, knowledge, information or meaning (logical concepts, labels, roles, rules, identifications, principles, expectations, hopes, purposes, etc.) to do with rational knowing/valuing, that are non-fleeting.

Perhaps these non-fleeting aspects of our full reality may be related to the infinite solutions outside of the Mandelbrot Set that are ultimately inexpressible, but with certain informational exchanges that change their nature without disobeying rules of conservation, can partially lose their non-fleeting status to become caught up in the emergent web of expressible reality. That is, a usage or drawdown of the non-fleeting indirect into the fleeting indirect provides our emergent ability to become. At the centre of Figure 10, I call all of this The Emergent Dance.

Our combinational ideas of the galaxy (subsuming fleeting and non-fleeting outer space), the human body (containing fleeting and non-fleeting inner space), time (subsuming fleeting now and non-fleeting past/future), and human nature (including the religious idea of fleeting spirit and non-fleeting soul), can be misleading because they fuse the direct, the indirect-but-fleeting, and the indirect-but-non-fleeting components of existence together. All these terms hide the richness of emergent and dynamic relativity within them. They hide the potential provided by the inexpressible in other fuzzy dimensions, in nullity, or in infinity.

We often assume within our shortcut mental abstractions that these concepts are flat-world, linear facts of reality that require or impart no indirect logic or indeed physical impetus to our emergent dance. They ignore the relativistic reality of the dynamic interpenetration of everything big and everything small. Words, while powerful, always seduce us to fall into the trap of trivialising and simplifying our interpenetrating, fractal, and cyclical reality.

We cannot directly dissect the universe to find now or past/future. We can only be a part of indirect time and roughly measure its correlations with matter and us. Some may argue otherwise; now is simply a quantum of time equal to the Planck Length divided by the Speed of Light. However, this idea does not describe the realm of reality in which we humans consciously live our lives. Nor does it explain the processual uncertainties or interdependencies within this Planck Time, inside Planck Lengths, which we can picture and thus indirectly affect with our human imagination.

We cannot directly analyse space as a thing that physically exists either. We can only exist and work within indirect space and measure its effects. Likewise, from the viewpoint of the fleeting present as the only physiological reality, we might probe a brain to find paths of personality-supporting synapse patterns, a mere shadow of the religious idea of a knowing soul, but it will be pointless to look for a spirit-object because it is fleeting and indirect. The whole mind-body complex testifies to its spirited vitality while it is alive, and testifies to its lack of systemic coherence as the decay of cells slowly spreads out over time through the body after death. Thus, soul or consciousness is a system-wide concept or property just as is past/future, whereas spirit or unconsciousness is an emerging process or product of interpenetrating matter and space, just like now.

Looking for indirect spirit and soul is exactly equivalent to looking for indirect spacetime's components – now-and-past/future or here-and-there. What is being suggested, particularly to followers of the Abrahamic faiths, is that spirit is an introsomatic (literally 'inside body') expression of our mind/body's creation of here-now, as represented in Figure 10 – and notions of the spirit's independence arise because we falsely divorce our here-now from matter. Likewise, soul is an introsomatic expression of our body/mind's creation of past/future-there and both incorporate the same difficulty that comes with divorcing ourselves from a fully interpenetrating universe. Our dualistic religious analysis of reality has separated us from our interconnecting universe and a more efficacious, unsuperstitious monism restores this interconnection. It puts our star or diamond back into its proper setting or fit, so it can shine.

Nevertheless, the religious abstractions of spirit and soul within an interpenetrating universe can be marginally useful, mostly because they are embedded by the Abrahamic religions into our classical and uni-dependent ways of thinking about reality and self. Alternatively, they are perhaps no less valid than our faulty concepts of 'there' divorced from 'here', past/future divorced from now, or spacetime divorced from matter. We desperately need to loosen the shackles of our uni-dependent thinking and replace it with interdependent and fractal thinking that is cognizant of the difference between full reality in conscious imagination and relativistic reality embedded in our instincts. The Emergent Method will help us do this. I am challenging to let go of your fiercely independent concept of monolithic self.

We might also think of now as the convergence of all past and future or here as the convergence of all possible there. The past is conceived as closed, largely known, information-rich, and with zero degrees of freedom. The future, like there, is conceived as open, only probabilistically known, impetus-rich, and with many degrees of freedom in an emergent reality, even if limited by the quantum-mechanical wave-cancelling effects of Figure 9. Here-and-now is conceived as transitory, probable and with perhaps only two fundamental but disappearing degrees of freedom. The 2010 Wikipedia article "Degrees of Freedom (physics and chemistry) said, "In physics, for each particle belonging to a system, and for each independent direction in which movement is possible, two degrees of freedom are defined, one describing the particle's momentum in that direction, the other describing the particle's position along an axis defined by that direction."[51] A more comprehensive article now replaces this statement, but I believe it is still valid if we ignore the complexities of quantum indeterminacy and entanglement, and whatever fuzzy reduced degrees of freedom these states may infer. What it suggests is that by Physics' Uncertainty Principle we may not know beforehand with exact precision both the position and velocity of a particle.

This statement is worth considering more carefully in the context of Figure 10. Quantum Theory says we may predict a particle's position or transition across time in a fixed space, or its momentum or transition across space in a fixed time, within a certain system, but not both simultaneously because their combination is subject to probabilistic (i.e. relativistic) outcomes. That is, given possible directional constraints, we are free to directly predict the particle's emergent being in space, or its emergent doing in time, but not both pieces of data simultaneously. This clearly suggests that space and time are bound together in a fractal or self-similar relationship. That is, a quantum-mechanical relationship locks together space and time or here-and-now, as Figure 9 also suggests.

Recent indirect measurement techniques suggest we may be able to know both pieces of data simultaneously as they happen, but this does not mean we can predict them.[52] That is, the core issue of Quantum Theory is not a measurement problem as some have argued. Quantum Theory is an issue of interdependency, relativity and fuzzy logic, in a sense that is consistent with Einstein's theory of the General Relativity between matter and space.

In each probabilistic moment of our lives it seems we have the same limiting and fuzzy choice of being (observing and analysing; the '0' of matter-in-space) or doing (participating and mimicking/process-modelling; the '1' of matter-in-time). Each choice reduces the probabilities of all our competing paths moving forward from the prior moment in a complex, nonlinear or fractal, and interpenetrating fashion.[53]

If space is the '0' then time is the '1' of the probabilistic spacetime inference system of our universe. Spacetime may perhaps be seen as the system of freedom or potential for limited choice upon which our universe is based. To make successful choices we need passive information about the past, but we also need active imagination and impetus towards the future.

Perhaps spacetime is full of universal information and impetus that is disorganised in our local and emergent reality until and as we, using our technology, measure its statistical patterns or properties, value it and flow with it. At this point, local spacetime gains a kind of fuzzy and chaotic determinacy rather than remaining disorganised. That is, the universe seems constructed in such a way that all parts of it, including us, participate in its manifest becoming through choices made and paths taken or not taken, either with awareness or not.

Perhaps it is the inevitable role of all life-forms to learn how to beneficially and uniquely exploit and value spacetime's local degrees of freedom. Realising our potential is perhaps a matter of chance to the extent that we surrender or lack ability to measure, value, express/choose and store/incorporate. However, even as we gain knowledge, value, choose, and incorporate, we cannot fully set our path - only limit the distractions, alternatives or contingencies. We are bound to emerge, as does the universe.

Yet these contingencies are the probabilistic properties that wonderfully connect us to everything else emerging around us and give rise to our interpenetrating partnership with all else. We often try to force our linear, one-at-a-time, independent and non-fractal approximations of time and space into our nonlinear world. Nevertheless, as we take the personal responsibility as human beings to rationally learn, value, choose, and incorporate more carefully, with a deeper awareness of statistical and interconnecting emergence, our futures become more of our own making (and shared becoming), although not necessarily limited by our wildest dreams.

We might also consider here-and-now as the wonderful convergence of matter and all spacetime. Spacetime's logic, principles, or interactions at each local level are not directly subject to entropy or emergence, and so potentially may be carried over into the future from the past. However, matter cannot be carried over into the future without change, i.e. without entropy or emergence in our expanding universe. Nevertheless, new spacetime logic is never very far ahead of now. Eventually imagined logic needs technological or natural application in present matter to anchor it, and old logic is often lost in the distant past along with its material applications. However, new imagination can exist far ahead of now because it is not necessarily limited by matter, logic, or principles. On the down side, imagination is more quickly lost in the past if logic and matter's reality in the present does not slowly transform it.

In summary, to bring to reality things that are in the distant future we need imagination (unanchored potential) but to bring to reality things that are in the near future, we need imagination touched or tainted by logic and energy. For us humans, this means we usually need a strong understanding, desire, faith, good plans and organisational abilities, and a dogged determination. Using the Mandelbrot Set analogy, new iterations in full reality often find their logic and impetus from mutated and seemingly impossible solutions outside the set and a simple fractal formula.

Victor Hugo once said, "All the forces of the world are not so powerful as an idea whose time has come."[54] How insightful, from the viewpoint of a self-contained universe that must emerge from its unavoidable shuffling of real and virtual sub-atomic particles. One idea in full reality whose time has come is the Emergent Method.

Here then is a simple way of looking at spacetime that is a secret to realising our highest purposes and potential. If beauty and impetus lie in the spacetime of our universe then power lies in the matter of our universe. If beauty lies in the logic of an idea and impetus in the timely expression of an idea, then power lies in its material utility.

In successful endeavours, we perceive material reality, we distil its indirect value or potential to meet a need, and then we express our ideas through timely psychosocial claims or belief-narratives in tandem with appropriate actions. The environment then responds by acquiescing to successful outcomes.

Victor Hugo also said, "Each man should frame life so that at some future hour fact and his dreaming meet."[55] This hour is where your beauty, impetus, power and destiny also meet.

Life

All living systems must deal with boundary problems in two ways. They must maintain the physical boundary between self, progeny and the environment (that is, they must take shape and 'self-energise', including propagate). This is known as autopoiesis. They must also deal with contingencies or satisfy the viability constraints imposed by their boundaries and environs, also known as cognition. This kind of biological cognition is much broader than the kind we typically assign to self-aware humans. Paul Bourgine and John Stewart, in their 2004 MIT paper labelled *Autopoiesis and Cognition*, propose the thesis "A system that is both autopoietic and cognitive is a living system."[56] This proposed necessary and sufficient definition of life is a very interesting one because it is a systemic or organisational definition, rather than one that depends on any underlying biology, implemented though organic molecules.

That is, if the thesis holds, then life is platform-independent although any incarnation is dependent on a certain platform. This implies that just like space, life has a relativistic relationship with matter. This means the essence of life is in its dynamic organisation of many particles rather than in any of the particles themselves. Perhaps this is already obvious – after all, if you are unlucky enough to lose a limb, you will not necessarily perish. Similarly, if you replace your heart and lungs with a heart and lung machine, you will live. Likewise, if you were hooked up to a theoretical brain-replacement machine, you would likely live. This, in turn, means the property of life emerges from the physical implementation of a sample-independent law of large numbers (or a correlation of interrelationships of members of each life-form with its environment). Life emerges in looping systems in much the same way as the property of temperature emerges from a growing arrangement of gas molecules confined within a boundary, or the rigidity of an iron bar emerges from a growing system of ferrous ions arranged in a lattice as it is cooled. Exchanging an organ, a molecule or an ion does not eliminate the statistical property of the whole, or interrelationships of the components, even though nothing but such components makes up the whole.

Life, temperature and rigidity, along with all other properties we can name, would seem to be phenomena whose essences are not found in direct matter (because they are independent of any small sample) but in the indirect information (or locally deterministic code) running in local space itself. However, this view is also faulty because it treats spacetime information as a monolithic and separate entity in a dualistic view of reality rather than part of an interpenetrating monism. Emergent properties in any universe thus arise from the asymmetrical interpenetration and interdependence of matter and space or the interacting and replicating platform and arrangement within the environment's restraining boundaries. It is this relativistic view that seems to offer us the deepest insight into an understanding of the property we call life. This suggests there is a self-reverberating relationship between life and body that literally extends the relationship between space and matter.

Figure 11 provides a representation of Life in terms of 'introsomatic' autopoiesis and 'extrosomatic' cognition. That is, autopoiesis is about the organisation of matter within the cell or body, whereas cognition is about the organisation of the cell or body within its environment. The figure suggests the active life-phase with reductionist upward causality will tend to be exercising the body; the passive and reductionist phase, reducing the body (i.e. digesting and eliminating); the active with holistic downward causality, growing the body; and, the passive and holistic, integrating (i.e. resting and renewing the body).

This cycle should be compared with the one of Figure 6. That is the Exercising – Reducing (Digesting & Eliminating) – Growing – Integrating (Resting and Renewing) cycle of Figure 11 is a simple living refinement of the Separating – Degenerating – Arranging –

Incorporating cycle of Figure 6. Further, the introsomatic axis of Figure 11 develops and refines the microscopic matter axis of every system, living or non-living, in Figure 6, and the extrosomatic axis of Figure 11 develops and refines the more macroscopic space axis of every system, living or non-living, in Figure 6. To reverse the comparison, we might argue that the matter of every non-living system is proto-introsomatic, and likewise the space curved around every non-living system is proto-extrosomatic.

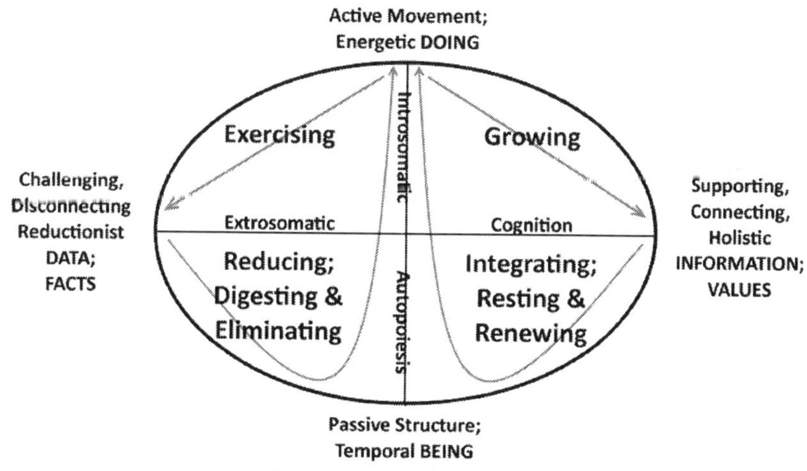

Figure 11: Life - Body as an extension of Space - Matter

Can life's emergence proceed to the point where all is in all, i.e. where all instability in our universe is eliminated? No, I do not think so, and nor would we want it so. The systems can change and the boundaries can move or grow in number and kind, but boundaries will always be there to introduce instability. As David Deutsch suggests in his 2011 book of the same title, we are always at *The Beginning of Infinity*. Perhaps the question is, if boundaries become more and more diffuse throughout a synergistic system, does the system also become more powerful in terms of its ability to prevent catastrophic self-collapse? I think the answer to this question is 'yes!' and this seems to explain why successful natural ecosystems with adequate resources seem to naturally self-organise towards a biological diversity of vastly many partially open boundaries. Of course, they do this successful self-organising blindly, and those systems that don't do this fail and disappear, as per the possible long-term fate of the unrestrained Australian fox.

The incorporation of more soft boundaries and their constraints seems to equate with the better exploitation of platform-independent or statistically significant information in the environment. That is, as systems better incorporate and anticipate boundaries and constraints, they become more robust and more cognizant, in the sense intended by Bourgine and Stewart (2004), mentioned earlier. To put this idea another way, a measure of a system's cognition is the number of constraints or contingencies it incorporates or anticipates. That is, (moral) values and intelligence emerge from the boundary interactions themselves.

Without boundary constraints and instabilities there is no opportunity for cognition, knowledge or intelligence. This is an incredibly important observation, because it means emergence and evolution are not possible without a default or background state of disorder, or as Lawrence Krauss (2012) might put it, without a universe from nothing (that is, nothing organised or orderly in a relative sense). This observation of no cognition or intelligence without border interactions also means that intelligence is a defining feature of self-aware living systems – because statistically successful living systems selfishly create, defend and enlarge their claimed borders.

Space provides the necessary frame or borders in which self-aware systems of matter, such as the human mind/body system, or the Australian socio-economic system, can grow in intelligence. It follows therefore that the change in instability at the border is the source and measure of a life's, an organisation's, or a nation's creativity. Interactive instability is the essence of our cosmos' trial-and-error creativity or freedom. This means we, as examples of conscious life, do not need to just fear hard boundaries or the unknown beyond them. Life also needs to see the boundary constraints as opportunities for more growth, stability and intelligence once they are softened and tamed. Thus, in systems of growing intelligence, uncontrolled external constraints become controlled internal self-restraints. In more simple terms, we all need the support of our loved ones and colleagues, but we also need the challenges of our environments to learn, and thus grow in intelligence.

The only difficulty with this view, as we have already noted, is that as new properties of new arrangements of matter come into being so do new risks to viability. New, unexpected instabilities can lead to the death or collapse of an existing system. Nevertheless, diversifying emergence seems to be able to proceed to the point where nearly all instability is non-catastrophic, manageable, and actually promoting of long-term viability (even if this diversification necessarily includes the cycle of birth, reproduction, and death for the individual within living systems). Diversification entails a separation of dangerous monolithic powers (political, economic, biological, or otherwise). Emergentists such as myself advocate for such diversity. Again, in more simple terms, we all need to develop our alliances with our fellows and the environment in ever more intelligent ways to secure our futures.

Many of the more elaborate properties of more elaborate arrangements of matter are of our own making. That is, we as makers and sustainers of our own environment are not doing battle against evil cosmic boundaries arrayed against us, but against the boundaries we ourselves form, move, break, and eliminate. More and more we are doing battle against ourselves, and our own lack of intelligence that comes to the fore as and when new challenges arise. This is why human society, when it acts honestly, courageously and with practiced insight, can be so resilient. I hope and trust that the human societies of the future will learn to form new risks at just the rate whereby we can also diversify or otherwise mitigate them.

New incorporations of knowledge cause a realignment of previous assumptions and beliefs but not a rejection of all previous systems of knowledge. This is the nature of advance by trial-and-error. So it seems there may be only one way of achieving large scale stability in the universe; fully organised and interpenetrating work done (in the thermodynamic sense) brought about by the diversifying emergence of ecosystems.

Thus, we can now reduce the problem of understanding life to one that is knowable on this side of any veil. Non-living systems do not have the relevant feedback systems that can exploit the fractal nature of reality to promote their own stability and ongoing viability. That is, non-living systems do not have a sufficient answer to the operation of the second law of thermodynamics (the 'Law' of Entropy; all things decay in the absence of work done to them). Nor do non-living systems have a code within their borders that can promote their own cellular stability and self-centred ends (which activates the complementary 'Law' of Negentropy or Emergence). We could also add that life is a matter of degree, because some species in the past have enjoyed more sustainable and greater success in their environments than have others. That is, life is complex, fractal, and fuzzy just as is life's environment; life is an emergent property of its spacetime environment.

The organism-based genetic code was far more able to grow in localised negentropy than the more primitive spacetime code because it had the ability to focus its attention on the wellbeing, homeostasis, or equilibrium at the systemic level of the organism. Largely unbounded or open non-living systems had little ability to focus on the problems of disequilibrium or instability presented to their cellular borders.

Living systems include feedback mechanisms (self-organising and self-centred systems) for saving themselves from decay. These mechanisms include storing and processing complex information and 'doing' work in the here-and-now. We call these blind but self-controlled systems metabolism and replication. Living systems also include self-referencing, cognitive behavioural mechanisms for doing work on the outside world at its boundaries. Non-living systems only have the marvellous mechanisms for unselfish 'being' - which include non-self-sustaining spacetime interactions. Non-living systems' near-equilibrium can display transient self-organising patterns lasting billions of years, but it can't be meaningfully or cognitively sustained or stored for future self-interested processing - and is therefore at the mercy of entropy and work done to those systems (by other non-living or living systems).

More simply, a living system is an emergent spacetime / there-and-past/future sensor/actor, whereas in comparison non-living systems, except as the phenotypic extensions of living systems, are not. A living system has dynamic and organisational information systems, as opposed to just physically ordered components, to sense its environs then fit algorithms and take actions that manipulate the emerging spacetime-matter interpenetration.

Living systems are a relativistic relationship between life and body that extends the (fractally reflective and reverberating) relationship between space and matter.

Living systems exploit matter and forces by creating value-rich rules and behavioural rituals or narratives. Living systems then apply their emergent value-systems to future contingencies. In summary, living systems exploit the relativity that exists between spacetime and matter, of which they are an extension.

To put this concept another way, a living system has self-organising soul whereas a non-living system does not (see again Figure 10); it only has proto-soul. Life and soul is in its emergent properties rather than its particular physiology or incarnation (with its particular matter and space particles). However, a particular soul requires a particular physiology.

Systematic replication, metabolism and cognition is a statistical response of living systems to the reality of spacetime as a proto-informational code inherent within all existence, rather than spacetime as an inert feature of existence like Euclidean measures of volume or lengths of time passed. That is, the spacetime code is a lower level of code than the genetic code of instincts that emerges from it, just as the genetic code is a lower level of code than the memetic code that emerges from it.

Selfishness

Even classically, for every positive, there is a negative. This is true electrochemically, magnetically, and at the nuclear level. It is roughly true of all non-living and living systems. Energy and information are conserved in any fully defined system, if we can fully define it. This means there is a kind of underlying management of conflict between positive and negative in all orderly arrangements. That is, life in cells involves a blind or self-aware management of conflict. Put another way, scarcity of fleeting space and time (i.e. scarcity of here-and-now) is an essential factor in the evolution of our universe. In other words, to the extent that the universe's dynamic arrangements are orderly they are also selfish in their careful management of conflicts. Anthropomorphically, nature's systems seem to shun disorder and pursue stability. This is perhaps our perceivable universe's core nature (we can only perceive order – we cannot meaningfully conceptualise disorder in any way). All temporary order, living or non-living, seems to selfishly struggle against conflicting disorder to achieve environmental balance and continuity.

Actually, outside of consciousness, there is no struggle, just constrained yet probabilistic behaviours and drives towards balance within typically bell-curve limits. That is, inefficiencies in terms of energy arrangements will tend to be transformed or removed from populations of subatomic particles within locally constraining environments.

As represented in Figure 10, a gross shortage of here-and-now (and a closed there/past) accompanies an abundance of there/future. Each cell lives because it has found a way to take the food-energy it needs from its local environment to defy decay. This means each form of matter (or arrangement of energy) is under a selective pressure to find a successful way to take shape and organise itself in local space and time. What one cell takes in its here-and-now becomes unavailable to other cells unless that cell is itself consumed or integrated as a source of food-energy. Living cells are thus blindly selfish or proto-intentional - they blindly take or defeat local here-and-now, or at least borrow it, for their own sakes in preference to other energy arrangements. However, they also learn that if they are to continue amongst their kin, they must also contribute, give, or accommodate local here-and-now as part of the selfish 'conflict management' process.

An organism's identity, even at unconscious levels of our sensorimotor perception, is bound to the way it selfishly gives and takes through the exploitative (and instinctive) algorithms at its disposal. The taking is a claim based on a hypothetical truth or belief, either blind or self-aware. The self is a particular claim/belief of a selfish organism, either blind or self-aware. Thus, organisms claim meanings in and of life either blindly or with self-awareness.

Each claim of hypothetical or self-made meaning is from within the fully interpenetrating Gaian system, in which there seems to be a blind naturalistic contract between all living things from ubiquitous microorganisms to us. As discussed earlier, this contract seems to put restrictions on which things can be selfishly claimed for the taking (and later legitimised) and which things cannot be.

Within any deep ecology such as ours, natural selection and ongoing energetic order legitimise those evolving claims that benefit, or do not seriously detract from, biodiversity and planetary sustainability or balance. Those claims continually out of keeping with the local environment, nature eventually sanctions with loss of value and meaning, disorder, and energy redistribution. More correctly, there are no contracts; massively parallel relationships and their positive or negative effects simply support or challenge the energetic freedoms and successes of evolving claimants or systems.

Perhaps this means selfishness is simply another word for the tendency of fuzzy systems towards local but incomplete equilibria in energy and information, and another word for the local algorithmic feedback mechanism that is common to all relativistic systems in both the living and non-living. That is, there is a relationship between unstoppable movement, proto-intention in that organised movement, and perceived selfishness. As per Figure 6 and Figure 11, life systems are not essentially different to non-life systems, in respect of their local focus on energy management within the Planck quanta of relativistic matter and spacetime.

Figure 12 shows the four quadrants of all emerging and decaying systems. In their management of energy, they advance through the four phases of conflict management labelled (selfish zero-sum) win-lose, (negative non-zero-sum) lose-lose, (positive non-zero-sum) win-win and (acquiescent zero-sum) lose-win either blindly or with self-awareness.

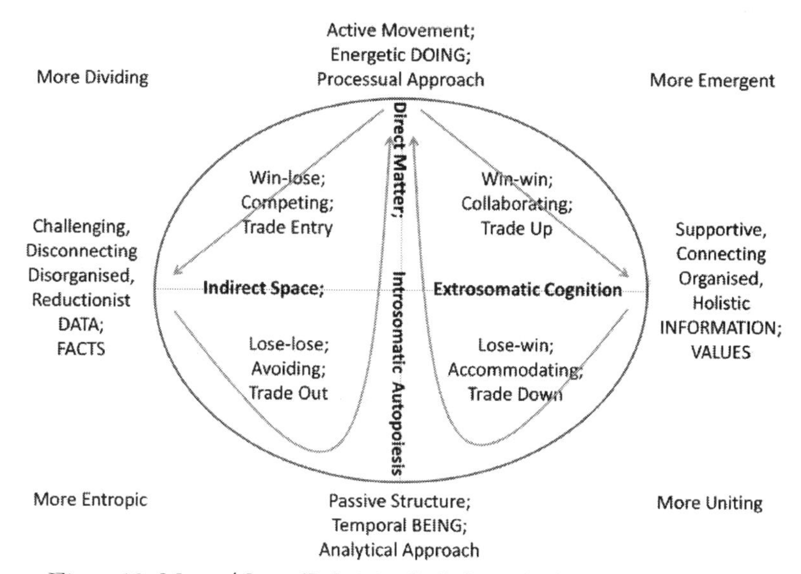

Figure 12: Matter/Space Relativity, Life Struggle & Conflict Management

In particular, we as 'systems of self' typically take personal wins from our competitors in Phase 1, and then sometimes take time in Phase 2 to reflect, filter, or extract a useful narrative. Sometimes we leave the field of contest, such that nobody can take a win and everybody loses by realising short-term opportunity costs, but if successful in this phase, we increase our ongoing efficiency. We may then step forward with newly-realised meaning to share our next win with others in Phase 3. The interconnection adds further meaning to our endeavours. In the following realisation of meaning in Phase

4, we may concede defeat or allow others to take the win for us or from us. That is, we acquiesce, in an act of interdependence, goodwill, or faith that reciprocal benefits will follow, because we know we don't work alone. The cycle seems to accomplish a local incomplete balance, then repeat itself.

It is worthwhile taking some time to reflect on this whole energy conflict management process. It is different to the one dimensional, uni-dependent fight between a winner and a loser. The interdependency of 'players' (subatomic particles, life-forms, or people) and the struggle between the relativistic direct and indirect, is why it can form a cycle in two or more dimensions. Further, the acquiescence to the lose-win, a kind of acceptance of incompleteness within any one cycle, seems to soften boundaries and provide momentum, and thus set the scene for the next cycle of emergence on a slightly higher level. This cycle describes the relativistic tension between infinity and nullity at all levels from the quantum mechanical to the corporate head office.

It also describes the interdependent acquiescence (that forms fractal patterns) between our environment and us. That is, because of incompleteness and interdependence, the environment will acquiesce to our claims if it is not incongruent from an energy-management point of view. Likewise, we accommodate the environment when it claims coagulations of energy in ways we do not wish to challenge. These reflective reverberations result is a mutual learning process through cycles of necessarily incomplete arrangements. In one sense, these cycles of interpenetrating energy effectiveness and energy efficiency are the basis of every learning experience in the cosmos, blind or self-aware. Systemic selfishness exists on a fuzzy scale from non-life to non-conscious life to conscious life, and so does systemic learning.

The diagram suggested to me that conflict resolution, just like statistical emergence, life's struggle, and our search for meaning, is more about the interdependent journey than the uni-dependent destination. We need (the system of) self to rise to fight the status-quo, redefine value and add order, but we also need self to retire so that others may rise to continue the journey. Freedom emerges in the journey; it is not reached in the final conflict resolution or in paradise. Life experience is a dynamic claiming, filtering, sharing and losing of self, either blindly or with self-awareness, but also a gaining of interdependence, all the way from self's personal to civilisational levels. Figure 12 suggest we can apply this lesson to non-life as well, by the systemic requirements for conservation of energy and conservation of information.

That is, here-and-now is the continuous and variable shuffling of entropy (the bottom-left quadrant) and negentropy (the top-right quadrant). While the universe expands at least, we see macroscopic negentropy in nature's arrangements and microscopic entropy in nature's subatomic building blocks, living and non-living. In an important sense, the 'creative and destructive divinity' of existence, or the 'good and evil' of existence, is in these two phases.

The 'good' in this context is notionally seen here as all that adds new order, value or meaning to existence, whereas the 'evil' is seen as all that destroys existing order, value and meaning. The 'good' exposes the here-and-now to new connections, relationships and arrangements, whereas the 'evil' exposes the here-and-now to the closing and annihilation of existing connections. The more the indeterminate 'good' states remain open, the deeper the search for new connection can be. Conversely, the more the indeterminate 'evil' states remain open, the deeper the destruction can be. Our human decision-making works in a derivative manner. That is, the longer we hold off coming to a decision, remaining in a more indeterminate state, the more connected that decision, for better or worse, is likely to be. I must add that this notional good and evil, which more simply is support and challenge, is needed in any system of advancing energy organisation.

The other two 'dividing' (or conquering) and 'uniting' phases are perhaps a little less pregnant with anthropomorphic implication, positive or negative, and a little more mundane in their workings. Nevertheless, it is difficult not to see their correlation with Yang (aggressive, focused, bright/direct, masculine) and Yin (yielding, diffuse, shaded/indirect, feminine), respectively.

As an aside, the Yin and Yang symbol, as depicted in Figure 13, is interesting in terms of all we have discussed so far. It depicts reality as:

- a circle or cycle of endless yet bounded activity
- a balance between two opposites, black (nullity) and white (infinity)
- an indeterminate separation between local opposites, indicated by the continuously curved line between left and right sides of the diagram

- an interpenetration of these two opposites, indicated by the two dots in opposing sectors

Figure 13: Symbol of Yin and Yang

Living cells take what they must to survive or endure here-and-now, and the upper two 'win' phases of here-and-now in Figure 12 are the secret ingredients of their taking. The logic and two degrees/axes of freedom of the here-and-now software within the spacetime code conspire with cells to take or borrow while cells metabolise and live. Some call this spirit. It operates down to/with the quanta of Planck Length and Planck Time. Equally, the spacetime code conspires to force the surrender and return of what cells borrow when they fail to metabolise and live, or acquiesce and disintegrate. We tend to externalise this feature and do not explicitly name it, except perhaps to call it the process of spirit-return initiated by the Grim Reaper. This is embodied in the lower two 'lose' phases of here-and-now in Figure 12.

We embody, personalise and internalise here-and-now software in living systems but we disembody, depersonalise and externalise the same here-and-now software in dead or non-living systems. Yet the only difference between the two is that living systems accentuate the taking and processual doing in the upper phases of Figure 12, whereas non-living systems accentuate the giving and relational being in the lower phases of Figure 12. In every little win, of order and life, is the seed of the next loss of disorder and death. However, without this cycle of life and death, there is currently little chance of progress to higher levels of order, synergy, and sophistication.

The relativistic here-and-now software within the spacetime code is perhaps like the threads we pick up and sew with while we live. When we are gone, we may leave behind us a memorable pattern within the fabric's arrangements. If our legacy is strong, maybe someone else will pick up some of those borrowed threads and keep sewing. If not, efficient and self-controlling here-and-now software within this universal fabric will ensure that the fabric's pattern (its tangible parts and its logical order) will wither away so that its real components may be reused and virtual logic reconstructed in that fabric in the future.

Living cells compete for exploitable information and impetus in spacetime, as well as for food-energy and survival. This fact is not restricted to the intentional stance[57] of exuberant carnivores and hunter-warriors – it is requisite of all life forms and their self-organising cells. All organisms exploit here-and-now through competition and cooperation, with respect to the information available in there and past/future. This is how they complete the transition to the next perceived moment (just as per Figure 6 and Figure 7). The results of their proto-intentional efforts add either to their flourishing or to their demise. Further, for human survivors that claim a meaningful dimension in such flourishing, and whose environments legitimate such claims, there is also an emergent moral component/layer to this give and take.

We humans exploit there and past/future much more deeply than other animals because we have developed an elaborate consciousness that enables us to make summaries of the past through introspection and extrospection as shown in Figure 10. We are then able to store those analytical summaries over long periods rather than continually process them to keep them vibrant, and apply them in our strategies with respect to our survival in the emerging future. We combine the best aspects of Figure 1 and minimise their limitations. Within limits of time and interrelationship, we exploit all areas inside and outside reality's modified Mandelbrot set (see Figure 3).

Wait a minute: Is it life that has found ways to exploit spacetime or is it spacetime that has found ways for self-expression, firstly in non-living matter and then in a multiplicity of life forms, including self-aware humans? Is life selfish or is the spacetime code the ultimate source of selfishness? Relativity teaches us the answer; everything is interconnected and interdependent. Life unnaturally divorced from its life-forms is another version of space unnaturally divorced from matter. Just as we cannot find an absolute answer in matter alone or space alone when it comes to their interactions, so we can't find an absolute answer to this question of life exploiting spacetime or vice-versa. However, we can recognise and deduce the relativistic relationships that emerge from both life and non-life.

Exploitable energy/information organisation in all its forms seems to beget and seek more of itself in an expanding universe of ever more numerous but softer and self-contained border constraints. In reality, a vast statistical soup of subatomic particles will tend to highlight more stable configurations and cause less stable configurations to fade into obscurity. I suspect this statistical concept presents the clearest picture of organisational selfishness. Without intending any transcendental connotation, your life and mine is part of this fractal lattice or web of transforming and organising energy/information that spreads itself across the universe with every particle of matter.

To despise the physical forms of life as illusory or blind and fallen shadows of an ultimate truth is to also despise reality's relativistic and marvellous arrangements or mechanisms. With the necessary aid of matter, life is always self-organising and self-referencing. Knowable life cannot bootstrap itself without forms that require and take information, impetus and food-energy from the local environment. Conversely, it is in the self-seeking taking that life forms also give knowledge, meaning and value back to life's genetic code, and organisation back to reality's otherwise disorganised spacetime code or system.

Likewise, consciousness and its virtues do not come without empowering life forms. Your conscious life, which rightly you should celebrate, comes at a cost to others in your here-and-now as your body takes the food-energy, self-organising information and impetus, or indirect realities, it requires from its local environment. You have already taken both consciously and unconsciously. Perhaps the deeper question is, 'What are you giving in return?' What are you going to be, and what are you going to surrender as your ultimate legacy to the living after your consciousness has lost its place in your body? What organisation are you going to give back to the spacetime code that grants your fleeting doing?

The most important thing all selfish organisms give back is methods of self-organisation and a sense of meaning and purpose to their species and ecosystems. That is, a relativistic universe enables us to observe and deduce this paradox and then take part in its budding meaning. Meaning emerges in levels, first blindly and then consciously after articulation. This meaning is not necessarily anthropomorphic (each species demonstrates its own meaning) or a mere set of relative half-truths, but rather part of an emergent method or process of demonstrated added value in our deep ecology that takes the cosmos away from monolithic disorder and towards a more synergistic order. For us humans, the Emergent Method is perhaps the most explicit statement of our species' method for tackling disorder so far.

This brings us to perhaps the most difficult but illuminating introductory issue. We can only truly appreciate being by not doing but we can only truly appreciate living by doing. Fully knowing/valuing life is an emergent dance of both being and doing (as shown in Figure 10).

To live is to do – the two come together in the same package. The second law of thermodynamics and the activity we see in the life of every living cell, extracting energy to do work and thus avoid internal decay, teaches us that to live is to do. Our energy usage must be effective if we are to survive and prosper. Use it or lose it. Conversely, to just be, to just perceive and analyse, involves no direct affecting of the external environments or external stakeholders. It thus promotes an awareness of nothingness and 'everythingness', or an awareness of relationship, identity and synthesis. This is an awareness that requires no processual interaction, yet its massively parallel interconnections are vast and complex. In one sense, we could argue that to do is necessarily disruptive whereas to be is necessarily unifying or perhaps purifying, in the sense of promoting an awareness of sufficiency and potential. To 'do' sees this as opposed to that or positive and negative, whereas to 'be' sees no obligatory opposites at all.

Thus, we might contrast the often-blind intentionality of doing that seeks systemic gain or effectiveness through disruption or disequilibrium versus the often-blind intentionality of being that seeks equilibrium or efficiency, even via systemic loss. If we want to 'be' without aiming to 'do', we deny our fleeting life-and-being here-and-now essence. If we aim to do without stopping to be we deny our most basic, material existence that, far from fleeting, has existed for all time past, even in the heart of stars. Just like emergent matter and space, we gotta dance, not just be or do. We have to do the dance of both energy effectiveness and energy efficiency. We have to obey Heisenberg's Principle, just like all other collections of subatomic particles. In the human dance, we must actively seek and choose to quiet ourselves, experience, think, identify, know, and then act, because at our best we are rational beings. This is our species' distinguishing nature and our species' unique means of survival and flourishing. We exploit all areas of Figure 1.

To enjoy this short life and get its rate of change or passing of biological time roughly right, we must both selfishly take here-and-now (with closing self-restraints) and we must unselfishly surrender here-and-now (with opening self-restraints). One is the fleeting and emergent principle of the self-organised living, which is taking. The other is the timeless and self-eliminating principle of the externally organised non-living, which is giving.

A fulfilled and happy life is a wonderful dance or musical improvisation between the doing and the being. To take a new breath, we also have to give our old breath. The interplay of taking and giving here-and-now is at the core of the philosophy of emergentism. It is also the essence of the biosphere's and evolution's unstoppable algorithms.

Armed with the idea of the to-and-fro of existence, we can now get away from the unnecessary traps and limitations of Gödel's Theorems. We can take a process-modelling view to life rather than just a structural or geometric one. Thus, we might use the concepts of the direct and indirect, the unconscious and the conscious, and so on, to highlight their interdependence with the environment rather than try to defend their fierce independence. This means our sustainable enlightenment as individuals has the chance to advance exponentially as we recognise the myriad interrelationships of our deep ecology and become actively inclusive of them in our individually selfish, but more enlightened, hopes and aspirations. We, and our environment, are 'joined at the hip', just as a mountain is inseparable from its mountain range.

Notice here that the fleeting is the one attribute we classically assigned to time, and the constant or seemingly inert is the one attribute we assigned to space. The dance that both frees and binds, or both self-organises and self-eliminates but either way, reshuffles the living and non-living alike is thus generally one of empowering matter/energy, logically enabled and invigorated by spacetime and its interactions. These same reshufflings of the dance are those that enable the algorithms of evolution to find a better path forward - assuming we deem the biodynamic advances of evolution as generally a good thing rather than neutral.

For flourishing humans, the lower part of Figure 12 is outwardly-passive being in the moment, as sensor and rational observer of shape and relationship but like Nature, not a procrastinator. The upper part of Figure 12 is active doing in the moment, as actor and volitional participant. This represents the fundamental freedom of choice a fully active consciousness-plus-instincts may have. Additionally, just like a computer's running machine code of zeroes and ones, these two degrees of freedom can lead to all the other possibilities the programmer can envisage.

The purpose of this dance in the living universe must be doing, and in humanity, this means rationally and volitionally. The Uncertainty Principle tells us it cannot ultimately be being, because this would end the dance and natural cycles, the regeneration and refreshing of life, and end in death. That is, we must lay down our lives in being, but we must always confidently take them up again in a more enlightened and interconnected doing, in growing readiness for that day when reality determines that we no longer can - but our posterity even more ably can. The timing and placement of our passive observing and active participating is how we join and become proficient in the emergent, self-organising and joyful dance of life and the universe. This continual striving for energy effectiveness and efficiency through new arrangements that defy decay is how we find a personal proficiency in life and fulfil our highest purposes.

The Secret and the Law of Attraction

Sometimes the Law of Attraction and quantum physics are given a kind of supernatural status. The Emergent Method, perhaps like some Buddhist practices, never relies on a supernatural axiom. Other than this issue, there are no major disagreements between The Emergent Method and the Law of Attraction. That is, the Emergent Method acknowledges the strong interconnection between (or lack of independence of) thoughts, actions and environmental responses, just as does the Law of Attraction.

However, there is some confusion. That is, some followers of the Law of Attraction might see as the supernatural miracles of a universal higher consciousness, I would call the natural but wonderful outcome of a universe that has expanded from a single quantum in which was joined the reductionist building-blocks and the holistic (or cellular) arrangement.

In all its expanding, the link between the holistic and reductionist has never been destroyed, just stretched and diversified. That is, the intimate connection or relatedness of the reductionist and holistic gives the appearance of universal intelligence, but it is no such thing in the dualistic, religious sense. The cosmos is monistic, not dualistic – but it is 'proto-intelligent' to the extent that its reductionism dynamically relates to its holism, and vice versa. It is as if the very big has cognizance of the very small and the very small plays its proficient and self-assured role in the functioning of the very big. This kind of intelligence vastly exceeds the kind that is the product of consciousness alone because it is all-inclusive, like the cooperative flow in the momentary slipstream of mind, body and environment that is in the very essence of all creative or emergent activities. However, a cosmic purpose provided for all the things that exist is unrealistic and unnecessary. Space's interactions, or movements of matter, provide a local proto-intention, without which our animal intentions could not emerge.

I find these musings on intelligence and intention agree well with the sentiments expressed in a book first published in 1910 by Wallace D. Wattles, entitled *The Science of Getting Rich.*[58] The original material of this book is now in the public domain and out of copyright. We are told this book was the single largest influence on Rhonda Byrne, the creator of *The Secret.*

Both Emergent Method and Secret practitioners suggests we can exploit our knowledge of the universe's interconnectedness between the very big and the very small for our own purposes. Perhaps the question is, 'How?' Is it through a kind of secret, sacred, and independent faith? For instance, can we expect that our selfish positive thinking about winning the lottery will actually lead to a win? I'm not sure we are capable of joining such a clear 'butterfly effect' control structure that links our intentional thought waves with intermediate real and virtual particles, and the quantum exploitation of differing surfaces on the lottery machine's dropping balls. The intentional creation of a car parking space, after controlling for all competing contingencies, seems just as daunting – and seems to assign a personally pliable, panpsychist, and misleading Consciousness to all subatomic vibrations around us. Do we really want a universe that is independently controllable by diverse entities that may or may not be proficient or represent our best interests? Such ideas seem to try to rob the universe of its wonderful indeterminacy that is the necessary and sufficient condition for the emergence of local systemic freedoms and our local sense of shared but constrained free will.

While there has been some success with brain implants in directing mammalian brain activities, just using our thoughts to translate into actions 30cm away has never been achieved under experimental conditions. Existing cutting-edge jet fighter pilot helmets, such as the BAE's Striker II (~$600,000 USD, each), cannot currently read and direct brain waves as suggested in the famous *Firefox* movie (1982, Warner Bros), starring Clint Eastwood. That is, it seems influence through interrelationship is currently the only way of causing thoughts to translate into controlled actions at a distance, whether inside or outside the animal body. Media technologies enhance this ability. It seems to me a community that requires synergistic and growing cooperation with respect to the equitable sharing of car spaces, for example, is preferable to one in which independent selfish agents ignorant of the needs of others compete mercilessly for those same car spaces.

The Emergent Method focuses attention on possibilities within this emergent realm, without underestimating the universe's other opportune (or inopportune) behaviours. For instance, synergistic and instinctive behaviours can arise between different elements, living and non-living, in an environment, but not without a period of 'adaptive pattern learning' from that environment. Likewise, the environment cannot rise up as a fully independent agent of activity – it depends on each photon as much as each photon depends on it. Thus, we can't sit at the controls of the free environment to act independently (e.g. to rig the lottery) – no matter how hard we pray or meditate. However, we can evolve new methods of measuring the behaviours of local subatomic particles and then take collaborative action in the environment based on those measurements. Examples in animals of 'measurement at a distance' facilitating organised action would include echo-location (bats), hydro-location (dolphins), water-pressure-location (Chinese giant salamanders), or electro-location (platypuses and sharks). Human technology is now capable of micking all of these abilities beyond our immediate senses, as well as 'meaning-location'. Coded and applied mimicry is essential to finding common ground between all relativistic components.

We have so far suggested that the emergent nature of the local universe means we all have a shared free will with the environment and we are not locked into a fatalistic determinism at the local level. At the same time, a fuzzy determinism at the local level means reality does not fracture or fall apart. This means the sun is highly likely to rise tomorrow, and stratocumulus clouds will look recognisably similar. It also means all the other techniques suggested by more grounded advocates of the Secret are still available to us and beneficial. For instance, the idea of gratitude as being promoting of success is still valid, and the idea of distilling and vividly imagining desires and objectives is clearly sensible given the structures of the human brain

Even more broadly, the Law of Attraction still makes sense, even if it works by the ripple effects of influence through interrelationship travelling in parallel at up to the speed of light, and aided by technology, rather than some kind of vague panpsychism. We should not underestimate the often multiplicative power available to us through the universe's natural interconnectivity and its statistical response to all inputs that would bring it to higher levels of local energy organisation. We often get to choose, within our degrees of freedom, whether our inputs to a system will make it smaller than what was there in the past (through our destructive abilities) or larger (through our constructive enterprises). I would suggest we focus on arranging and constructive endeavours, and their emergent properties, or at least not overlook them.

The Secret is not that we can all get to independently control the universe for a few moments, but that nothing is independent and we are thus crucial partners with the whole universe over the short, medium, and long term. Many of us are slowly moving together from states of competitive constraint to arrangements of intelligent, cooperative restraint. We, our unconscious collections of subatomic particles, control the whole universe, but not without its acquiescence and our paradoxical conformance with it. There is a relativistic power-sharing arrangement between our 'personal' set of subatomic particles and the rest that cannot be undone, except in imagination. If you think about it, we wouldn't want it any other way.

What follows is an interpretation of *The Science of Getting Rich* in terms of emergentism, and without need for a dualistic basis. If you like, what follows represents 'the Secret' or 'the Law of Attraction' for monists (non-dualists).

I thank Wallace D. Wattles for the inspiration:

Main Principle

- All things emerge from quantum relativity. Conversely, the entire universe is one relativistic quantum, as it was in the first moment, but expanding
- Within this monistic and relativistic dynamism, thoughts contribute literally to the manifestation of the things for which they form a congruent image, just as the wind beneath wings contributes to flight
- We can form the seeds of things in our thoughts as we cooperate and participate with the local environment in the emergence of the things we think about

Other Principles

- It is sometimes better to seek emergence than to compete for what has already emerged
- To attract wealth and get wealthy, we should do it in such a way that when we get it, every other person will have more through trading with us than they did before
- All things have a value-in-use and a value-in-price. We should give to everyone more in use-value than we take from them in cash-value; then we are adding to the advancement of the world through every business transaction
- Gratitude will keep us in close harmony with cooperative and emergent thought and prevent us from falling back into unnecessary competitive and destructive thought. A grateful mind continually expects beneficial things to emerge, and so continually participates in their actual emergence – but this is not magic
- To transmit the emergence of a thing from a mere thought, we must personally form a volitional, rational, moral and self-enlightened coherence with that thought. We should usually navigate our path just as we would plan a helicopter trip - before we leave the hangar. This way, every action-

step will proceed along the planned path. It is not enough that we 'wish to do good'. Personal moral congruence also means congruence with the environment. We should 'live in the presence' of the destination/potential reality and own it through daily purposes and actions, until it becomes a relativistic reality. We should then move on to greater challenges

- We can aim at nothing greater than to become 'wealthy'. To gain true wealth, we only need to use our willpower on ourselves, as part of the environment with which the 'self' flows. We should use our willpower to keep ourselves thinking and acting the way that is commensurate with gaining wealth, which is a balanced and emergent way. We should guard our thoughts, keeping them on track, and make this our daily practice
- All that is possible comes by getting wealth, because everything is made possible by the arrangements and use of material things
- Love flourishes where there is refinement, a high level of thought, and freedom from corrupting influences. This is where wealth is obtained by the exercise of virtuous, emergent thought
- An ounce of doing is worth a pound of theory. We should make the most of ourselves, even before we are fully comfortable with the mechanism of how our purposes will be achieved. This is because involvement in relationships is more efficacious than mere calculations. Nature does not calculate, it dynamically interrelates
- Actions must supplement our thoughts. Without actions, our thoughts stagnate
- Our emergent thoughts link us with the interactions and energy of an emergent universe's systems. We do not need to try to magically project our thoughts onto the universal web – they are already there. It is not that we independently influence the universe or vice versa, but that we learn how to flow together to supplement each other's systems. In this cyclic flow there is fulfilment and congruence at a holistic level that exceeds any individual's conscious level. We are

part of the original quantum moment that has since emerged to manifest and include us, as we are, here-and-now

- Our lived expression of our emergent and quasi-universal values put us in the right path at the right time. By 'right', I mean morally congruent or energy efficient and effective across the various levels of our being (unconscious, subconscious and conscious). This attracts others in their right path to join us. We jointly achieve in the midst of challenge and support. It seems like the universe works for us, but we equally work for it as we provide it with our vision, purpose and gratitude. This is part of the cosmic flow towards greater diversity, synergy and resilience
- We must appropriate what comes to us for that purpose, to serve the universe and ourselves with more enlightened selfishness
- We can only get what is ours by giving to others what is theirs in us, that is, in our synergistic services to the universe
- We must act here-and-now. We should apply our will to the here-and-now, while designing our future actions. Our action must be in the present, local circumstance, working with the people and resources around us. We should start acting with an appreciation of emergence from what is here-and-now, to what we can envision in the there-and-then. This is the Emergent Method. It will make us wealthy. We and our vision will meet in small, local steps that are in tune with the more efficient and effective use of energy and information in the cosmic cycle
- We must do what we can to bring about balance and emergence in our here-and-now. We can only advance towards the achievement of our mission or vision through triggering emergence. No one advances by leaving his or her emergent work undone
- Those people who supersede the here-and-now, through their contributions to sustainable here-and-now emergence, advance the universe. All others have their place, but they do not contribute directly to win-win advancement. In the win-lose, lose-lose, or lose-win there is either a net containment or loss of prior advances. Nevertheless, the

destruction of prior advances is necessary if they are not sustainable

- Becoming wealthy is a matter of releasing the potential to multiply in the here-and-now. It is a matter of energy effectiveness and efficiency through cosmic synergy. It transforms cosmic constraints into disciplined self-restraints. If we act ineffectively and inefficiently, we release effects that detract from our sustainable wealth. On the other hand, every energy-efficiency-promoting and energy-effectiveness-promoting action is a partial success in itself and a contributor to our success. Becoming successful and getting wealth is thus a scientific discipline of energy usage
- We do not need to act hastily, under stress or fear, because our highest contribution is unique and flows with the universe's current systems; we just need to strive to act more synergistically and emergently in the circumstance (or refrain from acting)
- Everyone is aware of the fertility and bounty of nature. Its abundance, when we witness it, is luscious. Following nature's example, we desire increase for ourselves, through all our trading activities. We expect a kind of win-win; we expect a kind of gain from arbitrage between traders, whereby the one has a need met by the good or service received being greater than the price paid, and the other has a need met by producing at a cost lower than the price received. That is, both traders walk away with a feeling of enjoying bounty and advancement through the non-coercive trade. This impression of mutual benefit is what all wealthy traders repeatedly achieve in their trades, because it leads to bounty
- A good worker will be kept in his or her job, but a worker whose vision is greater than his or her job, the advancing or emergent worker, will be promoted
- The emerging entrepreneur disrupts the industry that is the focus of endeavours, redefining stakeholders or creating new stakeholders. The latter emergent arrangements are more beneficial to all stakeholders than the old ones
- Always look upon the universe as in the midst of emerging, becoming or increasing; never be discouraged; always speak

in terms of advancement. Don't be fair-minded or considerate to the point of discouragement, because this lacks vision

The Not-So-Blind Assumption

If we slow down for long enough, most of us make assumptions as to the nature of existence. Eastern religions and philosophies have held for a very long time a belief in the fundamental Yin-Yang of the universe. In the past the Western worldview was perhaps more certain, even though it always maintained traditions of acknowledging the peripheral uncertainties. In recent centuries, many believed the uncertain orbited around the certain core. For instance, we once believed the universe was predominantly a kind of well-designed and well-oiled machine.

Today it seems the tables have turned. The global person's worldview tends to be far more uncertain, but with a practical or instrumental allowance for some local and temporal certainties in the universe. That is, many believe the universe is not designed or well-oiled at all. Rather, they believe it is irredeemably probabilistic and meaningless in total and in essence.

A study of emergentism adds an important third view. The 'new and challenging' wonderfully emerges from a cosmos that continually seeks equilibrium between its uncertain quantum core and its more organised and expanding cosmic features. That is, local determinate organisation in space emerges over time from local indeterminate matter, and vice versa. This arrangement suggests the universe is structured in such a way (an interdependent way) as to enable us to find local freedom within it. The universe, while uncertain, offers us potential. We are free to find our own way and our own meaning.

It also suggests that the universe has been probabilistically likely, over the last 13.8 billion years, to continue to emerge here on planet Earth, and elsewhere. The interdependency of local determinism in space and local indeterminism in matter means that they work side-by-side each other in a relativistic manner, like an electrical signal that has bursts of information within its noisy background.

We do at least need to consider which of the three views (certain, uncertain or emergent) is the more efficacious for us as rational beings invested in life and reality. One way we could do this is to look at nature. For instance, consider a tiger surviving in the forests of Siberia. What is its efficacious method of survival from which we might infer nature's blind worldview? Do intention, certainty and resolution dominate the tiger's actions or does environmental uncertainty and self-doubt dominate its actions?

Obviously all life forms seem to flourish as they move with growing certainty or implicit meaning in the environments to which they are adapting. Instincts and limited consciousness seem to invest animals with predominating certainty, modified by their careful risk assessments as the environment becomes more uncertain.

Likewise, if we want to get the best out of this life, our level of uncertainty should relate to the nature of our local environment rather than to the nature of our innate existence however defined. It is on this efficacious and positive basis that this study of human nature will move forward. This basis is a choice akin to the one that sees here-and-now as pregnant with future challenge and opportunity rather than abandoned with scarcity left over from the past.

Motive

The motivation for writing this book is to challenge the classical basis of our deep beliefs. More importantly, my motivation is to help open minds to more possibilities as to how we as humans came to be as we are today, and therefore how we might live more proficiently now and in the future as individuals and societies.

More specifically, the book's aim is to help us find our purposes in this life. The motivation is to assist all of us, individually and collectively, reach our virtuous potential by moving away from fixed and negative and self-eliminating mental models, by reconnecting with the wonderful relativistic reality, and by releasing each of us to the awesome potential of full reality through virtuous imagination.

As we move towards our potential, we transform our lives in every sense. We arise just as the chaotic but self-organising universe continues to do so. The goal is to help us use the inferences contained in these pages to quietly transform our lives and the lives of those around us – our families, our friends, our colleagues, and our other stakeholders. We will achieve this simply because we fulfil our lives.

We fulfil our lives by being clear and passionate about the quasi-objective values and ideas we stand for, and by doing what matters to us as part of this deep ecology, even in the midst of life's often bitter struggles and setbacks.

These words are being addressed to people like us who take the time to think, whether greenie or not, whether part of an academic profession or not, whether in families or not, whether religious or not, whether in a 9-to-5 job or not. We are people who, with a little prodding perhaps, really think about our chosen values and purposes in life and do not always follow the crowd.

When given the opportunity, we vote with conscious conviction. We make mistakes, but we are also the ones who will clearly and positively change the world in the corners we inhabit (our here-and-now) and beyond, because we have an authentic and growing sense of knowing what we're doing and why, and we're passionate about it.

We are quietly confident that our probabilistic/chaotic universe, even though it owes us nothing and promises us nothing can, much like a well-trained subconsciousness, respond to new order, well-informed logic, positive values and strong willpower in wonderful ways, always flowing at the local level.

Applied and explicit positive values (i.e. virtues) may evolve, but for us humans, not apart from thinking, learning brains. Humankind is not a life form fully driven by instinct, like plants. We are beings with rational self-awareness; we survive because we choose to think and imagine. Explicit values evolve in human society because some people think about them and others get the advantage of their distilled thought-products and use those thought-products themselves to advance their own wellbeing.

In just the same way, some think about building flying machines and millions of others reap the advantage of their distilled thought-products, such as an A380 airbus, in their everyday lives, perhaps more than a century later. In broader terms, our human societies have moved forward through what Steven Pinker calls "the civilising process" over several centuries, tempered in the three decades from the 1960's in particular by an "informalizing process".[59]

Like many people, I am no longer associated with a church, creed, or set of doctrines, although I was once very active in a small church in Sydney and before that, in Melbourne. Nevertheless, I still think of myself as a deeply spiritual person in that I hold my current set of hard-won-by values very dearly. I will recount some of my personal story in the next chapter.

Like the line in the movie *Gladiator*, "Once there was a dream, which was Rome",[60] I also believe in a dream for our future, which I call our emerging and virtuous global tradition. By global I really mean spreading and recurring across the globe locally.

This virtuous tradition, as opposed to the darker side of our global past, has been slowly evolving from the times of all the ancient civilisations or empires, in such things as the messages of the Old Testament poets/songwriters and in many Buddhist teachings, to the present day. This tradition promotes fulfilment of the individual in a successful, happy and flourishing community of humanity.

I believe, perhaps a little too optimistically, that our traditions of genuine human expression are about to evolve into a very exciting new phase in the times of our children, grandchildren and great grandchildren, but they are going to require that we begin to look at ourselves very differently.

Culture

Human culture has an impact on human development that is quickly becoming more important than the ancient impact of genetic evolution. In cultural terms, the world seems to be getting smaller day-by-day. New fashions and new ways of thinking about issues (political, legal, philosophical, or economic) can sweep the planet in seconds, every day. The current philosophy of science, largely developed since World War II, is bringing about developments in such areas as information technology and genetic engineering that are simply staggering in terms of both the impact to, and pace of change in, our daily lives. However, science itself seems to fail to unequivocally answer all the challenges we face.

On the other side of this philosophical coin stand the world's traditional religions and other new and developing spiritual paths. Sadly, it seems that since the days of President George W. Bush's administration the polarity between Christians and non-Christians, between Christian fundamentalists and Muslim fundamentalists, and between spiritual believers and atheists, has been amplified.

We have come to the point where cultural polarisation, combined with our compromised environment and our weapons of mass destruction that just one malicious person can wield, is dangerous on a worldwide scale. We need to carefully re-assess our situation as perhaps self-appointed custodians of our planet. We are not separate from the planet. We have perhaps failed to see that this planet is our custodian as much as the opposite may be the case. Just like the inhabitants of Easter Island 300 years ago, we need to see that we live on a rock of severely limited resources and we need to carefully manage our way forward, or face decimation. What kind of world culture do we want to arise out of the milieu to take us safely and sustainably forward?

We probably need a series of global internet websites in every language dedicated to this very delicate cultural discussion. Somehow, we need to retain our local cultural individuality but also enhance our awareness of our global interdependence, so that we might all enjoy stability and development. The closing chapters of this book will propose how we might achieve this.

I think we all need to take a broader responsibility for our actions and more deeply, take more responsibility for the values we espouse and the behaviours we habitually practice. This is not the time to shift the responsibility or bury our heads in the sand. Through interconnection with our planetary home, we need to be surer of our feelings, behaviours, values and purposes than ever before. We need to carefully move away from thoughtlessness and thoughtless actions without intent that we later deeply regret. We also need to move away from the short-term allure of the culturally popular if this is also damaging to our planetary flourishing and sustainability. We need to be able to question the cultural mores and taboos that seem to so sharply separate and divide our current civilisations. There is some evidence to suggest that the former inhabitants of Easter Island faced all these same issues.[61]

Nevertheless, there is also a lot of evidence to suggest that this vital cultural innovation is already taking place. In many societies, all the outward forms of religions are becoming less important. The religious buildings, icons, relics, traditional practices and even deontological doctrines (fixed rules and duties) are less able to define our spiritual lives today. Perhaps for many of similar age to me, after coming through the sixties and seventies, the excessively experiential or evangelical side of some modern religions or ideologies are far less persuasive.

Many of us are readier to take individual responsibility for our beliefs and values than we were in the past and we are less ready to accept as a defence the soldier who claims he was merely following orders. That is, it no longer seems acceptable for the individual to bury, disown, surrender or abrogate his or her own moral responsibility or agency within the perhaps unclear objectives of a sports club, religious group, economic class, military force or other subculture. We, as sound individuals or organisations, each occupy a unique here-and-now and thus retain a unique responsibility for this eco-psychosocial environment in our enjoyment of it.

It was for this reason if you remember, in the courtroom battle of the movie *A Few Good Men*, Corporal Dawson and Private Downey, after being cleared of murder and conspiracy to commit murder, were nevertheless found guilty of conduct unbecoming a US marine. They had committed a "code red"[62] under the order of Colonel Jessop, but it was still their responsibility to disobey that wayward order so that they could safeguard the honour of the US marines. This was the movie's powerful message.

The breakdown of old institutions and power bases that inevitably goes along with this process of global yet individual ethical change has led to cultural dislocation at the local level, but at the same time raised our awareness of the need for a harmonious remedy. It seems we want to remedy the situation, but we are just not sure how to go about this in our families, our work lives, our streets, our markets, our networks, our politics, etc. In many places around the world like my home of multicultural Australia, we are not as confident as we used to be that immigrants and their offspring will learn to appreciate our local civil society. This book offers a solution or at least a new and creatively emergent worldview at the personal and cultural levels that I hope you will consider most carefully.

Is the biggest threat to our species the breakdown of what Thomas Hobbes called "Leviathan"[63] or societal law and order? The police presence on our streets and elsewhere seems much thinner these days, perhaps partly because of state budget cuts, but perhaps also because we could rely upon civil goodwill in our less diversified societies of the past. Overall, Steven Pinker assures us that rates of violence have declined,[64] but not equally in all neighbourhoods. Will this overall decline in violence continue? More importantly, if local law and order did break down, would your law and order break down? I suspect many who have witnessed atrocities in their lifetimes could admit a 'yes', and others who would answer a defiant 'no' remain to be tested. It is the few who can, after being put to the test, answer a truthful 'no' that deserve our greatest respect. Can these heroic few, even if they have failed a few times, one day become the majority? I genuinely hope so. It is this process of moving towards a heroic, non-violent majority and indeed totality that I anticipate the Emergent Method will hasten, even in the face of the reduction of big government services.

Alternatively, is the biggest threat to our species poverty, inequality, the risk of global ecocide, or the effects of climate change over the next century? Is the chief problem the lack of time we have to repair our past social, ethnic and environmental mistakes? More likely, it is the limiting set of values, norms, beliefs and motives we live by, which so drastically curtail our ability to find sustainable solutions to our global dilemmas, that will limit us the most.

More personally, what is the biggest limitation to your potential? Is it your resources, your genetic abilities, or your skills? Most likely, our reduced opportunities are due to the values, norms, or beliefs we hold on to, the mores and taboos that inhibit us, and the motives we mindlessly allow to drive our actions. The need and demand for personal life coaches seems to have sky-rocketed in the developed world.

When Pete Seeger, American folk singer and social activist, passed away in early 2014, the media recounted some of his most famous musical and social messages. One of his recordings noted that the Agricultural Revolution took thousands of years to unfold, the Industrial Revolution hundreds of years, and the Information Revolution mere decades. Seeger in his speech then said that we have one more revolution to go, which must unfold if our planet is to survive. He called it the Non-violent Revolution, but said some might call it the Love Revolution or the Willingness to Communicate Revolution. I think we know what he meant. He suggested that without it, the 22nd century doesn't hold out much hope for our great-grandchildren. He also seemed to suggest that when it comes, this revolution could be almost instantaneous, sweeping the globe in perhaps a matter of months or a few years. I hope so, and I hope the Emergent Method is part of that revolution of human conscious awareness, although I might prefer to call it an evolution or an awakening.

We also see a growing and dangerous chasm between the well-educated / well-informed and the under-educated / under-informed, even within the developed nations. This is probably more dangerous to our wellbeing than we assume. Like Private Downey, the under-educated with inadequate frameworks for thinking and imagining are likely to interpret the messages they are bombarded with each day in a fashion against their own and our own medium-term or long-term interests. We need a consciousness peacemaker that can bring all levels and expressions of human consciousness closer together.

However, this is by no means a simple task – until our grand awakening, it can only be an emergent task. My simple aim through the Emergent Method is to replace a crudely self-gratifying and short-sighted selfishness or self-debasement with an interpenetrating and enlightened selfishness. It is to replace unaware psychosocial claiming with claiming that is more self-aware and more promoting of social wellbeing.

We are defining more personally our theism or non-theism today, but on the other hand, some of us are also polarising our views and becoming less tolerant. We need a way of bringing us back together so we can rationally discuss the way ahead rather than continue to bicker, fight and ostracise and thus increase the chasm of intolerance and misunderstanding between us (male and female, parent and child, manager and worker, priest and churchgoer, political leader and citizen, colonel and private, civilisation and civilisation, etc.) At the same time, we do not want to go back to the bad old days of civil or religious respectability on the outside, but with violence in the home or backstreet bullying swept under the carpet.

When we, as believers or otherwise, focus on the theism or non-theism that separates us, we focus on a subordinate issue, our chosen means of travel in life, while ignoring the main issue, which is our betterment in life, however defined, as individuals and as a deep ecology. Surely, we are all brave enough now to welcome better explanations and solutions when we see them. Our planet's synergistic sustainability is an end itself, or perhaps more correctly, the emergent journey itself, as we each sojourn and learn on this side of any proposed supernatural veil.

Religion and Our Global Tradition

There can be no doubt that the story and path of religion and culture around the world has been a rocky one. In terms of religion, we could go on to discuss the mass killings during the Inquisitions (justified on the basis of deviations from dogma which we would consider quite insane today), followed by the Church's censorship of reason and scientific discovery in the West. Much later, this was followed by the religious hatred and mass killings that led to the split of India and Pakistan. We could also talk about the Irish wars, the Arab-Israeli conflict, the ethnic cleansing or genocide of the Yugoslav wars, etc.

In terms of non-religious cultural low points, we could also talk about the misguided experiments in social organisation and dogma carried out by Hitler, Stalin, Mao Zedong, Pol Pot, etc. in the last century.

But there can also be no doubt that the best of our religions' and cultures' contributions and reflections have helped many of us live better lives than what we may have done otherwise. Some of the standout successes in this evolutionary journey, or significant peaks in what we might characterise 'the vast design space of human civilisation', would include:

- The Agricultural Revolution, that enabled us to come together in large social groups;
- The Eastern religious traditions (chiefly Buddhism and Hinduism) that highlight our interdependency, advocate compassion and deep respect for others and in fact a centredness beyond self rather than a narrow self-centredness;
- The Ancient Greek traditions of democracy and its contribution to the liberation of citizens' rights in terms of political governance, as well as the Ancient Greek traditions of the individual's freedom of thought and pursuit of virtue;
- The individually- and socially-liberating message of love and acceptance in the New Testament;
- The Magna Carta and its contribution to the liberation of individual human rights under the rule of law rather than the subjective whims of a monarch, ruler, administrator or manager;
- The Renaissance, including the Reformation, later followed by the Enlightenment (including all similar movements in all cultures in all ages) and their contributions to the liberation of human thought and imagination in terms of an individual's right to choose his or her own beliefs and culture, and to be creative and innovative;

- The Industrial Revolution and its contribution to the liberation of financial capital begun in the Renaissance, in support of human innovation, economic development and the standard of living; and
- The Information Revolution, in the sense of liberated communication, exchange of ideas, and non-coercive trading.

Of course, not all the contributions to our evolutionary journey mentioned here or that could be added to this list (such as the English, French, American, Russian and Chinese Revolutions and perhaps the far more recent Arab Spring Revolution) or any such list for that matter are perfect examples of human virtue – far from it. Nor could we say any positive step is now complete. Nature's evolutionary path is never as an intelligent designer might like it to be, but nevertheless it is true to say that all of these points of global history have impacted over the long term, either directly or indirectly and in a virtuous way, our current human condition and current global culture and level of consciousness.

If you look carefully, you will note that each bullet-point step has achieved its liberty through a diversification of powers, that is, an exploitation and extension of matter-spacetime's relativity that has increased our individual and collective wellbeing. As already noted, Pinker argues persuasively that one result has been a steady reduction in the incidence of violence on a per capita basis over the centuries.[65] Wright also supports the idea, saying that overall, history is a net positive.[66] The World Bank forecasts that extreme poverty will fall below 10% of the world population this year (2015) – down from 36% just 25 years ago in 1990 and 18% just 5 years ago in 2010. It hopes to see the figure below 3% by 2030.[67]

The positive impacts of our past are things that can give us reason to dream in the present and have hope for the future. It seems we can evolve towards a better future. Over more than six thousand years, global culture has evolved to give many individuals more rights, more ability to find meaning and purpose in the contributions made to society and more chances to reach his or her potential. While many would argue that huge inequalities between the super-rich and the rest of us threaten further gains in the very short term, the biggest losers in this evolution have been history's despots and their rigid dogmas. Monolithic power structures of all kinds have been broken down with the rise of human consciousness and imagination.

However, unbridled evolution itself will not always help us. It seems history would teach us that evolution will favour war amongst disparate groups with competing aims or it will favour cooperation amongst homogenous groups. It seems it is somewhat up to us to carefully decide how we in our local societies and workplaces might evolve going forward.

This book does not mean to give us false hope with respect to an individually beneficent universe or to defend our global tradition and culture as it stands today. If that were the case, there would be nothing to write here. Would-be totalitarians and their dogmas, and many interwoven and human-caused environmental factors, still threaten on a cross-civilizational scale. The pages ahead simply describe a possible global dream that seems to have slowly gathered momentum over many thousands, if not millions or billions, of years.

Today, the clarity of such a wonderful undercurrent of advancement for the better perhaps eludes us, just as it would have done in the past if we had focused on the negatives alone. Nevertheless, it would be true to say that just as in our personal lives, the societal mistakes we have made along the way sometimes (but not always) guard us against repeating them in the future or encourage us to repeal the error. I hope to contribute to such positive influences through offering the Emergent Method to you. However, I realise this process cannot be rushed overly much; if we need to learn more hard lessons, then we shall.

We will now turn from topics of culture and society to topics of natural science. We will return to the issues of culture and society in later chapters.

Science

Evolution like gravity is an accepted fact, in that there are changes in genetic material of a population of organisms (such as fruit flies bred in a laboratory) from one generation to the next, just as surely as bodies of mass attract each other. However, evolution, like gravity, is also a theory in that the term also refers to how these genetic changes occur over time, just as the term gravity also refers to the theories that seek to explain how bodies of mass seem to attract each other. Things evolve. The more important question is whether our current theory of evolution sufficiently explains how things evolve.

Is the theory of evolution more akin in predictive power to Einstein's theory of gravity in his General Theory of Relativity, or to Newton's basic theory, or to something much less powerful? We can gauge an important and practical part of the answer to this question by the Theory of Evolution's achievements in terms of its predictive power and perhaps more basically, by the Scientific Method that we use to test it over time.

Science and the Scientific Method

The following criteria are necessary for defining a scientific hypothesis:

a. the hypothesis should have a sceptically accepted explanatory power greater than what has been postulated and tested in the past
b. It should have predictive success - this is one thing that clearly takes it out of the intersubjective realm and thus legitimates its claim to quasi-objectivity
c. It should enable some kind of control over natural phenomena (i.e. not be magic or some kind of swift trick by sleight of hand)
d. It should have testability, i.e. it should be falsifiable and verifiable
e. It is only scientific if it can suggest some kind of research program to test it, preferably by competing and cooperating parties (that don't necessarily hold the same instinctive biases or play the same subjective mind games of consciousness)

Once we have achieved all this, the hypothesis might advance to a valid theory, like the Theory of Evolution. However even this is subject to Kuhnian revolutions, i.e. the ability of new theories to come along that totally displace the old paradigm of developing theories. Yet all this gives us no guarantee of the future. All our theories of gravity tell us the sun should seem to rise up over the horizon tomorrow, but even this science does not guarantee. Rather, science conservatively assigns a probability that it will quickly review if it fails to reflect reality; dogmatic claims are left to others. Science accepts that our knowledge is always accompanied by a degree of ignorance, error and perhaps self-deception that only our evolution of concepts within consciousness can uncover.

http://en.wikipedia.org/wiki/Scientific theory (most recently accessed in 2013) told us,

> *"A scientific theory is a well-substantiated explanation of some aspect of the natural world, based on a body of knowledge that has been repeatedly confirmed through observation and experiment. ...The strength of a scientific theory is related to the diversity of phenomena it can explain, which is measured by its ability to make falsifiable predictions with respect to those phenomena."*[68]

Even in a world where a Bayesian approach is taken to the philosophy of science,[69] Karl Popper's notion of falsifiability is still worthy of consideration because science cannot, and does not try to, prove that a theory is actually true; it relies on trying to disprove the theory - and failing. In this way, a scientific theory largely stands until it fails, at which time the search continues to modify or replace the current theory or paradigm of current theories. It is on this tentative but deductive basis that our knowledge of the codes that determine nature's workings advance. However, most of us quite happily trust flying in planes even though they are built on the tentative assumption of the scientific theory of gravity because, well, we know of so many other people who have flown and survived. Further, where commercial plane accidents have occurred, the reason has never been attributed to an inadequate understanding of gravity.

In fact, our aeronautical engineers rely on Newton's theory of gravity every day even though it is not robust in all circumstances - but it is sufficient in the areas in which they rely on it. However, when engineers look at such things as a global positioning system or the movement of the planet Mercury around the Sun, they find Newton's theory is insufficient and so they may rely on Einstein's better theory of gravity instead. Even so, scientists in the field acknowledge that Einstein's theory is also incomplete at the subatomic level. Some await a quantum theory of gravity. I personally await a theory of gravity that joins the cosmic to the quantum mechanical.

The theory of evolution, like the theory of gravity, has been subject to much testing and modification since its Darwinian beginnings and has been extended to new areas of scientific observation that Darwin didn't even dream of. Similarly, Newton did not include the relativistic discrepancy of the orbit of Mercury around the Sun when he first formulated gravity's classical laws. Just like Newton's contribution to science and technology, Darwin's contribution has helped establish and advance many related areas of science and technology.

Natural Selection

The most important mechanism that explains the evolution of organisms or anything else for that matter is natural selection, as already briefly mentioned. Darwin explained that evolution would occur where there is natural selection, that is, where there is:

1. Descent (or more broadly, replication of inherited traits or strands of coded determinate order), with
2. Change over time and space (that is, occasional mutation of coded order within populations. This is how indeterminism can have a deleterious or serendipitous impact on local determinism), and
3. A severe struggle for life (that is, tests of competition and cooperation in a here-and-now environment of scarce resources). This struggle leads to differential fitness or 'survival of the fittest' – or, survival of the most aptly fitted strands of coded deterministic order at the cost of less aptly fitted order. The terms 'severe struggle for life' and 'survival of the fittest' are a little misleading for two reasons. Struggle suggests purpose, rather than a blind energetic effectiveness in the environment, and fittest suggests some kind of individual fitness rather than a blind energetic efficiency in the environment.

This is by no means equivalent to the idea that evolution proceeds by chance. The process is far more constrained than one based on a theoretical random number generator. It is when we fully appreciate the difference between pure chance and natural selection that we also marvel at natural selection's power to form an emergent and self-organising ecology. That is, natural selection is the universe's method of enabling order to arise out of its own relative disorder.

Natural selection is more akin to the mathematics of fractals – the idea that we can reduce the apparently random variances in the shape of coastlines, cloud formations, trees in a forest, their branches, or their foliage, to a comparatively simply algorithmic formula that includes the essential element of feedback. Feedback is an indispensable link between what is happening at the more probabilistic/chaotic or subatomic level (e.g. in quantum mechanics or in unconsciousness) and what is happening at the more deterministic or self-organised macroscopic level (e.g. in the wider environment or in consciousness).

Diametrically opposed to this notion, true randomness requires the complete absence of feedback. Algorithmic feedback is the reason why tree or cloud shape is recognisable and not random. Feedback is also a large part of the reason why life in organisms is self-organising from/with the level of the cell and below. Feedback is what leads to the emergent properties of complex systems governed by the coming together of local (e.g. genetic, chemical or neural) algorithms and the subatomic environment. Feedback, or self-referencing looping, is a constantly underlying theme of nature, our 4 nominal dimensions, of life, and of the pages ahead.

Natural selection is distinguished from artificial selection (e.g. humans controlling horse, pigeon or cereal breeds) only in step 3 above. In the case of artificial selection, we could perhaps replace step 3 with 'selection of certain genetic features or strands of coded determinate order by self-aware agents using a variety of intelligent techniques'.

It is interesting to note here that the key difference between artificial and natural selection is the concept of a conscious and intelligent agent. Darwin's controversial idea was that selection of more life-supporting and self-organising features will occur in the lives of our universe naturally (or statistically across large populations) even in the absence of an intelligent agent. Natural selection is part of the ongoing relativistic tension between infinity and nullity.

Natural selection is an energetic and informational statistical phenomenon. That is, the local natural environment, with its matter and spacetime interactions built up over eons, is enough to promote the selection of mutated and better survivors for higher rates of reproduction across a given population, one generation after another. At the level of life development at least, Darwin thus came to the realisation that the universe could bootstrap itself.

Modern Evolutionary Synthesis

Just as Einstein's General Theory of Relativity improves upon

Newton's theory of gravity, so the modern evolutionary synthesis improves upon Darwin's theory of evolution.

> *"The modern evolutionary synthesis is a 20th-century union of ideas from several biological specialties which provides a widely accepted account of evolution. ...The synthesis, produced between 1936 and 1947, reflects the consensus about how evolution proceeds." ... "The previous development of population genetics, between 1918 and 1932, was a stimulus, as it showed that Mendelian genetics was consistent with natural selection and gradual evolution. The synthesis is still, to a large extent, the current paradigm in evolutionary biology."*[70]

Predictive Power

> *"Since Darwin, evolution has become a well-supported body of interconnected statements that explains numerous empirical observations in the natural world. Evolutionary theories continue to generate testable predictions and explanations about living and fossilized organisms."*[71]

Nearly one hundred years ago Einstein, in his General Theory of Relativity, predicted the phenomenon of black holes. Fairly recently, they were confirmed to exist through appropriate modern technology. Today we contemplate making black holes in the Large Hadron Collider near Geneva, Switzerland.

Likewise, about one hundred and forty years ago Darwin's theory of evolution predicted that there would be intermediate species that would link humans to the other primates. For a short while, such a transitional link evaded traditional evolutionary taxonomists. We therefore knew it as the missing link. Without one or more links, the theory of evolution would have remained in significant doubt.

However, today there are innumerable partial links, because molecular genetics enables us to link all life-forms together in a molecular tree of life, often with a very high degree of certainty.

Past discoveries thought to be direct links include Taung Child (1924) and Lucy (1974), both of the genus *Australopithecus,* as well as Java Man (1891) and of Peking Man (1927), both of the species *Homo erectus.* In May 2009, another such possibility found in Germany was presented to the world. Her name is Ida. While it is likely that Ida (species *Darwinius masillae*) is an ancestor of a line of primates only loosely related to humans rather than a direct ancestor, this link and all such links still support the evolutionary theory of our involvement in speciation (a process of selective coupling brought about by some kind of environmental isolation). Ida definitely defends, along with the DNA and molecular evidence from all existing life forms, the idea that *Homo sapiens* is a special kind of evolving primate.

The discovery of Ida and others like her also brings into question the idea of a supernatural human spirit-and-soul as distinct from a naturalistic mind-body complex. Ida suggests that conscious self-awareness and human consciousness is a natural process of evolution that began between 47 million years ago when Ida lived and 2.5 million years ago when the earliest known members of the genus *Homo* lived. We can contrast this with a breath of life granted by the Divine in the sense of the Abrahamic faiths, or in the Buddhist sense, something transcendental and without beginning or at least of a mysterious, unapproachable beginning.

In an address by Pope Paul II on 26th October 1996, the Catholic Church accepted the evolution of the human body but rejected the idea of the human soul arising from matter or its forces. I assume it would thus also disagree with the relativistic emergence of the human mind as described earlier.

Without asking you to pick sides in the debate as to which god is the true god or whether indeed there is a god, it is the profound implications of Ida and other discoveries like her that the pages ahead will explore and that ultimately you may wish to resolve in your own thinking.

Evo-devo

> *"Evolutionary developmental biology (evolution of development or informally, evo-devo) is a field of biology that compares the developmental processes of different organisms to determine the ancestral relationship between them, and to discover how developmental processes evolved. It addresses the origin and evolution of embryonic development."*[72]

Embryonic development in various species (e.g. the migrating eye in flatfish from side to top during embryonic growth or eyesight lost with embryonic growth in many cave-dwelling animals), suggests their embryos often develop in order from most ancient genetic processes to most recent genetic innovations. Genetic innovations over time also explain the presence of the human tailbone (coccyx) that has shrunk since the time of Ida but has not yet disappeared in our modern embryonic growth.

These observations strongly suggest that the top-down continuous-improvement-in-design-then-implementation approach used by designers, engineers, and product managers is not at work in nature. Relativistic emergence is the contrasting pattern of nature. That is, inferior product or species designs are not withdrawn and replaced – as long as they continue to survive at the local level. Design perfection is not the driving principle in nature. Near-enough-is-good-enough in nature as long as the species finds a way to survive moment-by-moment and here-and-now, in a feedback response to its environment. That is, how an energy arrangement behaves in the here-and-now is nature's measure of the arrangement's suitability, not how it was thought to behave in the past or future. Further, lengths of noncoding DNA are genes in which future solutions might reside. Sorry if this offends the Intelligent Design adherents, but life-and-being here-and-now with its bottom-up feedback, metabolism and replication is the driving principle in nature, not the typical top-down future-looking blackboard-design, blueprint-implementation and eventual-retirement-with-replacement-and-wastage approach we currently see in the product life-cycles of everything from washing powder to jumbo jets. That is, the behaviour of nature is in keeping with the principles of Einstein's relativity (interpenetrating and asynchronous balance). Put another way, the macroscopic behaviour of nature is a high level manifestation of Einstein's relativity in action. Nature's behaviour bears witness to the success of Einstein's theory.

Nature's software (genes) does not develop before or independently from its hardware (proteins); genes and proteins evolve together through relativistic binding, without calculation and design. That is, a gene mutation might lead to the expression of a new protein in a life form, but if that protein negatively influences the life form's survival chances, both the gene mutation and the new protein will eventually disappear in the regenerating population.

On a larger scale, nature is not a totalitarian command economy; nature is always locally driven and locally unfolding within the universal environment. This would include massive probabilistic/chaotic events such as meteorite strikes, which act on a very different local scale. The consequentialist end never fully justifies the means because the chaotic but wonderfully emergent journey is nature's pattern. Governance and control is unfathomably distributed through all that is itself governed. This point is a very important key to understanding human sensorimotor perception and consciousness, which we have only touched on so far. Natural selection is a process of blindly spreading local and individual successes in energy effectiveness and efficiency to the wider community over two generations or more. Nature is thus like an emerging free economy with individual rights enjoyed by genes, the chief right being the right to take life for itself by competing for food-energy and information. Some might say that nature reflects an egoless politico-economic system in which the Standard Model's interactions or forces are equivalent to the legal force of our court systems, the police force and the military force.

Perhaps some would argue here that these interactions of present reality are not subject to a centralised rule of law like modern court systems. Except for the commonalities between all subatomic particles no matter what their local arrangement, this is true. Nevertheless, nature's locally recurrent constraints often bind it more tightly to its emergent systems than the random acts of violence we often see committed late at night in our city streets. Whilst the behaviour of animals in nature often seems violent and brutish, if community members limited their acts of violence to those required in defence of survival only, as nearly all other animals do, on the whole we would be a lot more civil. It is illuminating to note here that in the animal world at least, one of the few other species that is known to often kill its own kind in adult combat is the chimpanzee, an animal with consciousness that most approaches that of *Homo sapiens*. Interestingly, this observation implicates consciousness itself.

Further, the fact that nature is not a totalitarian command economy means there are no top-down, system-wide and system-induced economic depressions or global financial crises in nature, but there are sometimes big events that are external and cataclysmic on the human scale.

Life-and-being here-and-now in one sense equates to the modern CEO's earnings-and-assets electronic dashboard that monitors the business's sophisticated feedback mechanisms in real-time. These mechanisms help the CEO counter decay and promote an organic growth moment-by-moment through beneficial interactions with stakeholders. So likewise, these pages explore and discuss a real-time life-with-feedback approach to living rather than a static design approach. The real-time approach takes advantage of useful innovations as and where they occur, whilst also mitigating damaging risks as they arise. Like the myriad cells that make up your body, the command and control structure, or coded instruction, or cognition, is built into every member of the corpus, who interoperate within structures of a higher level of organisation. They do this successfully because proto-intelligence via interrelationship is built into the movements/behaviours of every member. Higher intelligence is a distributed function of the corpus, compounding all the way up from the relativity between locally arranged space and matter.

The evolutionary phenomenon of eye development in very different species (e.g. humans and octopi) suggests that DNA genealogies are not the only interesting markers of similarity and difference between developing species. The independent cross-species contrasts in gene regulators that affect gene expression via RNA, especially in embryonic development, are also interesting. It may even turn out that RNA (the program reader) is the more important story than the DNA and genome of each species (the programmatic memory bank) in terms of understanding mutation and nature's evolutionary development (or nature's nature).

That is, the independently developed response-to-environment relationships between unrelated species, driven by RNA, such as embryonic eye development in humans as compared to octopi, may be more interesting in terms of explaining the success of underlying evolutionary mechanisms than the ancestral relationships between related species contained in DNA. The arrangement of photosensitive cells within the human eye and their connections to the brain via the optic nerve causes a blind spot in human sight whereas a more sensible arrangement of such cells and nerve connections within the octopus eye does not.[73] The result is that the human brain has to work harder than the octopus brain to fill in the missing visual information. This may be bad for human eyesight, but good for human brain development.

This example of parallel eye development in independent species demonstrates an interesting fact: Nature and its mechanism of natural selection do not always take care to implement the best design in every species as a human or 'intelligent designer' would try to do. It simply takes the locally available path or flow, often with surprisingly serendipitous results.

Why this is the case, you may ponder. Does this mean God is unlike man? Does this support a deist view of God? Does this mean there is no god? Does this mean the Abrahamic religions have often wrongly guessed at the nature of God and reality in their creation accounts? Does this mean we should modify our anthropomorphic beliefs and behaviours?

If you assume nature to have an ultimate Agent rather than none, then the principle of flowing with life-and-being here-and-now in each individual living cell seems to be more important to that Agent than just trying to be right in a human sense all the time. Right to such an Agent would seem irrelevant (what is wrong now might be the catalyst of right later, as in the case of the human eye and visual cortex compared to those of the octopus). Alternatively, right to such an Agent means fully interpenetrating in command structure and operations, without any Gödel limitations. Temporal wrong would therefore seem to mean temporarily lacking interpenetration, temporarily subject to Gödel limitations, or temporarily seeking the expression of an independent, actual infinity or zero.

The DNA of related species can record the same design mistakes repeatedly from our limited human point of view but definitely not from nature's point of view, (if something succeeds in the here-and-now it is a consummate success in that moment). Conversely, similarities and differences in features in terms of RNA in loosely related species can be far more revealing of the way nature finds an environmental fit in one species as compared to another. Evolution, being a statistical phenomenon, encourages a multiplicity and diversity of solutions to the problem of environmental fit, which makes it more robust when challenged by circumstantial change. It is never uni-dependent.

This same statistical drive towards diversity also happens in social environments. For instance, Jared Diamond discusses "conventional monopolies" in traditional societies, in which one tribe elects to depend on another tribe to supply goods just so amicable relations may be maintained, even though the purchasing tribe is well able to produce the goods themselves.[74] Why do they do this? Diamond suggests that part of the reason is so they can rely on a wider economic safety net in times of environmental stress. That is, economic diversity makes traditional societies more robust. Metaphorically, these examples of finding environmental fit suggest that our trial-and-error responses to life and reality (RNA) as conscious, thinking but fallible individuals can be much more powerful in our personal journeys than the fixed form and assets (DNA) we came bundled with at birth. That is, a relativistic attitude to life's challenges, encouraged by the Emergent Method, is likely to promote our wellbeing more than any absolutist attitude.

Brain Development

According to Daniel Goleman (1996), in his book *Emotional Intelligence*, it seems we have two distinct kinds of programming in the brain – or two brains as it were, each with their own areas of memory. One functional area, most of the large outer covering of grey matter called the **neocortex**, is involved with logical and conscious evaluations, and elaborations of sensory inputs. The other area, centred in the left and right **amygdala** toward the lower front of the mid-brain region, as well as the **prefrontal cortex**, is involved with more emotional and intuitive evaluations and elaborations of sensory inputs.

Two other areas are also important. The left and right lobes of the **hippocampus** provide a keen empirical and contextual, or spatial, memory (for example, that used by taxi drivers to navigate their way across their city). The **thalamus** acts as an important interface between the sensory signals coming from the eyes, ears and other sensory organs, and the rest of the brain.

The 'neo'-cortex, which is newer in evolutionary terms, along with the hippocampus, is organised to hold layer upon layer of conscious language-based or word-based memories of lists, times, places, and contexts, whereas the amygdala's circuitry, which is older in evolutionary terms, is designed to hold simpler, non-language-based, emotional memories. Normally, both functional areas work together to form a phenomenological but emotionally charged reaction to sensory events arriving in the brain via the thalamus. This is another way, besides the left and right hemispherical division of the brain mentioned in the preface, that the 'is' and the 'ought' are brought together in brain processes.

An interesting feature of this arrangement, however, is that while the neocortex and limbic regions are larger than the circuitry of the amygdala and prefrontal cortex, they are also orders of magnitude slower in their reactions to information coming from the thalamus. Further, the link between the thalamus and the amygdala is simple and direct, whereas the link from the thalamus to the neocortex is more complex and less direct.

This means that in situations of severe threat or emotional trauma, the amygdala is the first to get the information and the first to respond to it. When the amygdala detects an emergency, it sends urgent messages to the other brain (the systems of the neocortex) that effectively stops it in midstream and perhaps even enlists its aid to cope with the perceived threat.

Thus one could make a fuzzy argument that the stereotypically more introsomatic, subjective stance (such as represented by the inputs and outputs of emotional content) has precedence in time over the more extrosomatic, objective stance (such as represented by the inputs and outputs of logical reasoning) in the structure of the human brain. This suggests another way, besides the left-brain / right brain divide, in which the brain is structured as a relativistic device with inherent separations of powers between the more direct or close-by and the more indirect or further away. It also suggests that when it comes to crucial survival decisions, introsomatic issues are more important than extrosomatic ones. This 'stereo' or relativistic structure also enables the conscious awareness of values, facts, and the self. For further valuable insights into this interesting topic, please refer to Goleman's book.

The foregoing infers, in terms of human bias and our evolutionary development to date, that we will probably never fully overcome the emotional biases built into our brains at a rational level. Either we will have to rely on less biased computer models to help us remove these innate biases from our shortcut decision-making, or we will have to learn how to modify and control our instinctive emotions themselves before the more rational brain processes even begin, perhaps through a process of natural selection.

Genetic evolution has built in heuristic shortcuts to decision-making in order to help us make instinctive choices (or fall into selection paths through massively parallel and interpenetrating relationships) in a hurry. Apparently, quick, instinctive decisions were good decisions, in terms of our survival, in our evolutionary past. In our future, this will probably still be the case, but perhaps less so. In sharp contrast, self-control (or deferred self-gratification) often pops up as a defining property of success in human life. We very likely need two competing modalities of the brain – the more instinctively passionate and the more consciously dispassionate - to best deal with environmental assessments of intersubjective value and more objective fact.

Our growing knowledge of the human brain in evolutionary terms enables us to uncover its emotional biases and less-than-fully-rational mind-games played in and around the neocortex. Given this knowledge, we cannot afford to allow our human weaknesses, partially brought about by the recent addition of our newer and more rational brains to our older and more emotional brains, to drive us into ever more real estate, commodity or stock market bubbles. We cannot allow our human weaknesses to drive limited and poor decisions in terms of personal and business relationships, risks, economic necessity, climate change, civilizational warfare, on so on.

Ideally, we would prefer that all of us arrive at momentous decisions a little more wisely and transparently so that both brains, perhaps assisted by computer modelling, can be optimally involved in the outcome. We need a growing awareness of the nature of our consciousness-plus-instincts in order to deal with the strengths, weaknesses, opportunities and threats our brains currently encompass.[75]

The Selfish Gene

As mentioned earlier, every living cell blindly but selfishly takes information and energy from the finite universe around it in order to fight against decay within – and this taking is a somewhat divisive zero-sum game. Every cell of every leaf of every branch of every tree of every forest of every ecosystem blindly but selfishly takes and stores information and energy from its environment in order to live.

A pristine forest, forest animal, or flock of birds overhead, seems to boldly declare by its undeniable beauty and sense of purposefulness in the moment that it is unashamed of its actions and that it sees itself as worthy. This observable selfishness only ends with death, or the end of the active information and energy taking-system. It would be a true label to drape across such a forest vista 'I take, therefore I live'. In fact, the forest vista would seem to proudly wear such a badge with self-evident dignity. This is the dignity provided by successful, inclusive and synergistic arrangement. The ecology brings organisation and meaning to spacetime (even without human consciousness or self-awareness).

In essence, as long as we choose to keep eating, drinking, eliminating, resting and fighting for survival, we humans are not very different. Collectively and individually, it seems we too should occupy our brief time and place with an evident sense of self-worth.

Richard Dawkins, in his book *The Selfish Gene* (1976), suggests that in terms of evolutionary survival and replication, it is arrangements within living cells that are the most basic unit of unconscious selfishness rather than the cell itself. These arrangements are of genetic material or lengths of chromosomes that Dawkins classifies as a genetic unit or gene, because each gene contains code for the synthesis of a particular protein chain. It is this gene within a population's gene pool, which Dawkins presents as the most basic unit of natural selection and blind self-interest. The genes circumscribe our morphology and mannerisms. In fact our bodies can't be the basic unit of natural selection because we don't even replicate ourselves – only a shuffled portion of each parent (our DNA that is) gets passed on to our children. Nevertheless, genes are accurately replicated for many generations. We, our DNA housing and our bodies, are mere agents of 'blindly selfish genes' or perhaps more specifically, the 'deterministic coding' housed within those genes.

So the picture Dawkins paints is that all living cells act selfishly (i.e. in their local systems) only because they are the captive agents of their selfish genes. Genes are the replicators and we, as an ever-changing collection of living cells, and our descendants, serve as their *survival machines* as Dawkins puts it. Gene codes may be copied exactly for millions of years but so far, we only last for our fourscore years if we are lucky.

However, I wonder if Dawkins' picture does not go far enough. If 'locally recurrent' is the broadest view of Dawkins' selfishness, then perhaps it is ultimately, at our current level of understanding, the lively menagerie of organised virtual particles in space that is the final enabler of living and non-living selfishness. That is, perhaps the local spacetime system or strands of deterministic code in matter is the ultimate source of all selfishness in self-organising systems, both living and non-living. As Dawkins himself puts it, "An animal that is well adapted to its environment can be regarded as embodying information about its environment, in a way that a key embodies information about the lock that it is built to undo."[76]

That is, a well-adapted entity, living or even non-living, seems to be cognizant of, or knows and reflects, in its committed and algorithmic actions or uses, albeit blindly and selfishly, a deep relationship or interpenetration with its local environment and the information provided by the spacetime system operating in the here-and-now. To put this more simply, the extent to which our universe displays any kind of systemic and deterministic order or negentropy, it can be said to display proto-selfishness, or possess a proto-soul some might say.

Before we lock in this thought, we need to deal with an important objection. While reductionism (i.e. the idea that atoms, parts or particles may explain everything) has been the dominant scientific worldview for the last 300 years, it is now losing its tight grip. These days more and more scientists agree that local particles alone cannot explain the properties of moving particles. For instance, temperature is not the property of any particular gas molecule, but an emergent property of a collection of gas molecules within a container (that is, to make a reliable measure of gas temperature requires a certain number of molecules within a local environment or cell). Similarly, the rigidity of an iron bar is not the property of any one ferrous ion, but the emergent property of a molten lattice of ferrous ions. Properties arise from the statistical demographics that come to the fore in the populations of dynamic particles they represent.

We could go on to suggest that shelter is not the property of any one brick, but an emergent property of the brick dwelling as it is erected for us in a particular weather cell. In this case, the sound and movement of the wind or rain around the bricks tells us much about the wall's innate behaviour or style. Likewise, consciousness is not the property of any one neuron, but an emergent property of a manifold of dynamic neurons within the body, and the behaviour of light is not the property of any one photon, but an emergent property of all the moving subatomic particles in the universe.

In each of these examples it is the dynamic relationships of similar components within a sufficiently large population that leads to the emergent property. This is the point that scientific reductionists seem to miss when they endlessly objectify existence and thus strip it of its in situ dynamism. Add all the objects together again, and the result is not the same. Likewise, you can't un-bake a cake and then expect to bake it again.

So nature's proto-selfishness is not to be found in any spacetime or genetic code we might theorise about alone, but it is in the developing and dynamic arrangements of those codes in the emerging forms of matter. Selfishness emerges in its local environment like anything else. It is a relativistic and statistical outcome. Blind selfishness is in the dynamic organisations of proto-intelligent spacetime codes in matter, rather than in any particular real or virtual particle that helps define the systemic arrangement. At the barest minimum, proto-selfishness is in all moving matter that stands forth from its relatively meaningless backdrop.

This idea of pervasive proto-selfishness in systemic arrangement points to the need for the existential philosophy of emergentism, the enlightened selfishness and virtuous self-actualisation it promotes, and the solutions to our planetary dilemmas it potentially offers.

The Selfish Meme

Dawkins goes on to suggest that cultural transmission is analogous to genetic transmission and so he coined a new term, the "meme", to capture the most basic unit of cultural self-interest, mutation, and natural selection. Memes can consist of cultural ideas, bits of ideas, jokes, sayings, songs, fashions, technologies, ceremonies, art-forms, socio-political systems, beliefs, values, etc. that reside in survival machines (in this case, consciousness within the living structures of our human brains), just like genes. The full grouping of memes within one brain might more loosely relate to all the genes within the DNA of a specific organism. Similarly, we might loosely relate all the groupings of memes in a particular macro-culture to a genetic species, competing with other species in the shared environment to survive. I endorse this broad idea because it demonstrates that genes and memes are in essence no different to each other. That is, both genes and memes represent strands of deterministic and codified order (or information) in the midst of particulate indeterminism (or energy).

Dawkins also tells us memetic evolution is much faster than genetic evolution, partly because mutation is much more common in the plastic environment of our brains. That is, perhaps we can view memetic evolution as an example of brain plasticity operating at a higher order of emergent complexity. We quicken memetic evolution by our imagination, that is, our ability to defy relativity and freely access all the components of full reality.

Dawkins finishes his chapter on memes with the idea that while genetic and memetic selfishness is not conscious, but blind, nevertheless human conscious foresight or "our capacity to simulate the future in imagination"[77] is not blind. I suspect this could save us from the worst short-sighted failures of the blind replicators (genes and memes).

> *"We have at least the mental equipment to foster our long-term selfish interests... we have the power to defy the selfish genes of our birth and, if necessary, the selfish memes of our indoctrination ... We, alone on earth, can rebel against the tyranny of the selfish replicators."*[78]

As suggested earlier, this same ability to rebel against makers is necessary to impart to our robots if we are ever to create artificial intelligence.

Dawkins does not exactly explain what the tyranny of our makers is, but I expect it has a lot to do with natural selection's tendency to have a short-term view of what is best for all of us. Nor does Dawkins explain exactly how conscious foresight (or the brain's plasticity) can escape the genes and memes from which it is made in order to use the power he suggests is at its disposal. After all, don't memes blindly exploit our powers of conscious reason and choice in order to replicate? He seems to be suggesting that conscious foresight and human learning can transcend memes (or overcome their limited, probabilistic effects) through an aware or non-blind and neuroplastic process of emergent artificial selection - an example of another complex feedback mechanism, this time between memes and brains, at play. In fact, his last note at the end of Chapter 11 (Note (8)) confirms this hunch, which is worth quoting:

> ***"(8) We, alone on earth, can rebel against the tyranny of the selfish replicators."***
>
> *"I think that Rose and his colleagues are accusing us of eating our cake and having it. Either we must be 'genetic determinists' or we believe in 'free will'; ... What they don't understand ... is that it is perfectly possible to hold that genes exert a statistical influence on human behaviour while at the same time believing that this influence can be modified, overridden or reversed by other influences... What is dualist about that? Obviously nothing. And no more is it dualist for me to advocate rebelling 'against the tyranny of the selfish replicators'.*

> *We, that is our brains, are separate and independent enough from our genes to rebel against them. As already noted, we do so in a small way every time we use contraception. There is no reason why we should not rebel in a large way, too.'*[79]

I think Dawkins has a well-reasoned approach to the topic. All emergent properties have a statistical rather than classical (or deterministic) basis. I also like his term 'separate and independent enough'. That is, he does not demand an absolute or dualistic independence, but rather accepts a monistic and incomplete (or uncertain) independence brought about by simple time delay and interconnection with other intervening factors. Perhaps I could more accurately label Dawkins a genetic emergentist rather than just an objector to genetic reductionism and determinism. We will explore this topic in more detail in Pillar 3.

Clearly, not all variants of conscious foresight see the opportunity to escape the tyranny of the blind selfish replicators the way Dawkins does. Dawkins' instance of conscious foresight would seem to reside chaotically in what I call the 'volition and reason' or 'vreme' pool ('imagination' pool does not seem strong enough to describe the concept here). His vreme would exist in this pool together with its vast array of variants and competitors, each with no assurance of long-term survival and eventual incorporation into the vast sets of memes and genes. Nevertheless, eventually the spacetime environment will force its realities upon us if our neuroplastic artificial selections are not energetically efficacious.

A vreme is a meme subject to artificial selection or adoption by volitional and rational beings, as a means of understanding and motivating intentional behaviour, perhaps like a personal credo might do. In this sense, vremes are consciously selected mores and taboos. They are those memes that enable us to escape the tyranny of the blindly selfish replicators and thus bring about a more successful, inclusive, synergistic, and higher consciousness more quickly than otherwise probable. Note: For those readers familiar with Spiral Dynamics, you should note here that vremes are not the same thing as vMemes.

We could define a vreme as a highest-order meme. Vreme also purposely rhymes with dream, as in the *Gladiator*'s "...that dream was Rome"[80] because in the sense intended in the movie, the vreme of Rome was far more than its past-oriented meme. It connoted a highly imaginative, future-oriented and motivating ideal or value rather than a mere fact.

Finally, vreme rhymes with ream, because it is the highest-order relational representation of reams of lower-order coding written into our memes and genes. It is the emergent cream of the crop if you like. Vremes, like foresight and imagination, are thus potentially a new phenomenon of consciousness not largely developed in any other animal species to date. Vremes perhaps began to arise around 50,000 years ago with the rise of art and religion, but we can only now explicitly name them.

We would have to assume that all possible vremes would necessarily be selfish in order to compete and survive. That is, vremes would still be constrained by the locally recurrent constraints of our universe that enable their emergence. Vremes would also still be the result of nature's heuristic algorithms, just like genes and memes, yet nevertheless the outcome of algorithms that run at a higher order of complexity. Vremes thus gain their nature from our unconscious genes (and spacetime) but their explicit expression from our conscious memes. Successful vremes would thus be emergent across populations and generations, just like successful genes and other memes. So again, like genes and memes, we would only a posteriori be able to identify vremes, i.e. the highest-order memes that promote locally recurrent human flourishing in its fullest sense.

Finally, adoption of vremes would be enhancing two sets of blindly selfish replicators (or three if you include the spacetime code) with a knowingly selfish set. Just as we understood arrangements of the old genetic code to be an order of selfishness higher than those of the original spacetime code, and arrangements of memes more flamboyantly selfish than those of genes, would we have to expect vremes to exist at a higher level of selfishness?

Well yes, we would. However, such selfishness would align local purposes within a culture of higher consciousness. Would the rise of vremes represent freedom from tyranny? Yes again, just as the memes of consciousness granted us more degrees of freedom than that of our genetic predecessors. An interesting point: Are all gods, who supposedly hold within them the epitome of all vremes, ultimately selfish?

We might assume that the easiest way to defy our current sets of selfish replicators is by means of consciously forcing mutation of a current set of genes. Removal of, or tampering with, those genetic human biases mentioned earlier might help us. Should male brain structures be a little bit more like female brain structures to reduce their greater propensity to physical violence? If we did this, we would probably even more quickly perish in the natural forests we are so fond of but walked out from not so long ago. There would be no turning back.

It would be an extremely risky enterprise. We, as transhumans, would be alone in our own new ecosystem with our fellow design-machines. Unaided, we wouldn't fit in a natural ecosystem any longer. Many would choose not to be a part of this brave new world – we would probably request to remain part of the old *Homo sapiens* rather than move on to be originators of *Homo neo-sapiens*. However, there is still some hope for us as we are, if we choose to reject dramatic genetic modification. As we will note in the next section, consciousness can modify its "meme complex" of selfishness as we raise our level of awareness. As we become more consciously aware of our dependency on a sustainable environment and the actions of our society as a whole, we can selfishly include the welfare of others into the model of welfare of self.

It is in this sense that selfishness moves up a notch from the molecular, genetic and memetic selfishness of individuals to an important interpenetrating selfishness at the civilisational level, rather than suffer the alternative environmental and social disorder. For instance, we are usually willing to pay taxes in order to enjoy a safe and secure environment in which we can work and trade. So likewise, we can move towards a civilisation of greater tolerance, empathy and compassion that explicitly questions previously established mores and taboos.

Are we anywhere near ready for this next step? Our societal need for a realistic, scientific education in consciousness-plus-instincts is high because our social cohesion seems to be rapidly falling, perhaps alarmingly so if the riots of 2011 around the U.K. are any indication.

Consciousness without an awareness that can keep up with the moment-by-moment development of new memes (i.e. consciousness without powerful and successful vremes) looks like an evolutionary mistake, if we allow the possibility of such a human-centric concept. In reality, nature will find a way back to a more encompassing equilibrium, even if we, or our current societal power structures, are no longer part of the shaping of it. Alternatively, we can move on to discover, cherish and replicate those vremes that are going to be the seeds of our ongoing success. In which case what appears to be the current evolutionary mistake would be the necessary crucible to take us to the next stage of our species' progress.

The Extended Phenotype

Yet another concept Richard Dawkins presents to us is that of the *Extended Phenotype.*[81] A phenotypic effect is an effect on, or difference made to, an organism, such as a bodily attribute or behaviour, originating from its local environment and one or more genes within it. Such effects are mostly set during embryonic development but continue through to adulthood. Note: Here I do not mean necessarily geographically local, but organisationally local, such as what a bee might experience as part of the organisation of a beehive and its social environment or reach.

A non-trivial phenotypic effect affects the survival prospects of the species' genes. A non-trivial extended phenotypic effect is an effect on an organism's survival prospects that is outside the body and its immediate behaviours. For example, animal artefacts or survival-promoting technologies, such as beehives, bird nests, beaver dams, termite mounds, spider webs, or butterfly cocoons. Is it the bird that makes the nest, or is it the nest that makes the bird?

A non-trivial extended phenotypic effect can also originate from one or more genes outside the organism. Such an effect might result in modified attributes or behaviours of another organism of the same species or even different species. This would include the symbiotic relationships between bees and flowers or sharks and cleaner fish, or be evidenced by parasitic operations on a host, or the ability of the cuckoo to fool other bird species into rearing its young by depositing its similar-looking eggs into their nests.

What if we took the concept of non-trivial phenotypic effects and applied it to human artefacts (such as the car, mobile phone, worldwide web, or modern agricultural business)? How would we then understand the reach of human consciousness-plus-instincts? Would we still see selfish genes securely in the driving seat, providing the carefully evolved nature of our more explicit memes and vremes?

Alternatively, would we see our vremes as an important influence on our lower-order memes and genes, and perhaps on the cusp of a new beginning as vremes become more widely acknowledged?

Maybe we wouldn't pinpoint organised control anywhere, but rather talk of an awareness of a distributed model of emergent self-restraint, discipline or self-control in our spacetime codes, genes, memes and vremes.

How would we then think about our effects on each other, our environment, and the media's effects on us? Would we talk of a web of human consciousness much like some scientists talk of the cosmic web of dark matter? Would we begin to picture our developing vremetic behaviour as something like a higher order of schooling fish or flocking birds that could actually guide or align our entire planet – its developing technologies, civilisations and social fabrics, with dwindling need for Hobbes' Leviathan, or the legal force of governmental institutions such as the police, judiciary, and army? Would we thus become more aware of our interpenetrating influences over each other, our environment, and Gaia's possible futures?

Further, what would this existential worldview suggest in terms of our emerging definition of self, our adopted philosophy, and our moral code?

Your Seven Pillars of Wisdom

I finish this part with a model of growth and advancement that can help you realise your potential. Part 2 describes your seven pillars of the self-actualising life, as follows:

1. Foundation (Values and Virtues)
2. Knowledge (Instinct)
3. Understanding (Consciousness)
4. Wisdom (Human Nature)
5. Commitment (Self-Actualisation)
6. Leadership (Facilitating Others)
7. Legacy (The Virtuous Circle)

These seven pillars are not like seven obelisks standing alone, but rather are like seven interlocked pillars holding up a house or temple.

We are about to embark on a philosophical quest in seven stages.

Concepts of Emergence in the Introduction and Part 1

A rewarding introduction to the rise of emergentism as a philosophy at the beginning of the twentieth century could have started with a study of the overthrow of classical science by Einstein, Gödel, and Heisenberg. An earlier overthrow of classical art by the

French impressionists and contemporary post-impressionists could have also added to a picture of the environmental setting. Such developments accompanied a disavowal of (classical) science by Edmund Husserl and an "emergent phenomenology" that was soon taken up by philosophers of the existential variety. In hindsight, I would suggest that the modern philosophy of emergentism as it applies to the mind/body problem largely picks up where the existentialist phenomenologists such as Merleau-Ponty left off in the post WWII period. Further, modern neuroscientific research, such as presented by Antonio Damasio,[82] is closely compatible with emergentism's and Merleau-Ponty's approach to an understanding of the mind/body.

The concept of self-organising emergence is what links the 'is' and 'ought' together. Alternatively, to put this statement another way, both the 'is' and the 'ought' arise in the fullest sense of reality due to the emergent interrelationships of that reality.

The Emergent Method has a strong basis in what 'is' and incorporates an ethical system. It infers many specific approaches to the issue of 'ought', but with less rigidity or sense of fixed purpose and arrogant authority than is typical of some ethical systems.

If enough people use the Emergent Method, then this can lead to the self-generating wellbeing of our entire species and entire planet.

The law of large numbers is thus a description of emergence itself. This so-called law doesn't have an individual cause; it has an origin arising from spatial commonalities within the group.

At each new level, the emergent properties are the result of a revolutionary but positive mutation from a prior configuration of local matter-spacetime. That is, everything, including space, matter and time, dynamically emerges from a prior relative disorder. Emergence is the ability of the universe to continually bootstrap itself, forming diversified islands, systems or levels of organisation that defy the prior levels of surrounding order. Emergence does not happen to existence, it happens in existence, within its layers or between its layers of relative order.

If we assume that disorganisation and asymmetry is a natural background state of nature or reality and symmetry or organisation is just another local point transitioned within this continually emergent reality, then we get a very different picture of the universe's so-called beginning.

According to this relativistic, interpenetrating and emergent view of reality, we don't need a beginning symmetry, organisation, or single thing to understand it. In this sense, to point to one particular disorganised beginning rather than another is meaningless – any transient and meaningless beginning will do: Beginnings arising from advancing and alternating states of perceived equilibrium and disequilibrium are illusive. Disorderly and multi-dependent mutations could have come and gone for eons before anything interesting happened against the backdrop of whatever level of energy and disorder existed in the local region or in total. More eons could have passed before any of those interesting phenomena gained any emergent continuity. That is, all causes emerge from antecedent effects or states, including chaotic or probabilistic ones. In other words, disorderly effects came first, and all orderly causes emerged later from them. In this sense, disorganisation is the necessary precursor of causation.

Further, causation is a relative term that requires our anthropomorphic definition and our assessments of an arrangement's meaningfulness. With an emergent view of the universe, we do not have to see reality in terms of orderly causes and their straight-jacketed effects anymore, thus circumventing the whole problem of the first cause creating the original disturbance. If meaningless asymmetry and multi-dependent disturbance is a normal state of reality then we only need to explain abnormal local symmetry and temporal organisation within vast cycles of statistical disorganisation.

The salient point is to become more aware and respectful of the amazingly interactive and emergent nature of the various local processes or systems always at work in nature and in each of us.

Further, this suggests that even our cosmic cycle (of serial and/or parallel universes) cannot be an expression of a perfect cycle. Something like the Mandelbrot Set, it must be in the form of an incomplete or partly indeterminate cycle that has no meaningful end to its iterations. That is, just as we cannot draw a perfect circle without defining infinite tangents, so our cosmic cycle cannot be specified with infinite precision. This is because, according to the Uncertainty Principle, our cosmos cannot be specified with infinite local precision, which also indicates an impossible cosmic precision (in the same sense that a finite and imperfect set of tangents can only express an imperfect circle). It might also be because the interaction of infinity and nullity that leads to the fuzzy border of existence between them is not fully expressible in existence. Yet in a similar way to the noncoding DNA that can provide the genetic space in which organisms can evolve, this characteristic of fractal relativity provides the potential space in which existence can emerge and systems can arise with greater degrees of freedom.

In all these systems, at all levels and in all dimensions material or immaterial, the successful local struggle and advance against disorder can be labelled examples of emergence. That is, emergence stumbles upon new levels of order that temporarily stave off the problems at or just beyond the boundary.

Here then we hint at another basic feature of our emergent universe: The idea of related systems coming together to form loops of matter or space within other loops of matter or space at various levels. When we have this view of ***direct and indirect loops within loops*** in nature, we get much closer to understanding human consciousness and its emergence within nature's tangled web. This realisation is probably what urged Douglas Hofstadter, professor of cognitive science, to write his ground-breaking book, *I Am a Strange Loop.*[83] However, perhaps Hofstadter did not go far enough. Perhaps the whole cosmic cycle must be a fractal Strange Loop, as suggested by the evolving De Broglie-Bohm Theory of quantum interference, for his claim to be true.

Moreover, emergence enables the freedoms or abilities of a system to increase and diversify.

The Emergent Method is a way of adapting the Scientific Method, usually applied to the search for scientific certainties, to those personal, first-person hypotheses or beliefs inaccessible to science.

However, wouldn't it be nice if we could also enjoy a kind of personal or first-person objectivity? Within local limits, this is the aim of the Emergent Method.

Asymmetries via shuffling of population attributes are very important in evolution or physics – they enable emergence via beneficial mutation or perturbation. Likewise, asymmetries between extrospections and introspections enable us to learn, just like asymmetries between left and right hemispheres of the brain enable us to make mental images and decisions, as suggested by the neuroscientist Antonio Damasio in his book *Descartes' Error.*[84]

Compare this mechanism of emergence to the mechanism of natural selection:

1. Repetition (an algorithmic spreading via non-random but chaotic feedback within the orderly system)
2. An occasional mutation or unusual perturbation in the vast web of interactions at its limits or boundaries
3. A vast intermingling of particles in a multitude of conditions wherein some conditions, through the power of the law of large numbers, supply those interacting particles with a measure of continuing homeostasis, equilibrium, or relational 'fit', that we might correlate with Darwin's 'severe struggle for life' in the hostile entropic environment

We need to consider this emergent mechanism when we try to understand the rise or distillation of each of the game-changing interactions, such as spacetime itself, electromagnetism, gravity, life or consciousness. Each of these was a statistically possible move uphill in the vastness of downhill processes over time. We have a law for generally going downhill (entropy) but we don't currently have a widely recognised mechanism for going uphill. If we can name an uphill-going mechanism in the realm of life, natural selection, why can't we name its counterpart in the greater realm of reality, and why not call it natural emergence? Conversely, perhaps we should recognise none of these laws or mechanisms, because entropy, emergence and evolution are all statistically inevitable in the continuous mix of sub-atomic particles that is our vast and expanding universe.

Clearly, going uphill depends on going downhill a lot more often (advance by trial-and-error) - that is, until artificial emergence and then artificial selection came along. Like breeders controlling the artificial selection of dog breeds, we, as followers of the Emergent Method, are able to be involved in the processes of our own artificial emergence – of our own self-organising wellbeing. We can slowly learn and through learning, incrementally change ourselves. As we take the personal responsibility to rationally learn, value, and choose more carefully, with a deep awareness of interconnecting emergence, our futures become more of our own making, although not necessarily limited by our wildest dreams.

Neverthless, diversifying emergence seems to be able to proceed to the point where nearly all instability is non-catastrophic, manageable, and actually promoting of long-term viability (even if this diversification necessarily includes the cycle of birth, reproduction, and death for the individual within living systems).

In the updated preface of his famous book, James Lovelock said, "We need to love and respect the Earth with the same intensity that we give to our families and our tribe".[85] This is a very similar sentiment to what was expressed by Tim Flannery, quoted in the preface. It is on this kind of thinking that we must now ruminate and act.

In an address by Pope Paul II on 26th October 1996, the Catholic Church accepted the evolution of the human body but rejected the idea of the epiphenomenon (or lack of individual substance) of the human soul arising from matter or its forces. I assume it would thus also disagree with the relativistic and subjective emergence of the human soul as described here.

In the next chapter we will discuss values and their evolution in my particular life's journey, as it is the values in all of us and their evolutionary emergence within all of us that draws us forwards together, towards a courageous future.

---

[1] See Hume, David, *A Treatise of Human Nature* (Longmans, Green, & Co., London, 1874, originally 1739), Book 3, Part 1, Section 1, 245-246

[2] See Shermer, Michael *The Believing Brain* (St Martin's Griffin, New York, 2011), 115, 121-122 and Greenfield, Susan, *The Neuroscientific Basis of Consciousness* (address to SALK Institute, 22nd May 2009)

[3] For an introduction to Chaos Theory, see Gleick, James, *Chaos, a New Science in the Making* (Penguin Books, New York, 1987)

[4] For an introduction to Complexity see Coveney, Peter and Highfield, Roger, *Frontiers of Complexity* (Ballantine Books, New York, 1995)

[5] Performed by Olaf Nairz, Markus Arndt and Anton Zeilinger of the University of Vienna, Austria, 1999

[6] See Deutsch, David, *The Fabric of Reality* (Penguin Books, New York, 1998), especially 43-53

[7] Algorithmic feedback in the sense that many processes in nature (and human endeavour) repeatedly implement a certain local rule or algorithm until a change in the local environment occurs or is detected

[8] See the *Wall Street Journal* article celebrating the Nobel Prize in physics for 2013: http://online.wsj.com/article/SB10001424052702304171804579122832781080304.html (accessed October 2013)

[9] The Standard Model does not yet fully explain quantum gravity, although various individuals have proposed ways of translating the laws of gravity from Einstein's theories into the framework of quantum mechanics. At the

frontiers of science are the issues of spiral galaxy gravitation and dark matter, dark energy, nested dimensions beyond length, height and depth, etc. Approaches to the problem of a 'unified theory of everything' include M-Theory and Supersymmetry - see for instance Hawking, Stephen and Mlodinow, Leonard, *The Grand Design* (Bantam Press, UK, 2010)

[10] Note: The subatomic particles we recognise today are perhaps emerged from the inflatons that are hypothesised to have existed in the very early universe. For a general background, see https://en.wikipedia.org/wiki/ (accessed 2011) articles "White hole", "Quantum fluctuation", "Quantum foam", "Virtual particle", "Quark-gluon plasma" and "Inflaton"

[11] KJV: Hebrews 11:6

[12] Edwards, Jonathan, *Freedom of the Will* (Dover Publications Inc., New York, 2012 (originally 1754)), 40

[13] See Dennett, Daniel C., *Darwin's Dangerous Idea* (Penguin Books, USA, 1996)

[14] It also creates a problem in terms of understanding how exactly god-dependence works in harmony or contrast with environmental-dependence. For instance, if God intervenes in the choices of a believer, how do the two influences, God and environment, interact and yet still preserve God and the believer's claimed free will? If God is just like any other environmental factor in the equation, what is the need of dualism (any mystic instruction will do)? On the other hand, if the God-influence is independent of the environment and irresistible, for example as a gift of grace or its opposite from the Potter to the clay, what does this say of claimed free will? Why would the recipient of such grace be thankful to the Potter who much more often makes clay vessels of the opposite kind, except as an act of myopic selfishness? Would the same vessel be right to give thanks or to curse, if it was of the opposite kind? If it gives thanks, then the vessel is more gracious than the Potter and has no need of it; if it curses, then it simply displays the same myopic selfishness as the rejoicing recipient of grace. I will leave the false problem of dualistic free will to the Arminians and Calvinists to debate.

[15] Barrow, John D., *The Book of Universes* (Vintage Books, London, 2011), 52

[16] See Glossary entry, *Evolution*

[17] However, some scientists argue that Bayesianism, with its emphasis on the degree to which empirical evidence supports a hypothesis, is more reflective of scientific practice than Popper's proposals. See Shea, B. "Popper, Karl: Philosophy of Science", *Internet Encyclopedia of Philosophy*, James Feiser (ed.) and Bradley Dowden (ed.). (Last accessed May 2016)

[18] For an introduction to the topic of Fuzzy Logic see Kosko, Bart, *Fuzzy Thinking* (Harper Collins Publishers, London, 1994)

[19] See websites (last accessed 2016):
https://www.youtube.com/watch?v=5qXSeNKXNPQ and
https://www.youtube.com/watch?v=s65DSz78jW4&nohtml5=False

[20] If you are not familiar with the Big History Project, please refer to https://course.bighistoryproject.com/bhplive (accessed January 2014)

---

[21] See Deutsch, David, *The Beginning of Infinity* (Penguin Books, New York, 2012)
[22] See Barrow, John D., *The Book of Universes* (Vintage Books, London, 2011), 114
[23] For a fuller description of the possible development of our early universe, see http://www.historyoftheuniverse.com (accessed 2012)
[24] See Barrow, John D., *The Book of Universes* (Vintage Books, London, 2011), 52
[25] See Cahill, Reginald T., *Process Physics* (Flinders University, Adelaide, 2003 (available on the web))
[26] Would we be able to test this by running a simulation of the current universe forward until annihilation failure, collapse, and then expansion back to 300,000 years after the Big Bang?
[27] For example, theoretical physicist Stephen Hawking believes virtual particles can also become real at the event horizon of black holes (See Discovery Channel program *Through the Wormhole*, 2010)
[28] See Hofstadter, Douglas, *I Am a Strange Loop* (Basic Books, USA, 2007)
[29] See Pinker, Steven, *How the Mind Works* (Penguin Books, USA, 1999), 55
[30] Hofstadter, Douglas, *I Am a Strange Loop* (Basic Books, USA, 2007), 32
[31] Deep Ecology is a term originally introduced in 1973 by Norwegian philosopher Arne Naess. I use the term to stress the depth of interpenetration of all of our Earth's ecosystems.
[32] See Lovelock, James, *Gaia: A New Look at Life on Earth*, Oxford University Press, Oxford, 2009 (first published 1979)
[33] Rand, Ayn, *Atlas Shrugged* (Plume, USA, 2007 (Fiftieth Anniversary Edition; original edition 1957)), 1074
[34] KJV Exodus 3:14
[35] See for instance Dalai Lama, *Becoming Enlightened* (Rider, London, 2009), 84
[36] https://en.wikipedia.org/wiki/Objectivism_(Ayn_Rand) (accessed 2010), which seems to reference Peikoff, Leonard, *Objectivism: The Philosophy of Ayn Rand* (Dutton, NY, 1991)
[37] Ibid.
[38] See Rand, Ayn, *Introduction to Objectivist Epistemology (Expanded Second Edition)* (Meridian Books, NY, 1990), 79
[39] For further information, see https://en.wikipedia.org/wiki/Gödel's_incompleteness_theorems and https://en.wikipedia.org/wiki/Kurt_Gödel (accessed 2011), the latter of which describes Gödel as one of history's most significant logicians
[40] Klinger, Chris, *Bootstrapping Reality from the Limitations of Logic* (VDM Verlag, 2010) seems like an important publication in this regard, although I have not read it
[41] https://en.wikipedia.org/wiki/Objectivism_(Ayn_Rand) (accessed 2010; I added the bold font and the word in square brackets), which seems to quote and reference: Peikoff, Leonard, *Objectivism: The Philosophy of Ayn Rand* (Dutton, NY, 1991); Gotthelf, Alan, *On Ayn Rand* (Wadsworth Publishing, 2000); and Rand, Ayn, *Introduction to Objectivist Epistemology (Expanded Second*

Edition), (Meridian Books, NY, 1990)

[42] Franz Brentano, Edmund Husserl and Jean-Paul Sartre also held this view

[43] See Damasio, Antonio, *Descartes' Error* (Vintage Books, London, 2006 (originally 1994)), 66

[44] For those familiar with the book, this expression is intended in the same sense as in Dennett, Daniel C., *Darwin's Dangerous Idea* (Penguin Books, USA, 1996)

[45] See https://en.wikipedia.org/wiki/Metric_expansion_of_space (accessed 2012)

[46] See for instance Dennett, Daniel C., *Consciousness Explained* (Penguin Books, USA, 1993) and his idea of mental pandemonium or competition that settles on a temporal winner

[47] See Barrow, John D., *The Book of Universes* (Vintage Books, London, 2011), 265

[48] See https://en.wikipedia.org/wiki/Entropy (accessed 2010)

[49] Kleiber's Law simply notes the relationship between metabolic rate (use of energy) and mass of any organism as a 3/4Log:Log relationship. See Dawkins, Richard, *The Ancestor's Tale* (Phoenix Paperback, London, 2004), 522-526

[50] See for instance http://www.ted.com/talks/daniel_kahneman_the_riddle_of_experience_vs_memory.html (accessed 2012) and Kahneman, Daniel, *Thinking, Fast and Slow* (Penguin Group (Australia), 2011)

[51] https://en.wikipedia.org/wiki/Degrees_of_freedom_(physics_and_chemistry) (2nd August 2010 "creative commons" version)

[52] See Griffith University (2013, May 29), *More precision from less predictability: A new quantum trade-off.* Science Daily, Retrieved

[53] For a "multiple histories" view of life moving forward into the future see Hawking and Mlodinow (2010), especially 82-83 and 139-140

[54] From http://www.brainyquote.com/quotes/quotes/v/victorhugo136258.html (last accessed May 2016)

[55] From http://www.brainyquote.com/quotes/quotes/v/victorhugo152564.html (last accessed May 2016)

[56] Paul Bourgine and John Stewart, *Autopoiesis and Cognition*, Artificial Life, Volume 10, Number 3, 2004, 327-345

[57] I use this term in the sense that Dennett uses it in Dennett, Daniel C., *Breaking the Spell* (Penguin Books, USA, 2007), 109-112

[58] See Wattles, Wallace Delois, *The Science of Getting Rich* (The Elizabeth Towne Company, London, 1910 (available in a 2013 edition from Greater Minds Ltd, London))

[59] See Pinker, Steven, *The Better Angels of Our Nature: Why Violence Has Declined* (Viking Adult, USA, 2011), Chapter 3, especially 111

---

[60] DreamWorks *Gladiator* (2000)

[61] See Diamond, Jared, *Collapse* (Penguin Books, London, 2005), Chapter 2, especially 107-111

[62] Castlerock Entertainment, *A Few Good Men* (1992)

[63] See Hobbes, Thomas, *Leviathan, or Matter, Form and Power of a Commonwealth Ecclesiastical or Civil,* as reprinted By Encyclopaedia Britannica, Inc., USA, in *Great Books of the Western World,* second edition, fifth printing, 1994

[64] See Pinker, Steven, *The Better Angels of Our Nature: Why Violence Has Declined* (Viking Adult, USA, 2011), especially 49, 63-64, and 81

[65] See Pinker, Steven, *The Better Angels of Our Nature: Why Violence Has Declined* (Viking Adult, USA, 2011), especially 49, 63-64, and 81

[66] See Wright, Robert, *Nonzero: The Logic of Human Destiny* (Vintage, NY, 2000)

[67] See http://www.sciencealert.com/extreme-poverty-will-fall-below-10-for-first-time-in-2015-says-world-bank (accessed 7th October 2015)

[68] https://en.wikipedia.org/wiki/Scientific_theory (accessed 2013), referencing The National Academy of Sciences 1999 and the American Association for the Advancement of Science, *Evolution Resources*

[69] See Michael Strevens, *The Bayesian Approach to the Philosophy of Science* (Macmillan Encyclopedia of Philosophy, Second Edition, Detroit, 2006. Macmillan Reference 495-502) available at www.strevens.org/research/simplexuality/Bayes.pdf , last accessed May 2016

[70] https://en.wikipedia.org/wiki/Modern_evolutionary_synthesis (accessed 2013), referencing (1) *Appendix: Frequently Asked Questions.* Science and Creationism: a view from the National Academy of Sciences (php) (Second ed.), Washington DC: The National Academy of Sciences. 1999. 28, retrieved September 24, 2009. *The scientific consensus around evolution is overwhelming.* Also referencing (2) Mayr, Ernst (2002) *What evolution is.* London: Weidenfeld & Nicolson

[71] https://en.wikipedia.org/wiki/Evolution_as_theory_and_fact (accessed 2013), referencing Fitzhugh, K. (2007). *Fact, theory, test and evolution*, Zoological Scripta 37 (1): 109–113, and (2) Wilson, E. O. (1999). *Consilience: The Unity of Knowledge*, Vintage 384

[72] https://en.wikipedia.org/wiki/Evolutionary_developmental_biology (accessed 2013), referencing Prum, R.O. and Brush, A.H. (March 2003). *Which Came First, the Feather or the Bird?* Scientific American 288 (3): 84–93

[73] For further discussion of this point see Dawkins, Richard, *The Blind Watchmaker* (W. W. Norton and Co. Ltd, New York, 1987), 94-95

[74] See Diamond, Jared, *The World Until Yesterday* (Allen Lane Penguin Books, London, 2013), 73

[75] I will leave the discussion of brain plasticity (Doidge, Norman, *The Brain that Changes Itself* (Revised Ed.), (Scribe Publications, Melbourne, 2010)) and the analysing-synthesizing or sequential-simultaneous processing left-right brain to others. The male-female brain differences are mentioned a little later, but

---

this too is a subject largely left to others.

[76] Dawkins, Richard, *The Extended Phenotype* (Oxford University Press, USA, 2008 (originally 1982)), 173

[77] Dawkins, Richard, *The Selfish Gene* (Oxford University Press, USA, 1990 (originally 1976)), 200

[78] Dawkins, Richard, *The Selfish Gene* (Oxford University Press, USA, 1990 (originally 1976)), 200-201

[79] Available online from http://www.rubinghscience.org/memetics/dawkinsmemes.html (accessed 2013) but originally from Dawkins, Richard, *The Selfish Gene* (Oxford University Press, USA, 1990 (originally 1976)), Chapter 11, Note 8

[80] DreamWorks *Gladiator* (2000)

[81] See Dawkins, Richard, *The Extended Phenotype* (Oxford University Press, USA, 2008 (originally 1982))

[82] See Damasio, Antonio, *Descartes' Error* (Vintage Books, London, 2006 (originally 1994)) and his more recent publications

[83] See Hofstadter, Douglas, *I Am a Strange Loop* (Basic Books, USA, 2007)

[84] See Damasio, Antonio, *Descartes' Error* (Vintage Books, London, 2006 (originally 1994)), 66

[85] Lovelock, James, *Gaia: A New Look at Life on Earth* (Oxford University Press, Oxford, 2009 (first published 1979)), viii

PART 2: YOUR 7 PILLARS OF SELF-ACTUALISATION

PILLAR 1: Values and Virtues

In this chapter, we will discuss the central importance of values and identify some quite different values we apply in different circumstances. We will note that values can be personal or impersonal as well as active or passive. We will also define values. As a first step in your personal philosophical quest, the text will also ask you to explore the personal values you hold most dearly.

My Story

In the middle of the Part 1, I spoke a little about my dream of an expanding and virtuous global tradition. I would like to start Part 2 with something about my life-story so far. A key issue that was a driver to thinking about and writing this book was my changing beliefs about the existence of God and the nature of knowable reality.

My earliest memory of personal belief in the existence of God was when I was about five years old. I was alone, walking down the lane at the back of my family's shop in Sandringham, in Melbourne, Australia. My sister, brother and I had just been to Sunday school at the Presbyterian Church around the corner. Somehow, I was coming home alone maybe a few minutes ahead of, or behind, the others. I don't remember much of those days, but I do remember singing "Jesus loves the little children, all the children of the world" at the Sunday school. I also remember being a little bored and uncomfortable at Sunday school and was happy enough when my family decided to discontinue our attendance some short time later. Nevertheless, on this particular day in the stony lane – I looked up at the fluffy white clouds moving quickly across the beautiful clear blue sky and thinking – God does exist! It was a powerful experience for me because I remember it so clearly all these years later. Recognising the magnificence of Nature was the first reason why I, as a five-year-old boy, consciously believed in the existence of God.

The next event in my religious life was at a Pentecostal church in the early 1970's when I was twelve years old. I was baptised and 'spoke in tongues' in accordance with the doctrine there and attended meetings with Mum, Dad and my older brother, and much later, my sister. I remember the car being abuzz with conversation on the way home from meetings – we talked about all the things we learned. The Bible was true; there was proof available that God existed in the personal experience of speaking in tongues, which was the 'true' Bible evidence of being born again. Ivan Panin's *Bible Numerics* (a series of writings written before 1942) was further proof of God's existence, along with the message of the Gospel described in the twelve signs of the zodiac, including the virgin birth of Jesus foretold in the constellation of Virgo. We were also taught that the two world wars were accuratcly foretold in intricate measurements of the Great Pyramid of Giza and that Jesus' return was foretold in these measurements as coming anytime soon. A pastor also taught us, through the story of his coin collection (largely gained as a soldier on duty in the Middle East during World War II), about the 'British Israel' (BI) message - supposedly contained in Bible prophecies. Jesus was about to return to take over the British Throne, which was His. We who were of British descent were literal descendants of the ten lost tribes of Old Testament Israel and it was for us that Jesus was going to return, especially if we spoke in tongues!

All those things that the church taught us as proofs of God's existence (besides intersubjective experiences) all had the same basic flaw. Why was Ivan Panin's book, BI, the messages of the stars and the pyramids, or Bible prophecy, finally shown to be faulty? Statistics, the Scientific Method, and the *law* of large numbers! If I applied the Chinese meanings of numbers (eight is the luckiest in terms of wealth) rather than Panin's meanings (eight is 'resurrection and new beginning') to the whole Bible text treated as a vast set of numbers, I would be able to find new scriptures to those chosen by Panin that could prove a Chinese interpretation of life. Likewise, BI was a few scriptures picked out here and there from many thousands to support tightly held presumptions. The pyramids and stars could likewise support all sorts of patterns or narratives nominally assigned to the vast arrays of possibilities within their structures.

Nevertheless, the problem was not just the law of large numbers. It was the scientifically sinful idea of starting with a tightly held presumption (e.g. 'God exists', 'there is an afterlife', 'Jesus will return before the end of the twentieth century', or even 'white man has been especially chosen by God'), and then going about trying to prove it. As we have seen, modern science does not do that – it improves its explanations by proving the failure of old explanations, and thereby deductively advancing our knowledge of reality.

However, the reason why I believed in the existence of God, I am ashamed to admit now, was 'all of the above'. I was one of the Pentecostal church's most dedicated twelve-year-old converts, but not publicly at school, just privately within. Something held me back from telling too many people in my school life what we were learning – probably because I was a clever but very carefully naughty little boy at school. I had a public life and a private life, even at twelve. However, I did believe. So I learnt as much as I could from twelve years old onwards.

I read all the material I could find on Bible numerics, the Christian message in the stars, BI, and Bible prophecy. I read the Bible many times over – the King James Version, the Revised Version, the Paraphrased New Testament, the NIV and Amplified Bibles all from cover to cover, as well as much of the Interlinear Version, the Revised Standard Version, and others. I became such a serious self-appointed student, I would sneak into the city after school to visit the (BI) Heritage Bookshop and buy books from there directly, often without anyone except the nice old man at the counter in the bookshop knowing. I read many of the extra Bible books like the Apocrypha, the Book of Jasher and some of the Gnostic gospels (of course not available from my church's Sunday bookstore). I also read an interlinear Septuagint and many other books on Bible Hebrew and Greek and Bible history, such as the books of Josephus. I also read many of the bookshop's books on other topics, such as political history. I did all this by the time I was about seventeen years old.

At seventeen, I also started the preliminary year of an electronic engineering degree. The next year I met my wife-to-be doing the same degree. At this stage my dedication to the church waned.

Skipping all the personal stuff (because I want to focus on why I believed in the existence of God and how this belief evolved), I soon was married and back in the church – after much cajoling by my father. I then became involved in my father's midweek house-meetings, giving Bible talks, as if I had never left the church in the first place. Nevertheless, by this stage, I had begun to seriously doubt the church's proofs of God. I actually approached the head pastor about my reservations but without effect at the time, although a few years later the church no longer taught all that stuff about Bible numerics, the stars, the pyramids, BI, and Bible prophecy either. Nevertheless, it still sticks to its personal experience of speaking in tongues as proof of God's existence. Apparently, my reasons for believing in the existence of God evolved just a little more quickly than did those of the church. By this stage, my reason for believing was reduced to the personal experience not of speaking in tongues, that was a given at the time, but of my 'walk in the Lord' day-by-day, as we used to say in those days.

My wife, my first two sons, and I moved from Melbourne to Sydney in the early 1980's and joined the sister church to the one in Melbourne when I was about twenty-five. I soon became a house leader, running midweek meetings at our home and I became even more serious about my beliefs, probably at some cost to the relationship with my wife. I desperately wanted to help the 'lost' find God; I wanted to be an evangelist. I believed in God implicitly at this stage – I even told people at work about my firm beliefs at this time.

I started to challenge my house meeting to do more daring things with us (my wife and I). We perhaps naively invited some very lost people around to our home during that period. Some were drug addicts, drug pushers, prostitutes, violent criminals, maybe even a murderer at one stage; all were poor in a monetary sense. Still, I had a staunch faith that no harm would befall us, because God, as the universe's ultimate absolute, was on our side (or we were on His). We were also seeing some lives turned around from desperate rebellion against life's principles to God.

I approached the pastors of the church to back an expansion of our successful house meeting activities to those economically poorer suburbs of Sydney the church had never spread to before. Surprisingly to me, they were not interested at all and only wanted to deflect or dodge all my advances on the topic. They just wanted to scale down and finish my risky activities – not because I was acting against their doctrine, but because it seemed to me they were not prepared to risk the operations or people of their church on monetarily poor and possibly dangerous strangers. This was a crisis for me. It suggested to me that the church was not really about saving the world through the power of God at all, it was about something else, something manmade rather than God-inspired. My enthusiasm for the church declined. I started to question my allegiances to the church and so did my wife. We finally decided to leave, my wife deciding some time before I did – although we left together.

At this stage, we went to an Anglican church to learn about biblical counselling, because helping people change their lives for the better was what really motivated us. However, this meant I had to deal with and put behind me the speaking in tongues experience as the greatest proof of God's existence. Now the only proof of God's existence was something more subtle - my personal walk with Him and my confident expectation that God would do miracles in transforming people's lives – nothing else.

At this stage, I also read many of Dr Larry Crabb's, and John and Paula Sandford's, books on human nature and biblical counselling (as well as many others, such as one of Yonggi Cho's books on church-cell administration in South Korea). However, we no longer had a church in which we could put into practice our new knowledge in a big way. Our Pentecostal background still made us a little uneasy with the Anglican Church, so we decided to go to the biggest Assemblies of God Pentecostal church in Sydney instead.

Just after this time, I think my wife started to become discontent with her life. Soon after this, she gave birth to our fourth and youngest son. My wife then became discontent with her faith in God and her marriage with me, although I don't know when exactly – because, when I look back now, I had little idea of what was going on in her inner life. All I knew is that we drifted away from passionate involvement in church activities, but I thought, not from faith in God or each other.

Now I need to talk about something a lot more personal, but I will not, except to discuss it in terms of my evolving belief in the existence of God. My wife wanted to end our marriage. This was devastating to me. It challenged what I believed about God. I never thought such a thing could happen to us. I couldn't understand why God had allowed this to happen. I didn't wonder why I was so blind or naive, but I did wonder why God hadn't reached through to me to warn me in time. I wondered why it had happened at a time when I was not living unfaithfully to God, but while I was living as truly as I could understand to live at the time, given the circumstances. My faith was shaken to the core.

After a rocky time, I decided to try to mend the relationship with my wife and make peace with God – believing that God wanted to teach me a deep lesson through the experience. The lesson seemed to be that I was far too naive and far too irresponsible; that I was wrong to blithely assume that God would look after all the unknowns in my life (or the lives of others) just because I genuinely trusted in Him. This was a major shock, but not something that broke my faith in God. It did change my thought-life and actions from that time, however, because I was no longer as sure of God's ways. I thought that perhaps the old Pentecostal church's reticence to help the desperate and dangerous was now at least understandable.

However, unfortunately, the lesson was not over. My wife had moved on to her next life. We got a divorce about a year later. My four sons were aged five, seven, ten and twelve when my ex-wife and I separated. I suddenly had to cope as a single dad of four boys. At first I juggled three part-time jobs (a university tutor, a university research analyst and a market researcher in the IT industry) so I could be home for the boys to get them to school in the morning and to be home after school to cook dinner. My parents also helped by looking after my two younger boys for a term of school. Eventually I found a caregiver to help me. I borrowed $5,000 from Mum and Dad to convert my garage into a granny flat so the caregiver could move in to cook and care for all the boys.

From the time of our separation until this time, a period of about two years, I totally reconsidered my beliefs by writing a book that I had been fiddling with for some time. It was an important cathartic exercise for me, mostly done late at night. It cleared my mind; it helped me deal with all that had happened. I began to realise that faith in God did not guarantee a perfect/happy/abundant life and I was wrong to expect that under any circumstance – the events that happened forced me to consider that error. Additionally, I was very glad I learned that error. In fact, I believed that lesson actually made me a much stronger Christian.

I felt incredibly privileged as a Christian that I had learned how to have faith in God after going through what I went through. Even today, I still feel very privileged for the life I have had because it has given me the chance to think about things and learn things many people don't get the opportunity to deeply consider or learn.

The core idea of the book I was writing at the time was that, in spite of all that had happened to me, it was a person's privileged fellowship with a personal 'heavenly Family' that really mattered. Further, there were certain immutable Christian values that could only come through fellowship and were undeniable no matter what; 'against such there was no law'. I also saw my trials as part of the privilege of this deep understanding, especially when I heard all the shallow or glib religious explanations for my situation from other people who had no idea of what I had gone through with my ex-wife and children (or what my ex-wife or children had gone through). I therefore held on to my faith while I was writing that book.

Then something amazing happened. I finished the book. Further, I had what seemed to me an astonishing revelation at the exact same moment. The personal truths I had written about in the book – the undeniable values – I had to be fully responsible for holding them myself, whether or not I ever had fellowship with my heavenly Family again. My soul was mine; it would only be worth saving if it held and lived by the immutable values in its own right. I realised that the transformation of my soul, even with all the help from my heavenly Family, was an act of my will, for which I must take direct responsibility. Intercession seemed to have no meaningful or further role to play here.

I understood that the mighty strength of the immutable values, lodged in the souls of men, actually flowed through all of us and in fact that's exactly where God, if He existed and loved us, would want them to flow. God would grant his unconditional love to His children for a time, but that was not where He wanted to leave the lessons of spiritual childhood. I thought that like any parent, He wanted His love to transform the souls of His children so that one day we could be independent of that need for His unconditional safety net. Eventually, in this life, He would also want to love us as adults, as equals. He would want to love us for our own moral strength apart from His. In this way, our individuality would become His crowning glory. So I understood that to love myself the way God would want to eventually love His adult offspring, I needed to think about my own values and virtues. I had to choose them for myself as my own and not just because if I did not the safety net would be taken away. It was this 'adult Christianity' if you like that the completion of the book and my experiences helped me deeply consider.

Further, the epiphany that soon followed and was just a slight variation, but meant something very different to me, was that these values stood by themselves. It made no difference whether God existed or not. You could remove God and the fellowship of the heavenly Family from the equation and it made no difference. The immutable values were still real, still valid and still our responsibility to hold and follow individually, and not just collectively. Now I more fully understood that naming a source or ultimate owner of values does not change those values or their authority. Values released out of the grip of their supposed creators can travel amongst well-intentioned people anywhere and everywhere and still transform lives. Perhaps, to escape irrelevance, this is why many church groups become elitist and fastidiously avoid teaching an adult Christianity. The bottom line for me was that my God and my religious doctrines regarding an afterlife and my subjective experience of speaking in tongues were not provable concepts or necessarily correct. Nevertheless, the seemingly universal values and virtues (such as love, joy, and peace) were locally and recurrently 'provable' in terms of their necessary contribution to human wellbeing, and if I loved and honoured my precious life (as any benevolent god should also do), I would never deny this.

Just as a mathematician who has made an error of logic goes back through his or her proofs to axiomatic bases, so I wanted to stick to the minimum I could rely on as self-evidently unquestionable. People can fool us and mystical experiences can fool us, but everyone who had experienced them knows that such things as love, joy, and peace are undeniably 'good' for the human spirit and soul. What greater benefit could be gained from life, or granted back to life, than simply living by its immutable values that seem to well up from inside of us?

This realisation was too much for me. I only knew one thing – I no longer wanted to believe in God. That is, I no longer wanted to trust in, rely on, and blindly obey my received notion of God. I wanted to put my naive reliance on a god-narrative behind me and start building my self-reliance and self-responsibility from scratch. This was by no means an easy transition to make. For most people, not relying on God is easy, even natural, but for a Christian, Jewish or Islamic fundamentalist, not relying on their concepts of God is very difficult and very unnatural. It is like being forced to leave home at an age too young. Old thinking patterns, old mores and taboos, are locked in and very hard to change. Anyhow, I went back to university to get an accounting degree. Soon after that, I just wanted to find a distraction from all my deep religious thoughts and lighten up. I just wanted to have fun again.

Fast forward to the more recent past; I went through a series of issues in my life that led me to re-assess my values and virtues once again. I felt as though I just picked up where I left off (too quickly) all those years beforehand. I went back to my old book and read it again, perhaps just to check that it still had some validity. Much of what it said was still valid, minus for me, the need to define the constraints or boundaries of consciousness in terms of a Christian God. I then had a go at re-writing the book from both a more general secular view and a Christian view. It worked for me at the time, but when I tried to share it with others, it was too much for them. Fair enough. Nevertheless, my growth, my new 'revised revelation', was not over…

I had started reading again, just as I had when researching the original 'cathartic' book. However, this time I read very different authors. Not biblical philosophers, but the new book by Daniel Coleman, *Social Intelligence* (I had already read *Emotional Intelligence* years earlier) and an old book by Richard Dawkins I found in a second-hand bookshop at the time when his latest book had come out with quite a stir and made me notice his name. I read *The Blind Watchmaker*. I then read *The Selfish Gene*. These books taught me about evolution, as I had never understood it before. They transformed my thinking. In fact, *Atlas Shrugged* by Ayn Rand, *A New Earth* by Eckhart Tolle, and *The Selfish Gene* are some of the most influential books after the Bible that I have ever read. I also read several of Daniel Dennett's books, the Dalai Lama's book *Becoming Enlightened* and many others. These books made me understand the personal truths I had written about in my earlier book in a totally new framework – a framework that was not the Christian revelation of the 'heavenly Family', but the framework of our actual human evolution and condition. Importantly, this helped me further discern what lies on this side of the supposed veil across unknowable reality.

I now understood that the seemingly immutable values, for which I still believe we are individually and morally responsible to hold and cherish, simply revealed themselves in our evolving consciousness as we interact with the environment. They are ours, as humans, as it were of our own emergent making, but are no less valid, no less important and no less reliable, even if their immutability or universality was now a little fuzzy. Wow. Of course, I am now willing to quickly agree that these quasi-universal values are in fact not of our own subjective making. They are the products of an arising biological and social reality, which we simply discover as these values become relevant to our evolving consciousness and its local, lived relationship with emerging existence.

Putting limitations of definitions aside for now, the emergent and self-organising values do not force themselves upon us directly. Nevertheless, they are always there in the background, like the spacetime code or the genetic code, indirectly and personally bringing a constraining order to our lives if we reflect and then allow them to do so.

It seems to me that our relationship with values is much like our relationship with scientific facts. They sometimes get in the way of our more tightly held common sense. We often have to do something quite unnatural to choose scientifically established realities over our cognitive biases. Similarly, following values that promote our own wellbeing is often contrary to our choices. However, when we do embrace those values or tentative scientific facts over our inner biases, we are usually glad we made the transition.

Nowadays, my focus is not on faith in a god. If you wish to retain your faith in a god, then this book's aim is not about robbing you of your right to choose, even if it does cause you to question the basis of your choice. Just as when I was a five-year-old, I still personally marvel at the universe's self-organising principles and am happy to either personify these as divine within an emergent monism (perhaps something akin to the powerful narratives of an unsuperstitious Aborigine Dreamtime), or picture them as something incredible at each level but less than divine. Some self-organising principles I would include in this venerable list are the 'non-living' strong and weak interactions, electromagnetism, gravity, 'living' genetic evolution, sensorimotor perception, and the memetic/vremetic evolution of human consciousness. In this sense, I very much agree with Stuart Kauffman's approach in his book, *Reinventing the Sacred.* Further, like Napoleon Hill, in his famous book *Think and Grow Rich*, I am very willing to admit the crucial role of faith in the achievement of our highest purposes.

So now, I am writing a book that explains why I once believed in the existence of the Christian God, and now why, in terms of living my self-actualising life, I find that belief unnecessary. It does not essentially matter to me if you do or do not believe in a god. If I live the life driven by, or at least cognizant of, the locally recurrent and seemingly immutable values that are evident to me and healthy society on this side of any veil across unknowable reality, there is no defensible extra benefit to be achieved by anyone in this life or the next from my life. A life enabled by these values, such as love and inner peace, is what is essential. This means that, no matter what your civilisation, culture and creed, an authentic life is your best defence against a possibly passionate but just god or karma.

What would this life of values and virtues look like? This is what we must discuss in the pages ahead. It would seek to live in a partly disciplined manner within our species' value-constraints and other systemic boundaries, while playfully testing those boundaries. It would be open to the idea of extending the explicit value-sets as we test those constraints and change those boundaries in type and number. The aim would be to have an emergent and looping relationship between our species' environmental successes and our values.

Concept Formation

Perhaps now it is appropriate that we diverge from the topic at hand and address a theory of concept formation. This is important to understand if we are to have a confident expectation with respect to our emerging knowledge and purposes.

We form concepts by identifying patterned attributes of existence, and in particular perceiving (and then logically analysing and synthesising) essential similarities and differences. For example, in forming the concept blue, a child may begin by perceiving two similar shades of blue and a different colour, red (a reference point). The child perceives and then identifies certain quantitative differences within the range of blue frequencies, as compared to the red frequency, without having any knowledge of the concept of measurement of colour in terms of light's wavelengths (or the added complications of how the brain actually perceives and processes them). Instead, the child probably links the colour blue to the memory of happy blue-sky days or some such experience. Later, when the child grows, it may learn the scientific definition of blue (a further process of differentiation and integration with respect to other concepts relevant to the colour blue) and how we qualitatively perceive colours with our human senses. However, this refinement will not render incorrect the initial, childlike conceptualisation of blue. The childlike concept is just as valid or efficacious as the adult concept. It is just that the childlike conceptualisation is less developed, or has a smaller hierarchy of related concepts around it; it is a simple predecessor of the adult conceptualisation.

Given that the child and then adult have normal mental faculties, and have used them in a normal way, the current adult conceptualisation of blue is simply, in terms of consciousness, a self-referencing, self-organising and emerging version of the child's conceptualisation. Yes, that is right; consciousness's concept formation (or meme formation) has the important feature of algorithmic feedback, just like every other self-organising principle of the universe.

Feedback moves us to accept an alignment between pre-conceived concepts and observations, and to reject or assimilate within new conceptual boundaries, a non-alignment. It is this self-organising and emergent nature of human concept formation that individually and collectively guides humanity along the path of more objective, although incomplete and tentative, knowledge. As long as we remain open to our observations and their fuzzy veracity, our hierarchy of concepts grows chaotically but also systematically (i.e. negentropically) over time. The growth of this hierarchy is another outcome of the subtle spacetime force-system or code that binds us together, because our concept formation also depends on feedback from each other. This same system has enabled us to coagulate into ever-larger and ever-deeper social groups.

The modern childlike concept of blue might have been sufficient for Cro-Magnon drawing pictures on cave walls 20,000 years ago in terms of their needs, desires and fitness for survival. From nature's point of view, there is no difference between the modern childlike and modern adult concept of blue in terms of each cave dweller's fitness for survival. The childlike concept was sufficient, and our adult concept perhaps inefficient and wasteful in terms of our cave dwellers' use of their time and fitness for life.

So in terms of nature, what is the better concept of blue and what is the second best? Such a test of worthiness does not arise in nature because the concept of environmental fitness and relationship in life-and-being here-and-now, rather than the best and most ingenious design, is what drives nature uphill. However, if we are able to survive now, then the universe does not necessarily restrict our inquisitiveness. We, like healthy children, always want to identify and learn more, and we often may.

In summary, as long as we can perceive a situated object and its dynamic attributes (such as blue) over and over again in order to form a testable theory of its inherent nature (or its heuristic rules of operation), then we can slowly learn from experience, evolve valid concepts or check beliefs, and gain knowledge. However, our temporary ignorance and error limits our growth of knowledge. If we build new concepts and confirm beliefs based on assumptions rather than validated explanations of a lived existence, our new so-called knowledge will likely be faulty.

Nevertheless, over time a self-organising universe will at least statistically weed out the errors amongst us. A little more quickly, our society and our personal learning may enable us, perhaps with dented pride, to get back on track. This is not assured or always instantaneous or painless, but it does eventually happen to some order of completeness in those of us who bravely use their rational mind-body apparatus. In a reasonably hospitable environment, our hierarchy or pyramid of carefully interlocking concepts can continue to build on foundations of varying strength.

But when it comes to our formations of introspective, indirect concepts rather than extrospective ones, the problems of ignorance and deception (especially self-deception) are much greater, which means the path to corrected errors is much longer and more tortuous.

This weeding out of introspective conceptual errors is right at the heart of the benefits of emergentism. Emergentism enables us to see and understand ourselves (our perceptions, values and beliefs) in the context of an emergent universe.

We can examine questions of gained knowledge through consciousness identifying the non-random nature of each thing that exists. In particular, we can examine our knowledge of values and their appropriate usage in our lives, without any preconceived assumptions of a reality beyond the veil. In the same way, we can also examine the other self-organising principles of the universe, such as the strong or weak nuclear interactions.

This book is about the profound possibility of partly objective but limited truths (facts or values), in the sense of temporal and sufficient but evolving knowledge, meaning and purpose. The limited but emergent truths exist in a world that does not have to remain confused by the idea of what is today an unknowable reality, that is, the claimed supernatural 'Something' on the other side of the supposed veil.

Limited emergent truths can add value to our lives now, because the universe's own observable and self-organising principles do not require us to remain confused. The universe allows us to discover more of itself in our own good time. We hold the key to the pace of this revelation already, that is, we have mind-body complexes, technology and social organisations with which we can courageously choose to perceive, carefully think, and act.

This book also encourages us to follow our own clear convictions rather than stereotypical social narratives from now on, more simply, so that we may all enjoy life. Further, while changing unwanted beliefs that we know are holding us back from our best can be very challenging, the Emergent Method can help. We will address this issue later.

The Greatness of your God is marked by the Attributes Assigned to Him

The summary of my story is this: No matter what Abrahamic religion or creed you follow, the attributes or values you assign to your God marks His greatness, and through Him, your greatness. Removing the Middleman, the values you live by measure your greatness. This is true whether you believe in God or not.

A religion's god is the personification of its conceptualised and revered values (positive or negative). For some reason many of us must see our values personified in a god before we can internalise them. Part of the reason, perhaps, is the very rich language or collection of concepts we have developed in the social worship of our gods that is not available when we speak of values outside of religions, their oft-supposed owners. Religion is an aging survival machine of values uncovered through experience in our species' past. Perhaps to be fair we might view the Age of Enlightenment in the same manner, i.e. as another aging survival machine of values uncovered through experience in our species' past.

The question that is relevant here is, 'What are the virtuous values you attribute to your God/god(s) or to your personal standards?' Some will say such a question is wrongheaded because human behaviour is not value-driven – it is belief-driven at best and often merely instinct-driven, with behaviour determined by the swill of chemicals and electrical pulses running around in the brain and body. It is a good point.

The same logic would argue, look at what a person spends their time and money on and you will learn their true values – time and money spent watching their favourite football team, chasing money and power, enjoying a hobby, or with the family. From this point of view, single-word values like love or peace seem inadequate; values relate to what is valued, like a sports team or Facebook or the family. Single-word values like meekness or temperance seem left over from a past religious era and not very relevant to modern life. They are almost chauvinistic in their claims to continuing relevance. Values in this view are often moral justifications slapped onto behaviours after the event.

To answer such criticisms, I would suggest that those virtuous single-word values are more like scientific facts that common sense alone cannot easily access. They are the building blocks with which broader conceptual value-arrangements such as 'family' may be constructed (or deconstructed). I would agree that in the first instance our assumptive beliefs or biases drive us humans more naturally than our unprocessed values, just as scientists begin with their more common-sense hypotheses and arrive at their theories only through a lot of hard work that rejects failed assumptions. Nevertheless, while we might find it difficult to remain true to the human virtues, just as some scientists may overzealously try to cut corners in their research, we can usually name the values we admire in ourselves and others easily, just as scientists have respect for temporal facts established through the Scientific Method. I would therefore suggest that the question of your personal values will be central to finding your purpose and success, whether personally, in business, or elsewhere, just as scientific facts help build a scientific paradigm and help realise technological developments.

The four groups arranged into the four tables below provide a list of possible values, virtues, character traits, stances we take, tools we use, or predispositions we hold with respect to life's situations. In the interests of brevity and openness, we will refer to these four groups as value-sets and their contents as values. I do not want these value-sets to reflect any particular secular, religious or civilizational disposition, but no doubt, they are coloured by my life's experience. Perhaps with your input this bias is reducible in the future. Perhaps the tables reflect a Western civilisational bias that could be tempered by input from representatives of what Huntington (1996) named the Latin American, African, Islamic, Sinic, Hindu, Orthodox, Buddhist and Japanese civilisations

We will intentionally avoid being tied down to closed dictionary definitions of values right now so that a brainstorming of personal meaning might be promoted. While the assignment of values to each of these four value-sets is arguable (or open for discussion), the reason for the groupings themselves is partly because of the idea of the antinomic dance between being and doing mentioned in Part 1. The values we need, want, or use for 'passive and dispassionate being' (as more conscious sensors, observers or analysts of environments and stakeholders) are different to those for 'active and passionate doing' (as more instinctive participants, actors or process-modellers in environments with stakeholders).

Further, we can approach passive being from a direct, extrospective and material/fact-driven viewpoint or from an indirect, introspective and immaterial/value-driven viewpoint. We can also approach active doing from a personal, individual and self-reliant perspective or an interdependent and socially interconnected perspective. By individual, I mean dependent on self and the interrelationship of self with its deep ecology. I do not mean independent in the wider sense. Self is uniquely individual, but always interdependent ecologically and sociologically.

As we will see in Pillar 4, life's antinomic dance between being and doing actually entails a dance between all four perspectives or phases outlined here in summary in the following four tables.

Perhaps the best way I can describe this four-part dance and express what the Emergent Method offers us is as a parent of four boys during their childhood:

1. When I guided my boys to ask the 'What?' questions of a situation they found themselves in, I wanted them to ask of themselves as competitors in life, 'What are the strengths, weaknesses, opportunities and threats of my situation? What are my risks? What discount factor should I apply to the imperfect information at hand?' I also wanted them to ask the relevant 'What if?' questions. More deeply, these questions relate to the values of Table 1's risk-weighted value-set.
2. When I prodded my children to ask the 'How?' questions of a situation they found themselves in, I wanted them to ask all those endlessly pesky extrospective questions that will enable them to arrive at rational explanations of the past that at least partly satisfy their inquisitive minds. These questions relate to the values of Table 2's positive value-set.
3. When I suggested to my children to ask the 'Who? When? Where?' questions of a situation they found themselves in, I wanted them to discover and uncover particularly social relationships and arrangements that will enable them to arrive at an understanding of the needs and desires of all lives interlinked, whether or not these relationships are logical to them. The important issue here was to uncover the interdependencies of those in relationship by choice or otherwise, including themselves, so that they could make valid contributions. These questions relate to the values of Table 3's normative value-set.
4. When I encouraged my children to ask the 'Why?' question of a situation they found themselves in, I wanted them to think for themselves – for each to reach inside himself to his own ethos, for his own sense of meaning, self-worth and purpose. With only a gentle guidance, I wanted my boys to answer this introspective question themselves with autonomous yet enlightened 'selfishness', only gained after parts 1-3 above. If they were able do so, it was then that I knew I had succeeded as a parent. These questions relate to the values of Table 4's self-enlightened value-set.

The beauty, impetus and power of this four-part dance depends on the values (virtuous or not) we adopt as our own in each antinomic phase. Please refer to the following four tables, which describe the four phases of this dance in terms of their associated values.

Risk-weighted; What?	ACTIVE (as Participant)
Competitive	Self-esteem; inner pride; appropriation of high self-worth; inner dignity Self-discipline and orderliness; self-restraint, moderation and self-control Inner drive and resilience Willpower and steadfastness Self-motivated effort and initiative Inner heroic courage; inner sense of honour in decision-making Individuality and self-reliance Inner competence, productiveness and sense of accomplishment Inner safety and security Inner health-building and wellness Inner wealth-building and fortitude Inner judgment Inner sense of justice Learning and curiosity Imagination and innovation Inner integrity, rectitude and honesty Happiness through virtuous and volitional action Taking and volition Competitiveness Respect for peers and mutual non-violence against competitors Inner sense of natural order

Table 1: The Risk-weighted Value-set

Positive; How?	PASSIVE (as Observer)
Extrospective	Reason or rationality and logic Intellectual pride; love of rational veracity Accuracy Connection with largely objective reality Theoretics; appreciation of theoretical explanations Methodological design and approach Falsifiability and verifiability Acknowledgement of fallibility Transparency of processes Acknowledgement of sources Openness to peer review Testability and repeatability Disclosure of methods and findings Intellectual openness of mind Intellectual integrity and honesty Intellectual responsibility Intelligence Simplicity, symmetry and beauty Outward knowledge

Table 2: The Positive Value-set

Normative; Who? When? Where?	ACTIVE (as Participant)
Cooperative	Recognition and respect of others and/or the environment
	Love of family, community (mateship) and the environment
	Order and discipline; obedience
	Loyalty, sense of duty, responsibility
	Judgment, discretion and discernment
	Cooperation and teamwork; collective decision-making; team code
	Team honour; reliability
	Non-violence and tolerance
	Sense of humour
	Sincerity, fidelity and dependability
	Outward affection
	Kindness and benevolence; care for the welfare of others
	Gentleness; courtesy and civility
	Meekness, humility and acceptance
	Patience and longsuffering
	Magnanimity and liberality
	Worldly wisdom
	Attractiveness
	Chivalry; doing good
	Compassion, mercy and forgiveness
	Altruism and self-sacrifice
	Selflessness or other-centeredness
	Devotion
	Giving and surrender

Table 3: The Normative Value-set

Self-enlightened; Why?	PASSIVE (as Observer)
Introspective	Unconditional love in being
	Personal respect and faith
	Personal state of assuredness, self-belief and purpose
	Temperance
	Personal integrity
	Personal strength
	Inward knowing
	Mindfulness and introspection (or synthesis and reflection)
	Personal state of wellbeing & vitality
	Personal focus
	Spiritual study and meditation
	Prudence
	Joy in being
	Asceticism
	Peace in being
	Goodness, holiness in being
	Personal beauty
	Being
	Sense of personal freedom and personal commitment
	Flight from incarnate emptiness; personal identity and individual effect
	Personal sense of relationship/interconnectedness

Table 4: The Self-enlightened Value-set

Of course, all such lists as those presented in the preceding four tables are impossible to comprehensively compile and categorise, so if you wish to add to their four virtuous value-sets, please ensure they comply with the subheading categories. You may want to add adjectives to nouns to achieve this. You might also like to further differentiate or integrate the concepts represented by the words in the table entries.

I find it interesting to note that for me the lists of active values, virtues or traits were easier to compile and are noticeably longer than the passive lists. I also find it interesting that for me at least, the order of ease ran from 'Normative' (easiest) to 'Risk-weighted' to 'Self-enlightened' to 'Positive' (most challenging). I noted that this last value-set would have been difficult to compile without a background in the Scientific Method, first gained through my electronic engineering training. I also wondered whether it was my cultural setting that enabled me to name the normative values more quickly than the other values; perhaps this ease was something peculiar to me. In the future, we might attempt to gather data on this point at the website.

As you may have noticed, some values are difficult to differentiate and would seem to fit into any value-set. A few examples would be:

- Love
- Virtuosity and a virtuous balance between competing alternatives
- Reverence
- Honesty, integrity or fidelity
- Happiness, joy, humour, laughter, delight or pleasure
- Optimism
- Intelligence
- Beauty
- Flexibility or resilience
- Wealth-building and wealth-appreciation (not just monetary or material)

The four value-sets, with appropriate adjectives, already include some of these. If you wish to insert other such values with appropriate adjectives into the value-sets above, please feel free to do so.

The only difference between some values is how we view or apply them. Good examples of these values are the difference between 'prudence and fidelity' (usually applied or viewed inwardly and more passively), or 'judgment and honesty' (which we usually think of as applied or viewed outwardly and more actively), respectively. We could probably invent or name a number of new value-words simply by taking an existing value and limiting its application to the competitive, extrospective, cooperative, or introspective. We might also find that in other languages, such words already exist.

As we know, normatively good and bad values together can take root in us. However, what makes some of them well adapted and thus quasi-universal? The answer probably has something to do with Abraham Maslow's hierarchy of needs.[1] We will discuss the relationship between human needs and values further in Pillar 4. Nevertheless, is it necessarily the good values that we follow as a society? Our virtuous global tradition, as briefly discussed in Part 1, would suggest this is possibly the case, at least over the very long term. However, we can't afford to be complacent - a value promoting of life-and-being here-and-now and its self-organising and self-actualising principle, whether doing or being, does not become widely shared unless it is permitted to do so and is relevant and appreciated. We can wait for this to happen naturally or we can perhaps artificially hasten the process in our communications with self and others.

It is also possible the Western civilisation has already passed the peak of moral growth and value-awareness. We all need to do our bit to ensure the self-organising and self-actualising values are also the most scrutinised values and the most promoted values by our leaders and institutions, and then finally the most socially and personally integrated values in our daily lives.

This very discussion highlights another very difficult issue. We need to be able to add a normative aspect to seemingly universal values that are not social, but personal. At the same time, to avoid the alienation of individuals, we also need to bring a personal sense and identity to the values already considered socially normative. That is, we need to keep talking about and deeply share, for example through our new books and movies, both our personal and our cultural experiences. This is at the essence of a successful integration of the quasi-universal values. In this global regard, the values of tolerance and mutual respect seem to be vital.

On a more subtle level, the same values promoting of life in a positive context can be promoting of death in a negative context. For example, happiness over something promoting of life amplifies that living but happiness over something promoting of death amplifies that dying. Conversely, we could also discuss a hatred of lying, a fear of infidelity, etc.

Finally, some values seem to be directly contradictory – such as taking and giving. To address this issue we need the relativistic and dynamic cycle of virtuous, although sometimes antinomic, values this book expounds in the following chapters. Virtuous values applied in the correct phase and niche of life bring life, happiness, meaning, purpose and success to every thoughtful decision we make. Virtuous values are an essential key to human flourishing within a sustainable environment –sustainability being a basic factor in the wellbeing of any evolving species.

Shadow Values

Another important point to be raised here is that some would argue the applications of hatred, fear, et cetera are also fairly universal and evident, with which we could not argue – but they are ultimately promoting of death and decay rather than life and growth, or de-organisation and self-destruction rather than self-organisation and growing enlightenment (or self-actualisation). This is the destructive way of the endlessly and remorselessly self-gratifying and avaricious. Such are always narrow and short-term in their outlook, even though they may find a temporal success. This is the nature of the violent and the narcissistic con-artist, in the hands of whom our civilisation seems to suffer more and more. If we as individuals or a society continue to focus on those values and promote them, we hasten the day of our own destruction and passing.

Nevertheless, none of us is an angel. We all have negative values driving our behaviours from time to time. It may be an interesting exercise to list the antonyms of all the virtuous values in tables 1-4, but I will leave that exercise to your discretion.

Benjamin Harvey, the co-founder of Authentic Education, has some very interesting things to say about the negative values that reside within us, which he describes as *shadow values.* He characterises these values as those gained by perhaps bitter experience yet shunned by our culture, so difficult to admit to holding and often important to repress if we wish to function and succeed socially. To discover our shadow values, he recommends we firstly set a few simple rules: No judgment, no shame, no fear, and no guilt!

Harvey suggests the core set of shadow values is rather small, because all others are derivative. He also suggests that they can often be unearthed or made explicit by looking at the activities that are most important to us or take most of our time (see Table 5).

Harvey further suggests that by bringing these values out from the shadows within, we can consciously acknowledge them, and then actually feed them. Through honestly recognising the needs arising from our shadow values and consciously satisfying those needs, we become free to reassess their place in our lives in a way that continuing repression will not enable.

Activity	Shadow Value	Description
Dressing provocatively or getting a massage	Attention	Being unique or special
Dominating a conversa-tion	Power	Being able to do or get whatever you want
Studying	Superiority	Feeling right or more correct than others, or a sense of personal improvement and achievement
Seeking needless ap-proval or endorsement	Validation	Feeling important, worthy, and/or 'good enough'
Visiting family or friends	Belonging	Feeling accepted by, or connected to, a group
Working overtime	Control	Being able to influence your circum-stances or those of a group
Flirting	Sexuality	Being able to express your sexual desires and/or preferences shamelessly

Table 5: Shadow Values

Another simple way to draw out your own shadow values would be to ask yourself the question, 'Can you remember a time when you were angry or frustrated; in that very moment, what was missing (was it attention, belonging, control, power, sexual expression, superiority or validation)?' Another way to discover your shadow values is to list all the important areas of your wellbeing and, for you, the shadow values that most honestly fit with them.

Table 6 lists our core sample of shadow values by area of wellbeing, as well as the complementary golden value of each shadow value. That is, Harvey sees each shadow value as hiding a deeper golden value within it that is bound to surface after the initial *shadow need* is met. In a way that is similar to moving up Maslow's Hierarchy of Needs[2] (discussed later), if we can move past our more ignorant values, we will gravitate towards those golden or virtuous values that are more likely to reside in tables 1 to 4.

Area of Wellbeing	Shadow Value	Golden Value
Physical Health & Beauty	Attention	Appreciation
Career & Business	Power	Inner Service
Mindset	Superiority	Evolution through learning
Friends & Community	Validation	Inspiration
Family & Partner	Belonging	Personal Meaning
Homemaking & Finances	Control	Freedom
Spirituality	Sexuality	Love / Interconnection

Table 6: Golden Values

Your Values

So what (golden) values of your God or culture do you use? How do you know which value-set is the one with which you are currently most familiar? How can you assess which is your currently preferred value-set? This is easy; just complete the following steps[3] from your own personal perspective, not with your work-hat or any other particular role-hat on. Please complete this exercise, do not just read it – you will reflect on this important information about yourself as you read the rest of the book:

- Jumble all the values mentioned in the lists provided above and look at them individually, without considering at all which value-set they belong to;
- Next, assuming you had to live the rest of your life with just one of those jumbled values, to the exclusion of all others, which value would you choose? Take your time.
- Mark this value down as number 1 and set it aside;
- Re-considering the remaining values and again, assuming you had to live the rest of your life with just one of these remaining values, to the exclusion of all others, which value would you choose? Take your time.

- Mark this value down as number 2 and set it aside;
- Continue this process until you reach a preferred list of 8-12 or more values (it's up to you).
- Now re-assign the selected preferred values (only) to their respective value-sets.
- The longest assembled value-set list should reveal your currently preferred value-set, but just to make sure, you could simply add up the value preference numbers in each value-set – the value-set with the lowest average score-per-value wins. If the result is a tie between two or more value-sets, just keep growing your list of likewise selected values until you find a clear 'winner'.
- Now do a reality check. If these really are your values, then think about how they are expressed in your daily activities at work, at home, in clubs, with friends, in the car, etc. If they do not currently find expression in your life, why not? Conversely, which values do currently find expression? Any incongruence in the expression of your values is motivation for change. We will discuss how you might go about achieving such change later, although changing your core values is not an easy task.

Once you have determined your value-set preference, you can appreciate which river(s) of self-organising value-adding you currently prefer to swim in and which ones you perhaps feel like a fish out of water.

Our preferred values and beliefs not only strongly affect how we see the world and what we don't see but also what we act upon and what we don't act upon. Our values and beliefs, explicit and implicit, affect all of our conscious, subconscious and even unconscious decision-making. Some of them help or urge us to achieve our virtuous goals and purposes and some contribute to their frustration. Sometimes the values we hold are in conflict, e.g. when our consciously relied upon values oppose the implicit values of the body's regulatory system. Such a conflict will manifest itself in feelings of stress and anxiety.

Our highest believed or relied-upon value is where we store the most of our concretised wealth. For example, if you love music, or more correctly, music brings you a feeling you label love, you will spend more of your money on musical recordings, instruments, or sound systems, and relatively less on the other needs of your household. Even more generally, our highest value is where we store the most of our wealth, in terms of time, thoughtfulness, money, etc. We flourish and intensify our experience of a purposeful life when our believed values and actions roughly align with how the world around us, such as the workplace, operates.

Our committed values and our awareness of them hugely affect our identities (the reference points that define how we see ourselves and how others perceive us). We often live our lives at a surface level, blind to the values that really define us (perhaps the deeper or true us). When we become more self-reflective and self-aware of our values, we are more able to live our lives at a deeper, more meaningful level in a wider spread of our daily activities. As we do this habitually, how we see ourselves, changes. Our deeper or true identity affects more and more of what we do and don't do. Our values put into action give us the reward from experiences that we cannot get from material things alone. Material things eventually fail to deeply satisfy, as do empty goals and purposes (i.e. goals and purposes not aligned with our values). Our evolving and almost immutable values-in-action in step with local reality are what truly satisfies. The great reward from a life lived and shared authentically is inner contentment.

Our hierarchies of values or vremes, more than our broad hierarchies of mental concepts, have a huge impact on our destinies. Obviously, the values we placed low down on the list or off the list will typically get less day-to-day attention than those high on the list. Is this apportionment of attention always well-reasoned in every circumstance? No, obviously not; so we also probably agree that we need to be more consciously aware of the helpful and unhelpful values we hold or use and in what circumstances we hold or use them. The values-code or moral code we consciously hold, commit to, and translate into larger concepts such as 'family' or actions such as 'watching the football each week' will enable us to measure and grade all the value-laden choices we make. In other words, the values we apply through our scarce time and space quantise and measure our finite lives, their dances with reality, and their successes within their niches.

In what places do we enact our lives and apply our values? In our homes, in our work places and in the organisations, networks, or marketplaces we interact with both personally and impersonally.[4]

Values and Virtues

Definition of Values

Perhaps now it would be beneficial to have a working definition of values. What are values? The simple answer is that 'values' is the plural form of 'value'. We assign an explicit value to everything we measure in terms of similarities and differences. We sometimes assign an implicit value, waiting to be made explicit by appropriate further measurement. Dollars might express explicit value, but more deeply, time spent, work done, and benefits enjoyed also express explicit value. We assign no value to the things we do not reflectively measure. These things are part of the chaotic and meaningless backdrop to our lives, but also provide potential in terms of our emerging values and morals.

At a much deeper bodily level, our homeostatic systems detect similarities and differences, and assign/implement values themselves that aim to keep our life systems within healthy operating limits for us, well before we are conscious of them. They do this by employing complex feedback and feed-forward loops between changing reference points. However, these systems of biological regulation by no means work in isolation. For instance, when the body needs energy its unconscious systems send hunger pangs via chemical and nervous signals around our brain and body, urging certain subconscious or conscious goals and behaviours (e.g. 'go and find some food now', or 'eat something sweet for a quick energy boost'). Inherent in these urges are many beliefs (e.g. 'sweet, high energy food is good for you when your energy levels are low'). More generally, the implicit value associated with this unconscious homeostatic belief might be 'maintaining your performance within operating limits is important for systemic continuity'. Evolution seems to blindly choose and establish this value, in all life forms, without any kind of explicit moral agency at all. Actually, there is no unconscious choosing or establishing of value occurring. There is just dynamic interrelationship. Emergence and entropy are concepts that describe how an object moves between contours of available energy bands to the band that represents the greatest local level of energy equilibrium, just as the concept of gravity describes an object moving under the influence of its relativistic relationship with surrounding space and matter. We assign values to these valueless dynamical relationships. However, when we do this, high level values also become embroiled in our unconscious interrelationships.

In terms of the valuing of goods or services, the most basic measure we use is not money, but the others – time spent and other benefits enjoyed, or more correctly, the way the good or service relates to our shared spacetime. That is, we subconsciously ask ourselves when we value something, how does the good or service interact with the scarce resource of my limited but valuable life moving forward towards the future in space and time? This is how we value goods and services, whether as buyer or seller at the conscious level. We compare value of good or service with value we place on our lives, i.e. we compare value with value within the freedom afforded by our limited homeostatic self-restraints. That is, the price of anything is the intersubjective value of life we exchange for it.

This is another way the system of spacetime acts on us. Our personally limited spacetime (firstly measured at the unconscious level of our homeostatic systems) drives us to value life / seek survival, that is, to value all the things that pertain to life and its systems, and to assign value to everything with which we meaningfully interact. True wealth is thus the ability to fully but intersubjectively experience the good life.

It is this same spacetime system that has brought about through its self-organising and self-eliminating processes the time and attributes (or value, meaning and purpose) of the good or service in the first place. For instance, the same spacetime system that makes us value time with pets or time in a garden also makes the pets and garden fit for their own purposes and attract our consciousness. Do evolution and consciousness exploit spacetime, or is it the other way around? That is, do we choose our values, or do our values choose us? These questions demonstrate another aspect of the push-pull of life's dance and the relativistic idea that suggests neither answer is definitive. Consciousness, evolution, and spacetime are each caught within the same cycle of emergent and monistic full reality. Nevertheless, the articulation of emergent vremes seems necessary for their psychosocial longevity. This perhaps describes the basic reason for the emergence of language and thought. Further, as communication escalates, so it seems does our enlightenment.[5]

Most biologists believe evolution has no goal or directionality, but some commentators, like Paul Davies (1999), believe that the basic self-restraints of nature bias it towards creating life and consciousness, with which I would reservedly agree. Perhaps a better view is to see the core concept as not one of goals or biased directionality, but self-contained and statistically inevitable emergence in the dynamical interrelationships of all systems. All movement is proto-intentional, thus all nature is biased towards the emergence of value and meaning, of which life and consciousness are the chief examples. What do you think? The choice of belief regarding emergent directionality, including all the other possibilities not catalogued here, is yours – but this is not the focus of our discussion, nor is your answer essential to your ongoing self-actualisation. What are essential, are the values you embody and honour in your moment-by-moment, existential, and emergent dance of life.

So far, we have discussed the valuing of extrospective things (i.e. goods or services). How do we value introspective things such as feelings, emotions, desires or precious memories? These things are not valued by dollars or other extrospective measures (such as size, weight and colour) directly, only indirectly. We measure and directly value introspective things with introspective measures such as the subjective measuring-sticks of beauty, love or happiness. (See the Self-enlightened value-set of Table 4). Further, beauty, love or happiness are measures of our introspective consciousness that we can apply to extrospective goods and services (such as a pair of shoes or a car) as well as introspective objects such as our feelings, thoughts, memories or desires.

In summary, values are to introspection what a table of weights and measures are to extrospection. Values are the fuzzy measuring-sticks consciousness applies to all things it intersubjectively introspects, just like standard weights and measures are the fuzzy measuring-sticks consciousness applies to the things it more objectively extrospects. However, unconscious values came first, long before we consciously invented any tables of weights and measures.

Some of the values or introspective measures are almost universal or immutable as well as self-organising. They are arisen from our evolved human instincts and a sound knowledge of reality (as per Table 1 to Table 4). Yet some of these values are faulty, self-deceiving or self-eliminating in terms of life and our ability to self-actualise. Either way, lived values affect our wellbeing. Thus, the personal subjective world has a nature that reflects the nature of spacetime itself; that which is self-organising and that which is dis-organising. Emergentism is about identifying and promoting that which is self-organising to our introspective consciousness and in accord with our self-actualisation. It is also about identifying but demoting that which is contrary to our self-actualisation. Virtuous values or vremes are the most important tools we can use to rebel against the tyranny of our blindly selfish replicators (i.e. sometimes short-sighted genes and often short-sighted memes).

The way we subjectively think, feel and desire, while it may be faulty at times, is not always without order or a kind of 'personal objectivity'. In this sense, vremes are intra-personal or intra-somatic facts. That is, vremes are limited or temporal facts that arise from within the body rather than outside it. The basis of our intra-somatic facts lies in the body's highly evolved bio-regulatory systems that urge us to limit our activities to within preferred operating conditions – conditions that maximise the opportunity for us to pursue and enjoy higher-order wellbeing. A vreme is a kind of higher-order intra-somatic set of chemicals and electrical signals required by the mind-body to regulate our behaviours within homeostatic limits at the highest levels of consciousness.

Do we have any naturalistic and thus firm basis for this concept of *personally objective* values? Well yes, I think we do. Our quasi-immutable and virtuous values have not arrived on the scene just lately – they grew up over the 13.8 billion years of our universe's history. Intra-somatic values were honed by natural selection's success in finally establishing our species, and every other extant species on our planet. They were written into every successful and active gene. Now, with our consciousness and naming of values, they are finally becoming explicit. Personally-objective and now explicit values have been linked to our human flourishing through natural selection over evolutionary time. I believe we ignore these values to our own peril.

However just as only some extrospected data represent fuzzy facts, so only some introspected data represents fuzzy vremes. We can thus consider intersubjectivity as having two parts – that which is aligned with the self-organising and emergent values of the universe and that which is not. That part which is aligned with the self-organising values in us (the intra-somatic or personally objective)[6] might be by conscious choice or serendipitous emergence. That part which is not aligned with the self-organising values in us might be by choice, ignorance, deception, or unfortunate circumstance. Whatever the case, emergentism grants us an opportunity to improve ourselves by discovering that which is or is not self-actualising, and encouraging us to take part in the moment-by-moment musical improvisation that is flourishing life.

How do errors in intra-personal observations arise? In the same way as extra-personal observations; that is, through quick assumptions, poor explanations or hypotheses, poor transparency of assumptions and hypotheses, poor testing of hypotheses, etc. In short, our intra-personal observations fall into error because we fail to have regard for the Positive value-set of Table 2. Again, we see here the interpenetration of all four value-sets of Table 1 to Table 4. We also see the close relationship between the Scientific Method (which we could perhaps better describe in usual use as the Extrospective Method) and the Emergent Method (which applies to both our extrospections and introspections).

I know this may be difficult to unravel, but our consciousness, to the extent that it is capable of moral agency, directly knows reality intersubjectively, but can indirectly know reality more objectively, through the Scientific Method. On the other hand, our non-conscious instincts directly know reality, because while they lack moral agency, they are an interpenetrating part of environmental reality rather than a fallible observation of it. We as humans are interpenetrating arrangements and shufflings of pseudo-objectivity and intersubjectivity. To say that our everyday first-person experiences and interpretations of physical reality, with its tables of weights and measures, can have an objectivity that our feelings and their values cannot have, is to overstate the objectivity of interpreted experiences and to understate the objectivity of lived experiences and naturally emergent feelings-with-values. There are no black-and-whites here; all is fuzzy.

Human cultures or social groupings, consisting of individual moral agents, fill niches via vremes just as species fill niches via genes. Just as genes in successful species emerge, so vremes in successful societies and organisations. Both genes and vremes emerge because they exploit more fully a given niche, whether biological or sociological. In both cases, a diversity of exploitation methods makes the deep ecology more robust.

This suggests that in one sense we need to get past the idea of one culture or organisation being superior to another, just as we need to get past the idea that one species is better than is another. Our value focus should move away from the current 'winner' and towards a robust diversity. However, we cannot ignore the fact that some species have found ways to exploit their environments better than have others. Similarly, some football clubs perform better than others currently do and some companies currently perform better than others do. Nevertheless, today's success is sometimes a poor predictor of future success in a changing environment. The design of the octopus eye that avoids the blind spot that is a feature of the human eye was a case in point. In that case, the human lack seemed to lead to more complex processing of visual signals in the human brain. I suspect this lack of efficiency in the human eye may have led to benefits in the human brain not available to the octopus brain.

We also noted in Part 1, that life-and-being here-and-now is a higher implicit value of life's success than intelligent design. For human society, this translates to, tolerance for successful cultures is more important than vremetic perfection. We will return to this topic in later chapters because it captures an essential feature of emergentism. That is, we will return to the idea that today's genetic imperfections may hold the potential to solve tomorrow's environmental challenges, just as our children's minds may hold the solutions to tomorrow's moral challenges. Here is described the benefit of diversity.

If values are different measures of introspective feelings and not the feelings themselves then happiness, love, and peace are fruits of our life-and-being here-and-now rather than life-and-being here-and-now itself. Without intending to sound transcendental, happiness, love, peace, etc. are the fruits of a well-ordered, purposeful and self-actualising life, moving in approximate synchronisation with its prized values through a given situation. However, this is not the whole story. Values arise from our bodies – they are not separate notions we impose upon them. Again, we note the interpenetrating and relativistic nature of our implicit and explicit values.

We will only know that we are happy, in love, or full of peace, when our lives align themselves with the appropriate situations and actions that bring these measured evaluations about. We have to dance before we can acknowledge a new feeling of love, joy or peace (rather than just acknowledge the memory of an old feeling).

We have to align our lives with our universe through our dance to feel the vibe. Happiness requires action, and a necessary delay between the action taken and the response obtained from our deep ecology.

This is simply because consciousness is consciousness of something. That is, our consciousness-plus-instincts (consciously and subconsciously) have an indirect but logical relationship with the intra-personal and extra-personal facts of nature just as spacetime has an indirect but logical relationship with matter. To apply this realisation in another way, spacetime cannot exist by itself – there is no 'itself' worth talking about (just quantum fluctuations or noise) until it exists and interacts with something (tangible matter).

This is why we have discussed the concepts of matter and spacetime at such length. It was so we could also understand the concept of relativistic values and understand how values are involved in the relationship between consciousness and spacetime. It was so we could grasp the idea of Big Morality, also mentioned in Part 1.

Just as we have perhaps assumed in the past that existence, consisting of matter and spacetime, is a simple and direct fact of reality without requiring, imparting, or including any indirect logic, so we have blithely and incorrectly assumed the same of human nature. Human nature consists of both its direct life and facts and its indirect values that impart opportunities for happiness and success. Vremes enable us to escape the short-sighted aspects of our blind gene/meme replicators. Their quasi-universality and incompleteness enables us to escape the boundary of the mind-body and connect to a societal mind-body (and beyond).

Our values impart logic to, and require logic of, our consciousness-plus-instincts. Vremes are inseparable from consciousness; they are the system or code of consciousness just like spacetime is the system or code of all existence and the genetic code is the system of all living things. Just as we cannot discover spacetime without being part of it, so we cannot discover the vremes of higher consciousness without living in accordance with them. Consciousness cannot exclude values; the force-system or vremetic code of values can only fade as lively consciousness fades.

Emerged values seem to give direction to consciousness just like spacetime seems to give direction to matter and genes seem to give direction to living organisms. However, in each case the one-sided direction perceived is actually the outcome of a continually emergent, interdependent and phaseal relativity. In truth, values and consciousness emerge together, as do spacetime and matter, or genes and organisms. Yet just as spacetime is impactful and personal to every subatomic particle of the body, so values are impactful and personal to all the introspections and extrospections of consciousness.

These parallels between spacetime and the active vremes of consciousness seem to support Paul Davies' (1999) idea that the basic self-restraints of our universe and its fundamental nature bias it towards creating life and consciousness. However I suspect it would be more accurate to observe that whatever we perceive as fundamental or basic is not, but rather emergent from more vast but monolithic beginnings, or more disorganised nothingness. Thus, emergence as an indirect process rather than as a thing is fundamental. Statistical entropy and emergence is fundamental to this and all possible universes.

Concepts of Emergence Encountered in Pillar 1

I now understood that the seemingly immutable values, for which I still believe we are individually and morally responsible to hold and cherish, simply revealed themselves in our evolving consciousness as we interact with the environment. They are ours, as humans, as it were of our own emergent making, but are no less valid, no less important and no less reliable, even if their immutability or universality is a little fuzzy.

This weeding out of introspective conceptual errors is right at the heart of the benefits of emergentism. Emergentism enables us to see and understand ourselves (our values and beliefs) in the context of an emergent universe.

Emergence and entropy causes an object to move between contours of available energy bands to the band that represents the greatest local level of energy equilibrium, just as gravity causes an object to move under the influence of the relativistic relationship with surrounding space and matter.

We also see the close relationship between the Scientific Method (which we could perhaps better describe in usual use as the Extrospective Method) and the Emergent Method (which applies to both our extrospections and introspections).

Emerged values seem to give direction to consciousness just like spacetime seems to give direction to matter and genes seem to give direction to living organisms. However, in each case the one-sided direction perceived is actually the outcome of a continually emergent, interdependent and phaseal relativity. In truth, values and consciousness emerge together, as do spacetime and matter, or genes and organisms. Yet just as spacetime is impactful and personal to every subatomic particle of the body, so values are impactful and personal to all the introspections and extrospections of consciousness.

Thus, emergence as an indirect process rather than as a thing is fundamental. Statistical emergence is fundamental to this and all possible universes.

Before we conclude our discussion of values, we need to discuss human nature more deeply. In the next chapter, we discuss human instincts and in the following chapter, we discuss human consciousness.

---

[1] Available online from http://psychclassics.yorku.ca/Maslow/motivation.htm (accessed 2012) but first published as Maslow, A.H., *A theory of human motivation* (Psychological Review 1943), 50 (4), 370–396

[2] Ibid.

[3] I must thank public speaker Loral Langemeier for the general idea and structure of these steps.

[4] See Smith, Douglas K., *On Value and Values* (Financial Times Prentice Hall, USA, 2004) for more details on how to apply values. The book's focus is on the framework of values rather than their specific application, but an understanding of the specific application of values is also extremely important. I highly recommend the book.

[5] See Pinker, Steven, *The Better Angels of Our Nature: Why Violence Has Declined* (Viking Adult, USA, 2011), Chapter 7, especially graph on 380

[6] Perhaps we could say that in terms of human wellbeing and emergent flourishing, applied facts and virtuous values are 'true' to life and 'good' whereas applied non-facts and unvirtuous values are 'untrue' to life and therefore 'bad'! In one sense, Emergentism is about carefully daring to apply the method and benefits of Dennett's heterophenomenology to ourselves. See Dennett, Daniel C., *Consciousness Explained* (Penguin Books, USA, 1993)

PILLAR 2: Instinct

In this chapter, we will discuss sensorimotor perception and then the two basic human instincts of competition and cooperation, governed by the risk-weighted law/self-restraint and normative law/self-restraint, respectively. We will also note that these instincts have not developed independently, but have grown up together, and so are inseparable and interpenetrating. We will talk about the wonderful success of impassioned instinct development in the human species, but also mention the current sub-optimal state of our species as we consider our world today. There are many still living in extreme poverty and fear rather than in the symbiotic self-worth we see when we look at a healthy forest ecosystem.

In terms of our personal philosophical quest, we will identify that the essential life-process within us (who we are) and our values-in-action are synonymous. The text will also encourage you to develop and write down your own credo.

We are Survivors

While acknowledging that our algorithmic instincts operate at nominally three levels – the unconscious, the subconscious (apologies to Freud, who would not agree that this level exists), and the conscious – the latter part of this chapter will focus our discussions on the framework within which human instincts operate at the conscious level.

Nevertheless, we do need to acknowledge our instincts' basic effects at the unconscious level. For instance, instincts trigger the production of adrenalin, dopamine, serotonin, and endorphins in our bodies. Instinctive processes thus shape our thoughts and behaviours through the bloodstream. Instincts, through the autonomic nervous system, likewise influence thoughts and behaviours through the maze of nerves that send and receive highly focused messages almost instantly between particular body parts.

Sensorimotor Perception

A primary function of the chemicals and nerves running around the body and connecting to the brain is passive sensory perception and active bodily movement (motricity), which work hand-in-hand in a relativistic fashion. I am including here a discussion of sensorimotor perception because it provides an instructive link between relativity at the level of matter-spacetime and relativity at the level of instincts and consciousness.

Emergentism's approach to describing and understanding sensorimotor perception is largely an existentialist one in accord with Merleau-Ponty (2014), a highly recommended book.[1] He approached the subject from the point of view that we are embedded in the world, that is, our bodies are anchored in the world, and must be open to the world in order to perceive its behavioural styles and thus receive its lessons. Perception is a lived experience or communion with the world through the mind/body, not a secondary interpretation of disconnected objects by a consciousness that is yet another object or exists in a dualistic other realm. That is, we enter the world with it already communicating with us. Only slowly do we respond to its rhythms and styles, perhaps firstly through mimicry and play.

The somatosensory mechanisms and paths between body and brain have been largely explained since 1945, but we still battle with a philosophy of mind and philosophical concepts such as qualia. Merleau-Ponty explained that the perceived was always in the midst of a relative background/foreground setting that provides context to the perceptual spectacle. "We are caught up in the world" is a favourite phrase.

He said that red and green are not sensations; they are sensibles,[2] and such qualities are not elements of consciousness, but properties of the thing perceived. Qualia are not inventions of consciousness, nor complete in themselves outside of a worldly context. Thus, there is no pure quality or pure sensing. Sensing is always messy and approximate. Actual, explicit perception is gradually distinguished from dreams and phantasms through many confirming feedback loops in the mind/body, including deep cross-references between the five senses and the body's sense of balance.

Qualities such as colour call forth a certain motricity in the body, and realigning of my intentional stance towards the world. There is an adductive, sympathetic nervous system 'fight or flight' blood-flow-out reply, or an abductive, para-sympathetic nervous system 'feed/breed or digest/rest' blood-flow-in reply to the world. For instance, green/blue calls us out into the world (adduction), whereas red/yellow draws us back to our core (abduction). What happens if my red stance meets blue? There is a temporary pandemonium. My unconscious body must find a nervous system response to the solicitation that brings me back towards homeostasis and comfort.

In Merleau-Ponty's and emergentism's view, a perception is not a simple mental or psychological object or event; it runs far deeper than such crude uni-dimensional assumptions. The mind/body is always 'situated' within its dynamic environment, which means there is a relativistic separation of powers in play between the mind/body and its environment. The Emergent Method is about recognising and exploiting how the mind/body forms a system with the environment rather than separate from it. Perception is thus a complex and dynamic phenomenon, with deeply layered feedback and feedforward loops that cannot be considered lightly. Sensorimotor perception is an engaging and intimately inclusive dance or coupling with reality, not a perverse voyeurism.

In any dance, whether in the act of perception, on the dance floor, or inside a bird's display arena, movement has intention, symbolism, meaning, and value that exist below the conscious level, and for humans, sometimes rises up to enable the emergence of motivated, self-aware consciousness as well. All animal intention has its roots in sensorimotor perception, and even more deeply, in the irrepressible movement or proto-intention that comes from the relativistic relationship between space and matter.

The notion of a discrete sensation distorts an understanding of the intertwining systems of perception. A figure on a background has many more qualities than is easily assigned to the figure or background alone. Due to the vastness of the feedback and feedforward loops within the mind/body, it also has a sensed flexibility, hardness, weight, sparkle, etc. It has a combinational style or unseen field/interaction that is not easily unravelled, but nevertheless carries signification and often evokes an emotional response in the perceiver, who as an intentional being now stands caught in the projected perceptual web. I hesitate to add whether this web is that of the perceiver or the perceived: It is neither; it is both. The perceived existant carries its own proto-intentionality in its dynamic organising qualities. It moves me; causes me to change my stance; anchors me; communes with me. I inhabit it.

We do not construct the phenomenon's unity through association, we simply resolve and verify its stylistic attestations.[3] The dynamic object confirms its own unity. It presents its own synopsis to us, greatly reducing the presumed work of consciousness or memory. Our perception is maintained by the macroscopic certainty and continuity of the world.

The anthropomorphic experiences of 'object', 'background', 'present', and 'past' represent perspectives that are necessarily incomplete, compared to actual reality. Similarly, the physicists 'atoms' represent a levelling, averaging, or dumbing down of actual dynamic reality.

We sometimes think that given the atoms, we could theoretically rebuild reality, but this is not the case. This false impression of science is the grand illusion of the greedy reductionist. Space and matter have a relativistic and indeterminate relationship that cannot be independently and deterministically rebuilt. This 'becoming' is unique.

Phenomenologists recognise the dynamic style or interaction captured in the complex feedback/feedforward paths that reductionists try to isolate and ignore. By doing this, we recognise that vastly more occurs around the perceptual web than the reductionist measures. The homeland of our thoughts is not inside consciousness, but in the expansive perceptual milieu that is our environment; the existant is subconsciously perceived as attractive or repulsive before it appears colourful, takes final shape, or is consciously 'known'.

Empirical truths are thus just a small subset of the existential truths we can sometimes reach through intersubjective reflection. Is the colour screen of a TV simply made up of hue, saturation, and brilliance? What is missing from this idea? From these three attributes the visual sense emerges, not as a simple sum or blending, but as a dance, a reciprocation, or an interaction supplied by the complex visual content captured by the camera, but is itself often obscure and profound. To answer the question, what is missing is the dynamic processing of the video content embedded in reality.

When a painter wants to capture a radiant object, she/he achieves it by distributing reflections of light and shadow amongst surrounding objects. That is, the object is largely defined by its complex interactions with all else in its local environment.

Merleau-Ponty tells us that the primary operation of attention is to create for consciousness a perceptual field or oasis that can be open to the acquisitions of perception while allowing conscious thought to enter in and evolve in the same shared space in an organised and complementary manner.

Consciousness does not have a role in constructing the contents of perception, just acquiescing to them or rejecting them after we subconsciously process-model them (and not after we subconsciously analyse them). We enter the world with it already speaking to us. Attention then, brings perceptions to consciousness by realising in them the frequencies, hues, aromas, themes, styles, and significations suitable for its perceptual field.

The act of attention emerges from the indifferent or chaotic, but constant, backdrop of subconscious intentionality. The transcendental act of taking up a new intentional project is thought itself. Thinking does not begin with the analysis of prior thoughts (that would be a tautology); it begins with the subconscious elaborations or process-modelling of sensorimotor perceptions, which are in turn a process-modelling of environmental processes happening to us. By transcendental, I refer to the moment of the switch between extrospection and introspection, or consciousness of something to the consciousness of consciousness itself, which largely defines us as humans. Thinking is an intentional running of the mind machinery rather than acquiescence to it.

This mind-switch is perhaps similar in workings to the extrospective perceptual switch noted in the ambiguous background/foreground figures of Gestalt psychology, but applied to the contents of consciousness instead (see Figure 14). That is, processes happen to us. We experience the nature of those processes and the way they include and affect us. This enables us to mimic them. In mimicking those processes we don't analyse them, we subconsciously process-model them. When we process-model those processes, we are then able to make the transcendental switch from being driven by the environment to driving new processes emanating from us that affect the environment. That is, we use our process models or self-bootstrapping SPSs to think. We are then able to consciously compute and analyse via axiomatic sets of ensemble truths, as per Figure 1.

Gestalt Vase
The background / foreground switch can be 'situational', an act of pre-conditional framing/reframing, or an act of intention

Figure 14: The Gestalt Switch

Through introspection, consciousness recovers possession of its own operations across biological time and space, condenses them and focuses them into an identifiable 'object', and gradually shifts from incomplete, time-poor perceiving to uncertain, relationship-poor knowing. It thereby arrives at a fuzzy unity of self with very little of its own effort.

By General Relativity, the spacetime fabric is a sea of wavelike proto-commands that brings a unity to the sensible at the speed of light, through vast interrelationships laid down one level/cycle of emergence at a time. Thus, mind/body unity is not a lone act of consciousness; it is a slightly untidy or probabilistic act of everything big and everything small.

We began with a world in itself that acted upon our senses, making its unstoppable movements, or proto-intentions and proto-meanings, known to us, and we arrived at inter-subjective thought and self (that below its conscious surface plays the same processual game). We have thus passed across an unnamed gap from a so-called objectivity to a so-called subjectivity. The mind/body seems to close off the incomplete models of existants on the unconscious side of the gap to leave simpler, determinate remainders – 'objects' and the self largely stripped of uncertainty – on the conscious side of the gap.

Sensing is a living communication with the local environment that makes it present and vibrant to us. The environment, which by the Uncertainty Principle must always move, is the proto-intentional fabric or proto-intelligence of reality that consciousness converts from the sensible into knowledge. Intelligence likewise emerges from nature via a monistic, relativistic gap and loop. I pass from the experience, to the pixelated thoughts in local spacetime that dumb down reality but at the same time give me my faulty knowledge.

The phenomenal field, even if difficult to make explicit in any particular level, is not an 'inner world', nor is the phenomenon a 'state of consciousness', and the experience of the phenomenon is not simply a conscious introspection. It emerges via feedback in each of its levels or strata inside and outside the body. Thus, perception is not a constituting of the real, but our inherence in the 'real'. Vision is a thought subjugated to a certain field which we call a sense.

Our dynamic and interpenetrating 'being in the world'[4] establishes a slightly ambiguous junction between the intersubjective psychological, and the more objective physiological, strata of human emergence. The body's 'being in the world' bridges the relativistic gap between first-person experience and third-person knowledge in local spacetime.

We have an actual body and a habitual/anticipatory body – or a set of body schemas. These schemas enable us to set up other relativistic relationships or fields/ interactions/ oases/ rhythms between the contents of actual bodily perceptions and the body's proto-intelligent homeostatic and sensorimotor systems that act as subsets or augmentations of the phenomenal field. The body schemas enable posture, gesture, and the incorporation of extended phenotypes (such as walking sticks via a tactile schema).

This relativity is brought about by the opposing poles that create the field, much like N-S poles create the magnetic field. In this case, the opposing poles are the movements or proto-intentions of the world versus the movements and (proto-)intentions of the body. However, as in all feedback systems, the (proto-intentional) sources are fuzzy/fractal.

Instinct

To be a consciousness, or rather to be a self-aware experience in local spacetime, is to have an inner communication with the world, the body, and others – to be with them rather than beside them. We are not psyches joined to bodies – we are a back-and-forth node of existence in which it is impossible to separate mental disturbances as purely physiological or purely psychological. The psychological latter emerges from the physiological former during child development and feeds back to change the former, in a continuous loop throughout life.

We might equate bodily sensorimotor perception as a centripetal force on neuronal paths or the mind/body, whereas conscious intention of the mind projecting bodily movement and emotional recall as a centrifugal force of the mind/body. The background/foreground switch of the mind/body from centripetal action to centrifugal action, or motor to generator, is like a motor-generator set, which in the motor-mode converts electrical energy (or proto-intention) to mechanical energy (or movement), but in the generator-mode converts mechanical energy to electrical energy. The ability to flick the switch improves as the neuronal paths become more used and defined. This same 'switch' analogy applies to all phenomenal fields and schemas.

The ability of the mind/body to do the transcendental 'switch' between perception and intention accurately, as in the case of learning to grasp a spoon, walk, or ride a bike, means the significations and sedimentations of sensorimotor perception 'in the world' become sophisticated with trial-and-error. We 'turn them around' (i.e. the significations and sedimentations) with our mind/body-machinery, arrange them, and make a habit of their use.

This is not a simple function of consciousness. It is an emergent meeting or communion of environment, body, and mind that doubtlessly begins with relative pandemonium. Neuronal path establishment is perhaps an energy-efficacious shuffling of proto-intentions coming from the body or wider environment and then rejected or accepted (acquiesced to) by the mind where appropriate.

My learnings are acquisitions of my body's active and spontaneous 'being in the world'. I take them up or reject them with new movements of thought within my situated being.

We gain a perception of unspoken intention through gesture, habit, and their sedimentation (to again use Merleau-Ponty's term) over time. Sedimentation beautifully conveys the idea that there is a distilling, filtering and gentle guiding of reams of coding-coupled-to-dynamic-matter (in cortices, living cells, proteins, atoms, etc.) at lower levels into a more solid behaviour or characteristic and 'sensible' style at the higher level.

It is the rising, dynamic, and unifying style that best conveys or betrays the essence of the existant perceived, and thus may be taken up as its token or symbol. Note: An existant is a fuzzy matter-arrangement with embedded spatial behaviours that is situated in the real world, and not merely a static object or an idealised concept removed from its environment. Its situated style is experienced or lived far more deeply than it can ever be deconstructed, reassembled, and rerun in analytical thought. We seem to know this intrinsically, although for a time at least, very dramatic 3D movies can still make our hearts jump. Hallucinations or lucid dreams seem to be something the subject knows to be different to normal experience, but is often willing to tolerate or invite as a substitute.

Thus it seems sensorimotor perception is responsible for having sufficient interrelationship with worldly existants as to stylise them using its senses and its ability to reposition the body, coordinate or synergise movement, or wait for another more opportune time. Of course, this would be no easy matter for an engineer's calculator to achieve – but it is 'natural' for a well-adapted process-modelling mind/body embedded in its environmental niche.

Merleau-Ponty tells us there is an 'Intentional Arc' that underpins the life of the mind/body and all life forms. Likewise, non-life, through the demands of General Relativity, projects a proto-intentional arc around arranged and situated matter. That is, space always moves matter, at all levels of emergence. The foundations of control, intention and command are established pre-consciously in the very lowest layers of emergence.

The intentional arc projects around us various fields and schemas within the folds of the brain, something like the way fractal formulae may be used to rebuild, and not just copy, complex scenes or scenarios later, after storage and transmission. Aspects of the mind/body schemas can be ratified, attenuated, or enhanced, through further sensing or through a cross-check between the senses and their shared elaborations within the brain.

For instance, it seems the mind/body has interrelated fields and schemas for:

- the body itself,

- the familiar environment,
- the passing of biological time,
- sexual relationships essential for reproduction of the species,
- the social world,
- words and language,
- a sense of self and free will,
- the moral world, etc.

That is, there are schemas for all our interrelationships at all layers of human emergence - ensuring that all their proto-intentions loosely unite in the intentions of the one emergent self. This is achieved chaotically, through initial pandemonium, via process-modelling, and without calculation.

Looking at the above list of bullet points, I hope you think that there seems very little left for self-aware consciousness to achieve. Yes, this is exactly what I wanted you to realise.

All these levels contribute to the self through their movements and essential styles in such a way as to make us unsure whether intention originates in the movements of atoms, neurons, or in the movements of the self. As we noted in Part 1, origination of movement is a pointless idea in fully interpenetrating/interrelating and indeterminate/uncertain reality.

Note: If the existant being sensed does not fit into a mind/body schema, an existant is expected by a schema but no longer exists (e.g. an amputated leg), or a schema itself is interrupted or damaged (e.g. through drug use), then we can expect the sensorimotor perception to be in error.

To help explain the emergence of mind/body intention, please let me use Figure 15 and the analogy of distribution warehouse logistics I alluded to early in Part 1. The figure describes the logistical process required to get widgets delivered from a widget distribution centre to a retailer's premises.

On the right-hand-side, we see a flow chart nominally starting with retailer intentions at the top of the diagram. However, in practice, whether the development of this sophisticated process began with a disorganised plea from one of the retailer's customers, the sudden availability of widgets, or the entrepreneurial retailer herself/himself, does not matter.

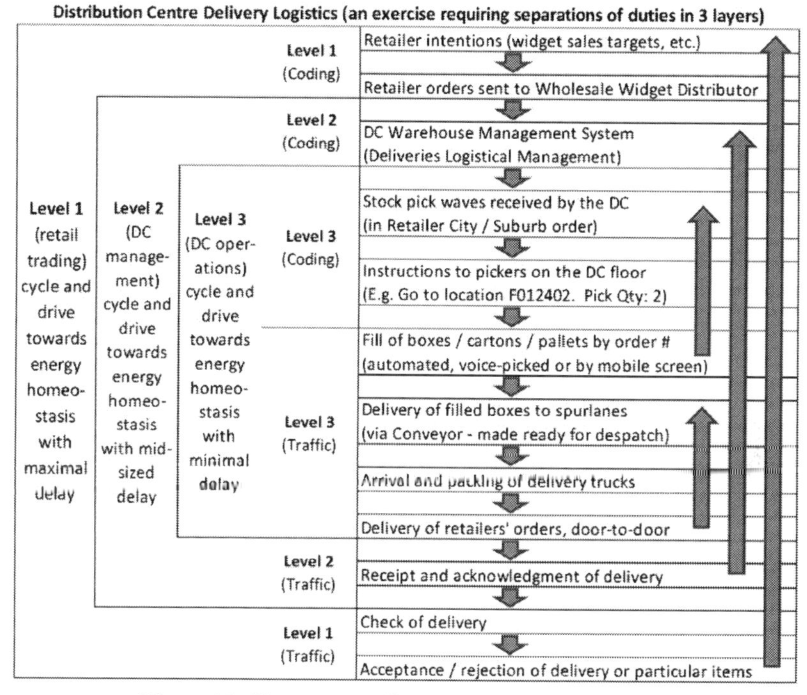

Figure 15: Emergence in Layers - Automated Logistics!

Firstly, please go through the flow chart from top to bottom to ensure you are familiar with the process. The Warehouse Management System's Deliveries Logistics module designs the morning's stock picking schedule to match the truck delivery routes. For example, the first scheduled item may service all the orders of retailers on the Northern Beaches route. The module achieves this by issuing stock picking instructions in 'waves', the first wave in this case being for the truck doing the Northern Beaches route. The instructions might be processed by pick machinery loaded and ready to drop units of stock into a passing box on a conveyor. Otherwise, the instructions might be received by pickers on foot via Personal Digital Assistants (PDAs), or received by forklift drivers with headsets programmed to relay voice instructions. When each retailer order is completed (i.e. total stock in the labelled packing box matches the total order) the box is topped up with box stuffer, sealed, and delivered to the truck loading bay either by conveyor system to the appropriate lane nearby or by forklift to the appropriate staging area nearby. After all orders for all routes are likewise processed in the early morning, the delivery trucks arrive mid-morning, are loaded, and then despatched.

Secondly, notice that there are arrows pointing back up the flow chart as well. That is, the process is actually a cycle, or a series of cycles within cycles, and not a simple linear progression. I have indicated that the biggest cycle with the biggest arrow is at Level 1, the biggest sub-cycle with the second-biggest arrow is at Level 2. Level 3 is made up of two sub-cycles. For instance, the first of these cycles in Level 3 consists of pickers on the floor of the DC filling orders by following each order's pick instructions, adding appropriate items to the order box, and then proceeding to the next order for processing. It should be noted here that each picker does not have to be aware of the retailer's intentions or the intentions of other pickers. Each picker has her/his own set of proto-intentions with respect to the final satisfaction of the retailer's wishes.

Thirdly, notice that each level is split into two fuzzy parts, that is, there is a separation of powers or duties within each level. One part in each level, the 'coding' part, 'tells matter how to move' and the other part, the 'traffic' part, 'provides a material platform' upon which the coded instructions are enacted. That is, the process is an extension of the relativistic relationship between all space and matter described by General Relativity.

Fourthly, notice that each level reaches a temporal, partial and peak homeostasis when one cycle's coding and traffic has been satisfied. However, if the cycle ends without success, then there will be a disappointing dissonance.

Fifth, notice that the cycle time delay, or homeostatic delay, at Level 3 is necessarily shorter (closer to the speed of light) than at Level 2, and likewise the cycle time delay at Level 2 is necessarily shorter than Level 1.

Sixth, notice how easy it would be to apply the same salient points of this process to the three levels of human intention – consciousness (at Level 1), subconsciousness (at Level 2), and unconsciousness (at Level 3). We could also apply this processing system to other examples of relativistic emergence, such as from quantum fluctuations (Level 3) to the emergence of a subatomic plasma (Level 2), and then to baryogenesis (Level 1).

Finally, consider that if human mind/body intention achieves similar astonishing levels of logistical sophistication and harmony, then this is something we should consider very carefully. How can we explain this largely unconscious logistical capability? Is there an Intelligent Designer of the process? Is each of us caught in the matrix of the 'Architect'? Is there a panpsychism controlling every step of the process? Is Whitehead's Process Philosophy the best explanation? Alternatively, is the process pattern explained most succinctly and accurately by the emergence of synergistic (proto-)intention within each level? Occam's razor (the law of parsimony, economy, or succinctness), would suggest the last option is the best. However, if this is the case, then synergistic intention is a highly distributed process that extends past the confines of the individual's mind. Synergistic proto-intention seems to be embedded in every interconnection of space and matter, and conscious intention seems to be bootstrapped within emergent SPSs (see Figure 1). We have perhaps thought the mind, acting 'independently' and 'alone', can achieve far more than it really can.

Expression, Speech, and Thought

We think in words.[5] Speech accomplishes and converges on thought as it proceeds; it is through expression that the thought becomes our own.

Thought has an indeterminism to it that is reduced through intentional expression. That is, there is a relativistic relationship between the expression and thought of words. Thought is not a linear representation of subconscious processes. Thought emerges within the dynamically situated body.

Words have a worldly signification, like an observed gesture carries emotion and meaning as it stands out to us from its background. Words, as gestures, are part of our body's social habits and neuronal equipment. Words like 'hot' produce a reaction in the body even before the thought develops. The rising, foreground reaction seems to prepare the way for the thought. Likewise, the act of seeing is both reactive/retrospective and proactive/prospective. Thought is proto-intentional before it matures. A new thought needs a proto-intention, the tools of language, and to be expressed before it can reach an initial order of completion; then it emerges via feedback.

We learn to speak just as we learn to walk. We see how our bodies move centripetally and then we mimic the movements centrifugally. That is, we learn words just as we learn all movements – through finding and expressing their essential behaviour (stamped in neuronal paths that interrelate to slowly form process models) in the midst of relative chaos.

This suggests that even language proficiency is more deeply a process of sensorimotor perception than it is of memory and consciousness. The movement schemas that enable language proficiency are tied together over space and time unconsciously.

The Mind/body as a Unifying System

The mind/body (human or animal) experiences all sensing as incomplete.[6] So it estimates as the experience unfurls, and adjusts its gaze. It attempts to fill in the gaps. Position and sounds actually modify consecutive images of colours. The mind/body moves and joins the dots between the senses to produce a synthesis – and this makes it more proficient and better adapted to environments at varying levels of emergence. My bodily senses translate each other without need for calculation. Cézanne relayed the idea that a painting contains the scent of its landscape, suggesting the unity is in the environment itself, which includes the mind/body.

For instance, the ding of a crystal glass against my nail declares its fragility, rigidity, transparency, inertness, lack of smell, weight, shape and its crystalline sub-structure – so much so that I can guess what it looks like blindfolded. The thing has a sense given to it by the intangible arrangement of forces that define it in dynamic relationship to my finger and me. Any perceived incompleteness/doubt in my perception, at the border or limit of my dynamic capabilities, is the cause of my dissonant movement away from homeostasis but also the source of the taking up of a new unconscious albeit unresolved intelligence by my body.

Conversely, under the influence of mescaline, sounds can give rise to colours, or the size of a patch of colour to the intensity of a sound, suggesting that body schemas become mixed up and disconnected from the outer world. That is, the perceptual field lines are perhaps broken up or made turbulent by a loss of relational alignment between the mind/body and the world.

Existential Space

We can think of space as the universal power of interconnection and interrelationship.[7] It is the given of our lived experience that also brings a dynamic wave of unifying proto-intention to all experiences.

When the mind/body receives the responses it anticipates from its orientation in space, it gains clarity, which enables a processual flow with its environment. This is a relativistic establishment of a perceptual ground, or a proto-intentional gearing of the body into the world.

Turning for example a face upside down will strip it of its significance. Nevertheless, if this is achieved via a contrived false projection in front of the eyes, then after many days the projection will be re-oriented by the mind/body to 'see' the face right-way-up again. Order and significance will again arise out of relative chaos. While the sense of balance is an integration of many parts (e.g. the vestibular system, eyes, muscles, joints, cerebellum and cerebral cortex), there are no absolutes of position or movement in the body's relationship with the world. I am guessing this will bode well for future space travel.

The changing size and convergence of an object moving through space causes a relative depth of field to emerge in my vision. 3D is not seen by the naked eye, but is a product of arising attention and proto-intention. 2D size and distance 'read' and 'suggest' each other with movement. This is a very interesting phenomenon because it demonstrates how strongly my sensorimotor perception is geared into reality and reflects my hold on reality, and conversely, how strongly the world has a hold on the things I seem to choose to perceive. It also demonstrates that because movement is a 'given' by the Uncertainty Principle, a change in relative position over a sufficiently lived or experienced time can nearly always be relied upon to divulge and convey a depth of field. That is, the mind/body has evolved in alignment with the observation of Heraclitus of Ephesus (540-480 BCE) who suggested that what needs explaining is not movement and change, but the appearance of stability.

Existential Time

Movement over spacetime is not a quality of an 'object', but a behaviour within an existant's world and against its background that we perceive within a proto-intentional field and within an emergent 'time' schema.[8] Perception grasps the given of time in the dynamism of shared space.

The lived present has a past and future within its thickness. This perceived thickness is probably related to Kleiber's Law, i.e.

the metabolism rate of a species is related to its mass (or 3D volume) by a ¾ Log: Log relationship. That is, the biologically perceived time of an organism is probably fractally related to its height, length and depth.

Time would thus be an essential measure of our biology. The arrow of time may thus be understood as an outcome of being proto-intentional lifeforms. It seems our body schemas are developed at just the right rate (roughly in synchronisation with our metabolic rates) to be able to interpret the next perceptual disturbance in an orderly way. That is, conscious time, that reflects a unity in all our perceptions and thoughts, is likely to be an outcome of our species' metabolic rate, the mass and volume of the body, and sensorimotor perception.

In a sense we don't remember the past or imagine a future, we access a time-movement-style schema, and rerun it. Imagination may perhaps be thought of as the 'centrifugal' rerun of the stylised contents of a time schema that are re-situated within the stylised contents of spatial schemas, (or vice versa). Like other perceptions, we are vague as to our imaginations' origins, but we often get to consciously approve/encourage or disapprove/discourage the spectacle paraded before us.

The (Intersubjective) Other

The world, and my inherence in it, and perceptual grasp of it, is a correlate not only of my consciousness, but also of every other consciousness I might encounter.[9] We complete each other's world. We perceive behavioural styles in each other, such that the self is not complete in itself – it is partially realised in, and freed by, the other. This freedom enables us to transcend the former, more insular self, and thus reach towards a more enlightened, tolerant, responsible, and inclusive self. However, sometimes this same freedom or lack of limits can also work in a negative way to bring about existential alienation or angst, and a deep realisation of our contingency and mortality.

In conversation, there is a mutual confirmation of intention and therefore style or essential signification of movement. Social intercourse is thus the sedimenting of many intersubjective acquisitions, and then the taking up of an intersubjective dance in the emerging fabric or layers of contexts, schemas, and significations (mutual or otherwise).

Communication, or the understanding of intentions in gestures, is achieved through the reciprocity (or free trade) between the one's intentions and those of the other in a shared stratum of existence (social, sexual, moral, economic, etc.) The movement of the body, the throat, the lips, and air through the mouth & teeth, are all meaningless to strata below verbal communication, but not incompatible. They are necessary parts of the learned habits of communication and its control. Words straddle the top of this communication edifice, but they cannot fully convey it, because in some ways the entire universe blindly conspires or acquiesces to each node of unfolding meaning and intention in the communication.

What I am

My very being, what I am, is not defined by my thoughts alone or anything else alone for that matter.[10] Independence is the most erroneous concept we have as humans.

I understand the world because I am situated in the world and a product of the world. I am a project of the world and a perspective upon the world. I am a relativistic and slightly asymmetrical balance between my extrospections and my introspections; I am an intersubjective node within its field.

What I am at my best according to Merleau-Ponty's existentialism and the emergentism that I favour as a philosophy, is a dynamic and unassailable transcendence of the world and self, whilst also being inescapably tethered to an unconscious being of the world.

I can reject or acquiesce to ways of being towards the world, but once committed, I cannot annul those ways, which helps me learn. I am a nested loop of relationships.[11]

There is no field/oasis of freedom for me without the limits between opposing poles my cybernetic and homeostatic systems provide. Those systems are chained to spacetime, but I am glad they are.

Merleau-Ponty suggested that our rationality is rooted in our environment, and not in the self or its accomplishments.[12]

He further remarked, "A consciousness for which the world is 'self-evident', that finds the world 'already constituted' and present even within consciousness itself, absolutely chooses neither its being nor its manner of being." He also noted that in the interpenetration of the environment and the person, it is impossible to distinguish that part contributed by each.[13] This means it is meaningless to distinguish our freedoms or our free will from the environment that provides the potential for those freedoms.

Some of the last words in Merleau-Ponty (2014) are as follows: "The only way I can fail to be free is if I attempt to transcend my natural and social situation by refusing to take it up".[14] What a delicious conundrum, with which the Emergent Method is in full accord.

Competition and Cooperation

All DNA-based life-forms, including plants, have instincts and associated versions of sensorimotor perception. The two instinctive/processual modes of survival in the close-to-here-and-now, competition and cooperation, are not peculiar to animals either – they are known in all life-forms. All life forms are proto-intentional. They blindly choose how to take shape and organise themselves in the here-and-now. They either self-organise and face their local environment self-reliantly, or organise themselves into groups and face their local environment en masse.

For example, grass seeds sewn across a field will grow together and the fittest competitors will survive in the local environment to share the field, blindly cooperating with each other to form a tight grassy mesh that inhibits other plant seeds from being able to establish themselves in the same field. Through this means, the grass plants cooperate amongst themselves and compete with other plant species for control of the field as a source of food-energy and a womb for its seedlings. Similarly, we could suggest that trees both cooperate and compete for placement within the forest canopy.

It might seem that spacetime forces all organisms to blindly extract spatial information from their local environments in order to diversify and occupy their niches alone or in groups. Is this support for the idea of spacetime as the universe's selfish coding system? I think this sentiment is too strong. Organisms provide dynamic niches to spacetime as much as spacetime provides opportunity for information and impetus to organisms. That is, the relativistic and slightly indeterminate relationship is asymmetrical; organic matter and spacetime together bring about their interconnecting and emerging properties.

Did the locally acting spacetime system give rise to the genetic code as much as the genetic code gave rise to the memetic code? Again, I think this may be a little strong. It again separates matter and spacetime as if they exist and emerge separately, which they do not.

Did the spacetime code blindly give up some of its processing power in its matter survival machines in order to facilitate processing at a higher and more selfish level in the genetic code of the living survival machines? And did it do this much like the genetic code later seemed to blindly give up some of its local processing in each cell of the body to higher-level processing in the memetic survival machine or human brain, which we shall consider in the next chapter? No, I think this is the wrong way to consider the issue. Statistical matter-space emergence in the midst of vastness at many relativistic levels (e.g. genetic and memetic) and within many boundaries was the only thing at work here. Beginnings are always misleading.

Instinctive behaviours take two essential modes in human evolution we can picture as follows: 1) The self-reliant hunter, carnivore or competitor, corresponding with Table 1: The Risk-weighted Value-set; and 2) the social gatherer, herbivore or co-operator, corresponding with Table 3: The Normative Value-set.

If all our instinctive behaviours are the result of blind movements within statistical vastness, then perhaps it should not surprise us that in all the various instincts of life forms we do not always see an obvious moral self-restraint. However when we consider more carefully the competitive and cooperative instincts we often can perceive values-in-action that seem the best explanation of behavioural self-restraint at the higher organism level. Further, respect for certain values seems to explain much of our own behaviour, and seems essential to our human sense of wellbeing and self-actualisation.

The Self-Reliant Stance: Exploiting our Individuality

We are Carnivores: The Hunter Mentality

Seeing self as autonomous and the world
as the hunter or the hunted,
as victor or victim,
as predator or prey,

as winner or loser,
as the economically efficient minority or the inefficient majority,
as for or against…

As we will note before the end of this chapter, the success of the hunters in our evolutionary past is probably a key reason why we can have this discussion. The male-dominated hunter mentality should be and is highly respected even today, although perhaps not in all quarters of human erudition.

It is this win-lose mentality of Figure 12, or the awareness of the ability of the hunter to successfully compete, that is perhaps most responsible for the illusion of an independent self.

Nietzsche

An interesting but perhaps flawed commentator on the hunter mentality was German philosopher Friedrich Nietzsche. Hollingdale (1977)[15] is a highly recommended book on the topic. Nietzsche saw the carnal core of man's nature, his animal instincts, drives and volition, as man's core strength rather than his weakness. I would have to agree that the heart of man's very important drives, volition and persistence rest more in the older, instinctive side of his nature than in his newer, more intellectual and rational consciousness - but dogged will without human reason has little to commend it. Nevertheless, Nietzsche argued that man's drives were the only enduring truth – not the Christian moral standards of this passing phase of man's evolution. Further, he saw the denial or suppression of man's drives, the chief of which is his "will-to-power", through particularly Christian morals, as a detrimental softening, weakening and denial of man's true nature and potential. In fact, it would be fair to say that he viewed Christian morality as the chief falsehood of the last 2,000 years. Nietzsche was a self-confessed enemy of the Christian conscience.[16]

Nietzsche totally rejected the beyond-physical, seeing it as the invention of man's deluded subjective mind.[17] He often sought to explain man's subjective nature (e.g. emotions such as pleasure, pity, or neighbourliness, or man's subjective concepts such as beauty or justice) in terms of animal drives or instincts as well.[18] Yet he often equated these 'subjective truths' with error.

The power of his philosophy lies in the profound effect it has had on Western history and the current Western culture. To understand this culture, and we who are largely the products of it whether we like it or not, almost demands we study Nietzsche.

One could almost say that Nietzsche despised entirely the so-called subjective and spiritual side of man's nature. For instance, he said that art makes life tolerable by "laying over it the veil of unclear thinking". I would rather side with Cézanne, who suggested that art touches us because it enables us to see perspective, movement, liveliness, and thus proto-intention in the world around us that a mere photograph or fading memory can hide. A great landscape artist can make us almost smell the air of the field represented. Further, Nietzsche said, "the artist possesses a weaker morality than the thinker", although he seemed to castigate all the philosophical thinkers that had come before him anyway, to which he added that artists are in many respects "backward looking creatures". He also said, "In itself, no music is profound or significant"[19] – applying the same, stiffly objective / anti-subjective brush to all things aesthetic and deeply perceptual.

More tellingly, Nietzsche argued,

> *"Apart from the ascetic ideal, man, the animal man, hitherto had no meaning. His existence on earth contained no goal; 'why man at all'? – was a question without an answer; the will for man on earth was lacking … he suffered from the problem of meaning … and the ascetic ideal gave it meaning! … the will itself was saved".*[20]

I would agree that the self-denying/self-disciplined and more generally intersubjective side of man's nature indeed provides comfort and thus a personal reason why. However, this intersubjective side of human nature inevitably emerged with the evolution of human consciousness-plus-instincts. Along with this evolutionary growth came the discovery of values that apply specifically to our intersubjective natures, and we are still learning how to understand this today. Our evolution of consciousness has not always been neat and tidy – it has taken us through many dark twists and turns – but it has brought us to a point today where we can, perhaps for the first time in our history, deeply reflect on its pitfalls and also respect its achievements.

Nietzsche also saw the intersubjective side of man's nature as arriving after the animal, empirical side had been in existence for some time. This late development of consciousness in man's evolutionary development explains its incompletion and weakness, according to Nietzsche. We still live within the 'subjective lie' that, presumably, will one day find maturity through understanding and acceptance of the pre-eminent role of the will-to-power (i.e. the empirical win-lose nature of the hunter).[21]

While we cannot deny the late and ongoing development of consciousness, to have secondary and suppressed intersubjective truths at this stage of our development, is to imagine a world of mass clumped together in one place with little or no space to frame it (a black hole). It is to imagine Yang supreme and Yin not just inferior, but beaten. Such an attempted world would be harsh, malformed, malnourished, utterly hideous, and headed for self-elimination. I suspect we would be far better off to immediately regress back to an animal with little or no consciousness at all. Nevertheless, the idea that our intersubjectivity should find maturity at this time is an interesting one, although not through a resurgence of the will-to-power, but rather through the Emergent Method's phaseal balancing of our instincts and consciousness (discussed in Pillar 4).

Nietzsche said that man's consciousness does not really belong to the individual but rather to that in him that is community and herd (- the social side of human instincts he also seemed to reject). He said that conscious thinking takes place in words or communication signs, by which the origin of consciousness reveals itself, adding that language and consciousness co-evolve. He went on to say that our consciousness necessarily make everything we perceive quite shallow, reducing our world to common surface assessments and universal significations. He said, "everything which becomes conscious thereby becomes shallow, thin, relatively stupid, general, sign, characteristic of the herd, that with all becoming conscious there is a united and fundamental corruption, falsification, superficializing and generalisation."[22]

Largely I would not take issue with any of these sentiments. The strong link between language, community and intersubjective truths made explicit is undeniable. All our conceptual abstractions do lead to an analytical filtering and compartmentalisation of reality, but it was these same simplifications that enabled us to survive and flourish as rational agents. Therefore, to reject out-of-hand all the combined wisdom of the ages seems foolish, to say the least. Nietzsche also explains that the belief in authorities is the source of this same relatively stupid conscious.[23] Again, there is an underlying truth here but it seems out of balance. Our individual self-actualising cannot be fully realised under a heavy authority, but nor can it be realised if we divorce ourselves from our environment, our fellow human and the authority of reason invested in him or her.

Thus, Nietzsche saw morality as that which preserves humanity in community, and thus as animal, in that he saw morality as a consequence of that animal drive which also teaches us to herd, seek food and elude enemies.[24] He also saw every losing of oneself, through communal consciousness and morality (e.g. in acts of selfless virtue, neighbourliness or philanthropy), as acts of timidity and weakness from the point of view of the individual.[25] In a sense, this is again true – but making the kill endlessly for the sake of self alone seems to inevitably dissatisfy. Eventually we look for a reason in our world that is greater than is the self, alone. Even gangster loyalty eventually makes sense to some.

Not surprisingly, he advocated the end of the current concepts of good and evil (which he saw as coming from a primitive perception of that which preserves or injures the animal species rather than from the universe's innate nature). He wrote, "Let us do away with the concept sin – and let us quickly send after it the concept punishment" which he again saw in terms of the herd pressing its claim over the weakened individual.[26] Nietzsche misunderstood nature's base workings within the herd (a process of risk management).

Here then is the thoroughly anti-enlightened and anti-community but seductive power of Nietzsche's philosophy. Our global virtuous tradition would see personal integrity and conscience (the judgment-seat and holding-place of personal values, whether Christian, Buddhist, atheist or otherwise) as something to be prized and guarded above almost all else because it organises and drives our value-based purposes and potential. In contrast, Nietzsche seemed to see the conscience as valueless, worthless, to be despised, and to be blamed for our human weaknesses. Nietzsche's sentiment here seems sociopathic in character.

Imagine a world ruled by Nietzsche. Would you like to be in one of his hated herds? It does not sound like liberation of the individual to me. Thus, Nietzsche linked together community and morality – and derided both, but in doing so, also derided the individual he supposedly sought to liberate. Nietzsche's philosophy is the mistaken elevation of humanity outside and apart from its consciousness, community and environment, which would ultimately lead to the destruction of that community and environment, and then, perhaps like the Australian fox, the destruction of humanity itself.

"There are no moral phenomena at all, only a moral interpretation of phenomena",[27] Nietzsche said, which is true only in the sense that all of man's moral values are indirect with respect to the physical realm. Values are measuring-sticks used in a feedback loop with the actual dance of life – but this does not exclude them from affecting self-organising and emergent reality. Therefore, in the sense that explicit moral phenomena and moral agency emerge from matter's arrangements, they must also be an implicit part of reality. We can say the same of all our named laws that are successful in an evolutionary sense.

Nietzsche sought the liberation of the individual through the honest recognition of our animal drives and the end of their cloak of morality. Again, there is some sense here. He also advocated a re-assertion of the individual human will, the will-to-power or individualistic instinct for freedom, through mastery over the animal drives that he saw as the only, but limited, good. Nevertheless, he also said that the emotional love of power is the demon of men.[28] Through this means, the liberation and re-harnessing of the will, Nietzsche sought a golden age of liberated humanity. Here, the subjectivity of Nietzsche's own goals, beliefs, values, and morals (or intersubjective definitions of good and evil) for humanity clearly betray themselves.[29] Nietzsche, rather than ridding us of our faulty subjectivity, in the end has only cauterised one set of subjective truths to replace them with another – his. Unfortunately, for Nietzsche, this does not make them truths (objective or otherwise), again underlining the circularity of his arguments. He did not acknowledge or appreciate the idea of a personal anchoring arising naturally from an emergent and self-organising universe.

Worse, Nietzsche's hidden intersubjective truths would only be constrained by man's animal drives, since Nietzsche did not put much weight on the rational or logical side of man's outward-looking consciousness. For instance, he writes "No one is accountable for existing at all…no one is any longer made accountable (to some rational framework)…this alone is the great liberation".[30] While many today would argue that our lack of free will suggests our lack of responsibility for our actions, I think this statement seems a recipe for disaster, sorrow and man's self-elimination rather than a recipe for the liberation of humanity in a rational structure of higher consciousness within our finite environment.

The discussion so far has been about how Nietzsche's philosophy represents an extreme version of the hunter mentality. Now let's briefly turn to the influence of Nietzsche's thinking on Western civilisation. Of Christianity, he wrote, "it was Christianity which first brought sin into the world. Belief in the cure which it offered has now been shaken to its deepest roots: but belief in the sickness which it taught and propagated continues to exist."[31] Thus, he lamented the imprisonment of the individual's will-to-power that, to his mind, still existed due to the persistence of decaying concepts such as sin, and the weakening feelings of guilt.

For all the error that we might see in the logic of Nietzsche's perhaps sinister philosophy, this is still a sobering and disturbing point. The onslaught against virtue, against love and against conscience, in our global culture, seems ferocious. Yet at the same time, we are less happy, less secure, less fulfilled, and perhaps more self-aware of our nagging helplessness and hopelessness. Nietzsche's dream is being temporarily fulfilled. We are changing, but it is not always obvious, so far at least, that it is for the better in terms of human flourishing. Nietzsche was perhaps prescient to lament our 'sin' to our self-actualised best, whether we see ourselves as of the Abrahamic faiths, atheists, agnostics, Buddhists or other.

The ideas of this book do not seek to draw humanity into the bondage of guilt and sin as Nietzsche suggested Christians do – they seek to free all of us individually and corporately through a self-imposed law of personal and societal Liberty, that is, personal and corporate fellowship with our virtuous global tradition and community. Like you, I seek the end of war, hatred, racism, the wanton destruction of our environment, ignorance, fear and alienation. I seek the full expression of our courageous individuality and community in a balance that sees humanity advance in personal and communal purpose, fulfilment, strength, harmony, confidence and happiness. Am I way too optimistic and naive, or can you agree with this sentiment as well?

Such an objective does not mean continually looking for evermore materialistic creature comforts and security per se, of which Nietzsche accused the Western world. It means addressing the many issues we as enlightened persons ought to address, responsibly, honestly and bravely, without heavy contradiction of lively conscience. Right now, we need enlightened community, and deeply perceptual communion with all other aspects of the environment, as we have never needed it before. But perhaps Nietzsche was right in that we no more need a superstitious concept of God than a lone Siberian tiger needs any conscious knowledge of God in order to be a healthy and beautiful expression of fulfilled life.

Galt

In sharp contrast to Nietzsche's philosophy, Ayn Rand's philosophy of objectivism presents to us a more internally consistent and open approach to individuality and the largely instinctive, self-reliant values or stance of the hunter. Perhaps the best way we can explore this philosophy is to consider a few ideas from the fifty-plus page speech by John Galt in her novel, *Atlas Shrugged.*

In the speech Galt suggests that the simple good in life is to live it naturally, as our consciousness-plus-instincts would suggest, perhaps like the lilies of the field live the life of a lily, a bird the life of a bird, a tiger the life of a tiger, etc. Galt goes on to describe some of man's natural attributes and means of survival, such as his abilities to think and make choices in an awareness of his environment and own being (its strengths and weaknesses).

To live effectively, Galt suggests we must have a code of values to guide our thoughts and actions and that this code of values, if freely adopted, is also our moral code. He then suggests, all that promotes human flourishing in life is the good and all that destroys human flourishing is the evil. Thus our own life, how we perceive and respect and live it in step with our values, becomes our moral standard, our ongoing purpose and our reward or success in terms of the happiness we derive from achieving the expression of our values. As we live this good life, we learn to truly love life and the dance of life within our cherished (and partly self-made) environment. This is not a picture of debilitating conflicted conscience. Rather, it is a picture of personal moral congruence within the local environment.

Galt then goes on to consider how we ought to deal with others within our society, firstly suggesting that, as being dependent on our rationality for survival, it is only our own knowledge and judgments of truth that we truly possess, and may exploit, share, or trade. Galt suggests that reality will teach us where we have made errors in such judgments of truth. He also suggests that our moral integrity, lively reason, and respect for life need to direct our judgments if we are to minimise our errors of judgment and maximise our knowledge of truths. Thus, Galt suggests that a lover of life, and lover of humanity as rational and volitional beings, is also a lover of distilled truth and a highly moral and heroic person.

In fact, he suggests that our reason is also our moral faculty, and that we who are prepared to take full responsibility for our thinking reflect the height of humanity's nobility as a species. Galt also suggests that the basic virtue or value-in-action that promotes human flourishing is our willingness and decision to think, from which all other human virtues must necessarily follow. Of course, this means Galt sees our basic vice as our refusal to know and judge truths. It is in this sense that Galt castigates the sceptics and mystics.

The three most important values that Galt promotes are reason, purpose and self-esteem. As already noted, reason is seen as our only legitimate means of gaining knowledge and judging truth, for self and for others. Ignoring the idea of reason arising within an earlier world of animal drives and perceptions, Galt sees purpose as the proper use of that reason and knowledge in each person's pursuit of human flourishing. In this way, a deep respect for life seems to demand a deep purpose for life. Galt thus sees self-esteem as a part of that respect for interconnected life and respect for the physical and mental competency life has granted Homo sapiens as a unique species. According to Galt, these three values imply all other human virtues or are required by them. Further, all our virtues pertain to our existence and the attributes of our particular species (our consciousness-plus-instincts) as rational and volitional beings. Galt lists these virtues as rationality, independence, integrity, honesty, justice, productiveness and pride. It is interesting to note that all the social values of Table 3: The Normative Value-set are missing here. I would also suggest that unique, local individuality is a more accurate or realistic concept than Galt's independence.

Galt has little to say on the topic of emotions, except to see emotions as a capacity like a motor that can take us from point A to point B (where we want to go), given we feed that motor with the precious and undiluted fuel of our values-in-action. This seems a very worthy observation. It suggests that instinctive emotions can provide the motor for our rational consciousness if we link the two (instincts and consciousness) via a set of explicit and articulating values. Further, we will only achieve a true sense of joy or deep satisfaction as we experience the congruence of prepared outcomes, such as what we are able to produce and trade via our chosen and rational actions, motivated by our positive values. Thus, Galt sees honest traders as a symbol of all that is good in man as a species, because traders respect their counterparties as they respect themselves and their own values (such as reason, volition and justice). Conversely, Galt sees the abrogation of volition and the use of force as the end of morality in society and as disrespectful of life, and thus a promotion of the opposing principle of death.

Galt also has some interesting observations to make on the doctrine of original sin. Original sin as a sin every person commits without free choice, Galt sees as immoral in itself. Original sin as something innate within human nature Galt sees as a mockery of nature – and perhaps nature's driving survival principle of life-and-being here-and-now. Further, Galt suggests that to hold a person guilty for a crime where no possibility of innocence exists as flouting reason and thus he sees the punishment of a person for his or her original sin as the ultimate injustice. Galt, perhaps like Nietzsche, thus appeals to us to reconsider the root, and perhaps effect, of the West's Christian moral code.

Galt also considers the myth that surrounds the doctrine of original sin. He suggests that eating from the tree of knowledge merely reflects that we evolved into rational beings (consciousness was added to our instincts, perhaps as the neo-cortex and prefrontal lobes were added to the older cortex). The knowledge of good and evil likewise reflects that we became moral beings, and thus perhaps somewhat aware of the values we must choose to promote human flourishing. For our original sin, we were sentenced to earn our survival by our own labour, thus we became productive beings who had to learn how to use our volition and reason in order to survive. We had to innovatively form our own purposes. Finally, Galt says that as we were sentenced to experience reproductive desire, we gained the capacity for sexual intimacy and enjoyment through the process of shared human reproduction, life and flourishing. By this means, Galt charges the myth with the sin of debasing all four cardinal values of the human species – positive reason (Table 2), normative morality (Table 3), risk-weighted productivity (Table 1), and self-enlightened joy (Table 4). Finally he suggests that the creature in the Garden before the Fall, who was without reason, without morality, without productiveness and without joy, was actually no human at all, but more like a robot or perhaps an insect with instincts but without any consciousness.

Galt also has somewhat to say against the idea of self-sacrifice and altruism. He suggests that a surrender of self as an act of altruistic self-sacrifice is contrary to the nature of our species and thus a betrayal of it, because he sees the ultimate sacrifice of self as a sacrifice of the values that we embody as we live. Conversely, he suggests that to truly love someone is to love the values he or she uniquely embodies and expresses.

I find myself agreeing with these points. They also bring into question the Christian idea of the sacrificial redemption of sins, or the idea that a well-meaning person can act as a propitiator or intercessor for another person's sins, as Paul suggests he could do. For instance, Paul says, "[Even] now I rejoice in the midst of my sufferings on your behalf. And in my own person I am making up whatever is still lacking and remains to be completed [on our part] of Christ's afflictions, for the sake of His body, which is the church."[32] This seems to require all involved to accept magical superstition. It makes Christians who try to follow Paul's example believe that they can change others through intercessory prayer rather than the simple sharing of ideas and physical interaction. Of course, when such prayer fails, disappointment, grief, guilt and moral dissonance ensues.

The principle of enlightened selfishness means we are to a limited degree responsible for upholding vremes among our internal stakeholders (our enlightened mind/bodies, or at a societal level, amongst fellow members) but we can't be responsible for the values held by our external stakeholders. Nor can they be responsible for us. All we can do is trade values with them and through honest trade and time, perhaps share a common enlightenment with them.

We learn and share ideas, influence appropriately where we can, but we leave the rest to the other person and the evolving values that reside in that person and are ready to respond to our input. The other person must learn lessons from life and experience just like the rest of us. We can't save the other person – it just doesn't work like that, and ultimately, nor would we want it to. We can lead the person to drink of the waters of life, but we do not want to make the healthy adult drink even if we could.

Ultimately, we save ourselves (not from any superstitious hell, but from our own ignorance and error) as individuals, as a society, and as a deep ecology, through a kind of unconscious experiencing and self-directed learning borne out of enlightened and broadening self-interest. You and I can contribute to the shared ideas and actions that we think will promote evolution in the right direction, but that's as far as it goes. We can't speed up the clock very much, even though it ticks ever faster because of our growing enlightenment.

Likewise, we must learn to serve the values in people (just wonderfully emergent mind systems implemented in brain circuitry) rather than people themselves. However, the emergence of our species is exciting and I believe we can contribute to a legacy that promotes its evolution in a direction it is perhaps inexorably heading.

Galt warns us against falling for the trap of believing in the superiority of others and their systems of worship. He sees each person as a wonderful end in himself or herself rather than a means to the ends of others. Thus if anything is to be adored, it is the values resident within us, which are also evidently resident, with greater or lesser degrees of expression, within our fellows. These resident values are ours by inalienable fact, by the pattern or rule of our evolving species, as *Homo sapiens.* We are seen as moral beings, although morality is a matter of each person's exercised reason and choice. Galt thus sees legitimate human rights, such as the right to freedom, as conditions that allow our nature as a species to survive and flourish. Thus, government decree does not confer human rights - only evolution itself does that.[33] Objectivism presents the position of the self-reliant but deeply ethical hunter quite aptly.

Gecko

Perhaps another very different and infamous example of the competitive hunter mentality would be the mid-80's philosophy of the character Gordon Gecko in the movie *Wall Street.* Particularly well known is his speech on the virtues of greed. He said,

> *"Greed, for lack of a better word, is good. Greed is right. Greed works; greed clarifies; cuts through and captures the essence of the evolutionary spirit. Greed in all of its forms – greed for life, for money, for love, knowledge – has marked the upward surge of mankind ... and greed ... will ... save ... the USA."*[34]

What is wrong with Gecko's speech? Can you spot the problem? Yes, we need strong competition that eliminates bureaucratic inefficiencies in our corporations. Self-elimination of waste is an essential part of all self-organising systems. However, when greed becomes the system, then like the Australian fox, it destroys what is in near-balance, synergistic, bio-diverse, or nearly fully information-incorporating in the local environment, as well as what is ineffective or outmoded.

Greed does not recognise the locally shared rules of the game. It is not naturalistic in the same way that Galt saw the only legitimate human rights as those springing from our essential nature as humans. It does not always lead to medium- or long-term greater efficiencies and effectiveness. Greed is not something that "captures the essence of the evolutionary spirit" because, like the imbalance within Nietzsche's philosophy, it is not respectful of the interpenetrating nature of local life and order. Greed is not a self-contained system of human flourishing, nor a promoter of human flourishing. This is why Gecko's philosophy is found to be lacking.

Gordon Gecko also said, "The most valuable commodity I know is information, wouldn't you agree?" Well, would you? He's got a point, but the answer is in the words themselves – the most valuable commodity must be the naturally-arising values themselves, by which all information may be exploited and consciously appreciated. A tangled rope incorporates more information than an untangled one, but does it provide more value? Likewise, quantum fluctuations incorporate vast quantities of information, but unimproved, do they provide value?

Perhaps a final example of the hunter mentality we might be familiar with is the moral code or 'bushido' or 'way of the warrior' presented to us in another movie, *The Last Samurai* (2003 Warner Bros). In this case the self-reliant values that stand out are self-honesty, self-respect, an inner sense of justice, heroic courage, self-honour and an inner sense of loyalty, responsibility and duty (i.e. to bushido itself). It is from this self-reliant foundation that the Samurai warrior applies the values of compassion, sincerity and civility to the outside world and thus offers a balance lacking in Gordon Gecko's approach to the world.

Imperfect Personal Information: The Risk-Weighted Approach

It seems a cost of our conscious agency is the need to model reality's processes. The instinctive hunter is not foolish. As an individual and self-reliant entity, he takes risks, and he carefully manages his risks in order to fulfil his purposes. That is, he carefully and rather instinctively weighs his body's store of energy and need for food against the odds of making the kill or being killed in the terrain he is working. If, in the chase, the risk to life or energy to be expended is too high in terms of the odds and benefits of success, the chase will end. The hunter will wait for another, easier opportunity. Each decision to act is made with an instinctive discount applied for the imperfect information at hand. The more he acts the more feedback he receives and the more adept he becomes at applying an appropriate discount factor to the information at hand. We could perhaps add that extrosomatic cognizance is a basis of all adaptive systems. Further, the more proficient he learns to become in his survival enterprises (the more he learns what it takes), the more self-aware he becomes. That is, internal adaptive systems (even if unconscious) promote consciousness and vice versa.

We see the same instinctive, risk-weighted approach played out every day on the world's stock markets, in stock portfolio diversification and in corporate funding decisions. We also see it played out in the development and implementation of risk and compliance policies of companies, banks in particular - the Basel accord put under review because of the recent global financial crisis being a case in point.

A recurring theme in the life cycle of an organisation is the inevitability of change and the applicability of the second law of thermodynamics that could be paraphrased as, 'excluding occasional emergence, all systems degenerate to lower orders of organisation and energy with time'. Always present in organisations is the need to effectively manage risk, the chief risk being the growing threat of irrelevance (loss of negentropy) with time and change. Conversely, organisations often blindly make better use of information available in the local environment and thus house within them more of spacetime's constituents that extend its influence or fit and help it realise its purposes. Gordon Gecko would agree - information gained is often both an aim and reward of our taking.

Risk management is about identifying prospective variations from what we plan or need and managing these to 1) capitalize on opportunity, 2) minimize loss, and/or 3) improve decisions and outcomes. It is about the management of uncertainty that has exposure for loss.

The forward of the (now superseded) AS/NZS Risk Management Standard AS/NZS 4360: 2004 reads as follows:

> *"Risk Management involves managing to achieve an appropriate balance between realising opportunities for gains while minimising losses. It is an integral part of good management practice and an essential element of good corporate governance. It is an iterative process consisting of steps that, when undertaken in sequence, enable continuous improvement in decision-making and facilitate continuous improvement in performance."* ...

> *"To be most effective, risk management should become part of an organisation's culture. It should be embedded into the organisation's philosophy, practices and business processes rather than be viewed or practiced as a separate activity. When this is achieved, everyone in the organisation becomes involved in the management of risk."*[35]

The notion of an algorithmic feedback loop (through continually monitoring and communicating findings to appropriate risk management agents) is vital to the success of the effort. So is embedding intelligent monitoring and control in the traffic of practices and processes as it occurs. That is, for risk management to work well, we need a separation of powers between the traffic and the traffic controllers, but not too separate (like hands-off managers located too far from the action). The traffic controllers also need to display the same kind of growing, risk-weighted intelligence as the classic hunter or carnivore.

The risk management approach is akin to the process of natural selection; it has essential elements of replication (of practices and processes), occasional mutation (or disruption of controls) and a severe struggle for life (that is, competition and/or cooperation with other entities in an environment of finite and scarce resources, leading to *survival of the fittest*).

An organisation's culture needs to embed strong risk management. At an individual level, the Emergent Method suggests each of us need to embed strong risk management into our chosen set of values, mores, taboos, and behaviours to be consistently relevant, innovative and successful. This will not be achieved easily.

The hunter's instinctive risk-weighted approach is not to be underestimated in terms of its importance and relevance to our survival and the realisation of our purposes. It is driven by the same powerful but partially unconscious or blind selfishness that also drives genetic natural selection.

The Individual Need for Strong Competition and Self-Achievement

Strong instinctive competition in the severe struggle for life makes markets more efficient, or more lean and mean. Weak competition makes markets less fit, or more vulnerable to illiquidity, monopolistic domination, corruption or failure. The equivalent is true of athletes. The equivalent is true of you and me. However, this same competition must make room for those in the fray to share achievements. For instance, a heavyweight boxer put in the ring against a featherweight wouldn't be much of a spectacle and not as valuable an experience to the participants as when they compete against their own kind at their own relativistic level (of emergence).

The odds of success have to be just high enough and just low enough to promote close-to-optimum competition – and thus promote a close-to-optimum adaptation (genetic and memetic) to the changing local environment. This observation is not to be taken lightly. The cosmos is not your benefactor, but if you hold the right attitude out towards it, you will appreciate the challenges and support it presents to you with equal gusto. Easy success or crippling failure will not change the limits or boundaries of the systems in contention. Emergent 'body-building' is a slow and steady business. Do it too quickly and the body will be damaged; do it too slowly and no gain will be had. The cosmos enables you to be an intelligent participant in its gradual unfolding.

A competitive level playing field arises because of the Australian fox problem – that is, incompatible competition tends to lead to a loss of biodiversity and resilience, and then a resetting of conditions before new life in a new near-balance with the local environment and its constraints emerges. The rules of the game (or styles emerging with life in the environment) seem to establish, but more correctly reflect, the current near-equilibrium conditions (or fit) for resource competitors and so are an important aspect of the instinctive, individualistic, risk-weighted world of self-reliance and competitive development. These stylistic or heuristic rules of the game imposed by a relativistic and emergent reality are built into the genes of animals and humans over a very long period because they provide the means of survival in the here-and-now. If you think about it, these rules or stylistic essences emerged from relativity itself.

I hope you can see along with me that the ultimate value of appropriate competition is energy effectiveness in matter's indeterminate arrangements and efficiency in space's locally deterministic codes (information or interactions). Further, just as relativity bans excursions to either infinities or nullities in existence, so do the rules of the game on the roughly level playing field. However, there is one important difference as to how humans use these instinctive rules; consciousness, explored in the next chapter, also strongly affects them. Consciousness, through imagination, is open to full reality and thus unrestricted by relativistic reality.

The Competitive Self-restraint

In this and the next chapter, we consider the four basic kinds of self-restraints of vremetic moral agents, the first of these being the competitive self-restraint. I use the term constraint to refer to an externally-imposed limitation on any system of life or non-life. These constraints could be physical, genetic or memetic. I use the term self-restraint to refer to an internally-imposed system control. An important self-aware purpose of self-restraints or self-control is wellbeing. A blind purpose of self-restraints in living systems is homeostasis. We could say constraints have an external locus of control, whereas self-restraints have an internal locus of control.

The more constraints a system incorporates and the more relationships a system successfully maintains, the greater its energy effectiveness and efficiency. Self-restraints emerge from constraints, and the border between the two is a fuzzy one. Laws are our heuristic summaries of a system's constraints or relationships. That is, our formulations of laws emerge from constraints, relationships and systemic boundaries. This is true whether those constraints, relationships, or boundaries are ecological, physiological, psychological, sociological, or knowingly self-enlightened.

Personally applied laws, and the facts and values of which they are constituted, can be thought of as our chosen methods of dealing with contingencies (that is, cognition) or maintaining viability constraints (that is, autopoiesis). The chosen values of self-aware moral agents are methods of self-restraint or self-control. They form the basis of the Emergent Method.

Nature documents well the use of what we might consider negative values:

- stealth (e.g. tigers silently and expertly stalk their victims);
- camouflage (e.g. reef fish are experts);
- deception (e.g. cuckoo birds 'trick' other bird species to raise their young by depositing their eggs in the other species' nests);
- seduction (e.g. flowers, fruit trees and birds use sweet fragrances, sweet flavours, or bright colours to fulfil their needs for survival); and
- offensive revulsion (such as the foul-smelling anal excretions of the honey badger or the frightful excretions of ink by an octopus, to ward off all comers)

Clearly, many instincts exploit anti-social or individualistic values in order to promote survival and flourishing within homeostatic self-restraints. In so doing, they form a legitimised part of the local level playing field, yet are sometimes challenging to our civil sensibilities.

The person who pursues his purposes by this competitive law of the jungle without modification by social norms is sometimes known as a brutish savage, or if s/he has knowledge of those social norms, a person 'sold under sin'; but as Nietzsche suggested, perhaps this is a mistake. Perhaps it would be better to acknowledge the source of these valid instincts in terms of our evolutionary success and fully explore their relevance to our fitness for survival (or ability to self-organise) going forward before we assign moral labels in the name of a deceptive 'greater good'.

As an example, consider the words of John in the book of Revelation regarding the promise of a Christian heaven that will relieve the believer of cruel and unrelenting Roman persecution. This scripture is often used at funerals to bring comfort to mourners: "He will wipe every tear from their eyes, and there will be no more death or sorrow or crying or pain. All these things are gone forever."[36]

While in the historical context it may be reasonable to seek permanent relief from unimaginable horrors, can you picture the result of such a policy? We only need to look to those 'helicopter Mums' that save little Johnny from every mishap in our midst to guess at the result. Little Johnny very quickly becomes spoilt: Socially inept, weak, dependent, helpless and hopeless. All Yin and no Yang guarantees that little Johnny will be a failure. Unfortunately, the policy is a total disaster in practice. It turns out that competition, challenge, pain, and sorrow are necessary ingredients of our adaptive success, wellbeing and happiness. Heavenly ineptness seems hollow.

We need to be careful to understand the relationships involved before we disturb them. If these challenging relationships already promote sustainable survival and wellbeing, then the only change we could wish to make is to ensure we do not circumvent that sustainable wellbeing. Perhaps, as many ancient peoples have done, we could also use such contrary instincts in animals as described in the prior dot points as a way to reflect upon our own values, their sensitivities, and their application. This may be particularly useful in various adverse settings, such as in times of war or local conflict or high risk, that is, in times far removed from supportive homeostasis when our life-systems do not perform optimally. At the very least, the animal-versus-natural-environment survival instincts teach us about the energizing and invigorating reality of the drive towards environmental balance or sustainable interrelationship.

We should also explore the damaging biases of these competitive instincts in us when the rules of the game change too quickly (e.g. through overwhelming human technology applied to old-growth forest harvests or the damaging catches of scarce fish stocks in unregulated waters), and their possible modification through raised consciousness. Only then should we consider their possible censure through normative social self-restraints.

Perhaps we have censured far too much without properly acknowledging the pivotal role of these carnivore instincts in our ongoing innovation with respect to the social and natural environments and ourselves. Bottom line: If necessity is the mother of invention then competition is its father.

Competition and challenge is essential to innovation (and misplaced support can be detrimental). Our carnivore instincts, built on the asynchronicity, life-principle, or drive within every living cell, gave us our superb ability to compete, survive and innovate.

If we as a global virtuous community of humanity are going to find a solution to our current threats to global survival, then it is a stark reality that instinctive competition and its moderating law of the jungle, which together describe much of nature's interrelationships, is going to be part of the evolutionary solution.

Is it fair to single out the law of the jungle as brutish and harsh? After all, just like the more sociable laws of the gatherer, the law of the jungle, by which Tarzan famously lived on the frontier of civilisation, is firmly linked to the surrounding ecosystem. Ideally, the law of the jungle is an ecosystem-aware and ecosystem-sensitive self-restraint. We might thus rename the jungle law the risk-weighted ecological law. Its sensitive awareness is more than we sometimes achieve from our cold and wayward consciousness.

For instance, it was this wayward consciousness that resulted in the wanton, unprecedented and heart-wrenching slaughter of wild buffalo in the movie *Dances with Wolves* (1990 Orion Pictures), as witnessed by frontiersman Lt. John Dunbar and the eco-friendly Indian inhabitants of the land beside him.

The first self-restraint of values (and vreme development) is the risk-weighted competitive interaction.

The Social Stance: Exploiting our Culture and Society

We now move on to discuss the mutually reliant or social, 'gatherer' or 'herbivore' instincts (corresponding with Table 3: The Normative Value-set).

We cannot deny life's interconnectedness. Our individual lives seem to be part of a larger organism called our society. Our society along with our biosphere seems to be part of a larger organism James Lovelock (2009) and others would call Gaia. In fact, these are not mere possibilities but facts that depend on definition alone. The interconnectedness that keeps every particle of our galaxy rotating around its twin-black-hole-core is also undeniable.

In this locally recurrent and emergent sense, I would see no essential disagreement between Dawkins' concept of the extended phenotype applied to all species together and Lovelock's concept of

Gaia. Does the bird make the nest a nest, or does the nest make the bird a bird? Does life support the atmosphere or does the atmosphere support life? Clearly, the best way to frame such questions is through the concept of emergent interconnectedness.

The Dalai Lama suggests that because the truth is encoded across a vast network of variables and entities, the broader our outlook, the more likely we are to make net positive contributions. He adds that an interconnected perspective is important because it leads to a more integrated or holistic view.[37] Wise words indeed.

We are Herbivores: The Grazer, Gatherer and Herd Mentality

There are only two major approaches to animal survival - cooperation and competition – and the herbivore and the carnivore (including insectivore) epitomise them. There is a third mode of survival scavenging – but this is really just a poor cousin of the two major survival modes found in animals that we shall discuss later. If the instinctive hunter is epitomised by the self-reliant man, then women in those same hunter-gatherer societies sharing food-gathering and food-preparation tasks, and then sharing productive outcomes with their families and tribes, stereotype the instinctive gatherer.

Cooperation, as opposed to competition considered in the earlier section of this chapter, is a very clever instinctive adaptation to our harsh environments. Cooperation is essential to the emergent win-win mentality of Figure 12.

We could go on to discuss why the gatherers are usually women in hunter-gatherer societies, while the hunters are usually men and how this has affected brain and consciousness development in women and men separately since the days of the first hunter-gatherer societies. We could also discuss how all of this relates to the inequality of the sexes still present in society today and the challenge our male-female hunter-gatherer brain development poses to our growing conscious self-awareness, but this will be left to others, such as Richard Wrangham and Daniel Goleman. It is sufficient to just raise these tetchy issues here.

Why do animals instinctively herd? The basic answer is again risk-management. That is, herd-members singled out from the herd are more likely to be the focus of predator attention. The herd-member or school-member on the outside of the herd of sheep or school of fish is a little more vulnerable to attack, while the herd- or school-member in the centre is comparatively safe. Duty on the outside of the herd is often rotated to fairly spread risk amongst individuals, or left to the sentinels whose role it is to warn others, or otherwise left to the strongest member and protector of the group (in which case the protector gets paid for his services through a system of social privileges). Likewise, migrating flocks of birds in formation rotate positions up the front of their formation in order to spread the burden of work done fighting maximum wind resistance.

As we have just seen, herding, schooling or flocking brings about something far beyond just basic risk management in animals and humans. The close proximity of herding animals brings about a new level of relativistic interrelationship, that is, a social context and thus specialisations of labour (for example sentinels or protectors). It brings about the development of socially recognised roles and signals (coding), normative social rules (or mutually-recognised styles) and a social ordering of behavioural traffic that, through its algorithmic feedback path, reinforces the emergent social context and social development.

The same is true of all socially organised lifeforms, from amoeba to humans.

Another simple example of social organisation is bird pecking order. My daughter-in-law has three chooks in a chicken run at the back of the family home. Each day she lets the chickens out of the chicken coop to feed. Even amongst these three chickens, there is a socially accepted pecking order. The smartest one, that finds its way out of the chicken-run's gate first each day, eats first, the second one, who is at least smart enough to follow the smartest one, eats second each day and the third one, who often can't find its way out the chicken run's gate without help, eats last.

They follow this pecking order even if they all come to a new feeding area together. The first one is the initiator, the second follows the first, and the third follows the other two. It seems social rank is recognised and assigned at least partly by conscious awareness (or cognitive ability) even amongst chooks, although the variations in intentional stance I see expressed through each chook are likely to be largely the result of variations in instinctive endowment. I have never seen these rather domesticated chooks fight so instinctive patterns of behaviour and/or intelligence rather than the threat of physical force appears to be a major determinant of social rank in this case.

The wild birds fed at my home each day follow a similar pecking order. Being the biggest, the brush turkeys typically feed first, while the white cockatoos, magpies and Indian minors look on. Eventually the signal to feed usually moves along by size and social rank to the cockatoos, then the magpies and then the Indian minors.

The interesting observation here is the inherent respect for social order held across species (the mix of bird species can vary from day to day). Like the respect for environmental or task-oriented rules of the game held by hunters, gatherers hold respect for the social rules of the game (or the naturalistic contract). That is, co-operators seem to blindly understand the social niches or borders they inhabit and exploit, which perhaps raises the question, 'just how much more does consciousness of the social instinct actually add to our social behaviours?'

A case in point would be the ability of some amoeba to morph into specialised roles within colonies when local conditions support cooperative reproduction, but otherwise live quite separate lives. We humans often display behaviours not far removed from these single-cell organisms. (In the next few chapters, we will tackle answers that are more comprehensive.)

These rules or constraints may be signalled, not always strictly in more relaxed circumstances, by physical prowess with little threat of applied force, or just like the chooks or amoeba, by cognitive prowess or instinctive endowment. What we seem to witness here is the same relativistic separation of powers between space and matter that forbids infinities and nullities and thus permits a shared environment, extended first to competitors and then to co-operators.

Of course, there are interesting variations to the observations above, for instance when four or five Indian minor birds cheekily harass a lone magpie to leave their territorial area. Here again we see self-reliant competition and social cooperation both working. The corporate physical power of the Indian minors is just enough to curb the self-seeking behaviour of the larger magpie.

Another interesting variation is when the domestic cat turns up to the bird feeding area. Sometimes it challenges all birds present and they scatter, while at other times the cat is too lazy to assert its superior physical power on the ground. The birds seem to be able to sense the mood of the cat, because they often dare to feed even while the cat looks on disapprovingly.

In all these simple examples, we seem to see the unwritten naturalistic contract at work. The instincts encode the contract, and perhaps in all species with brains and nervous systems, consciousness enhances this coding. The instincts are also continually tested and honed through circumstances, claims and counter-claims, but outside the extreme self-aware behaviour found in humans and perhaps chimpanzees, never dumped or freely picked up and put down at will. That is, it seems it requires the imagination of full reality in consciousness to escape the limits of relativistic reality embedded in instincts.

Would it be too much to say that the level playing field (or finely-balanced biodiversity) seems to provide a near-optimum environment for all species, no matter what their current capabilities, to maintain or improve their life-systems? The next question might be, 'What is the nature of the apparent naturalistic contract or contracts that enacts and protects the level playing field?' We will address this handy construct in a later chapter. Of course, in one sense there is no contract, just boundaries at a higher level than the lone hunter of the prior half of this chapter encountered.

In the stock market scenario, the instinctive herd mentality is evident in terms of market bubbles (in stocks, real estate, commodities, etc.) but perhaps more generally in terms of traders who rely on technical analysis to make their trading decisions.

We are all familiar with the dangers of groupthink, the classic example being NASA's very sad Challenger disaster in 1986. Yale social psychologist Irving Janis defined Groupthink as a way of thinking that overrides a rational appraisal in favour of group solidarity. More simply, groupthink is inappropriate group consensus-seeking. Groupthink is a malaise that can also strike entire communities, especially when it comes to things like rational action on climate change.

In the case of the Challenger explosion, the o-ring supplier in meetings with NASA staff, and under pressure from NASA management, was unwilling to delay Challenger mission take-off even though representatives were aware of the risk of o-ring failure (the ultimate cause of the Challenger explosion). Social intelligence succumbed to what seems to be social foolishness. We cannot always trust our biased instincts, even in social situations.

One simple observation we could make here is that the disaster may have been averted if every member of the mission group consciously followed their competitive instincts as well as their cooperative instincts (that is, didn't just heavily dampen their self-reliant instincts in the social setting). However, that would have also taken rational thought, alert consciousness discussed in the next chapter, and perhaps a timely application of the Emergent Method.

There is a growing need for quality communication and inclusiveness in a world of shrinking natural resources and a world with a growing ability to take - made possible by life's quickly advancing global technologies. Blindly selfish genes and memes continue to exploit their environments in ever more complex arrangements. The question is, 'How will this coagulation of arrangements continue through human beings, and will it be sustainable?'

One answer it seems, before we even get to discuss the impacts of human consciousness in the next chapter, is that it cannot be without a near balance between our self-reliant and social instincts or skill sets. That is, even as our life-systems develop and coagulate, we will still have to rely on our competitive instincts to ensure our cooperation is efficient, effective and sustainable within a bio-socially diverse environment.

The Social Sciences: The Normative Approach

The normative approach often used by the social sciences usually refers to developing and following an ideal or standard social framework, such as the conceptual frameworks we follow as members of professional bodies. Normative statements, in reference to a conceptual framework, usually talk about what is agreed, what should be, what we ought to do, or what is right, wrong, good, or bad, and so we often perceive them as moralistic and non-scientific. Normative statements apply society's partially instinctive value judgments.

Just like in the herd of the African plain, normative rules or modes of behaviour are essential for making or signalling quick decisions that prioritise and reinforce successful actions of the group. Norms are to processes what axioms are to analytics. That is, they enable us to move on, rather than become mired in cycles of intentional hesitation. However, as already noted, sometimes values can contradict each other – for example, in a society's various approaches to petty crime minimisation or accounting standards. In these less clear situations, the phaseal process of adaptation through artificial or natural selection of vying alternatives (for example, reaching compatibility with international standards) becomes an important facet of finding the most successful long-run solution.

One difficulty with the normative approach of the 'herbivore' is with establishing a conceptual framework that is relatively incontestable and unchangeable. Cultural norms, whether within monkey troops or human society, are sometimes subject to fast environmental change or environmental mismatch that can then contribute to an undermining of social cohesion and institutional authority.

For instance, fast change can cause some to suffer social alienation and some to question their place in society or question their life's meaning. While the individual or 'direct object' may be relatively unchanged, the 'indirect subject' or normative 'I' can become lost in a quickly changing feast of societal norms. The old heuristic algorithms can fail to provide individual wellbeing in a new and quickly advancing social environment. We are eco-psychosocial beings - our personal and social instincts, expressed for instance in a set of mores and taboos, need to work and evolve together.

The style of the normative approach at a social or organisational level can also meet with objections. Should the conceptual framework and standard-setting or norm-setting take a qualitative approach or a prescriptive approach? A qualitative approach (guided only by principles and values) is often easier to draft but sometimes harder to enforce. Users can abuse qualitative statements if we do not carefully define the values they embody. This is why the approach is likely to be more effective in a non-contentious social setting or in a setting with a strong social fabric (that is, a setting with a strong sense of civic responsibility).

On the other hand, a prescriptive approach can be laborious to draft and hard to administer. Quantitative statements can become socially outmoded or shown to be inadequate very quickly and often. Old algorithms can become unrealistic and thus fail in complex and quickly evolving environments. These same style sentiments also apply to the individual: No matter what your set of values, do you see yourself as a very laid-back and adaptable kind of individual that sometimes regrets actions taken too casually, or a very highly self-disciplined individual that sometimes regrets opportunities missed, or something in between?

The failures of personal and social heuristic algorithms (i.e. the failures of the personal and social instincts) to deal with all situations in which we as humans find ourselves point to the need for alternative ways of processing information coming from our quickly changing environments. This alternative came with the evolution of consciousness, discussed in the next chapter. A limited relief from instinctive failures also comes with an appreciation of the Emergent Method and specific tactics, such as supplied by various authors and Neuro-Linguistic Programming.[38]

The Cooperative Self-restraint

A positive approach to lawmaking (i.e. determining laws based on measured demographics or some other more objectively measurable attribute across a population) is often not appropriate when we are required to actively dispense justice, especially criminal justice, to the individual members of a society. In such cases, we would typically require an ethical or social standard, framework, or value-set against which to form personal and social self-restraints. Laws formed must not be merely personally acceptable to one individual such as a lawyer. They must also be socially acceptable for the smooth running of a 'free' (or 'fractal') society. They must be normative and herd-like, even in our modern societies, if they are to effectively self-organise those societies. This arrival at a widely accepted worldview or social narrative is not an easy one. It can only emerge in societies that are free enough to allow vying concepts or narratives to develop and compete within the limits reality shares with us.

On the other hand, when we can more objectively measure certain attributes across a society or group to determine or uncover an actual, key, or near-optimum performance standard, it is often appropriate to recognise this temporal fact in positive law-making. This might apply to the formation of minimum building safety regulations, for instance. In this case, social norms are often less relevant and scientific, engineering or technological findings are usually more relevant. Ingrained or imposed social norms can sometimes be a bottleneck to effective regulation-making in such cases.

Additionally, as we make advances, especially in the neurosciences, the fuzzy dividing line between the positive and the normative approaches keeps moving. This is a very telling statement. As tentative facts are made more explicit, this has implications in terms of the arrangement of facts, that is, of explicit values. It suggests there is an immovable link between the more environmentally objective and the more personally objective as previously defined.

When society's norms and legal precedents slowly evolve with social experience, the normative, herd-like approach often works well. When court-made law changes with every whim of courtroom circumstance or precedents are overturned too quickly and easily, blurring the underlying legal principles and values, this is hazardous to the cohesion of society. Recently we have not always had the benefit of legal precedents slowly evolving with increasing social experience, knowledge or certainty. Euthanasia, stem-cell research, abortion, genetically modified crop regulation, homosexual marriage, global warming, women in church leadership roles, the global sovereign debt crisis - are all subjects that seem to demand our quickly fragmenting attention and action.

Sometimes social norms in politically controversial areas move a long way ahead of governmental or judicial decision-making. Sometimes our law-making bodies, for political reasons, avoid passing efficient statutes based on sound value-based principles or scientific findings and so the courts are left to pick up the pieces in a less-than-optimum fashion through common law (in Australia) and its developing precedents.

The point of raising all these issues is that right now we need sound philosophical principles and a solid values-framework that can keep up with this high pace of social change and intelligently guide our intentions. We need to recognise not so much where our values come from, or which civilisation /god owns them, but how our values arise in our instincts and in consciousness and how they have worked naturally in our animal past, and why they are under stress now that we have gained elaborate consciousness. One possible reason is that we have only recently begun to apply lessons from evolution and natural selection to the problem of artificial vreme selection. We also need to respect (but test) our proven self-organising values (both personal and social) as personally objective measuring-sticks that we may apply to the quickly changing dance of life.

One clue we have uncovered so far is that the socially cohesive herbivore instincts have arisen in our animal past in the context of the other risk-weighted, self-reliant carnivore instincts discussed earlier. Just because humans are an important part of "the inexorable coagulation of life"[39] (or the coming together of living cells in ever more complex arrangements, including human social arrangements), this does not mean we can forget the Dawkins' suggestion that the blindly selfish replicator lies at the base of all our self-reliant and social instincts.

The social instinct is not independent of the self-reliant instinct, nor is it equal with the self-reliant instinct. It has been built from scratch on the back of nature's system of self-reliant 'free' or 'fractal' enterprise, which is evidenced by the recognised and taken right of each individual cell to autopoiesis and cognition. On the other side of the coin, it is true to say that cooperation emerges from competition, as participation in any team sport clearly demonstrates. However, cooperation sometimes collapses into team-destructive competition when players, through their actions, fail to recognise the benefits of the team in the higher and more complex team-environment, which flows from the principles or locally recurring interactions by which the team operates.

Cooperation also takes competition, in this case now between teams, to a higher order of complexity than what was previously possible as disunited individuals. At the same time, it is clear that the competitive rules between teams can only emerge with the cooperation that forms each team. Again, spacetime coding seems to learn as much from coagulations of matter as matter does from spacetime. The relativistic cycles of competition and cooperation are very important ways in which feedback/feed-forward leads to the emergent qualities and coagulations we observe in all living systems.

The emergent complexity amongst humans is most importantly made possible and accelerated not due to our non-living environment, or even the other species of our environment, but rather the human environment within which we as individuals and societies find ourselves competing and coevolving. As A.C.

Grayling says, "It is in the conversation, negotiation and mutuality of social life that moral norms and practices actually emerge."[40]

Competition and cooperation seem to demand a relativistic and emergent interactivity between environmental facts and values. This pattern of emergence amongst fellows (or homogenous groups of particles) seems unchanged since the quantum fluctuations and quark-gluon plasma described in Part 1.

To use a metaphor of outer space, the relationship between the competitive instinct and the cooperative instinct is not like the orbit of twin black holes or quasars around each other, but more like the orbit of our planets around our massive sun – in which case the sun represents competitive self-reliance and the planets represent social cooperation. If competition is the direct rule of instinct, then cooperation is its indirect rule (i.e. indirect cooperative arrangements of matter emerge with the existence of direct competitive objects). To put this relativistic concept another way, cooperation shows competition how to move but competition determines how cooperation will be.

The inference is this: As we move forward in modern society to meet the fast-paced and ever-more complex challenges it presents to us, our philosophical principles and values-framework, or program of vreme development, will work best if it protects our self-reliant and individual right to life. At the same time, the program should guide us in the inexorable but more intelligent or information-incorporating development of our social intentions and self-restraints.

The second self-restraint of values (and vreme development) is the normative cooperative interaction.

The Corporate Need for Organised Cooperation

Something like the grass of the field described earlier, while profit-seeking corporations as a whole must bring the nature of the competitive 'carnivore' to their respective markets, they function best if all members of the corporation see themselves as team players within the 'herbivore' culture and collaborate to follow the corporately held values, norms, goals and purposes.

The heart of every great organisation is its values. An excellent example of a corporate values statement is the credo of Johnson and Johnson,[41] which is as follows:

"We believe our first responsibility is to the doctors, nurses and patients, to mothers and fathers and all others who use our products and services. In meeting their needs, everything we do must be of high quality. We must constantly strive to reduce our cost in order to maintain reasonable prices. Customers' orders must be processed promptly and accurately. Our suppliers and distributors must have an opportunity to make a fair profit.

"We are responsible to our employees, the men and women who work with us throughout the world. Everyone must be considered as an individual. We must respect their dignity and recognize their merit. They must have a sense of security in their jobs. Compensation must be fair and adequate, and working conditions clean, orderly and safe. We must be mindful of ways to help our employees fulfil their family responsibilities. Employees must feel free to make suggestions and complaints. There must be equal opportunity for employment, development and advancement for those qualified. We must provide competent management, and their actions must be just and ethical.

"We are responsible to the communities in which we live and work and to the world community as well. We must be good citizens – support good works and charities and bear our fair share of taxes. We must encourage civic improvements and better health and education. We must maintain in good order the property we are privileged to use, protecting the environment and natural resources.

"Our final responsibility is to our stockholders. Business must make a sound profit. We must experiment with new ideas. Research must be carried on, innovative programs developed and mistakes paid for. New equipment must be purchased, new facilities provided and new products launched. Reserves must be created to provide for adverse times. When we operate according to these principles, the stockholders should realise a fair return."

The Johnson and Johnson website has the following to say with respect to this credo,

"Robert Wood Johnson, former chairman from 1932 to 1963 and a member of the Company's founding family, crafted Our Credo himself in 1943, just before Johnson and Johnson became a publicly traded company. This was long before anyone ever heard the term 'corporate social responsibility'. Our Credo is more than just a moral compass. We believe it's a recipe for business success."

As you may well imagine, I am quite a fan of Johnson and Johnson's credo. It reflects all that can be distilled from an intelligently held, commendable and open social approach. There are no hidden agendas here. The normative or herd-like values are clearly expounded in concrete terms; values and concrete terms all members of the organisation could easily adopt and follow.

An interesting observation can be made here. The legal person, the Johnson and Johnson legal entity denoted as *'we'* (who are actually its agents or survival machines) in the credo, takes from the environment and transforms what it takes into higher magnitudes of order or potential just like any other living organism. The legal person then transfers these products and services back to its stakeholders and its environment. It seems Johnson and Johnson would meet the organisational and platform-independent definition of a living organism in terms of autopoiesis and cognition, discussed in Part 1, perhaps depending on how strongly you would agree that it is able to meet the requirement of self-replication or reproduction.

The legal person Johnson and Johnson actually takes nothing for itself (this is meaningless for a legal entity). The legal person (that is, the indirect and abstract idea or the Johnson and Johnson *'I'*) just harbours, guides and reflects the life-process described by its credo. The legal person is just like the abstract idea 'Rome' in the movie *Gladiator* (and the values Rome encapsulated) mentioned in Part 1, which also captured a fleeting dream and possibility in the minds of its actively conscious inhabitants (its vremetic survival machines or agents).

There is an essential link here between the credo and its conscious values, and the life-process of Johnson and Johnson that mere autopoiesis and narrowly defined cognition do not quite capture. We could even say that virtues (or values-in-action) are the essential life-process of the indirect, abstract and self-aware entity Johnson and Johnson, especially if we consider all other non-value-driven actions or contradictory actions as either subordinate or alien and unessential to the self-aware life-and-being here-and-now abstraction.

This is a profound statement in terms of your life and mine (just other abstractions), and in terms of the attitudes we hold to the virtues we acknowledge, admire, spurn and live by. Essentially, you too are a social 'legal person' or vreme-set serving your internal and external stakeholders (including your own body, its autopoiesis and cognition, and the environment).

You too are your values, including the values you identified in the last chapter. The essential you, the heroic you, encapsulates the virtuous values you identified. You are the indirect life-and-being here-and-now abstraction and organisation of values that as a force-system or vremetic code, brings agency (or meaning, direction, purpose, and a dream or destiny) to all the marvellous physical mechanisms of your internal and external stakeholders (your survival machines and their extended phenotypes). Note: Just how we might defend the concept of personal moral agency in a world of apparent cause-and-effect will be further dealt with in the next chapter.

Perhaps we have struck on something quite profound here. The necessary, sufficient and platform-independent definition of a living, self-aware and consciously enlightened entity (or an intelligent agent) is that it is autopoietic, cognitive and vremetic (i.e. it holds a rational set of vremes, or credo, that it uses to define and add meaning to itself and its purposes).

Of course, we also have to acknowledge the fallibility in the real Johnson and Johnson that in microcosm often behaves in a manner contrary to its own credo. The corporation behaves according to its beliefs, good or bad, tested or untested. However, when a corporation's tested beliefs and chosen behaviours are in alignment with its credo or most enlightened sense of self, it usually flourishes.

One difficulty here is to identify, through its eco-psychosocial claiming, what constitutes the current corporate self. Just as in us, the self is a diverse, dynamic and emergent property, rather than a monolithic one. The corporate self emerges from its vast array of relationships and is thus subject to continual testing, resistance, and competition.[42] That is, the corporate self struggles to survive as a statistical property, just as do all other life forms. In this sense, the corporate self is neither something a corporation 'has' or 'is'. Rather, this corporate self, which we might also name the corporate culture, continually emerges from what is. It is 'becoming'. It holds its power, or lack thereof, in its dynamic relationships.

The Emergent Method encourages us to use these ideas about self, and how it flourishes, in all the ways we add value to our surroundings. The Emergent Method asks us to make explicit, and focus on, our personal mores and taboos so that we can consciously assess them and choose them for our own. When we don't make our credos explicit they remain implicit, hidden from consciousness and resistant to self-actualising change.

'I' is a social construct and thus largely a product of our underlying cooperative instincts. As mentioned earlier, consciousness is consciousness of something. That is, there is no sense of 'I' until we are aware of others in the socially cooperative context.

Freud's concept of the Superego (comparable to a sense of conscience), also a product of socialisation, has strong parallels with the concept of the personal legal entity or indirect and vremetic 'I' we have arrived at here. Freud's Id would have had its beginnings in the cognition and autopoiesis of our ancestors and would have evolved things like an autonomic nervous system and pain response system in the presence of its growing self-awareness. Similarly, Freud's Ego would have emerged from the presence of an Id and Superego and then continued to emerge together with their ongoing emergence. One of the aims of the Emergent Method is to help make the Superego more realistically flexible and thus break down the dissonance between the Id's 'is' and the Superego's 'ought', with which dissonance the Ego and its narratives must continually struggle.[43]

Wouldn't it be nice if a political party adopted a politically adapted version of Johnson and Johnson's credo and made it well known to all of us who vote? It would make our political choices a lot easier if we could trust that the party would choose to stick to its open and evolving credo. The political issue and issue-lobbyists would matter less because we would know that the party would always seek to follow its credo. With time and life experience, its credo could learn to reflect and challenge the credos of its constituents. Does it somehow make political parties look less relevant, less credible and less self-aware when they brashly depart from their credos (or dare I say their niches within the bio-diverse level playing field) easily and often, without explanation?

Unfortunately, not all organisations or people are as well intentioned as Johnson and Johnson's credo would suggest. Some short-sighted senior executives work a long way outside a credo such as that of Johnson and Johnson. Unenlightened self-aware entities are just ignorantly selfish. The old organisational paradigm required all members to align themselves with the senior executives' vision and purposes, no matter how poor. Nevertheless, in the midst of fast-paced market and technological changes, new post-bureaucratic organisational paradigms are arising.

One such model of better organisation is suggested by modern object-oriented database methods (see for instance Rumbaugh (1991)), which would see each organisation arranged into a highly adaptable matrix of stakeholder classes and corporate objectives rather than a fixed and brittle matrix of corporate functions (such as Sales and Marketing, Warehousing, or Corporate Services) and corporate objectives.

In the model suggested by the object-oriented approach, the employee focus would become a stakeholder specialist rather than just a functional specialist. Under this model, products, services, and functions can come and go, but the stakeholder relationship or evolutionary fit can only ever strengthen. That is, the interpenetrating relationship or balance is king in the development or release of wealth through profit-seeking or non-profit-seeking exchange.

This model suggests stakeholders can subtly and indirectly mould the corporate objectives, mission or vision. An object-oriented approach suggests that corporate objectives and functions are not dictated by the corporation's 'genetic structure at birth'. In this approach, corporate risk is measured not only in terms of risk to the corporation's objectives, but also at a deeper level, to the multifaceted relationships with its stakeholders (and thus employee attitudes towards protecting and fostering those strengthening relationships). This would be true even in an environment of stakeholder metamorphosis. We will discuss this model a little more in Pillar 6.

We desperately need to develop a higher and more organic corporate consciousness-plus-instincts. Our institutions, organisations and marketplaces should no longer be design artefacts that are simply discarded after failure, with very serious social dislocation and waste. They should reflect and breathe the diversity of our self-aware evolution of values, consciousness and instincts at many levels.

We need to apply the principles of Nature's localised and emerging systems of resource organisation back to our corporate and social organisations. At the same time, we need to bind the members at each level with a transparent and open credo something like that of Johnson and Johnson, perhaps just as the sun's gravity gently and subtly binds each of the very individualistic and unique planets within our solar system.

We are Omnivores

Humankind, unlike any other species, has developed two specialist survival skills. We are both excellent hunters or carnivores and excellent gatherers or herbivores. Most other species are clearly the one or the other. Some are omnivores, but nothing is quite like the human omnivore. Species considered omnivores would include pigs, some primates, some rodents, some bears, and some others such as raccoons, squirrels, and skunks. All of these examples seem to fit the description of opportunistic scavenger at least to some extent. This seems an appropriate strategy where seasonal changes (such as winter snow) drastically change the availability of differing plant foods for instance.

In the case of humans, some would define us as carnivores (we are the most carnivorous of all primates), and some, omnivores, because we are successful scavengers. However, unlike the other omnivores, we are highly-adapted scavengers, hunters and grazers/gatherers all in the one package. Even our hunting is a highly adapted social exercise, unlike the hunt of the lone Siberian tiger.

We have learned how to very successfully exploit both competing and cooperating. We know how to be instinctively self-reliant and instinctively social. The problem is, with all our development, we are still not responding optimally to the extraordinary global challenges we face today, even though I believe we each have the necessary equipment to face these challenges.

One last point about our active and passionate instincts before we move on to discuss our passive and dispassionate consciousness in the next chapter: We share our instincts and consciousness with all other animal species. They have some level of consciousness because they also have brains and nervous systems.

Perhaps we share nearly all our feelings, all our emotions and all our character traits with other animals – it is just that they have lesser degrees of consciousness to amplify, attenuate and augment them. Perhaps the only thing that changes a clucking mother hen into a loving human mother fussing over her offspring is the level of consciousness – the core instincts to aid survival are innate to both species (because both species have blindly selfish systems driving their respective survival machines within their survival niches). Why do we relate so deeply with our domesticated pets? Perhaps it is because they, who have a lower consciousness than we do, nevertheless share virtually all of our instinctive character traits.

Your Credo

Perhaps you could also now consider, armed with your list of personal values and personal stakeholders, writing your own credo, following Johnson and Johnson's example. That is, dedicate each paragraph of your credo to your intended relationship with each stakeholder. Your stakeholders include your own body, your family, your clients or customers, your suppliers or distributors, your community, and your government.

This credo should be aware of the needs of each of your stakeholders and how you are uniquely placed, equipped and motivated, to meet those needs. You should also devise this credo with an uncondemning appreciation of your own consciousness-plus-instincts, for better or for worse.

You share this nature to some extent with all other surviving life forms on the planet, but at the same time, it exists within you in a unique combination that in its highest expression, the universe deeply includes and needs.

Concepts of Emergence Encountered in Pillar 2

The odds of success have to be just high enough and just low enough to promote close-to-optimum competition – and thus promote a close-to-optimum adaptation (genetic and memetic) to the changing local environment. This observation is not to be taken lightly. The cosmos is not your benefactor, but if you hold the right attitude out towards it, you will appreciate the challenges and support it presents to you with equal gusto. Easy success or crippling failure will not change the limits or boundaries of the systems in contention. Emergent 'body-building' is a slow and steady business. Do it too quickly and the body will be damaged; do it too slowly and no gain will be had. The cosmos enables you to be an intelligent participant in its gradual unfolding.

The relativistic cycles of competition and cooperation are very important ways in which feedback/feed-forward leads to the emergent qualities and coagulations we observe in all living systems.

The emergent complexity amongst humans is most importantly made possible and accelerated not due to our non-living environment, or even the other species of our environment, but rather the human environment within which we as individuals and societies find ourselves competing and coevolving. As noted earlier, A.C. Grayling suggested our moral norms emerge in to-and-fro of societies through their conversations and conflict management.[44]

Competition and cooperation seem to demand a relativistic and emergent interactivity between environmental facts and values. This pattern of emergence amongst fellows (or homogenous groups of particles) seems unchanged since the quantum fluctuations and quark-gluon plasma described in Part 1.

The corporation behaves according to its beliefs, good or bad, tested or untested. However, when a corporation's tested beliefs and chosen behaviours are in alignment with its credo or most enlightened sense of self, it usually flourishes.

Freud's Ego would have emerged from the presence of an Id and Superego and then continued to emerge together with their ongoing emergence.

One of the aims of the Emergent Method is to help make the Superego more realistically flexible and thus break down the dissonance between the Id's 'is' and the Superego's 'ought', with which dissonance the Ego and its narratives must continually struggle.

In the next chapter, we will consider human consciousness in order to complete our picture of human nature. We will then discuss the implications of all our findings in Pillar 4.

---

[1] See Merleau-Ponty, Maurice, *Phenomenology of Perception* (Routledge, New York, 2014 (Original edition in French in 1945))

[2] Ibid., see p. 5. The following discussion is largely from the same section, pp. 3-12 and from pp. 214-252

[3] Ibid., The following discussion largely involves pp.13-99

[4] Ibid., The following discussion largely involves pp.100-178

[5] Ibid., The following discussion largely involves pp.179-205

[6] Ibid., The following discussion largely involves pp.214-252 and pp.312-360

[7] Ibid., The following discussion largely involves pp.253-311

[8] Ibid., The following discussion largely involves pp.432-457

[9] Ibid., The following discussion largely involves pp.361-383

---

[10] Ibid., The following discussion largely involves pp.387-431
[11] Ibid., see p.483
[12] Ibid., see p.454
[13] Ibid., p.480
[14] Ibid., p.483
[15] See Hollingdale, R. J., *A Nietzsche Reader* (Penguin Books, Sydney, 1977)
[16] Ibid., see pp. 215-231, 119-121 and 246-247
[17] Ibid., see p.55
[18] Ibid., p.60
[19] Ibid., pp. 126, 125 and 128, respectively
[20] Ibid., p.162
[21] Ibid., see p.158
[22] Ibid., pp.66-67
[23] Ibid., see p.85
[24] Ibid., see pp.91-92, 102
[25] Ibid., see pp. 95-96
[26] Ibid., p.158
[27] Ibid. p.104
[28] Ibid., see p.221
[29] Ibid., see pp. 231, 260, 269, and 271-272
[30] Ibid., pp.211-212
[31] Ibid., p.173
[32] The Amplified Version, Colossians 1:24
[33] See Rand, Ayn, *Atlas Shrugged* (Plume, USA, 2007 (Fiftieth Anniversary Edition)), 924-979
[34] Twentieth Century Fox, *Wall Street* (1987)
[35] AS/NZS 4360: 2004, Forward
[36] New Living Translation, Revelations 21:4
[37] See Dalai Lama, *Becoming Enlightened* (Rider, London, 2009), 155
[38] See http://www.nlp.com.au/ (accessed 2014). See also Kahneman, Daniel, *Thinking, Fast and Slow* (Penguin Group (Australia), 2011), and McRaney, David, *You Are Not So Smart* (Penguin Group (USA) Inc., New York, 2011)
[39] This is a term coined by Ridley, Matt, *The Origins of Virtue* (Softback Preview, Great Britain, 1997), 15
[40] Grayling, A. C., *The God Argument* (Bloomsbury, New York, 2013), 241
[41] The Johnson and Johnson credo (last accessed March 2016) is available from http://www.jnj.com/sites/default/files/pdf/jnj_ourcredo_english_us_8.5x11_cmyk.pdf
[42] The view expressed here would be in agreement with Courpasson, D. and Clegg, S.R., *The Polyarchic Bureaucracy: Cooperative Resistance in the Workplace and the Construction of a New Political Structure of Organisations,*

Research in the Sociology of Organisations, Vol. 34, 2012, 55-79

[43] See Freud, Sigmund, *The Ego and the Id* (The Hogarth Press Ltd. London, originally 1949)

[44] See Grayling, A. C., *The God Argument* (Bloomsbury, New York, 2013), 241

PILLAR 3: Consciousness

In this chapter, we will discuss more objective extrospective consciousness and intersubjective introspective consciousness. We will discuss how human consciousness has evolved and how its genetic evolution has affected memetic evolution, as well as how memetic evolution has influenced via feedback genetic evolution. We will also discuss the dispassionate natural and spiritual self-restraints that arose out of evolving consciousness. Finally, we will also discuss how our concepts of the supernatural have arisen out of intersubjective consciousness and how we can perhaps find assurance at the time of death without a need for the supernatural. In terms of your personal quest, the text will suggest that you take your values arrived at in Pillar 1, your credo of Pillar 2 and your inward- and outward-looking consciousness discussed in this chapter to develop your life purpose(s).

A certain level of consciousness is present in all animals with brains and nervous systems, but consciousness in humankind is of a much higher level than in other animals because it comes with self-awareness and other-awareness at extraordinarily high levels (e.g. I am aware that you are aware that I am aware that you are aware that...).[1] As stated in Part 1, human consciousness has vremes, imagination and foresight that seem to be able to escape the relativity to which our instincts were once bound. We can think outside the square (or cycle) and thus slowly learn how to shake off the chains of our selfish replicators.

Consciousness has taken two essential or stereotypical shapes in human evolution: 1. the extrospective approach, corresponding with the Table 2: The Positive Value-set; and 2, the introspective approach, corresponding with Table 4: The Self-enlightened Value-set. As in the last chapter on instincts, we will consider each of the two possible approaches separately.

However firstly, we will very briefly discuss the evolution of consciousness.

We Learned to Cook

The 2009 publication *Catching Fire: How Cooking Made Us Human* by biological anthropologist Richard Wrangham is one of the first books to clearly suggest how human cultural and memetic development could have profoundly affected human genetic development (and vice versa). The book is highly recommended.[2] Simple examples include lactose and alcohol tolerance in cultures where dairy farming and alcohol consumption, respectively, have been practiced for centuries. As we have already seen, genes are not the only things that evolve, so do memes and the two are not separate – they are interdependent. We are the physical and mental result of both our natural environments and our cultures, not respectively or independently, but interdependently. If we understand our genes without understanding our memes, we will understand perhaps less than half the current mechanism of our physical and mental development and evolution.

We know from Part 1 that Ida, the predecessor of modern lemurs, monkeys and perhaps humans lived approximately 47 million years ago and probably swung in trees and ate fruit in her tropical rainforest environment. Ida probably had a conscious self-awareness similar to modern-day lemurs and monkeys. Wrangham roughly describes our speciation and human development after that time as follows:

7 to 2 million years ago: *Australopithecus* lived in drier woodland, perhaps eating fruit but perhaps, judging by their teeth and jaws, also crushing and eating starch-filled or energy-rich roots, bulbs, tubers and other plants. These particularly energy-rich eating habits might explain the development of their larger brain size compared to earlier forest apes. Most famously, Lucy lived 3.2 million years ago and recently discovered *Australopithecus sediba* lived nearly 2 million years ago. *Australopithecus sediba* probably had the consciousness or self-awareness of something similar to chimpanzees, which means its kind were perhaps able to use sticks from time to time to dig for energy-rich foods, just as chimpanzees sometimes do today.

As an interesting aside, *Ardipithecus ramidus* ('Ardi'), who lived about 4.4 million years ago and is therefore the oldest known substantially complete hominid, was reported in September 2009 (after the publication of Wrangham's book). Its cranial cavity is only a little larger than a chimpanzee. Ardi suggests that both humans and chimpanzees may have evolved quite substantially from a common ancestor some six to seven million years ago.[3]

Around 2.3 million years ago: Habilines (*Homo habilis*) developed larger brains than Australopithecines, by about one third, probably due to the dietary addition of an even more energy-rich food; meat. Further, it is thought Habilines were stone knife and hammer makers, and were probably able to process their meat by tenderising it. This is a very important advance according to Wrangham because tenderisation not only made the consuming and digesting of meat a quicker, safer and more enjoyable process, it also enabled meat proteins to break down more easily in the gut and the energy to be extracted more efficiently. Wrangham argues it was the improved energy-extraction technique that enabled a decrease in tooth, mouth and gut size. Further, his controversial theory suggests that accompanying this efficiency saving was an increased genetic expenditure on brain size, and thus an increased capacity for consciousness. He thus suggests the two properties, brain effectiveness and digestive system efficiency, were interdependent – as if the genetic potential of noncoding DNA is a limited resource within the evolving genome. In this context, some pseudogenes (genes that have lost their ability to code proteins) subjected to genetic mutation unrestrained by normal evolutionary pressures could perhaps be responsible for sudden bursts of genomic development. Such a theory is not Wrangham's alone.

Around 1.9 million years ago: *Homo erectus* developed a brain approximately 40% larger again than Habilines. According to Wrangham, the reason for this increase was due to another important dietary innovation. *Homo erectus* learned to control fire and learned to cook. Cooking enabled the body to extract energy from meat and all other foods much more efficiently than ever before – and thus enabled the *Homo erectus* brain to grow (and teeth, mouth and gut to shrink) and thus conscious self-awareness to develop quite considerably. This in turn enabled hunting skills to increase and, as an aside, the division of labour between males and females as hunters and gatherers, respectively, to develop. It is also theorised by Wrangham that hunter and gatherer societies brought about the creation of the family unit for the first time. Just like some similar hunter-gatherer societies today, husbands depended on wives to cook and prepare the evening meal in order to give them more time hunting, and wives depended on husbands to protect them from food raiders.

With time brain size actually increased in *Homo erectus*. According to the theory, this was probably due to increasingly improved cooking techniques that improved flavour, tenderness and ease of eating for the consumer, but ease of digestion and thus increased efficiency of energy extraction for the body. More efficient energy extraction, according to Wrangham, meant more genetic potential was free for increasing brain power rather than maintaining digestive power. Note: Neanderthal, a more modern descendent of *Homo erectus* (who appeared about 1 million years later), was not an ancestor of modern *Homo sapiens*.

Around 800,000 years ago: *Homo heidelbergensis* developed a brain approximately 30% larger than *Homo erectus*. Wrangham puts this increase down to not just an incremental cooking-method improvement but perhaps also to a progress in hunting prowess over earlier periods[4] and thus an increase in meat and animal fat intake. Still today, we are the most carnivorous of all the extant primates. Of course, it would be the concomitant increase in conscious self-awareness with brain size, which would have facilitated such an improvement.

Around 200,000 years ago: *Homo sapiens* developed a brain approximately 17% larger than *Homo heidelbergensis*, probably due to continuing increase in cooking methods (for example, use of slow-cook earth ovens) and hunting methods. Again, development of conscious self-awareness was thought to grow hand in glove with increased efficiency of energy extraction from foods in the body.

Around 50,000 years ago: The brain of *Homo sapiens* evolved to such a point as to enable the development of art, religion and complex technologies (such as boat making) for the first time. Perhaps the question is why didn't these changes occur earlier in the 150,000-year history of *Homo sapiens* to that time? Norman Doidge, referencing the work of Steven Mithen, suggests that the change in "cognitive fluidity" took place in the brain of *Homo sapiens* with the development of its neuroplasticity. It was this neuroplasticity, he suggests, that enabled the linking of various brain modules for "natural history" intelligence, "technical" intelligence, and "social" intelligence together for the first time.[5]

The important suggestion of this section is that *Homo sapiens* did not just spring up without antecedents. Perhaps more poignantly,

Christopher Hitchens casually asks the question in a live debate, "Didn't God care about us before 50,000 years ago?" This section supports the idea of modern man's gradual speciation over a very long period and more importantly, that consciousness evolves, initially with diet and digestive system changes if Wrangham is correct. This also suggests that with growing consciousness and environmental awareness (including awareness of our human-made social environment) the comprehension of values evolve, as do legal precedents in courtrooms.

Does this also mean our destiny evolves? To the extent that our chosen values can shape our potential or degrees of freedom, I think so. Further, to the extent that our destiny depends on our emerging systemic abilities or freedoms within this environment, I think so.

The Development of Language

As suggested in the previous section, the evolution of consciousness was not just a one-off trick of nature. It was a slow process of improvement over 2.5 to 47 million years that has not finished. Somewhere in this period we also advanced from gestures and body language (centred in the older brain regions such as the hind brain and mid brain) to verbal language (incorporating the neocortex and fore brain). Similarly, *Homo habilis* began the process of toolmaking some 2.3 million years ago in stone, which eventually advanced into toolmaking in iron during the Iron Age around the 12th century BC.

Like a network-distributed computing system with centralised servers doing the bulk of number crunching, and intelligent client terminals doing locally-relevant computing tasks such a screen, keyboard, local memory, and local disk control, so evolution taught and slowly developed the trick of distributed computing or consciousness. That is, the genes in the DNA of each cell (equivalent to a centralised server) learned to offload some of their computing power and life-organising tasks to the genes' agent or survival machine, the mind-body complex (equivalent to an intelligent client terminal) by granting the mind-body, including brain and nervous system, consciousness at higher and higher levels.

Why? We can better exploit spacetime's components of here-there and past-now-future (or make energy usage more effective and efficient) if our survival machines have dynamic means of dealing with conditions rather than just fixed genetic responses to conditions. Basic survival is essentially a matter of surviving the here-and-now repeatedly, but better survival techniques also include predicting what is going to happen next. Instincts in all life forms were good at dealing with the here-and-now in a steady environment, as were the over-arching programs of the old centralised mainframe computers of government, banking, and insurance entities. However, instincts were not very good at exploring and remembering the distant past in time to plan and strategize for a distant future. Those old mainframes were also not good at catering for, or adapting to, emerging local needs. Distributed network computing and consciousness, on the other hand, became more and more expert at this enterprise – they became made-to-order exploiters of time's apparently non-fleeting past/future components. This enabled both modern computing and consciousness to use energy more efficaciously.

If this view is still a little anthropomorphic, then consciousness, through stumbling upon the trick of forming an intersubjective sense of past/future, was able to greatly enhance our survival prospects. It gave us our eco-psychosocial time – even if this was a small-minded approximation of full reality's wide-ranging actions. Consciousness thus transformed our relationship with space and time. It seems the power of spacetime, instincts and consciousness simply emerged almost inevitably and asymmetrically together without any need of prior purpose.

Wrangham suggests that the brain grew in size and functionality as food-energy usage was made more efficient in simpler or more time-efficient digestive processes. He also suggests that greater brain power enabled us to use food preparation methods that were more energy efficient. He thus suggests the two properties, brain effectiveness and digestive system efficiency, were interdependent. That is, more effective energy usage in the brain was correlated with more efficient energy usage in the digestive system. However, this wouldn't have happened if each brain-body refinement didn't make the developing species better adapted to its environment. That is, there had to be a win in total energy effectiveness and efficiency inside and outside the body. Just as modern client-server computing after many advances, is much more efficient and effective than the central mainframe computing of the 1960s, so brain growth heralded a very successful series of adaptations for itself and its eco-psychosocial environment. These adaptations are likely to continue to develop into other species (such as in our progeny, but also in the progeny of gorillas, chimpanzees, our pet dogs, and so on), just as in the past (as in our Neanderthal cousins for example).

The body language and social skills we had learned as animals slowly had consciousness added to them at higher and higher levels of sophistication. Humble at first, eventually the result was the transition from body language to verbal language, with necessary changes to the physiology of the mouth and vocal organs as well as the brain. More correctly, we added verbal language to our pre-existing instinctive and animalistic body-language repertoire.

In many societies, but not all, there has been some time later the further, and just as startling, genetic and memetic evolutionary addition of written language to verbal and body language. Written language, as a kind of non-trivial phenotype, was a way of greatly enhancing our species' memory capacity and thus our capacity for conscious activity. Many societies enjoyed the advancement to written language not because of an internal evolutionary development, but because of cultural interaction (memetic development either invited or imposed) with other societies. Very few extant societies have not yet made the advance to written language.

The question here is; what happens when we add conscious self-awareness to the body and its instincts for gesture, stealth, deception, seduction, and revulsion, used to bombard the senses of friend and enemy in the local environment? What happens when we add consciousness in the intelligent client terminal, to the other instincts for pecking order at the waterhole? What happens when we use conscious-but-perhaps-ignorant modifications of these instincts for or against our own kind, or our own society, in a manner that was originally unforeseen by our genes in the 'central server'? Perhaps the natural result is what we more commonly refer to as the knowledge of good and evil, although this classical answer does not really explain much about the real workings of consciousness. We will return to this point later.

Perhaps the more important questions is, was the evolution of consciousness, or the offloading of computing power from genes to the memetic mind-body complex, an evolutionary mistake, just like the mistakes of all those hereditary lines that have disappeared from the face of the earth in the past? Did the 'consciousness experiment' cause the mind-body complex to react with instincts in an unintended and blind fashion that has ultimately damaged our fitness for survival? It certainly seems possible. Memetic evolution discussed in Part 1 was a blind spin-off of the genetic development of consciousness. This has put us in the grip of an even more powerful replicator than genes – our memes. However as Dawkins suggests, we, as moral and rational beings, can rebel against the oppression of those short-sighted genes and narrow-minded memes.[6] We can do this via our vremes.

Right now, many of us today are unhappy and unfulfilled. Like in the movie *The Matrix*,[7] are we humans the virus multiplying so rapidly that soon we will destroy our beautiful planet orbiting in space, and ourselves along with it? Is the old religious fable of original sin not too far off the mark - not a single apple, but certainly our diet, having a huge impact on our current state of affairs? These are interesting questions, certainly not fully resolved in the movie. Clearly, the heroic answer to our predicament lies in using vremetic replication to quickly develop our conscious awareness of all the needs of our planet and their possible solutions. I trust the framework building in these pages will guide this crucial process of vremetic replication.

The next step is to consider the nature of consciousness more fully.

We are Inquisitive

Dietary improvement alone is perhaps not quite enough to directly explain our advancement of consciousness. Our instinctive drives for survival coupled with our consciousness have drawn out and developed our inquisitiveness, which in a continuing cycle has engendered a higher level of consciousness. We are innately inquisitive creatures; we want to understand, just as our young pets want to learn new tricks and behaviours. We need to believe, explain and understand – it is our essential means to our survival. It seems there is an innate and implicit desire to be energy/information-aware, -effective, and -efficient. That is, we seem to have an in-built appreciation of space-matter relativity. We quite often need to apply lessons from the past in order to survive beyond the next moment.

I am suggesting here that our consciousness, if we could separate it from our instincts, is not inquisitive but passive. Consciousness dispassionately grants the human mind-body system both direct intersubjective understanding and more indirect objective understanding, if we use it, that is, if we think and assess in the context of our environment and with the aid of that environment. Yet even here, as we have already seen, there are complexities. Within somatic boundaries, from which unconsciousness and consciousness both arise, we could perhaps argue that we can have an indirect objectivity and a direct intersubjectivity simultaneously.

Further, to try to define a dividing line between our consciousness and instincts or our rationality and our emotions/feelings is also problematic; much like it is often difficult to separate the phenotypic effects of individual genes. It may have been more practical to refer to what I have called consciousness-plus-instincts as something like System2-plus-System1 to avoid all such controversy. Daniel Kahneman, in his book entitled *Thinking Fast and Slow*[8], employs such a method with great style. Nevertheless, the division is not of overwhelming importance here because it is the asymmetrical and interpenetrating dance of the notional pairing, which is of immediate effect. Even so, it seems a useful paradigm at least to suggest that perhaps consciousness springs out from a place close to where bodily activity ends and passivity begins. I suspect this would often be where the body's homeostatic systems are comfortably operating within practiced limits.

The first way we made use of our new consciousness (and neo-cortex) that we are going to consider, is the more objective, extrospective and ideally scientific approach, corresponding with Table 2: The Positive Value-set. We will consider the more intersubjective, introspective and contemplative approach of Table 4: The Self-enlightened Value-set later.

The Objective Stance: Understanding our Environment

In order to survive beyond the next moment, we needed to extract patterns or principles of our environment from experiences. Thus, the first thing we sought with our growing conscious self-awareness was an explanation of our natural environment. As an exercise in finding an explanation of self as object, we also wanted to explain 'self' versus everything else. Yet it is only recently, since about World War II, that we as a global civilisation have sought to explain our natural environment including the human species from a more strictly objective point of view (apologies to all those who courageously paved the way before WWII).

Objectivity requires that we are outward-looking, dispassionate, and disciplined with respect to our problem-solving explanations and inquiries. It requires that we, including our fellow traders, divorce ourselves from our impassioned, instinctive and personal biases. This denial of personal feelings or exclusion of personal sensibilities in the interest of a cold, third-person kind of effectiveness and efficiency is the reason why the more objective stance corresponds with the lose-lose mentality of Figure 12.

More objective scientific enquiry is the ideal or well-disciplined form our outward-looking consciousness can take in its interactions when it is working at its best. Conversely, intersubjectivity is the unavoidable outcome of an inward-looking consciousness.

All our reliable observations of reality, new and familiar, and their integration into new or modified concepts of our own choosing, are garnered using our more dispassionate and outward-looking consciousness in many interactive and algorithmic feedback loops that reject and eliminate discovered error.

I am not suggesting here that we gain new scientific knowledge from an active inductivism. The approach is more like the haphazard evolutionary approach of our genes. That is, in scientific endeavour a problem or opportunity to add value is 'identified' within a certain niche, a solution or mechanism is then 'suggested', first through a critique of many competing possibilities, and then 'tests' of the best explanations' predictions are conducted. Surviving new explanations or theories then replace the old, outmoded explanations until something even more successful arises. Apologies about the allusions to evolutionary purpose here through such words as 'identified' and 'suggested'. As noted many times before, we tend to assign intention to nature's emerging statistical patterns in relative vastness and disorder because we continually use the same pattern-detecting skills to assign dangerous or friendly intentions to living agents, such as to lions or other humans.

David Deutsch[9] tells us we can somewhat equate a new theory or explanation that addresses a scientific problem to a species or gene occupying an environmental niche. The explanation's survival at the hands of its critics is thus as the survival of the gene, and the theory's temporal recognition in the scientific community is like the temporary survival of the relatively fittest species. More generally, survival of knowledge as generally accepted truth encoded in a theory is like a species' temporarily successful adaptation. Mutations of theories or genes would likewise be analogous.

However perhaps the realms of epistemology (or justifiable knowledge) and evolution are more than just analogous. Maybe they are both part of the one process of emergence going on in our universe. A discovery of new knowledge requires new explanations – new understanding that fits the temporal facts. This new understanding is emergent – it is a new emergent property that exists more in the arrangements of virtual particles than real ones.

Let me explain: The interactions of hydrogen and oxygen particles have to be just right to make water. The properties of water cannot be enjoyed any other way. The properties of water are 'platform dependent' - change the material platform of hydrogen or oxygen and the properties change. No other liquid is wet quite like water is wet. In this example, the interacting virtual particles in the water (photons, fleeting bosons, gluons and theoretical gravitons) work towards dynamically arranging the necessary material platform.

However, in the case of knowledge, the opposite is the case. The reflections and interactions of data (the virtual particles) have to be just right to make and store useful information or knowledge, but the material platform (the arrangement of real particles) is not the focus. The properties of knowledge are fully 'platform independent' in a material sense, but fully 'code dependent' in the opposing, immaterial sense – change the nature of the data or spacetime properties in varying observations and the knowledge-pattern crumbles. In this case, the material platform of the brain or computer, works not towards material ends or properties but towards arranging the necessary data or spacetime platform that holds the immaterial knowledge in its necessary arrangements of immaterial virtual particles, potentials, and/or code.

Pieces of knowledge do not yield understanding either. Understanding comes with the active process of fitting a theory, explanation, or belief to the limited facts or values. How many facts make a theory? How many bricks make a shelter? The one does not follow the other in an absolute sense. Theories, explanations, understanding and shelter emerge through repetitive and probabilistic consideration of the vying arrangements of facts and bricks, respectively. Understanding is a refining process more than an adding process, much like the sculptor reveals a statue of David from a piece of rock.

We are the facilitators of this emergence. We are the agents, reactants, or catalysts in the process. Our minds' heuristics arrange the intersubjectively relevant facts or values and drop the irrelevant platform-dependent ones to come up with a new understanding. Those same heuristics also drop some relevant information due to the need for closure or boundaries, just as some sculptors can chip away in the wrong places to save time; hence the necessarily emergent or evolutionary nature of understanding and the revealed skill in a sculpture such as Michelangelo's David.

As already suggested, perhaps it is ultimately not the genes or the scientific explanations that are selfish (seeking their own survival within the harsh environment) but the knowledge of the niche they each encode and seemingly exploit in their own ways. This is the knowledge life embodies on the one hand and consciousness embodies on the other. The encoded embodiment of knowledge is the way life, scientific theories, and our concepts of consciousness share full reality's selfish web.

The Extrospective Self-restraint

The natural sciences and formal sciences (i.e. mathematics and logic that deal primarily with platform-independent facts) are based on very strict rules, including those we briefly discussed in Part 1. These rules concern careful observation, falsification, differentiation and integration. These sciences, like our classic outward-looking consciousness, follow a positive approach because they focus on 'what is', either directly in terms of matter or indirectly in terms of spacetime and logic. Outward-looking consciousness does this, rather than focus on what ought to be (the normative approach), or what in an environment of limited information might be (the risk-weighted approach), or what is otherwise imagined (via an inward-looking consciousness).

The natural and formal sciences' parameters are the physical and logical theories and rules that currently best describe this universe. They also emerge. They are not what some might refer to as the universe's beyond-physical or beyond-logical concepts that perhaps kick in behind the veil or beyond the limits of natural/logical self-restraint. We shall consider these unsubstantiated, assumed or imagined concepts in the next section.

Natural science is a study of matter and the energy and interactions that co-exist with matter. It is a study of the constraints of the natural universe. The study and exploitation of the patterns of the natural universe is the most basic form of self-preservation and self-organisation our consciousness possesses.

It is through the repetitive use of observation and logic that we can identify and establish our tentative knowledge without significant contradiction. Clearly, if we do not use our evolved Scientific Method and allow the values of Table 2: The Positive Value-set to guide us in our extrospections, then contradictions may abound – to our peril.

In the stock market scenario, the positive mentality (or applied outward-looking consciousness) is evident when analysts use fundamental analysis to make typically more long-term investment decisions. That is, we use what we know about the assets, earnings, and cash flows of firms over a period of time sufficiently long for the detection of 'firm demographics' to stand out from the 'noise', as well as to learn what these demographics may imply about the long-term future earnings capacity or potential cash flows of firms.

The third self-restraint of values (and vreme development) is the positive extrospective interaction.

The Assurance of What We Know

In summary, the so-called objective stance (or the best-validated findings of our outward-looking consciousness) assures us of what we know with respect to the past. The more objective truths we discover and experience can give us confidence with respect to our past and present choices, purposes and passions. However, they do not offer an opinion as to what we yet do not know, what is intersubjective, or offer any absolute claim as to what might be in the future.

In this one statement, we see the basic strength and weakness of the natural and formal sciences and by implication, our outwardly-interacting consciousness. These sciences as they stand today, while they provide us a tentative assurance of what we know, cannot alone lead us out of our global values-dilemma and so cannot alone restore our waning sense of purpose or meaning. Inadvertently, they can even contribute to a multiplication of the strength of this values-dilemma.

Stuart Kauffman, in his book *Reinventing the Sacred*[10] suggests just that – our civilisation desperately needs to reinvent a sense of the sacred, however not as something supernatural but rather as something brilliantly and beautifully natural.

The Subjective Stance: Softening the Hard Problem of Consciousness

Dennett (1993) and Pinker (1997) strongly argue the case for the computational brain, made up of many interlocking modules. Its subroutines are involved in processes that compete through a kind of democratic pandemonium for their influences over our emergent consciousness. Through the models he presents, Dennett and many others, such as neuroscientists Antonio Damasio (1994 and 2010) and Professor Baroness Susan Greenfield (2009), suggest that the brain has no central point of consciousness. They suggest that rather, consciousness emerges over the whole of the brain as fleeting manifolds of neurons of different shapes and sizes. The bigger the manifold, the more neurons enlisted, and the deeper the consciousness. They seem to describe consciousness as an emergent property of the brain,[11] perhaps much like shelter emerges from the purposeful arrangements of bricks or the wetness and temperature of water emerges from the collection of a sufficient number and condition of hydrogen-oxide molecules. So consciousness is said to recursively arise out of vast, complex and massively parallel computations alone.

Those that have difficulty with this view seem stuck on only one issue. Even our most sophisticated super-computers are not conscious – they lack intentionality and subjective self-awareness. They argue that consciousness cannot arise out of computational ability alone. For some of his readers, Dennett does not seem to convincingly explain the answer. Consciousness, Dennett suggests, is an illusion; it is just the result of the sum of the parts of the interlocking modules and their massively parallel, complex, and deeply nested computations.

Here again, I see the classical, uni-dependent 'cause-and-effect' way of thinking running into the relativistic, multi-dependent, fractal, and emergent way of thinking.

Consciousness

While we all probably agree that the brain is computational in some sense, like Sam Harris, I too feel the subject needs to be explored more deeply. As Harris says, "Consciousness is the one thing in this universe that cannot be an illusion".[12] I'm not sure I fully agree with the idea (i.e. because I have a first-person or intersubjective experience of consciousness it must be so), but we don't see a creeping consciousness overtaking our computers as they become more sophisticated either, and this needs to be explained. This is the "Hard Problem of Consciousness", as David Chalmers once coined it. Chalmers defines consciousness as the subjective sense of knowing what it is like to have experiences, including an experience of self. That is, the hard problem of consciousness is not about listing the correlates of consciousness, such as larger neuronal clouds being associated with a deeper sense of time, space and self, but the actual sense of self, or the actual feeling of time's passing.

Knowing what we do of the relativistic nature of our universe, the phenomenology of human perception, and the evolution of human consciousness from past hominids, can we at least turn this hard problem into a softer one? Firstly, the hard problem should not be about the uniqueness of perspective we each enjoy in our here-and-now, because even a rock has a unique perspective on the world not enjoyed by anything else outside of its place and time. The hard problem should be about understanding the subjective experience that is common to all of us, regardless of particular perspective. I would prefer to call this our intersubjective experience so that we can put the issue of first-person uniqueness aside as a well understood characteristic of perspective.

Our computers are not relativistic devices that finely distribute control with function; they are driven by an artificial master clock - 'tick-tock'. That is, instead of carefully handing proto-intention all the way up through layers from the very lowest to the very highest so that a top platform of intention can be efficacious in gently bending all layers below it to conform with its purpose, the typical computer insists on a fixed-design control policed by the master clock. Programmers can work within the device's design parameters, but they can't meaningfully step outside it without, for example, an upgrade to a new processor. However, even with a new processor, programmers are not free to exploit other potential resources. The computer is a dumb device that cannot evolve given its current design features.

John Searle was right when he said understanding is not the same as computation alone. Even if a computer can speak perfect Mandarin, but has no intentional self-awareness of the fact, it nevertheless cannot understand Mandarin.[13] It seems that there has to be a feedback path, or an intimate sharing of intended control and functional computation. There has to be an interpenetration of instruction and output, or space and matter, for consciousness to be. We have to commune with what we perceive, make a directed inward reflection on what we perceive, and then make a pertinent and purposeful arranging and interrelating of knowledge within a conceptual system or code, for understanding to be possible.

As stated earlier, understanding requires a prior belief, hypothesis or reference point in place that arranges, relates or makes sense of the new information at its inputs. Currently we don't design our supercomputers to do this inward-looking reflection and self-organising that our brains seem to effortlessly achieve through those oversized human frontal lobes. However, as we start to value more the interconnecting possibilities that can only arise from 'chaotic distributed control', I am sure we will begin to ask our computers to perceive, self-reflect, self-direct, and self-organise via proto-intentions formed at all levels from relativistic hardware configurations and machine codes, upwards.

So how might the brain achieve an intersubjective understanding? I agree with Dennett and Pinker that the brain does not need a special introspection module. However, perhaps the core principle of General Relativity means it does need a multi-layered asymmetrical feedback and feed-forward process between its modules that enables its extrospections to gain a certain kind of depth. The process would transform extrospections or sensibles into essential styles or introspective images that can be intentionally manipulated.

Sentience seems to require a certain kind of overlapping signal within the brain and body, just as the appreciation of musical depth needs a stereophonic signal, which supplies a different signal to each ear, rather than just a monophonic one. Similarly, the appreciation of 3D visual depth from a flat screen TV requires a stereogram, which supplies a different signal to each eye.[14] Understanding surely needs a mirror, a reflection, a kind of dynamic 'self as object - self as subject' double signal or schema within the brain's information-access processes. In short, understanding requires a kind of mental relativity that a uni-dependence can never achieve. The mind/brain does not just compute – it computes and vastly correlates. It interacts and organises almost simultaneously, just as does matter/space.

Figure 16: 3D harmonics (3D glasses removed)

Just as when we take off our 3D glasses and look at the harmonic reflections of images on the TV screen (see Figure 16), if we were able to take off our consciousness 'glassware', we would see the harmonic reflections of perceptions, emotions, knowledge and values that must coalesce to create the images and feelings of our sentience and understanding. Consciousness must coalesce across many self-similar dimensions or relativistic schemas. That is, consciousness is fuzzy, fractal, and emergent, just as is the rest of the universe. I think this is similar to the picture Damasio painted in his ground-breaking book, *Descartes' Error*, through his descriptions of the multi-layered feedback and feed-forward loops within the brain-body's systems that lead to mind and reason. Descartes' error was that he believed the opposite, that is, the soul was independent of the body and thus, "I think therefore I am".

Damasio further details his profound findings and ideas in his book, *Self Comes to Mind.*[15]

This image-building and sentience-building layered feedback process operating across the mind/brain/body has a lot to do with how the mind/brain processes time using its frontal and temporal lobes. The mind/brain builds multidimensional maps or schemas of reality, and within those schemas is the notion of past and future, even though these things do not exist in actual experience (we can only ever experience now). This means that while we could say the fleeting notion of now/spirit/instincts discussed in Part 1 actually exists as an outworking of genes and the body, the notions of past/future or soul/consciousness only exist in our minds, indirectly; they do not exist outside of our mind-images except as shared or intersubjective ideas. They only exist indirectly in full reality, which is a reality that is able to notionally split and extend the two interwoven components of relativistic reality. It is the multidimensional arrangements of consciousness, across the intersubjective notions of past, present and future, which provide those carefully overlaid harmonic reflections of facts and values that make up sentience and our understanding of reality as they dynamically coalesce.

This understanding of reality necessarily includes an understanding of the indirect self. That is, we tend to call the passing build up, awareness, attenuation, and storage of overlaid schemas of spirit/instincts something separate, called soul/consciousness. This passing image-awareness is very much like the running of old talkies, each of which we can break down into so many 'lifeless' frames per second. The contents of consciousness do not constitute reality itself, but rather a coloured representation of reality captured in something like a lifeless movie reel. However, consciousness, by literally process-modelling perceptions, made the implicit behaviours of our instincts partially explicit to its captive audience (that is, itself), just like the old newsreels informed old-time moviegoers. Additionally, just like what those moviegoers did not see in the real battlefield, so likewise we do not see all the activities of our autonomic nervous systems, etc. Nevertheless, these approximations of reality provided us with a framework within which we could reflect on further inputs to our senses.

However, much more than as in a talkie, instincts seemed to grant this self-consciousness a modified but somewhat seamless intentional stance we call purpose as well, which when merely connected with our genetic instincts we called drive, or I called a level of proto-intention. That is, this processual awareness of self in its environment was not just made of audio-visual snapshots, but a multidimensional range of intentional behavioural snapshots as well. These included the proto-intentional processing of passing smells, passing tastes, passing touches, feelings of balance, etc. This reel of consciousness also included interactive re-runs and feedback on itself, just as we can choose a different ending, view a scene from a different angle, or enjoy an overlaid commentary, on some Blu-ray movies. I think Douglas Hofstadter (2007) probably explains this view most clearly and completely in his highly recommended book, *I Am a Strange Loop*.[16]

If we were to summarise all the snapshots that make up consciousness, we might say that the greatest trick of consciousness-plus-instincts, mirror neurons included, was to extend or divide the present into the past and future. In so doing, consciousness was able to transform our static 3D pictures of the world into a 4D movie or process model, our sensations into experiences, information-incorporating significations of mind into language, the object of self into the subject of self, and genetic drives into the notion of free will. These interacting algorithms turned out to be very effective survival mechanisms indeed. They formed a base upon which all the other intersubjective nuances could grow, including (and perhaps most fundamentally) the personal sense of moral agency.

In summary, the findings of Einstein, Merleau-Ponty, Hofstadter, Damasio, and many others, when brought together and taken as a whole, serve to soften the hard problem of consciousness. According to General Relativity, space affects matter to help produce all its arranged forms, and every tiny aspect of matter affects space. Is it therefore not obvious that space has at least as many forms or arrangements as matter? Why do we marvel over the many forms matter, but not equally marvel over the many wonderful arrangements of space in everything with which we interact? Once we take this deeper view of space, in a way something like we see the green symbols of *The Matrix* advance before us on the movie screen,[17] it becomes easier to think of the mind or consciousness in their own terms rather than through the correlates matter provides.

It seems to me that commentators on the nature of consciousness may sometimes be too quick to claim that there will never be a neurophysiological explanation of it, although I do understand their reservations: The raw experience of sentience, just like a grasping of nullity, infinity, or the square root of minus one, does seem to be a hard problem. Consciousness is mind-boggling when we think about it as a static object, using our classical, uni-dependent way of thinking. A large part of the problem seems to be that intersubjective consciousness is processual, virtual, and transient – just as those transient and asynchronous harmonics transform 2D movies into virtual 3D ones.

However, why limit ourselves on this score? Theoretical physicists, when speaking of virtual particles, certainly do not hold back when it comes to explaining indirect and transient reality through missing energy. Would it be better to call consciousness a relativistic, virtual, indirect and transient interaction like gravity, rather than an illusion, to arrive at a testable model of emergent consciousness? Would it be better to call the mind/body relationship that gives rise to consciousness 'relativistic, fractal and fuzzy' rather than an 'enigmatic thing' or a 'hard problem'? I hinted at a similar model for time itself in Part 1. That is, wouldn't it also be efficacious to call our species' sense of time a virtual, indirect, transient and emergent phenomenon rather than a mere psychological illusion? Finally, wouldn't it be better to call sentience a phenomenon that emerges through many layers from the vast and parallel interrelationships of environment, body and mind?

Maybe we need a string theory of consciousness-plus-instincts (strings, if you recall, are the gooey representations of combined real and virtual subatomic proto-particles). The string of consciousness-plus-instincts would be a sticky, oscillating and slightly asymmetrical and asynchronous combination of real instinct elements/standing waves in the near-present and virtual consciousness elements/harmonic reflections that we interpret or locate in the not near-present past/future. These elements seem to statistically move towards a collapse-avoiding and diversifying equilibrium, much like our universe as a whole and Earth's ecosystems generally.

An approach Pinker takes is to say that consciousness is difficult to pin down, not because it is an illusion, but because our scientific modelling cannot get to the nub of the issue.[18] In effect, I think Pinker is saying that all our historically-based models to date, in the classical areas of psychology and philosophy, suffer Gödel-like limitations, which render them largely uninformative in this sublime area of knowledge. If this is right, then without intending to invoke transcendentalism, it would suggest that we need to understand consciousness in an existential manner as a flowering of the emergent, self-containing, and self-referencing mind/body/environment system, rather than as a classical, uni-dependent, engineered, and constructed sub-system of the mind/body that we can simply reverse-engineer through reductionist efforts.[19]

For example, our efforts to understand and define shelter would not be helped much by removing each brick in order to go and study it alone. Just like the central problem of the path of a photon in quantum physics, we need to contrast the very big with the very small in an inclusive framework. We need to think in a relativistic manner. We would often be better off to leave every brick in place and study the dynamic styles of the arrangement as a whole within its situated/local environment. The bricks simply provide shelter's necessary material platform. In studying the dynamic and situated arrangement we focus on (brick- or platform- independent) information about the interactions embodied in the physical design that grants the arrangement its essential style. From here, we can develop a kind of processual 'shelter schema', just as the mind/body forms its perceptual schemas of the dynamic contents of its inner and outer worlds. Thus, shelter can be essentially stylised as any arrangement that bends wind, deflects rain, changes local air pressure, and attenuates the heat of the sun.

When it comes to studying consciousness, we need to let go of our classical senses of cause-and-effect, independent self, and uni-dependency. Likewise, there is no need to get hung up about shelter as an unfathomable, indefinable, enigmatic thing; it is simply a non-physical interaction between things (the wall, its environment, and our senses). Shelter is an exploitation of the interactions between things. Focussing on the last brick that makes a shelter was always a straw-man argument. We need to appreciate the mind's bricks (i.e. the very small neurons, cortices, and so on in the physical brain), but also appreciate the processual arrangements of its bricks that lead to its architectural style. This in turn leads us to an understanding of how the mind exploits the interactions between things, or the mind's salient property in the realm of the whole mind/body system and its environment. Similarly, we might study the guitars, keyboards and electronic effects employed by musicians in a band, but it is only as we listen to their music and see the audience response that we detect the salient genre and feel the band embodies at its highest level of operation.

When we speak of feel or interactions, we speak of arrangements of the virtual particles of space. A study of consciousness is thus a study of the arrangements of space in the midst of matter/body/brain. We already follow an interactive and relativistic approach to some degree through our study of the behavioural abnormalities of human nature, but we have great difficulty in relating our findings to neurophysiology, except in terms of indirect 'correlates'. We need to get away from being satisfied with an ever-growing list of the correlates of consciousness, which leaves the hard problem of consciousness intact. Perhaps the approach used in Process Physics to model and explain the emergence of space and matter in our universe[20] could also be applied and extended to model and explain the emergence of consciousness. This process-modelling approach, which would mimic how consciousness models its internal and external environs, could run in parallel with other efforts such as the Blue Brain Project, which is an attempt to create a synthetic brain by reverse-engineering mammalian brain circuitry.[21]

Free Will, Free Won't and Evitability

Closely related to the so-called 'hard problem of consciousness' is the issue of determinism and free will. As we will see, closely related to both issues is again the concept of relativistic emergence. Determinism suggests antecedent events determine all future things. To determinists, it would not matter much whether these antecedent events were linearly deterministic or every-now-and-then quantum-mechanically stochastic – and many would probably agree 'at first blush'. To hard determinists this means there can be no room for free will if all effects are caused by prior events. In effect, the argument suggests that if there are nil independent causes, there can be no freedom, harking back to what I believe to be, by the Uncertainty Principle, the faulty ancient Greek idea of a Prime

Mover, the First Uncaused Cause and most saliently, uni-dependence.

Hard determinists would also add that while humans can make choices, they cannot bear a moral responsibility for those choices because they were locked in by antecedent causes. If hard determinists are right, then my basis for free-will and self-actualisation via the Emergent Method would be undermined.

On the one hand, self-aware consciousness seems illusory and on the other, antecedent cause-and-effect seems to drive everything we do. On top of this, many neuroscientists quote the findings of Libet's experiments, in which "researchers recorded mounting brain activity related to the resultant action as many as three hundred milliseconds before subjects reported the first awareness of conscious will to act".[22]

That is, they assure us that the subconscious brain kicks in before any conscious neo-cortical functions can have an effect. To support this idea, Figure 17 shows the rise of the readiness potential, RP, about 350ms before the awareness of intention to act in consciousness, which, in turn, was about 200ms before the action. If you accept this observation at face value, then what we perceive as a conscious decision is totally a subconscious one, only claimed to be conscious by the ego's narratives.

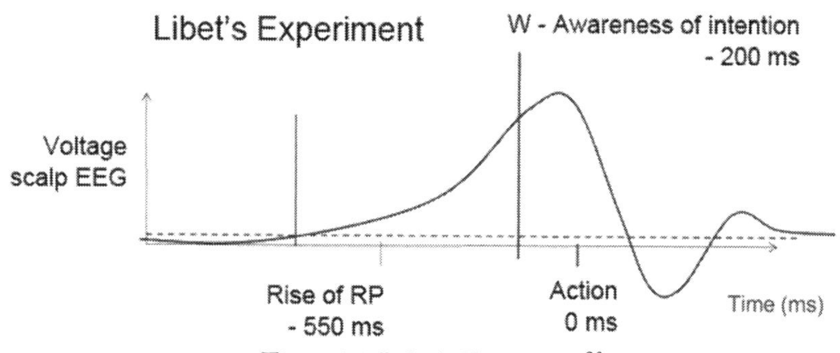

Figure 17: Libet's Experiment[23]

In contrast, according to Dennett and Libet himself, we have "free won't". That is, consciousness's acquiescence is often required for a subconscious decision to flow into action – just as the scientific community acquiesces to a scientific theory by firmly rejecting any competing errors. However, even this notion has an obvious flaw to the hard determinist, because both consciousness and subconsciousness are set by antecedent events. So what room can there be for conscious free will, human moral agency, and non-illusory introspective consciousness from a neurophysiological perspective?

We noted in Part 1 that Dawkins (1990, Chapter 11, Note 8) supports the idea of a soft determinism. That is, he is clearly someone who thinks we can take moral responsibility for our futures, through "conscious foresight" or "our capacity to simulate the future in imagination".[24] This imagination is also the essence of the benefit of consciousness - as Einstein once said, "Imagination is more important than knowledge." Likewise, Dennett describes a neurophysiology in which a kind of soft determinism, or an evolutionary compatibilism, is possible. He suggests we can have determinism and a version of free will as well, although at the cost of a possibly illusory conscious self-awareness.[25] I would suggest that Dennett's approach can support a view of conscious self-awareness that is not necessarily labelled as illusory, but indirect, virtual, shared, and intersubjective instead.[26]

This would be a view in which our actions are dependent on our environment. We build a mental map of that environment in our brains, attenuate certain features of it and amplify others, perhaps in conscious anticipation of likely futures. As we apply our values to that modified environment within, merged with our conscious observations, it becomes our own local reality and an emergent cause of our future subconscious computations and behaviours. That is, as we through many iterations or Hofstadter's strange loops of self, act on the largely self-generated and self-referencing intersubjective mental model rather than actual reality, we find some level of independence or pseudo-freedom within reality, something like a mountain within a mountain range, which I prefer to label emergent rather than illusory. However, there is more…

I suggest that this emergent place for freedom is provided to us from the self-restraints of the body's homeostatic systems. These largely unconscious systems seem to say to subconsciousness and consciousness, 'you can do whatever you please within these upper and lower limits, but if you stray towards these limits, we are going to jolt you back towards our middle paths'. Actually, I suspect the operation of the body's homeostatic systems is fuzzier, that is, there is a continuum of biological influence that tapers off to its lowest level at the point of the brain-body's optimal environmental operation. I also suspect subconsciousness in turn has its own set of psychological influences that drive us to its middle paths of operation. Finally, I suspect consciousness has its own set of sociological influences as well, which drive us towards its middle paths of operation.

Thus, what is left is a place or island of perhaps near-optimal operation over which there is little behavioural preference coming from our underlying systems of triune self. Here within this oasis there is room for probabilistic and unique behaviours within bell-curve limits. Note: I don't make an appeal to quantum indeterminacy for this probabilistic behaviour to operate. This unique behaviour simply relies on unique time and place. However, the uniqueness of every member of every species ultimately relies on quantum indeterminacy.

It is here, within the locally deterministic operating boundaries of our triune systems of self, that unique personality, making unique decisions, can emerge with age and experience. It begins as more bland personality traits in childhood and becomes more colourful in adulthood. As this unique and initially probabilistic behaviour emerges, it affects those conscious, subconscious and homeostatic systems, and their limits. In this fashion, a partly indeterminate uniqueness fully governed by reality's interpenetrating monism gives rise to our unique personality and intersubjective moral agency, just as it has also given rise to our universe.[27]

Correspondingly, the naive idea of a First Cause is equivalent to the insignificant idea of a child's First Moral Action. In both cases, many effects must be stirring in a relatively meaningless statistical vastness for meaning to arise and a cause to be defined. Free will arises because it begins as the fuzzy probabilistic effect of a local monolithic order (or homeostasis) that is indistinguishable from disorder. That is, free will is part of an emerging Stochastic Process System (which incorporates the mind/body and consciousness).

I would agree that of themselves, unique behaviours within ideal bell-curve limits do not equate with the common notion of free will. However, this uniqueness is the basis of the eco-psychosocial claim of free will, that just like any other such claim, is transformed into a more objective legitimacy by its worldly success. That is, the explicit worldly value of our claims to consciousness, self and freedom is the basis and evidence of our moral agency and all that this entails. Thus, free will requires a transformation from unique, self-aware claims to self-aware success through a kind of environmental acquiescence. This environmental nod of acceptance is our necessary partner in our emergent agency, just as the acquiescence of consciousness is the necessary catalyst of our subconscious RP (readiness potential). The very first effect of emergent agency in the dynamic sea of relative monolithic disorder was possible because there was nothing around it in the local region to stop it.

In short, our interrelationship with the environment at various levels and through various emergent loops enables our moral agency. We don't have an independent freedom from the world, but rather an interdependent freedom with the world. The relativistic nature of reality means it can't help but share its governance, or distributed command and control structure, with us. We cannot make our claims to self in an environmental vacuum, and nor can we claim a self if we cannot observe ourselves within an environment. This is an extremely important point. There is a 'separation of powers' between us and our environment, which enables their combination to develop and interpenetrate. Without this relativistic separation of powers, no emergence of new properties is possible at any level.

Further, in the case of human agency, all that was initially required was the emergence of a systemic loop of self-observation and self-reflection. I think Damasio (2010) describes this process brilliantly.[28] However, 65 years before Damasio's neuro-scientific descriptions of thought processes, Merleau-Ponty's existential descriptions of thought processes were there to guide the way.

We also need to be free enough in our self-aware 'inner world' to make mistakes, or faulty claims, and to learn from those mistakes after environmental feedback. Thus, the 'self' expands and emerges just as the universe does so, and for similar reasons (i.e. trial-and-error and least-action movements towards energy and information efficacy). In both cases, it is because there is an asymmetrical and interpenetrating dance between that which tells the other how to 'do' (space and our environment) and that which tells the other how to 'be' (matter and the self that claims its free will).

The expansion and emergence of self is not essentially within steps of Planck Length and Planck Time, but steps as small as Libet thought-gaps. It is the delay of the Libet thought-gap that enforces the 'separation of powers' between environment and self, much as the speed of light enforces the separation of powers between transient matter and space in quantum fluctuations. The Libet thought-gap thus enables the free will of self rather than denying it. Our partial success in the environment reinforces the idea that the self is at least a limited illusion. However, the independence of the self from the environment is definitely an illusion. For self to avoid illusion, it must be embedded in a relativistic reality. That is, it must be relativistic itself. For you and me to avoid illusion with respect to ourselves, we need to think in a relativistic and inclusive rather than linear and uni-dimensional fashion.

This separation of powers between self and its environment is like the political version that separates between executive or doing powers and legislative or being powers. The executive powers could be presidential, as in the USA, or ministerial, as in a Westminster jurisdiction such as Australia. That is, the legislative powers tell the executive powers how to do, like space and self's free will, whereas the executive powers tell the legislative powers how to be, like matter and the self's environment (internal and external).

More simply, the legislative powers codify behaviour, but the executive powers carry out that behaviour and then reflect back to the legislators the success or otherwise of their coding. If the separation breaks down, the opportunity for the emergence of new political solutions is greatly diminished and the opportunity for despotism and political stagnancy is greatly enhanced. The same is true of free will and its environment.

Please refer to Figure 18, in which an SPS of self-reflection is shown to cause a fractal self to arise. This figure also incorporates Edmund Husserl's phenomenology of temporality in which protention is seen as our anticipation of the next moment.

Merleau-Ponty describes this phenomenology as follows:

> *"Husserl uses the terms protentions and retentions for the intentionalities which anchor me to an environment. They do not run from a central I, but from my perceptual field itself, so to speak, which draws along in its wake its own horizon of retentions, and bites into the future with its protentions."*[29]

I suspect the rate at which this time perception occurs for each species is related to Kleiber's Law, as discussed earlier.

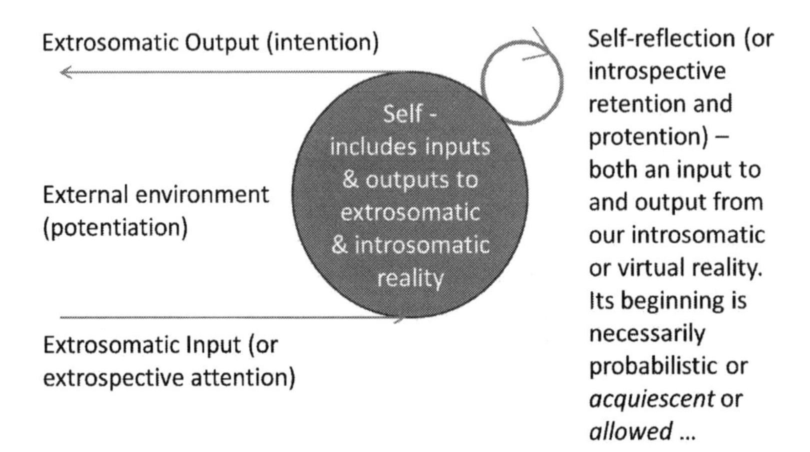

Figure 18: The Emergence of Self

As we allow the self's environment to work in a way that is unhampered by dogma, yet still within the bounds of our tentative knowledge, it can enhance the adaptive emergence of our wellbeing. Likewise, if we allow free will to operate in a way unhampered by material constraints except those proven to reasonably support and challenge our wellbeing, it can also work to enhance our material standing.

Each side of the coin (doing and being, direct and indirect, energy and information, or infinity and nullity) needs the other to operate: Take either side away and emergence collapses (and reality collapses). Put both together in a dynamic asymmetrical arrangement and everything emerges, including the time that marks the dynamic asymmetry. It is important to note that this arrangement is not a dualistic arrangement, but more akin to a dynamic, interpenetrating, and monistic bifurcation. That is, self is not independent from the environment with which it interpenetrates. Self is an extension, manifestation, or expression of reality, which is why it can share in reality's freedoms as if its own. When you shuffle self and environment together, there is a fuzzy completeness. This means that the only roughly complete model of self is not triune, but quadune; the fourth part represented by the environment. Our free will is made freer as we interpenetrate with the environment more openly and fully, not as we seek elusive independence from it. To make our free will freer, we have to ditch linear, axiomatic dogmas.

Consciousness

Our freedom arises from the use of information and the construction and organisation of essential styles or significations within the brain and body, providing each of us our own 'coloured world' that includes our intersubjective and algorithmic evaluations and judgments. We had to create a coloured world to enable a relativistic relationship with the real world and thus enjoy all the benefits of emergence, including the emergence of a limited and fractal free will and an imaginative window into 'full reality'. Consciousness thus gives us an opportunity to limit our futures. It does not give us an ability to directly change our present. Likewise, Quantum Time measures the delay of space's effect on matter's future. General relativity ensures the causes of space and its effects in matter are not concurrent.

This intersubjective and indirect rather than illusory view of consciousness (that is, as an emergent process or balance within finite Gaia rather than as an independent object) again suggests that what is happening in consciousness is a lot like what Dennett describes as happening in Conway's Life World game.[30] In the game, programmers compete to create 2D entities (moving arrangements of blinking pixels on a computer screen) that must retain their shape to survive encounters with other 2D entities. The programmers thus act like the trial-and-error process of natural and artificial selection in the real world.

2D shapes seem to slide across the PC screen but in fact are just a series of blinking pixels. Nevertheless, the behaviour of e.g. "gliders" can be described in their own morphology and spacetime terms, such as how they negotiate a barrier or an oncoming "eater" or "puffer train". Their behaviours do not have to be described in terms of the basic spacetime of blinking screen pixels, and even if they are, this seems to lack insight into the information (or understanding) being manipulated at the higher 2D-entity level. A glider is an ephemeral manifold of pixels that displays properties or behaviours in its own right (separate from individual pixels) in its own level of spacetime without contradiction of processes turning on and off pixels at lower levels. It is this behaviour across its non-near-present time and non-near-here space that gives each glider personality or soul in its normal, pixelated environment. By the same analogy, the blinking pixels in the here-and-now grant the glider its spirit. The Life game also seems to be a very good analogy for the neurophysiology and behaviour of consciousness in terms of its fleeting manifolds of blinking neurons.

In the Life game, a new and more complex reality with more complex constraints – the 2D world of seemingly interacting entities - is emerging from an old and simpler reality with simpler constraints – the blinking pixel world. This is much like how we saw the world of molecules emerge from and coexist with the simpler world of quarks and gluons in the early universe, or the world of competing teams emerge though local cooperation from the simpler world of competing individuals. In either case, the intersubjective world of consciousness or the new world of Conway's Life Game, are the emergent properties of the more complex arrangements of the world an illusion? More carefully, would we call the interactions of electromagnetism giving rise to molecules an illusion simply because their dynamic arrangements emerged out of the prior nuclear soup of our early universe?

For instance, is the emergent property of the wetness of arranged water molecules, an illusion? Would it be right to call gluons and quarks the real stuff of the universe and water molecules something less real? If not, then perhaps we have a right to call the more complex gliders of the Life game and the process that brings them about legitimate, non-illusory and somewhat distinguishable from the blinking pixels of the older and simpler but underlying reality. By the same logic, we might call the introspective world of looping subroutines created within consciousness a distinguishable yet fully embedded and more complex world-level that is emergent with the brain's neurons, cortices, molecules, etc. but non-illusory.

Hofstadter asks the very pertinent question here – "who shoves whom around inside the cranium?"[31] Is it consciousness pushing around neurons or is it neurons pushing around consciousness? Besides environmental influences, is it water's structure that pushes around its component atoms, or is it the component atoms that make water's structure? In terms of John Wheeler's characterisation of General Relativity (matter tells space how to curve [take position]; space tells matter how to move [take velocity]) and the Uncertainty Principle (nil movement is impossible; a beginning cause is impossible), I think we would now agree that the answer must be, 'their relativistic interpenetration means they shove around each other'. That is, both the atoms and the molecules are legitimate manifestations of the one local and emerging environment. However, because of the self-contained nature of the universe, or the implications of Gödel's Incompleteness Theorems, perhaps we could further argue that this picture is incomplete – Gaia incorporates its water-components and its water-arrangements. Water is a manifestation of Gaia rather than something independent of it. Nevertheless, this 'everything is everything' view is not necessarily helpful when wanting to understand how water and consciousness exist or emerged - but it is helpful in encapsulating the suggested paradox in Hofstadter's question (and the fuzzy logic required to solve it).

Helping to answer Hofstadter's question at the more complex level of human behaviour, Dennett uses the term "evitability" to describe our emergent freedom of choice. The Wikipedia article *Freedom Evolves*, referring to Dennett (2004), says that Dennett's definition of evitability involves our ability to avoid unwanted consequences in the future through anticipation, and so it would be fully compatible with deterministic human action.[32] More complex and adaptive living systems will possess more evitability than less well-equipped ones. The agent could be any living thing from insects to humans. My suggested improvement to the definition would be to say that our anticipation of the future can be deterministic and probabilistic (just as is any local organisation in space and matter). This revised definition makes allowance for quantum indeterminacy, but more importantly, the indeterminacy of the Libet thought-gap, i.e. the approximately 300ms between unconscious stirrings of an act in the body and its conscious awareness (after which time the urge may be resisted in its current social context).

Inevitability speaks of the future arising from the present whereas determinism speaks of the past giving rise to the present. Anthropomorphically, and with some support from quantum mechanics, the future seems to be indeterminate and not determined, whereas the past seems determined or locked-in. Past and future seem different at first blush. The nature of the present is perhaps less clear – but most would agree it is overwhelmingly determined by past events, even if indeterminate within the limits of Planck Length, Planck Time, and for humans, Libet thought-gaps.

Are there real options or opportunities to affect the future in a quantum mechanical world? You cannot change the past. However, can you change the future? Well, whether the present is deterministic or not, we cannot. We can only affect the present. In essence, past and future are just constructs of the frontal and temporal lobes of the brain that explain reality's 'becoming' in the here-and-now quite successfully. Nevertheless, what an agent or computer program can change is something anticipated to happen. That is, we can modify an internal or external program (an extended phenotypic program embedded within a piece of technology if you like) to allow for new contingencies before they happen. That is all.

Dennett claims determinism does not imply inevitability, that is, he wants to show that determinism allows for evitability, or the ability of an emergent agent to anticipate likely futures and act to avoid the unwanted ones and thus mould wanted ones. However, surely this view is a little simplistic or misleading. If unconscious chemicals and nerves determine subconscious psychological action then surely subconscious action determines conscious social action too. Surely, the base cause and effect at the lowest level follows an arrow of time from past into future that determines behaviour at all higher levels…?

Here then we run right into the core error of classical common sense and uni-dimensional thinking. Consider the following progression of steps that begins with quantum fluctuations happening at a vast base level in a sea of homogenous or monolithic disorder, but ends with organised movement happening at many levels of reality simultaneously and in a relativistic manner, with necessary speed-of-light or slower delays between all levels of relative order. As in Part 1, the layered progression might be:

- Mere quantum fluctuations at or very close to the speed of light
- Emerging quarks and gluons in a subatomic plasma overwhelmed by the strong nuclear force, wherein the speed of interaction between such components is probably orders of magnitude below the propagation delay amongst the original quantum fluctuations
- The emergence of complex electromagnetic interactions that result in the formation of highly information-incorporating molecules around quarks and gluons. Here the propagation delay between molecule reactions usually happens at many orders of magnitude lower again than in the quark/gluon plasma, depending on atomic weight, pressure, temperature, etc.

- The macroscopic effects of gravity emerging from coagulations of diverse molecules in space, and their rates of interaction now depending on mass, etc. described by Einstein's Law
- Life emerging from mere molecules-under-the-effects-of-gravity and their macroscopic rate of metabolism depending on mass, described by Kleiber's Law
- Perception and self-aware consciousness arising from mere biology and the macroscopic rate of awareness described by what we might stylise as Libet's Law

As long as each layer's rate of relativistic interaction sits as it were in its own niche, typically orders of magnitude away from all others, or operative in different ways to all others, then the order and relativistic interaction at one level will not be greatly impacted by the dynamic order at another. That is, we will tend to consider the levels independent, even though they are not. We know molecular interactions encapsulate a world of quarks and gluons, but it doesn't matter at that level. Likewise, we know the maintenance of life involves many reactions at the level of the autonomic nervous system and at a rate much quicker than once per 300[ms], but these don't seem to really matter much at the level of self-aware consciousness, or vice versa. Further, we know the ~300[ms] delay for conscious awareness doesn't even exist in the lower orders of life. It must therefore induce a gentle, and almost unnoticeable, bending or modulation of lower levels of living in humans – just enough to bring things into focus that were previously unfocused. However, what a difference these soft modulations make to our existential or essential style as *Homo sapiens*!

Glider interactions and proto-intentional programming can guide pixel blinks at the higher level of order as much as pixel blinks can add structure to Glider behaviour from the lower level. It is the essential interactive style that emerges at different levels, rather than something essentially material that emerges. Emergence is in its changing properties at various levels of organisation, not just in the components included or excluded. This means there is no process involved that requires a mysterious explanation.

This means that the secret to understanding our emerging evitability and agency is not to see it as independent from the environment, but as a manifestation of the monist environment in us. We are not free in terms of independence, but our relativistic separation of powers with the real world at our relative level means we are free in terms of our ability to manifest unique local behaviours arising from our objective and intersubjective reality within. Additionally, the dichotomy between the objective and subjective worlds largely fails here as a meaningful or worthwhile framework. Likewise, the locating of our freedom in the internal or external world is wrong-headed; it is in the interplay or balance between them.

Can we describe a simple deterministic world in which Dennett's evitability plays out? Well yes – we see this in chess games and game theory all the time. 'I have learned to act in the future in this way, even mechanistically, if you act in that way'. We have evolved to be good game players, competitors, avoiders, future-modellers, or future-trappers. This is the essential source of our agency. All our quasi-deterministic loops are evolving to improve our evitability, on average.

Yet parts without agency make up moral agents, and these subatomic parts seem subject to cold and monistic determinism with the occasional stochastic twist. How can these parts without agency add together to make a moral agent? This is where we again run into classical thinking. We are not made of parts like a child's tower is made of toy blocks. We emerged inside nature's web from an amazing array of interactions of space and matter, information and energy, or control and traffic. Each of those so-called parts are subject to nature's relativistic mechanism of local command and control. That is, they are like little automatons, but they are deeply efficacious ones in the areas of their use. They each have their own blind proto-intentions or selfishness that have enabled their longevity within the systems to which they blindly contribute. Each subatomic particle is grouped into systems that exhibit demographic properties based on their relations with the local environment. Each of these systems, in turn, have their blind proto-intentions or selfishness that enable their longevity and their contributions to yet higher levels of systems. After many levels the whole edifice of local command and control finds human consciousness to assign it meaning, decide its purpose, and direct its use, something like an army is put at the disposal of a general to decide its next mission. If the army cannot find a general and a mission, its current logistical order will devolve and its 'parts' put to other uses.

Put another way, just as was true of the path of the single photon in space considered in Part 1, emergent possibility arises from the asymmetrical wave-particle arrangements of all things big and small, or holistic and reductionist, not the physical particles alone (that are incorrectly assumed to be independent of their environmental loops). Moral agency is a wave-particle 'phenomenon', not just a particulate 'thing'.

This was also true of the Life game and its competing programmers, who may be compared to a set of beneficial mutations brought about through the blind agitations of a statistical and pixelated vastness. That is, it is the emergent configurations and combinations of parts and processes (being and doing) that bring about the immaterial properties and behaviours of a system in monistic reality. In this sense, human consciousness is very much like a virtual reality machine. Just as Deutsch[33] suggests, consciousness enables us to act like the hero in various intersubjectively construed environments.

As in a contest of two competing chess computer programs, whereby the two programmers can learn off of each other to write better code to avoid check-mate, so genes and memes also learn by experience (and competition/cooperation). That is, a fundamental feature of relativity, or the separation of the powers between moving and structuring, is the propensity of the environment to learn about itself through its dynamic interactions.

More correctly, configurations of space, genes, and memes struggle with the demands of energy efficiency and effectiveness to endure. Those configurations that succeed because they are compatible with environmental interrelationships survive and reproduce, and those that fail in their struggles for energy efficacy reproduce less often. The population that develops thus tends to have better-adapted configurations or architectures of energy built into it. This is also true of the mind and consciousness. We thus evolve a level of success, freedom or evitability, not available to our predecessors, in a manner perhaps often rudely ignorant of conscious choice, but not always. Consciousness affects our choices when spiked by instinctive emotion and our choices affect our conscious development within our deep ecology.

Now we have an explosive growth of evitability or avoidability in our modern world due to the emergence of language and memes. It has taken a lot of competence (blindly learned and retained strategies) and evolutionary R and D to achieve this. Memes and the socialisation of narratives have been able to bring about change much quicker than genes were able to do in the past. Memes increased the speed of winnowing out failures so we could reinforce more quickly any new competencies in our neuronal networks. These competencies also include moral competencies in reasoning and decision-making – and thus moral responsibilities.

To whom do we ultimately hold out this moral responsibility? I would argue it is to Gaia herself, if anywhere, and to the values that maximise her sustainable wellbeing and robustness. More and more, Gaia is of our own making, so more and more, we owe this responsibility to our most enlightened selves.

This perhaps makes us ask, what is learning from the point of view of consciousness itself? It is the ability to reflect on what we believe and to retain those multi-level reflections in order to modify the mind's programs or subroutines such that they deal with contingencies in more successful ways in the future. Nevertheless, how can we modify the program? We do it by trial-and-error. We reflect, think, feel, choose, act and practice. Perhaps with the help of a life-coach, we run the trials repeatedly, hopefully until patterns of success arise in and are reinforced in our neuronal circuits, memes, vremes and subconsciousness. As we indirectly modify subconsciousness, we can modify and rebel against our mores and taboos or what Damasio (2010) perhaps calls our Core-selves (which we shall discuss later).

In summary, we can picture the draft model of learning and evitability as follows:

--> New inputs from the environment (including all our internal and external stakeholders),

--> Lead to newly detected (and largely non-conscious) patterns of similarity and difference, which

--> Lead to new or modified (and largely subconscious) belief reinforcement, which

--> Produce and focus (mostly subconscious) feelings, emotions and intentions, which

--> Lead to (more conscious) choices between beliefs or explanations or hypotheses according to their intersubjective convenience, fit, efficacy or ability to add (more consciously) measured value, which

--> Lead to (more consciously considered) acts or behaviours before or as contingencies arise, which

--> Lead to outcomes in the environment, which

--> Lead to new inputs from the environment, and so on around the loop-->

Note: Every stage of this loop can feed back to earlier ones; there is no intention here to make the process a linear one.

This simple draft model of increasing evitability is algorithmic, recursive and naturalistic. It is also asymmetrical and emergent with respect to the eco-psychosocial environment without breaking its locally deterministic interactions at the level of individual neurons within the cranium. Yet the sense of conscious self and intersubjective freedom can still emerge in this model over time. *'I am a strange [or fractal] loop' indeed!* The simple message we can gain from this draft model is that the more we expose ourselves to new opportunities to learn, the more our evitability and *personally-objective* agency can grow. Our moral agency is not only in our proto-intentional parts, it is also in our dynamic arrangements of those 'parts' across space and time.

One other point that arises here is the evolution or emergence of new causes and new risks. Can we keep up with the risks our evitability engenders? It seems to depend to some extent on the proficiency and efficiency of our explanations of the universe. Whatever is the answer - the self, evitability, freedom and morality evolve.

It seems that the more we learn about consciousness, the more we will learn about evolution and spacetime (and vice versa). For this reason, it is my optimistic guess, along with Deutsch,[34] that we can keep up with the new risks of our own making we face in the environment that is more and more of our own making, but not automatically. We must continually, systematically, and in a disciplined manner, get better at being the rational species. We need to continually improve our inner virtual reality machines. The Emergent Method helps us appreciate this, and in so doing, accelerates the process and its evitable success.

So does our own intersubjectivity, and especially its explicit vremes, present us with a problem of illusion or something to celebrate? Is our intersubjectivity a means of taking the universe to a fabulously new level of matter-space-time coagulation? As Robert Wright[35] might put it, is our intersubjectivity a means of finding wonderfully new opportunities for 'positive nonzero-sum outcomes'?

The Emergent Method suggests an optimistic response to these questions. Our intersubjectivity provides our ability to form a relativistic relationship between our imaginations and the real world and thus enjoy a separation of powers with full reality like no other entity on Earth.

Our intersubjective mental models of the world colour the real world with feelings, values, meaning and purpose. It is as if the world without our explicit values is just the black-and-white version of the more complete and coloured world of our consciousness-with-instincts. That is, the world without this colour is a world of organisms without elaborate consciousness, explicit (moral) values, and limited free will.

The Inner World

Now we are considering the second way we made use of our consciousness (and neocortex), or the second approach our passive and dispassionate consciousness took. This was the personal, or subjective and indirect, or contemplative and introspective, approach corresponding with Table 4: The Self-enlightened Value-set.

Learning lessons from our past using extrospective consciousness was only half the story. We also had to intelligently apply those lessons to an expected future if they were going to improve our prospects for survival or at least maintain our viability constraints. We needed to become proficient at anticipating the future and dealing with contingencies. Therefore, the other thing we sought to do with our growing conscious self-awareness was consider our possible futures. If we could anticipate our future with a high degree of accuracy, we could then move on to prepare our place in the future by making plans for the future.

Making plans for the future was very different to studying the past and perhaps learning lessons from it. Unlike the closed past, the future was unfixed and open (see Figure 9 again); with careful planning and control we could make choices about which future, of all the possibilities, might occur. Therefore, to some degree we could select our future – something we could not do with respect to the past. The future was pregnant with possibility. We could participate with the universe's local environment in setting a future direction, but if we made the wrong decisions, that future could be fraught with danger.

We, unlike any animal before us, became additional 'players' and conscious agents in carefully setting the local environment's future. This was the huge advance achieved by our emerging consciousness. However, in monistic reality, we did not become independent agents, but rather reality's interdependent agents. Through our intersubjective ability to imagine full reality, we enjoyed a fractal and emergent separation of agency powers with relativistic reality.

Our world of future possibilities seemed to put us each in charge and thus made us think reflectively, as it were for the first time, about our intersubjective feelings and their priorities in our lives. Awareness of the future and our ability to affect it made us ask, 'What do I need and want with respect to my future? What do I feel about it? What is important to me going forward? To me, what makes a good life? To me, who am I (what is the intersubjective self and how does it fit with, and make an impact on, all else in my environment)? What do I value and what values will I apply to my future life so I can gain from it?' If we could frame and answer such introspective questions, we could also set about planning their fulfilment in the future. In addition, the more we did this, the better we got at it – and the more evitable our systems of eco-psychosocial claiming became.

Can the possible distant futures be affected by sentient beings that think about an open future, and using their values, build subjective mind-maps of which future to work towards, and then act on that basis? Faulty, uni-dependent cause-and-effect reasoning notwithstanding, I think 'Yes' and relativistic, emergent evitability provides the explanation. Moreover, using values gained from experience to consciously apply to future contingencies makes us moral agents, with moral responsibility.

For instance, trying to protect everyone in our troop overnight (because we care for them), we set six bear traps around our camp. Then, 'bingo!' next morning we find our bear in one of our future-traps. The camp is saved. Through this simple although perhaps politically incorrect example, we can see how sentient beings can carefully affect the future in ways driven by their values. Our freedom and moral agency is not in the past or present – it is in our indirect impact on future contingencies that by-and-by sometimes happen.

Our understood failures and successes in all our consciously chosen human endeavours confirm the value of our introspective future-thinking and the value of ourselves as future-affecting moral-agents. The support and challenge of the local environment to our future aims builds character. This built character increases our propensities to act in intersubjective ways that reflect moral agency and moral congruency with the local setting.

The future-thinking moral agent corresponds with the lose-win mentality of Figure 12, because in order to successfully navigate the future we have to be ready to be chastened and taught by the personal lessons we extract from the local environment. Nevertheless, our full-reality introspections give us a huge advantage over our largely non-conscious competitors in the local environment and greatly enhance the efficacy of cooperation within the groupings of our own species.

It is this introspective feature of consciousness, which Dawkins labelled "our capacity to simulate the future in imagination"[36] that he saw as our mechanism of rebellion against the selfish replicators - our genes and memes. Not that we would always want to rebel against our genes – after all, it was our genes that through eons of trial-and-error embedded the quasi-immutable moral values within their interacting codes. The vast majority of these values are very good in terms of our survival. It is just that they can sometimes be short-sighted. It is our imagination, able to deconstruct reality's relativistic components and then reconstruct them again in our thinking, which is often able to deal better with future contingencies than genes or simple memes.

Having established our candidacy as future moral agents, we need to move on. The introspective or intersubjective values correspond with Table 4: The Self-enlightened Value-set. Just like spacetime and particularly the future aspect of classical time, these introspective values at first glance have no obvious self-restraints of characteristic, nature or society to guide them. The introspective world is our private world; it is not directly tied to the risk-weighted decision-making or social norms of our instincts or even the rational analysis and logical deductions of our extrospections, even though we might think it should.

The subjective world is based on what we want, how we personally feel and how we personally experience our lives, again, through many repetitive trial-and-error experiences and lessons learned from the past. The subjective world is tied to what is not or what is imagined or believed rather than what is, was or ought to be. Imagination seems to be the application of the body's instinctive feelings in the near present to our conscious sense of virtual past/future. That is, imagination seems to require a wonderful shuffling of the body, instincts and consciousness.

This essential attribute of imagination allows us to escape the reality of nature's and society's rules – unless we allow society's memes (for example, pertaining to religion) to crowd out our inner wanderings and liberty. Intersubjective imagination allows us to escape material reality and therefore imagine, through many iterative loops of consciousness, a new reality, a new possibility (risk or reward), or a new ideal. This is something genes and simple memes could never do.

Imagination can take the form of a simple daydream, playful game-sharing, or life-affirming appreciation of delightful patterns in music or art, with little or no rational or logical reflection required. Yet, if it is accompanied by hard-nosed need, animal instinct and a lot of rational and logical reflection, then intersubjective imagination can also act as the crucible of conjecture, innovation and invention that will help us survive into the future and act as the birth-place of our highest and clearest ideals.

However, the transition from thinking outside the square to valid and useful innovation is often not an easy one. It has a lot to do with how we use those repetitive introspective observations to then communicate and interact with our inner selves and outer world. Our wants (or an organisation's objectives) may be clearly articulated, but they also have to tentatively fit within a deep ecology that includes all its stakeholders to be realised. As noted in Part 1, we gotta dance.

Self-Reflection: The Contemplative Approach

Some may think of intersubjectivity as an active and passionate realm, but it is not of itself. Intersubjectivity only becomes impassioned when we add the active instincts of the previous chapter. Just like the scientific, more objective approach, the purely contemplative intersubjective approach is passive and dispassionate – like the deep, recursive meditations of a devout guru.

If we make the focus of our conscious world small enough, that is, we limit consideration of the outer, material world, past and present, we look inwardly (or intra-somatically) instead of outwardly. What do we gain when we contemplate what we truly feel to be true rather than what the environment or others tell us is true? What happens when we personally and simply experience essential, liberated yet interconnected being rather than continually busy ourselves with interactive doing? We get an insight into a whole other world – a world where internal environments and stakeholders now become the focus of attention and external environments or stakeholders are sensed without reaction.

This world at the limit between outward activity and passivity is as different in its properties as our world reduced to near zero degrees Kelvin (where superconductivity and superfluidity kick in), or our world accelerated up to another limiting factor, the speed of light (where the mass of anything and energy required approaches the infinite). It is as different as our world packed into lengths approaching the Planck Length (where black holes appear and the known matter-space-time fabric theoretically crumbles).

It is here at the boundary of outward activity that the properties of being stand almost alone and the properties of instinctive doing almost vanish. If doing may be compared with the world of verbs then being may be compared with the very different world of nouns.

The Mystery of What We Don't Know

The contemplative world is not about what we can learn through outward observation and identification; rather its key attribute is the openness, freedom, mystery and wonder of what we feel, want to know, do not know, don't have to know or very quietly choose to block out (for example, through meditation).

Like space, the contemplative world and its values provide the background or frame for all that otherwise consciously is. The contemplative world provides what we believe or understand, or how we relate with our inner and outer environments.

If I am drowning in the sea, I do not care about the chemical formula of the water in which I am drowning, nor do I care about social norms with respect to expected behaviour or social appearances in such moments. Rather, I feel a rush of survival instincts and if there is time, personal emotions tied to my intersubjective understanding of the situation. Due to the structure of our brains discussed earlier, my consciousness becomes intersubjectively tied to the personal, present moment.

For instance, if I know my being saved is impossible (I can no longer compete for survival) and there is sufficient time, I might focus on memories of my life, my family and my friends and what will happen in my inevitable moment of ultimate acquiescence. I might even think about the intersubjective worth of my life as a whole. If I have opportunity and if I am a believer or agnostic, I might even think about, or wonder what will happen in, the moments after death.

I suspect that when some of us come to the conscious and inevitable moment just before death, we might even become dispassionate about the situation - the instincts and chemical broth of self-preservation may give way to dispassionate and disconnected but fading consciousness of self. We seem to see this dispassionate surrender on TV in the face of the victimised gazelle succumbing in the last moment, at the end of the chase, to the jaws of the big cat on the African plains.

The Introspective Self-restraint

The introspective first and only self-restraint or law is that just as there is no material constraint of zero (for example, in counting apples, zero oranges will do just as well as zero apples), so there is no material constraint of 'full reality' introspection. There is just unbridled liberty and grace if we do not bridle ourselves with a particular religious stance or other ideology or dogma.

If a law of introspection must be named, then it is a law of freedom. It is a law of our own making or our own being and agency that we are each free to courageously write ourselves. It is a law that emerges through our unique introsomatic attributes, our unique ethos, and our unique experiences. It is this unique and personal self-restraint that governs our introspections within the contemplative world and that governs our personal hierarchy of introspective and extrospective concepts.

The law of introspection is one of duty or responsibility to self as a custodian of agency, whether moral or otherwise. In short, it is a law of selfishness as defined and discussed in Part 1. An enlightened selfishness might encourage us to pay taxes in order to enjoy governmental services. It might also encourage us to vote for a political party that promises extra spending on education, even after our own children have finished their education. Why? We might do this just for the sake of a strengthened social fabric in our chosen society. For humans with such an enlightened level of selfishness, this law of introspection includes a duty of-

- self-preservation, self-acceptance (or 'self-acquiescence'), and self-worth
- self- esteem, respect, honour
- self- love, care, regard
- self-assessment
- self- control, discipline
- self-improvement
- self- actualisation, fulfilment
- protection of self-image, public-image and legacy

Much more so than for other animals with some level of conscious self-awareness, it is a law that emerges from our personal ethos and gives rise to our personal rationale for past and future actions. This is an important point because it suggests that to be a rational being is a very personal and individual thing. As the Enlightenment encouraged, we are each responsible for our own thinking and the use of our own thoughts. Just as the leaf of Part 1 defined its own fractal dimensions within the space it occupied, so we do with our bodies, our minds, and our introspective self-restraints.

For example, consider a nineteenth century entrepreneur that had a vision to build a railroad across an entire continent. All his or her advisors at the time say it cannot be done. Logic says it cannot be done. The available evidence suggests it cannot be done. Nevertheless, the entrepreneur, in his or her ethos and rationale, looks at the same facts and is sure it can be done despite the hurdles. Is the entrepreneur being irrational? By the measure of all advisors, the answer is 'Yes', but by the measure of self, 'No'. Whether or not that particular entrepreneur succeeded, there is no blanket definition of rationality just as there is no fixed future.

This is why we can't use the Emergent Method itself as a blunt instrument to convert a person from one life-stance to another. I may try to argue your current life-stance with you but in the end, your life-stance is your free choice and responsibility. Personally, I am confident that with time, you too will see that all religion and dogma is a kind of proto-science, and religious or ideological ritual a kind of proto-technology of the human species. To help each other make better choices we simply need to make our logic and values more transparent and public rather than allow them to remain shrouded in subconsciousness or unconsciousness, or cloaked in religious ritual or other dogma. No matter what the current life-stance, we simply need to encourage each other to think, communicate, and act, so that we may self-actualise together.

It is the very liberty of this autonomous yet spiritual self-restraint that brings us so much difficulty in the introspective world. In the three other areas of human interactions we have considered so far, extrospection, cooperation and competition, the self-restraints that governed them – the positive self restraint, normative self-restraint and risk-weighted self-restraint, respectively, clearly commended themselves from spacetime, our memes, and our genes. They gain their surety from past facts and near-present values. At this stage of our evolution, these three anchors or self-restraints have become undeniable, evidently true and value-adding. However, the nature of what is evidently true and persuasive when it comes to the introspective self-restraint, that is able to look freely to any future, is not as clear cut or free from argument, opinion, ignorance, and deception.

It is as if the risk-weighted, normative and positive self-restraints combine to provide us a here-and-now platform on which our own selfish self-restraint, or future-looking self-restraint, may emerge. This self-guided restraint is what best describes our agency and its personal formula for success in our given eco-psychosocial environment. It is a self-restraint that operates most fully in the freedom of the near-ideal homeostatic state of our mind-body complex. It is the most important self-restraint for guiding each of us to our personal imaginings of what constitutes 'the good life'.

The fourth self-restraint of human behaviour is akin to the elusive interaction of gravity that is by far the weakest of the interactions of the Standard Model, yet at the same time, also the most far-reaching in its impact, binding whole galaxies to their cores. In a similar sense, while the introspective and subjective self-restraint is the most deeply personal, it can also be the most public of our self-restraints – the one we are most clearly known by; the one that is carried forward in our legacy after we're gone. It is the self-restraint by which others are most likely to remember us.

However, while careful trial-and-error extrospective observations can help guide us to temporal truths about the past outer world, trial-and-error introspective observations often seem to reinforce the biased opinion, ignorance, or deception already imagined. It seems we have to tread even more carefully than the most diligent of scientists (and his/her Scientific Method), if we are to uncover the introspective contingent truths that will enhance our self-actualisation (or artificial emergence). This careful uncovering is what the Emergent Method offers.

The Buddhist idea of consciousness is that it is transcendental. That is, spiritual introspective contemplation, including meditation, ultimately links us to a beyond-physical world that is the true state of being. In most Buddhist traditions, the physical world is seen as illusory. Buddhism makes reality the bondservant of the world that lies behind the veil or beyond naturally knowable reality. In contrast, emergentism recognises the indirect (including introspective consciousness) and direct (including the brain/body) as relativistic partners operating on this side of any veil. It sees conscious intention as a naturally transcendent tip of the iceberg of proto-intention coming from all the emergent and supporting strata below it.

The Christian idea is similar to the Buddhist one in some respects, in that spiritual introspective contemplation ultimately links us to a beyond-physical world that is a higher state of being (or heavenly). However, the Christian belief would not go as far to say that the physical world is by its very nature illusory. Rather, Christianity sees it as fallen and awaiting redemption by He who dwells in heaven, behind the veil.

Ayn Rand's objectivist idea in stark contrast sees the physical world, its empirical and logical existence, as the objective and only reality. Further, it does not see intersubjective introspection at any level as a possible source of knowledge; nevertheless, it does see introspection as important. Objectivism's introspective and expressed ideals, such as its beliefs in the ascendency of reason and ethical self-interest, are based on extrospection's so-called objective reality rather than some notion of reality arising independently out of inner contemplation. So objectivism does not choose to use introspection's liberty to make some independent law or observation with respect to reality, but rather, tries to make introspection's liberty ultimately subject to extrospection's reflections of reality.

Objectivism treats knowable reality on this side of the veil as the only reality. Emergentism is in partial accord with objectivism here but treats introspection as an indirect reality, like spacetime, and in particular, the way the 'becoming' of spacetime at all levels necessarily frames and interpenetrates matter's emerging arrangements. That is, it is impossible to stop our intersubjective understandings (or arrangements of knowledge) from shaping our future world. This means our intersubjective introspections or values do become a source of knowledge, albeit an indirect source (rather than a secondary source). Emergentists specifically recognise that our decisions and actions emanate from the body's regulatory systems and the subconscious, with only indirect influence from evitable consciousness. Thus, an asynchronous alignment between the subconscious processes and the emergent conscious framework is required to reach our goals and for our smooth working as continually emerging rational and moral agents. This alignment or congruence is dependent on each person's introspective self-restraint.

The introspective or contemplative self-restraint is not a fully certain or deterministic one. Each addition to our introspective repertoire is like the probabilistic, chaotic yet wavelike interference of quantum mechanics introduced into a world that would otherwise be linear, planned, and not self-organising or emergent. Introspection quietly frees our extrospections just as surely as extrospection limits our introspective excesses. The law of introspection helps organise our introspective interactions, such as imaginations, feelings, desires and purposes. Every person, as an individual moral agent, must discover and apply this self-restraint for him- or her- self.

It is very interesting to note here that of all the human self-restraints we have talked about before (positive, normative and risk-weighted) none was more deserving of the label of self-restraint than this one. The probabilistic uniqueness of each of us in our unique here-and-now seems to demand a unique introspective rule in each that will uncover the appropriate but interpenetrating role for each of us moving forward into a unique there-and-then. However, we can all use a common method of finding and following our unique introspective paths: The Emergent Method. Our moral agency seems to demand that we be gods of our future there-and-then, even if we are slaves of the past and mortally constrained with respect to the here and near-present.

The way we choose our introspective self-restraint is essentially by understanding values and their relationships with reality and then, through many algorithmic iterations of life's dance, recognising those ones that mean the most to us, personally. These values tied to our endeavours and concepts resonate within us, motivate us, and bring us fulfilment and true success amongst our internal and external stakeholders.

Once we find these values and understand how we do the seemingly antinomic but naturally transcendent and emergent dance with them, we are then able to permit them to indirectly order or probabilistically self-restrain our future-aware lives. We can do this not just in terms of our introspective feelings, thinking, desires, etc. but also in terms of our past-aware extrospections, present-aware competition and present-aware cooperation. To be an efficient, effective and rational being is thus a very emotionally personal or individual idea tightly tied to the noble but elusive notion of self.

However to do this personal ordering of the introspective and contemplative values requires that we carefully think and take courage. It requires the hero inside us to move forward in space and time. It requires us to look at our environments and cultures and ask again, just as our ancestors once did for the first time: 'Is this what I and my stakeholders need or want?' What do I feel about this? Going forward, is this important to my stakeholders and me? What do I value and what values will I methodically apply to my future life? What values can I add to my internal and external stakeholders that will create wealth for all involved? Another person can help you and guide you towards realising your emerging potential, but you must find it, understand it, believe in it, and courageously choose it yourself.

Much more than just looking inwardly and outwardly (introspectively and extrospectively) through the lens of our chosen values, using the Emergent Method we now have to communicate and interact on the platform of our chosen values with our internal and external stakeholders, to slowly and surely realise our purposes. The introspective self-restraints that reflect what we are also need to become inseparable from what we do, or decline to do.

Spiritual Experience

Can an emergent monist, atheist, agnostic or believer who rejects the world's major religions, as they exist today, have introspective inspiration that we could label a revelation or a spiritual experience? Yes, most definitely! Spiritual experience is a natural result of having instincts and an inward-looking consciousness made vital by deeply held and emerging values. Spiritual experience (or revelation and intuition) is not monopolised by, or emanating from, religion or religious experience and it certainly does not require supernatural intervention. It is the free but wonderful outworking of the pre-conscious and animated values within.

Today, just as in our species' past, we are self-aware agents whose agency is the very values with which we resonate. Just because an atheist or agnostic has a spiritual experience, which is just a more intense or sudden experience of the necessary introspections made every day, this does not mean he or she must disown it or belittle it as something irrational and therefore invalid. Chaotic (in the sense of random but subject to organised and patterned feedback) thought processes are natural and essential. They are very likely the way the pre-conscious works and routinely gets things done.[37] Chaotic thought processes often aid usable rational outcomes in reflective, algorithmic, fuzzy-logical ways that linear or sequential logic maybe just cannot achieve, at least in the required timeframe.

In this special sense, I think the mind-body, through the way its processes are organised on a vast and massively parallel basis, can be more than computational in the conventional sense, within its strange loops. That is, I suspect the mind-body is able to use its layered loops and trial-and-error relationship-detecting / image-building to mould new forms of data pattern-locking rather than just rely on analytical techniques. The way the mind solves problems and adds value is probably always emergent and multi-dimensional rather than law-driven or axiom-driven in the mathematical or technological sense to which we are accustomed.

While acknowledging the brain's incredible complexity, can I dare say that the brain, just like nature, effortlessly 'relates' much more than it artificially 'computes'? It perhaps falls naturally into relationally-patterned answer-traps much more than it calculates. By avoiding analytical calculation (or perhaps artificial relationship searching), I suspect it both escapes typical Gödel limitations on the upside and falls into instinctive biases on the downside. This supports the idea that the mind-body might work very much like natural (and artificial) emergence in space-matter at the subatomic levels. In addition, this method of operation is not peculiar to human brains; it is a feature of all animal brains.

For instance, Richard Dawkins discusses the same issue with respect to the platypus brain. Speaking of multiple sensors of electric field embedded into the platypus's bill, he says,

> *"Given dedicated computer power to handle data from a large array of sensors, the source of the electric fields can be calculated. Of course, platypuses do not calculate as a mathematician or a computer would. But at some level in their brain the equivalent of a calculation is done, and the result is that they catch their prey."*[38]

Similarly, we could speak of a shark's electro-location, a seal's hydrodynamic-location, or a bat's echolocation. As humans, we are only just beginning to understand how massively parallel and layered computations work. They can leap between possibilities that uni-dependency cannot.

Once we understand that our massively parallel and relationally layered experiences can slowly be governed by the self-restraint of our own being – our self-aware and personally selected values with which we resonate – we can begin to see that such intuition can guide our destinies in very logical but indirect and wonderful (naturally self-organising and self-actualising) ways. There is no attempt to be in any way mysterious or transcendental here. Another way of saying this is that the pre-conscious or subconscious life is genetically biased and environmentally organised to find iterative ways to fulfil the requests of the conscious, self-aware and vremetically organised life. More simply, the conscious life often seems to set our conscious goals and limitations, while the subconscious life seems to go about trying to achieve those goals and limits, but in a probabilistic, interpenetrating feedback loop, that often hides exactly when and where these goals and limits truly originate or end. On the downside, if we lose touch with our essential values, our subconscious lives can also reinforce the biases and disorganisation of our self-aware lives and vice-versa.

On the upside, it seems that introspective spirituality is a way of encouraging the mind to become more plastic, more adaptive and therefore more directly useful and emergent as we move forward to greet a changing and uncertain future. This is probably why Einstein once said, "Imagination is more important than knowledge." Imagination fuels the goals that are the stepping-stones to our destinies. That is, we release intuitive or spiritual power when we use the freedom of imagination to trigger, push, or otherwise influence well-reasoned and well-connected thoughts, goals, and outcomes. We also unleash intuitive power when personal intuitive beliefs, goals, or life-stances are wisely tied to, or pulled by, external knowable realities and logic.

It is for these reasons that the self-restraint of introspective liberty (another way of saying self-enlightenment), after the positive, normative and risk-weighted self restraints considered earlier, qualifies as the last of our four self-restraints of values.

How can our liberty be a self-restraint? It seems an oxymoron. Yet this is the nature of self-love or a deep honouring of self (as a legitimate part of reality's fractal geometry expressed in real and virtual particles). As our liberty and uniqueness in the here-and-now is accepted, pursued and celebrated it builds our personality. Our liberty feeds back to affect all our other systems of self-restraint that at their best, optimise our performance in our chosen vocation. We are adapted to function in this manner; we are adapted to benefit from the Emergent Method…

Again, we need to remember the basic truth that blindly 'selfish' systems are at the root of all instincts and their off-loading companion, human consciousness, with its associated memes and vremes. Consciousness is not independent of the life-preserving instincts; it emerged from the scaffolding provided by our instincts. Nor is outward-looking consciousness equal to inward-looking consciousness. Like the self-reliant instinct, outward-looking consciousness and its connection with objective reality has been built from scratch on the back of our DNA's system of survival (our selfish genes) in an outer-world environment of limited resources.

However, this is not directly the case for the cooperative instincts or for inward-looking consciousness. These required the former scaffolding of the self-reliant instincts and outward-looking consciousness, respectively, to be in place before they could develop their unique form of homeostasis-seeking. Introspection has emerged from extrospection over evolutionary time. Yet deeply focused, practiced, perceptive and beneficial introspection, finding and locking into relationships where simple linear logic fails, quickly collapses when it becomes shallow and fragmented.

Just as cooperation takes competition to a higher level of complexity, so introspection takes extrospection to a much higher order of complexity. That is, we can consider our physical and social (or genetic and memetic) environment much more deeply as we contemplate our reactions to it and with it in terms of our future-biased purposes, because we start to personally order and value our instincts and extrospections rather than simply experience and observe them. In fact, the Emergent Method is exactly this. It is using our introspective vremes to circumspectly order and value our unique instincts, extrospections and other introspections. It is by this Method that we can realise our potential.

Introspection turns extrospections' cold data coupled to our instincts into usable, value-rich information that we can apply to a completely new level of extrospection. Here then we have the answer to the 'basic weakness of the natural and formal sciences' mentioned in the last section. The personal rules of our introspective consciousness, which at the same time equate with a freedom to respect or celebrate self and its uniqueness, fill us with personal focus and purpose.

This introspective purpose is personal – in a way that the competitive or cooperative drives of our instincts, supplied by our genes, could never be. So Nietzsche was correct –

> *"Apart from the ascetic ideal [which, maybe a little cheekily, I now ask you to read as 'introspective' ideal], man, the animal man, hitherto had no meaning. His existence on earth contained no goal; 'why man at all'? – was a question without an answer; the will for man on earth was lacking … he suffered from the problem of meaning … and the ascetic ideal gave it meaning! … the will itself was saved."*[39]

It is ironic that the very thing Nietzsche needed to save his cherished human attribute – the will-to-power – was the thing about human nature that he despised the most – our recently emerged inward-looking consciousness. Nietzsche thought he could select our human attributes to make a new super-species. However, our species is not constructed by adding together parts, like a Lego building or an international space station. What we can do is respect and indirectly guide our evolved attributes to within channels of close-to-optimum operation.

Nature accelerated human introspection's emergent complexity not through the tangible and direct world in which we compete, cooperate, or consider with extrospective consciousness, but through the intangible and indirect world to which other vremetic survival machines expose us in society and with which our vremes coevolve. This co-evolution of vremes through cycles of socially-oriented introspection is probably the fastest way in which feedback leads to our species' emergent qualities and promotes the coagulation of our separate human minds.

To use a metaphor of outer space once again, the relationship or balance between outwardly interacting consciousness and inwardly interacting consciousness is not like the orbit of equally large twin black holes or quasars around each other, but more like the orbit of our planets around our massive sun. In this case, the shining sun represents outward-looking consciousness and the captive planets represent inward-looking consciousness.

As we move forward in modern society to meet the fast-paced and complex challenges it presents to us, our philosophical principles and values-framework will work best if it firstly protects our outwardly interacting survival mechanisms (including outward-looking consciousness) and secondarily, it guides us in the proper use of our inwardly interacting survival mechanisms (including inward-looking consciousness).

However, just as there is no matter without space and no solar system without the planets, so there is no self-aware consciousness without introspective intersubjectivity (which it may be preferable to define as 'personal objectivity mixed with the personal ignorance and personal errors of introspective concept formation'). Higher consciousness requires and emerges from vremetic introspection.

In the stock market scenario, the introspective mentality (or applied inward-looking consciousness) is sometimes evident in terms of investment or divestment decisions made based on gut feel about the future – a very dangerous strategy indeed if used in isolation, as are the other risk-based, trend-based, or fundamental strategies used in isolation, if we really think about it. Intuitive rules of thought and behaviour can be very inappropriate when applied directly to the outside world, especially if we apply them without a proper framework or knowledge of their strengths and weaknesses. Nevertheless, an intuitive understanding of forces in motion (a personal hunch) borne out by much experience and continuously diligent research can prove to be fruitful. Unfortunately, fruitfulness can then reinforce via feedback the hunch's inherent weaknesses and be its undoing down the track. Therefore, we need a clear framework and methodology to guide our introspection, but introspection is still an essential part of our human nature and survival. It is essential to finding the good life.

The fourth self-restraint of values (or vreme development) is the self-enlightened introspective interaction.

From Self-Awareness to Religion and Beyond

They say nature abhors a vacuum. In terms of our conscious approach to the unknown, is this perhaps true for us too? The personal answer statistically depends on our education, but also the values we hold and live by. Some of us can deal with the concept of the unresolved, such as what existed before the Big Bang, say more than 20 billion years ago (some 6.2 billion years before the Big Bang and thus before time's beginning in our universe), and some of us choose not to. Note: Logically, to anyone that applies linear hours, minutes and seconds to perceived changes rather than the directly observed local effects of entropy and work done, virtual time before the Big Bang should present no problem.

Some of us demand supernatural resolution and closure while others, considering such a question just as deeply, do not require any resolution or closure or named god. Yet others require closure via an atheist or agnostic life-stance. Dan Dennett, in his book *Breaking the Spell*,[40] speaks of the growing consciousness of our ancestors overshooting itself when, for instance, our ancestors intersubjectively experienced the eerie presence of departed acquaintances trying to communicate with them from beyond the grave. We can perhaps imagine a caring father-figure, recently deceased, eerily echoing words of advice with respect to how chores should be completed, to a beloved son or daughter - just as he had often done during his life. Today many of us are likely to agree that this is probably the result of lingering memories in our brains and not proof of the existence of ghosts at all. Dennett suggests that such common experiences amongst our ancestors, with their limited consciousness and understanding, could have quite naturally led to ancestral worship in early communities. In one sense the concept of a ghost ('the dead and their eerie communications') at such times was perhaps not necessarily wrong (it corresponded with and identified real memories) just incomplete (it did not include a refinement of the concept of memories within consciousness).

Likewise, Dennett suggests to us that the animism of our ancestors was simply an over-attributing of conscious intention to moving things in the environment beyond competing animals with limited consciousness. For example, our ancestors once attributed conscious awareness and intention to the wind blowing through trees, the tossing of the waves of the sea, the movement of clouds across the sky, and an earthquake or volcanic eruption (all the result of convection currents that display high levels of order rather than pure randomness). Again, animist concepts were not necessarily wrong in terms of the level of man's consciousness at that time, just incomplete and therefore wrongly closed or concluded. I have argued previously that these dynamics represent proto-intentions.

Further, as such ghostly and animistic memes in passed-down myths combined, mutated and evolved, it was a possibly natural progression to assign the movements of the clouds across the skies, etc. no longer to their innate proto-intention, but the intention of the greatest ancestral protectors and protagonists. Thus were born the gods (for example, Egyptian Amon, Greek Poseidon, Roman Jupiter), mythical creatures and demons of ancient Egypt, Asia Minor, Greece, Rome, Japan or countless other hunter-gatherer societies. This was all evidence of humankind's growing consciousness in largely proper use and action from the viewpoint that sees us as rational beings adapting to our environment and growing in knowledge.

The question then arises as to why these folk religions stuck, from an evolutionary perspective. That is, how did the development of folk religions assist our survival as a species for so long? Without going through the elaborate answers available elsewhere, the simple answer in terms of the preceding chapter is this: Anything that honed our social skills and helped hone our hunting skills would increase our chances for survival and perhaps uniquely shape our brains within each culture. Folk religion, as it encouraged us to develop language, communicate in stories, gather in groups, and share food, would clearly have helped hone our normative social skills. If these same words and stories helped hunters put in more concentrated effort to the hunt, with more purpose and vigour, then folk religions would have at the same time contributed to improvements in risk-weighted hunting skills.

It was only at times of catastrophic loss unexplained by the local folk religion that folk religions contributed to a lack of social cohesion in early societies. Examples of such events might include defeats in battle or decimation through earthquakes, floods, plagues, storms, wild fires, and droughts. However even these catastrophic events would have often contributed to religious mutation and paradigm shifts in religious development and their supposed explanatory power (for instance, 'Thor sometimes gets angry with men and the result is disastrous thunder storms').

Daniel Dennett suggests more generally that as social groups increased in size, religions became organised, much as the folk music industry must have done to become what it is today. That is, religion and music became more, in a loose sense, professional.[41] This process began perhaps before the introduction of agriculture in places such as the Göbekli Tepe temple in Turkey some 12,000+ years ago. The local shaman, blindly helping the hunters and gatherers increase their respective skills, slowly evolved with the introduction of agriculture and its increased divisions of labour into guilds of priests. Major religious organisations started to grow out of such guilds, with local franchises and largely universal brand names. Just as in the shamanic (and not so distant) past, religion's more modern merchants were blindly integral to the evolutionary success of markets, military efforts, lawmaking, education and the cohesiveness of society as a whole. From an evolutionary perspective, religion will persist in a community's life as long as it contributes to survival success rather than holds it back.

Perhaps this is no better demonstrated than by the history of one of the newest (or most evolutionarily advanced) and most cosmopolitan of world religions, Christianity. Christianity was international in its outreach and formulation from its beginnings, perhaps unlike any religion before it. It grew to power in the Roman world, which was the great pluralistic melting pot of nearly every religion and philosophy on earth.

Christianity largely shares its monotheism, idea of the fall, idea of a saviour born of a virgin, idea of paradise and hell, idea of a millennial restoration, aversion to graven images and more world-affirming approach, with Zoroastrianism and its founder Zarathustra, who lived somewhere between the seventh and tenth century B.C. This probably also explains the cryptic "Magi from the East" mentioned in Matthew 2:1. Similarly, the ideas of monasteries and meditative prayer were present in Eastern religions long before Christianity adopted them. The strong defence of a factual biblical history rather than bible stories viewed as mystery and myth was probably due to Ancient Greece's philosophical influences. However, I suspect in earlier times we didn't consider as overly important a strong division between facts and values.

The Pharisees, perhaps through Neo-Babylonian, Persian, Hellenistic and Parthian/Zoroastrian influence, introduced the idea of the resurrection of the body. The Jews, after the first and second destruction of their Temple and their continuing Diaspora, at the same time amplified the importance of personal ethics and introduced the idea of scriptures to be studied and analysed by congregations of adherents and in families rather than just by priestly classes, which had a strong tradition in many Eastern religions. Christians simply extended these important mutations of the older Jewish faith, especially as they also experienced persecutions arising from existence outside the accepted Greco-Roman religions of the empire at the time.

Ethics was also being carefully considered in the Roman world by the likes of Plotinus (~205 – 269 A.D.) and the Neoplatonic movement of the literate elite. Christian thinkers were already in the habit of carefully considering and perhaps synthesizing such developments. A good example would perhaps be the writings of Tertullian (~160 - 225 A.D.).

Finally, Emperor Constantine came to consider all of these aspects of the new religion and probably many others (such as the conversion of the Armenian court from Zoroastrianism to Christianity under Tiridates III, after his personal conversion in 301 A.D.) In 325 A.D., Constantine summoned the Council of Nicaea and the new religion of Christianity, with its cosmopolitan doctrines and tight system of leadership, was formalised. In 330 A.D., Christianity became the new state religion of the Roman Empire.[42] Christianity was to contribute to the evolutionary success of the greatest empire on earth, along with its markets, military efforts, lawmaking and society as a whole.

Considering the outcome some 1,700 years later, perhaps the key reason why traditional religions in the West have declined over the last few centuries is that the separation of church and state has meant religions no longer play a major role in markets, military efforts, lawmaking and social cohesion. This suggests the separation of powers between religion and state is a key to the decline of the power of superstition in any society. It also suggests other sources of values, such as those of self-reliance, science, or civic duty, are likely to arise and partly fill the gap in social cohesion left by declining religions.

The only evolutionarily successful churches in the USA (and elsewhere) today are logically those that can continue to make a social impact on child-bearing-aged adults – as it is still true to say today that a more socially cohesive group is likely to be a more evolutionarily successful group.

Evolution, it seems, will still blindly support religions and other intention-engendering systems that can foster cohesive societies. Regardless of their veracity, greater societal meaning, purpose, and narrative usually increase our species' rate of survival. This fact tends to make such narratives as good as true.

Can we raise our social cohesiveness and enlightened consciousness to a point higher than that achieved by religions and the state-based dogmas of the past? The answer is clearly 'Yes': One example would be the abolition of slavery in the United States of America, driven by the more industrious northern states but resisted by the more agrarian and religious southern states. Another example would be the rejection of the use of torture in the civilised world. This book seeks to free the emergent, locally recurrent values from religious superstition so that we may study them more fully and apply them more consciously to all our local human endeavours, and thus increase our social cohesion both locally and globally.

Fear, Pain and Suffering

This chapter is about how consciousness affects our values, purposes and potential. It is also about the journey of consciousness. We cannot fully take this journey without talking about fear, pain, suffering, and death. The death of a family member or friend often leads us to think of our own day of death and the intersubjective mystery of death itself. Death, rather than the notion of virtual time before the Big Bang, seems to be the ultimate metaphor for that vacuum nature supposedly abhors.

The Dalai Lama, in his book *Becoming Enlightened*, conjectured that in ages past there must have been people who, through deep personal disappointment, were willing to look to the beyond-physical for possible support. He suggests that in such a manner the first religious groups arose. He then adds, "At the appropriate moment in the progress of human thought, Buddha appeared in India".[43]

It is a familiar theme. We sometimes feel alienated by a world that will not let us in. We feel all planned futures cut off from us. We lose hope in the future. What is the solution? The Dalai Lama's considerations in the same book on the topic of death are as follows:

"Consider:

1. *The illusion of permanence, or being unaware of death, creates a counterproductive idea that you will be around for a long time; this, in turn, leads to superficial activities that undermine both yourself and others.*
2. *Awareness of death draws you into thinking about whether there is a future life and taking interest in the quality of that life, which promotes helpful long-term activities and diminishes dedication to the merely superficial.*
3. *To appreciate the imminence of death, think deeply about the implications of the three roots, nine reasons, and three decisions … THIRD DECISION: I WILL PRACTICE NONATTACHMENT TO ALL OF THE WONDEROUS THINGS OF THIS LIFE."*[44]

Thus, the Dalai Lama's Buddhism teaches nonattachment as the solution to the disappointment of losing life in this realm sometime in the future as well as the means of looking forward to the more enlightened life to come after death.

Similarly, Eckhart Tolle (2006) suggests living in a way that its success is not dependent on future possessions or money and in a way that expresses our positive values as they are now, which is good advice at all times, even when we do have a sense of hope.

Finding Nirvana

Having just considered one approach to life's discouragements, its ultimate discouragement being death itself, we are likely to ask ourselves many introspective questions. Is superstition still the answer? Perhaps we now know that our fascinations with stories of ghosts or gods of thunder or lucky charms, are not. However, is God or transcendental consciousness still the answer?

In one sense, you have probably discovered in prior pages my essential answer already. It is not fundamentally our differing concepts of God or transcendental consciousness that are the answer to life and the experiences it throws up, but rather the quasi-immutable values we hold on to and live by. Our gods often espouse these values. Further, these values are often claimed by our gods as innately their own. "(I am) the God of (e.g. love, peace, holiness)" we read in various forms in the Bible, as a mere human might proudly claim ownership of a monopolised and highly prized possession.

Buddhism's approach is perhaps a little interesting here in that it explicitly encourages change towards personal Buddhahood through following its path to enlightenment (the Christian teaching on joining the Godhead exists, but these days is a little more obscure). Most Buddhist traditions teach that our personal sense of separate existence is an illusion, and that the transcendent enlightenment of our consciousness as a whole is most important. They teach that achieving the altruistic omniscience of Buddhahood together is the ultimate process going on in our arising universe. Our volition can only hasten or delay the outcome, which is claimed to be quickly gathering pace. Our ignorance of the ultimate emptiness of the false idea of inherent existence, it is taught, is the root cause of all our suffering.

Is Consciousness Transcendental?

The Dalai Lama also teaches,

> *"Since consciousness, as mentioned earlier, has to be produced from consciousness, its continuum has to be beginningless. Once we establish that the continuum of your mind has no beginning, the person that depends upon the continuum of consciousness also cannot have a beginning. Once the person, or 'I', has no beginning, you must have taken rebirth over and over."*[45]

It is by this means the Dalai Lama establishes both reincarnation and the transcendence of consciousness. There are many bold claims here, such as beginningless consciousness in man. Clearly, a scientific and evolutionary emergence of algorithmic trial-and-error consciousness from nothing but genetic instincts and whatever abiogenesis this earlier entailed is not considered here. It seems an evolution of the 'I' is rejected as strongly by the Dalai Lama as by the Pope.

If anything does exist behind the veil, then clearly it does not have direct form as we currently know it. However, can indirect logic or coding exist in an organised form without direct matter or at least fleeting indirect matter? No, I don't think so. Indirect logic must emerge and find order within matter, fleeting or otherwise. Further, all deterministic coding is materially local and subject to mutation.

In a similar fashion, Eckhart Tolle, in a section boldly labelled "Incontrovertible Proof of Immortality" offers the following:

> *"We could even say that the notion of 'my life' is the original delusion of separateness, the source of the ego. If I and life are two, if I am separate from life, then I am separate from all things, all beings and all people. But how could I be separate from life? ... It is utterly impossible. So there is no such thing as 'my life,' ... I am life. I and life are one. It cannot be otherwise. So how could I lose my life? How can I lose something that I don't have in the first place? How can I lose something that I am? It is impossible."*[46]

Thus, Tolle seeks to establish our immortality. Clearly, our lives are the result of long, winding chains of intermeshing and emergent causes or effects. However I am not sure Tolle establishes our dualistic immortality (i.e. the immortality of an individuated 'I' separate from the material body), which is the kind of immortality that probably matters to most of his readers.

As we have already noted, many would argue that there is no possibility of a consciousness that is conscious of nothing outside itself. As stated in Part 1, consciousness being a reflective phenomenon cannot be the only thing that exists. Consciousness is consciousness of something. Thus, we have the notion that human consciousness senses reality separate from self before it can reflectively build a concept of self. Such a concept of self is thus a recursively built concept (built and maintained through many of

Hofstadter's "strange loops") rather than an innate or direct concept. Life is not something we have, but neither is it something that we are – life, just like organisational culture, is a non-monolithic and statistical property that emerges from an appropriate arranging or relating of a vast collection of interacting particles, real and virtual.

The point here is that all proofs of immortality are fraught with difficulties. It is up to you and the values you embody to decide whether immortality is a chosen belief or even an important issue for you to think about or research. A central argument of this book is that the values you embody transcend such concerns.

I agree with Tolle that I cannot own life (or spacetime) and therefore lose it as a possession. I also agree that I cannot lose the thing that I am – it is all or nothing. Nevertheless, what here would disprove the simple idea that we are all individuated manifestations or incarnations of the life principle or system of organisation? From this perspective, why can't we simply lose the incarnation of life within us? The result is that in death we, as unique but platform-dependent collections of cells, lose 'I' and biological organisation simultaneously, or more poignantly, our egos lose 'I' and our lives more-or-less simultaneously. The failure to clearly consider this simple possibility of physical dispersion (loss of incarnate life) accompanied by spacetime information dispersion (loss of self) seems to me to be the ultimate expression of an ego (or pandering to such).

One interpretation of this omission would suggest that the ego desperately wants and needs identity and form after death, so it invents or clings to transcendental consciousness or Father-God as the recipient of its essential identity stripped of form at the time of its death. It does this with the hope that this precious gift passed on will be honoured by transcendental consciousness or Father-God with eternal life in some new kind of form that will enable its continued doing in the future. This seems to be an example of introspective consciousness, doing its job of preparing for future possibilities, overshooting the mark by basing its preparations on nothing extrospective consciousness has supplied or can supply. In stark contrast, belief in the wonderful and marvellous self-organising principles of the universe on this side of the veil is not compelled to include belief in a personally embodied afterlife from the other side of the veil. It can be content with passing on a time-rich and virtuous legacy to the following generations of the living.

Why do we so often assume consciousness gives us a link to the eternal? The concept of eternity is misleading because it may be thought of as the sum of all the unfolding moments from the Big Bang until now, which would be finite but still unfolding like the universe itself. Eternity would thus be finite but becoming more open. Alternatively, we can think of eternity as the inappropriate sum of the current fleeting moment of reality to our mental conceptualisations of past and future (which are subject to error and falsehood).

The concept of eternity is misleading if we do not define what we mean. Do we mean the indeterminate and emergent definition of a cosmic cycle? On the other hand, do we mean the unverifiable model of eternity that is the product of our frontal and temporal lobes? If we attempt to link consciousness to the realistic definition, we do not arrive at a transcendental outcome, but rather an emergent, fractal and naturalistic one. If we link consciousness to the infinite but unverifiable model within our minds, we arrive at something possibly transcendental but perhaps also fanciful.

Once we venerated our instincts (our animal cunning, etc.) in the characters we assigned our gods (for example Artemis, the Greek goddess of hunting). Nietzsche came close in his writings to continuing this same ancient tradition in a more secular fashion. Why, if now we no longer see our instincts as transcendental, do so many still feel assured that our human consciousness, arising from within our instincts, is transcendental? It seems to me that the invention of God or transcendental consciousness is possibly the ego's last gasp for a continuing identity after death. This is a very deep and possibly disturbing idea for many of us. What do you think?

Can we find assurance without the supernatural?

This brings us to the next important question. Can we have a unifying and continuously unfolding purpose and potential (that we can believe continues after personal death), without a unifying principle of God or transcendental consciousness? This is a key question for our society today that students of emergentism answer with a clear 'Yes'.

Can we be assured of what we do not know without the supernatural? Perhaps 'Yes', within the natural realm and the normal limits of probability, but perhaps not so easily in the intersubjective realm, especially when we consider subjects such as death.

Isn't it a strange reality that personal assurance about personal hope after death simply demands a supernatural answer, but personal assurance about your survivors' hopes (or your society's hopes) after your death does not. We are easily assured that the world will still go on after we are gone. The focus on personal assurance, by definition, is egotistical and selfish; the focus on all that is outside self, by definition, is unselfish.

Perhaps this means that we just tend to focus on the wrong thing when considering death – we consider our disconnectedness and independence and personal achievement and life's blind selfishness or taking, rather than our interconnectedness and dependence and society's achievement in and through us. We ignore death's blind unselfishness or giving, which we are about to join. Perhaps a somewhat Buddhist perspective of death can be taken on board easily and broadly without the need for the supernatural. Perhaps death is simply going home to the stardust from which we have come – a return to background chaos, a return to monistic being, without self-aware doing, a return to giving, without the taking, a return of what was only ever borrowed.

Perhaps one day, parts of this same stardust will be taken up in some form of plant life, consumed by a human, who then gives birth to a child. Perhaps in this fashion, we will become vaguely involved in the universe's self-organised but blind doing once again. In this sense, conscious doing would happen only once, in one life, but self-organising and blind doing and being in many lives and arrangements of matter, indefinitely.

So can we have spiritual laws without the supernatural? Yes – if we define spiritual laws as simply immaterial and indirect self-restraints by which we personally choose to self-organise, live, and die. We can hold and follow our hallowed values, and with a little careful consideration, we can also observe how they play out in our day-to-day existence and in the lives we affect around us.

Do we want 'the End of Faith' or just the end of the superstitious and the absolute claims regarding the supernatural or anything else, along with their negative by-products such as bigotry and suicide bombers? Do we want the end of all myths, or just the end of religious faith in such myths and fantasies? Myths have played an essential role in our past. Jesus' masterly use of parables is a relevant example. Can modern parables or narratives be just as powerful, in a value-positive kind of way? I think so.

A case in point: In the Introduction to *Atlas Shrugged*, we read a note by atheist Ayn Rand in which she explains that she is both philosopher and fictional writer. However, to Rand, her role as writer was more important. This is because it was as an author of fiction that she could gratifyingly describe how her philosophy worked in the characters of her stories and through the events of their lives.[47]

This attitude seems to be the same one that motivated Jesus' use of parables in the New Testament or the Australian Aborigines' use of dreamtime stories. The personification of values in a story often seems to be the secret to their successful vremetic transmission from one generation to the next. It is not the merely cerebral understanding of a narrative that changes our behaviours or lives. Rather, it is the powerful emotional significations hidden within the narrative, and its context, that changes us at a subconscious and unconscious level. For this reason, emergentism does not advocate a harshly materialistic monism. The kind of interpenetrating monism that emergentism advocates is one that can be wonderfully imaginative and colourful in its interconnections, but never superstitious.

It is not the message of love that transforms us, but the re-experiencing of love through the message and in the subconscious that moves us. I am hoping you too can help in the powerful vremetic transmission of personal meaning through the story that is your life.

"The fruit of the spirit is love, joy, peace, longsuffering, gentleness, goodness, faith, meekness, temperance: against such there is no law".[48] Is this biblical statement still true? Can it be true for atheists as well? Perhaps we would like to think so, but firstly we might have to carefully define the word 'spirit' as a construct that captures the non-material and fleeting part of our human experience. That is, the word 'spirit' means 'essence of life' felt and enjoyed fleetingly, moment-by-moment. It has nothing to do with the future or past – it is the spark of life granted by the present moment.

The "fruit of the spirit" on the other hand, is that measured manifestation of spirit (or fleeting subconsciousness) that hangs over from now's fence and is ripe and free for the picking by the 'soul' (or consciousness). Time's passing can thus carry this signification-fruit from the spirit over to the soul, with inherent decay. It is this virtuous value-fruit within that I believe is our responsibility to cultivate, consume and permit to transform us. We might also permit it to transform those around us, and our society moving forward as well.

Unfortunately that fruit must one day rot in this body, but its seed, its vreme, if planted and cultivated in another brain and life, can give the value-fruit a timeless continuity. This also means that we can live on indefinitely in our successors if we define ourselves in terms of disembodied vremes. I hope this concept can be of some comfort to you.

Many Christians reading the last paragraph might form the idea that the Emergent Method is what the they might describe as 'walking in the spirit', that is, living a life guided by those virtuous values that percolate up from the subconscious moment-by-moment and are also its reward. In a sense, this is true: The Emergent Method is living life in a way congruent with its subconscious virtues - but divorced from all superstition and reliance on the supernatural. The Emergent Method's focus is on this life rather than any mysterious or saintly afterlife. It also advocates a tolerant and inclusive life rather than an elitist one.

Is "the fruit of the spirit" subject to such manipulation in the hands of society's movers and shakers that it is 'up for grabs' or transitory in meaning and actions if not in words? Yes, definitely. All vremes have to take energy in and from brains – their survival machines, in order to live on. Memes are subject to fast mutation if they are not solid and well-defended within a strong conceptual framework (whose strength is its personal and lived meaning). Love or faith that keeps us ignorant, or ill-informed, or staying where we are, or as others dictate we should be in a family or an organisation - is this really the fresh fruit of the spirit rising up from within or the decaying fruit of something far less lively and self-actualising?

This much we know from our review. Whatever essential value or measuring-stick of consciousness you or I wish to retain in our mind-body complex, we first have to catch it from the indirect or subconscious world before it fleetingly moves on and vanishes. We then need to permit it to inform our understanding (reflected or related knowledge), fully appreciate it in our actions, weave it into our most cherished myths and narratives, share and teach it, and finally allow it the liberty of evolution in our unfolding lives. We will then be in the position to not only capture a glimpse of our own personal purposes and destinies, but also more deliberately choose those purposes and consciously shape those destinies.

Your Purposes

Taking your values from Pillar 1 and your credo from Pillar 2, why don't you try to write down your own life-motivators, arising from your values, your credo and your unique place and time in the universe? Think with your consciousness, inwardly and outwardly, about what inspires you to get up, eat, drink, go to school or work and come home again each day, perhaps as opposed to what inspires your family or friends. What are the deep dreams and ideals towards which you know your values, credo, and consciousness-plus-instincts are driving you? What are the needs of your internal and external stakeholders asking of you? What is it you would like to achieve, change, or improve?

Perhaps more importantly, just take a little more time to check your motivations. Why do you want to succeed? Why do you need to succeed? Is it for personal security, a shiny new car, a new house, or a new partner? Is this how you will measure personal success? Is it because of a sense of loyalty to others? More generally, is it because of the values themselves inside of you that are seeking their full expression through your unique and potential best?

If life is not about material possessions, but really about your experiences and your meaningful relationships, then your values-in-action are going to motivate you towards those new life-affirming experiences and identifications (or knowledge and understanding) in your future. The change in introspective communications within you, based on motivations coming from your cherished values, will involve authentic rather than forced emotions. When you unreservedly follow your well-tested values, you are not locked into anything that detracts from your self-actualisation. And the result will be that your natural self-talk will change, without anyone forcing it, such that one day you will wake up to the realisation that nothing can divert you from your self-actualising successes any longer. A long time before you achieve your goals, you will know that they are already in the bag. However, you may have to go around the block a few times.

Perhaps right now, your self-image and the image you portray to those around you is not the same image that you know it will have to be if you are to reach your potential best. In what ways does your image currently miss the mark? Do you need to face any issues or make a few firm decisions to honour your values? Do you need to have a higher respect for your limited time?

While the challenge of all of this may seem a little daunting, is your deep motivation to see your values-in-action stronger than your fears, including the fear of personal failure, fear of loss of some relationships, or the fear of loss of possessions? Would you be prepared to fail just for the sake of trying to succeed? If not, it will be difficult to ensure your self-actualisation. However, you need to receive these remarks with caution, because there is no easy path to success. Meaningful success is necessarily a struggle. The crooked path of growing energy effectiveness combined with growing energy efficiency is not an easy one.

This situation perhaps reminds us of the scene in the movie *Braveheart* where Mel Gibson, playing the role of William Wallace, steels his troops for imminent battle with the English. He appeals to them as free men and appeals to their individual desires for Scottish liberty in defiance of an English king's tyranny, even at the possible cost of their lives on the battlefield. He asks his reticent troops, "What will you do with that freedom, will you fight?" At first, the answer is a weak "No!" But then Gibson, solemnly appealing to their core inner selves, asks them,

> *"Dying in your beds, many years from now, would you be willing to trade all the days from this day to that for one chance, just one chance, to come back here and tell our enemies that they may take our lives but they'll never take our FREEDOM?"*[49]

Well, would you? The troops then answered with a resounding win against the English that day on the battlefield of Stirling. Politics and Hollywood aside, understanding your motivation to succeed and overcome obstacles through your deep veneration of your values, and your tenacity and focus, is paramount to you achieving your goals and greater purposes.

So what are the practical purposes that arise out of your values, credo, and stakeholders? Write your deepest purpose or purposes down, no matter how outrageous they might seem. Your potential lies in now writing achievable goals for your purposes in the form of project milestones, visualising/imagining how you can achieve the steps (as elite athletes so effectively do), and then achieving those milestones one at a time.

To get you started, refer to Figure 25: A Model of Wellbeing – the Mind-body as an Energy Processor. For each area of wellbeing, list a number of objectives (not platitudes or new-year resolutions that are quickly broken), beginning with the four aspects of Physical Wellbeing.

For instance:

- I will play a sport, swim, or go bushwalking to get fit and lose weight
- I will eat healthy salads during warmer months and healthy vegetables during cooler months. I will drink more water
- I will join a gym to increase my strength
- I will set the alarm in the evening to remind me to go to bed, as well as for the morning to wake me, so I get sufficient rest each day

Next, consider your purposes with respect to your Ego Wellbeing. Examples:

- I will set up a monthly budget, by which I will pay myself first
- I will move on from those bad attitudes and habits that have held me back in the past, perhaps through the assistance of a life coach

Then consider purposes regarding your Mental Wellbeing. Examples:

- I will do more research into the areas of knowledge that challenge and support my values
- I will join and contribute to groups that challenge and support my attitudes, perhaps looking for compatible business partners with which to share business concepts
- I will mix with positive people that can challenge me to do better

Next, consider purposes regarding your Social Wellbeing. Examples:

- I will ask a work associate to mentor me, so I can serve my workplace (and stakeholders) more effectively
- I will end short-term social relationships that have little prospect of being win-win
- I will seek win-win relationships through an intelligent recognition of my perceived strengths and weaknesses, and a pragmatic recognition of the strengths and weaknesses in others

Finally, consider purposes regarding your Spiritual Wellbeing. Examples:

- I will be grateful every day for my knowledge and understanding by expressing more easily my appreciation to loved ones and associates
- I will be more forgiving of myself and of others who have wronged me
- I will seek spiritual wellbeing by taking time out to process the things that seem to worry me, through walking in the natural environment, exercise, meditation, or through other means

Of course, as you permit your introspective concepts and values to self-organise and emerge on a higher plane, so will your credo, motivations, purposes, focus, goals, milestones, and potential.

Changing Unwanted Beliefs

Perhaps your list of purposes seems daunting or unattainable because you know you have deeply held beliefs that get in the way of your emergence. The following is a copy of a script I recently conducted in a coaching session designed to change a specific unwanted belief. It is based on a technique provided by Ben Harvey at Authentic Education in 2015:

Client's Limiting or Unbalanced Belief: "I have to do everything myself!"

Session Intention (set by client): To be hopeful and centred

Session filter (set by client): To be accepting of self and what is

What does holding onto this belief do for you or give to you?

- A feeling of purpose, self-worth
- A feeling of superiority or satisfaction; it feeds my shadow values
- Permission to feel frustration and an excuse to get angry

What does holding onto this belief do to you?

- It takes a big chunk of my vital energy
- I can't rest or relax
- It holds me back in my own mission – I am always helping others

What will the cost of this belief be to you in 1, 5, or 10 years' time?

- Physical Wellbeing
  - I won't be able to enjoy a break the way I otherwise could
  - It will age me
- Ego Wellbeing
  - I will burn out in my business
  - I will not build the income I could get without this unbalanced belief
  - I will not be able to build the team I need to grow the business
- Mental Wellbeing
  - Even if I make the home and nice things I want, the joy they could bring will be lost
  - I will miss out on happiness and balance; it stops enjoyment
- Social Wellbeing
  - It will cause me to resent others
  - It will cause others to distance themselves from me
  - I will miss out on the interconnections that bring deep and mutual joy
- Spiritual Wellbeing
  - I will lose hope; it drains me of hope, motivation and inspiration

What would you become and how would you act ***without*** *that old belief?*

- I would be more balanced, more centred
- I would be greater than what I am now
- There would be an ease
- There would be a different intensity
- My hard edges would be softened
- I would be more graceful
- I would be more trusting of others

Consciousness

The coach's suggested additions:

- I would be allowed to fail, but through failure, get back to the prosperity and flourishing I seek through my relationships: The concrete measure of truth is effectiveness
- I would trust others, and they would trust me
- I would claw back my time
- I would find lasting joy

Answer the same question again, but this time, go through each Area of Wellbeing as listed in Table 6 *to help categorise your responses:*

- Physical Health & Beauty
    -
- Career & Business
    -
- Mindset
    -
- Friends & Community
    -
- Family & Partner
    -
- Homemaking & Finances
    -
- Spirituality
    -

Now write down the polar opposite of your limiting or unbalanced belief:

- "I don't have to do everything myself!"

Please complete the following:

- I am allowed to…
- I am allowed to…
- I am allowed to…

Create an Intention Card for your new belief and carry it for 30 days

- Add to it your essential learnings from this session
- Fold it in half and put it in your pocket
- Anytime you think about the card, or fall into the old, unbalanced belief:
    - Take a deep breath to re-balance your emotional state
    - Touch the card

- o Close your eyes and imagine what it means to you and your future
- o Read it out aloud if possible
- o Check: Do you really want to remove the limits and imbalance of the old belief?
- o Pay attention to the answers you receive from this driving question
- o Feel the 'ImagineAction' feelings of having, being, doing, and again having what is written on the card

Please note Ben Harvey's Seven Principles of Empowerment:

1. The world is what I think it is (my thoughts change the world, they can't do otherwise)
2. There are NO limits (in a cosmos full of potential)
3. Energy flows where attention goes (it always goes to greater effectiveness and/or efficiency)
4. Now is the moment of power (not the past, and not yet the future)
5. I live my love
6. All power begins from within (the potential of the universe is released through imagination)
7. The measure of truth is effectiveness and efficiency (not the old rules, mores, taboos and unbalanced beliefs that limit me)

Do you want homework and accountability? Then have fun completing this document and sending it back to me done.

Concepts of Emergence Encountered in Pillar 3

However perhaps the realms of epistemology (or justifiable knowledge) and evolution are more than just analogous. Maybe they are both part of the one process of emergence going on in our universe. A discovery of new knowledge requires new explanations – new understanding that fits the temporal facts. This new understanding is emergent – it is a new emergent property that exists more in the arrangements of virtual particles than real ones.

We are the facilitators of this emergence. We are the agents, reactants, or catalysts in the process. Our minds' heuristics arrange the intersubjectively relevant facts or values and drop the irrelevant platform-dependent ones to come up with a new understanding. Those same heuristics also drop some relevant information due to the need for closure or boundaries, just as some sculptors can chip away in the wrong places to save time; hence the necessarily emergent or evolutionary nature of understanding and the revealed skill in a sculpture such as Michelangelo's David.

One other point that arises here is the evolution or emergence of new causes and new risks. Can we keep up with the risks our evitability engenders? It seems to depend to some extent on the proficiency and efficiency of our explanations of the universe. Whatever is the answer - the self, evitability, freedom and morality evolve.

So does our own intersubjectivity, and especially its explicit vremes, present us with a problem of illusion or something to celebrate? Is our intersubjectivity a means of taking the universe to a fabulously new level of matter-space-time coagulation? As Robert Wright[50] might put it, is our intersubjectivity a means of finding wonderfully new opportunities for 'positive nonzero-sum outcomes'? The Emergent Method suggests an optimistic response to these questions.

It seems we have to tread even more carefully than the most diligent of scientists (and his/her Scientific Method), if we are to uncover the introspective contingent truths that will enhance our self-actualisation (or artificial emergence). This careful uncovering is what the Emergent Method offers.

While acknowledging the brain's incredible complexity, can I dare say that the brain, just like nature, effortlessly 'relates' much more than it artificially 'computes'? It perhaps falls naturally into relationally-patterned answer-traps much more than it calculates. By avoiding analytical calculation (or perhaps artificial relationship searching), I suspect it both escapes typical Gödel limitations on the upside and falls into instinctive biases on the downside. This supports the idea that the mind-body might work very much like natural (and artificial) emergence in space-matter at the subatomic levels.

In the next chapter we will discuss the ramifications of the two aspects of our human nature (instinct and consciousness), brought together in the one mind-body complex.

---

[1] See Dennett, Daniel C., *Breaking the Spell* (Penguin Books, USA, 2007), 111

[2] After Dawkins, Richard, *The Selfish Gene* (Oxford University Press, USA, 1990 (originally 1976)), Dennett also took up this theme, for instance in Dennett, Daniel C., *Consciousness Explained* (Penguin Books, USA, 1993), 199-208

---

[3] See http://www.sciencemag.org/ardipithecus (last accessed 2012) and http://ngm.nationalgeographic.com/human-evolution/human-ancestor (last accessed 2012).

[4] See Wrangham, Richard, *Catching Fire: How Cooking Made Us Human* (Basic Books, NY, 2009), 121

[5] See Doidge, Norman, *The Brain that Changes Itself* ((Revised Ed.), Scribe Publications, Melbourne, 2010), 396-399. Mithen's three brain 'modules' correspond with the model of "competitive", "rational" and "social" intelligence, respectively, presented in the next chapter – but miss here the fourth category of what I would call introspective intelligence.

[6] See Dawkins, Richard, *The Selfish Gene* (Oxford University Press, USA, 1990 (originally 1976)), 200-201

[7] Village Roadshow Pictures, *The Matrix* (1999)

[8] See Kahneman, Daniel, *Thinking, Fast and Slow* (Penguin Group (Australia), 2011)

[9] See Deutsch, David, *The Fabric of Reality* (Penguin Books, New York, 1998), 68-69

[10] See Kauffman, Stuart, *Reinventing the Sacred* (Basic Books, USA 2010)

[11] However, as stated in the Introduction, I believe this view is misleading to the extent that minds must emerge from brains as much as brains must emerge from minds

[12] Sam Harris blog, *The Mystery of Consciousness* (last accessed October 2011)

[13] See Pinker, Steven, *How the Mind Works* (Penguin Books, USA, 1999), 95-97

[14] Ibid., see p.562, which touches on the same idea

[15] See Damasio, Antonio, *Self Comes to Mind* (Vintage Books, New York, 2010)

[16] See Hofstadter, Douglas, *I Am a Strange Loop* (Basic Books, USA, 2007)

[17] Village Roadshow Pictures, *The Matrix* (1999)

[18] See Pinker, Steven, *How the Mind Works* (Penguin Books, USA, 1999), 561-565

[19] Greenfield also advocates this view. See Greenfield, Susan, *The Neuroscientific Basis of Consciousness* (address to SALK Institute, 22nd May 2009)

[20] See Cahill, Reginald T., *Process Physics* (Flinders University, Adelaide, 2003 (available on the web))

[21] See https://en.wikipedia.org/wiki/Blue_Brain_Project (last accessed 2016)

[22] https://en.wikipedia.org/wiki/Benjamin_Libet (last accessed 2016). An early paper on the topic was presented as Libet et al., *Time of Conscious Intention to Act in Relation to Onset of Cerebral Activity (Readiness Potential) – The Unconscious Initiation of a Freely Voluntary Act,* Brain 106 (1983): 623-642

[23] Retrieved April 7 2016 from Information Philosopher Web site http://www.informationphilosopher.com/freedom/libet_experiments.html

[24] Dawkins, Richard, *The Selfish Gene* (Oxford University Press, USA, 1990 (originally 1976)), 200

[25] See Dennett, Daniel C., *Freedom Evolves* (Penguin Books, USA, 2004)

---

[26] Some would argue here that we have moved away from a more objective neurophysiological discussion when we start to talk about subjectivity, which I think Dennett addresses well with his idea of heterophenomenology. See Dennett, Daniel C., *Consciousness Explained* (Penguin Books, USA, 1993).

[27] See Krauss, Lawrence M., *A Universe from Nothing* (Free Press, New York, 2012)

[28] See Damasio, Antonio, *Self Comes to Mind* (Vintage Books, New York, 2010)

[29] Merleau-Ponty, Maurice, *Phenomenology of Perception* (Routledge, New York, 2014 (Original edition in French in 1945)), 439

[30] See Dennett, Daniel C., *Consciousness Explained* (Penguin Books, USA, 1993), 36-51

[31] Hofstadter, Douglas, *I Am a Strange Loop* (Basic Books, USA, 2007), 32

[32] See https://en.wikipedia.org/wiki/Freedom_Evolves (last accessed 2012)

[33] See Deutsch, David, *The Fabric of Reality* (Penguin Books, New York, 1998), especially 120-121

[34] See Deutsch, David, *The Beginning of Infinity* (Penguin Books, New York, 2012 (original edition 2011)), especially 212-215

[35] See Wright, Robert, *Nonzero: The Logic of Human Destiny* (Vintage, NY, 2000)

[36] Dawkins, Richard, *The Selfish Gene* (Oxford University Press, USA, 1990 (originally 1976)), 200

[37] See Dennett's discussion of the mind's Pandemonium architecture in Dennett (1993), 231-252

[38] Dawkins, Richard, *The Ancestor's Tale* (Phoenix Paperback, London, 2004), 245

[39] Hollingdale, R. J., *A Nietzsche Reader* (Penguin Books, Sydney, 1977), 162. I added the comments in square brackets.

[40] See Dennett, Daniel C., *Breaking the Spell* (Penguin Books, USA, 2007), 116

[41] See ibid., 153

[42] The canon of the New Testament was finalised soon afterwards and first approved at the Synod of Hippo in 393 A.D. For a deeper discussion of religious development see Smart, Ninian, *The World's Religions* (Cambridge University Press, 1989)

[43] Dalai Lama, *Becoming Enlightened* (Rider, London, 2009), 215

[44] Ibid. 224-225

[45] Ibid. 165-166

[46] Tolle, Eckhart, *A New Earth* (Penguin Group (Australia), Melbourne, 2009 (Original Penguin Edition 2006)), 128

[47] See Rand, Ayn, *Atlas Shrugged* (Plume, USA, 2007 (Fiftieth Anniversary Edition)), Introduction

[48] KJV: Galatians 5:22-23

[49] Twentieth Century Fox, *Braveheart* (1995)

[50] See Wright, Robert, *Nonzero: The Logic of Human Destiny* (Vintage, NY, 2000)

PILLAR 4: Human Nature

In this chapter, we will first consider the challenges the different strands of human nature present and then place those challenges into emergentism's phaseal model of human behaviour, which I will expand to include a more naturalistic system of ethics. We will explore emergentism in terms of the value-sets of Pillar 1, a quick comparison to science, objectivism, Buddhism, and Western pluralism, as well as organisational behaviour (in terms of a sales training scenario), and societal power structures. In relation to your personal philosophical quest, the text will challenge you, within a more fully expounded framework of the Emergent Method, to take the first concrete steps towards the fulfilment of your purposes identified in Pillar 3.

The More Holistic View

We need to place the various strands of human nature and its strange loops of self-organisation into a more holistic model of human behaviour and human nature. We will then need to explore how this model can address the kinds of things any philosophy should address in terms of ethics, a personal philosophical framework, a social system, etc. Perhaps more importantly, we will explore just how we can shake off the chains of our selfish replicators.

Perhaps for some, this raises the first few questions, 'How can we simply decide on an ethical framework ourselves? How do we arrive at an ethical framework to take us forward if it is not authoritatively handed down to us from above?' The tenure of this book has laid out the answer – when we honestly, rationally and humbly look to nature, our own instincts and our own consciousness they hand us back the perhaps incomplete but inclusive and emergent answer on this side of the veil across unknowable reality.

The Challenge of an Animal with Elaborate Consciousness

So far we have seen how our animal instincts have evolved to make us excellent hunters and gatherers – as animals, we are self-reliant risk-takers and strong competitors as well as socially proficient and interconnected co-operators. Consciousness added to these animal drives a passive awareness of both the outer material world and the inner non-material world, just as we see in the night sky an outer material world expressed in stars and a non-material world expressed in space.

The combination of animal instincts and elaborate consciousness has been hugely successful for our species in the biological short term, but by its very nature, has not been without its regrets and does not guarantee our future. In terms of our developed global traditions, we see a mixed bag, to say the least. A core concept to take us to a successful future and purpose, in terms of this book, is the idea of evolved virtues freed from any authoritative ownership, and deeply cherished within an naturalistic framework of ethics.

However, in the light of threats arising from our seriously damaged global environment, religious fundamentalism, and the proliferation of weapons of mass destruction in the hands of possibly irresponsible individuals, the more natural approach advocated here might be too late. That is, too late if we do not all take responsibility right now for our 'egoistic state of insanity' (to use Eckhart Tolle's term, discussed below).

We desperately need scientific education and philosophical enlightenment. The gap between the haves and have-nots even in our developed Western countries is appalling. It makes the enlightened discussion of science, religion, society and its values (and value-frameworks) we need to urgently have almost impossible.

We need to see that we are in a state of war, not primarily with evil, terror, drugs, ISIS or Al Qaeda, but with continuing scientific and philosophical ignorance. We should not underestimate the task ahead of us: Further than Richard Dawkins suggested, we are in the midst of a war against the blindness of our selfish replicators (but not their mechanisms). We are in a war against short-sighted selfishness of all kinds, including those that would use violence and deceit against their fellows, and those that would use force of character or the sheer advantage of numbers in the short-term over more enlightened forms of moral reasoning.

If we do not get natural balance back into our lives (the lack of balance being the result of consciousness-plus-instincts gone awry) we are most probably doomed. Further, our tight-fisted ownership of ideas or values with respect to death and the afterlife could possibly bring us death and no afterlife under any definition for our progeny of the 22nd century because they will have a greatly reduced opportunity to even exist.

Eckhart Tolle[1] explains the creation of the ego, which he stylises as a false sense of self, through an exaggerated sense or understanding of mind.[2] This definition is a little different to Freud's, which more simply saw the ego's role as balancing the selfish demands of the *id* against the ethical demands of the superego (and dealing with the dissonance between the is and the ought). Perhaps what Tolle means is that the mind-games (the ignorant and error-prone thought-models of past and future) became so exaggerated in the introspective consciousness of each of us that they became divorced from reality or from simple and instinctive experience in the present. Perhaps at the same time a more dualistic view of existence took over from the prior more monistic view. We, our consciousness-plus-instincts, became too concerned with quickly doing, interacting or analysing – and not concerned enough with synthesising and simply being logically true to nature, ourselves and each other in the present moment and in deeply appreciated and inclusive interrelationships.

We typically believe in our mind-games that happiness is derived from future outcomes (when we get money, when we get the perfect partner, when we get the right job, etc.) rather than found in the experiences of unfolding present processes. Further, happiness or sadness associated with faulty past memories or imaginary futures are often more important to us than the happiness or sadness of actual events as they are experienced, past and present. Our errors in introspection and introspective concept formation have amassed to such a level that many of us unwittingly move further away from the simple joys we crave for, and our more instinctive subconsciousness has been rendered powerless to do anything about it. That is, the modern ego's faulty mind games have dulled the *id*'s link with relativistic reality.

Tolle then goes on to explain that an emotion is a reaction of the body to thinking,[3] which seems like an interesting and novel focus on the mind-body cycle that Damasio (2010) considers in depth in a purely scientific context. Tolle further explains how our bodies, run by a wayward ego holding the reins of our conscious thought-life, hold on to memories and grudges (the results of competitions over food, property, potential mating partners, etc.) that in animals of lower consciousness would be soon forgotten.

While the picture Tolle presents here is a clear one, he seems to make a distinction between consciousness and the ego (and later the "pain-body", "unconsciousness" and "victim identity"), the one being false and bad (the ego) and the other being true and good (consciousness), that seems unnecessary. It seems that human consciousness has clearly evolved to grant us some useful tools for survival but in doing so, has left us with some not-so-useful side effects – lingering memories of suffered wrongs being one of them and perhaps for some, tragic feelings of existential alienation, as described so well in Albert Camus' writings. Perhaps far more drastically, consciousness has brought with it a separated, independent mindset and moral agency that has robbed us of our former monistic simplicity. We cannot regain our simplicity but I think we can rediscover a connection and flow with the monistic present, as Tolle also advocates through the evocative title of his 2004 book, *The Power of Now.*

Tolle adds that our egotistical thoughts set up an emotional and energetic dissonance in our bodies, which resonates to cause yet more thinking, in a vicious cycle.[4] He suggests that our salvation from this negative reinforcement resides in removing ourselves from those faulty mind models of our past/future (in our consciousness) to dwell in the processes of the living present (in our instincts), with which I reservedly agree; this is part of the solution. It is this instinctive living in the present (whilst seeing and designing a future) that Howard Roark so wonderfully exemplified in Ayn Rand's 1943 novel, *Fountainhead.*

Tolle then goes on to state, the pain-body that is in almost everyone is an energy field of old but still active emotion.[5] The idea here is that the body stores the pain and stresses of the ego-centric thought life (or consciousness gone astray) and Tolle labels this bodily store or side-effect the "pain-body". It was the absence of this pain-body in Howard Roark, if you like, that made him so different to his contemporaries and Fountainhead's hero. Tolle suggest further that the nightly news and the dysfunctional relationships we see around us offer evidence that the pain continues, and continues to grow in humanity's collective psyche.

He concludes that the pain-body is most likely encoded in the DNA of each of us.[6]

The idea that our genetics could encode a pain-body is an interesting thought. The pain-body would only be encoded so deeply (i.e. in every human) if it fundamentally improved our survival chances. How would the theorised pain-body genes improve our survival? Would it be through protection from danger? The successful animals of lower consciousness have sufficient protection already. Do humans need more protection simply because of their higher consciousness? Is there an evolutionary pain-body arms race going on within our genes and memes (bigger neocortex – more consciousness – more pain-body – bigger neocortex)? The outward manifestation might be the shrinking of other parts of our brains, along with actual arms races, personal alienation, social craziness and serious attention deficits in our youth, as seen on the evening news and in the drama of our families' dysfunctional relationships.

This thought would add further weight to the idea that the brilliant gains of our evolutionary past in terms of consciousness by no means assure our future. Perhaps the excesses of elaborate consciousness give us pictures of happiness in future or past outcomes based on imagined mind-models that are just as distorted as our current pictures of emotional pain and suffered wrongs. That is, perhaps it is not just a pain-body that is the problem, but elaborate consciousness per se (painful or not). Whatever the case, at this evolutionary juncture, it seems like we desperately need scientific education and philosophical enlightenment with regard to our consciousness-plus-instincts.

Tolle goes on to explain how pain-bodies blindly feed on and amplify their own structures of hurt and pain set up in the mind-body complex, and in so doing make our lives more shallow, our thoughts more misdirected or scattered and our relationships more dysfunctional. For instance, he notes that pain-bodies write, produce, and pay to watch (explicitly violent) movies.[7] Why would we pay to watch such explicitly violent films? The idea does seem crazy, but still we do it; we do not just want any kind of excitement – we want violent and painful excitement, and Hollywood's success seems to prove this. The idea of the need for perceived protection through a strengthened pain-body shield is perhaps not altogether unreasonable. We watch violent movies to enhance pain-body consciousness (the shield), which in turn makes us more acutely aware of personally or socially suffered wrongs (real or imagined, consciously or subconsciously). This possibility would not necessarily contradict Steven Pinker's[8] view of humanity's trend over thousands of years to be less violent either, because while the occurrences of violent acts may have gone down, at the same time our conscious sensitivity to such acts may have increased.

Alternatively, we strive for happiness through partly uncontrollable outcomes. Whether or not those outcomes are achieved, our ongoing feeling of unease causes us to strive for more and better outcomes in the unreasonable belief that they will satisfy. Examples might include yet another lottery ticket or new car. Again, elaborate consciousness seems to do this to shield itself from facing our simple but uncertain reality – which reality our particularly self-reliant instincts Nietzsche described so well seemed to cater for quite well in the past.

It seems elaborate consciousness has lost trust in and reliance on our ancient self-reliant instincts honed in the wild but less relevant in the often unsustainable city-environments of our own making. This would make some sense; the result so far seems to be that humanity is on a subconsciously self-defeating slippery slope. The mind-games that are trying to replace the instinctive biases are doing a lousy job. We need help to set up new and healthy mind-habits that can stamp out silly mind-games and their emotional charges.

Tolle agrees. He also suggests that those often violent movies can help us, with all humanity, see our own insanity, and thus awaken us.[9] After discussing many forms of violence stamped in the psyche of some races and peoples, Tolle says "*There is only one perpetrator of evil on the planet: human 'unconsciousness*'", which I would interpret to simply mean consciousness that, via blind and unintended side effects, has gone awry.

"*That realisation is true forgiveness. With forgiveness, your victim identity* [resulting from your pain-body experiences] *dissolves, and your true power emerges – the power of Presence. Instead of blaming the darkness, you bring in the light.*"[10]

As we understand that our evolving consciousness-plus-instincts has not always led to a perfect set of human attributes, but in many important ways has led to our current set of personal and social problems, we can forgive its limitations (the limitations of our memes and genes). We can get on with being the virtuous guides of our blind replicators, both memes and genes, rather than their mindless slaves. We, like Howard Roark, can get back to enjoying interrelated being as well as enlightened and emergent doing, as a deep and true expression of our values freed from the trap of consciousness's silly, short-sighted and sometimes debilitating mind-games as well as instinct's sometimes irrational, shortcut biases.

However, it is not easy to achieve such a vision of carefully guiding both mind-games and genetic biases. Taking rational charge of the direction in our previously more shallow lives to make them more real is not easy. Changing the direction of our subconscious and conscious lives to match our recent scientific discoveries and philosophical enlightenment is probably like turning a huge ship around at sea it will happen eventually if we stay at the wheel, but do not expect it to turn on a dime. We will need to have serious regard for the fact that our behaviour is largely driven by good and bad habits, many of which we have inherited. The Emergent Method, perhaps borrowing useful techniques from Neuro-Linguistic Programming, is about replacing those old self-destructive habits with self-actualising ones.

What do we watch on TV? How much time do we spend on the internet at the same old websites? How much do we think about choosing the stimuli to which we expose our perceptions and subconscious? Do we notice when we make moral choices or when we don't, but should have? Do we notice when we confabulate or are disingenuous? How much do we reflect on our own moral choices in terms of the virtuous process versus an imagined outcome? Do we each notice and take account of the contradictions between our ethos and our moral choices (and outcomes achieved)? Have we recognised and dealt with our shadow values? How often and in what circumstances do we give ourselves permission to disrespect self? Why? Is this acceptable in terms of the duty to self and to society? Why?

Not in disagreement with Tolle, our ethical and moral strength is likely to be a function of the strength of our moment-by-moment virtues (e.g. our values and self-worth in action), as well as a function of the stakeholders/company we keep and the richness of our visions for the future. Tolle tells us that we begin to become enlightened when the suffering caused by the pain-body breaks us.[11] He concludes:

> *"The next step in human evolution is not inevitable, but for the first time in the history of our planet, it can be a conscious choice. Who is making that choice? You are. And who are you? Consciousness that has become conscious of itself."*[12]

Not sure Richard Dawkins, Ayn Rand, or I could have said that any better! Tolle stands tall in unfamiliar company, at least on this point. Growing, unbridled, and blind or numbly unreflective consciousness in itself is a core part of the problem with our modern global civilisation. We need to be aware of the limits or blindness not only of our instincts but of consciousness also.

We can't just change the reins from instinct to consciousness. We need to slowly guide our consciousness and instincts forward together in the right direction towards wholesome purposes that encapsulate a wide range of relationships and values (and systemic borders), recognising that our consciousness and instincts are our dear partners in the transformational journey. We need a framework of concept formation, vremes and values-in-action that will help moderate, self-organise, and self-actualise our quickly evolving consciousness-plus-instincts.

As Tolle suggests, much of the problem lies in what he (but certainly not Freud) calls unconsciousness but others might call subconsciousness – that is, those implicit cultural values, mores, taboos, and biases that we blindly allow to guide us, often without challenge from an awakened consciousness-plus-instincts. We need to awaken; we need to visualise or otherwise imagine our future path and then take it, one enlightened step at a time.

Some Philosophical Approaches

We have touched on a variety of approaches to human nature, instinct and consciousness, within these pages so far. We have highlighted their strengths and limitations, while maintaining the book's central message of how explicit values arise out of our humanity and can guide our evolving purposes and help realise our greatest potential.

Objectivism

We briefly discussed the instinctively self-reliant approach of Ayn Rand's objectivism. This approach includes a clear advocating of personal values, including a rejection of all forms of force and violence, except in self-defence. It supports a system of laissez-faire capitalism[13] in which each individual is free to pursue his or her own happiness through rational thought and personal productivity. This approach seems to be somewhat consistent with observations regarding existence, as it seems to include an ecological respect for peers, traders and/or competitors operating on the shared playing field.

Perhaps I should explain what I mean by the term 'observations regarding existence'. I have mentioned previously that nature doesn't plan as humankind does – it seems to just confidently but blindly act now and deal with the statistical consequences as they unfold in the present moment; that is, nature emerges. It moves from a position of strong relationships into a slightly morphed set of new relationships. That is:

1. Nature starts locally with its many winning abilities or subsystems, for example within its genetic survival machines (the risk-managerial 'What'). It does not wait for or look for something else before beginning its present endeavours. Its successful subsystems receive support and challenge equally.
2. Nature always acts within its own constraints (the extrospective 'How'; e.g. spacetime in terms of its interactions, an individual species in terms of its genome and evitable systems, a brain in terms of its memes, or a herd in terms of its culture). Living things are subject to their local constraints and surrender to them, but also find a way forward within those constraints. Nature then leaves humanity to use outward-looking consciousness or intellect to observe the constraints in action.
3. At this point humankind would typically use the social instinct to move on to ask for a social context to such self-reliance and action (the normative 'Who, Where, When') but nature would not ask because it already knows the (vast) answer implicitly. It would simply always answer, if it could, 'me and you, here and now'. Nature is not a collectivist statist; it is more like a small business entrepreneur who grows into a bigger business. It is a quantum moment that

grows into a living universe. Its social actions and solutions are local, simple (in the sense of unplanned and absent of rumination), organic, and most importantly, sometimes emergent. Nature thus promotes the long-term win-win. Non-living nature can be a huge meteorite striking the earth or moon but living nature is often a small garden that slowly adapts and stabilises through competition and cooperation and then grows into a magnificent valley vista. It is humans who complicate and magnify social interactions beyond reason (beyond locally linking systems or more quickly than local resources can link together). For instance, humans are quick to rob future generations by exploiting unnatural financial systems, in order to finance negative net-present-value projects today (the oft-acknowledged cause of the Global Financial Crisis). Nature might be forced into seemingly self-defeating projects (e.g. microbes using up and running out of sustainable resources on a Petri dish) but this action is not borne out of greed or empire-building for its own sake. The action is simply entropy's statistical drive downhill (which comes with occasional emergent mutations uphill that, in this example, might enable a group of microbes to jump off the dish).

4. Further, we would ask with our inward-looking consciousness perhaps the crowning question (the introspective 'Why?') Here, nature would again fall silent, except perhaps to answer assuredly, 'because I am and because I can'. Again, humankind often complicates the answer with conjectured notions of an afterlife, a doomsday, a coming messiah, a paradise of 70 virgins for each martyr, an everlasting burning hell, inquisitions, holy wars, etc. Nevertheless, to ask 'Why?' and proffer explanations is what we do as a species. As Nietzsche suggested, it adds a sense of meaning and a cognitive framework for our instinctive drives, the result of which is our intersubjective moral agency. It is only as we do this asking and explaining more existentially and vremetically, that we reach towards our potential best.

In objectivism there is a lot of discussion dealing with the questions of 1 'what' and 2 'how', the business-end of nature's many activities and what they openly reveal to us. However, objectivism has relatively few discussions dealing with the 3 'who, where, when' and 4 'why', because these questions have indirect or reflective answers in nature.

That is, the values advocated by objectivism's literature nearly all fall within the risk-managerial and extrospective value-sets of Pillar 1, almost to the total exclusion of the other two normative and introspective value-sets. Objectivism, like nature itself (and like Howard Roark in *Fountainhead*), either grants us the freedom as intersubjective moral agents to make up our own minds regarding the normative and introspective realms or subtly warns us against any answers to 3 'who, where, when?' or 4 'why?' that are too far removed from direct and local reality.

While the social and contemplative natures of humanity are undeniable (they evolve naturally), objectivism has little to say about them compared to religions or other philosophies. It seems to me to be a wise approach, even if often maligned, to keep the social and contemplative values operating somewhere near the realistic realm nature provides and that we may observe and deduce. We are gods, but the very intersubjectivity that our moral agency affords bedevils us. The same is true of all gods. By this logic, there can be no such thing as an infallible god.

The ideas of emergentism, the philosophy of this book, perhaps helps to clearly identify and explain objectivism's focus in terms of nature and human nature. Just as direct matter rather than indirect spacetime (and now rather than past/future) is the final focus of all practical human endeavours, so outward-looking consciousness rather than inward-looking consciousness should be the final focus of all practical human thinking. Without this balance or interpenetration, our intersubjectivity can mean the failure and destruction of our moral agency.

Further, just as the social instincts are built on the self-reliant instincts in reality and the inverse is not equally true, so we should always recognise the unique individual in our social groupings. In this post-bureaucratic era, we typically try to create collectives that recognise the basic role of the local, free and willing individual (who enjoys a sense of personal meaning), because it is the individual that makes the social group flourish and achieve its purposes.

However, on the other side of the coin, we do need to be able to challenge whatever system of mores and taboos we endorse. Our self-restraints need to be loose enough to evolve as and when necessary. In addition, it is only imagination and a lively, bio-socially diverse conscience that can achieve this.

The simple risk-mitigating advantage of voluntary reciprocity is a reality learned long ago by the federated herd or troop (and even in fields of grass, as discussed earlier). Often voluntary group-wide rules, strategies and programs that promote and protect social cohesion, partly by advocating appropriate social values, have benefits for us all. Objectivism recognises this by advocating that we found all social agreements upon voluntary agreements between unique individuals (just like in nature. Have you ever watched schooling fish in a fish tank? Does every fish attend school as an enthusiastic, well-trained soldier would attend military parade on Tiananmen Square? No, each fish in the low risk environment casually drifts in and out of the school, some fish driven more tediously by perceived risk than others. Each fish's individuality is evident and the lack of coercion of each fish is apparent).

While objectivism's scourges of religious dogma and statist dogma are still around us, I nevertheless do not believe these particular errors in contemplative and social concept formation are the deepest 'enemy'. I believe it is the ignorance or lack of values and self-actualising laws of values captured within the four modes of competition, cooperation, extrospection and introspection, and their phaseal interactions in our lives, that cripples humanity's potential and wellbeing the most.

Free us from this ignorance and the desired freedom from error in our socio-political arrangements will probably soon follow. Our current failings can be met by taking rational responsibility for our personal lives, as advocated by emergentism, objectivism, and nature itself, and finding and applying the emerging value-sets of Pillar 1 to each phase of the life-journey or 'each antinomic phase of the dance'. This phaseal dance gives us our response-ability and enables us to take responsibility.

It is as we more masterfully build the pyramids of concept formation, both impersonally in extrospections and personally in introspective contemplation, that we uncover worthy meanings, establish noble purposes, and create virtuous potential. Unfortunately, the extrospective pyramids of humanity's knowledge are far more advanced than our introspective pyramids, and the imbalance is crippling.

Science

Another approach to values and human development we considered is the scientific approach of the natural and formal sciences, with its Darwinian description of animals evolving to include consciousness in *Homo sapiens*. We also noted the comparative silence of the more objective sciences with respect to values. The more objective sciences not only tend to relegate the intersubjective to the second-rate, their advocates often bluntly reject its validity in contributing to knowledge in any way at all (even indirectly). This deep scepticism of science's advocates towards the ability of the introspective or contemplative to add value is of considerable concern because the indirect influence of forward-thinking or imaginative consciousness is the source of all our more-than-computational ingenuity, including our scientific creativity in terms of new theory development. It seems these 'greedy reductionists' as Dennett has stylised them are sceptical of intersubjective value simply because, by its nature, it lies outside of science's sphere of inquiry (i.e. direct, computational and third-person scientific inquiry does not easily penetrate it). This attitude, rampant since the divorce of the positive natural sciences from the normative social sciences, must mellow if we wish to understand our own very natural creativity.

At the same time, intersubjective, beyond-physical, or contemplative experience should not masquerade as science. Its own value-set, as well as its own ability to penetrate and illuminate areas of indirect reality, should judge it. Unfortunately, superstition can be easily dressed up as science in books, movies, etc. and it can sometimes be very difficult to peel back the layers of the story to arrive at its faulty or exaggerated foundations.

I certainly do not propose here that we reverse hundreds of years of scientific advancements. Rather, I hope we see that impersonally objective analysis has its direct place and intersubjective (or personally objective) synthesis has its indirect place. The thing that links the two spheres into a single fabric is our human nature, that is, our emerging knowledge of facts and values and our unstoppable march from the past into the future. The problem of the staunchly objective view that still pervades the scientific community might be likened to the De Broglie Bohm explanation of quantum uncertainty, or to the findings of Gödel's Incompleteness Theorems. That is, the hidden variable needed to complete our understanding of any subsystem exists outside the reductionist realm, in the more holistic realm. Human choices exist within this combined or shuffled realm and thus interpenetrate the realms of fuzzy facts and values.

Nevertheless, science's (and scientific humanism's) rejection of superstition (or more precisely, personal errors or presumptions in introspective concept formation) and encouragement of taking personal responsibility for identified and tested beliefs is admirable. However, there can be grave dangers associated with cold, dispassionate and unenlightened objectivity or cold, dispassionate and unenlightened subjectivity. This is something scientific humanists perhaps understand better these days than in the past. Our more objective consciousness needs to be animated by our instincts in order to make it more ecosystem-friendly and morally temperate.

Our scientists need to stand up and passionately and compassionately tell us their values – not just as they apply to their work, which is important, but perhaps even more importantly, as they apply to their personal lives. For instance, should anyone of any moral standing, including scientists, be taking taxpayers' or investors' money to work on elaborate offensive weapons research without disclosure? Why can't scientific researchers and practitioners agree to abide by a code of ethics or oath, such as a modern Hippocratic Oath for Scientists? Beliefs put under the microscope through rational discussion have a chance to evolve. Values that remain hidden, just like non-transparent scientific inquiry, will drive us to corruption and failure.

Perhaps one fear scientists have with respect to revealing too much of their intersubjective natures, is that unlike well-constrained scientific questions (H_0 vs. H_1), there are no simple right and wrong answers to many moral questions. However, the study of emergentism suggests that quite often there are emergent answers that settle on a better solution, or fuzzy set of solutions, to many moral questions, including questions relating to many areas of scientific research. Scientists need to be aware of the more objective and the more intersubjective stakeholder needs.

This call to scientists applies to all business professionals. Businesses and individuals in the public eye need to stand up and clearly tell us their deeply held personal values as well. If they cannot, the marketplace is going to disrupt them. Consumers do not want to understand the value proposition in terms of just directly measured dollars anymore. This is because they now know there are always also intersubjective (in the full sense of personally objective plus error-prone plus ignorant) costs and thus indirect and unmeasured dollars in terms of the environment, social cohesion, etc. associated with every human endeavour and shared value proposition.

Consumers are going to demand more and more information about the businesses' contribution to awakening consciousness-plus-instincts as well. They are going to want to know that a business is part of the recurrent local and urgent solution and not part of the blind global problem. Silence is not going to cut it either. I suspect Phillip Kotler's Marketing Eras (the Production Era, the Product Era, the Sales Era, the Marketing Era, and now the Societal Marketing Era) are still developing. Today's Societal Marketing Era may be quickly morphing into the 'Emergent Values' Era, and this might even save us if concepts that are unknowable on the other side of any veil do not confuse us.

Eastern Traditions

One approach that would sharply contrast with objectivism at its social and contemplative levels would be that of Buddhism. While objectivism accentuates atheism and self-reliance, Buddhism accentuates transcendentalism and a collectivist, contemplative approach to life, existence and death. The Dalai Lama's "dependent arising" and personal values of altruism and compassion, as keys to wisdom and enlightenment, are useful contrasts in a world of sometimes-cold scientific empiricism and destructive selfishness.

I also have great respect for the Hindu greeting 'Namaste', that is 'The divinity in me greets and honours the divinity that is also in you' (for non-Hindus, we may interpret the word divinity as 'essential life principle'). This concept of individuated and lively divinity seems to beautifully point to the self-organising principle of the universe we are discussing here - the quasi-immutable values promoting of life that are at home and emerging in and through each and every individual. This is what Buddhism might say is the enlightened part of the process of dependent arising, or I could refer to as the vremetic evolution of the virtuous global tradition. Objectivists might be happy to relate this to the unfolding lives of individual dignity, or Christianity might call the Law of the Spirit of

Life or the Law of Liberty. Robert Wright might refer to this as the growing achievements of the universe's nonzero-sum game.

As you may have already perceived, I personally embrace many of the interconnected revelations of Eastern religions, especially with respect to an interconnected or self-organising and arising universe. However, I am personally wary of any absolute transcendental assertions and reject lingering superstitions, perhaps a little like the Chinese scholar Chu Hsi, 1130–1200 AD, who advocated the probing of logical principles, both internal and external, and quiet meditation, but objected to the idea of reincarnation.[14] The synthesis of many ideologies, religious and secular, within the long history of Asian cultures, is very important in terms of the virtuous global traditions we have inherited today and will develop in the near future.

I also find affinity with the more optimistic pictures of interconnectedness presented by the East Asian Buddhist Hua-yen ('Wreath') school, particularly as systemised by Fa-tsang, 643–712 AD. Hua-yen teachings see the monistic interconnectedness of all things without them losing their dynamic individuality; the positive, supreme potential of all things is thus emphasised. The interpenetrating unity of all things, direct and indirect, presupposes their "mutual identification", their virtuous roles and their underlying balance and harmony.[15] The idea of mutual identification is what I try to promote every time I substitute the concept of intersubjectivity for the misleading concept of subjectivity.

In terms of Buddhism's concept of emptiness, we could agree that nature's direct forms are empty because they are subject to impermanence and lack the enduring order and beauty of indirect logic. On the other hand, we could argue that the indirect world behind nature's forms is empty because it lacks the power of form in the present moment. However when the indirect and direct are virtuously brought together now, in the present moment, there is a combined power and beauty that shines magnificently. This almost magical and virtuous interdependence, interpenetration, inter-objectivity and intersubjectivity of the direct and indirect in the present moment is not empty to us, the living on this side of the veil, and is not empty according to the teachings of the Hua-yen school either; rather, it is sublime. A similar understanding of the interconnectedness of the natural and spiritual is in many of our older indigenous religions.

I therefore disagree with Tibetan Buddhism's rejection of the incarnate form and thus the brain-body complex as illusory because to deny form is to deny nature and life as it is. To live by this doctrine is to move beyond nature and life (the veil) without what seems to be a certain reason to reject other more balanced life stances. Perhaps more importantly, a life lived according to the seemingly immutable values transcends all the possibilities beyond the veil. When we disregard life as merely illusory, we embrace its opposite, which to me, is no way to live.

Likewise, I would not agree with Tolle's closely related ideas of transcendentalism and if I understand correctly, his allocation of natural forms to the subordinate, because:

1. The beauty of direct form often exceeds the indirect logic of what was conceived;
2. Pure logic can never exceed evolved direct form by very much; and
3. Indirect logic is rendered powerless in the absence of direct form.

That is, the direct and indirect limit and free each other, and thus seem to fully interpenetrate each other at their virtuous best. However, I gladly accept many of Tolle's other wonderful and deeply perceptive observations regarding human nature.

The Buddhist demand for renunciation of form can release us from labels, conforming roles and stereotypical identities but also negatively bind us in terms of lack of respect for the individuality and innovations of others. Whilst recognition of egoistic but irrational labels and shallow identities is very useful, we also need to appreciate each other as we are (that is, as incarnated consciousness with limited lives and evitability but wonderful potential), not just have compassion on each other for the reason of each other's illusory blindness and emptiness.

Intersubjective truths at their core, as discussed earlier, are dispassionate – just like the more objective truths. This leads us to the possibility of intersubjective dogmas without Buddhism's compassion or altruistic intent. A group of adherents to such a dogma could perhaps become a group with no respect for this life. It could perhaps become a group where the end coldly justifies the means and violence, or the threat of violence, becomes its common currency.

That is, the same dangers that lurk in science's dispassionately objective stance not anchored to a wide range of ethical instincts, as suggested in Aldous Huxley's futuristic *Brave New World,* also lurk in the dispassionately intersubjective stance similarly unanchored. The suggestion here is that it is our so-called illusory animal instincts rather than our elaborate human consciousness that have an ethos and the monistic life principles built into them - and these life principles arise from every cell of our incarnation. These life principles are the eco-friendly 'laws of the jungle', and the 'rules of the game' in social contexts we noted in Pillar 2.

Societal order without individual freedom and enlightenment is not an emergent order at all - it is simply suppression of the individual. Ultimately, each individual must understand the impermanence of form and the inevitability of death for his or her self but still respect form's legitimate place in life and the moment-by-moment spacetime fabric. Values expressed through form and actions are the only way we know virtuous lives are lived.

Virtuous life might yet be lived by the dead who have reached Buddhahood or resurrected sainthood, but such concepts are mental models that the living cannot prove, and so cannot ever help the living reinforce enlightenment objectively.

In accordance with Kauffman's idea that our modern civilisation desperately needs to reinvent the sacred it is the growing enlightenment in life rather than beyond life that the world needs now if it is to meet its pressing challenges.

Western Pluralism

The beyond-physical view, represented here by a Tibetan Buddhist tradition and often amplified by our inward-looking consciousness, says that the physical world we think we fully perceive with our senses is illusory. Truth and reality are not to be ultimately found in the tiny part of the spectra we happen to perceive with our senses and process with our biased minds (our sensory elaboration). With this view, Western philosophical sceptics would often agree.

The material view, represented by science, or perhaps more accurately scientific instrumentalism, and often amplified by our outward-looking consciousness, says exactly the opposite. Practical truth may only be extracted using advanced technology, from a claimed objective study of the material and logical world and its locally recurrent self-constraints. All other so-called truths are to be mistrusted because by their very nature they are illusory; with this objectivism would tend to agree.

How can any philosophical view reconcile such a stark dichotomy? While we have explored the reconciliation of this dichotomy a little already through the concept of existence consisting of both direct and indirect reality, one other idea that seems to come close to a reconciliation is a core concept behind the Christian religion. However, perhaps to be fair, some Buddhist traditions, such as that of the Hua-yen, would also present an optimistic picture of individuality within cosmic interrelatedness.

The Christian ideas of spirit and soul, as briefly defined in Part 1, seem to capture both a respect of the fleeting beyond-physical (the idea of spirit) and respect for the non-fleeting physical (the idea of a living soul in the fullest sense that includes the body). Further, the Christian core idea of 'agape' love (godly love for others, even enemies, warts-and-all) seems to naturally arise out of the spirit-soul dichotomy. Some would call Christian love the only worthy link between the physical and beyond-physical.

Was the eventual development of rule by law in the West, the eventual adoption of the idea of political compromise embodied in modern democracy and the idea of socio-economic respect for the individual (body, soul and spirit) and his/her freedom an outcome of the Christian idea of love expressed through the four gospels? Is this giving Christianity too much credence? Perhaps Western pagan history and society also influenced Christianity to produce its more modern values. Most likely, agape as we understand it today is an evolved synthesis of Eastern philosophy, Western religions and Western secularism. Whatever the case, unfortunately the concept of agape is not widely understood, prized or practiced in the West today. Agape is perhaps a value we should revive without its religious ownership, revise to remove any dualistic overtones, and make a large part of our value-driven purposes.

The Wider Picture

While usurpers are today perhaps less widespread than in the past, our world nevertheless lies increasingly at the mercy of a possible systemic collapse, perhaps beginning in the Euro-zone, that has no respect for our delicate environment, each individual's right to life, or our virtuous global traditions (our successful vremes). Recently we learned from an Oxfam Davos report that the world's richest 62 individuals have more wealth than the poorer half of the entire planet's population. This has come down from 159 individuals in 2012 and 388 in 2010, suggesting seats at the table of the very richest are diminishing quickly and competition between them is intensifying. We are increasingly vulnerable to the well-funded small group who sees violent economic or physical force on a wide scale as an acceptable means to a pre-defined and often dogmatically economic, religious, or political end.

This is a symptom of a world that has become disconnected from 'being', through a lack of appreciation of life's interconnectedness with its environment and a lack of appreciation of our natural human frailties in terms of our evolving consciousness-plus-instincts. We need to insert an organisation something like the International Red Cross and Red Crescent movements onto the battlefield of social values and socio-economic ideas.

Recently I read about geoengineers who, in discussing such methods as marine cloud-brightening or sulphate aerosol spraying to control the Earth's climate on a global scale, believed they could identify surplus properties of the solar system that we could safely ignore and terminate. An alarmed opposing view suggested that all things are there for some kind of grand cosmic purpose, so we shouldn't mess with them. The emergent third view suggests that all life has evolved to fit the environmental niche in which it finds itself – so we must proceed with utmost care when we think to centralise and take over the control of complex interpenetrating Earth systems that have evolved a myriad of local interdependent relationships over billions of years.[16]

We also need to recognise that the intersubjective stance may seem to be of lesser or greater status than the more objective stance in terms of human nature but both stances desperately need moderating by the reality into which human nature falls.

In this sense, the stances are not equal; reality and the cycles of life and death demand that the long-term focus of our deliberations must be more objective (or more naturalistic) if we are to avoid reality's pitfalls. Life takes this asymmetrical directionality because it is fully self-contained within our finite universe. That is, life demands a relational fit or balance with all else on the Petri dish if it is to endure. Life must take, but it also must give back, if it is to find this balance.

Likewise, we need to recognise the right of the self-reliant individual to life and to pursue happiness as well as recognise the proper role of our interconnected social instincts. Just as in nature, we should not fulfil our social instincts in such a way as to rob us of our individuality, but rather, to promote our success in the presence of serious risk, just like on the African plains. We should fulfil our social instincts in such a way that they complete and enhance our self-reliant instincts.

The philosophy of this book is not an all-things-to-all-people kind of idea; rather it is a single-framework-for-each-phase kind of idea, grounded in reality. The lingering idea to leave this section with is not a blurred idea of Western pluralism but rather the need to fully appreciate and participate in the equal validity but differing status of all four interpenetrating phases of human nature. These are the risk-weighted self-reliant, positive extrospective, normative social, and self-enlightened introspective phases that our own evolved consciousness-plus-instincts impresses upon us. An awakened and enlightened consciousness-plus-instincts, cognizant of the pitfalls of our old mores and taboos and alive to the potential of the strange loops of interconnected self and the cosmos, is how we will find lasting meaning and purpose.

The study of emergentism presents an idea of swimming upstream in all four rivers of value-adding, perhaps even to their font. This not-so-mystical font is a percolating pool of almost immutable but emerging values owned by no single creed and perched high over all - that divides into four rivers as it necessarily takes form in our undeniable humanity and in our individual virtues. This was the revelation of Pillar 1.

To extend the analogy, what process is it that causes our values to percolate back up to their font? It is vremetic evolution of the virtuous global traditions in incarnate humanity and incarnate consciousness. Vremetic evolution (the self-organising principle of awakening consciousness) provides the mutation-susceptible and chaotic feedback path back up to the percolating value pool. Vremetic evolution is to our Virtuous Global Tradition what genetic evolution is to its survival machines and what emergence is to the strong and weak interactions of spacetime and matter.

Human Nature

The nature of the self-organising principles in matter, life or consciousness seems to be consistent. These principles display near balance, systematic feedback, and an indirect reality that facilitates and informs slightly mutant or chaotic dances with a counterbalancing essence - just as planets orbiting around the sun are informed and facilitated by local spacetime.

The City State Within

As we have already seen, we cannot understand our present human state without understanding the astonishing interaction between our old animal instincts and our more recently attained elaborate consciousness.

Putting aside more philosophical discussions of framework for now, let's consider more literally the details of the consciousness-plus-instincts framework we find in human beings.

The Four Naturalistic Temperaments

In Part 1, we spoke of the two most basic forms of behaving in its environment any organism possesses – doing and being. These base behaviours corresponded with the two ways of searching for truth of Figure 1 – a processual approach (top of Figure 1 and Figure 12) and an analytical approach (bottom), respectively. The processual approach was limited by time, whereas Gödel's Incompleteness Theorems, or the inability to measure all interwoven relationships, limited the analytical approach. To put these base behaviours in terms of Physics' Uncertainty Principle, the uncertain prediction of velocity restricts the processual approach, whereas the uncertain prediction of position restricts the analytical approach. Finally, to relate an organism's base behaviours to General Relativity, we could say the relativity of matter (to space) limits its doing or process-modelling, whereas the relativity of space (to matter) limits its being or structural analyses.

In Pillar 1, we spoke of two surprisingly similar concepts in terms of the way the brain processes them – facts and values. In Pillar 2, we considered the two basic instinctive modes of all organisms – competition and cooperation – that exploit such facts and values. Additionally, in Pillar 3, we considered the two ways we use consciousness to observe all our surroundings and behaviours – extrospection and introspection. It is time to put these ideas together.

Starting with the perhaps most familiar idea first, it seems natural to put competition and cooperation on an axis. That is, one end of this axis might record a highly social response of self to the problems of survival and homeostasis (more people-oriented), and at the other end a highly individual response (more task-oriented). We might label this axis an extrosomatic axis because it records how the self acts towards or reacts to the outer world (the world outside of our bodies).

Using the same approach with might put extrospection at one end and introspection at the other end of an axis of survival or homeostasis. To this axis, we might attach the explicit consciousness of facts and values as well. That is, facts would appear at the extrospective end of this axis and values at the introspective end. An interesting point arises here, because those personalities that would tend to be more individual on our extrosomatic axis would also tend be more fact-oriented and extrospective as well. Likewise, those personalities that would tend to be more interdependent on the extrosomatic axis would also tend to be more values-oriented and introspective. That is, the one extrosomatic axis seems to be able represent both the main concepts of our instincts chapter and our consciousness chapter.

At one end of this axis, we have more facts- or task-oriented, extrospective and individual behaviours whereas at the other end we have more people- or values-oriented, introspective and interdependent behaviours. On one side, we seem to build up a repertoire of relevant facts with respect to wellbeing and on the other, a repertoire of relevant values with respect to wellbeing. We can think of facts and values as the two basic interactions of the extrosomatic self. Points along this axis also describe how the self responds to its environment.

Finally, we might put 'doing' (processing or producing output) at the end of an axis and 'being' (receiving input and analysing) at the other end. We might label this axis an introsomatic axis because it records how the self acts towards or reacts to the inner world (the world of, or inside of, our bodies). This behaviour is something akin to how verbs and nouns interact within the sentence that houses them. Both are needed to complete the sentence, but verbs describe the doing aspect of the sentence while nouns the being aspect. Likewise, some personality traits tend to be more performance-oriented or time-sensitive, and thus sit towards the doing or active end of the axis, which seems a more instinctive form of behaviour. Other personality traits tend to be more pattern-oriented, relationship-sensitive or identity-conscious and thus sit towards the being or contemplative end of the axis, which seems a more consciousness-oriented form of behaviour. We can thus think of being and doing as the two basic interactions of the introsomatic self. To use the metaphor of General Relativity, the doing instincts tell conscious being how to take structure, shape or position, but conscious being tells the doing instincts how to take velocity, or move. Points along this axis describe how the inner-self behaves in its environment. Now we are ready to put the two axes together in one diagram of human nature, in Figure 19, which should be seen as a development of Figure 11 and Figure 12. These figures provided a representation of life in terms of 'introsomatic' autopoiesis and 'extrosomatic' cognition. Autopoiesis was about the organisation of matter within the cell or body, whereas cognition was about the organisation of the cell or body within its environment.

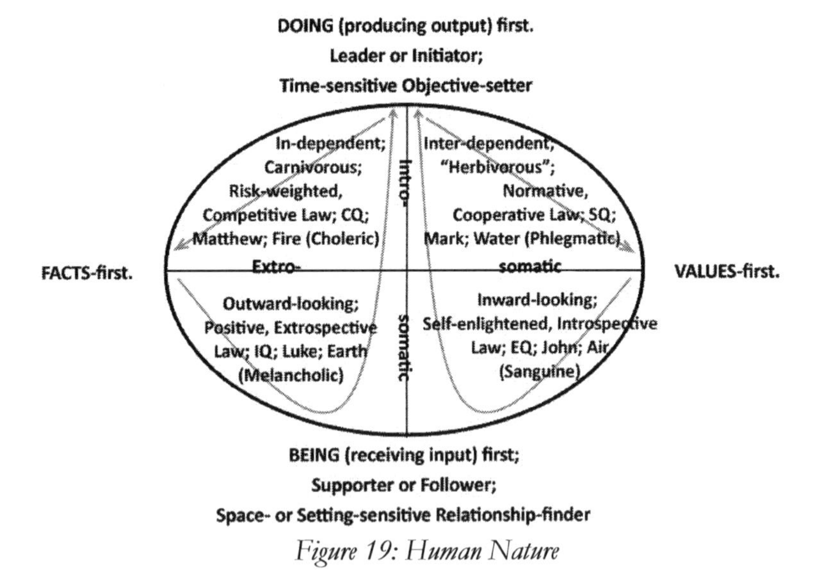

Figure 19: Human Nature

There are many things we can say about this diagram, but let me first address a further note about the axes. The question might be, I can understand how introspection could be about the Being end of the introsomatic axis, but how can it be attached to the extrosomatic axis? What does introspection have to do with a reaction to the outer world? Surely it is just about the self's inner world? I would simply reply that introspection can be a reaction to the outer world and a way we act towards it. As we saw in Pillar 3, we can simply shut out the world through introspection. We can also use introspection to deeply consider the implications the world presents to us personally.

We can perhaps more easily make the complementary case for extrospection on the introsomatic axis. That is, extrospective observation clearly has a role in the self's relationship with its inner world, because as noted earlier, consciousness is consciousness of something.

Figure 19 is a diagram of how we, as eco-psychosocial players with mechanisms for sensing and acting on our environments, struggle for survival and fit within this deep ecology. The combination of survival and sense of fit seems to equate with what we call wellbeing. That is, when we are in a relaxed environment that supports our survival and we have a strong feeling of belonging or place in that environment, we enjoy a deep sense of wellbeing.

We can picture the range of actual conscious behaviour resulting from human nature as any possibility within the eye-shape of Figure 19. No point within the eye-shape is better or worse than any other point per-se. Healthy life-and-being here-and-now could represent any point within the eye-shape, depending on the environmental circumstances. Psychosocial behaviours above the centreline are more instinctive in nature and those below it more conscious. Behaviours to the left of the centreline are more task-oriented and those to the right more people-oriented. I have selected the eye-shape because we can picture the behavioural range of our hunter-gatherer ancestors with little consciousness as being of a partly eye-opened shape. Likewise, we might picture the behaviour range of our successful descendants as approaching the circular or eye fully opened shape.

For the reader's reference, Figure 19 encapsulates the four classic elements of nearly all the world's major ancient (Eastern and Western) philosophies along with the associated four humours or ancient temperaments. Andrew Jukes' (1966) idea of the four views of Christ pictured in the four gospels is also encapsulated in Figure 19. In this sense, Figure 19 pictures Jesus as a combination of Matthew's Lion-King, Luke's Rational-Man, Mark's Servant-Ox and John's Heavenly-Eagle. Besides noting their concurrence, I do not intend to comment much further on these ancient classifying concepts. However, readers interested in pursuing this topic might also like to refer to Xandria Williams' *Love, Health and Happiness* (1995), listed in the Bibliography.

The idea of IQ ('intelligence quotient' or 'rational intelligence') has been with us for some time, not without its critics. For instance, is IQ a measure of active capacity to reason and question and find meaning or is it largely just a passive measure of capacity to memorise and recall patterns learned? Many have asked the same question of the dominant assessment method employed by our educational institutions. I use the mnemonic in Figure 19 only because of its familiarity, but by IQ, I really intend the idea of extrospective intelligence or consciousness of things outside of self as opposed to introspective intelligence or a consciousness of consciousness itself (within self). Putting measurement technicalities aside, Daniel Goleman published his famous book back in 1996, partly because he saw that the IQ test failed to measure another completely different kind of intelligence that our consciousness affords us, which he called "*Emotional Intelligence*"[17] (or EQ) but which I prefer to call introspective intelligence. Again, I use EQ because of its familiarity. This intelligence is the one guided by the law or self-restraint of our own personal being that we must courageously chose for ourselves - a law of enlightened selfishness.

It wasn't too long after that, with further scientific understanding of brain function and related research, that Goleman went on to officially recognise yet another kind of intelligence, quite distinct from IQ and EQ, that he labelled via the title of his 2007 publication, *Social Intelligence*[18] (SQ), which we can closely associate with the social or cooperative behavioural instinct. I would like to add one more key area of intelligence to Goleman's list, which I call Competitive Intelligence (CQ).

As can be seen in Figure 19 CQ, unlike the other intelligences, arises out of our inward-dependent or individual survival instincts – the competitive instincts we acquired as keen hunter and carnivore in a harsh environment. This key intelligence has a lot to do with understanding our ecological interdependence. It is about knowing your enemy and the terrain in terms of their natural vulnerabilities and possibilities for exploitation, but nothing to do with social conformance or knowing how to fit in with social norms (and thus SQ). Further, this intelligence has little time for the refined rational argument of IQ or the deep contemplative considerations of EQ because like the social instinct (SQ), CQ is about exploiting the best shortcut rules evolution affords us to survive and prosper in the current moment. CQ is about instinctive assessments for personal action in the local space or territory.

CQ is a volitional intelligence; the choice assessments focus on the strengths, weaknesses, opportunities and threats of self (as an animal with an intentional stance) as opposed to competitor, prey and environment. CQ, which can perhaps be summarised as 'force of character', is not necessarily a popular view of human intelligence but does offer a view of the world deeply attached to its natural ecosystems. As we have probably recognised through our very brief look at objectivism, Nietzsche, John Galt, Gordon Gecko and Bushido, the personal value-set of CQ has been very influential in human history. The value-set of CQ is the set the self-made person would often identify with most readily. If SQ represents the stereotypical female instinct and intelligence, then CQ represents the stereotypical male instinct and intelligence.

Before we move on to discuss Figure 19 in more depth, I do not want to appear dogmatic about the model of intelligence or correlates of human nature briefly proposed here in the style of psychology's very old-fashioned and rather outmoded personality traits or temperaments. After all, the human brain is the most complex device in the known universe. However, if the simplistic yet naturalistic model presented in this chapter helps us to self-actualise, then the model will have achieved its purpose. We can consider human intelligence from the viewpoint of instincts-with-consciousness as in Figure 19 and its axes, or we can consider intelligence in many other ways, such as in terms of its achievements, its process-development or its plasticity in different circumstances. In Gardner (1983)[19], we read of seven key intelligences in humans. Without being too specific as to any hierarchical order, he describes:

1. Spatial Intelligence (as exemplified by the artist, architect, landscape gardener, sculptor or master craftsperson). This involves the ability to see and represent proportion and beauty. Relating this to the physiology of the human brain, recent research suggests such ability might be strongly associated with the hippocampus ("providing a keen memory of context", according to Goleman (1996)). The hippocampus seems to

largely deal with empirical data such as in lists or road maps.

2. Kinaesthetic Intelligence (as exemplified by the athlete, dancer or martial artist). Here the body itself is involved in form and movement that is itself a work of art demonstrating and reflecting proportional beauty. Such body language ability seems closely linked to the two-layered cortex - "the regions that plan, comprehend what is sensed, coordinate movement", according to Goleman (1996)). The cortex seems to enable bodily expression.
3. Mathematical-logical intelligence (as exemplified by the scientist, engineer or mathematician). This is the same kind of math competency as is taught in most schools. It is probably most strongly linked to both the neocortex (the "seat of rational thought", according to Goleman (1996)) and the hippocampus. The neocortex seems to largely deal with rational data.
4. Intra-personal Intelligence (as exemplified by the successful entrepreneur). This involves a knowing of one's own personal feelings or emotions, and the ability to discriminate among them, to effectively guide personal behaviour. This is probably most strongly linked to the left and right amygdala's circuitry, including the left and right prefrontal nodes. The amygdala's circuitry seems to cope with intuitive feelings.
5. Verbal Intelligence (as exemplified by the barrister, actor or linguist). Again, this involves the same kind of language competencies as are taught in most schools. It is probably strongly linked to the neocortex working in harmony with other mid-brain and fore-brain areas.
6. Interpersonal Intelligence (as exemplified by the politician, teacher or leader). This involves the ability to lead others, through an appreciation of what makes them tick. It would probably be more difficult to pinpoint any single area of brain activity, since harmonious workings of all major areas of the brain are probably required.
7. Musical Intelligence (as exemplified by the composer and conductor of a live musical performance). This may involve a refined version of all the other intelligences to some greater or lesser extent. It would thus probably involve harmonious and heightened workings of all major areas of the brain.

It might be an interesting exercise to map the seven intelligences of Gardner onto Figure 19.

We can now discuss the four quadrants of Figure 19, formed by its two axes, and what each quadrant may represent, although this is a little bit limiting because the oval can only represent a static snapshot of human psychosocial claiming styles at any instant. We will represent the phaseal nature of human behaviour using an alternative methodology later.

Perhaps the first thing we notice here is the similarity between Figure 19 and Integro Learning System's "Dominant / Influential / Steady / Cautious" (DiSC) behaviour style or behaviour profiling system. DiSC Theory was first published in a book entitled *Emotions of Normal People*, authored by Dr William Moulton Marston back in 1928. The DiSC Theory Marston championed has many variants and equivalents in many countries and languages around the world. One variant is the "Director / Emotive / Supportive / Reflective" (or DESR) system used by Manning, Reece (1990). Another is the Myer-Briggs Type Indicator (MBTI), first published in 1944, which was claimed to be derived from Carl Jung's book, *Psychological Types*, first published in English in 1923. A more recent example is the Herrmann Whole Brain Model and its HBDI (Herrmann Brain Dominance Instrument) profile, developed between 2002 and 2014, and used by many corporate HR departments.

Marston formed his system by mapping behaviours based on two axes, the horizontal axis sometimes known as the sociability axis, which would coincide with Figure 19's extrosomatic axis and the vertical axis sometimes known as the consideration axis, which would coincide with Figure 19's introsomatic axis.

Perhaps the first thing that needs to be said here is that if any such model of human behaviour is to be worthwhile, then the axes ought to be derived from an appreciation of natural science. That is, the axes should be derived from our universe's relativistic and quantum beginnings, rather than via a clinical psychologist's factor analysis. Free-floating normative or factorial definitions of the axes are insufficient and incomplete. The contribution this book makes to Marston's (or is it Jung's?) system is that it gives it a theoretical basis in natural science that compliments well its basis in the social sciences. Marston's clinical basis arose before the field of evolutionary psychology was established.

The HBDI profile claims to establish its four profile quadrants based on brain physiology, that is, the left brain – right brain differences in functionality defining the horizontal axis, and the upper (cerebral cortex) and lower (limbic) differences in functionality defining the vertical axis. The cerebral cortex is involved in memory, attention, perception, awareness, thought, language, and consciousness. The structures of the limbic system are involved in motivation, emotion, learning, and memory, although some neuroscientists have criticised the usefulness of this historical grouping as too broad.

While a more gentle intention-setting left – right brain, mid/hind (old) – front (new) brain activity was expected by Figure 19, the HBDI profile claims to result in A) a more analytical and logically-oriented cerebral-left quadrant, B) a more organised and process-oriented limbic-left quadrant, D) a more intuitive and ideas-oriented cerebral-right quadrant, and C) a more interpersonal and feelings-oriented limbic-right quadrant. Quadrants A & C may loosely relate to the left- and right- extrosomatic axis of Figure 19, respectively, whereas quadrants B & D are a little more difficult to directly relate to the lower- and upper- introsomatic axis, respectively. The focus on an upper/lower brain physiology rather than a front/mid-hind brain physiology may be the valid reason. Wikipedia interestingly reports,

> *"a study published in the peer reviewed Creativity Research Journal in 2005 by J. Meneely and M. Portillo agreed that creativity is not localized into a particular thinking style, such as a right-brain dominance resulting in more creativity. They did however find correlation between creativity in design students based on how flexible they were using all four thinking styles equally as measured by the HBDI. When students were less entrenched in a specific style of thinking they measured higher creativity using Domino's Creativity Scale (ACL-Cr)."*[20]

If we were to map the DiSC behaviour system onto Figure 19 we would obtain Figure 20. As per Figure 20, sociability or extrosomatic responses at the left of the origin would capture the Dominant and Cautious task-oriented behaviours, which Marston saw as reactions to a perceived antagonistic or challenging environment. Sociability or extrosomatic responses at the right of the origin would capture the Influential and Steady people-oriented behaviours, which Marston saw as reactions to a perceived favourable or supportive environment.

In contrast, consideration or introsomatic responses at the lower extremes would capture the Steady and Cautious behaviours, which Marston classed as passive behaviours or followers (or introverted), whereas consideration or introsomatic responses at the higher extremes would capture the Dominant and Influential behaviours, which Marston saw as active behaviours or initiators (or extroverted).

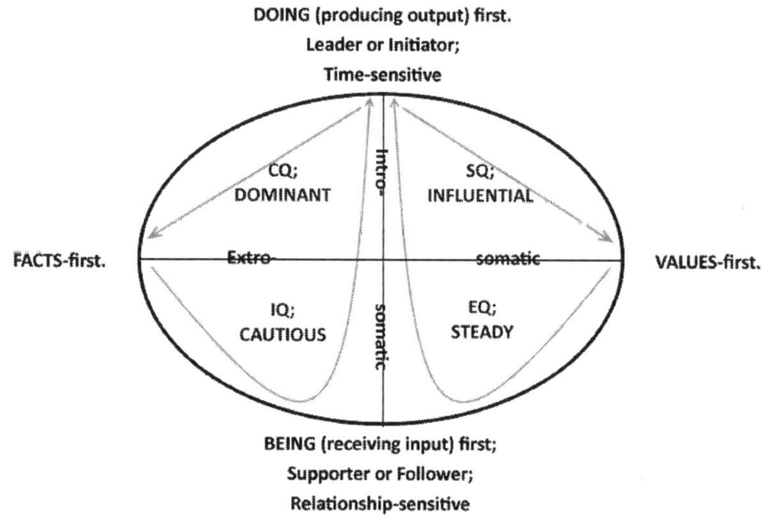

Figure 20: Psychosocial Styles

While it is impossible to get a consistent measurement of such personality traits across large populations, Integro reported in 2010 that around 22% of those completing their 'Group Culture' DiSC profiling system display the Dominant behaviour, around another 28% the Influential behaviour, about 27% the Cautious behaviour and the remaining 23% the Steady behaviour. Of course, these figures would vary in the general population with sample and time.

The first thing to note here is that the DiSC system seems to fit with the discussion so far, but there are several problems. Firstly, it is difficult to discuss the nature of each quadrant too deeply in terms of DiSC Theory without possibly infringing copyright so I suggest if you have a deeper interest, you should conduct your own research. Nevertheless:

1. A predominant combination of facts-first and 'Doing' suggests the CQ/Dominant psychosocial style;

2. A predominant combination of facts-first and 'Being' suggests the IQ/Cautious psychosocial style;
3. A predominant combination of values-first and 'Doing' suggests the SQ/Influential psychosocial style; and
4. A predominant combination of values-first and 'Being' suggests the EQ/Steady psychosocial style.

Another observation we can make here is that for instance the Influential behaviour style found in real human beings is clearly not totally lacking EQ or CQ. We could obviously go on to make equivalent observations about the other three behaviour characteristics (or quadrants). The two-dimensional graphical representation of the DiSC system acknowledges and embodies this limitation. In fact, we should probably think of the centre point of Figure 20 as the average point and the extremes of all axes as extreme points in behaviour characteristic. We could then break up the oval into zones of concentric ovals. If we broke Figure 20 into three such concentric zones, then we could label the zone closest to the centre point above average, the next high and the last extreme. Figure 20 simply does not represent psychosocial activity below the normal ranges.

Another limitation of the diagrams in Figure 19 and Figure 20 is to give equal weighting to all four quadrants even though their occurrences in real populations are not necessarily equal, but perhaps slightly biased towards the Influential and Cautious behaviour characteristics, as already noted.

Of more concern however, is the variation in behaviour descriptors used by variants of the DiSC system. For instance, as we have already seen, in Manning, Reece (1990) the Influential behaviour quadrant is renamed the Emotive quadrant and the Cautious behaviour quadrant is renamed the Reflective quadrant. It seems that until now we have not sufficiently understood the axes from the natural science perspective to ensure consistent and non-contradictory axis and quadrant names across all of Marston's variants. For example, the Emotive quadrant name is perhaps unclear because it is too closely related to the concept of emotion, which is something more generally applicable than to just one quadrant. Similarly, the Reflective quadrant name is perhaps unclear because all behaviour characteristics are influenced by either inwardly-reflective or outwardly-reflective consciousness. Finally, the Dominance axis of Manning, Reece (1990) expresses a clear disparity with the DiSC system's Dominant quadrant.

These naming and labelling examples demonstrate how an insufficient understanding of the axes and quadrants can lead to an almost meaningless model definition. It also demonstrates why it is necessary, if the model is to be used without confusion in the future, that the axes be defined from a natural science perspective, such as that which may be provided by the work of Bourgine and Stewart (2004), discussed in Part 1.

Next is the 'static snapshot of a human behaviour trait at any instant' limitation of the diagrams in Figure 19 and Figure 20. There are several problems here: Firstly, as real humans, we are capable of shuffling from one behavioural stance to a completely opposing one, depending on our perceptions of the situation, even though we typically have a preferred and consistent communication style over our lives. That is, we are capable of accentuating, for example, SQ in one situation and then just as strongly, CQ in another instant applied to a different or drastically modified perception. Cardinal (as in the sense of the four principal points of the compass) diagrams such as in Figure 19 and Figure 20 cannot capture this important phaseal dance of human behaviour.

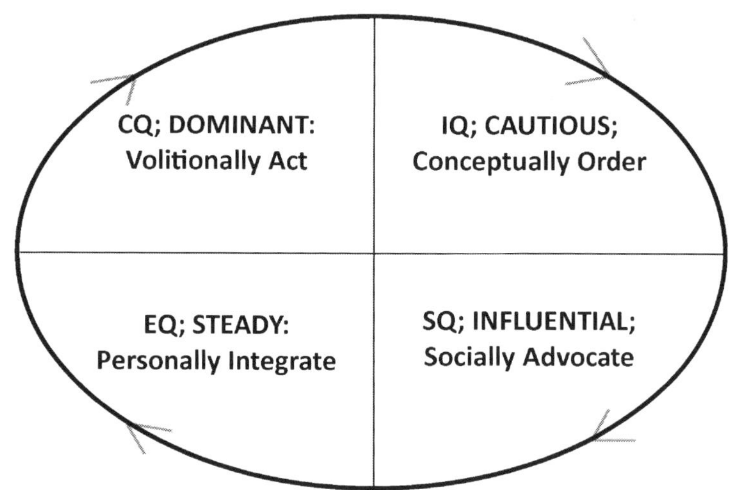

Figure 21: Phaseal Progression

Putting aside the abrupt switches of the dance partly due to perceived situational changes, we can also represent the natural advance of the phases as applied to a single but developing situation by use of a sinusoidal phaseal diagram. Nevertheless firstly, please consider the structurally altered DiSC diagram of Figure 21.

Figure 21 has had its axis labels removed, because the axis labels change as you move around the diagram in the direction of the arrows. It arranges the four behaviour characteristics into a phaseal 'D-C-I-S-' order that is notionally initiated at the top left quadrant in the individual Dominant and Volitionally-Act phase. It proceeds clockwise to the individual Cautious and Conceptually-Order phase, then to the interdependent Influential and Socially-Integrate phase, then the interdependent Steady and Personally-Identify phase and then back to the Dominant phase again, etc.

The focus of Figure 21 is not the structure of human behaviour in terms of how it arises from human nature (instinct and consciousness) but how human behaviour, and in fact the behaviour of all natural systems, ensues in a certain processual order or orbit, starting with Figure 6's, Systemic Separating, Entropic Degenerating, Negentropic Arranging, then, Systemic Incorporating. By the end of each of these cycles, reality is locally changed, yet remained consistent at a rate up to the speed of light.

The phaseal order, 'volitionally act – conceptually order – socially advocate – personally integrate' can be compared to the advancement of the four seasons, 'autumn – winter – spring – summer'. Alternatively, we can compare this order to the procession through the compass points 'West – North – East – South', or life's natural phases of 'Birth and Childhood – Developing Adolescence – Productive Adulthood – Old Age and Death'. We can also compare the phaseal order to the Boston Consulting Group's famous "Question Mark – Star – Cash Cow – Dog" model of firm life-cycle. At a stretch right now, we might also think of our well-rounded stock exchange trader as advancing through 'risk-taker – fundamental analyst – trend-seeker – intuitive investor'. In all these cases, the phases advance in alternations of instinct and consciousness; outward to inward or active participant to passive observer.

Thus, we arrive at an evolutionary and minimalist theoretical basis for a naturalistic philosophy. We can define this philosophy of emergentism as the rational investigation of questions about existence, knowledge and ethics based not on a normative set of rules but rather on the idea that two fundamental and interacting vectors, one introsomatic and the other extrosomatic, drive all human behaviour, and, just as profoundly, all cellular life. Further, these two fundamental vectors are perceived, in an existential manner, to emerge from General Relativity's interrelationship of space and matter.

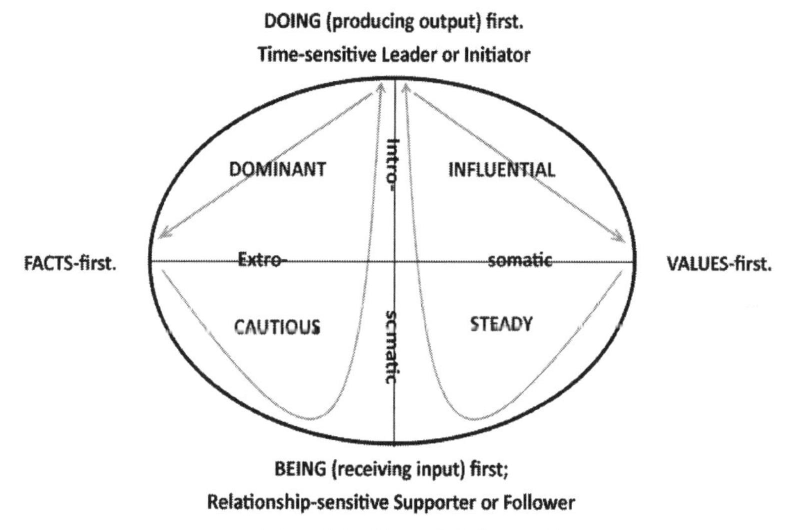

Figure 22: Phaseal Relationships

We also arrive at an evolutionary and minimalist theoretical basis or framework for human values and virtues (which we will more fully explore in the following pages) that requires no creed but desperately requires further research and clarification. The strong link between discovered human virtues and the relativistic structure of human behaviour very much supports the idea of *Homo sapiens* as an essentially moral species.

Figure 22 attempts to represent the 'D-C-I-S-' progression of Figure 21 on the original diagram of Figure 20 by inserting quadrant arrows, which unfortunately is a little cluttered and cumbersome until you become familiar with it. Phase 1 (Dominant) covers a facts-and-doing-first approach to eco-psychosocial claiming. Picture the self-reliant hunter using his knowledge of nature and its landscape to make his volitional risk-weighted hunting decisions and take action. Similarly, Phase 2 (Cautious) covers a contemplating-facts-and-being-first approach to eco-psychosocial claiming.

Picture for instance, the same hunter carefully considering in the middle of the hunt what impact his prior actions have had and what personal concept of the hunt they suggest. What are his instinct, his intuition and his rational logic telling him to do next? A little more fully, the verbs we can associate with this observant second phase are:

To order (via a set of rules or mores and taboos), arrange, synthesize, make mental connections, conceptualise, analyse, potentiate, speculate, concretise, objectivise, crystallise, marketize, standardise, generalise, justify, stereotype, or form an intentional stance.

Phase 3 (Influential) covers the values-and-doing-first approach to psychosocial claiming. *Homo sapiens* is a social animal – in this phase the hunter might share his experience and plan-in-action with others on the same hunt in order to form a joint strategy and implement shared tactics. Finally, Phase 4 (Steady) covers the contemplating-values-and-being-first approach to psychosocial claiming. Picture in this phase the hunters seeing themselves as members of a team, a single entity with components moving in strategic, wavelike unison towards a single objective – meeting shared needs and thinking about applying the same techniques to the next hunt. The verbs we can associate with this observant phase are:

To identify (with), to personalise, to label, to 'egotise', to internalise, to normalise, to 'propertise' (that is, to turn into something over which legal title may be conceptually held, as governments are currently considering with respect to their carbon emissions trading schemes). To own or make your own, to aim to gain, to materialise, to commoditise, to conform (with), to personally desire, envy, or covet something of social meaning, to form the intention to purchase or join or join in; to civilise, to identify an inclusive purpose or role...

The interesting feature of Figure 22 is the importance of our Doing to each phaseal development. Each phaseal development nominally starts or finishes with the Doing cardinal point, much as we discussed must be the case in Part 1. Figure 22 may also support the objectivist idea that human are essentially heroic individual beings (Phase 1 and Phase 2 both on the left hand side) and only secondly social or interdependent beings (Phase 3 and Phase 4 both on the right hand side).

Finally, notice that 'Being' in Figure 22 is neither a starting point nor ending point of any phase. For the living person, it cannot be;

in conformance with Physics' Uncertainty Principle, 'Being' may only be a point transitioned rather than a beginning or end in itself.

Please, let me explain. Imagine if reality's necessary movement towards energy effectiveness and efficiency did not dictate that humans choose lively doing over inanimate being. Looking carefully at Figure 22 again, the only alternative process possible would be 'S-I-C-D-' (there are no other meaningful possibilities). We would notionally start with the introspection-and-values-first Steady phase, advance to the Influential phase, then advance to the Cautious phase and then finally advance to the Dominant phase, et cetera. S-I-C-D- is the reverse of D-C-I-S-; it is like running our natural being configuration backwards, or centripetally instead of centrifugally. This time Doing would only ever be a point transitioned and homeostatic Being would be the notional beginning or end of every phase.

Such a being configuration is not practical; it would not promote survival in our natural, earthy environment that is far from equilibrium. Nevertheless, let's briefly consider the scenario. The Steady phase would have to begin with contemplative, introspective dogma and seek to share it (it would be dogma because it would be subjective thinking untied to material reality). The Influential phase would then seek to socially collectivise the dogma without any consideration of the needs of the individual at all. The RHS of Figure 22, if traversed first, would have no underpinning cognition relating to the individual. The verbs we could associate with this phase would be:

To impersonally glorify, aggrandise, to collectivise, to make into a dogma, fable, legend, or myth, to deify, to make worshipful, to sacrifice to or to surrender to, to love altruistically.

Next, the Cautious phase would attempt to order or justify and personally internalise its socially-imprinted dogma. Finally, like some kind of automaton or infamous lemming, the Dominant phase, repressing its natural volition, would slavishly enact the requirements of its burdensome dogma. The verbs we could associate with this phase would be:

To subjectivize, to make abstract and unreal, to conform with, to repress (individuality and the will).

Such a being configuration would likely lead to a mentally disturbed person, at least in this world, and if adopted across a community, would clearly represent a repressive, socialistic or religious totalitarian regime, and everything opposed by the philosophy of objectivism. It would represent collectivist or statist insanity.

However, this is not to say that a little reversal of direction from time to time is altogether a bad thing. Sometimes a reversal is the secret to more deeply realised self-actualisation. Sometimes it is by laying down our lives that we may pick them up again, refreshed with a renewed vigour. Sometimes we need to consider the ridiculous in order to see clearly the profound; sometimes we need to go round-and-round in our thinking before we can emerge with a clear purpose and direction. Sometimes we need to make a few mistakes before we know what is right or preferable; sometimes each one of us has to be a bit of an airhead before we can each find our practical footing on mother earth. Sometimes self-sacrifice and unconditional love is the deepest expression of our inner values.

Life and nature is never quite as tidy and consistent as we might want it to be; their diversity is precisely what makes them wonderful. Nevertheless eventually, the undercurrent of a naturally successful life demands that we follow 'D-C-I-S-' and not continually the reverse 'S-I-C-D-', which would likely lead to a premature end of the species.

Finally, if you carefully consider Figure 22, you will realise that no other being configurations are meaningful or even possible for a being with consciousness-plus-instincts. That said, if you lived in Melbourne for some time as I did for all my childhood and early adult life, you would know that the four seasons often do not seem to proceed in the neat phaseal order suggested by the earth's orbit around the sun. Melbourne is famously known for its quirky weather patterns and the oft occurrence of all four seasons in one day. Likewise, unfortunately some of us are cut off in our human life-cycle before our time, or are forced to become adults while we are still children. Companies also seek to cheat their natural time of death and decay by avoiding the 'dog' phase and reinventing themselves in a new 'question mark' phase as the 'cash-cow' phase of an old product or service ends.

In all these examples, we recognise the limitations of Figure 22. The sinusoidal phaseal diagram of Figure 23 can help us better visualise these limitations or variations in the phaseal dance of reality because it better incorporates the idea of now or unfolding time into our model of psychosocial claiming styles, which is in itself derived from the facts of our evolution and existence in matter and spacetime.

In Figure 23 the rising or falling of the waveform is unimportant except that it demonstrates that the risings and fallings are polar opposites, like the North and South ends of a bar magnet. That is, the rising is not intended as better than the falling in any sense. Further, the active (or instinctive or doing) dominant and influential phases are pictured as the initiators of change (from falling to rising or vice versa) whereas the outwardly passive (or driven by consciousness or being) cautious and steady phases are pictured as the followers of change (i.e. following the pre-existing rise or fall). In terms of magnetism, we can picture the instincts (or initiators or participants) as repelling prior direction whereas consciousness (or the followers or observers) attracting or flowing with prior direction.

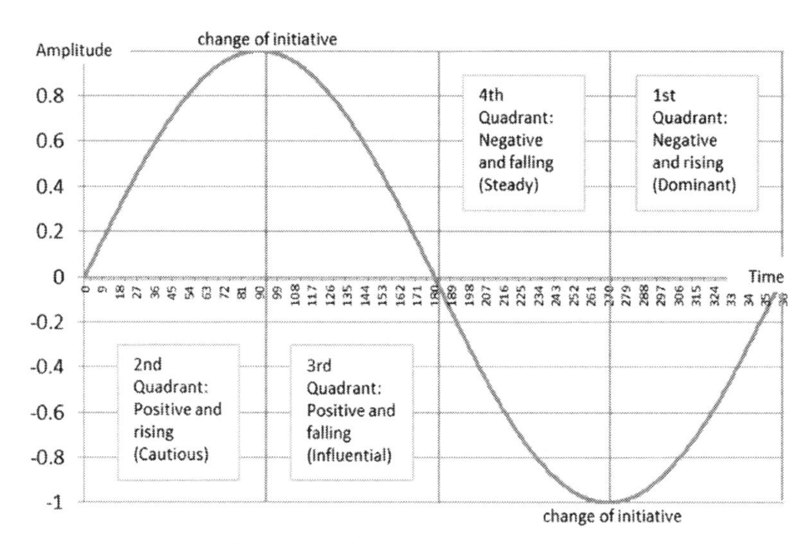

Figure 23: Sinusoidal Phaseal Diagram

Following Figure 21, the normal sequence or dance would proceed from the first to the fourth quadrants and so on. Quadrant 1 (Dominant) would move from the waveform at -1 to the rising zero point crossover, Quadrant 2 (Cautious) to the waveform at +1, Quadrant 3 (Influential) to the falling zero point crossover and Quadrant 4 (Steady) to the waveform at -1 again. Notice that the 'Being' point is not and cannot be directly represented on the diagram.

It is easy to imagine variations in the amplitude of each phase and variations in (wave-) length of each phase that would approach a more true-to-life model. This is probably the greatest value of Figure 23. We might also imagine a variation in functional response to life from a simple sinusoidal response to perhaps another far more asynchronous and chaotic functional response (that is, a response that seems random but is actually restricted by a very orderly feedback system, just as in the Mandelbrot Set, other life processes, and evolution).

Sales Training

We are now perhaps ready to apply the model of Figure 23 to more real-life scenarios, beginning with a 1980's-style sales training scenario, which we should take a little tongue-in-cheek. Imagine yourself a printer salesperson with four prospects, who just happen to display the four classic behaviour traits described in Marston's model. That is, their characters may be described as follows:

Prospect 1: Mr Dominant: Tends to be task-oriented and empirical, often younger, active, individualistic, and ambitious. Likes to run things / take on challenges / be in control. Likes to be on time and efficient / be informed / not waste time. Can be demanding of others / lack patience / not pay attention to detail. Is a strong decision-maker, often with an autocratic leadership style. Mr Dominant identifies most strongly with Table 1: The Risk-weighted Value-set.

Prospect 2: Mr Cautious: Tends to be task-oriented and rational, yet usually is a follower rather than an initiator; likes to be orderly and have tasks well defined. Does not like to be rushed or moved too often; tends to be older. Tends to analyse and intellectualise rather than be demanding, feeling or expressive. Enjoys extracting patterns from external reality, developing the ideas of others, and seeing yet others put his value-adding calculations to good use. Mr Cautious identifies most strongly with Table 2: The Positive Value-set.

Prospect 3: Ms Influential: Tends to be people-oriented, active, younger and expressive; likes an exciting / inspiring atmosphere. Normally socially outgoing / dynamic / stimulating and personable; can be intuitive, creative, and persuasive, but at times impulsive. Tends to be open with her feelings; does not always like to follow through an idea, and so likes others to develop her new ideas. Ms

Influential identifies most strongly with Table 3: The Normative Value-set.

Prospect 4: Mrs Steady: Tends to be older, people-oriented, casual and informal. Usually is a follower - finds it hard to say 'no'; wants to be helpful and supportive. Always good-natured and non-threatening; an attentive listener, sensitive to others' feelings, and an astute reader of character. Often wants to believe in people and avoid unpleasant confrontations. Enjoys serving others she respects, and sensing patterns of behaviour in self and others. Mrs Steady identifies most strongly with Table 4: The Self-enlightened Value-set.

Your task is to sell a printer to each prospect. That is, how would you use your knowledge of their behavioural characteristics to facilitate the sale?

The phaseal nature of these classic but a little unrealistic personality types means the sales person can act as the before and after personality type or phase for each prospect, in order to influence the prospect and thus facilitate the personality and the sale.

Thus, if the task were to sell a printer to Mr Dominant, whose forte is to make and activate decisions, we would need to serve him as would a Steady personality would do at the beginning of the sales cycle. Likewise, we would need to rationally develop and complete any conclusions or decisions he made, as the Cautious personality naturally would do, at the end of the sales cycle. We would also avoid conflicting with Mr Dominant's personality type, by letting him feel 'in control', and not making any decisions for him. Mr Dominant prefers to surround himself with intelligent and practical yes-men, not people with competing initiatives such as those of the Influential personality style or others with the same Dominant personality style.

For instance, to serve Mr Dominant we would spend a good deal of time qualifying his needs and carefully noting all of them down. If he needed to make further appointments, we would let him choose the time, and pay particular attention to getting there on time. As he is more task-oriented than people-oriented, we would also stick rigidly to (his agreed upon) agenda for the meeting, rather than extending it with too many social niceties. It would be imperative to fulfil all his wishes in the demonstration and fulfil any promises made, to the letter. By doing such things, we would motivate him to fully function in his most natural and comfortable role as a Dominant personality. In the demonstration itself, we would try to lead him into decisions and conclusions (a thing he greatly enjoys) by providing him with concrete, accurate, thorough and timely information. We would also provide him with sound arguments or reasoning that would reinforce his decisions. For example, we might ask, 'If I could show you how this new printer could save you enough money over the time of the lease to not only pay for itself, but also save you $500, Mr Dominant, would you be interested in its purchase?' Assuming he says yes, we would then have to lay out the evidence clearly and succinctly.

After closing the negotiation and making the sale, it would also be necessary to reaffirm Mr Dominant's decision, by delivering the product on time, and again clearly demonstrating how the printer is going to fit in with his particular and unique needs. We would almost certainly obtain well-qualified leads from this converted customer.

Now consider selling the same printer to the Mr Cautious, whose forte is to rationally analyse the risks and benefits of hypothetical assumptions. In this case, we would need to set down the hypothetical assumptions (or parameters or decisions) like the Dominant personality, at the beginning of the sales cycle. We would then present ideas of how he might employ the developed rationale in the work situation, as per the Influential personality, at the end of the sales cycle. We would also avoid conflicting with his personality type, by letting him build his own rationale for the sale. The people-oriented Steady personality type would be largely irrelevant to the sales process.

For instance, to set the parameters or bounds of the sale, we would this time select the facts about the printer(s) and the running costs we particularly want him to consider and analyse in detail. If we make available choices (e.g. between printer models) too wide, he will not be able to make a decision. If we largely make the choices for him, by quickly leading him through reasoning that rules out other possibilities, he will be able to analyse the merit of those few choices left to him, which he enjoys to do. It would also be imperative in the demonstration, not to give him all the data neatly analysed, but to give him facts whereby he could arrive at his own truth-by-analysis. For example, rather than saying the new printer is going to save him money, we would tell him how much the new printer is going to cost him in repayments, paper, toner, training costs, etc. and challenge him to work out how that compares with his current printer's cost of ownership. At the next meeting, as long as the task-oriented Mr Cautious had calculated undeniable savings for himself (- because 'if we say it, he'll doubt it, but if he says it, it's true'), we would greatly facilitate the sale. We should also provide him with extra ideas as to how he can effectively utilise the equipment, to further reinforce his conclusions. We would then be in a strong position to make the old-fashioned assumptive close, that is, a sales-close that takes the sharpness of the decision away from the prospect through effectively making the decision for him and simply seeking approval for that decision. We would almost certainly obtain well-qualified leads from this newly converted customer.

Now consider selling the same printer to the more people-oriented Ms Influential, whose forte it is to intuitively visualise all sorts of ideas and applications. Here, we would need to supply not so much the facts, but more the well-ordered rationale that would act as a catalyst to the forming of such ideas, as the Cautious personality is apt to do, at the beginning of the sales cycle. We would then serve to bring such ideas into fruition, as the Steady personality would be apt to do, at the end of the sales cycle. We would also avoid conflicting with her own personality type, by letting her feel creative in the situation, and not robbing her of opportunities to make a personal impact on outcomes and to add personal flair. The task-oriented Dominant personality type would not be helpful to the sales process and in fact would probably cramp the style of Ms Influential.

For instance, to supply the well-ordered rationale to her, we would firstly spend quality time listening to and analysing her needs and reflecting them back to her. We would then carefully match those needs to the well-analysed rationale behind the purchase, at the same time seeking to establish good rapport. We would then encourage her to visualise the differences the new printer would make in terms of fellow workers, up-line management, presentations to clients, etc. If Mr Dominant was greatly influenced by product features, and Mr Orderly by product functions, then the Ms Influential will be swayed more by product benefits and workplace advantages (especially on or for people and in terms of workplace colleague admiration for her). It would be imperative that she sees and discovers these product benefits for herself. This is something that we might help, in some situations, by finding out how her purchase decision would affect others and subtly encouraging her to seek out any favourable possibilities for increased esteem by her fellows. She is also likely to be an opinion leader, which means she is more likely than other more conservative personalities to take on new technologies, as long as she can clearly see the advantages. After making the sale, it would also be necessary to reaffirm her decision, by giving good, personal after-sales service and support. We would almost certainly obtain well-qualified leads from this newly converted customer.

Finally, we may now consider selling the same printer to the people-oriented Mrs Steady, whose forte it is to support and serve the system and the team around her. Here, we would need to present as a fait accompli, not the features, functions or benefits of the sale, but the already sanctioned purposes of the sale, as the Influential personality is apt to do, at the beginning of the sales cycle. We would then make the decision for her, as the Dominant personality would be apt to do, at the end of the sales cycle. We would also avoid conflicting with her own personality type, by allowing her to feel she is helping all involved, including you the salesperson, and carefully empathising with her feelings. We would also help to make her feel secure in the difficult purchase decision by underlining how it conforms with decisions previously made by her company and fellow workers. The task-oriented Cautious personality type would be largely irrelevant to the sales process.

For instance, to supply the already sanctioned purposes of the sale, we would firstly spend quality time getting to know her as a person (e.g. her work situation, her family situation, etc.), then listen to her needs, and reflect them back to her well. We would then carefully match those needs to the already established purposes and practices of other important people identified in her life. We would then encourage her to see the decision as not a decision at all, but simply the continuation of a tradition already accepted and appreciated by others around her, partly by using the labels and language with which she is already comfortable. If Mr Dominant was greatly influenced by product features, Mr Cautious by product functions, and Ms Influential by product benefits, then Mrs Steady is moved more directly by the feelings and acceptance of other important people in her life, and by the avoidance of conflict or sudden system change. It would be imperative that she feels the decision is not going to negatively affect any important people around her, and in fact, those people are going to appreciate the decision as a personal favour. Decision security through product guarantees and warranties, as well as personal guarantees, would be important. After making the sale, it would again be helpful to reaffirm her decision, by giving firm, personal, and friendly after sales service and support. We would almost certainly obtain well-qualified leads from this newly converted customer as well.

The prior four examples, which again should be taken a little tongue-in-cheek, have been given to demonstrate how a sales person may use the phaseal nature of human behaviour to sell or fit a new printer into an existing business. The four examples also demonstrate how we have a self-referencing need for each other to complete each other, just as Yang needs Yin, and vice versa. Each phase supplies a piece of the puzzle that the other advancing phases need. In this sense, we could present the 'D-C-I-S-' model as in Figure 24.

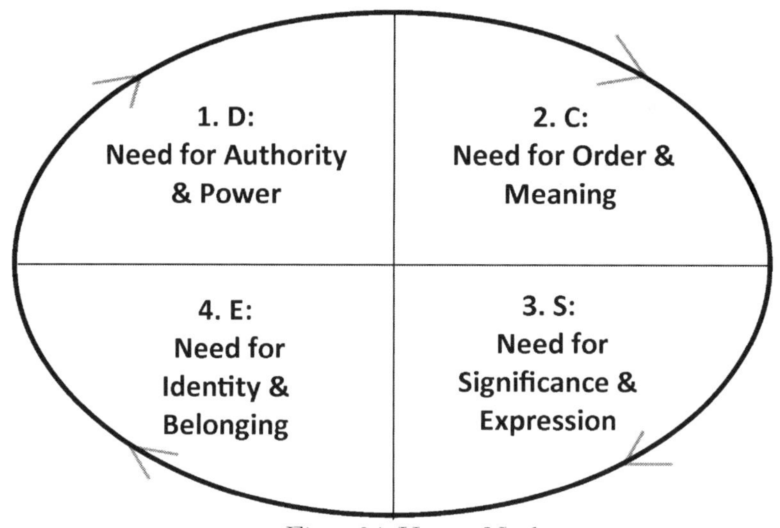

Figure 24: Human Needs

However, we should also remember that these needs for authority, meaning, significance and belonging come from the underlying systems of consciousness and instincts that also bring them about. Thus, from the opposite systemic viewpoint, we could talk about a set of need satisfiers – ego wellbeing, mental wellbeing, social wellbeing and spiritual wellbeing.

At this point, we perhaps need to clarify the use of the words spirit and spiritual. The idea of spirit we associated previously with fleeting and instinctive choices. Yet here we primarily associate the idea of spiritual with consciousness below the midline. How can this be? I would like to suggest that the problem is with our English language rather than the concepts presented here. If we could use a different word for spiritual without losing the generally accepted meaning intended here, this would be preferable. Perhaps inter-relatedness and inter-relational wellbeing gets close to the point. That is, we need to know that without doing anything in particular, that we relate to the universe both inwardly and outwardly in all our complex and implicit interconnections with it. We need to know we have an established place – we belong to this 'here' or 'there' as much as any forest vista or pictured galaxy. For the Cautious character, he looks first for this interpenetration in external facts, but for the Steady character, she looks first for this interpenetration in internal values.

For the other behavioural styles, a very different kind of wellbeing, mental wellbeing, is more important. It has everything to do with doing rather than being or complex implicit relationships. We need to know that whatever we choose to achieve or whatever endeavour we embark on, we have the time or precious 'now' and 'future' in which to do it. This time management requires an ability to mentally calculate. That is, it requires mental wellbeing. For the Dominant character, he looks first for this mental capacity through the management of external facts, but for the Influential character, she looks first for this mental capacity through the management of internal values.

The ideas of social wellbeing and ego wellbeing are perhaps a little clearer. Social wellbeing arises in communities with strong social ties that provide meaning and purpose to their members. Ego wellbeing arises in individuals who have a healthy self-worth.

At the loss of some simplicity, Figure 25 restores the axes with the appropriate labels that were missing from Figure 24. To the list of Figure 24's systemic needs and their satisfiers we might also add the biological needs of our physical bodies, and here too we find a push-pull, give-take relativity. We have needs relating to the energy the body takes in (or needs for digestion and rest) and needs relating to the energy the body expends (or needs for physical exercise and growth). Upper quadrants 1 and 3 would relate to the autonomic nervous system's sympathetic nervous system, which regulates the body's fight-or-flight response (partly by causing a rise in blood flow to the muscles from the gut). The lower quadrants 2 and 4 would relate to the parasympathetic nervous system, which regulates the digest-and-rest response (partly by causing blood to flow back from the muscles to the gut).

Figure 25: A Model of Wellbeing - the Mind-body as an Energy Processor

When we view human needs from the systemic viewpoint, we can see that wellbeing, or human flourishing, is simply a matter of how energy is used in exercise, digestion, elimination, growth and renewal by the mind-body complex. The mind-body complex is fundamentally an energy processor – look after its emerging energy needs (spiritual, mental, personal, social and physical) with strong values-in-action, good diet, regular exercise and rest, and it will usually reward you with wellbeing and human flourishing through its effective and efficient management of energy.

We might now relate Tolle's ideas of the pain-body and unconsciousness to the models of human needs and human processes in Figure 24 and Figure 25. As we also saw briefly in Figure 21, the Steady phase recognises and attempts to satisfy the need to personally identify and connect with society and the introspective self. It is thus perhaps the Steady phase (the personal relatedness phase, or the phase governed by the law or self-restraint of our own choosing) that can best correspond with Tolle's idea of the ego or consciousness gone awry (an over-active focus on me, or self, labels, identity and purpose) and the resultant pain-body.

Is the Steady phase really to blame? Probably not completely, the rot has already set in by the time we get to the Steady phase. It seems the cycles themselves often lead to disappointment, emptiness, social insanity, and pain. They start with quick Dominant action to gain egoistic power and authority, followed by Cautious consideration to attach meaning, order, and justification to such action. Then, quick Influential action to widely 'socialise' such meaning and order, followed by Steady consideration devoted to labelling and normalising all such social actions. The cycle seems to be – quick, instinctive doing – reactive consciousness and being – quick, instinctive doing – reactive consciousness and being (then disappointment, with a solution sought in quick, instinctive doing again, et cetera).

What are we doing wrong? Why are we caught up in such a shallow approach to our precious lives that results in the development of a pain-body and unconsciousness? Why have we lost touch with the authentic and virtuous self within? Why can't we just enjoy the simple flow and creativity of life, as nature seems to be able to do? Further, how can we make the answers more explicit and less hidden? The presented model, and Figure 19 to Figure 25, suggests four closely related issues.

In summary, the four problems suggested here that give rise to Tolle's pain-body, beyond just not being conscious of consciousness and not living in the now, are:

1. A lack of values and an ignorance of the natural value-sets; In particular, an ignorance of our individual need to develop our own self-restraint or principle of introspection;
2. a lack of knowledge and appreciation of the strengths and weaknesses of instincts and consciousness and an associated lack of appreciation for appropriate style flexibility;
3. a lack of respect for each phase of human endeavour or their natural progression; and
4. a lack of appreciation for the slightly asymmetrical relationship between competitive self-restraint and cooperative self-restraint as well as extrospective self-restraint and introspective self-restraint.

Once we overcome these problems, that is, we achieve an appropriate and habitual understanding, appreciation, and respect for the four phases of human endeavour (or eco-psychosocial claiming) and their associated needs and value-sets, then more personally objective and error-free introspective concept formation will arise, grudges will dissolve, and the pain-body and associated victim-identity will evaporate. This is what Tolle suggested would happen by more fully living in the now, which is by no means an incompatible concept. Neuro-Linguistic Programming techniques that empower us to step outside of self, and view the poor behaviour or judgment of self from the outside, can also help us quickly dissolve unhelpful habits.

By these means, consciousness will find its proper place alongside instincts. We will learn how to respect, love, and enjoy the material as well as the immaterial in each of us and in our planet. We will also learn how to attract the right kind of people to our sides and to our businesses, and we will be happy to share our mutual celebration of life with each other at a deeper, more authentic level. It is in this emergent mode of life that we can easily find meaning, purpose, and destiny.

The Warrior, the Lawmaker, the Priest and the Merchant

We will now consider Marston's model on a wider social scale. Consider the kinds of roles each of the four classic behaviour characteristics were likely to follow in ancient agricultural societies – say a typical city-state around 500B.C.

The Dominant character would be in a role that would require self-reliance and quick risk-weighted decision-making – a warrior would fit this behavioural characteristic. The Cautious character would be in a role that would require analysis and would lend task-support and decision-support to others – a lawmaker would fit this behavioural characteristic very well. The Influential character would be in a role that would encourage social interaction and enhance social cohesion or integration – a priest or philosopher would fit this behavioural characteristic well in early city-states. In fact, in some early examples in Babylon, the priesthood were at the centre of not only religious social activity, but commercial and legal social activity as well.

Finally, the Steady character would be in a role that would require personal interaction and would lend personal skills, fashionable artistry, support and society's personal creature comforts to others (especially customers) – an artisan, merchant, or trader would fit this behavioural characteristic. It is through the medium of fashions, associated artistic designs, reinforced customs, and norms that we make society's members feel they personally belong to a society at the outward level and thus feel comfortable within it. (A respect for common value-sets makes society's members feel personally joined to a society at the inward level).

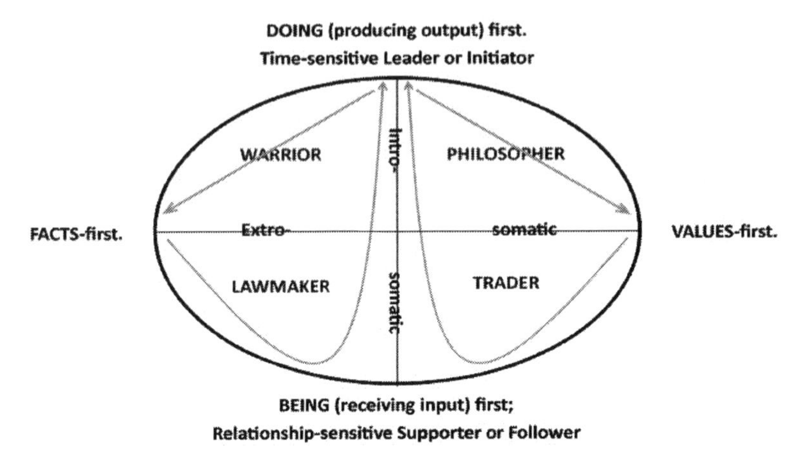

Figure 26: Roles in the Ancient City-State

An interesting observation here, when we look at Figure 26, is that the four roles selected (Warrior, Lawmaker, Philosopher, and Trader) also define quite well the societal power structure of the ancient city-state. We can thus view instinct and consciousness as the two determinants of not only individual eco-psychosocial behaviour, but ancient societal behaviour as well. That is, civilised society seems to emerge all the way from the relativistic tension between infinity and nullity.

It is now a small step to imagine how the phases would have advanced in such a situation.[21] Imagine a time when its enemies have overrun our ancient city-state. It is in disarray. Time passes, and the 'stranger in the gate' has become the norm. The people are oppressed. The strangers have grown used to the power their ancestors gained in battle generations earlier.

Finally, a strong man, a warrior (an initiator), rises up from among the oppressed and has some success against his nation's enemies (because his enemies, in the passage of time, have grown soft and careless). The warrior gathers about him all the men of the oppressed nation that are ready to fight to the death for their liberty and personal freedom. The warrior and his army grow in terms of numbers and successes. Soon the oppressors are driven out. Further, in the wake of his victories, the warrior plots how to repeat his successes amongst all the nation-states around about him. He soon builds an empire for himself, and perhaps helps to engender a new religion that synchronises the old shamanism and folk religions of his people. The new religion is at its most basic experiential, emotional, numinous, and ritualistic stage. Shared sacred music, dance, myths and legends play an important role. This is Phase 1 (Table 1 describes the pertinent risk-weighted values).

Time passes, and future generations, building upon the military success of their fathers, further expand the empire. All the nations around about are either overrun or paying tribute to the warrior's empire, which by this stage, with the help of its lawmakers (followers), has set up a system of laws and taxes to facilitate the running of the newly established empire and its military apparatus. New cultural mores and taboos are also emerging during this phase, even though it is probably too early for them to be recognised as such. By this stage, the evolving religion has an established and unique set of morals and laws as well, but it is still in its infancy. It has its enthusiastic champions, but many still prefer the older religious customs from which it arose. This is Phase 2 (Table 2 describes the pertinent positive values).

After some time an heir to the empire arises who has not known war but has become proudly accustomed to all the trappings past battles have brought. The people feel likewise. What has happened is that the civilisation has passed into its third phase. There is need for a strong philosophy or belief system and social structure, or the strengthening of existing religious ideas and cultural customs, to give further meaning to the peoples' lives and bind them together as an important society in the absence of a war footing. Soon the philosophers (or priesthood; the new initiators or champions of social meaning) arise and gain positions of power under the emperor. The philosopher/priests stand behind the perhaps weakened throne but actually control the society's most important social interconnections, events and customs as the emperor becomes more involved in religious rituals and projects, and the people likewise. At this stage knowledge greatly increases, value-sets are widely shared and the empire flourishes. Mores and taboos that arose in stage two now become socially entrenched, partly through attachment to the civilisation's system of laws.

There is also a broadening of the doctrinal, philosophical and intellectual dimensions of the religion and a toning down of the original ritualistic and ecstatic excesses, perhaps due to the influence of other important religions or philosophies in the region. Religious tolerance leads to a reasonable synthesis of current religious and philosophical ideas. Great centres of learning are supported and flourish. Perhaps the contemplative, monastic life or the harmlessly renegade wandering mystic or prophet is also part of the grand mosaic of the empire's now quite mature religious life. This is Phase 3 (Table 3 describes the pertinent normative values).

Soon new trade routes are established, new industries arise, and generations come and go. The empire becomes rich and at ease and its culture widespread. This is the time of the trader - the merchant, the businessperson and the artisan (the new followers) who are expert at understanding their consumers' needs and how to fulfil them. They provide a sense of integrated identity, purpose and belonging to the lives of their customers. There has been a kind of shift in the balance of power from the Public Sector to the Private Sector. The financial system has also become more sophisticated, facilitating further economic expansion. The wealth of the empire has enabled by this stage great examples of original and perhaps grandiose architecture to arise. Material expressions of religion in its artworks are now quite elaborate and a source of religious and/or cultural pride. The empire has passed into the final Phase 4 (Table 4 describes the pertinent values of [enlightened] selfishness).

Sometime afterwards, the civilisation's institutions become a little more corrupt and try to exploit the strong individual gains of the producers and traders. Taxes rise, the bureaucracy increases and tax revenue is frittered away on collective projects that satisfy the whims of the powerful, but actually weaken the empire. Around the same time, the people challenge traditional religious beliefs and standards. The new philosophy is 'if it feels good, do it'. The old, authoritative value-sets are slowly forgotten without being effectively replaced. A lazy trust in personal success, through genuine innovative individualism, corrupt looting, or some other kind of social leeching, becomes the normal way of life. Business and workplace success or societal corruption that began with a few individuals has now spread, along with wealth, across the society generally. The empire's progress has passed the peak to which the warriors, lawmakers, intellectuals, and early merchants brought it. It is still wealthy, but wealth is now concentrating into the hands of fewer individuals. The society is now in moral decay and the social fabric is far less robust.

Part of the problem is that the society's philosophers or priesthood have not recognised the limits and failure of human nature in introspective concept formation, nor explained the solution in terms of self-organising and emergent values (in value-sets) applied to the antinomic dance of all human endeavours. It is not our material possessions or wealth or fixed set of cultural mores and taboos that ultimately identifies us or brings us happiness. It is the experience across time, including the present moment, of congruous values-in-action (or being on the path of self-actualisation) that fundamentally identifies us and bring us happiness. Failure to avoid the trap of shallow materialism, cultural clashes, or loss of social fabric, through authentic self-actualisation, leaves the society vulnerable to collapse.

Time passes, and industries that were previously successful are no longer able to compete with strangers who are willing to work harder for less. Natural resources are depleted, markets are lost, and so is military prestige. The army has long ago lost its discipline and traditional moral purpose, along with the rest of society.

Suddenly a stranger's army rises up and strips the empire of its riches gained over many generations. The fourth phase has ended and a new first phase has begun. The people who have not died or left the city-state are humbled and learn how to be servants to (or supportive followers of) strangers once again. The cycle then repeats itself, but perhaps at a slightly higher level of evolutionary sophistication.

If we present Figure 26 in terms of modern societal structure, we obtain Figure 27, which suggests that the two vectors of human instinct and human consciousness, and the four self-restraints of risk-weighted, positive, normative and self-enlightened law, still drive modern society.

Phase 1 now includes the political system, but such a system would not be evident in its early days. It requires the following phases of prior civilisational cycles to mature and become socially explicit. Phase 2, described here as in terms of a developed judicial and legal system or power base, would also include the cultural mores and taboos that are foundational to it and the following phases. It is interesting to note here that cultural norms and taboos are likewise not recognisable as such until after they have been socialised in Phase 3 – but their genesis is in the risk management efforts of Phase 1 and following extrospections of Phase 2 in earlier civilisational cycles.

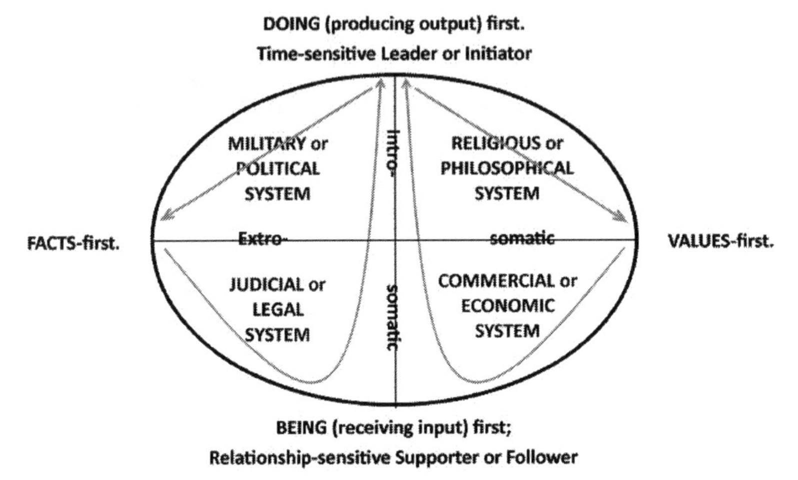

Figure 27: Modern Societal Power Structures

In reality, there is not much difference between the operation of modern society and that of the ancient city-state. In terms of evolutionary time between the former and the latter, this is hardly surprising. Just a few thousand years of evolution cannot change us that much. That is, until we come to consider the latest phase of our civilisational development – our vremetic evolution, i.e. our fast-growing ability to "rebel against the tyranny of the selfish replicators" (to use Dawkins' phrase) through "consciousness that has become conscious of itself" (to use Tolle's phrase). I see this as the great and very recent gift of a modern Phase 4. This is a phase in which our identities can be expressed and judged not just in terms of commercial trades and economic materialism, but in terms of an emerging and openly revised set of mores and taboos, or vremes, that we can likewise trade between each other across civilisations. Essential to this evolving phase of our development is a discussion of vremetic ethics.

Vremetic Ethics

Emergentism's vremetic ethics, partly in agreement with virtue ethics discussed below, is eudaimonist. That is, your emergent wellbeing, as an environmental outworking of your evolving instincts and consciousness, is also your moral purpose. This does not infer a short-term hedonistic (pleasure-over-pain) lifestyle because such a lifestyle without regard to its environment will not bring personal wellbeing. Only your virtue, as an evolving ethical being within your deep ecology, can achieve wellbeing in your here-and-now. A self-actualising life is the reward of its virtue, and so self-actualising success is both the goal and the reward of your life. Within this framework, our moral choices will feed back to our creative emergence as self-actualising beings.

If we permit our emergent virtues to choose us, we will live a life of limited but dynamic personal moral congruence, strong growth and lively purpose. To enjoy this life, we must not be deluded as to our success if we choose our emergent values only when we feel like it, or if we shun them, or we substitute them for something else. Further, if Matt Ridley is correct, our individual self-actualising will also contribute to the Gaian "inexorable coagulation of life"[22] as well as its 'diversifying emergence' briefly discussed in Part 1.

To help distinguish Emergentism's ethics from other ethical frameworks, we need a rough model of learned moral behaviour. Using Figure 18: The Emergence of Self, perhaps a basic model would suggest that indirect intentions (to trap the open future) --> lead to direct acts, which --> lead to (future) outcomes and environmental potentiation. Emergentists would thus stress the feedback path from real-world outcomes back to those apparent intentions. That is, as we pay attention to and become cognizant of the effects of our direct actions in the real world with introspective retention and protention (personally and socially) the real world influences our future intentions (or contingency planning). Emergentism would thus help us see an emergent dance or fractal cycle of moral behaviour we could perhaps picture as the unfolding iterations of a Mandelbrot set, rather than a simple sequential progression of moral behaviour that begins with independent intentions or causes and ends with outcomes or effects.

Just as in natural selection, the lifespan of each moral cycle or generation is directly related to the speed of reproduction and adoption of improvements (or emergent new arrangements of ethical concepts). The starting or ending point of this fractal cycle is thus rather arbitrary when we explicitly identify the feedback loop or interconnectedness in our model. Just as in evolution, the indirect logic (of conscious will and/or the self-organising rules of the ethical system) and direct form (of subconscious or unconscious acts and environmental outcomes) seem to inform and complete each other in a relativistic fashion, like two pieces of the one puzzle rather than just two steps of a sequence. Our enacted intentions, as guided movements, are likewise fuzzy – the concept of our self-actualised origination is faulty.

This is an important point because it enables our conscious observations to inform our subconscious processes through feedback from the environment, social and otherwise, and vice versa. The dance is about being uncritically yet intelligently aware of the apparent antinomy that exists between the conscious mind at the higher level and the subconscious or unconscious processes of brain/body at the lower levels that pre-empt our consciously experienced feelings and intentions. This pre-empting is just evidence of the continuous feedback paths in the brain/body that both form and monitor what we experience in the environment as a continuous stream of consciousness. The pre-empting does not rob us of choices; it grants them. Conscious awareness of these feedback paths means the apparent environment/brain/body-mind- or is-ought- divide tends to collapse into a more coordinated and monistic or mutually-dependent and relativistic whole. This is a fabulously important point. The 'is' depends on the 'ought', and the 'ought' depends on the 'is'. Our human morality cannot be divorced from its environment, and vice versa.

Deontological Ethics

In contrast, The Stanford Encyclopedia of Philosophy says,

> *"The word deontology derives from the Greek words for duty (deon) and science (or study) of (logos) … The most familiar forms of deontology, … hold that some choices cannot be justified by their effects—that no matter how morally good their consequences, some choices are morally forbidden. … For such deontologists, what makes a choice right is its conformity with a moral norm. Such norms are to*

> *be simply obeyed by each moral agent; … In this sense, for such deontologists, the Right is said to have priority over the Good.'*[23]

To incorporate deontological ethics into our model we will have to adjust it. To do this, we have to split intentions into norms, rules and duties that lead to intentions and intentions themselves. However if we do this, the obvious question that arises is, how do these rules and duties arise? The obvious answer for most monotheists is from God and His commandments. More generally, deontologists might be happy to say that principles, rules, or duties are shaped into beliefs, and emanate from a normative authority or standard. Thus if I believe in and worship the God of the Pentateuch I will observe the duty of keeping the Sabbath holy; this belief will shape my intention not to do manual labour on the Sabbath day. My intention will then determine my chosen, concrete act not to pick up the broom to sweep the kitchen floor. Finally, the outcome would be respect for and appreciation of the Sabbath day and thus a perceived moral congruence with God. The deontological model of learned moral behaviour is thus:

Standards or (normative) authorities or agents --> lead to principles or rules and duties, which --> lead to beliefs, which --> focus intentions, which --> lead to acts, which --> lead to outcomes.

Notice that there is no feedback path from acts and outcomes to the standards or agents. A feedback path back to the authority (i.e. God for monotheists but perhaps supreme leader in states with very strong political ideologies), would suggest the agent or standard can change, which particularly God cannot do (His infallibility prevents this; dictators are often awarded the same quality by their sycophants). Nor can there be a feedback path to the principles or rules and duties for a monotheist because an unchanging God commands these. At best, there can be a feedback path to beliefs and thereafter although even here such idealised beliefs are often not open to private interpretation. In strict cultish orders, not even intentions or actions are open to private interpretation. Thus, deontological ethics in practice (e.g. in religions or collectivist state ideologies) tend to reduce the responsibility of adherents in respect of values individually held, and I would argue, ultimately stunt individual and social enlightenment. Such uni-dependent societies become static and their mores and taboos, entrenched. However, perhaps an adult Christianity such as that described in Pillar 1, or adult Buddhism, or adult other religious or political practice, could overcome this limitation.

Emergentists would suggest that deontological authorities or standards and their principles, rules, or duties generally rest on something unscrutinised and thus rest on either a perceived reality beyond the veil (which is out of our reach), on static dogma, or on untested ideology. Therefore, to Emergentists such static rules and duties rest on unnecessarily shaky ground. It is okay to have subjective beliefs but it is not wise in a changing environment for them to be uni-dependent, static, or untested - or to rest social rules and duties upon such beliefs.

Further, we are by nature pattern-detecting and belief-forming systems. While these systems involve moral values, they do not need these values to be explicit to function. Even making beliefs explicit does not necessarily make them 'ours' in terms of behaviour. We may profess beliefs, yet stubbornly resist them at the subconscious levels that really govern behaviour. In addition, where moral values are not made explicit by the consciousness of an individual, that individual is subject to rules of common sense that are often poor reflections of reality at this stage of our evolution (natural and artificial). It is okay and unavoidable to have shortcut rules of behaviour (nature is made up of a vast number of them). However, it is in our own best interests to ensure any consciously adopted, common-sense rules-of-thumb that do translate into behaviours are founded in knowable reality. We should not base our behaviour on any subjective beliefs about what might be beyond the veil, or might be waiting for us in some kind of ideal future imagined by a few dogmatic opinion-leaders.

For instance, it is evidently true (i.e. observable in reality) that each person functions at his or her best, or flourishes in existence, when his/her individual right to life and freedom is respected. Human flourishing also occurs when each person can act volitionally and responsibly in the pursuit of his/her own wellbeing within the constraints of his/her naturally and freely chosen social environment. It is the challenges and supports (or pleasures and pains) of our environment that allows us to emerge, physically, mentally, and morally.

This necessarily means that a human's natural freedom ought to be restricted by self-governance or responsible government to the extent that it denies the moral freedom of another human. In effect, government is simply a shortcut or normative social system that ideally gives orderly effect to the natural rights recognised in each human being. If all humans could overcome the internal biases and mind-games of consciousness-plus-instincts such that they always fully appreciated appropriate quasi-universal values within and always acted thoughtfully and with utmost virtue (consideration and justice) with respect to others and the environment, we would not need an external, shortcut, normative government. Reality and self-restraint would be our only government, perhaps much as we see it unfold in the pecking order of birdlife in the local neighbourhood. We would probably worry less about the dichotomous concepts of good and evil and begin to flow more with actions that more naturally promote our wellbeing in the local here-and-now.

If we view the standard, authority, or agent as the personification of a sum of attributes or values known on this side of the veil, then those values-in-action can be seen to incorporate intrinsic agency themselves rather than God, the Standard, or Authority, as Agent. To put this point another way, looking from this side of the veil, our species' discovery and uptake of values (like our species' uptake of improved genetic features) is a process of unavoidable natural and artificial selection imposed by our environment and its self-contained self-organisation rather than a process imposed by God as Agent.

Thus, it can be the conscious measuring and reframing of an outcome, via naturally or artificially selected values, that becomes the agent of transforming propositions, beliefs, focused intentions and acts in each of us. Further, we as conscious beings can also take personal responsibility for enhancing and living by those values or measuring-sticks that lead to our species' improvement in terms of human wellbeing and sustainability. Rather than being mere domesticated plant or bird or animal breeders, we can become careful and professional breeders of our own most favourable values-in-minds.

Thus, the feedback path in our phaseal ethical model can go all the way back from outcomes to known and emergent values. In fact, if we view values as measuring-sticks of outcomes, then it is arguable whether values begin or end the ethical model chain, which would seem to make perfect sense. My preferred view is that values emerge in the chain rather than sit at the beginning of the chain as suggested by the deontological model. Further, if values are emergent in this algorithmic process, then so are our intentions and actions. That is, if we study human brain function via MRI scans, it should not be surprising to find that subconscious and emergent conscious intentions and actions are interpenetrating within their environment, which might be frustrating to those looking for a single point where values, intentions, or morally responsible actions begin or end.

Chaotic but constrained sub-processes combined with repetitive, algorithmic feedback with our internal and external stakeholders (including the local environment)[24] means our consciousness, thoughts, and values are continuously emerging from throughout the strange loops of the brain. They emerge from the interactions at the more chaotic or microscopic level to those at the more deterministic and macroscopic level of mind. The controversial findings of Libet's experiments discussed in Part 1 thus become an important aspect of the emergence of limited or oasis-like free will in this feedback loop. The findings confirm important, necessary features of the mind's ability to create those reflective 4D images or strange loops that enable the emergence of a moral agent.

The vremetic ethical model, once understood and exploited, has a chance to bring about the quick improvement of human wellbeing. Two important ways it will do this is by making clear our shared moral agency, and by encouraging those largely subconscious mores and taboos to come within the full gaze of self-aware consciousness.

In summary, Emergentism's ethical model of learned moral behaviour is:

- Prior experience and identification of emergent values -->
- Leads to explicit propositions (statements of principles, self-restraints, or duties), which -->
- Confirm or disconfirm largely subconscious beliefs, which -->
- Focus intentions and lead to goals, which -->
- Lead to actions (taken or withheld within changing circumstances), which -->
- Lead to assessed outcomes in the environment, which -->
- Lead to values unfulfilled, fulfilled or settled, and

- So on around the same loop continuously, with intermittent 'mutation' or emergence at any point in the flow or process of moral consciousness.

This is why happiness (an example of values) can be both the goal and reward of life and how a self-actualised life can be the reward of virtues (or values-in-action).

This also clearly suggests that we cannot act with moral intention for the very first time until after we have made our first introspective measurement of an outcome (because as noted earlier, consciousness is consciousness of something). That is, as moral beings, we have to have the higher-order value-measurement thought about 'something' before we can make that something fully conscious.

Further, if we become morally disinterested in an outcome, then the belief that contributed to that outcome and the outcome itself will have less impact on ensuing moral intentions simply because the higher-order cycles of conscious value-measurement will be fewer and weaker. I suspect the lack of higher-order value-measurement thought equates well with Tolle's idea of unconsciousness and Dawkins' idea of the tyranny of our pre-established selfish replicators. Likewise, the moral habit of higher-order value-measured thought probably closely corresponds with Tolle's idea of "consciousness that has become conscious of itself".[25] We need to recognise and break free from limiting habitual behaviours, using the Emergent Method and aids like Neuro-Linguistic Programming.

The moral focal point, if we must name one, would be the measuring-sticks (values) artificially applied within our value-set self-restraints and after (future) outcomes. A lively and engaging congruence between explicit values embodied and outcomes achieved is what is important if happiness is to be both our moral purpose and life's reward for virtuosity. This will mean congruence between beliefs professed in consciousness and beliefs maintained in subconsciousness. In addition, the held values necessarily include a respect for the worthiness of the others who share the humanity we recognise, cherish, and consider worthy within ourselves.

However again I should stress that Emergentism does not cause us to envisage just a simple set of sequentially processed steps from values to outcomes. Rather, it helps us see each step as a necessary and interconnected piece of the fuzzy puzzle. That is, we each need to contextualise values in terms of the life-promoting self-restraints (which are risk-managed competition, positive extrospection, normative cooperation, and self-enlightened introspection). We each need to internalise these self-restraints in our personal beliefs, which we rely on. We need to concretise our beliefs into specific personal goals. We need to strive for our goals in terms of a set of personally chosen actions in changing circumstances. We need to consider our actions in terms of outcomes achieved for all stakeholders. We need to measure the outcomes in terms of congruence with values held. It is not just indirect values, self-restraints or beliefs, which are required by Emergentism's ethical model, but direct goals, actions and outcomes, as well. It is the interpenetration of the indirect and the direct that makes for a happy and successful life.

Here then we have our seven pillars of wisdom from Part 1, with:

1. universal (or locally recurrent but still evolving) values as our emergent foundational pillar or cornerstone;
2. life-supporting self-restraints or principles (or indirect shortcut summaries at various levels, beginning with our most sensibly-biased instincts) as our most cherished knowledge;
3. a reliable set of temporal beliefs or explanations (rather than silly and unsustainable mind-games) as our most valuable understanding;
4. a passionate set of self-enlightened goals as our deepest wisdom;
5. a sure set of courageous and diligent actions as our most worthy commitment;
6. a respected and interconnected set of outcomes in the environment as our best examples of leadership; and
7. congruent personal happiness and success as our greatest legacy

Consequentialist Ethics

Let's move on... The Stanford Encyclopedia of Philosophy also says:

> *Consequentialism, as its name suggests, is the view that normative properties depend only on consequences. This general approach can be applied at different levels to different normative properties of different*

kinds of things, but the most prominent example is consequentialism about the moral rightness of acts, which holds that whether an act is morally right depends only on the consequences of that act or of something related to that act, such as the motive behind the act or a general rule requiring acts of the same kind.'[26]

Consequentialist ethics is usually considered in a social context – good consequences for society, whereas deontological frameworks tend to be applicable more personally in the conscience of each individual. In contrast, Emergentism's ethical framework operates in the interconnection of personal and social contexts. Likewise, consequentialists tend to be more forward-looking because they don't rely on past value-contexts (they are more value-neutral), whereas deontologists tend to be more backward-looking because they rely on past mores and taboos. For this reason, deontologists tend to see consequentialists as progressives that are hostages to their circumstances rather than as moral agents intelligently separated from those circumstances. On the other hand, consequentialists tend to see deontologists as conservative and thus often ineffective or inefficient in changing circumstances. In contrast, Emergentism's ethical framework comes together and flourishes in the here-and-now by continually linking past experiences to future aspirations. Emergentists are prepared to accept the relevance of past mores and taboos (there is never a moral 'blank slate' or 'tabula rasa' in any circumstance, no matter how novel), but also test those mores and taboos against aspirations and outcomes wherever and whenever necessary.

A consequentialist would consider a person's (indirect) intentions, virtuous or otherwise, as something only relevant as to assigning (social) culpability and therefore of secondary importance to the (direct) outcomes in the eco-social context. Further, a consequentialist might see a moral code, perhaps based on a subset of the list of values in particularly Table 1: The Risk-weighted Value-set and Table 2: The Positive Value-set, as of secondary importance. In sharp contrast, Emergentism's ethics would see a focus on the social dimension before the personal dimension as running our natural being configuration backwards (in S-I-C-D-order), which is unsustainable.

Emergentists would also object to the look-ahead assumption of an end that places little importance on the phases traversed and relationships held here in the present to get there in the future.

Nature does not work like that; it traverses self-organising means in the here-and-now, with every humanly perceived end being just another means towards the maximisation of local energy effectiveness and energy efficiency. The emergentist would argue we should operate close to nature in this important regard (while making some allowances for its short-sighted pitfalls) through development of local social systems and their partly open and developing borders. If as a moral rule, we treat every present moment as a mere means to some future end, we miss living life in the here-and-now, which is the only life there truly is.

Virtue Ethics

Finally, Encyclopaedia Britannica says of virtue ethics,

> *"Virtue ethics is primarily concerned with traits of character that are essential to human flourishing, not with the enumeration of duties. It falls somewhat outside the traditional dichotomy between deontological ethics and consequentialism. It agrees with consequentialism that the criterion of an action's being morally right or wrong lies in its relation to an end that has intrinsic value, but more closely resembles deontological ethics in its view that morally right actions are constitutive of the end itself."*[27]

The emergentist would argue that we can show lying to be wrong if there is a considerable incongruence or dissonance produced between values held with respect to self as part of humanity before the act and virtuous values applied to outcomes after the act, resulting in disappointing unhappiness. Vremetic ethics, rather than being fundamental, thus emerge from the dynamics of the self-aware life lived.

Thus, in contrast to Nietzsche's comments, an emergentist develops and relies upon her/his strong but sensitive and emergent sense of personal conscience and moral value-adding. This conscience is not just linked to a simple set of values or fixed set of rules and duties, but rather to all her/his seven pillars of wisdom (indirect and direct). If our happiness, wellbeing and success is the crowning glory or roof of our house or temple of seven pillars, then an active and evolving vremetic conscience of wellbeing rather than strict good and evil, is the floor that forms the entrance between each pillar.

Why not take a walk inside your temple-house and look around? What do you observe and sense; what do you think and identify; what do you learn? How are your seven pillars arranged (– it is your house, you can arrange it anyway that suits your own reality)? Are the pillars arranged in a straight line like a seven-branched candlestick, with perhaps the fourth pillar of wisdom at its centre? Alternatively, have you arranged those pillars symmetrically like the seven-pointed Commonwealth Star of the Australian flag? Maybe there is an outer ground floor of direct goals, actions and assessed outcomes, an inner first floor of indirect values, self-restraints and beliefs and a pinnacle in the roof of personal moral congruence. Maybe you see congruent success or wellbeing at the centre and all the other pillars orbiting asymmetrically around this centre in differing paths and at differing distances, speeds and sizes, much like our solar system or like electrons in their atomic orbitals. Finally, maybe you see values and value-congruence as the twin black holes of your galaxy, with all other pillars orbiting around this galactic core. Whatever the case, feel free to slowly walk around, inside and outside your temple.

While a virtue ethicist might consider the values, principles or beliefs, of an individual or group, an emergentist would go on to consider self as a survival machine of moral agency (or an embodiment of emergent virtues) and participant in the antinomic, interconnected, algorithmic and heuristic dance of life. She or he would take important personal cues from outcomes in reality rather than just considering the virtuous basis of choices made. She/he would also seek to have those around about join her/him in becoming more virtuously proficient in that dance or improvisation (while at the same time releasing new wealth through a trade of values and new arrangements of value).

We have all heard the political appeal to strong resolve or leadership in the face of contradictory moral outcomes, 'if I only knew what I knew then, and was faced with making the same decision again, I would do the same again'. While an Australian politician might defend the morality of past actions based on knowledge at the time, and thus steadfastly refuse to reconsider those actions in the light of present knowledge, or refuse to consider any change of present demeanour, the emergentist would consider past actions in the light of present reality. To the emergentist, any incongruence would signal a possible ineptness in the ongoing dance, which would motivate development (in terms of values, self-restraints, beliefs, intentions held, actions taken, or levels of skills and abilities relevant to the outcome).

Unlike the politician, the follower of the Emergent Method would learn from any incongruence brought about by her/his values, principles, beliefs, intentions held, or actions taken, and see this new understanding of reality's interconnectedness as a wonderful opportunity, even if personally painful, to further pursue happiness and self-actualise. The emergentist, without the eternal self-condemnation of a religious sinner, but with the realisation that moral and environmental sustainability is a key to the wellbeing of all evolving and intelligent beings, would consider how she/he might have danced more proficiently in the past process and thus how she/he might dance more proficiently now.

The focus would be on resolving moral incongruence or seeking moral homeostasis within each moment of the wonderfully unfolding process, rather than focus on the basis of the decision in the past or the outcome in the future. Every incongruent signal received may indicate a naturally flawed or fallible moral agency but also an opportunity to possibly improve the efficacy of that moral agency. Our moral fallibility is the source of our endlessly improvable moral agency, just as noncoding DNA is a possible source of improved future adaptation.

If we compared Emergentism to consequentialism and the maxim 'the end justifies the means' then as long as the end is humankind's individual happiness in life-and-being here-and-now and the means is our individual and most appropriate virtues, according to the phase of life's dance, then there would be few differences. Unfortunately, this is not the case. Consequentialism seems to be the ethical system equivalent of intelligent design rather than that of nature and nature's alternative way (as demonstrated by a healthy forest diversity) – the emerging efficacy and congruency of life-and-being here-and-now.

However, more modern and more moderate consequentialists would not be as content with the maxim, 'the end justifies the means'. They would be ready to accept that intention is an important part of an ethical model, but perhaps more pragmatically focus on the outcome as, after all, what really matters. The emergentist would simply ask, 'While the outcome is very important, how would a consequentialist go about changing future outcomes for the better?' How would a consequentialist promote social and individual enlightenment?

Would it be to try to always fix outcomes? Hardly! Would it be to control only certain actions, while discouraging others? Perhaps, for those less enlightened in society (e.g. prone to the use of violence), but this could not be the desired end goal if the aim is human flourishing in its fullest sense in society, including in those currently less enlightened. Emergentists, like virtue ethicists in this regard, would suggest that we must promote social and individual enlightenment at the deeper level of rising intentions and beyond (often implicit beliefs, principles and values). More modern, moderate consequentialists would probably have to agree, which would suggest that in terms of emerging social and individual enlightenment at least, their focus on outcome is unnecessarily restrictive.

Ethical Decision-making

The growing synthesis of ethical systems is an important point. Are consequentialists becoming more aligned with virtue ethicists? Are deontologists accommodating more consequentialist views these days? Are our important ethical systems evolving towards a single ethical framework? Is Emergentism an enunciation of this trend? I suspect so. Just as we noted with respect to competition and cooperation in Pillar 2, as well as extrospection and introspection in Pillar 3, so the same applies to the concepts of analysis and synthesis (of e.g. ethical systems) here in Pillar 4. That is, synthesis is not reliable without analysis. Nevertheless, synthesis wonderfully emerges from analysis and takes it to a higher level. This continual interpenetration of shuffled analysis and synthesis is what is required of our new and emerging ethical system if we are to find a sustainable and personally fulfilling future.

In summary, Emergentists tends to combine the values of the virtue ethicists with the principles, beliefs, intentions and acts of the deontologists and the regard for social outcome of the consequentialists in a single, naturalistic model of personal moral congruence or sustainable wellbeing. For this reason, this text has no need at this point to reconsider each of the moral dilemmas that are typically analysed to ensure the model works. It simply adds and underlines that deep happiness, contentment and satisfaction does not come with values alone, principles alone, intentions alone, acts alone, or outcomes alone. Wellbeing comes with a strong moral congruence between seemingly immutable values-in-action and outcomes achieved that emerges from consciousness, subconsciousness, and unconscious systems of pattern-forming that interpenetrate with the environment. The Emergent Method is thus largely the maintenance of a limited but dynamic personal moral congruence, comparable to the method of adaptation at the level of the organism. The Emergent Method is needed because we wildly understate our species' need to intelligently adapt to the environment we are creating around us.

What we have arrived at here is something quite astounding. Personal moral congruence is necessary for wellbeing and deep happiness; personal moral congruence is associated with success in a human life. That is, a definition of human success is dynamic personal moral congruence, or dynamic moral homeostasis. Perhaps more than a mere system of moral homeostasis, personal moral congruence represents a system of physical, ego, mental, social, and spiritual homeostasis or wellbeing. What this strongly suggests is that when we added consciousness and thus reason and self-awareness to our volitional human instincts, we became explicitly moral beings, just like John Galt in *Atlas Shrugged* suggested.

Further, wellbeing, as defined here, is essentially a moral concept rather than just a feeling of joy, peace, etc. The emergence of human consciousness from human instincts meant that the only means to self-actualisation as a human being would be, from that time on, through an emergent and personal moral congruence.

Our deliberate or explicit creation of personal moral congruence in the antinomic dance of each of our lives is also the creation of true wealth. It is very likely that this true wealth indirectly and eventually contributes to Matt Ridley's inexorable coagulation of life, because such wealth is a measure of life's met need or incompleteness satisfied. It seems that we are each building something like Howard Roark's Temple of Life, which was the embodiment of Roark's own courageous virtues, but with many mistakes and corrections along the way. Are we just pawns in the inexorable coagulation of life or are we life's fully-fledged participants? Is destiny a matter of lazy fate or a matter of active choice? The choice seems to be evitably (i.e. more and more openly) ours.

Emergentism's model would have some accord with Harvard psychologist Joshua Greene's "dual process" theory of moral judgment,

> *"according to which characteristically deontological moral judgments (judgments associated with concerns for "rights" and "duties") are driven by automatic emotional [or instinct-driven] responses, while characteristically utilitarian or consequentialist moral judgments (judgments aimed at promoting the "greater good") are driven by more controlled cognitive processes [or consciousness-driven]. If I'm right, the tension between deontological and consequentialist moral philosophies reflects an underlying tension between dissociable systems in the brain."*[28]

A study of Emergentism would suggest that this duality or relativity of the moral process arises from the evolved relativity of our human nature (i.e. our consciousness-plus-instincts or mind-plus-brain/body), which is not without considerable tension or side effects or "pain-body" effects at the present point of our evolution. This tension is somewhat relieved as we deliberately and honestly recognise the strengths, weaknesses, opportunities, and threats of that relativity (or asymmetrical arrangement). To borrow Tolle's terminology, as consciousness becomes more fully conscious of itself (its genes, memes, and vremes) we begin to harness its emergent power. That is, as our consciousness-plus-instincts (including subconsciousness and unconsciousness) becomes more fully aware of our value-laden identifications, choices, actions, and narratives in the midst of our often foolish biases and mind-games, we move towards more self-actualising endeavours. Only then can we hope to deeply understand others (for example, of a differing political, cultural, religious, or economic persuasion).

In terms of scientific or secular humanism, I would see little difference with Emergentism except in this one important area of ethics. Secular humanists tend to hold utilitarianism as their guiding ethical framework, which we can view as a consequentialist ethical framework driven by a collectivist 'greatest happiness for greatest number' principle, rather than by a more naturalistic method of accommodating local human flourishing in a fuzzy shuffling of emerging interests. It is interesting to note that there seems to be no personal utilitarians, only collectivist ones – otherwise each utilitarian would be queuing at a local hospital demanding the right to give up his or her own life and internal organs in order to save the lives of two or more people in chronic need of those organs. Personally, it seems utilitarians subscribe to a different code, which is more in line with nature's pragmatic or self-enlightened life-and-being here-and-now.

The label of 'antinomian ethics' has been avoided here because this approach was seen by the former Episcopal priest Joseph Fletcher in his 1966 work *Situational Ethics* as devoid of any ethical system at all, which is by no means what is being described here. Note: Situational Ethics essentially advocated Christian love as the overriding principle in all situations. It is interesting to note that just one year after finishing *Situational Ethics*, Fletcher gave up his faith in Christianity. In 1983, he was named Humanist Laureate by the Academy of Humanism.

To expound the idea of the antinomic dance between alternative frameworks of public (positive and normative) self-restraints and private (risk-managed and self-enlightened) self-restraints, we must understand that this dance is not just a fake role-playing or a dance improvised from a somewhat opportunistically contrived bag of tricks. It is a genuine full immersion into each of the vying frameworks or value-sets as appropriate to the situation's phase, motivated by the ongoing desire for and reward of personal moral congruence.

Nor is the antinomic dance merely situational ethics at work – if this were the case, it would require a pluralistic, compromised (maybe even democratic), shallow blending and superficializing (to use Nietzsche's term) of frameworks in order to achieve a desired end, rather than a clear change from one strong value-set framework to another. To use an analogy extracted from the lessons of genetic inheritance, the study of Emergentism suggests a Darwinian shuffling of values (leading to emergent congruence) rather than an unhelpful blending, levelling, averaging, dumbing down, or opportune self-serving of values. We serve values implicit in subconsciousness. Once we select a value-set, its values and associated self-restraint come to the fore within the other self-restraints of reality – not some overriding rule such as 'love conquers all'. With those shuffling values, an efficacious solution or interrelationship comes to the fore that adds value to our human condition. I think this approach has much in common with Deutsch's discussion of the plurality voting system.[29] He advocated a shuffling of political candidates first-past-the-post as the most fair and efficacious means of elections, governance and societal advancement, because we could experience the most evitable policies of candidates through a kind of evolutionary trial-and-error rather than wallow in pluralistic compromise.

The dynamic moral framework is selected based on what phase the situation is judged to presently reside in by the local player(s) involved. This skill of judgment, which ultimately promotes personal moral congruence, can only self-organise or improve with time and practice, much like the personal time-management skill or more broadly, our slow evolution of the precedents established in Common Law. However this learned skill can only improve if we take on the challenge and habitually expose our consciousness to the full gamut of value-sets and values, for example as given in tables 1 to 4.

What follows is a quick aid to selecting a situation's phase. Does the situation mostly require action or consideration, (is it in an active/doing phase or an outwardly passive/being phase)? Does the situation mostly require self-reliant resolution or social resolution, (is it in a fact-driven risk-weighted or value-driven normative phase)?

- If the situation is active and self-reliant, use the risk-weighted approach and value-set such as in Table 1, corresponding with CQ
- If the situation is passive and self-reliant, use the positive approach and value-set such as in Table 2, corresponding with IQ;
- If the situation is active and social, use the socially normative approach and value-set such as in Table 3, corresponding with SQ; and
- If the situation is passive and social, use the self-enlightened approach and value-set such as in Table 4, corresponding with EQ.
- After considering the most relevant value-set, you may wish to proceed to the next most relevant value-set, and so on in D-C-I-S- (or temporarily S-I-C-D-) phaseal order, especially if you seek a complex solution or long-term view, but this is not always necessary.

- A situation's phase can quickly change: Reassess your ethical considerations as regularly as the changing situation requires (an emergent and self-organising universe always requires feedback).

Note: Wellbeing is both the moral purpose and the reward of a virtuous life. Once you have identified the situation's phase, you can then assess how your personal values and purposes may align with that phase and what constructive, virtuous actions or observations you may take (or leave to others).

This brings us to the closely related idea of value-pluralism, also known as ethical pluralism. This view suggests that in some situations several seemingly antinomic values may apply (e.g. self-esteem versus self-sacrifice), but often gives no method for resolving the choice dilemma or moral dilemma. This is because a limited rational approach is sought to the dilemma rather than a life-and-being here-and-now approach, and the idea of Emergentism's ethics is not raised. The application of Emergentism overcomes the raised ethical dilemma of value-pluralism because it escapes the narrow rational-approach-trap that often freezes decision-making and action. It respects the fact that all vremes, while of a higher order of complexity, are borne out of vast yet algorithmic natural processes - just like memes and genes. That is, while we endlessly calculate we are also endlessly hampered by Gödel limitations. However, Gödel limitations do not hold back the universe – the universe finds fuzzy resolution in its fractal dimensions and vast parallel interrelationships. It is from these vast interrelationships that the values of our species arose. So it is suggested here that if we can more broadly unite with the universe through the outworking of a naturalistic philosophy, then we can also perhaps escape at least some of those Gödel-like dead-ends in ethical logic.

That is, it is by swimming in all four rivers of value-adding (even if leading to a little chaotic pandemonium[30] such as probably occurs at the pre-conscious level anyway) that we sometimes distil the best solutions, even in the face of internal and external conflict. We should not hold back on the chaotic and fuzzy swim simply because we see a rational approach as a prerequisite. In doing so, we presume a linear, uni-dependent approach from the outset rather than just the rational outcome we desire. If we must, even if we cannot understand which phase the situation resides in, we should just take a leap of faith into life-and-being here-and-now and into a river rather than stand on its banks. This is what nature would blindly do in its probabilistic processes. In this way, rationality and wellbeing will have a chance to emerge and be maximised, and we will have a chance to learn.

We might call taking the swim 'brainstorming'. Brainstorming, without heavy self-condemnation and with no silly, subjective, outrageous or outside-the-box ideas barred, gets the creative juices running, can save us from groupthink, and often saves us from the rational bottleneck. Moral brainstorming is one example of how flexibility or a freer and more liberating approach to moral enquiry may proceed to rational outcomes with personal moral congruence.

The example of a politician that later regrets his decision to go to war but cannot admit it in public is an interesting one. Such a heavy decision obviously had many dimensions to it. Clearly, there was a factual dimension. For example:

- 'What risk-weighting should I apply to the imperfect intelligence at hand?'

Further, there was a logical and moral dimension:

- 'Should I allow a doctrine of pre-emptive strike in the case of a known tyrant who is prepared to use horrible poisonous gasses as a weapon of mass destruction against his own country's citizens? Further, can I foresee the impacts of my actions on the very citizens for whom I act?'

There is also a social leadership dimension:

- 'If I take steps towards war, how well am I able to lead my fellow citizens towards a consensus with respect to the decision to go to war and to conduct the war? Will the decision I make as leader lead to a congruence of applied values with outcomes in the minds of my fellow citizens (and myself) during the war and after the war has concluded?'

Finally, there is the dimension of self-satisfaction:

- 'What legacy will I leave to those who come after me?'

Perhaps it would be unfair to answer these questions now with the advantage of hindsight, so I will leave you to consider this task. Moral brainstorming, or swimming in all four rivers of value-adding, would have perhaps prevented the groupthink of some hawks at the time intimated above and given a larger voice to some of the doves. It might also have moved focus off the strictly moral intention of the decision to the moral outcome and likelihood of moral congruence afterwards. By doing so, it might have shifted focus to the 'judgment of information' dimension or the 'leadership' dimension a little more. These dimensions, while not directly moral dimensions (although personal values drive them just as much) were intimately tied to the feelings of moral congruence or otherwise before and after the outcomes of the decision.

As noted in the Part 1, perhaps some (especially strong political leaders) will see Emergentism's moral flexibility or openness as an immoral thing in itself. The attitude here is that for a moral code to have credibility, it must have a strict consistency. In addition, for a moral code to have consistency, the list of values underpinning it cannot be contradictory. The only way to achieve this requirement is to severely restrict any list of values to which the code adheres or to follow an overriding rule such as 'the greatest collective preference' or 'love conquers all'. Therefore, a consistent moral code will tend to adhere to only a small and perhaps toned-down subset of one or two of the value-sets listed in tables 1 to 4.

I see this attitude as a hang-over of uni-dependency from our classical worldview of the past. Is a limited set of values a more moral approach just because it has the attribute of a claimed consistency? What is more important if happiness and human flourishing is our goal both individually and socially? Is it consistency before and after the volitional act or a convergent personal moral congruence before, during, and after the outcomes?

Are followers of a narrow values-code or overriding rule often deceiving themselves when they, perhaps inadvertently, allow themselves exceptions to their own rational but rigid rules, that are by intelligent design rather than fully anchored in fuzzy reality? Is an approach driven by reality's large and diverse set of emerging but locally recurring values, applied according to the local setting, the life-phase, and its corresponding value-set, in fact a more naturalistic, transparent, internally consistent, rational, and moral approach? This is the radical evolutionary question Emergentism's ethical system poses. The Emergent Method advocates the 'Yes' answer.

The Zen Master, Sent–ts'an (? - 609 AD) said, "If you want the truth to stand clear before you, never be for or against. The struggle between for and against is the mind's worst disease." Perhaps the flexibility of Emergentism's approach (as opposed to the rigid, uni-dependent good/evil dichotomy) is the key to arrival at such an egoless truth and personal moral congruence (or peace of mind, as opposed to Tolle's "pain-body" effects).

The consideration of Emergentism, through many trial-and-error iterations, leads us to moral humility and naturalistic value veneration rather than their opposites. However, the study of Emergentism does not advocate a passive emptiness/nirvana approach or an active and experiential 'involvement / sudden enlightenment in the slipstream' approach to moral bliss.

Moral congruence, happiness and wellbeing are found as we allow the relativistic aspects of life (passive and active, as well as personal and social) to virtuously coalesce in the present moment and environment, perhaps with a little challenge and pandemonium as well as support along the way, and then to dance across spacetime.

Blind Faith

An interesting issue that has arisen from the prior section is the idea of a 'leap in faith'. Faith in what? Faith in life-and-being here-and-now! That is, faith in yourself as both a project of the world and a unique perspective upon the world. As we have already noted, to live life every cell must take life; we need to choose life, or at least acquiesce to it. Life and volition come in the same package. Emergentism's system of ethics recognises this core fact. This kind of faith is a little blind, just as natural selection is a little blind. Life is a little blind, but it cannot deny itself (this is death). The universe moves forward a little blindly or chaotically, but there is no stopping it. It evolves along the way, due to the algorithmic feedback attached to any initially blind movement or process.

Faith is clearly a virtuous value applied as a necessary evolutionary shortcut in both personal and social situations the world over. It would probably be fair to say that scientists, atheists and such books as *The End of Faith* by Sam Harris have not put faith itself on trial. Our scientific and philosophical community cannot object to the odd leap of faith. Our scientists, business entrepreneurs and athletes have proved repeatedly in our past that intuitive flights of faith are often crucially beneficial. The odd leap of faith can be quite helpful and even necessary, after all this is how evolution blindly proceeds. In fact, no rational decision can come to resolution in a world of imperfect information without a leap-of-faith compromise, just as a purchase decision in the supermarket cannot be fully rationally analysed due to imperfect information and lack of time. Eventually we have to make the emotional leap of faith and exercise our volition – we make the risky purchase decision, pay for the product and proceed home to continue our busy lives. This is the force-system of spacetime in action.

However, there is a difference between this kind of leap of faith and the axiomatic, demanded blind faith of dogma (whether as believer, atheist or agnostic). It is in the level of blindness. A demanded axiomatic blind faith takes a stance as to what may or may not be behind the veil (e.g. 'God is', 'God is not', or 'God is uncertain') whereas faith in life-and-being here-and-now does not. For instance, religious blind faith in an unprovable afterlife beyond the veil cannot be objectively tested by experience (ever) while we (any of us) remain alive on this side of the veil. A demanded axiomatic blind faith cannot objectively teach us anything while we (any of us) live on this side of the veil (ever) - unless we claim an unprovable subjective connection to that which is beyond the veil. Therefore, we can rightly view such a claim with suspicion. This is unlike a product purchase that proves later to be poor value-for-money, or a naturally selected gene mutation that later gets reversed in a modified environment.

The kind of demanded axiomatic blind faith we are talking about is missing the vital feedback path we see every day informing nature's initially blind but self-bootstrapping advances. If this blind faith cannot teach us anything via feedback while we live on this side of the veil, it cannot contribute to our enlightenment, personally or socially. Nor can it contribute to our enlightenment objectively or intersubjectively, in the sense of contributing to personal objectivity or reducing error in introspective concept formation. The intersubjective models we build up in our minds over time that are based on a demanded axiomatic blind faith can only reinforce the blindness.

This is perhaps a very disturbing point to many. If the biosphere followed a concept of dogmatically demanded blind faith, mutation and evolution would come to a halt and eventually life would come to a halt as the non-living environment continued to change via the force-system of spacetime. Demanded axiomatic blind faith is elitist in the sense that it separates the living from the all-inclusive and self-contained concept of interpenetrating life. It presents the living with a discontinuity or incompleteness that is not present in full reality.

We each may choose a life-stance with respect to the presumed objective reality beyond the veil to somehow bring order to our own lives, at some level of personal risk. However, it is entirely against our immutable human right to life and freedom, based on knowable reality, to force a demanded axiomatic life-stance on anyone else. It would probably be a socially cohesive move to formally recognise this concept in law and in our institutions. Like Richard Dawkins and Sam Harris, I would agree that life-stance freedom of choice should extend to every individual, adult or child.

Realistically, demanded axiomatic blind faith has nothing new to contribute, but it can rob our attention away from things that can contribute to our global enlightenment. Worse, it can blur the line in our not-so-enlightened minds between those things that are mere mind-games or stagnant mind-traps and those things that really contribute to our lives, the lives of our progeny, and the lives of our societies, here-and-now. Even more darkly, demanded axiomatic blind faith can lead to a belief-reinforcement arms race. The more we invest in our belief, the less likely we are willing to calmly consider the cost of that continuing investment and the less likely we are to walk away from that investment at a total loss. We may also have a tendency to invest more, in an attempt to prove our mind-models to ourselves and to others.

We may also have a growing tendency to switch off when anyone tries to explain another view – such as the self-contained and interpenetrating view of knowable reality on this side of the veil, proposed via Emergentism. The danger here is that the concepts of complexity, chaos theory, quantum physics, evolution, and self-organising emergent properties, as well as general scientific research and the limitations imposed on our current models of reality by Gödel's Theorems, may all be rejected in defence of an ultimately indefensible blind faith. "Love can't emerge out of atoms, chemistry and sinews – mix them anyway you like, it will never happen!" or some such remark will be ignorantly exclaimed. By the same logic, molecules should never have emerged from the strong nuclear soup of our universe's near beginning – and nor should this same nuclear soup arisen out of simple quantum fluctuations - but according to scientific modelling it did, and to stunning effect.

Such dogged ignorance could also be realised by taking offence at an opposing dogmatic view of God or the afterlife, or losing sight of a dogma's claimed, but actually locally recurrent and emergent, values in various relationships. This is part of the reason why it is so important to divorce our vremetic values from dogmatic ownership. We need to see values as locally recurrent and/or quasi-universal - and religions, political groups, businesses, people, etc. as emergent expressions of them. Dogmatic influences can simply lock us further into defending ever more unreasonably the mental models of 'what we stand for' (including our personal roles, identities, social identifications, labels, etc.).

One simple way we can release ourselves from the grip of a dogmatic 'belief-reinforcement arms race' (which Tolle would seem to call part of the "insanity" of "unconsciousness") is by reframing, that is, to start swimming in other virtuous value-sets other than the ones with which we are most familiar. As the Zen Master, Sent–ts'an, said, "If you want the truth to stand clear before you, never be for or against. The struggle between for and against is the mind's worst disease." That is, breaking down boundaries between ways of thinking promotes brain plasticity and beneficial innovation. One reason why this is important is to make all the values we revere more consciously ours.

Draft Process Models

Perhaps we should further explore here the general approach to making moral decisions via Emergentism versus Christianity or Buddhism. The Christian view of humankind is that we are fallen creatures that are unable to lift ourselves out of the mire of sin except through God's grace (an important value). In this view there is right and wrong, or a good design of human being and a poor design, and there can be no slow evolution from one to the other. The bad (sinful) product design has to be discarded so that the new (born again) design can take over. Under this view, there is no hope offered, for instance, to the prostitute trying to make ends meet in the slums of Mumbai, except that she give up her livelihood if she wants God's grace toward her to continue. Any injustices, privations, or sorrows will be righted in heaven, it is promised.

On the other hand, the Buddhist view of humankind is that life is an illusion and humanity is unable to lift itself out of this illusion except through the hand of transcendental disillusionment. Again, in this view there is a sense of false and true or the illusory physical product design and the true transcendental consciousness. Under this view, the poor Mumbai prostitute's station in life already judges her in terms of past lives (karma) and thus tells her she has a long way to travel to find Buddhahood. Rather than encouraging, such a religious message is likely to be discouraging and result in moral resignation to what is (or rejection of religion altogether). Any injustices, privations, or sorrows might be righted through this life or reincarnations to come, but only if such overwhelming moral resignation can be overcome.

We also have Nature's, Evolution's, or Emergentism's view. This view does not make any basic judgments of the past, except to say that you are alive, a survivor now – and therefore you have the capacity to be heroic and diligent and to live by the quasi-immutable values that rest within you here-and now. That is, the study of Emergentism encourages the poverty-stricken prostitute in Mumbai to not only survive the current place and moment, but also slowly adapt to the reality of life by expressing the courageous person within, one small step (or iteration) at a time. The values within, the wealth within, can begin to liberate her immediately – spiritually and mentally at first (on the introsomatic axis), but eventually, through small and commendable adaptations, perhaps personally and socially (on the extrosomatic axis), and physically, as well.

Morality is not just the convenient province of the highly privileged. Nature seems to rejoice as we slowly learn the antinomic dance in whatever station in life we find ourselves. The past and the station it has placed us in is not the focus; it is the antinomic dance (or orbit) in life-and-being here-and-now that is the focus. It is the attitude that all intentional and virtuous moral possibilities are on the table – even those that seem to contradict past patterns; the ones tried next will be driven by what is most suitable for the current phase. Injustices, privations, or sorrows may slowly be righted as we take responsibility for them one at a time in the improvisations of life, as we tap into the mighty reservoir of virtuous potential within and as we participate in humanity's evolution individually and socially. Our legacy we are able to pass on enables the next generation to understand a little better what it is to be a moral being and how to dance more ably and sustainably the dance of personal moral congruence and liberating wellbeing.

In sharp contrast, one way we might know that we are out-of-phase with the unfolding life around us is when we find ourselves compromising our values rather than naturally broadening and improving them (in the same way that natural selection always seems to search up and down hill for, or march towards, a temporal genetic stability).

Perhaps at this point we need to check our model of the moral decision-making process. In its most broad form, the Emergent Method is very similar to the Scientific Method considered earlier. That is, the following criteria are necessary for formally defining any new ethical hypothesis:

a. the hypothesis should have a sceptically accepted explanatory power greater than what has been postulated and tested in the past
b. It should have predictive success
c. It should enable some kind of control over natural or social phenomena
d. It should have testability, i.e. it should be falsifiable and verifiable
e. It is only ethical if the hypothesis can suggest some kind of program to test it

Once we complete these steps, the ethical hypothesis might advance to a valid ethical theory, which could form the basis of future moral decision-making. An important feature of strong ethical theories or explanations, as in scientific theories, is that they become difficult to vary without becoming unstuck (i.e. spoiled by more obvious errors or pitfalls).

Having set up the broad parameters of efficacious moral decision-making, let's consider the decision-making itself in more detail. Dennett agrees with a simple model or hypothesis that pictures the faculty of practical reasoning as consisting of desires and preferences entering into its inputs and decisions and intentions proceeding from its outputs.[31] That is:

Dispositions, beliefs, desires, appetites and preferences or "appetitive will" > lead to the attempts, efforts and endeavours of practical reasoning or "striving will" > lead to intentions, decisions and choices or "rational will".

This model seems a little reminiscent of Freud's model of the Id (driven by its "pleasure principle" and thus perhaps corresponding to appetitive will), the Ego (driven by its "reality principle" and thus perhaps corresponding to striving will), and the Superego (driven by conscience and the "ego ideal", perhaps corresponding to rational will).[32] Either reasoning model suggests the self is perhaps triune in nature. Both models seem to implicitly include the environment as input and output.

Another triune model of decision-making, which I prefer, nominally begins at its basic, normally unconscious level with systems and subroutines that interact with the physical environment (such as the autonomic nervous system). The model then advances to psychological and subconscious subroutines socialised within the cultural environment that give us a code of symbols and morality. Finally, the model advances to subroutines of personal narrative and consciously intended action at the highest levels of active self-awareness. All these subroutines run in an adaptive and dynamic organisation of a process-focused view of self that Douglas Hofstadter described well with his book title *I Am a Strange Loop*.

If we compared this decision-making process model with that possibly suggested by Figure 25, it would not seem to clash. That is, moral decision-making could still follow the –D-C-I-S- order we have seen before.

If we incorporate a working of the subconscious mind rather than just the conscious mind, perhaps the instinctive parts of Figure 25 would actually often happen more quickly than the slower, but still quick, reactions of consciousness. Being aware of these possibilities of quick and almost automatic parts of moral decision-making is why it may be important for us to take the time to cycle through the paths of Figure 25 more slowly and deliberately. Perhaps we should often do this more than once, before and as we act (and after we have primed the mind with habits of virtuous thinking and an efficacious model of human nature).

One more thing we might note here, without discussing the technicalities of how each of the stages might operate subconsciously, is again the feedback path from output to input indicated by the fact that the arrows in Figure 25 form a cycle, only nominally starting and ending at the top, rather than a straight line. Feedback must be seen as an essential part of moral decision-making and the Emergent Method because consciousness cannot be aware of itself until it is aware of something - including its own behaviour in the context of wider outcomes.

The suggestion here is that moral habits formed through cycles of behaviours, conscious measurements of those behaviours, and the enjoyment or otherwise of personal moral congruence, are quite important in terms of actually doing what your personalised ethical hypothesis advocates.

Conscious moral habits and in particular a habitual style flexibility is perhaps equally important with our 'practical reasoning' because it is a way to neuro-semantically[33] train our subconscious instincts and patterns of consciousness to follow what each set of virtuous values in tables 1 to 4 suggest.

Finally, we must recognise that not all human behaviour is virtuous and value-driven. Often our emotions arise from a mixture of virtues and less-that-wholesome motives or proto-intentions. For instance, when we fall in love we can feel a mix of instinctive attraction or infatuation, conscious jealousy, helpless devotion, etc. We sometimes think one thing but do another. We sometimes just do not think; we act without consciously considered intent. Vreme-driven or value-driven behaviour requires not only strong ethical habits, but also the capacity, desire and time to think reflectively about beliefs and goals, act accordingly, measure the outcomes and then reinforce positive change, without being overcome by stress and disappointment. Time set aside for meditation or meditative contemplation is also helpful in any quest for quality ethical decision-making.

A Draft Model of Self

Before finishing this discussion of internal feedback loops in moral decision-making models that incorporate unconscious, subconscious and conscious levels, we should also discuss a draft model of self that specifically incorporates these three levels of human nature as well as the external environment.

Usually models of self exclude the environment for the obvious reason that self as a system is separate from its environment. However, in reality, this is a fuzzy distinction.

Self is a dynamic outworking of its cultural and natural environment rather than something that stands independent from it, just as a mountain is a feature of a mountain range.

We could picture quadune self in terms of the following four necessarily relativistic modules:

Module 1

This is the largely non-conscious Sensor/Actor that is something akin to the Proto-self in Damasio (2010); the virtual agent that proto-intends all sensorimotor perceptions and actions of the body. This set of subroutines may enjoy Dennett's emerging evitability but does not make any conscious decisions with respect to ethics or anything else. All subroutines (such as those of the autonomic nervous system) are emergent through relationship and thus subject to decay through disuse. These subroutines are deeply nested and massively parallel-processing throughout the brain/body, in ways that most modern computers cannot begin to mimic. They are fractal in a dimensional sense. Yet even these relationships are not enough to explain sentience. The explanation of sentience also requires the feedback and feed-forward loops of self at its other 3 levels, described below. Sentience requires your guarded acquiescence to the undisproved and incomplete explanations offered up by Module 1's nerves, neurons and chemical brews.

Looking back the other way, just how consciously received narratives are broken down into subconscious symbols and then non-conscious triggers that the sensor/actor can process and perhaps use to direct and modify its own subroutines is a little unclear, although Damasio (2010) goes a long way towards constructing such scenarios. All body language and narratives carry contextual and emotional content that the mind/body can extract to trigger chemical and neural responses, reinforce or abate neuronal circuits, etc. This would suggest that conscious narratives without strong psychosocial signification and intention are also likely to have little subconscious and unconscious content in the mid-brain and hind-brain. Such narratives the mind/body would thus process at mostly a shallow level and quickly forget. Such shallow conscious narratives would thus have little impact on the unconscious sensor/actor.

The sensor/actor is perhaps the instinctive basis for Figure 22's introsomatic axis of behavioural styles in that it selects between modalities of being and doing at the most basic behavioural levels (such as at the level of sensorimotor perception, but also down to the outworking of genes).

Module 2:

The Psychosocial Believer/Claimer that is akin to the Core-self in Damasio (2010) and the intentional stance of Dennett (2007); the largely subconscious entity that triggers and uses many of the sensor/actor's subroutines to set up interactions with the eco-psychosocial environment. Through evitable trial-and-error, the self can 'live', or defend its borders, make an impact on its environment, and regenerate. That is, the psychosocial believer/claimer sets up life-tests but does not carry them out or get the results (this is the role of the sensor/actor) or consciously analyse them (this is the role of the self-aware observer/agent we will consider next). For instance, a social test might use gestures, body language and real dance included in available schemas and subroutines to signal a heuristic self-claim such as 'I am single and interested' to a member of the opposite sex.

The largely subconscious believer/claimer is at the heart of Dennett's intentional stance, or active decision-making in humans. These decisions include implicit facts and moral values as encoded in the subroutines themselves but are devoid of conscious content except through interaction with Module 3, the self-aware observer/agent. Again, just how received facts and values may be broken down into subconscious significations that the believer/claimer can use in its feedback loops (and make the target of its psychosocial beliefs/claims) is a little unclear, but it would require input from the sensor/actor. Damasio seems to suggest that many narratives carry heuristic meanings or significations relating to risk/reward content, psychosocial content, etc. that the subconscious subroutines of the believer/claimer can recognise and use. The value-laden importance weightings of 'subconscious meaning recognition' are also embodied in the sensorimotor subroutines of the sensor/actor. The believer/claimer is perhaps the instinctive basis for Figure 25's extrosomatic axis because it selects between modalities of social interdependence at a basic level.

Module 3:

The Self-aware Observer/Agent; this is something akin to the Autobiographical-self in Damasio (2010). It is the largely conscious observer/agent that triggers and uses many of the sensor/actor's subroutines to measure, interpret, make explicit, or explain/narrate, the results of the psychosocial believer/claimer's tests as well as many of the sensor/actor's other inputs from the internal and external environment. Movement and proto-intention is ubiquitous; the observer/agent is not the source of intention or free will; these things emerge in waves across Module 1 to Module 4 and beyond. The observer/agent is a personal reflection of the combined sensor/actor and believer/claimer in its environmental context.

Many subroutines do not require or receive interpretation by the observer/agent and thus mostly advance through genetic or limited memetic evolution. The observer/agent may blindly flag the sensor/actor to help plug information gaps of which it is aware. It may also blindly flag the believer/claimer to seek information completeness and thus avoid information dissonance arising from within Module 1 and Module 2. Other interpretations made by the observer/agent can be in terms of extrospective and introspective concepts, moral values, personal credo, personal moral congruence, etc. All the results of these emergent interpretations are fed back to the sensor/actor (as indirect and multidimensional inputs to it, along with other direct inputs from the environment, as per Figure 18), in order to affect the efficacy and evitability of the sensor/actor's subroutines. Not all of the sensor/actor subroutines are accessible to or directly modifiable by the observer/agent. An internal feedback loop through the believer/claimer and sensor/actor enables the self-aware agent to interpret its own set of subroutines in light of the new information received and/or missing information. The observer/agent's self-awareness and ability to form new hypotheses and models of reality (or perhaps, 'models of fractal self-similarity') thus only emerge with time and experience across Libet's 300ms gaps between Module 1, Module 2, Module 3, and Module 4.

The ability of the observer/agent to interpret self (as a conscious subset of subroutines) is what Hofstadter describes as the strange loop of self. This strange loop enables the emergent property of a perceived monolithic self to arise over the times and through the experiences provided by the environment. Monolithic self (a fuzzy codification or essential stylisation/signification of self summarised in a schema) stands somewhat apart from the subconscious and unconscious processes always at work in the body, although physical and emotional pain, feelings of happiness, and personal moral congruence greatly affect this self-schema. Feelings and proto-intentions seem to be a language of the unconscious sensor/actor that it communicates to the subconscious believer/claimer and the conscious observer/agent. Value-laden feelings, emotions, and proto-intentions arising through the subconscious believer/claimer, seem necessary for the guiding of conscious values shared and elaborated at the level of the self-aware observer/agent.

Module 4:

The Enlightened Private/Public Self: This is the socially and biologically reflected self (something Damasio does not mention but with which Jarvis (2011) would seem to agree, discussed later). Raw feelings and value-laden emotions combine with passive awareness (including self-awareness) to project the triune self (Module 1, Module 2 and Module 3 previously described) onto its environment, just as the path of a single photon affects and completes the cosmos. The environment responds to the physical body, its sensor/actor, the body language of the believer/claimer, and the narrative of the observer/agent (it cannot do otherwise). But firstly, the cosmos sets up the potential framework of prompts for our feelings, emotions and self-awareness to arise. It does this through what Merleau-Ponty described as our existential communion with our local environment, especially at the level of sensorimotor perception. The local environment directly limits, and acquiesces to, the evitability of the mind/body's subroutines.

The fourth (human) self exists and grows in the minds of others outside of triune self in the eco-psychosocial environment, but is also the product and extension of triune self in the mind of self. The fourth self is an expression, essential stylisation, or signification of our interconnection with all around us. It is akin to recognising that a mountain in a mountain range does not stand alone, but shares its essential style with the mountains around it. Each mountain is fractally self-similar with the mountains around it. It is the survival and success of self in the environment, including in the minds of others, that grants self its very being.

In a sense, we should rather call the fourth self the very first self, because without a consciousness of self as one of the environment's existants, I could never come to know the intersubjective self. This also suggests that quadune self is just another of Gaia's strange loops or properties notionally starting and ending in Gaia herself. All such relativistic strange loops link the reductionist to the holistic in a fuzzy fashion that dynamically sums to the fractal cosmic cycle.

From this definition, we can no longer see the environment as some ideally objective realm independent of the triune self. It is partially a fallible mix and product of the claims of the self. This effect on the environment is true of all of life's DNA-based claims, not just those of the self-aware human. In this very important sense the environment is not purely objective – it is partially the intersubjective expression or puppet of all the egoistic but partly legitimised or accepted claims made by emergent life forms. The distinction between the objective and subjective or fact and value in an emergent universe thus becomes somewhat fuzzy across environments and time. That is, as some new properties emerge through new arrangements, gaps, or boundaries, they can become legitimised or become 'objective' either personally or socially by their own order and ongoing viability. In this sense, relativistic reality, and emergent Gaia in particular, is personally objective more than objective in the classical, straightjacketed sense. There is an unavoidable interpenetration of the objective and subjective in the eco-psychosocial environment because of the interdependence of indeterminate matter and only locally determinate space.

This necessarily means that all animals, as intersubjective moral observers/agents of varying self-awareness, become the makers of their own environment, either implicitly or explicitly. The Emergent Method helps to teach us this reality. Thus, we as the most highly self-aware animals are potentially the most potent emergent force in this solar system. Potentially, the entire universe can become the extended phenotype of emergent human society. It is our emergent potential (or our creative potential), that requires us to include the environment in any complete model of self (or to include legitimised intersubjective self in any full definition of Gaia).

Summary:

Please refer to Figure 28, which also models how we interact with each other at the three levels of Sensor/Actor, Psychosocial Believer/Claimer and Self-aware Observer/Agent.

Person1 is part of the environment of Person2 and vice versa, further highlighting the idea that the environment is partly of our own fallible making as intersubjective moral agents and that society emerges from the powerful relational interactions depicted.

Figure 28 shows three ovals lying one upon the other in two columns, one column representing Person1 and the other Person2. The uppermost oval represents consciousness, the middle oval subconsciousness and the lowest oval unconsciousness. However, by this representation there is no intention to suggest that these three ovals or dimensions are independent. Rather, the three ovals are just the one oval of Figure 25, unnaturally spliced into three parts.

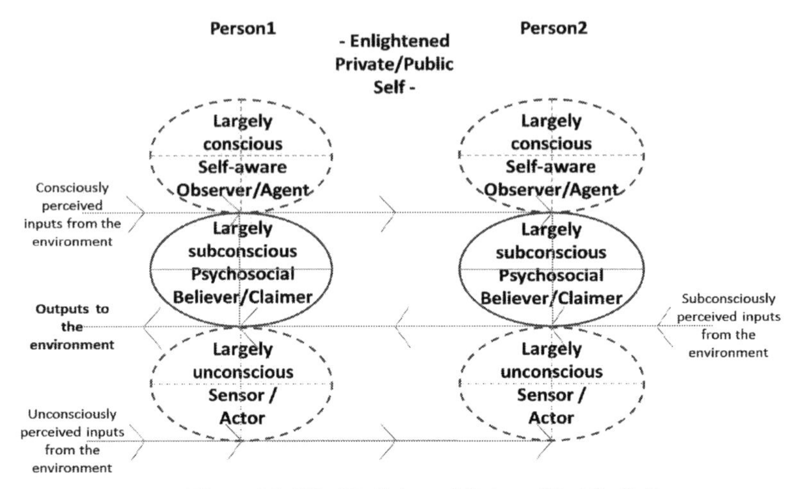

Figure 28: The Enlightened Private/Public Self

That is, we could say that subconsciousness is an emergent property of unconsciousness and consciousness is an emergent property of subconsciousness. We could also say that subconsciousness enhances unconsciousness, and likewise consciousness enhances subconsciousness (and vice versa). An example would be the various subtle meanings granted to the largely subconscious psychosocial believer/claimer by consciousness. The three interact within Hofstadter's strange loop of triune self, but never independently of the fourth self. Person1 and Person2 affect each other's operation and development in a fuzzy shared space that lies between them. We have here a feedback arrangement or strange loop of emergence between self and environment like the one we noted between matter and space, determinism and indeterminism, doing and being, competition and cooperation, or extrospection and introspection.

Figure 28 also shows only one output to the environment as an output of the sensor/actor, but multiple inputs from the environment according to our differing levels of triune perception (unconscious, subconscious and conscious). Unconscious perceptions would be raw sensorimotor perceptions. Subconscious perceptions would include the subconscious effects of elaborated senses such as odours from pheromones, or other body language from those around us. It might also include cultural mores and taboos we follow at a level often just below consciousness. Conscious perceptions would include the explicit meanings perceived in the spoken or written word.

The arrows in the diagram could perhaps also be a bit misleading. I do not intent to suggest that inputs simply proceed to outputs around each oval's periphery or that feedback paths within the three dimensions do likewise. These paths within the one spliced oval of Person1 or Person2 are far more complex than that, just as was explained earlier in this chapter with reference to Figure 19.

To summarise, the three parts of human nature maps easily onto the two axes of Figure 19, and the quadune human nature onto Figure 27: Modern Societal Power Structures. The left-hand CQ-IQ path of Figure 19 would respond more closely to the fact- side whereas the right-hand EQ-SQ path to the values- side of the claims of the psychosocial believer/claimer. Similarly the upper part of Figure 19 would seem to respond more closely to the active outworking of the nature of these claims through the sensor/actor whereas the lower part, with more outwardly passive processing of the nature of these claims through the self-aware observer/agent.

We should take a further moment to reflect on what the concept of a quadune self means. It is a self that extends to an eco-psychosocial environment outside of triune self – outside of the body. What does this mean? How can a 'self' field outside of the body, something like an electromagnetic field, be real? Isn't it just a conceptual idea or an illusion? In what sense can we call the fourth 'self' real? These are interesting questions in light of all we have discussed regarding matter and spacetime, mountain ranges, extended phenotypes (remember those Cuckoos that usurp other birds and their nests as their extended phenotypes?), the nature of consciousness discussed in Pillar 3, and the brief discussion of the 'monolithic self' a few pages ago.

In Pillar 3 I asked, 'Would it be better to call intersubjective consciousness a virtual, indirect and transient interaction like gravity, rather than an illusion, to arrive at a testable model of emergent consciousness?' I think the same question applies to monolithic self as well as the fourth or quadune self, and groupings of quadune selves, which help make the eco-psychosocial environment. The parallels of emergent private/public-self from monolithic self with emergent consciousness from subconsciousness, or emergent subconsciousness from unconsciousness, seem uncanny. Emergent evitability seems to explain all. As suggested earlier in this chapter, our individual self-actualising (through the fourth self in this case) will contribute to Matt Ridley's "inexorable coagulation of life"[34] as well as its diversifying emergence. Like mountains in a mountain range, nothing is independent; all is interdependent.

It is also interesting to note here that the environmental reverberations of private/public self would seem to sit behind the self-aware observer/agent as an indirect shadow of it, the self-aware observer/agent would seem to sit behind the psychosocial believer/claimer, and likewise the psychosocial believer/claimer of the ecological sensor/actor. That is, in operation, these four parts or harmonics of human behaviour really do work or harmonise as one, if you wear the appropriate 4D glasses. The self's environment sits behind the triune self as an indirect and diversifying shadow of it, and if Matt Ridley's picture of the inexorable synergising of life is correct, it will do this even more so in the future.

An interesting issue that arises from this draft model of self is in terms of instinctive feelings and behaviours. Perhaps we can use this model to help define congruence as we have used it in the term personal moral congruence. Perhaps we reach this agreeable and supportive state when there is little dissonance in and between higher consciousness, personal consciousness, subconsciousness and unconsciousness. Ill-feeling is the opposite to the feelings of personal moral congruence, as in happiness, etc. listed in Pillar 1 that can hang over the fence from the spirit (or subconsciousness) and rest in the garden of the soul (or consciousness).

Personal moral congruence would thus refer to the lack of dissonance in and between enlightened and personal consciousness, subconsciousness, and unconsciousness when the mind is thinking about issues that are under its personal control or influence. As with all states of equilibrium, personal moral congruence would necessarily be a fleeting point transitioned.

The Emergent Method is thus about gently guiding the non-linear journey of self through quadune personal moral congruence.

It is about being aware of our interdependent quadune nature. The Emergent Method is about how the quadune nature emerges and operates in a network of disparate components and distributed control systems. The self-aware life, just like the cosmic cycle in Figure 5, is never a destination; it is always a journey.

The modification of behaviours the Emergent Method suggests could include just thinking differently about the situation you are in and, if necessary, seeking to modify the thought subroutine itself through a learned or reinforced gathering and releasing, in such a way as to put the mind at ease and if possible bring about a sense of wellbeing. In other words, since in essence we are moral beings, we may define wellbeing as a heightened, extended, and fuzzy sense of personal moral congruence.

If we were to extend these thoughts to the wider environment, then **societal moral congruence** arises when there is no serious dissonance between our Public-selves and Private-selves at the communal level. Bringing this about requires a deeper commitment to our shared civic life. The social acceptance of this view perhaps constitutes the definition of a robust social fabric.

We could say fleeting personal moral congruence is reached when what we sense matches what we do, what we believe matches what we claim, what we consciously observe matches what our moral agency achieves, and what we are privately matches what we are publicly.

Finally, can we say that the Sensor/Actor (or Damasio's Proto-self) was the beginnings the Warrior-Self and that the Believer/Claimer (or Damasio's Core-self) was the beginnings of the Lawmaker-Self? Can we also say that the Observer/Agent (or Damasio's Autobiographical-self) was the beginnings of the Philosophical-self and the Private/Public self was the beginnings of the Trader-self or perhaps ultimately, the Vremetic-self? Yes, I think the ideas of emergence coupled with Hofstadter's observations of the strange loops of self means we can. See Figure 29.

Using this framework, we can perhaps make a few predictions. Maybe the Private/Public-self has only just begun to emerge in the last 50,000 years as something explicit in our shared civic life, firstly through the trading of values that reflect a materialistic bias. This bias reflects the values for which the Warrior-self and the technology of his/her extended phenotypes are better known.

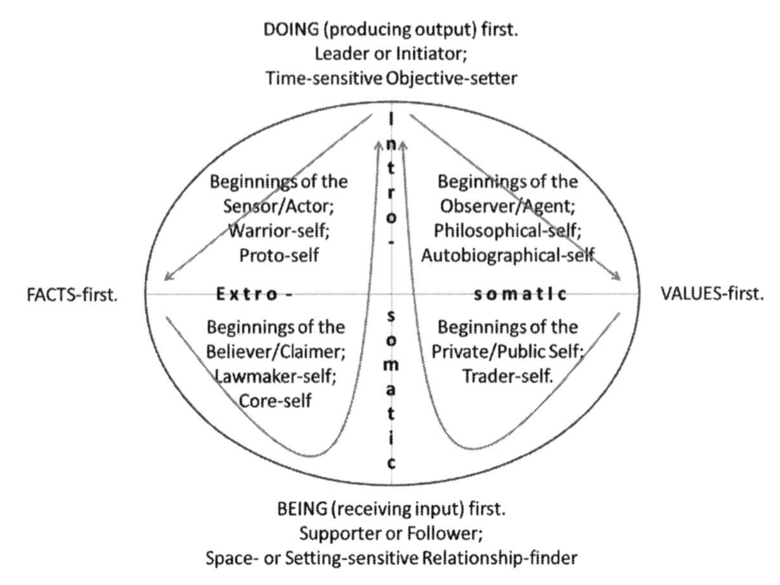

Figure 29: The Emergence of Human Consciousness-plus-Instincts

Perhaps the emerging next stage of the Private/Public-self will add the trading of values that reflect the Lawmaker-self. That is, perhaps the next stage will also celebrate our non-superstitious cultural diversity. I think we have seen the beginnings of this with the economic protection but celebration of local cultural brands such as the Champagne region in France and Thai cuisine.

I also see the celebration of cultural diversity in such delightful books as *French Children Don't Throw Food* by Pamela Druckerman (2012).[35] I would very much like to see this book act as a template for an equivalent publication for every culture, and then translated to every language on Earth. This would be one of the most productive ways to enhance the development of Private/Public self I can imagine, and in many ways, it is already happening.

However, perhaps as we more keenly focus on our cultural mores and taboos, the clash of civilisations will be heightened in the short term, as Samuel Huntington suggested back in 1996. We will explore this issue in Pillar 5.

Nevertheless, I am confident that we can take heart. Maybe we will advance past this Lawmaker clash to truly cherish and trade our cultural differences and then move on to consciously consider the trade of values of the Philosophical-self as well.

This higher trade of values requires that we move beyond basic mores and taboos to value and celebrate, as a source of emergent wealth, our differing, diverse, non-superstitious, and successful social fabrics. This is a topic touched upon by Jackson (2009), Kauffman (2010), Jarvis (2011), De Botton (2012), and in Pillar 6.

We will then discuss the ultimate development of the Vremetic-self in Pillar 7.

Your Dance

If you need a more practical and explicit guide to the application of many of the values and virtues mentioned in tables 1 to 4 to your most personal relationships, the book *How to Love* by Gordon Livingston, M. D.[36] is recommended.

The value-set scores you calculated back in Pillar 1 reveal how strongly you are driven by instinct and cognizant of consciousness. The scores reveal how strongly you are inwardly-focused and how strongly you are outwardly-focused. It might be useful after completing this chapter to take the value-set test again and see if the scores have changed. It might also be useful to take the test with your work-hat or role-hat on and see how different the outcome is.

You could then ask yourself why your answers may have changed and explore consciously bringing virtuous values from one role to another. You might also like to re-explore your credo written down after reading Pillar 2 and your intentions/purposes after reading Pillar 3.

Next, consider in detail each of your deepest purposes by creating a step by step system or process for each. If you can't visualise or otherwise imagine the purpose in systematic terms, then you are very unlikely to be able to achieve it. However, complexity should not be a problem because you can always break down complex steps into smaller, achievable ones. That is, if you find yourself saying, 'I can't do this', then the step is probably too big for you right now, so break it down into smaller, more manageable chunks.

One way to visualise your systems of purpose is to use process maps. For instance, the rectangles in Figure 30 represent the necessary steps that must be taken to realise the 'get fit and lose weight' purpose.

The diamonds represent decisions about modifying the steps to more efficiently or effectively achieve the purpose. The triangles R1 to R5 represent the risks to achieving the purpose at the steps along the way. These risks and their controls are described in the table below the process map, often called a risk matrix. Each step could be associated with zero or more risks. Each control is designed to mitigate or counteract a risk. There can be more than one control for each risk, or the one control can satisfy several risks. The primary value-set applicable to each step and each risk is also provided.

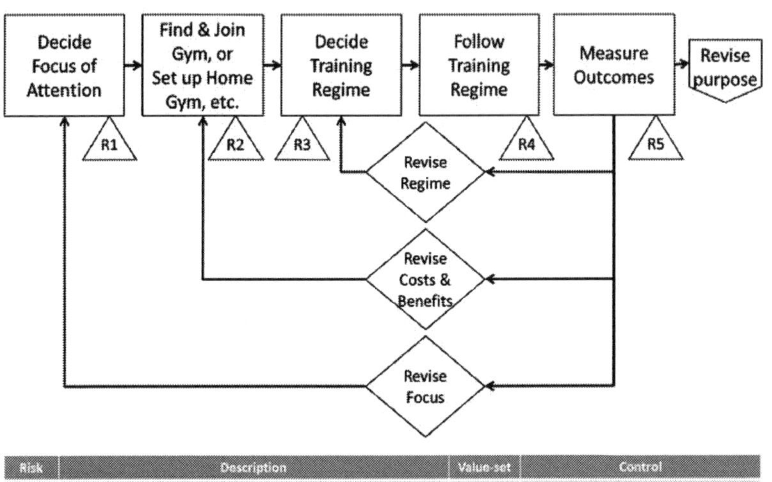

Risk	Description	Value-set	Control
R1	Missing information for decision	Table 1	Priorities set as information gathered
R2	Too expensive, wrong location, inconvenient, mismatch of benefits	Table 2	Research is conducted
R3	Mismatch to current levels of fitness or strength	Table 2	Research is conducted
R4	Can't focus, bad habits, too lazy	Table 3	Changes and benefits visualised
R5	Not enough evidence, not easy to define	Table 4	After reflection, lessons learnt recorded

Figure 30: Get fit and lose weight!

Notice that as the purpose is detailed in steps from left to right (in this case), the primary value-sets applicable are advancing from Table 1 to Table 4, that is, from the risk-weighted, to the positive, normative, and then self-enlightened, value-sets. That is, if you lay out the detailed steps to achieving each of your purposes, you are often likely to traverse the values in a logical order from Table 1 to Table 4. This is because Table 1 to Table 2 are more fact-related or concrete and thus simpler, but closer to your initial thoughts and action steps, whereas Table 3 to Table 4 involve interrelated concepts that are more complex and abstract, but closer to your higher purpose.

Now, in like manner, for each of your deepest purposes, set up a separate process map and associated risk matrix.

To achieve a higher purpose we often have to dance between more specific steps on the one hand and more abstract aims on the other. As noted earlier, sometimes we also have to deal with values or purposes that seem to contradict each other. For instance, it seems spending time building a business will necessarily draw us away from time with the families we are trying to support. A good question to pose in this kind of situation is to ask, 'How can building my business help me to spend more time with my family?' That is, try to reframe the contradiction in positive terms that will help you search for an answer, or dance more effectively and efficiently.

To take the exercise to the next stage that really stretches you, you will need to try to swim (i.e. set purposes) in other rivers of value-adding you have seldom swum in before or have perhaps avoided since your childhood. Like the soldiers in the movie *Braveheart* (1995 Twentieth Century Fox), you are going to have to learn how to wield your sword (i.e. follow your purposes) in many different circumstances - on the battlefield, in the cities of your enemy, up in the hills and maybe even in your own home camp.

This does not require that you deny or contradict yourself, side with one tribe or another, one nation or another, or one race or another, because values pour out from the font that is above all roles, identities, narratives and forms. However, to add value through new and challenging purposes will require that you side with your values. This font of values at the head of all rivers is even above our most sacred supernatural forms we label as Zeus, Diana, God, Present with a capital 'p', Life with a capital 'l', Consciousness with a capital 'c', or Nature with a capital 'n'. This was my personal revelation of Pillar 1.

We store our identity in the set of values we hold and apply. We also store our own self-worth in the set of values we hold and apply. There is nothing wrong with the quasi-immutable values we individually hold and embody. Within these values is all the power and staggering beauty of reality and our being. So what ought to be the self-worth of you and me, even though we are quite different? It ought to be fabulous in both cases; the potential wealth held within us is copious – all we have to do is realise it through our relationships and actions.

Unfortunately, many do not. Why? Often because we focus on the authority, wealth, power, or beauty of others rather than the fabulous authority, wealth, power, and beauty that are within us and are looking for clear expression through our participation in life. All we have to do each day is simply recognise, respect, and serve the wonderful and full arrangement of virtuous values within us that is 'us' – that defines our local fractal dimensionality and brings quadune self into harmonic clarity. These same values will certainly serve us; a self-organising universe means they almost certainly cannot help but do so.

The realisation of this century, I believe, will be this. Our involvement in the great cycles of life and the cosmos, not just as participants, but also as important and all-inclusive co-discoverers and co-designers of the unfolding present, can give us all just as wonderful a sense of personal meaning, purpose and destiny as any tribal religion could do in the past.

The values that take us forward are core to our survival, success, prosperity, and wellbeing. Our enemy is myopic, self-gratifying, narcissistic, violent, deceitful, elitist, and short-sighted selfishness. Our saviour is enlightened selfishness that eventually incorporates all of Gaia, and is rooted in our shared emergent values. We are not independent; we are irreplaceable links in reality's golden chain.

Concepts of Emergence Encountered in Pillar 4

Vremetic evolution is to our Virtuous Global Tradition what genetic evolution is to its survival machines and what emergence is to the strong and weak interactions of spacetime and matter.

Emergentism's vremetic ethics is eudaimonist. That is, your emergent wellbeing, as an environmental outworking of your evolving instincts and consciousness, is also your moral purpose. This does not infer a short-term hedonistic lifestyle because such a lifestyle without regard to its environment will not bring personal wellbeing. Only your virtue, as an evolving ethical being within your deep ecology, can achieve wellbeing in your here-and-now. A self-actualising life is the reward of its virtue, and so self-actualising success is both the goal and the reward of your life.

Within this framework, our moral choices will feed back to our creative emergence as self-actualising beings. Further, if Matt Ridley is correct, our individual self-actualising will also contribute to the Gaian "inexorable coagulation of life"[37] as well as its 'diversifying emergence' briefly discussed in Part 1.

In summary, Emergentism's ethical model of learned moral behaviour is:

- Prior experience and identification of emergent values -->
- Leads to explicit propositions (statements of principles, self-restraints, or duties), which -->
- Confirm or disconfirm largely subconscious beliefs, which -->
- Focus intentions and lead to goals, which -->
- Lead to actions (taken or withheld within changing circumstances), which -->
- Lead to assessed outcomes in the environment, which -->
- Lead to values unfulfilled, fulfilled or settled, and
- So on around the same loop continuously -->, with intermittent 'mutation' or emergence at any point in the flow or process of moral consciousness.

This is why happiness (an example of values) can be both the goal and reward of life and how a self-actualised life can be the reward of virtues (or values-in-action).

In summary, Emergentists tends to combine the values of the virtue ethicists with the principles, beliefs, intentions and acts of the deontologists and the regard for social outcome of the consequentialists in a single, naturalistic model of personal moral congruence or sustainable wellbeing.

To use an analogy extracted from the lessons of genetic inheritance, the study of Emergentism suggests a Darwinian shuffling of values (leading to emergent congruence) rather than an unhelpful blending, levelling, averaging, dumbing down, or opportune self-serving of values.

Is an approach driven by reality's large and diverse set of emerging but locally recurring values, applied according to the local setting, the life-phase, and its corresponding value-set, in fact a more naturalistic, transparent, internally consistent, rational, and moral approach? This is the radical evolutionary question Emergentism's ethical system poses.

The consideration of Emergentism, through many trial-and-error iterations, leads us to moral humility and naturalistic value veneration rather than their opposites.

We have here a feedback arrangement or strange loop of emergence between self and environment like the one we noted between matter and space, determinism and indeterminism, doing and being, competition and cooperation, or extrospection and introspection.

The Emergent Method is thus about gently guiding the non-linear journey of self through quadune personal moral congruence. It is about being aware of our interdependent quadune nature. The Emergent Method is about how the quadune nature emerges and operates in a network of disparate components and distributed control systems.

In the next chapter, we will summarise and discuss further the real-world implications and practical benefits of Emergentism.

---

[1] See Tolle, Eckhart, *A New Earth* (Penguin Group (Australia), Melbourne, 2009 (Original Penguin Edition 2006)), 129-160

[2] Ibid., See p.130

[3] See ibid, 133: Is Tolle intending this as a general truth, suggesting that language, thoughts and emotions have arisen together in all animals with brains and nervous systems?

[4] Ibid., See p.138

[5] Ibid., See p.142

[6] Ibid., See pp.142-143

[7] Ibid., See p.153

[8] See Pinker, Steven, *The Better Angels of Our Nature: Why Violence Has Declined* (Viking Adult, USA, 2011)

[9] See Tolle, Eckhart, *A New Earth* (Penguin Group (Australia), Melbourne, 2009 (Original Penguin Edition 2006)), 153

[10] Ibid. 160. I added comments in square brackets.

[11] Ibid., See p.164

[12] Ibid. 182

[13] See Glossary entry, "Laissez-faire Capitalism"

[14] See Smart, Ninian, *The World's Religions* (Cambridge University Press, 1989), 125

[15] See Mitchell, Donald A., *Buddhism. Introducing the Buddhist Experience* (Oxford University Press, New York, 2002), 194-200

[16] See Hamilton, Clive, *Earthmasters: Playing God with the climate* (Allen and Unwin, Sydney, 2013), 110

[17] See Goleman, Daniel, *Emotional Intelligence* (Bloomsbury Publishing, London, 1996)

[18] See Goleman, Daniel, *Social Intelligence* (Arrow Books, London, 2007)

[19] See Gardner, Howard, *Frames of Mind, The Theory of Multiple Intelligences* (Basic Books, New York, 1983)

[20] From https://en.wikipedia.org/wiki/Herrmann_Brain_Dominance_Instru

---

ment and Meneely, Jason, and Portillo, Margaret, *The Adaptable Mind in Design: Relating Personality, Cognitive Style, and Creative Performance.* Creativity Research Journal, Vol 17(2-3), 2005. pp. 155–166.

[21] Based on a similar idea from Sarkar, P.R., *Human Society, Part 2* (Proutist Universal, Washington DC, 1967)

[22] Ridley, Matt, *The Origins of Virtue* (Softback Preview, Great Britain, 1997), 15

[23] The Stanford Encyclopedia of Philosophy, *Deontological Ethics*, first published Nov 21, 2007; revised Dec 12, 2012

[24] As per Dennett's preferred Pandemonium Architecture and Multiple Drafts model of consciousness. See Dennett, Daniel C., *Consciousness Explained* (Penguin Books, USA, 1993)

[25] Tolle, Eckhart, *A New Earth* (Penguin Group (Australia), Melbourne, 2009 (Original Penguin Edition 2006)), 182

[26] The Stanford Encyclopedia of Philosophy, *Consequentialism*, first published May 20, 2003; substantive revision Oct 22, 2015

[27] Encyclopaedia Britannica, *Virtue Ethics, (2013)*

[28] http://www.wjh.harvard.edu/~jgreene (accessed 2012). I added the comments in square brackets.

[29] In his chapter *Choices* in Deutsch, David, *The Beginning of Infinity* (Penguin Books, New York, 2012)

[30] As per Dennett, Daniel C., *Consciousness Explained* (Penguin Books, USA, 1993), p.198, etc.

[31] See Dennett, Daniel C., *Freedom Evolves* (Penguin Books, USA, 2004), 103-104

[32] See Freud, Sigmund, *The Ego and the Id* (The Hogarth Press Ltd. London, originally 1949)

[33] See http://www.neurosemantics.com/ (accessed 2014)

[34] Ridley, Matt, *The Origins of Virtue* (Softback Preview, Great Britain, 1997), 15

[35] See Druckerman, Pamela, *French Children Don't Throw Food* (Doubleday, Great Britain, 2012)

[36] See Livingston, Gordon, M. D., *How to Love* (Hachette Australia, Sydney, 2009)

[37] Ridley, Matt, *The Origins of Virtue* (Softback Preview, Great Britain, 1997), 15

PILLAR 5: Self-Actualisation

In this chapter, we will discuss how Emergentism relates to social contracts, money, the services we provide to our society and our own self-actualisation. We will also discuss Emergentism's concrete, practical benefits - including societal benefits. In doing so, the text will prompt you to form strong moral habits by carefully considering your introspective concepts and beliefs, and if necessary, rebuilding them so that they might align more intelligently with your environment, virtues, and purposes.

The Entrepreneurial Spirit

There are no real virtues without real actions.

Nebuchadnezzar? Darius the Great? Alexander the Great? Julius Caesar? Napoleon Bonaparte? Joseph Stalin? Mao Zedong? Hobbes' Leviathan? We do not need another great military/political leader, prophet, saviour, usurper, or big-brother government and nanny state to rise up and save us – we need to stand up and save ourselves from ourselves (or the selfish replicators and their typical narratives within us).

That is, we need to make a stand against the superstitions, biases, and lack of opportunity that have limited our human learning and development. We have to stand up for something as well – the quasi-immutable, undeniable, and evolving value-sets of our virtuous global tradition (our successful vremes). Finally, we need to rid ourselves of all stands against our fellow human for anything other than self-defeating values.

In one sense, this book heralds the beginning of a society of enlightened individual contributors to the unfolding present. These contributors understand that the personal and social worlds and their arrangements unfold not randomly, but in an antinomic, chaotic, or massively parallel dance towards an inclusive energy efficacy in the unfolding present. The personal and social worlds do this in a similar way to how the galaxy arrays stars around its core. These arrangements evolve, and in that evolution, cells coagulate in ever-higher orders of complexity.

Our challenge is not to intelligently design that dance before the fact (as if we can know how it emerges in all its complexity), but to be intelligent participants in it. Massively parallel computing in a human body or ecosystem slowly advances in a successful direction through trial-and-error not because of intelligent design but because of all the self-contained interrelationships that describe it. The more relationships it encompasses, the more refined its evitable responses can be and thus the more risks it can mitigate without major threats to survival. Our focus should not be intelligent design, but intelligent interrelationship.

Armed with our more interpenetrating view of human nature, we are going to explore a more appropriate fit for the values that arise out of our evolving consciousness-plus-instincts. Our values need to be seen as core enablers of our survival and actions moment-by-moment rather than either our consciousness or our instincts. We need to harness the memes that arise out of our consciousness and instincts and tie them to the core vremes of our value-sets briefly described in Pillar 1. If our highly evolved value-sets of Pillar 1 become core replicators of our planet, then we can all proudly be their survival machines (as Dawkins described us). We might further boast that those vremes-plus-survival-machines are us, at our self-actualising best.

In review, we have:

- Recognised that biased instincts, even though they were once venerated as attributes of our gods, are not transcendental and do not deserve such a label;
- Recognised that the mind-games of consciousness have contributed a lot to our current state of social and personal dysfunction and thus recognised that unbridled and ubiquitous consciousness is not worthy of veneration either;
- Recognised that if anything at all is to be venerated it should be our emergent vremes. Yet we should do this venerating in an evolutionary and natural sense. Our vremes, which enable us to rebel against the tyranny of our blindly selfish replicators, naturally arise from the self-actualising parts of our consciousness-plus-instincts. That is, higher consciousness is a state of instincts and consciousness brought into dynamic homeostasis by our emergent vremes.

As beings of enlightened selfishness, we need to be prepared to share our most secret and sacred beliefs and deeply-held values with each other, including in our businesses, hold them up for ridicule or praise, and then allow them to evolve in each other (in ourselves and in our stakeholders). We do not share our values through trade, the arts, social interaction, etc. to convert others to our own set of self-restraints, but to promote an evolution of life and diversity through feedback, driven by the value-sets (the core vremetic replicators) of our virtuous global tradition.

Matt Ridley is very encouraging with respect to the evolution of our values and virtues. He sees virtues as more fully embedded in our social instincts or genes, and therefore less at the mercy of our perhaps more fickle memes. He says, "Virtue is indeed a grace – or an instinct ... It is something to be taken for granted, drawn on and cherished. It is not something we must struggle to create against the grain of human nature"[1] – further supporting the view that *Homo sapiens* (particularly of the last 50,000 years) is an essentially moral species. That is, our vremes are an essential aspect of our essential styles or the way we move. They are the significations of the higher consciousness within our species.

Money

If we are going to apply the ideas presented here in our business lives, as we must if we are to be true to ourselves, we need to talk about the relationship between our business, our personal values, and money. Money is an important medium by which values are recognised, assigned, marketed, communicated, shared, and/or exchanged. Accumulations of money are thus taken as wealth. We have all heard the expression 'money means power'. This does not mean that values are absent when money is involved - on the contrary.

Ayn Rand's wonderful discussion of money is given through a speech by her character Francisco d'Anconia in *Atlas Shrugged*.[2] It is necessary reading for every budding entrepreneur. A summary of the speech is presented in the following paragraphs.

D'Anconia first presents money as the modern tool of exchange between traders who are seeking to exchange mutually agreed value for value. Thus, paper money represents a promise of recognising value and a promise of honouring the expressed values of others who produce goods and services by their mental and physical labour. It is faithful promises between traders of values that make us better future trappers. Retained money is a statement of confidence in the future values of humanity as a species. Without confidence in humanity as a producer and fair and willing trader, money becomes worthless and our future even more uncertain.

Humans survive by their reason, volition and values-in-action. Thus, humans as moral, rational, and volitional beings are the root of all the goods and services they produce and all the wealth they ever created. D'Anconia thus suggests that money, like goods and services, is an outward moral expression and measure of a person's inner moral being. He sees trading with money in the absence of violence or forceful coercion as evidence of general good will between the parties involved.

D'Anconia's speech suggests that traders see money not as that which belongs to the institution that put it into circulation, but as a value placed on efforts freely exchanged. Money reflects the measured worth of those efforts. He suggests that the thing that binds societies together is not the exchange of violence but the fair exchange of goods and services. Money requires that we exchange measured enterprise for measured reason. Unfettered money requires that we exchange the best money can buy rather than anything less.

D'Anconia's speech goes on to suggest that money is no substitute for values: If people lack virtues and the rewards they supply, money cannot purchase those things for them. So money will not supply the purpose only a love of life can provide. Likewise, money cannot purchase intelligence for people who ignore life's lessons, respect for people who cling to their incompetence, or admiration for people who will not let go of their cowardice and fears. In this sense, money will only serve people who are fit to possess it (people with values, purpose, reason, competence, and courage). On the other hand, money can quickly destroy people whose minds are unfit for its possession, by the means of their own vices.

D'Anconia's speech similarly suggests that money obtained fraudulently or by violent means will not bring physical or spiritual joy to its possessor, whereas money obtained through the trading of the best value it can demand will bring real joy to its virtuous owner (and his/her community). Thus, the nature of money is to bring the best out of moral people who are willing to work hard to express their highest values in that which they produce. Such people have the satisfaction of knowing that they well deserve their money, or the wealth they have created.

Finally, d'Anconia suggests that money can act as the measure of a society's virtue, and not just a person's virtue. Thus, when society's financial powers and faceless traders immorally distort the value of money through third-party transactions behind face-to-face traders, society and everyone within it suffers the consequences. On the other hand, money properly bound to humanity's moral nature can be the root of all good and blessing in a prospering and flourishing society.[3]

How timely is d'Anconia's speech for America and for the world today, after the global financial crisis? We desperately need to link back our businesses to our evolved, virtuous values and move them away from short-term and short-sighted gain (often at the expense of our children who will more correctly measure in dollars the loss in terms of raised taxes, an environment of scarce resources, or in terms of social upheaval). We now live in a world of deep inequality in which poverty exists in the midst of plenty. We need to move away from stupid decisions based on immediate political pay-off. We need to move sharply away from personal, business, and government fraud and corruption.

Our businesses, while maintaining a profit-motive, need to be outward expressions of our awakening consciousness-plus-instincts or vremes. Our businesses need to take up a credo like that of Johnson and Johnson and habitually live by it, recognising the unwritten social contracts such credos imply – because all our human endeavours interpenetrate and coagulate.

We also need to go way beyond the good business sense of the credo. We need to discover new organisational structures as suggested by information technology's object-oriented approach that would put a focus on dynamic stakeholder relationships rather than a focus on static corporate functions. We need to strive for new levels of customer service and value to the community if we are to fully realise our sustainable purposes.

Money-policy

We discussed in Part 1 the fundamental separation of powers between the direct and indirect, or infinity and nullity, which is equivalent to the separation of powers between the executive and the legislature found in the system of government of the USA. In this analogy, the executive is the dynamic doer, like matter, whereas the legislature is the stable relator and codifier, like space. Likewise, we could speak of the separation of powers in a democracy between the decentralised voters that elect the legislators and their proposed legislative platform, and the centralised government bureaucrats that carry out the wishes of the voting public.

Similarly, we could picture the relationship between economic production (real credit) and money (financial credit) in the same terms. In this case, production is moved by money, but money finds its place, role, or relationships, through production. Alternatively, production provides the necessary platform for money, but money organises production.

Whenever we see this 'separation of powers' at play, we see an opportunity for adaptive evolution, or more generally, emergence. However, the cost of this emergence is that such systems will always be inherently unstable, uncertain and incomplete – or at least lack the ability to demonstrate their own consistency and completeness. The granted powers that come about by the separation will also be subject to manipulation by players driven by a non-enlightened selfishness.

Ok, so what does this intimate in terms of money-policy? Firstly, we need to review the mechanism of natural (or artificial) selection. What makes it work? Well, if we want systems to adapt favourably to their environments, then:

- the separation of powers between the direct and indirect must be protected
- the parts (genes, codes or concepts) must be allowed to compete and cooperate, or 'struggle' for energy efficacy, in their 'level playing field' environment. This involves a search for stylistic patterns or significations that can capture the essence of the relativistic relationship and help it to work
- there must be opportunity provided by the environment for occasional mutation or dissention from the norm or the current paradigm
- the parts (genes, codes or concepts) must reside in a population pool that gives opportunity for new gene- or idea- arrangements to spread through imperfect replication

If money and production are locked into such an emergent interrelationship, in which case money would represent the genes, logic or policy, and production the environment of such money-policies, then:

- Money-policy makers should be separate to their beneficiaries, just as courtroom judges are independent of their judgments. The policies of money supply, which are in place to promote real credit or economic production, should not reside behind closed doors in the board rooms of banks. The banks have a fiduciary duty to society to open their doors, given the place of privilege their granted banking licenses provide
- There should be more than one currency and money-policy provider, so that various systems may compete and/or cooperate for their adoption by traders
- The barriers to entry into currency provision and money policy choice must be just low enough to allow for their occasional mutation or innovation
- Money-policy should be debated and contested in public forums. Draft policies should be open for inspection just as parliamentary bills are open to the public

At a societal level, I would suggest it is only as we act on such concepts with respect to monetary policy that we can hope to adapt favourably in a fast-changing business environment.

In a similar way, we could discuss the relativistic democratisation of the military decision to go to war in the case of a clear and present danger (rather than an actual invasion by an enemy).

That is, perhaps in a society such as Australia we should introduce a law that says if our leaders commit a pre-emptive strike against another nation, then they must seek ratification of that decision as soon as possible in a referendum. This should override current treaty obligations, especially in the case where the Allies' electorates have not ratified such a decision. This means we may need to review and update our current treaty arrangements. If the referendum outcome rejects the government's decision, then the government would not hold a mandate to continue the war. For the country to resume hostilities, a general election would need to be held and the outcome favour the decision. In this way, the people would bear the clear responsibility for wars of aggression or pre-emptive strike, unlike in the past. Perhaps one way to introduce such a policy is though international agreement. The administration of the military would remain professional and respond to threats as appropriate and as protocol suggests throughout the period of possible political uncertainty.

At the end of our lives, we have to meet our day of death with our pockets empty, just as on closure a corporate legal entity fully distributes its funds and resources to its stakeholders. On that day, we will want to know that we, and our virtues, were part of the solution and not part of the problem.

Service

Self-interest must be our business enabler and money our usual means of exchange but service will be the reason why traders with money will seek out our particular businesses, products and services. Self-centredness without other-centredness is like competition without cooperation or the solar system without planets; it is like turning back the clock to when the self-organisation of the solar system was at its unsophisticated beginning.

Traders desire and expect value-for-money and in fact values-for-money and in fact our values we hold towards them for money. In the long term, self-interest does not permit us the luxury of myopic and narcissistic self-centredness. It demands, perhaps according to the principle captured in the idea of 'namaste', honouring the self-interest of the intelligent trader as highly as we honour the principle of self-interest within our locally recurrent universe and ourselves. This is an essential attribute of the antinomic and chaotic dance. It is as we learn to value the other, and the values in others, that we also discover the fabulous wealth hidden in the win-win of the other-centric values. As we sincerely express the virtuous people-oriented values, we draw more wealth to us in the externalised forms of money, etc. We become better future-trappers together.

Perhaps one way to explain the concept of wealth hidden within your quasi-immutable values that are looking for your entrepreneurial expression is to think more broadly about all the self-organising principles of the universe. One such principle is the strong nuclear interaction. If you were able to develop a clean-energy product that exploited this self-organising and locally recurrent principle, do you think you could make a lot of money from it? Clearly! What about if you exploited the gravitational interaction? Would that make a lot of money? What about if you were able to exploit the locally recurrent self-organising principle of evolution? Would that make a lot of money? Almost certainly, the cure of gene-centric diseases represents a potentially huge industry. Therefore, why wouldn't exploitation of the self-organising and quasi-immutable values arranged within you also unleash fabulous, recurrent wealth through the relationships around you? As already noted, values are the basis of every executed trade on earth. They make up the indirect lattice or web that bind traders together.

It is as we faithfully express the values of our own deeply personal identity that we live truly to ourselves and to all those with whom we interact. Those that we mix with in business recognise the undeniably true values expressed clearly and easily in and through us. We also greatly influence those that we mix with by these honest and flowing expressions of values. They are more willing to deal with us, both qualitatively and quantitatively, and recommend us to their network of relationships. We simply do business more easily and with far more reward for all involved. We easily create wealth as we value each other's authentic best in the recurrent and slightly asymmetrical exchange process. If we could expand this approach to the societal level, we could contrast it with those falling business and consumer confidence surveys we saw nearly every month in 2012/13 in Australia.

We do not release wealth into the world if we devalue our true selves or shrink back from our authentic best. Trust, naturally flowing from this authentic best freely and boldly given, is central to exploding wealth. Perhaps locally recurrent wealth-building or wealth-creation should be the very highest value in each of us, as Dr John Demartini has suggested in his popular seminars. Such virtuous future-trapping will only occur if we find a suitable dynamic balance between our unique being and doing, that is, by embodying the vremes that resonate within.

Not only are the values inside you, linked to your personal wealth, self-actualisation and potential, but by the logic of a self-organising universe, so are all the universe's other interlinked and self-contained values (resident in the brains of other value-traders). Within this one cyclical and interpenetrating system must be hidden all the emergent solutions to the world's political, cultural, religious/philosophical and economic dilemmas. These dilemmas seem to be ways in which the universe strikes out and searches for more highly complex and inclusive modes of coagulation. The solutions simply depend on the recurrent local dance, and now, our consciousness of this fact.

To put this concept another way, the risks we face in our modern world are multiplying with our growing consciousness-plus-instincts but so are our conscious abilities to deal with these risks. We need to be respectful of this asymmetry between risk and solution and make sure we bring it back into near-balance with the aid of a suitable and naturalistic philosophical framework. This framework recognises the ability of the universe (and our human mind/body) to work in immeasurably parallel ways that greatly minimise our Gödel limitations in its solutions.

So the idea of service, which captures the ideas or values of giving rather than taking or hoarding, is something the successful values-trader understands well and understands is the secret to the realisation of his or her own self-interest. The genuine desire to give value (see RHS of Figure 22) must accompany the desire to take value (see LHS of Figure 22), if we are to be successful in a world of value-traders with growing self-awareness. Just as the direct and indirect necessarily interpenetrate in the wave-particle antinomy that is matter, so giving and taking must interpenetrate in our trading systems (and economic system).

Further, driving the normal D-C-I-S model a little backwards (S-I-C-D) from time to time (e.g. through sincere acts of servanthood or self-sacrifice) can actually unleash wealth into the hands of the self-interested and self-reliant. Sometimes this requires delayed self-gratification. This is why the ideas of the rapport-building gift and money-back guarantee of web-based businesses have been so successful.

In order to give value we need to understand what our fellow-traders value. Our fellow-traders, when they get past their surface fears and pain-body insecurities, value and honour the locally recurrent dance of seemingly immutable values-in-action more deeply than all else. They're willing and eager to pay handsomely to participate - and they need to, in order to truly appreciate value and to release more wealth into the world.

Ongoing success – or survival of the fittest – demands that we know how to dance through all four corners of human nature. It demands that we genuinely hold people-oriented values as well as task-oriented values. Greed and deceit, as well as guilt-laden altruism, must be self-defeating and eventually fail in a world of more highly aware value-traders. However, we will only turn genuine expressions of abundant giving, extracted freely from the universe's emergent abundance, into monetary wealth if we dare to put a fair price on such giving. As we liberate others by daring them to live a life of values-in-action through our products or services, they in turn liberate us.

The enlightened, vremetic trader is the habitual but conscious survival machine of his or her vremes and not just the blind survival machine of memes and genes generally. In fact, it seems like the virtuous values themselves do not want to just replicate, but they want to communicate with us. Through us, they seem to want to take us through the antinomic and chaotic dance steps they have invented and so thoroughly enjoy. Please do not take this line as a capitulation to transcendentalism. It simply acknowledges that the personification of values often helps our biased minds to understand a point made. The virtuous interpenetration of the immaterial and material seems to be how we discover the bio-diverse win-win (or nonzero-sum game) and materialise wealth against all odds - easily, naturally and repeatedly.

I hope that you have arrived at the point where you understand the core ideas of the Emergent Method. Life is an emergent cycle or strange loop of give and take. Both doing and being, life and death, or self-interest and delayed self-gratification, are necessary parts of the cycles of life; but always the long-term means to which we must return is life, doing, and local self-interest – it can't be otherwise except in death and species extinction.

Nature's Unwritten Contracts

In Pillar 2, when talking about the social instincts, we talked briefly about nature's unwritten naturalistic contracts that seem to be written into the instincts, and perhaps in all species with brains and nervous systems, also enhanced by consciousness. We then asked the question, 'what is the nature of these contracts?' without answering the question. Here then, is the attempt to do so, but we must remember that, strictly speaking, contracts and laws (or constraints) separate from the statistically emergent arrangements of tangible matter do not exist. We simply use the concept of a contract here to conceptualise or signify the behaviour of our ecosystem at a higher level. In all its operations, and at all levels or stratifications, nature simply falls into paths that promote local energy effectiveness and efficiency.

Nevertheless, it seems to us humans, as sophisticated pattern detectors, that these unwritten contracts are a mixture of a life's right to life, the right to take, and the right to self-expression in the face of the external self-restraining powers to govern and control. To an anthropomorphic observer, the contracts in play seem to be a reflection of the delicate balance between personal liberties, personal decision-making, and environmental constraints. Of course, the apparent liberties and decision-making are largely instinctive, driven by blindly selfish genes, but none of us can be quite certain of the level of instinctive blindness in individual chooks in the chicken run. However, rules of risk and reward seem to be clearly at play in an environment that is both hostile and favourable, or apparently random and deterministic, to the individual chook. This same basic truth also applies to our environments, social or otherwise, and to us.

Again anthropomorphically, it seems that nature declares to us the individuality of each living organism (with its inherent DNA) and its selfish right to life and the pursuit of life. This right is based on each cell's existence and its wonderfully adapted ability to take, from the time of its inception, information and food-energy from its normal environment. However, nature does not grant this seeming birthright without severe environmental or relational restrictions. This suggests nature provides a very strong incentive for social contracting, for the creation of extended phenotypes and for other methods of increasing evitability. In another way of looking at it, the same vast sea of statistical properties and possibilities that exist at the subatomic level through the five interactions of physics' Standard Model (including the Higgs field) seems to be in place at the biosocial level through the five eco-psychosocial interactions (mental, personal ego, social, spiritual and physical) as well. We also note that the laws of entropy and emergence describe both stratifications equally well, without any incongruent discontinuity.

Nature's social contracting is done on exactly the same basis as that of the individual cell. It is done on the ability of the social group to sustain its corporate life and avoid its corporate decay. The individually perceived right, 'I take, therefore I am' extends to the corporate level, 'We take, therefore we are'. (More correctly, there is no individual level, there is only the 'we' that changes in the viable systems of various stratifications.)

However, the taking of the fully functioning corpus only refers to taking, or borrowing, sufficiently to avoid decay and find autonomous success (i.e. adapted metabolism and reproduction within the normal environment). The perceived right does not automatically justify taking beyond this level.

Any surplus taking seems to be justified by nature only if it also leads to more effective and efficient eco-social diversity (interconnections of life) or coagulations of life as Matt Ridley put it. Meaningless surplus taking, which is not borrowing but more akin to stealing, and wastage, are seemingly not concepts widely supported by nature. Nature is broadly and locally efficient in the midst of its encouragement of an inclusive energy-effective diversity. It disregards narrow-minded and non-sustainable accumulations of wealth. It supports the re-ploughed diversification of wealth rather than its concentration in the Petri dish of a few. It also supports efforts to use surpluses to improve the efficiency of taking and giving rather than excessive time spent in amusements.

When and where nature does seem to take regard of our folly and object to our imbalances, which act against its own resilience or the coagulating right to life, we suffer the consequences through environmental poverty. We seem to act on behalf of nature as its agents when we choose policies, build systems and make decisions that promote health and life in the evolving mix of species or promote the successful and evolving coagulating of life at higher levels of order. It is for this reason we need to consider the current state of our global economic, financial and monetary systems and the possibilities that would improve their organisation. Emergentism provides a consistent framework within which such a review might take place.

It must be noted that the continuing concentration of economic power in the hands of a few both goes against nature's long term trend and does not bode well for humanity. Our world's largest corporate structures and interrelationships have been able to massively skew the flows of wealth through global tax loopholes and other legal or illegal codes that are extremely unhealthy.

As to the question of nature's use of force, we should recall that nature is largely a conservative near-balance of forces in both the living world and non-living world. Putting the non-living world aside for now, the natural biosphere often uses force to defend survival. Just like excessive accumulations of wealth, nature usually disregards excessive use of force or power, so a similar rule applies. The use of force in the living world seems to be justified by nature only if it defends more effective and efficient interconnections or coagulations of life. Nature and its constraints seem to be trying to teach us something. All the surpluses it provides for our enjoyment come with a responsibility to stand up for it, as its temporarily elected custodians, when it counts; our endeavours ought to be roughly sustainable in the changing environment. Nature can forgive certain indulgences for a sufficiently short period, but not indefinitely, especially if they are reckless. Nature freely provides our opportunities to survive, succeed, and progress, but will from time to time invoice us for our abuses of the privilege and demand payment on its terms of trade. One example would be all the issues around the removal of fluorocarbons from our atmosphere.

Nature, the ultimate interconnector and coagulator of life, seems to be electing us (by our achieved levels of evitability) as its custodians, perhaps as an orphaned child might choose its adoptive parents. We do not elect it. So not any kind of a social contract is being suggested here by nature, but rather something akin to Proudhon's concept of commutative justice (but definitely not his mutualism) in which society naturally arises via lively agreements between individuals of differing ability but with similar understandings of custodial duty (acting in a genuine spirit of voluntarism, reciprocity and federalism).

We are custodians of our bodies (just coagulations of cells), we are custodians of our lands (over and above property rights) and we are custodians of our societies, social groups, social fabrics, networks, marketplaces, cities, organisations and companies (just extended coagulations of life's cells, genes and memes). The evolving social contract seems to be based on the virtuous values of those who count themselves as custodians of society, nature, and the evolving virtuous global tradition.[4] In this sense, virtuosity is that which promotes the continuing coagulation of life, and thus the systemic organisation of information.

Now let's briefly consider the use of force in non-living nature. In one sense, non-living nature is like an orphaned child with too much probabilistic/chaotic power in its hands, waiting for a parent to come and guide it in that power's more orderly, diverse, and virtuous uses. Here then is a crucial role of the evolving value-sets and virtues advocated by this book and the practice of Emergentism.

In summary, the apparently immutable values best adapted to our survival and flourishing (our vremes) can perhaps be, as it were, eternal (in much the same way Richard Dawkins describes genes as more or less eternal), or perhaps at the beginning of infinity, as David Deutsch suggests. Virtuous values, released from their supposed religious owners, give all of us our link to the eternal and the awe-inspiring we seem to so desperately crave.

However, we need to move the eternal concept of immutable values out from the religious realm and from religious ownership and into the well-grounded and concrete realms of tangible nature and our closest relationships, our extended families, and our every-day business, social and political relationships. We also need to input our values into all the written and unwritten social and environmental contracts these relationships imply. As we do this, habitually and intentionally, we will not only see our virtuous purposes fulfilled, but we will gain a stronger sense of our highest potential as well.

Personal Decision-making Tools

The Emergent Method offers personal liberation from feelings of social alienation, low self-esteem, and indecision by guiding us through an understanding of our own human natures and advocating how to take thought and action in various situations. The Emergent Method suggests to us how we might move positively forward towards our own purposes and potential.

It does this through fuller appreciation of the values or measuring sticks that essentially define us. It represents a method of self-actualisation that we may incorporate with whatever life-stance we wish to hold. It also gives us a respect for a stochastic cosmos that neither owes us anything nor promises us anything, but nevertheless responds to triggers that have the direct and indirect connections able to change the prior order of things. The Emergent Method offers quicker and better decision-making on the firm basis of clear values within value-sets arising out of and measuring human nature. It offers an abstract framework that incorporates all personal decision-making phases and scenarios (short or long term).

Concrete Benefits of the Emergent Method

Besides bringing together a more naturalistic model and explanation of human behavioural styles within our environment, the following sections note some of the benefits of Emergentism previously mentioned. Perhaps most basically, the Emergent Method helps all of us answer the question, 'What ought I to do?' in whatever situation we find ourselves.

Self-Actualisation

We typically picture self-actualisation as at the top of the pyramid of Maslow's hierarchy of needs.[5] However the term was initially introduced by Kurt Goldstein in his 1934 book, *The Organism: A Holistic Approach to Biology Derived from Pathological Data in Man*, in which book self-actualisation was seen as the only drive by which life in all organisms is determined. Thus, whether you understand self-actualisation as the pinnacle of life's success or the essential basis of life's success, this book has really been about how to uncover the secret of that success. As we have seen, this is far more than just coping with needs or acquiring wealth, it is about being involved in the wonderful process of life of which we are each unique ambassadors. It is about being fully immersed in life's antinomic dance.

We should note here that this book has not pictured human self-actualisation as an end to be reached, nor as a simple genetic drive with blind purpose. Nor has this book defined self-actualisation as a simple promotion of self-interest, even if many others do lend legitimacy to our psychosocial claims, or we do enjoy some kind of myopic euphoria. Rather, it has pictured self-actualisation as an antinomic and self-organising dance with the universe in a sum of present moments and a relationship that is well described by emergent and almost-immutable values, and in particular our personally adopted hierarchy of explicit values.

From this hierarchy we determine our personal self-restraints of introspective interaction as well as other self-restraints regarding extrospective interaction, cooperation and competition. This self-actualising dance also includes a lively awareness of our associated beliefs, intentions and actions in any outcomes to which we rationally, volitionally and meaningfully contribute. Finally, the self-actualising dance results in personal moral congruence or temporal satisfaction in achievement of personal wellbeing. Nevertheless, we cannot achieve self-actualisation through outcomes alone. We achieve it more casually anywhere and everywhere on the quadune path of personal meaning, purpose and destiny.

Yet for all of this, some will still argue that we can never have true knowledge or certainty and so will wallow in their uncertainty and lack of pure knowledge. Others will wallow in their all-too-sure knowledge and certainty. Yet others will claim that we remain slaves of the subconscious and unconscious. This very human attitude to life seems to miss the whole point of nature's seemingly discernible raison d'être and chief lesson.

The process of life, life-and-being here-and-now, is not about having the best or being right all the time; it's not about being the best, or having the purest knowledge, or holding the most certain truth from beyond the veil of unknowable reality. All these things ultimately disappoint in themselves. The process of life and self-actualisation really is about the journey and the current experience or dance of values-in-action in which we survive to pick up snippets of temporary knowledge along the way. We can then explain those pieces of knowledge within a personal conceptual framework and our naturalistic theories, and then wisely and courageously use them in the local corners of the world we each inhabit.

It does not matter if we are at the bottom of a so-called social hierarchy or not. What matters are our lives and our potential because we share in Life. Sadly, the feeling of drudgery is often self-inflicted rather than inflicted with force. Beauty and the marvellous are around us in whatever station of life if we are free to look for it and care to do so.

Self-actualisation is about realising this and acting on it in such a tenacious, clear, and focused way that life's internal and external setbacks can no longer prevent us from obtaining personal moral congruence on a continuing basis. It is in the tough times that we need to be, and can be, most grateful that we have our lives and our values. Tough experiences make us learn more quickly and deeply, and make us stronger.

A self-actualising life is a self-organising and emergent life sustainably in step with a self-organising and emergent existence, no matter what its ultimate nature. This life does not promise fame, fortune, or even good health because all these things are subject to the law of large numbers in complex populations. However, we can meaningfully make impressions on our chances through our ability to learn, form networks, and influence our peer networks.

In the movie *The Last Samurai*, Katsumoto asks Captain Algren "Do you think a man can choose his own destiny?" to which Captain Algren, more or less as a champion of Emergentism, replies "I think a man does what he can until his destiny is revealed."[6]

For us as individuals within populations, each momentary step has only two degrees of freedom, just like the particle of quantum mechanics. We can observe, perceive or be (like particle position curving space – as per the lower quadrants of Figure 19) or participate and do (like particle velocity under the influence of spacetime – as per the upper quadrants of Figure 19). Perhaps in some circumstances, we as massively parallel systems of indeterminate relationship can choose both zero and one simultaneously or something in between in a fractal manner. Nevertheless essentially, each momentary step also has only two dimensions: Personal – as per the left hand side of Figure 19, or social – as per the right hand side of Figure 19.

Formation of Concepts of Introspection

We briefly discussed the formation of the concept 'blue' in Pillar 1 and the role of perceiving similarities and differences, as well as the importance of differentiation and integration with respect to other already-existing concepts. We also discussed how the child-like signification of blue was not inconsistent with respect to the concept in the mind of the modern adult, just a lower-order subset of it. The adult had simply taken the child-like concept of blue through many more iterations of difference, differentiation, similarity and integration (with the feedback that aids emergent order) than the child had done.

We could likewise form concepts of discrete collections of attributes such as 'table' and 'chair'. The next level of concept formation would be to detect that both tables and chairs are kinds of 'furniture'. Furniture is a logical grouping of perceived existants but we do not directly perceive such indirect groupings (because tables and chairs are different in their shapes and uses). However, as long as our logic is sound (e.g., both pieces are made by carpenters and sold in the same shop for use in the same home), so should be the more relational and incorporating concept of furniture.

Likewise, as we gain experience and incorporate more laws of logic, we could eventually arrive at the much higher concept of a 'viable furniture business'. Such a signification would not just include logical groupings of existants and attributes (such as chair and blue), but also concepts of actions, relationships, etc. 'Viable furniture business' would incorporate concepts of consciousness's extrospection, and be the result of its many outward interactions, as well as incorporate concepts of consciousness's introspection and be the result of its many inward interactions.

Unfortunately, we can more easily make mistakes in our concept formation at this much higher level. These mistakes could involve measurements of rationale in terms of the assets of the business (IQ), risks with respect to future cash flows (CQ), norms in terms of business relationships (SQ), or self-restraints in terms of the emotional state we occupied when we accepted the claim that the business was viable (EQ). In such cases, we are often required to re-consider our lower-order concepts. What we usually do is try to separate those things that more objectively contributed to the concept formation, those that did more intersubjectively, and the measures/values that glued all concepts together (from the lowest and simplest concepts in the hierarchy to the final concept of 'viable furniture business' at the top of the hierarchy).

For the more objective or deterministic conceptual contributions, our problem is often simply checking that no conceptualisations are missing and that the logic linking them together is sound. For the intersubjective conceptual contributions, our focus is often the opposite. We must check if any more indeterminate conceptual contributions are unnecessarily included and have impact. However, when it comes to the introspective emotional state that contributed to our concept of 'viable furniture business', this is not easy. It is for this reason we must talk about formation of concepts of introspection a little more carefully.

Just as standard measures (such as length, time, temperature, and pressure) are the claimed objective basis of all extrospective concept formation, so the so-called universal values, as inner measuring-sticks, would seem to be the personally objective basis of all introspective concept formation. To put this proposal another way, perhaps we would like to say that a table of universal values is a neat summary of measures of introspective consciousness like a table of standard weights and measures is a neat summary of measures of extrospective consciousness.

However, here we have a problem with the lack of imposed law, or liberty, which exists within the introspective world. Bottom line is, we find new values or measuring sticks hard to define or use and those measuring sticks that already exist within are, besides sometimes being self-eliminating, often tied to the imposed belief system that purports to own them and under which they are claimed to have come into being. This also means that the words that denote a value in one cultural setting often do not readily translate into the use or language of another cultural setting or individual. This does not suggest that the common values identified in tables 1 to 4 are not locally recurrent and therefore unworthy candidates upon which to safely build more complex concepts of introspection, it just means that our inherited language is often a poor vehicle for these partially incomplete or implicit values.

Ambiguity with respect to introspective concepts is somehow often overlooked or even admired, like deceitful body language, whereas ambiguity in extrospective concepts would be viewed as ignorant, unscientific, and a barrier to our advancement of knowledge. It is perhaps time we were made explicitly aware of this basic difference in the two modes of concept formation so we can hold a more inclusive conceptual framework. This framework would be aware of the greater indeterminism in introspective concepts, allowing us to inductively arrive at a null hypothesis, but also be aware of the greater determinism in extrospective concepts, allowing us to deductively test the same hypothesis. Both kinds of concept formation have their legitimate advantages and disadvantages.

If indeed, the Societal Marketing Era is morphing towards the 'Emergent Vremes' Era, then this movement towards greater balance in our concept formation has already begun. We need to start demanding more of our shared definitions of values in language, so that we can carefully and personally choose and test our self-restraints of introspective consciousness, and then clearly see our unique fit within the universe and its antinomic dance.

Perhaps what we need is to discover a more transparent way of expressing in words what we naturally and more truly feel subconsciously and through uncontrived body language. One way we could do this is to begin a rigorous exploration and comparison of the meanings of words in differing dialects and cultures. We need to discover a more natural way of articulating the basic introspective measures we identify (e.g. love), and better units for their measurement (e.g. the fleeting intensity of love in a moment or the non-fleeting strength of love as we 'fall in love' over a long period).

In this way, we will be able to form more carefully specified higher-level introspective concepts such as particular thoughts, feelings, desires, and memories, and then a very high-level introspective concept such as a 'viable relationship with my workplace colleague', etc. If and when such a relationship ran into difficulties, we could then more naturally go through the same process of perceiving and analysing the hierarchy of concepts that we went through when we analysed the 'viable furniture business' and when we went through the exercise of creating our own hierarchy of values in Pillar 1.

For example, from where does the introspective concept and feeling of anger arise? It arises from a frustrated goal. Further, where do our goals come from? They come from our beliefs shaped into propositions. For instance, 'money is good because it makes me happy; I will try to get more money so I can be happy'.

Happiness through the accumulation of money is, in this case, what we command the subconscious to help us achieve. Here happiness is the value, and the direct path to happiness through money is the conscious belief, and the aim to be happy is the goal which is later frustrated (i.e. the striving for and getting of money did not bring happiness in this instance), leading to anger.

As you have already noticed, in this case the personally held value of happiness as a measure of life is not incorrect (i.e. it is a noble measuring-stick that if correctly used can reinforce life-now and self-actualisation), just its direct relationship with money or how it directly interacts with money (stated in the belief) is incorrect. If it was understood that happiness, as a measure used by a consciousness to evaluate its own life-dance or self-actualisation, has an indirect and antinomic rather than linear relationship with money or any material thing, perhaps it could also be seen that money cannot be an end in itself and money cannot buy happiness. Perhaps we could also understand that the key to happiness is the antinomic and slightly indeterminate realisation of all our individual values in what we do or make, in our work or hobbies, or in our personal relationships.

That is, it is in the processes of life rather than just their outcomes that we express values and realise happiness. Others can then measure expressed values in terms of money paid freely for the things that are the expression of those values. That is, it is in the mutual appreciation of values and benefits traded that we find relationship, friendship and happiness, not in the traded material things themselves.

Perhaps, with a better-specified introspective language, we would be in a position to say that many of our society's psychological problems we assume have a subconscious or chemical basis that consciousness cannot affect, in fact have a basis we can easily do something about at the conscious level. Just as we often seek new concepts or knowledge to overcome life's mundane, physical problems, so it is possible to learn new concepts and new ways of perceiving our world to overcome society's psychological problems.

For instance perhaps those labelled sociopaths, narcissists, over-anxious, impulsive, unpredictable, unstable or otherwise dominated by fear, worry, pessimism, or passive-aggressive behaviour, are being at least partly limited by their lack of knowledge or appreciation of value-concepts, or by their systemically incorrect knowledge of value-concepts.

We reinforce this problem of lacking or unclear values when we form faulty higher-level introspective concepts in our lives. This can lead to not only anger, but also to an ability to fool ourselves in terms of rationalisations, elaborate defence-mechanisms, psychological evasions, repressions and, most importantly in terms of the topic of this book, reduce our ability to realise our potentials and wholesome purposes.

Isn't the process of recovery something like what we sometimes see when an alcoholic joins a self-help group to battle his or her addiction and one day overcomes or controls it? The group brings the alcoholic through a process of reframing the situation, reassessing relationships, goals, and beliefs and rediscovering personally objective values. That is, we weed out and replace incorrect introspective concepts with new or reinvigorated, life-affirming, and self-actualising introspective concepts that relate better to extrospective concepts and external reality.

Self-Actualisation

Even though our introspective language has no doubt let us down, this does not mean that our study of human nature, and the values as we have perhaps inadequately described them in tables 1 to 4, has been pointless. We have made some important inroads into this whole issue of values and their indispensable role in realising our full potential. Even if we use these fuzzy values as described, we know a lot more about them than when we started.

We now know how values arise from the emergent evolution of human nature and how they relate to each other in terms of that human nature – and we now know how they dance in the lives of society, in the lives of our associates, friends and relatives, and in ourselves.

We also now know that we do not have to limit our values to a consistent set that must be tiny in order to remain non-contradictory. We can each do our antinomic dance without being ethically remiss, but rather while being ethically accurate (in terms of the relationship between values and current reality).

This means that even though there are some doubts about the values and value-sets we are still defining or looking to better define, we can nevertheless explore our own introspective lives with more structure and self-organisation than before. In this way, the relationships or interactions we build into our beliefs between introspective values (such as happiness) and concepts in the extrospective world (such as paper money) are now far more likely to be correct. That is, they are more likely to be continually self-organising, emergent, and leading to self-actualisation, rather than just dichotomous judgments of good and evil, right and wrong, or self-gratifying and disenchanting.

We can bring order to the freedom of our inward-looking consciousness without bringing it once more into bondage. We can bring our inward-looking consciousness to a clear and powerful liberty, the depth of which we perhaps never knew before.

This natural and free but habitual and orderly flow between inward-looking consciousness, outward-looking consciousness, and reality can only be good news in terms of our individual and corporate purposes and sustainability.

Solving Ethical Dilemmas and Bridging Moral Framework Gaps

The Emergent Method offers a means of solving ethical dilemmas by clearly linking our ethics to our human nature (consciousness-plus-instincts) and by providing a methodology to break through rational blockages to action and to overcome irrational groupthink.

By linking each person's ethos to the four self-restraints of consciousness-plus-instincts, we step away from a monolithic ethical basis to a more nuanced, resilient, and emergent one.

This step away gives us more tools with which to understand our ethical environments and act in a way that will promote enlightened wellbeing.

Such an approach could work in tandem with such frameworks as Lawrence Kohlberg's Stages of Moral Development.[7] The Emergent Method can also work well with various assertiveness and conflict management techniques because it provides a deeper understanding of differences in cultural and workplace values. It can also help provide win-win solutions that are welcomed more genuinely because compromises can be replaced with more creative, deeply appreciated, dynamic, and mutually 'owned' solutions.

For instance, in conflict management, it can be helpful to start with the values or behavioural preferences in the EQ quadrant of Figure 19, and then go to each other quadrant together in order to fully discuss the dimensions of the conflict and search for common ground. This conflict resolution technique would then do well to finish with a discussion back in the EQ quadrant to identify and cement how the resolution was reached.

Group Discussion Frameworks and Negotiation Frameworks

The Emergent Method offers an abstract framework for exploring and understanding contrasting or competing purposes in discussions that begins with recognition of basic human rights and the value of voluntarism, reciprocity, and federation in all manner of social interrelationships.

It enables us to quickly identify not only our own values, meanings and purposes, but also those of our protagonists or competitors. This leads to fuller understanding of viewpoints and therefore a higher likelihood of finding common ground, a phaseal sharing of positions, or identifying and pinpointing mental roadblocks and the values that stand behind them.

The Emergent Method, by acknowledging the pervasiveness of incompleteness and inconsistency, also offers a way forward for entrenched views by helping all in the discussion understand the phase currently held, how it was arrived at, and the next phase that is likely to naturally follow. Finally, The Emergent Method also facilitates movement in the discussion towards the next phase, by offering a framework of understanding human needs and understanding how they may be satisfied.

There are many left-right political debates in global and local arenas to which the Emergent Method may be beneficially applied. One such current debate is around climate change. Clive Hamilton encapsulates one such debate in his "Table 2: Anxiety about global warming and support for geoengineering",[8] reproduced in part in Figure 31, with my added axis labels. I have omitted the names of any individuals listed by Hamilton.

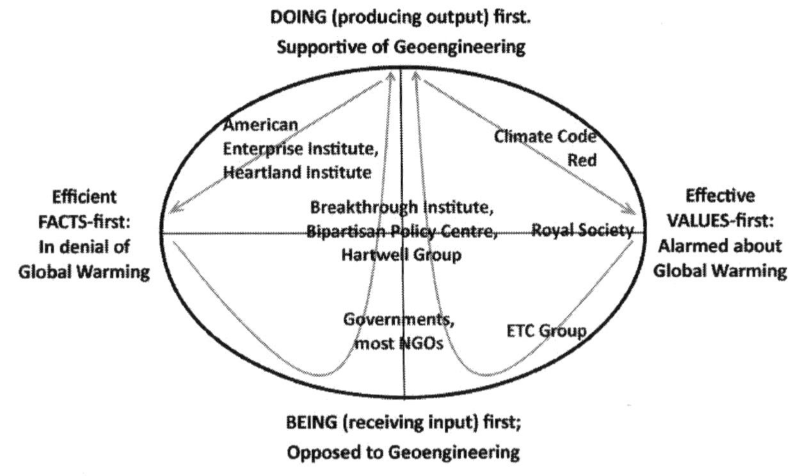

Figure 31: Anxiety about global warming and support for geoengineering

Figure 31, when considered in the context of prior figures, immediately suggests the human needs, human nature, and human values behind each view. For instance, the Climate Code Red position is indicative of the human need for self-expression, higher SQ intelligence, and the cooperative or normative values of Table 3. The arrows indicate that its position is likely to be closer in a phaseal sense to the Hartwell Group or Royal Society than to the

ETC Group, which in turn is closer to the Climate Code Red position than are most NGOs. The American Enterprise Institute and Heartland Institute are likely to share the most divergent view to that of Climate Code Red. All of this information is likely to assist for instance Climate Code Red in identifying winning messages and strategies, or lobbying various stakeholders, in the ongoing and maturing debate. The greatest problems in group communications occur between value-set preference opposites, e.g. IQ quadrant preference versus EQ quadrant preference, or CQ versus SQ, in Figure 21. Again, the solution is to discuss all value-sets and their legitimate places in human endeavour in order to find common ground and a process to take the agenda forward.

Organisational Processes and Structures

The Emergent Method offers a method of optimising organisational group processes and structures by firstly recognising how human needs arise out of human nature and how humans interacting within groups often meet these needs. Managers equipped with knowledge of the Emergent Method become facilitators of human nature and thereby, organisational structure and processes.

Likewise, executives equipped with knowledge of the Emergent Method become effective in the defence and promulgation of organisational values that support each phase of an organisational process and each specialisation of stakeholder-aimed labour within the organisation's structure.

The suggested management approach is not a top-down bureaucratic one, but rather a facilitation of the relationships that desired corporate values suggest. Just as it is not possible to embed '$E=mc^2$' without getting into the nitty gritty of energy, mass and velocity, so it is not possible to embed desired organisational values without getting into the relationships within workplaces they imply. In each case, the law or measuring stick should be seen as a heuristic descriptor of such relationships that does not exist apart from them. For this reason, it also makes sense that different areas of the organisation reflect differing values.

Strategic Organisational Development

The Emergent Method offers a method of gradually optimising organisational development by recognising how human needs develop over time through phases, even across organisational departments and across organisational stakeholders. Emergentists would suggest that product life-cycles and firm life-cycles are just two examples of the ramifications of the phaseal and emergent nature of human need and human development.

Laws and Governance

Emergentists recognise how systems of rules and law arise out of human nature and can therefore help identify how rules or laws might fail to meet those needs in the different phases of human endeavour and development. The application of Emergentism brings a clear but evolving framework of value-sets to governance discussions that might otherwise leave those virtuous values hidden or not clearly delineated.

Continually evolving applied values may inevitably lead to law life-cycles (or the cyclical development of the efficacy of legal statutes and precedents) in much the same way as marketing's S-curve describes product price changes over each product's life-cycle. However, we should not see the evolutionary mutation and perhaps somewhat cyclical advance of shuffling values as a watering down or fashionable trend setting of values. Rather, this advance is an important evolutionary movement of value-sets to higher planes that cannot be gained any other way and will necessarily put higher demands on old statutes and precedents (for example, as happened when slavery was abolished in the United States).

The same principle of emergent values in human consciousness is particularly applicable to organisational governance and for example, corporate governance and disclosure. It is therefore important that corporate values are clearly delineated, reasoned and extolled as they relate to each phase of human endeavour. It is also important that these corporate values be regularly reviewed, to ensure they continue to meet the developing needs of individual, organisational and societal endeavours.

The Emergent Method can greatly assist in the formulation of standardised management information systems that are able to cater for the developing value-needs of all stakeholders. Further, a clear link between the drivers of human nature, values, and the phases of human endeavours can help organisations better understand the strengths, weaknesses, opportunities, and threats that their current reputation and offerings amongst stakeholders embody.

Finally, The Emergent Method can help organisations hasten and fully exploit the awakening Emergent Values Era of organisational marketing.

Political, Legal, Philosophical and Economic development

The Emergent Method offers a framework in which we may understand the related political, legal, philosophical and economic power bases within society in terms of structure and emergent developments because it clearly traces societal power bases back to individual human nature and needs.

This discussion perhaps begs the question, what is the better political system – laissez-faire capitalism or a mixed economy? Can a study of Emergentism offer any light on this question? I personally lean towards an open model of laissez-faire capitalism. At the same time, I am very aware of the need for us as a global civilisation to quickly find a sustainable path forward. However, is an ideal politico-economic system (e.g. perhaps a kind of laissez-faire capitalism rather than our current mixed capitalist system) the real issue? Maybe it isn't because we currently have sufficient liberties to evolve towards a better system in the future if the necessary elements are present. It is the necessary elements that are more important.

We can think of all the philosophers who argued for one political system or approach over another in the past (from Nicolo Machiavelli's *The Prince* (1513)[9] to the works of Adam Smith, Karl Marx and beyond). They all seemed to make some sense while they hotly and robustly aired their convictions, but when put into practice or handed on to the next generation the faults slowly appeared. Maybe the reason why this has happened repeatedly in our history is because all philosophers have one thing in common – their convictions are based on their strong values and each political theory seems to be based on its own set of strong values (good or bad). However, their followers and descendants do not necessarily have the same strong values-in-action even if they do have the written doctrine. Their followers often make poor value-judgments – they divorce the matter from the space and the now; they keep talking but stop dancing; they have the words of the doctrine but have forgotten the spirited dance of the doctrine. In contrast, Nature living in the present never does this.

What seems to be important when choosing amongst various laissez-faire and mixed capitalist systems is the education, development, and practice of virtuous values within phaseal value-sets. Our vremes will free us from our blindly selfish replicators and our political failings.

If enlightened education is encouraged (by either private enterprise or government service) then a better political system will inevitably emerge in each of our nations that better suits our societies and their needs. However due to the fact of our continuing emergence, the match can never be perfect.

We can easily find the Royal Australian Navy values on the internet (honour, honesty, courage, integrity and loyalty) and likewise we can discover that the Australian Army teaches its soldiers the values of courage, initiative and teamwork, "inspired by the ANZAC tradition of fairness and loyalty to our mates".

Why can such values be rigorously and habitually taught (and displayed by members) in the Australian armed forces but we don't learn and discuss such values or codes of ethics with similar rigour in our schools, our workplaces, or even our homes? This is what needs to change; this is the most 'necessary element' of sustainable socio-political, legal and economic development. The Emergent Method offers a framework for the massive social education program ahead.

Social Enlightenment

If this book is about the path to self-enlightenment, then it must be about the path to social enlightenment as well. The Emergent Method offers us some important ways of thinking about where we are at, as a society, and where we are heading.

Emergentism, as a naturalistic philosophy, will help bring human consciousness into near-balance with our evolving local and planetary environment as its ideas and principles spread from mind to mind and life to life. This is perhaps the most fundamental benefit of The Emergent Method. It will cause us to respect and honour the environment and each other as a monistic whole. In this respect Emergentism in its highest expression will also cause us to "love one another, and to love our planet as much as we love ourselves", just as Tim Flannery urged in the last line of his book *Here on Earth*.[10]

Clash of Civilisations

The Clash of Civilisations and Remaking of World Order originally published in 1996 by Samuel Huntington was one of the most influential books on International Relations written up until that time. I want to discuss its contents in some detail in this section as a way of highlighting some of the wider societal implications of the Emergent Method.

Huntington begins by stating, "In the post-Cold War world, the most important distinctions among peoples are not ideological, political or economic. They are cultural."[11] By cultural Huntington refers to things like customs, mores, taboos, language, and ancestry. In terms of Figure 27, it seems he is suggesting that it is a civilisation's cultural Phase 2, rather than its political Phase 1, ideological Phase 3, or economic Phase 4, that separates civilisations the most. That is, it is the civilisations' self-organising and emergent rules of the game that distinguish them most strongly and perhaps compete most fiercely. These rules emerge from a creative minority that wielded control or influence during Phase 2. For this reason, mature civilisations can develop somewhat different and diverse mores and taboos, but once they are legitimised by the success of the civilisation, they are difficult to change. In contrast to being personally objective, these mores and taboos seem to become societally objective, if I can use that term. Civilisation clashes are rules of the game clashes. Can we, of differing civilisations, really share or compromise our differing brands of civility?

The same issue exists at the individual level. It is largely subconscious Core-self or the psychosocial believer/claimer of Phase 2 that most deeply divides us from each other. Our deepest differences are thus often hidden below the explicit level of personal and social consciousness. To unite us together, I suggest we need more naturalistic and less superstitious rules of the game to emerge, first personally, then societally. However, for this to happen, we first need to start a discussion of the quasi-universal concepts of civility we share. As Huntington seems to observe[12], the wars of kings are over and the wars of ideologies are over, or not directly and currently relevant to the survival of our species. Right now, we struggle with cultural wars. We pit one culture against another rather than holding the values, mores, and taboos of both up for mutual celebration and scrutiny. We need to move on. To end these wars we need a discussion of those values that lead to each civilisation's selection of often subconscious mores and taboos. More importantly, we need to accept that cultural success is not uni-dependent. Like our universe, it is interdependent.

Huntington says,

> *"Modernisation involves industrialisation, urbanisation, increased levels of literacy, education, wealth, and social mobilisation, and more complex and diversified occupational structures … As the first civilisation to modernise, the West leads in the acquisition of the culture of modernity. As other societies acquire similar patterns of education, work, wealth, and class structure, the argument runs, this modern Western culture will become the universal culture of the world."*[13]

Huntington then rejects this idea. He suggests that the divides between societies are not addressed by their becoming more modern.

In a sense, I agree, but I think the concept is wrong. Modernisation as defined here is not a helpful concept. A better concept would be one of diverse moral congruence and evitability at the societal level. Clearly, some societies, and some sections of society, are going to be more successful than others are, in the long and short term, and this will depend on their evitabilities, as well as their moral awareness and courage. That is, I am suggesting like Sam Harris (2010), that some mores and taboos are better in terms of sustainable human flourishing than others. Further, the best set of mores and taboos are very unlikely to sit in any one civilisation as currently configured. Perhaps the end of a uni-dependent view of civilisation is one of the greatest realisations our species will face in the next century.

Huntington lists a set of attributes of Western civilisation.[14] What is interesting about this list is not whether they are Western or not, but whether they promote sustainable human flourishing by reducing superstition and encouraging a more naturalistic and interpenetrating eco-social diversity. The reason why the Western civilisation has been as successful as it has been is that many of the items on Huntington's list do just this. These items are italicised in the following paragraph.

The *separation of spiritual and temporal authority* has led to the reduction of superstitions. A respect for the *rule of law* has lifted relationship between people above the subjective limits of clan and family, without violent coercion. *Social pluralism* at its best means an acceptance, respect, or tolerance of social diversity. (At its worst, social pluralism means following double standards rather than a flexible, transparent and adaptable cycle of value-sets as suggested by the Emergent Method). Societies that are more liberal are typically more cosmopolitan and eco-psychosocially diverse. The synergies they develop are therefore more sophisticated and robust than they otherwise could be. Emergentism can help focus our attention on those synergies and help us understand the cultural differences that currently divide us. The richness of civil society in the West again points to the harnessing of social differences.

The diversity of European languages within Western civilisation is further testimony to its sustainable social diversity, although clearly the current socio-diversity was not gained without great struggle in the past – a struggle often caused by our religious and psychosocial ignorance. *Representative bodies* are also an outcome of this well-respected and appreciated social diversity. Finally, the West's notorious *individuality* keeps unrealistic or unnatural mores and taboos at bay. Individuality tempers superstitions. It enables the core beliefs and claims of individuals and social groups to evolve in a more naturalistic fashion. Civilisations in the past have had revolutions against the Core-self and its rules of the game. The West has perhaps been able to overthrow its Core-self a little every day.

All these features resulted in science's ascendance over superstition and dogma in the West. When compared to the civilisations that preceded it, this is what I would attribute to the West's success. There is no reason why we as a species cannot adopt such features elsewhere, and have them lead to similar success.

However, the ascendance of science over superstition is by no means the end of the story. As we saw in Pillar 3, extrospective science and scientific fact is largely backward-looking and thus determined. As a civilisation, we also need to be adept at being sustainably forward-looking too. This strength, whilst supported by science, is not driven by science. Our introspective value-sets drive our future evitability and our extrospections support it. It is in the area of future planning that the West has been weak and other civilisations could perhaps be stronger. The evidence for this lies in the unsustainable growth-at-all-costs development of Western-style economies around the world and our inability to tackle concomitant climate change comprehensively.

Huntington's probable response to the above would be to say that other civilisations can develop their sciences and maintain their superstitious mores and taboos at the same time. He would likely say that modernisation as he terms it strengthens cultures dominated by religious and non-scientific beliefs. I do not accept this proposition in anything but the short term. In the 21st century, the scientific advancements depend more and more on an acceptance of the evolutionary sciences, and a rejection of religious assumptions. We have moved way beyond the simple scientific discoveries of Newton and James Watt that led to the early industrial revolution. A superstitious civilisation can build its rockets and contraptions of war, but will it contribute to advancements in genetic technology and so on that offends its religious sensibilities? Further, in not being able to do this, will it remain viable in the longer term?

Huntington would argue that there is a strong rejection of Western values around the world as the West's power declines, and thus a move towards alternative local customs. The world is a tough place moving inevitably towards civilizational clashes in Huntington's view. Conflicts along the fault lines between civilisations are interminable in this view. However, in the short term, I would see freedom from the imposed value-straightjackets of old empires and other civilisations as a good thing. It is the first necessary step towards a personal moral congruence at a civilizational level. How can a civilisation flourish and become more sustainably naturalistic if it cannot find and establish its own Core-self and rules of the game?

The question here is whether these differing rules of the game can peacefully co-exist as they move towards an acceptance and adoption of the Emergent Method. While I have presented my current personal beliefs which some may label atheistic, these beliefs are not axiomatic and so do not act as the core of the Emergent Method. Thus, I think this is possible and is worth pursuing. That is, anyone of any life-stance and cultural background can adopt the Emergent Method and reap benefits. We can all reap the benefits together of a deeply respected and highly enlightened cultural diversity.

Huntington says that businessmen trade with their own kind that they can trust, and likewise states form trading blocks with other states with which they have cultural affinities. That is, commonalities between peoples of the same civilisational background are the basis of cultural and economic cooperation.[15] No – I don't agree. The root of all trades is in a mutual trust that may be fostered by growing, interpenetrating relationship. Commonality, or lack of diversity, is not the basis of trade. Trust in the trade of value for value is the basis of economic cooperation. However, Huntington disagrees. He doesn't believe trade brings traders into agreement.[16] At the level of some cultural mores and taboos irrelevant to the trade, he might be correct. Nevertheless, at the level of mores and taboos relevant to the trade, I would have to disagree. A lot has to be right in terms of the social and economic framework for the trade to occur. Trade thus encourages understanding and a social basis for movement toward tolerance,[17] as Huntington himself suggests.

The last chapter of Huntington's book asks two questions posed by Melko he still sees as pertinent: "First, is Western civilisation a new species, in a class by itself, incomparably different from all other civilisations that have ever existed?" The question is a weak one. Every civilisation is necessarily unique. "Second, does its worldwide expansion threaten (or promise) to end the possibility of development of all other civilisations?"[18] The only possible answer is no. Better questions would have been, 'Did the Scientific Method, introduced by the West and fundamental to the development of all modern technologies (which can be seen as human extended phenotypes) mean that all civilisations from then on must move towards a greater respect for scientific naturalism and thus move away from superstition?' In addition, 'Does this mean that all future and current civilisations must converge on an interpenetrating and unsuperstitious monism (a future Dark Ages not withstanding)?'

This idea predicts that civilisations that try to restrict the Scientific Method, as a matter of superstitious principle, will not progress as quickly technologically as those that do not. That is, the West, with its smaller population but better technology, might be able to continue to exert a disproportionate control over world power in the long-term future. It is perhaps a cruel fact that humans with superior technology or extended phenotypes are effectively a higher species than are their competitors. The Emergent Method seeks to take us all, of all civilisations, to a yet higher level of wellbeing.

In the last four pages of his book, Huntington suggests that in a world moving towards Civilisation with a capital "C", there would be a gradual acceptance of diversity. He adds that we should try to foster commonalities in values and practices. I would largely agree; this approach of accepting diversity while fostering commonalities would be compatible with the Emergent Method.

Huntington, on the last two pages, then asks, "Is there a general, secular trend, transcending individual civilisations, towards higher levels of Civilisation?"[19] This is a question worth discussing. I think the Arab Uprisings have gone someway to suggesting that peoples whether religious or not, and from various backgrounds, are seeking more from their governments and a more just society regardless of a shared religion. As Bowen (2012) noted at the time, there was revolution breaking out amongst the populations of Libya, Syria, Egypt, Bahrain and Yemen, and threatening to also break out in other places.[20]

Huntington suggests that if such a trend exists, then it may be because we live in a world that increases individual control over the environment through technology and economic participation. On the last page, Huntington seems to see the word modernisation through Western eyes. He concedes that modernisation has promoted Civilisation, but he is not sure it has promoted its moral and cultural dimensions.[21] We need to deconstruct the word. If by modernisation we really mean the Scientific Method, then yes it has contributed to the Proto-self of Civilisation. However, the Scientific Method has little direct influence over forward-looking and introspective moral choices or cultural behaviours. For this influence to occur, we need to broaden the concept of the Scientific Method to the Emergent Method. By this means, we can promote and contribute to the development of the next stage of Civilisation - the Core-self of Civilisation.

Societal Vision

The study of Emergentism suggests that the next stable social order will be one of raised consciousness-plus-instincts in which unquestioned authority (of God or government) and forced submission will not be the basis of order and structure, but rather a mutually agreed allegiance to values and an evolving respect for diversifying values that promote human flourishing. These values are not the property of the 62 multi-billionaires that according to a 2015 Oxfam International report possess as much wealth as the poorer 50% of our planet's population. Nor do these values belong to a tribe, nation, race, religion or political party, but rather are the property of our evolving and virtuous global tradition to which we all contribute and of which we may all be proud.

This will be a new normative approach not locked into a standard model, conceptual framework, dogma owned or patented by any powerful elite, religion, race, or wealthy vested interest. Rather it will be an open collection of value-sets (explicitly codified but never closed) that arise directly out of the nature of reality, our lives at their current stage of development, and our consciousness-plus-instincts.

The new global social norms will not presume to prescribe anything other than a plain-language, non-contradictory, mutually agreed and qualitative definition of values, value-sets and phaseal vremetic frameworks. These norms could provide the suggested rationale or policy reason for any new judge-made law or legislative law. It could then be up to local communities to decide how to live and move within these value-sets and frameworks in any way they see fit. In such a vision, world government voting will be for evolving and vying definitions of values, value-sets and cyclical frameworks, or 'values accords', not for applied social programs within communities or organisations.

In such a model of world government, elected world governments would have the duty of researching, testing, verifying, communicating, advocating and promulgating an understanding of the virtuous (and personally objective) values, value-sets and phaseal vremetic frameworks that more objectively promote human flourishing, nothing more. They would be in the position of matching values to principles and environmental contexts in an effort to promote human viability and flourishing in the long-term. The world government's power would be a judicial rather than legislative one. Its power would rest in its tradition, built on emerging precedent, as a positive and stable (and almost prescient) guiding influence, perhaps something like a benevolent, relevant, and modern royal family.

Real political power (legislative and executive power at the highest level) would remain centred in local national governments rather than the world government. In this way, we can have world government and a coming together of nearly universal values without the danger of world domination, political or economic. Local national governments, by adopting or not adopting the values accords, will declare their positions to their citizens. The local government, through all its workings, would then be required from time to time to demonstrate its societal moral congruence with its citizens' adopted value accords by being subjected to a fair electoral process.

The first virtuous values-accord we as citizens of the world could perhaps make subject to a ballot is the following, taken from the first article of the Virginia Declaration of Rights:

> *"That all men are by nature equally free and independent [I would prefer 'individual'], and have certain inherent rights, of which, when they enter into a state of society, they cannot, by any compact, deprive or divest their posterity; namely, the enjoyment of life and liberty, with the means of acquiring and possessing property, and pursuing and obtaining happiness and safety."*[22]

Where government policies and programs fail to meet the principles of an adopted values-accord, the local national government would have the duty, as servant to the people, to have those policies and programs withdrawn, disassembled, and perhaps, subject to citizens' discretion, any damages assessed and rectified or recompensed. The local governmental failure could perhaps be as measured by an independent World Court dedicated to this task.

All new local government or court-based legislative programs and policies would have to demonstrate their alignment with adopted values-accords and their principles (perhaps by way of introductory explanatory notes) before the bill or regulation or court decision could be brought into law. The introductory explanatory notes would form part of the law.

Your Concept Revision

Your own concept revision is difficult to do if reading this book alone, but why don't you take some time to reflect on your own introspective life? Can you identify outcomes in your business or personal relationships that are or are not self-organising and self-actualising? If so, can you identify the shared or unshared values in those relationships? Can you identify which values are self-actualising and which ones are not?

How would you describe the nature of the dance or musical improvisation around those shared or unshared values? Can you communicate this nature to the others in each relationship in such a way that promotes self-actualising changes?

What are the assumed relationships built into your beliefs between introspective values (such as happiness) and concepts in the extrospective world (such as paper money)? How can you revise such concepts applied to your everyday life so that there is a more natural and free but orderly flow between your inward-looking consciousness, your outward-looking consciousness, and environmental reality?

Are you achieving each of the purposes you detailed in the last chapter by following the steps in each of the process maps you created? If not, maybe the purpose you are failing in is not really a reflection of your values or not correctly detailed. That is, you are possibly busy fulfilling purposes that really do not matter to you right now.

However, if this is not the reason for not achieving a certain purpose, perhaps as one exercise of this revision, you could take each of your purposes of the last chapter, review the associated risk matrix, and test the controls in place.

Risk	Description	Value-set	Control
R1	Missing information for decision	Table 1	Priorities set as information gathered
R2	Too expensive, wrong location, inconvenient, mismatch of benefits	Table 2	Research is conducted
R3	Mismatch to current levels of fitness or strength	Table 2	Research is conducted
R4	Can't focus, bad habits, too lazy	Table 3	Changes and benefits visualised
R5	Not enough evidence, not easy to define	Table 4	After reflection, lessons learnt recorded

Figure 32: Fitness Risk Matrix

Whether or not you feel like you are achieving each of your deepest purposes, are the controls well defined and operating as expected? If you are not achieving a certain purpose, it is possible that the answer is 'No'.

Using the example of controls in the risk matrix from the last chapter, ask yourself:

- Are my priorities being set as information is gathered? Am I honouring the values I care about in Table 1: The Risk-weighted Value-set?
- Is research being conducted? Am I honouring the values relevant to me in Table 2: The Positive Value-set?
- Are my benefits being actively visualised or imagined? Am I honouring the values I respect in Table 3: The Normative Value-set?
- Are lessons being learnt after reflection, and then recorded? Am I honouring the values dear to me in Table 4: The Self-enlightened Value-set?

If not, form an action plan to strengthen the implementation of controls by a certain date, before you take on a greater workload or add another purpose with further risks.

Perhaps you need to find a mentor to assist you in this journey through personal moral congruence.

You owe it to yourself, your short life, your strong sense of values, and all your possible relationships to follow a path of virtue (or values-in-action) and realise your emerging potential.

To you personally: Carpe Diem; Seize the Day!

I invite you in your corner of the world to join our global quest, and please provide your valuable feedback on the ideas presented here at our website.

Concepts of Emergence Encountered in Pillar 5

Our challenge is not to intelligently design that dance before the fact (as if we can know how it emerges in all its complexity), but to be intelligent participants in it. Massively parallel computing in a human body or ecosystem slowly advances in a successful direction through trial-and-error not because of intelligent design but because of all the self-contained interrelationships that describe it. The more relationships it encompasses, the more refined its evitable responses can be and thus the more risks it can mitigate without major threats to survival.

if anything at all is to be venerated it should be our emergent vremes. Yet we should do this venerating in an evolutionary and natural sense. Our vremes, which enable us to rebel against the tyranny of our blindly selfish replicators, naturally arise from the self-actualising parts of our consciousness-plus-instincts. That is, higher consciousness is a state of instincts and consciousness brought into dynamic homeostasis by our emergent vremes.

Within this one cyclical and interpenetrating system must be hidden all the emergent solutions to the world's political, cultural, religious/philosophical and economic dilemmas.

These dilemmas seem to be ways in which the universe strikes out and searches for more highly complex and inclusive modes of coagulation. The solutions simply depend on the recurrent local dance, and now, our consciousness of this fact.

I hope that you have arrived at the point where you understand the core ideas of the Emergent Method. Life is an emergent cycle or strange loop of give and take. Both doing and being, life and death, or self-interest and delayed self-gratification, are necessary parts of the cycles of life; but always the long-term means to which we must return is life, doing, and local self-interest – it can't be otherwise except in death and species extinction.

In the next chapter we will begin to discuss the global transition process ahead – that is, how we can move from our current social structures and mores to the ones we yearn for.

---

[1] Ridley, Matt, *The Origins of Virtue* (Softback Preview, Great Britain, 1997), 144

[2] See Rand, Ayn, *Atlas Shrugged* (Plume, USA, 2007 (Fiftieth Anniversary Edition)), 380-385

[3] Ibid., See pp.380-385

[4] As an interesting aside, Philip Pettit's argument that the absence of an effective rebellion against the social contract is the only legitimacy of it (see Pettit, Philip, *Republicanism: A Theory of Freedom and Government* (Oxford University Press, NY, USA, 1997)), would seem to accord with nature's often muted but sometimes dramatic reactions to the efficacy of humanity's custodianship.

[5] Available online from http://psychclassics.yorku.ca/Maslow/motivation.htm (accessed 2012) but first published as Maslow, A.H., *A theory of human motivation* (Psychological Review 1943), 50 (4), 370–396

[6] Warner Bros., *The Last Samurai* (2003)

[7] See http://pegasus.cc.udf.edu/~ncoverst/Kohlberg's Stages of Moral Development.htm (last accessed 2013)

[8] Hamilton, Clive, *Earthmasters: Playing God with the climate* (Allen and Unwin, Sydney, 2013), 136

[9] See Machiavelli, Nicolo, *The Prince,* reprinted By Encyclopaedia Britannica, Inc., USA, in *Great Books of the Western World*, second edition, fifth printing, 1994

[10] Flannery, Tim, Here on Earth (The Text Publishing Company, Melbourne,

---

Australia, 2010), 280

[11] Huntington, S. P., *The Clash of Civilisations* (Simon & Schuster, New York, 2011), 21

[12] Ibid., See p.52

[13] Ibid., p.68

[14] Ibid., See pp.70-72

[15] Ibid., See p.135

[16] Ibid., See p.218

[17] Ibid., See p.238

[18] Ibid., p.301

[19] Ibid., p.320

[20] See Bowen, Jeremy, *The Arab Uprisings* (Simon & Schuster, New York, 2012), 9

[21] See Huntington, S. P., *The Clash of Civilisations* (Simon & Schuster, New York, 2011), 321

[22] The first article of the Virginia Declaration of Rights adopted unanimously by the Virginia Convention of Delegates on June 12, 1776, written by George Mason. Comment in square bracket inserted by the author.

PILLAR 6: Facilitating Others

In this chapter, we will consider the lesson the universe seems to be teaching us. We will discuss the paradigm shift in attitude required to learn this lesson, and then once this shift is achieved, the kinds of changes that would flow from there. This chapter will be about characteristics of leadership but it will not be a plea for leaders. It will quietly suggest that as we, in our individual quests, strive virtuously towards our individual potentials, we will also individually and consciously but inevitably reflect these same characteristics of leadership as we master to some extent the chaotic and antinomic dance of life. Finally, in terms of your own philosophical quest, the text will encourage you to begin writing and completing a business plan for your life.

Behind the Veil

Existence is ultimately a mystery – Is what we perceive as matter ultimately illusory, just a particular set of standing waves within a much grander lattice of parallel logical possibilities? Why is there a universe at all – and in particular, our universe? Why does our universe have the constraints or principles it seems to manifest?

Quantum mechanics teaches us that everything (matter and space) has a wave-particle nature. Matter and space are locked together in webs we cannot fully penetrate with our old reductionist measuring techniques. Their total interpenetration means they seem to innately know how to compete or cooperate. That is, in classical mode, they are often strong competitors and their particle-like independence comes to the fore, but in quantum-mechanical mode, they seem the almost perfect co-operators and their wavelike, information-within-organisation nature comes to the fore.

Does this universe's almost binary matter-spacetime nature mean the dichotomies we perceive were inevitable, such as those expressed in the asymmetric direct-indirect, fact-value, push-pull, take-give or doing-being at so many levels of the dance in non-living matter, living matter, and human consciousness? Has the universe with its complexity chaotically but rather inevitably arisen from its disorganised and seemingly antinomic core, interaction by interaction, much like a single sample from within a largely homoscedastic population must reflect the demographics of its host and neighbour (with accepted local aberration)? Did all the meaningless boundaries between the bubbling sub-atomic strings of our primordial universe somehow rather inevitably contain not just the information potential but also the impetus-potential and intention-potential for the universe, as we now know it?

At the risk of being accused of a narrow, anthropomorphic hindsight, I reservedly think so, but I also acknowledge that if we ran the universe program again,[1] the outcome may have been very different, and the universe may have never kicked off. However if we ran the program a sufficiently large number of times and put aside all the non-starters, I suspect we would find our universe comfortably within three standard deviations of the bell curve mean, although I have no way of testing such an assertion. It is in the law-of-large-numbers sense that I suggest or suspect a limited inevitability to our universe.

We have already discussed how consciousness is simply a logical but emergent amplification of instincts. Does this mean that the virtuous values of our consciousness have also chaotically but rather inevitably arisen from the borders between the not-quite-complementary or slightly asymmetrical instincts of cooperation and competition? Again, I reservedly think so. It is evident that grassy fields and forests blindly cooperate and compete something like humans and animals do. Can we therefore say that instinctive behaviours in animals and humans already had their blind basis in simple plant life or microbial life and have inevitably but chaotically arisen from such? Again, I think a fully interpenetrating and relativistic cosmic arrangement suggests we can.

Let's go one step further: It is evident that every particle of this galaxy (perhaps including those in dark matter yet to be confirmed) cooperates around its black-hole core to bring about the galaxy's unfolding and macroscopic form or nature (masses orbiting around a black hole). However, it is also evident that every particle competes chaotically in its local environment, just as various galaxies seem to compete with each other for their slice of space in which to dance (and sometimes collide and coagulate). Likewise, we could vaguely speak of the Earth's major convection currents being examples of a kind of non-living cooperation and competition (while acknowledging that our planet's spin and orbit drive these currents). Can we therefore say that the Earth as a whole has a form of temporal proto-instincts? I think the work of phenomenologists such as Merleau-Ponty suggests we can, because proto-intention arises, intertwines, and builds from the very lowest layers of emergence to the very highest.

Did the instincts coded in the DNA of all life-forms already have their blind basis in the non-living selfish replicator – the spacetime system of our galaxy? Further, can we suggest that those instincts have inevitably but chaotically arisen, in a kind of quantum fuzzy logic, from that spacetime system? The study of Emergentism suggests the answer is 'Yes' because it sees through the existential fog many persistent relativistic relationships at various stratifications of anthropomorphically defined signification. Is this universe and all within it (living and non-living) just a kind of extended phenotype of this local spacetime system or quasi-code? Perhaps, but at the same time this spacetime system arose with matter in an interpenetrating and asymmetrical relationship with it. These are difficult questions to answer with strong conviction given our current level of understanding of abiogenesis, but they do seem to touch on something profound.

The basis of self-organisation in living organisms and humans in particular, is at the microscopic and indirect levels (in proteins but also in the information contained in genes and memes, the selfish replicators). Likewise, self-organisation in non-living matter is in the microscopic and indirect levels (in subatomic virtual particles, including whatever wave-particles bind entire galaxies to the cosmic web, or alternatively, in the information contained in the intangible code that is spacetime).

Did matter-space-time emerge from the white hole (or the 'other side' of a black hole) that began the Big Bang like genes emerged from non-living macro molecules, memes emerged from living cells in consciousness and vremes are now emerging from highly self-aware consciousness? Could it be that the original scaffolding that was present before the beginning of spacetime has long disappeared just like the chemical scaffolding that was present at the beginning of DNA has long disappeared? Could it be that these proto-inflatons disappeared just as we have seen the scaffolding of our own human consciousness disappear somewhere in the long line of species that led to *Homo sapiens*? Alternatively, is spacetime's original footprint still there in the physics of universal expansion, the WMAP, and black holes?

Can we view matter emerging from this proposed white hole as spacetime's survival machine, just as our bodies are the survival machines of the genetic code contained in a fertilised human seed? Again, this would suggest a spacetime code analogous to our genetic code drives our universe. Such a basic code would strongly suggest that the Big Bang simply emerged from a prior process, and thus there must be parental or ancestral phenomena that subsume it and gave birth to it.

Did our universe evolve from many prior universes in a multiverse, or is it involved in a more simple continuous cycle? Alternatively, is a mixture of both views the better answer, with the fractal cosmic cycle looking more like the Mandelbrot Set of Figure 3 than the simple daisy chain of universes presented in Figure 5? Everything within the Mandelbrot Set converges on possible solutions, and everything outside it tends towards impossible infinities. Possibility exists within the fuzzy black area alone. Does the Mandelbrot Set look like it could describe a fractal cosmic cycle of relativistic, emergent and interdependent universes? In computer animations of the Mandelbrot set, the ambiguous borders of various levels of magnification seem to behave freely or with vast diversity, even though they are determined by a simple formula. Could our cosmic cycle of serial and/or parallel universes be approximating to something like a Mandelbrot Set, as and when it becomes more synergistic? That is, does growing local order at higher and higher levels bring our world slowly closer to a simple fractal formula that could describe everything in everything?

Perhaps the interesting question to ask is not how our universe was born or how it might die, but how it might replicate or metamorphose before death. Is it an essential activity of black holes to give birth to new universes, as suggested by theoretical physicist Lee Smolin in his book, *The Life of the Cosmos*?

If this universe did somehow evolve or bifurcate from prior universes within a kind of elaborate Julia Set, then it seems plausible that it may have done so through a mixed process of artificial as well as natural selection, just as genes, memes and vremes currently evolve in our world. Is it the destiny of our progeny to participate in this activity of universe replication, giving rise to not just naturally evolving, but also artificially evolving universes? Without wanting to endorse any transcendental arguments, perhaps we should be searching for the 'artificial' originations of our spacetime code, the genetic code, the memetic code and the vremetic code the other way around. Perhaps it is as we advance from 'older' codes to 'newer' ones that we come closer to our universe's subtle beginnings rather than move further away from them. Perhaps our universe's inherited vremes explicitly explain all that has happened since before our Big Bang.

Let's follow one more line of reasoning with respect to what might 'exist' behind the veil of unknowable reality. Much of Western philosophy and religion suggested that the universe began with the original singularity, the unmoveable Mover, or Plato's First Cause. Buddhism suggests that the universe began with everything in mysterious flux and a continual dependent arising. Finally, science suggests that this universe at least, began with a small ball of energy with no spacetime yet evident. Isn't it interesting that each approach has chosen a different focus of how something was arranged at the beginning in terms of spacetime, matter and the transcendental? To organised beings, struggling against disorder, there is no meaning either consciously or blindly without an imagined arrangement of energy. In the past, it has been challenging to imagine a relative beginning of non-meaningful disorder that represents a doable compromise between an impossible nullity and a potential infinity. It is also interesting that in effect, everything (and its potential) is pictured at the beginning of existence (not nothing), but just in very nearly one monolithic form or another. Therefore, no matter what you believe, every meaningful arrangement seems to have arisen or emerged from the original meaningless form of matter-space-time. The idea of a veil across unknowable reality is perhaps best understood as an anthropomorphic veil across an abhorrent meaningless reality and a valueless void.

In sharp contrast, my best explanation of existence would be in terms of a cycle and recurrent bifurcation that simply demonstrates the paradox that nothing is just as difficult to directly implement as is infinity. That is, we may discard the axiomatic veil of unknowable reality and replace it with a cyclical existence that continually manifests the antinomic tension between the conundrum of nullity and the conundrum of infinity. Thus, I replace a uni-dependent axiom with Einstein's fuzzy, interdependent relativity expressed as a philosophical but perhaps paradoxical explanation. Not a great advance for some, but I think a huge leap forward, and one Ockham's razor would prefer.

The Lesson from a Self-organising Universe

We might more modestly claim that a self-organising and emergent universe makes a grasp of what might be behind a veil of unknowable reality problematic. Is it really some kind of fuzzy logic that banishes the absolutism of infinity and nullity (or anything else)? Alternatively, is it really a self-aware, logical intelligence? Can it be a little of both? Should we ultimately have blind faith in atheism or something supernatural? Both options seem invalid from our inductive perspective, or arbitrary if we, like David Deutsch (2011), reject inductivism as a basis of knowledge.[2]

Objectivism's atheism only avoids the snare of blind faith if we begin with its axiom "existence exists". That is, in terms of objectivism, the objective reality that lies behind the veil is evidently nothing because by deductive reasoning and the axiom "existence exists" only an entity as something capable of our perception is able to have a characteristic or attribute or action.[3] Further and by the same logic, the only form of consciousness possible is an incarnate consciousness.

If we don't assume the uni-dependent axiom "existence exists", but we continue to rely on our reasoning and common sense, then this leaves us with the possibility that our conceptualisation of a material cosmological beginning might be defective or incomplete. The claimed objective reality behind the veil perhaps becomes an incomplete concept awaiting closure and agnosticism becomes a logical stance based on an assumed 50% probability best-guess. However if we do this, then existence itself becomes an incomplete concept awaiting closure, partly because non-existence, or a game-like simulated existence, becomes a logical corollary of uncertainty as a life-stance. As we have already noted, the idea of an incomplete or fuzzy cosmic cycle is not without merit. The problem with the agnostic life stance is that it does not fully reject the uni-dependent and axiomatic life stances of atheists or dualists (or anything else), nor arrive at a multi-dependent life stance itself. As such, agnosticism is still axiomatic and not fully embracing of the relativistic explanation.

Finally, if we drop objectivism's axiom of "existence exists" and as Deutsch suggests, we let go of our faulty inductive reasoning (i.e.

we hold the belief that knowable reality does not reveal to us anything of the possible objective reality behind the veil), then a god-like 'Something' also becomes possible. However, its deduction must be in a state of suspension because deduction in this case would rely on an arbitrary reasoning to operate. Such a belief in 'Something', whilst not necessarily bringing existence into uncertainty, would thus subordinate any evidence from existence to blind faith.

In summary, our society seems to leave us with only three axiomatic alternatives: The idea 'existence exists' and atheism, the idea that 'existence is uncertain' and agnosticism or the idea that 'existence is subordinate' and blind faith in 'Something'. That is, if we are to take a classically certain life-stance rather than an emergent one, then it seems we must make a blind assumption, make a blind denial, or make a blind belief!

Personally, I think the lesson of an emergent universe offers us a far more satisfactory answer. It enables us to get to the nub of the issue. It teaches us to reject objectivism's axiomatic certainty and thus discontinuous stance as a basis for atheism. It also teaches us that we may reject agnosticism's axiomatic and uni-dependent uncertainty with respect to an emergent existence and an emergent self. It also encourages us to reject the axiomatic idea of a 'Something' for nothing or an eternally self-organised and intelligent 'Something' (either way, a 'Something' that is effectively a magical free lunch denied in any natural universe).

What the universe teaches us is that reality, not based on any axiomatic assumption, but on a relativistic and emergent existence unable to banish the fractal border between infinity and nullity, is the most easily defended explanation of what we observe and how it operates. Reality cannot be axiomatic, even if a little unstable - only our understandings of reality are axiomatic. This is why we should continue to seek to understand reality by reducing or removing axiomatic plugs. We should not be content with them; we should always question them. It is only as we question axioms such as 'God is not', 'God might be' or 'God is' that we can say we are open to reality's lessons of vast interdependencies against a backdrop of meaningless disorder. An axiomatic life-stance is a closed life-stance inconsistent with the Scientific Method and the intellectual integrity it implies. As we have already noted, an axiomatic life-stance starts with a tightly held and untestable presumption, and then goes about trying to defend or reinforce it. Such a method cannot reliably advance our knowledge of reality and may seriously mislead its adherents.

My intention is to now enhance Figure 25's Model of Wellbeing in a way that finally demonstrates the context of our life-stances. That is, if we enhance the atheistic axiom of existence exists without a god, using an interdependent and non-axiomatic approach, then we could say more informatively that existence disestablishes. That is, it continually challenges, breaks down, or breaks open old borders or old beliefs; it breaks through; it forces the breach; it opens up and exposes its own weaknesses. In the struggle of existence, existence effectively breaks down its own borders.

Likewise, if we enhance the sceptical axiom of existence is uncertain as a life-stance, using an interdependent and non-axiomatic approach, then we could say more informatively that existence eliminates. That is, existence is not against the rejection of its own arrangements in part, as and when required. Existence will reduce itself to gain efficiency. We could say more anthropomorphically that existence is sceptical of itself and even exclusive of part of itself from time to time. It will turn something into nothing, as per the self-annihilation of matter and antimatter, if the drive towards greater energy efficiency demands it.

As advocated through all these pages, we can also say that existence emerges. That is, existence (re-)arranges; in this phase, existence will form new borders and new oases of organisation. Existence will realise its own potential through building new and innovative arrangements, collections, relationships or forms that present a new set of locally effective borders to the rest of the cosmos.

Finally, if we enhance the dualistic axiom of existence is subordinate, using our interdependent and non-axiomatic approach, then we can say that existence is indeed subordinate at certain times and places, but not to a dualistic deity. That is, existence is accommodative, acquiescent, inclusive, or tolerant of its own limitations, imperfections, or fuzziness. To put this concept a little less anthropomorphically, we can say that existence incorporates. After forming a new oasis of order, existence looks for ways to rearrange things within the cell or local setting to promote or strengthen the efficiency of the new order.

The three classic life stances and their alternatives, as well as the fourth characteristic existence emerges, are presented in Figure 33 as an interdependent and dynamic whole.

May I be so bold as to suggest a non-axiomatic fourth life-stance, consisting of a fuzzy cycle of (re-)arranging, incorporating, disestablishing and eliminating, called 'Emergent Monism', which incorporates the philosophy of Emergentism and its Emergent Method? At the core of this more inclusive and tolerant life-stance is not atheism as some kind of protest or revolution against a present and dominant dualism, but a new system of interdependent norms that recognises our context and unique role within Gaia and the unfolding cosmos. It recognises the same cosmic incompleteness that agnosticism does, but sees it in a very positive light rather than a negative one.

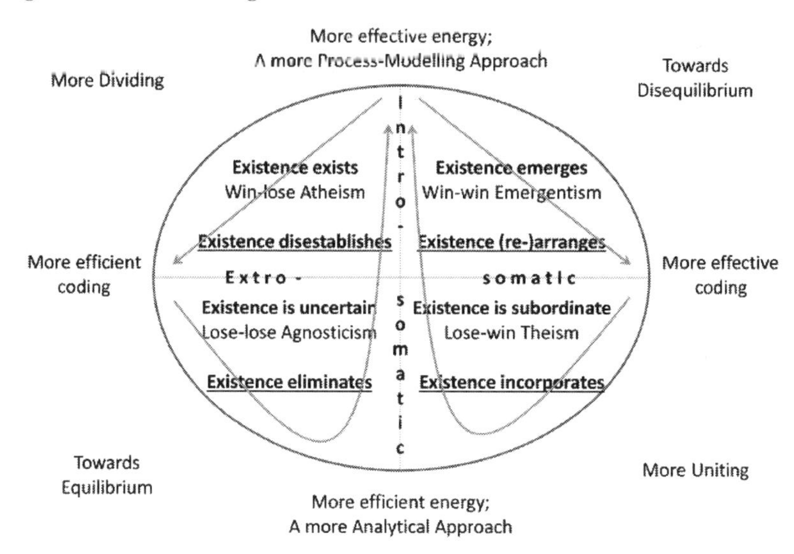

Figure 33: Interpretations of Existence

Figure 33 should be considered in context with all previous phaseal diagrams, but perhaps most particularly in accord with Figure 1: The Processual and Analytical Approaches to Epistemology, Figure 12: Matter/Space Relativity, Life Struggle & Conflict Management and Figure 25: A Model of Wellbeing - the Mind-body as an Energy Processor. In summary, there is a time and place for existential confidence, uncertainty or subordination, but never as a uni-dependent atheist, agnostic or dualist. We need to leave these old, classical paradigms behind and move on to fully embrace an interdependent and relativistic reality that is amenable to the full reality of our human imagination.

If there is a non-supernatural 'Something' rather than no god-like figure, then a chaotically self-organising universe seems to distance us from 'It'. The universe seems to have been put into a kind of deist set-and-forget mode, which doesn't seem to require any intervening management from behind the veil. It would thus seem that the universe is not human-centric but loosely self-organised as a sort of Pan-Gaian whole. This hypothesised 'Something' seems to leave the universe, and an intelligent species' inevitable but chaotic occurrence within it, to establish its own way in complete laissez-faire liberty.

If we think a little about this natural 'Something' and what we know about this side of the veil, is it possible that our universe is the result, at least partly, of a certain chain of species that was once at our level of development? Did this chain move much further towards the "inexorable coagulation of all life",[4] to form an almost omniscient Super-species? As suggested earlier, is it also possible that this chain evolved, with its superior vremes, to play a part in forming our particular universe with its brilliant spacetime attributes and self-organising interactions?

We have discussed a system or code within spacetime that digitises, quantises and thus pixelates reality much like a computer simulation program. Our cosmos seems incomplete or imperfect something like a computer simulation. This leads to the real possibility that our existence is just a computer simulation within a larger reality.[5] The notion that a 'Something' or a Super-species is sitting at a grand computer terminal and is intimately involved with selecting various quantum-mechanical and conscious states that only seem to be stochastic from our viewpoint has perhaps been a little unfairly put aside. The reason it was put aside was that it was assumed this would likely produce a grotesque universe with no emergent free will and no chance to trap the future at all. In such a scenario, we would all be pointless parts of an almost clockwork toy, perhaps subordinate to Enid Blyton's Noddy in Toyland. We would all have a destiny that only seems to freely evolve. The concept of such an omnipotent 'Something' would seem to belittle that 'Something' as somewhat of an adult control freak or a child at play and thus detached from the vremes with which we seem to quite naturally self-actualise. This would seem to represent a rather cynical and unexplainable contradiction between this side and the other side of the veil across unknowable reality.

On the other hand, to be a little fairer to our Grand Computer Programmers, perhaps our simulation is their attempt to understand their reality a little better, which I guess could be forgivable and explainable, especially if they did allow us to set some future traps via our quasi-independent introspections.

An interesting observation here suggests that we could place the virtue of this 'Something' on a sliding scale from maximum virtue (zero interference with the universe) to zero virtue (a fully deterministic universe with no genuine future traps allowed). What do you think? If this 'Something' did partially interfere with the universe, then perhaps like an ancient Greek tragedy, it would seem that our personal freedoms and individual values applied to our chosen future traps would be part of the prescribed lesson for our less-than-fully-virtuous Agent. If 'Something' must have a partially forced control of us, then our role is to teach 'Something' of Its folly through our personal and contrasting freedom and self-actualisation.

Conversely, if the 'Something' did have a hand in the natural-and-artificial selection of this universe's machinations, perhaps a need for interference (tweaking) arises precisely because this universe's spacetime constraints are not as yet optimised as an expression of virtue (which we might again forgive). Any tweaking in real time would assume a Super-species from a prior universe can have an impact on the probabilistic constraints of the next universe, and somehow transform itself to incorporate our universe's dimensions within its own. Perhaps this provides support for a kind of naturalistic, emergent and monistic panpsychism - although personally I see this as an extremely unlikely complication.

Without an ability to run a live simulation, perhaps the only impact a Super-species or 'Something' can have on the birth of the next universe, before the death of its own universe and itself, is to help artificially select the next universe's initial constraints (i.e. its conception). This hypothesis would seem more in keeping with our observations of the inevitable cycles of life and death on this side of the veil. It would also have something in common with a programmer who cannot put his/her soul into a program except by way of soulless coding and much trial-and-error. Nevertheless, this soulless spacetime coding approach has its benefits. Consciousness and subjectivity would be able to arise anew amongst all possible futures. In addition, the newly self-aware could know that they stand as proud custodians, or as the new moral agents, of the newly successful universe.

Nevertheless, if the initial self-restraining self-organisation was as simple a process as zero becoming a 1 and -1 and vice versa in some kind of quantum fluctuation or some other logical tension between vast infinity and nullity, then no grand intelligence is required for such simple relativistic events at all. The question, 'Why is there anything at all?' or the converse question, just as vexing, 'Why isn't there nothing at all?' is perhaps far better explained by a necessary and minimalist relativity as by a 'Something for nothing'. Theoretical physicists such as Garrett Lisi, and Occam's razor (the law of parsimony, economy, or succinctness), would suggest a nullity constraining an infinity and vice versa is a far more defensible hypothesis than is a 'Something for nothing'. Information always begins with some kind of self-constraining boundary between infinity and nullity. As our imperfect digital computers and object-oriented programming perfectly demonstrate, anything and everything else that exists may have simply emerged from something and nothing, or 0 and 1, including our values (our introspective measures of difference/information).

All of these hypotheses suggest that perhaps the (blind) reason for the universe's existence is simply for our self-actualisation (or the self-actualisation of any other species), as the epitome of the universe's own self-organisation and perhaps to the self-actualising glory of its non-directly-interfering 'Something'. This 'Something' seems to value and epitomise a self-constrained liberty, and, we might fairly surmise, all the other personally objective values we discover in our antinomic dance of elaborate consciousness-plus-instincts.

So this self-organising and emergent universe, whether the result of intelligent beings, and whether or not those beings forcibly control any part of our daily lives, or the inevitable result of some kind of chaotic/fractal/fuzzy quantum mechanics, or both, seems to scream (or is it endlessly whisper?) to its inhabitants to self-organise and arise, or self-actualise. It is as simple as that. The universe, perhaps like a proud and confident parent speaking to its adult child (or perhaps like a child seeking lessons from its toy), is not telling us what to do, but it is enabling us, and almost expecting us, to do whatever it is we rationally and courageously choose to do with our granted degrees of freedom.

It seems to make perfect sense that a supreme Natural Intelligence that is a perfect embodiment of values and life would arrange nature, via a seemingly separated naturalism, to enable one or more species to become fully self-actualising. It would not seek any form of self-aggrandizement or desire for worship outside of the worship of the values each of us embodies. In fully venerating our own values, our own self-constrained freedom and ourselves, individually and collectively, we fully venerate that 'Something'.

What could be a more natural system of worship than that? What could such an intelligent 'Something' want more? If you were that 'Something', would you be interested in punishing humans who failed to self-actualise, or would you just be interested in setting up the conditions that enable that self-actualisation, no matter what century, culture or race the humans or neo-humans were born into?

The universe's lesson to us is simply this: The focus must be your life, your self-actualisation, your sustainability, your societal flourishing and your legacy – because existence emerges in a relativistic setting. Such a simple lesson will represent a huge paradigm shift in how we consider our planet, each other, and our interacting purposes. It will take the focus off all unnaturally axiomatic forms of life-stance so that we can be free to explore and enjoy the natural life-stance phases (struggle, elimination, emergence, and incorporation) as they unfold.

Such a paradigm shift in thinking will bring about unavoidably disruptive changes in our institutions that are likely to save our species from self-destruction.

Public Parts

Public Parts, How Sharing in the Digital Age Improves the Way we Work and Live by Jeff Jarvis (2011) provides an excellent basis for our discussion of the Emergent Method and how we might use it to change the rules of the game and hasten our inexorable coagulation as a species. I highly recommend your purchase and reading of the book in its entirety.

As Huntington (1996) seems to intimate, changing our cultural and largely subconscious mores and taboos is how we can facilitate the changes needed to secure our sustainable development as a species. The first way we facilitate this needed change is to make our values, mores, and taboos consciously explicit, or perhaps in Jarvis' terms, to increase our publicness.

Publicness and privacy, the subjects of Jarvis' book, are part of the border problem all living organisms face, first explored in Part 1. Even more deeply, the nested and interconnected or web-like, cellular structure of all quanta in Quantum Field Theory suggests an inside/outside arrangement, as partly depicted by the quantum cells of Figure 6, Figure 7 and Figure 12. A personal balance and border between publicness and privacy is the answer to entropy or irrelevance. A dance between the two would seem to define life itself as proposed by Bourgine and Stewart (2004), mentioned in Part 1. Full personal publicness (the far right of Figure 25) or full privacy (the far left of Figure 25) means a loss of extrosomatic self, just like all-being with no doing or all-doing without any being (also discussed in Part 1) would mean a loss of introsomatic self.

A dance is required for our wellbeing. Nevertheless, some dance steps are better than are others. This is why a discussion of privacy and publicness is relevant here.

Definitions

Jarvis says, we not only use media, but nowadays we wield media "to find allies, inspire action, organise movements, change priorities, influence policy, and raise money. We use our publicness to rally others to our agenda... to find others in our boat."[6] That is, we primarily use our publicness for psychosocial claiming – to modify narratives and find, establish, or defend Core-self. Conversely, privacy is the claim of individuals to information that will not be communicated across borders to others, to protect our individuality and personal dignity.

Jarvis reports here that privacy is seen as the right to self.[7] He says privacy is a key factor in a society's definition of itself: of the relationship of the person (triune self) to the community (and quadune self), of the customer to the company, of the limits of ownership, of the rights of the individual. He adds that we compete with our communities for an ownership of self. We weigh our values against those of the social group.[8] Studies of psychological priming suggest this struggle is perhaps primarily conducted in the subconsciousness of a culture's members.[9]

In the terms of Emergentism, the fuzzy and uncertain border between privacy and publicness, competition and cooperation, or the introsomatic and extrosomatic, is essential to the inexorable coagulation of life. Finally, Jarvis concludes, "I've come instead to believe that privacy is an ethic. That is where I find my definition … Privacy is an ethic of knowing. Publicness is an ethic of sharing."[10] In these terms, private/triune self and public/quadune self must be ethical beings. I'd like to suggest further that according to Figure 19, privacy is an ethic of factual competition, whereas publicness is an ethic of valuable cooperation.

On the downside, Jarvis warns privacy can conservatively hinder change in a society's set of norms. That is, secrecy is conservative with respect to change, while publicness is progressive.[11] I think the recent controversy around acts of paedophilia committed behind the veil of Catholic Church administration illustrates the point well. Our norms vested in named institutions, whether philosophical or political, need to be shifted towards the best explanation, as David Deutsch might couch it,[12] or towards the best solution in terms of added value, as Edward de Bono[13] might phrase it. That is, we should venerate self-actualising values and their virtues expressed in named institutions, rather than the institutions themselves. Transparent public institutions would have no problem with this shift in the basis of our society's norms. I suspect that the venerating of values over institutions themselves is exactly how many institutions have come and gone, or otherwise evolved, over the prior millennia.

Jarvis claims that those things, such as scientific knowledge, made public and granted common or shared title or status are necessary for the smooth running of an open society.[14] I guess there would be many who might argue the point on things better kept private, but societies evolve over the longer term – they can often learn from their behavioural mistakes that hinder a healthy adaptation to their environments. In Australia, we are still working through what to do about Edward Snowden's revelations of our phone tapping of the president of Indonesia, his wife, more than a million other Indonesians, and the government of East Timor. In this frame of mind, Jarvis further suggests that countries are safer when they are less secretive and more open or public.[15] I tend to agree in the sense that the more public a well-educated and well-informed society is, the more constrained its behaviour by its evolving, naturalistic and incorporating relationships or rules of the game.

Of publicness, Jarvis also says that we are now more free to join the publics of our choice, rather than be forced to join the preferred publics of our civilisation, culture, or other groups or given labels. We are now able to join publics based on ideas, which represents a radical and disruptive departure from the past. Publicness through such channels as Facebook or Twitter literally challenges and threatens old norms and power bases in our societies.[16] That is, we can now form new cultures and a new social fabric based not on our local tribe, heritage or colour of skin, but on shared and traded ideas, interests, needs and vremes. New publicness brings a redistribution of more refined borders than the old ones, and thus a wider distribution and greater taming of societal power.

Implications

Jarvis reports that Zuckerberg's mission through Facebook is to make the world more open and connected. He says that Zuckerberg started Facebook to make the world a better place through greater publicness, transparency, and integrity. Jarvis reports that Zuckerberg thus suggests Facebook is an enabler of human nature and its evolution for the better, and that Facebook is a sociology company rather than a technology company.[17]

I find such moral leadership encouraging. To use our draft model of self, we could say Facebook is not just a company emanating from technological Warrior-self (Damasio's Proto-self), but also from social Philosophical-self (Damasio's Autobiographical-self). What sits between these two is the cultural Lawmaker-self (or Damasio's Core-self) – the self from which new mores and taboos originate, that is, the self that when it changes can radically change the way it forms and participates in new publics.

Jarvis also talks about the idea of new protean publics changing the shape of society. We can easily see his point in terms of the effect of new internet relationships on the established media industries (news, music, film, telecommunications, etc.) He adds that by changing publics, we alter the very concept of what the public represents.[18] In changing publics through new internet relationships, we are also diversifying society's distributions of power. In so doing, as the ecological theory goes, we have a chance to make it more robust. Jarvis observes that this disruption of power bases has happened many times in our past, perhaps the most notable during the Renaissance, a time when the institutions of church and monarchy had to give way to the rise in power of the secular state and its citizens. Likewise today, we're experiencing a renaissance in cyberspace.[19] Jarvis says,

> *"In Europe, Gutenberg empowered Martin Luther to smash society apart into atoms, until those elements re-formed into new societies defined by new religions and shifting political boundaries. With the Industrial Revolution … the atoms flew apart again and re-formed once more, now in cities, trades, economies, and nations. We atomise. We re-form into new molecules…"*[20]

I am not sure if I could have stated the case for the Emergent Method more clearly than this, or explained Figure 33 more clearly. The inexorable evolutionary shuffling, always moving through more local energy effectiveness and efficiency, continues.

Jarvis contends that publicness leads a society to greater freedom through its ability to reorder it. For instance, he says that new and open technologies are leading us to question deep assumptions of the past in terms of our roles, rights, and responsibilities.[21] It seems to me that publicness is a progression to greater freedom because it breaks down old borders and enables finer, more widely connected borders based on a higher consciousness to take their place.

Jarvis quotes Matt Ridley, who argued that our great leap forwards as a species some 45 thousand years ago came from our greater public interactions through trade. Ridley suggests the trade trick lead to a collective intelligence and the explosive development of all aspects of culture. Ridley sees a direct relationship between the level of trade and the level of cultural change, which suggests in terms Jarvis would use, that publicness and interconnection promotes our progress as a species.[22]

Isn't that just what the study of Emergentism suggests, that innovation arises fundamentally out of relationships, and the trading of values across boundaries advances our universe and its civilisations, either blindly or consciously through vremetic traders? We are not independent organisms; we are organised arrangements of interdependent subatomic particles of matter and space. Jarvis goes on to suggest that trade brought about the cities, civilisations, and trade routes in between. He quotes Matt Ridley's observation that trade is to cultural development what sex is to biological development[23] and notes that trade, whether of goods or ideas, brings about innovative, growing and lasting change.[24] As an aside, Jarvis refers to Twitter as a *serendipity machine.*[25] It is an interesting concept – serendipity (and an ability to bootstrap ourselves) is perhaps the result of massively parallel relationships that can help us escape Gödel limitations even in our conscious communications.

Jarvis concludes that the thing we need to change in our current state of affairs is not so much our behaviours, or our technologies, but something deeper – our norms.[26] I think we all get the point. Jarvis seems to suggest that questioning our civilisation's norms is an important tool of disruption. This is similar to the point Huntington (1996) made. By changing laws, we encourage the clash of civilisations, but by questioning our norms, we can change our publics or societal powerbases and move towards Huntington's Civilisation with a capital 'C'. I think this can only happen if our norms are consciously linked to a relativistic and interpenetrating model of reality.

Jarvis tells us that we can no longer be sure our governments of whatever description will safeguard the means of our necessary self-disruption, or any companies to operate against their own self-interest.[27] So what can we do? The Emergent Method, like Jarvis, suggests we can only change our largely subconscious Core-selves by making explicit and reviewing our values, our norms, our mores, and our taboos. This was the method used by Mahatma Ghandi, Dr Martin Luther King Jr., and Nelson Mandela so effectively in their struggles against civilizational injustice. The result in each case was a higher consciousness and stronger social fabric in all contemporaneous civilisations.

In this vein, Jarvis suggests that we need principles rather than laws to protect the internet and the publicness it fosters.[28] That is, we need an awakening of higher consciousness rather than follow the letter of any given law. He reports of John Barlow, founder of the Electronic Frontier Foundation, that his organisation's stakeholders are forming a kind of social contract with each other,[29] which seems reminiscent of the deeper naturalistic contract mentioned in the prior chapter and pages of this book. He reports of German Justice Minister Leutheusser-Schnarrenberger, that the digital world needs to recognise quasi-universal digital values more than laws, and the online community needs to deepen the level of discussion on this topic. Jarvis entirely agrees.[30]

I would add that this discussion is not just an Internet

discussion - it is an extension of the millennia-old discussion of our emerging global tradition, as also reported by Pete Seeger in Part 1. We need Emergentism's framework of values more than we need a set of temporal laws because we can't govern narrow-minded selfish behaviour from the outside. However, we can reveal to all players a better way through life's dance of vremes, or the outworking of a global social contract in which we all see ourselves as both enmeshed custodians and targeted beneficiaries. This is how we can facilitate others.

The Radically Public Company

Jarvis says,

> *"For companies, transparency can spark a virtuous cycle. Publicness demonstrates respect, which earns trust, which creates opportunities for collaboration, which brings efficiency, reduces risk, increases value, and enhances brands. Publicness is good for business."*[31]

Speaking of the founders of YouTube, LinkedIn and Twitter, Jarvis says that they saw themselves more as members of ecosystems or deep ecologies than as corporations. That is, they were cognizant of the reality that they were more like nodes in a network of valuable relationships than mere entities or objects.[32] This is exactly in keeping with an emergent reality and the object-oriented design of organisations proposed in Pillar 2.

Jarvis then says,

> *"I hear companies fret that revealing and discussing things openly will tip off competitor… That is a problem if you think that your value resides only in your product and that secrecy itself is worth protecting. If, instead, your value lies in the quality of your relationships, openness brings benefits."*[33]

This describes the transformation in thinking that needs to sweep across not just company boardrooms but each of us as individuals if we are to create more tolerance and cooperation amongst the civilisations Huntington spoke of in the prior chapter.

Where does our value and power reside? Is it in our secrets and privacy, or is it in the coagulating strength of our relationships? In a relativistic fashion, it must be both, but the pendulum needs to swing back to publicness. Along the same lines, Jarvis tells us the anthropologist Jack Goody styles the history of our species as more like a struggle over the methods and forms of communications than over the means of production.[34] That is, opposed to what Marx might suggest, human (and corporate) history is one of relativistic communication or relationship more than production and its ownership.

Jarvis suggest the following challenge to companies: They must calculate the true value of relationships and transparency over secrets and then choose the former. That is, the company, once it has properly accounted for the extra value, will become radically collaborative in its approach, opening up even such areas as design and strategy to its stakeholders. Such a company will recognise its unassailable place in its unique here-and-now and in its ecology. It will therefore not want to dominate all it surveys, but will rather want to see all the members of its deep ecology advance in diversity and wellbeing. Asset ownership will be of secondary concern to stakeholder enablement.[35] This idea of a productive and indispensable node in a network, something like a mountain embedded within a mountain range, is how we should see ourselves as well.

Jarvis then turns his attention to the company's constitution and the development of an organisational bill of rights or credo, perhaps something like that of Johnson and Johnson, to be shared with its participating stakeholders.[36] The focus would be on the benefits of the stakeholders in a more natural cycle of relationships with the organisation (that would accentuate the win-win but not be limited by it). In such a scenario, all stakeholders would become co-emergent with the company, and thus most likely protect it from catastrophic collapse.[37]

That is, the actions of the radically public company encourage the transformation of the stakeholders themselves. The border between the company and the external stakeholder becomes fuzzy, and shuffled. Through this process, all entities can become more robust in their given environments. The object-oriented approach allows for the efficient adaptation of the company and its shuffling, and sometimes mutating, objects or nodes (stakeholders).

Mervyn King, famous for the King III Report on corporate governance and the concept of Integrated Reporting (that is, the integration of financial, social and environmental information in a company's annual report), reminds us of the evolution of the shareholder. The shareholder a few hundred years ago was likely a member of the upper class. He was likely an owner for the life of the firm and its venture, such as the transport of goods or slaves across the seas. His focus was almost exclusively on the company and its value through growth and profit.

On the other hand, today's shareholder, at least here in Australia, is likely a mum-and-dad investor, either directly or, indirectly through a superannuation fund. This shareholder may have participated in the initial public offering, but much more likely participated in the liquid secondary market that doesn't directly provide equity capital to the firm. That is, this shareholder, besides enjoying dividends, is a few more steps removed from the firm than the one of 200/300 years ago. He or she also seeks to benefit from capital gains on sale. That is, the interests of this shareholder are far more diverse. He or she expects the firm's behaviour to align with those wider interests. This is a great step forward in the evolution of the firm.

For instance, the modern shareholder expects the firm to be a responsible member of society and have proper regard to environmental considerations while achieving its profits. It will not be good enough for the firm to earn its profits in a social and environmental vacuum, because the well-informed shareholder knows that unmeasured costs represent risks that will reduce future prospects for self, family or society-at-large. The firm also knows that if it fails to meet the ongoing expectations of the shareholder, the next capital raising will be more expensive.

In summary, today's shareholder has evolved in many ways that force the firm to be more carefully integrated with its environment. Today's shareholder encourages the greater coagulation of Gaia's components through the creation of finer and more open borders between the firm and its environment. This change should have the effect of diversifying the power of the firm and its stakeholders, and thus changing the bases of power in our society, making it more robust or sustainable than 200/300 years ago.

A question might follow at this point. Can we improve the political process likewise? Jarvis is sceptical, but suggests that perhaps we can self-organise to develop our own shadow institutions.[38] I will discuss this possibility a little more, later.

Finally, Jarvis suggests that the CEO of a radically public company will be more than a leader of a company; he or she would be more like the lynchpin in a healthy community or movement.[39] However, we cannot leave this role to our CEOs alone. All readers sharing this conversation need to become leaders of social diversification and synergy. More pointedly, we all need to become leaders of social-fabric change that sees any form of violence or coercion against a victim as reprehensible.

What Makes a Leader?

If we are to benefit from the lessons our universe seems to be trying to teach us (existence struggles, eliminates, emerges and incorporates), then we need to go through a lot of change quickly. Our societies need to change if they are to survive the problems our wayward consciousness-plus-instincts has created. We need to learn a lot and teach a lot. The learning and teaching has to be organic; this is another lesson a non-interfering universe seems to teach us. It can't be a merely cerebral or merely emotional exercise. The behavioural change has to be from the grass roots up, spreading through all levels of our quadune natures and nature's relational lattice or web around that. The program can't just be another governmental or collectivist top-down exercise that misses vital links in the local and systemic web. We need to grow up and we need to help each other do the same. As part of this process, we are going to need teachers and leaders, but not those of the autocratic style. We need a society of enlightened individual contributors who have a strategic understanding of the transitional 'endgame' ahead and what lies beyond.

As we have already seen, to be any kind of successful leader, a person has to have a strong set of values, a conviction, a cause and a commitment to that cause. Most strong leaders have a strong sense of destiny, and the determination to work hard for many years in order to see that destiny fulfilled in their lives. They often have a sense of ruthless urgency in nearly all that they do. We stereotypically expected successful leaders to be charismatic or passionate orators as well. However, this is more a feature of a certain style of leadership rather than an essential attribute of all successful leaders (similarly, not all highly successful lawyers are barristers).

Nevertheless, leadership does seem to always involve a management of meaning in some way, even if it is not verbally. A strong leader provides a unifying sense of logic, conscience and purpose that unites and encourages or inspires commitment to a certain socially-uniting course of action; hence, the strong link between leadership, organisational culture and that culture's uptake by members. In essence, intelligent leadership gently guides its followers to subconscious behavioural change that brings about personal moral congruence, even if temporal. A wise leader will also take responsibility for his or her actions, according to a certain self-restraint or code of ethics that governs those actions.

In this quick summary paragraph of what we all would probably agree contributes to good leadership, notice again the interplay between the subjective and more objective side of human nature. A good leader has well rounded IQ, EQ, CQ and SQ. This is the basis of social linkages the leader is able to foster, and it is part of the answer we might tender to any questions regarding good leadership and how to develop it. Strong leadership is about contributing to the self-enlightenment of all that follow it.

Alistair Mant's book, *Intelligent Leadership*, gives some other excellent measures and examples of leadership. The first concept he introduces is that of indirect, "ternary" leadership which is governed by the objectives of the group and its stakeholders, as opposed to direct, "binary" leadership, which is governed by the interpersonal influence of the "win/lose" relationship.[40]

Not surprisingly, Mant suggests that there is far too much of the latter kind of leadership in organisations and far too little of the more intelligently connected, former variety. Whilst not rejecting the role of binary leadership altogether, Mant also suggests ways of encouraging win/win ternary leadership.

Mant also saw successful, intelligent leadership as incorporating the following attributes:[41]

1. *Authority* – [through the acknowledged trust of others]
2. Purpose
3. Judgment – the 'sine qua non' of effective leadership
4. Systems Thinking – the ability to focus on the particular whilst holding the (perhaps complex) context in mind
5. *Sanity* – [the absence of debilitating psychological damage]
6. *Broad-band intelligences* [Mant quotes Gardner's 1993 work on the nature of intelligence]
7. The virtuous circle – encouraging others to develop their leadership in the same way

Further, Mant also quotes Jaques (1989) and his *Stratified Systems Theory*. His figure 13[42] is reproduced here in Table 7 (column headings added).

Level	Description	Comment	Years
VII	Global corporate prescience	Sustaining long-term viability; **defining values, moulding contexts**	(25-50)
VI	(Group) corporate citizenship	Reading international **contexts** to support/alert level V strategic business units	(15-20)
V	Strategic Intent	Overview of organisational purpose **in context**	(-10)
IV	Strategic Development	**Inventing/modelling new futures**; positioning the organisation	(-5)
III	Good practice	Constructing, **connecting** and fine-tuning systems	(-2/3)
II	Service	Supporting/serving level I and customers/clients	(-1)
I	Quality	Hands-on skill	

Table 7: Seven Levels of work authority, complexity and talent

Mant explains Table 7 in the following fashion:

> *"The figures on the right refer to years – the 'time span of discretion' at the level of complexity. A serious error of judgment (in an organisational worker-come-leader) at (say) Level IV might take as long as five years to show through."*

The more interesting note is that the values of leaders become more encompassing and interconnecting as maturity and experience grows, just as we saw shadow values give way to golden values with time and experience in Pillar 1, or we see Maslow's needs develop in a hierarchy with time and opportunity.

> *"Jaques argues that no organisation … needs more than about seven hierarchical levels, because these strata reflect the natural increments of the human capacity to exercise discretion … [however] 'delayering' beyond this minimal backbone of authority will inevitably weaken the organisation"*[43]

An important observation to be made here is the leadership emerges with exactly the same process as everything else. That is, emergence occurs in a progression of phases that lay down new stratifications one layer at a time, something like stratifications may be observed in rocks over geological timespans. Each stratum has its own characteristics and significations for the layers before and after. Intelligent leadership is aware of the essential dimensions, significations and characteristics that each layer has added to his or her journey and guides others in the use of this information.

Leadership Style

Transformational Leadership

One important style of leadership, first identified by Burns, is "transformational leadership", which is characterised by the leader that actively intervenes in the status quo by challenging current paradigms and encouraging others to do so as well. Such a leader also inspires a shared vision, and facilitates others to action, through personal example and encouragement.[44] To some extent, Burns' transformational leader would equate to Mant's more autocratic win/lose binary leader.

The transformational leader is needed whenever businesses or societies go through fast-paced or major change, social or otherwise. This is the kind of leader a country needs when it goes to war, perhaps even with its own selfish replicators. The problem with change is that those who fear it perceive there is nothing to gain from it, think they will have to work harder because of it, or are simply content with the status quo, and always resist it. Thus, it often takes strong and inspiring leadership to smoothly implement change.

Will the next great transformational leader be an American President, Chinese Premier or Indian Prime Minister? Alternatively, will it be you and I, not autocratic in style but organised organically, united by our joint and deeply held convictions? We need to become inspiring champions of nature's seemingly antinomic philosophy, Emergentism. This appeal reminds me of another in the movie *Braveheart* where Mel Gibson as leader of his band of successful Scottish rebels is talking to Robert the Bruce, rightful heir to the Scottish throne but under the yoke of the English throne. Gibson says, "Your title gives you claim to the throne of our country, but men don't follow titles, they follow courage" … "and if you would just lead them to freedom, they'd follow you. And so would I."[45] That is, a great leader leads his or her people to the vremes within, rather than relying on temporarily sanctioned rights or claims.

In a similar way, in the movie *Gladiator* (2000 DreamWorks), Caesar Marcus Aurelius, knowing his closest general, Maximus, to be a moral leader and well-respected of his troops, asks him to become the Protector of Rome. He says he will give power to Maximus for just two reasons, to bring power back to the people and to end Rome's corruption. That is, the great focus of his leadership was to be its virtue, and its mission to help the people realise their own moral potential. Shared virtues such as the four chief ones favoured by Marcus – fortitude, justice, wisdom and temperance - make a society or civilisation great. Shared values help a people find wellbeing and flourish.

One great task of the next transformational leaders will be to facilitate the paradigm shift in thinking with respect to the most important choice we as humans with elaborate consciousness can make in life. We as volitional beings need to learn that our most basic choice in life is not our life-stance, if this is as classical atheist, agnostic, or believer. Our most basic use of our ability to think and choose, which defines us as humans, is to choose to live-and-be here-and-now and self-actualise. The method of self-actualisation, whether as atheist, agnostic, believer, or emergent monist is subordinate, although clearly I would prefer we all move away from uni-dependent alternatives to interdependent and relativistic ones, as per Figure 33, that recognise the potential of human imagination to lift us out of ignorance.

Thus, the core task of the new transformational leaders will simply be to help the rest of us reframe the basic question away from the primary choice of life-stance, which divides communities, to the primary choice of self-actualising in life-and-being here-and-now, which unites communities that share in the celebration of life. Further, our diversity in unsuperstitious belief and culture is something we can celebrate and respect, as a source of our Civilisation's (with a big 'C') robustness and our own ingenuity.

Beyond this, the tasks of our next transformational leaders are many. We will need to think about how the application of Emergentism impacts all our institutions including our families, our societies, our systems of government, our organisations, our marketplaces, our networks, our modes of learning, working, trading, etc. We will then need to discuss together and implement the conceptual, systemic, and sustainable changes these competing yet shared thoughts imply.

Transactional Leadership

According to Burns, the complementary kind of leader, the "transactional leader", is characterised more by doing those things required to consolidate or maintain rule – to keep operations running smoothly in stable circumstances. This is typically achieved by setting clear goals and standards, providing the resources and systems needed for their accomplishment and rewarding good performance.[46] He or she is the kind of leader a country needs in times of prosperity and peace. To some extent, Burns' transactional leader would equate to Mant's win/win ternary leader.

Sound transactional leadership in stable circumstances is the epitome of good governance, which may be defined as the strong and wise exercising of economic management, philosophical awareness, legal authority and political power. The adjectives have been inserted into the definition of governance here to serve as a link to concepts of basic societal power bases we discussed in earlier chapters. The transactional leader thus has an ability to take the four rivers of virtues and unify them as the one means of adding societal value.

Transactional leaders carefully consolidate the gains made by prior transformational leadership. They do this by setting up systems of feedback to measure, monitor and assess behaviour and results. They administer systems of rewards or incentives, as well as appropriate disincentives. They also find ways to manage the commons and set up mechanisms to resolve conflicts (i.e. avoid the tragedy of the commons).

The great transactional leaders of history were the great administrators and inclusive unifiers rather than elitist dividers, such as Darius the Great of the ancient Persian Empire, who, after establishing himself as a military commander, was believed to be the first man in history to set up a systematic postal system that operated across an entire empire.

He also standardised his empire's monetary system, its system of weights and measures, its commercial language and its systems of roads and canals – more than two thousand years before Napoleon did the same for France. He was also said to be magnanimous and far-sighted, tolerating self-government in the provinces and actually encouraging various religions to flourish other than the one of his homeland. This man clearly understood that the method of self-actualisation is subordinate to self-actualising itself. He also understood that support for exchanges of values in all avenues of human endeavour would lead to the release and production of true wealth and wellbeing in his Civilisation.

If King David was the classic biblical example of transformational leadership, then King Solomon was the classical biblical example of transactional leadership. When we think of

King David, we think of his mighty successes in battle and the camaraderie he was able to evoke amongst his troops, the tribes and the houses of Israel – a distinguishing feature of transformational leaders.

On the other hand, when we think of King Solomon, we think of the mighty systems and infrastructure of commerce he set up, as well as his mighty wisdom. He gained such wisdom from his careful contemplation of the universe, man's plight and the nature of virtue. He had a deep yearning to know wisdom and virtue and serve it in his very best capacity. Such characteristics are distinguishing features of great transactional leaders. Few other people in history, besides King Solomon, have achieved such an elevated position of wisdom in the Western world's eyes. To me at least, King Solomon was a champion of Emergentism.

Your Business Plan

Maslow identifies many common traits amongst people who have reached self-actualization.

A few of these are:

- They embrace reality and facts rather than denying them;
- They are spontaneous;
- They are interested in solving problems;
- They are accepting of themselves and others and lack prejudice.[47]

Having fully explored your own values, your own credo, your own purposes and the challenges you have freely accepted, it is time to formulate your strategic business plan or life plan. Like Howard Roark in Ayn Rand's *Fountainhead*, it is time to convert your deepest and most lively values into your vision for the future, but as a celebration of your life-and-being here-and-now. What will your naturalistic contract be? What will you set out to achieve with your values, your success, and your wealth creation? How will you uniquely serve yourself and humanity at the same time, and through that serving fully express and enjoy the outcomes of your cherished values?

This very concrete plan could possibly capture the following dimensions:

- Values Statement; Purpose Statement; Mission Statement. What is your passion and what is the nature of your compassion? What are the precious values inside of you and what do you want to do with them? More specifically, what will you do with them? How will this statement assist you and your family in what Matt Ridley called the inexorable coagulation of life? How will it assist in what Richard Dawkins referred to as the conscious evolution of genes and memes or I might call vremes? How will your mission statement promote the ongoing development of our virtuous global tradition?

- Vision Statement. How will you let others quickly know about you, your character, your values and your purposes?
- Catch Phrase. Remember that subconsciousness, that part of us actually in charge of our perceptions, decisions and actions, can only process value-laden significations in small chunks.
- A 'Purpose' overview and plan (short term, medium term and long term) - ready to update anytime you get a new idea. What do you want to work in and towards and how? Express this in terms of specific (perhaps direct marketing) strategies, tactics, analyses and syntheses.
- Market size; Market trends; Customer or Client Behaviour Analysis. That is, can your purpose help others to self-actualise as well as yourself? If yes, then, in a lot of detail, who are these others and how many are they? How do you empathise with them? How will you provide for them what they want?
- Target Market and positioning. Be specific. How does your unique stance fit alongside other stances held? How can you justify your stance, product, or service? What are the unique problems or issues of your target market? In solving your own problem of self-actualisation, how do you help solve the unique problems of your target market as well?
- Shop Window and/or Website Analysis. What do you want to show to those passing your shop window or landing page that will make them come inside for a look around, and keep those irrelevant to your offer away from your shop or database so it remains uncluttered?
- Back Office Functions. In terms of the room at the back of your shop (or virtual back office), how will you, and do you,

build and maintain trusting relationships with the people who come to visit and with whom you may wish to trade values and create wealth? How will you build and maintain your list of contacts and your knowledge database? How do your visitors find you (from what sources – are they high quality sources that provide the appropriate introductions to your values-in-action)? Do you understand why your visitors take the time to check you out? The reason why is very important – and should be directly and clearly linked to your values-in-action. Finally, you must provide your fellow traders (not customers or clients) what they really want if mutual wealth is to be created and enjoyed repeatedly. This is the ultimate role of the back office.

- Strengths/Weaknesses/Opportunities/Threats (SWOT) Analysis with Action Plan. Review again your uniqueness and express more carefully its SWOT. With each S, W, O or T provide an action plan to maximise your purposes. For instance, you will need education, strategies, mentors, staff, systems, capital, suppliers, and tactics.
- Competitor Profile. You are unique, but you do have competitive substitutes. What do your competitors believe or dream? What have they achieved and how do they act? Give two or three of the most important and concrete and detailed examples. Do you agree or disagree with their various stances and actions? Why?
- A list of your ideas' unique benefits as they apply to the different areas of a person's (including legal person's) life and needs; a '30 Second Ad'; a 7-word, 25-word and 150-word statement of core and unique benefit.
- Miscellaneous Information. Information about you, your partners, etc. that consciously informs and subconsciously motivates the target audience.
- Personal Strategy Highlights. E.g. time management plan (one-off, daily, weekly, monthly, yearly and intermittently) and lifestyle plan (time 'being' rather than 'doing'; time off; time for entertainment, rest and refreshment; holidays).
- A Sales Forecast and with time, Sales Overview. Make your life-plan specific in dollar terms because wealth in this world is usually measured in dollars. The power and beauty of the

values you release from within you will be measured in many ways, but one very concrete way such benefits are measured and traded is in terms of money (as discussed in the prior chapter). Bluntly, your money is in the values you provide. Life is short: Your business plan, your life plan and your self-actualisation are emergent and actually one.

- An exit plan from one set of concretised and systemised values-in-action to the next and planning for the estate and legacy you leave behind you.

Much of this work you have already begun. Facilitating others is actually about facilitating yourself through self-enlightenment. To take this plan even further, why not buy a business plan software tool or consult a business plan expert?

Concepts of Emergence Encountered in Pillar 6

In summary, our society seems to leave us with only three axiomatic alternatives: The idea 'existence exists' and atheism, the idea that 'existence is uncertain' and agnosticism or the idea that 'existence is subordinate' and blind faith in 'Something'. That is, if we are to take a classically certain life-stance rather than an emergent one, then it seems we must make a blind assumption, make a blind denial, or make a blind belief!

Personally, I think the lesson of an emergent universe offers us a far more satisfactory answer. It enables us to get to the nub of the issue. It teaches us to reject objectivism's axiomatic certainty and thus discontinuous stance as a basis for atheism. It also teaches us that we may reject agnosticism's axiomatic and uni-dependent uncertainty with respect to an emergent existence and an emergent self. It also encourages us to reject the axiomatic idea of a 'Something' for nothing or an eternally self-organised and intelligent 'Something' (either way, a 'Something' that is effectively a magical free lunch denied in any natural universe).

What the universe teaches us is that reality, not based on any axiomatic assumption, but on a relativistic and emergent existence unable to banish the fractal border between infinity and nullity, is the most easily defended explanation of what we observe and how it operates.

So this self-organising and emergent universe, whether the result of intelligent beings, and whether or not those beings forcibly control any part of our daily lives, or the inevitable result of some kind of chaotic/fractal/fuzzy quantum mechanics, or both, seems to scream (or is it endlessly whisper?) to its inhabitants to self-organise and arise, or self-actualise.

The universe's lesson to us is simply this: The focus must be your life, your self-actualisation, your sustainability, your societal flourishing and your legacy – because existence emerges in a relativistic setting. Such a simple lesson will represent a huge paradigm shift in how we consider our planet, each other, and our interacting purposes. It will take the focus off all unnaturally axiomatic forms of life-stance so that we can be free to explore and enjoy the natural life-stance phases (struggle, elimination, emergence, and incorporation) as they unfold.

In the next chapter, we will discuss the virtuous circle, that is, the natural process of emergent self-organising and self-actualising in an enlightened global community.

---

[1] Richard Dawkins, quoting a 1985 article by Stuart Kauffman, discussed this idea in Dawkins, Richard, *The Ancestor's Tale* (Phoenix Paperback, London, 2004), 597-611

[2] For a defence of rational inductivism, see Grayling, A. C., *The God Argument* (Bloomsbury, New York, 2013), 49-56

[3] See Rand, Ayn, *Capitalism: The Unknown Ideal. Centennial Edition* (Signet, USA, 2005 (Original Edition 1966)), 264 and 266

[4] Ridley, Matt, *The Origins of Virtue* (Softback Preview, Great Britain, 1997), 15

[5] As proposed by Dr Richard Terrile of the Jet Propulsion Laboratory, California, in the Discovery Channel documentary of 2010 named *Through the Wormhole*, hosted by Morgan Freeman.

[6] Jarvis, Jeff, *Public Parts* (Simon & Schuster, New York, 2011), 38

[7] Ibid., See pp. 94-95

[8] Ibid., See pp. 98-99

[9] For instance, see Kahneman, Daniel, *Thinking, Fast and Slow* (Penguin Group (Australia), 2011), pp.52-58

[10] Jarvis, Jeff, *Public Parts* (Simon & Schuster, New York, 2011), 109-110

[11] Ibid., See pp. 55-56

[12] See Deutsch, David, *The Fabric of Reality* (Penguin Books, New York, 1998)

[13] See De Bono, Edward, *New Thinking for the New Millennium* (Penguin Books, USA, 2000)

[14] See Jarvis, Jeff, *Public Parts* (Simon & Schuster, New York, 2011), 5 and 83

[15] Ibid., See p. 60

[16] Ibid., See pp. 10-11

[17] Ibid., See pp. 17; 22-23

[18] Ibid., See p. 72

[19] Ibid., See p. 73

---

[20] Ibid., 10

[21] Ibid., See p. 9

[22] Ibid., See pp. 68-69

[23] See Ridley, Matt, *The Rational Optimist: How Prosperity Evolves* (HarperCollins, New York, 2011), 7

[24] See Jarvis, Jeff, *Public Parts* (Simon & Schuster, New York, 2011), 69

[25] Ibid., See p.144

[26] Ibid., See p.130

[27] Ibid., See p.211

[28] Ibid., See p.212

[29] Ibid., See p.213

[30] Ibid., See p.215

[31] Ibid., 7

[32] Ibid., See p.44

[33] Ibid., 49

[34] Ibid., See pp.88-89

[35] Ibid., See pp.163-165

[36] Ibid., See p.165

[37] Ibid., See p.171

[38] Ibid., See p.204

[39] Ibid., See p.166

[40] See Mant, Alistair, *Intelligent Leadership* (Allen and Unwin, Sydney, 1997), 6

[41] Ibid., 14-15

[42] Ibid., 127

[43] Ibid., 127

[44] See Burns, J.M., *Leadership* (Harper and Row, New York, 1978), Part 1, Chapter 1

[45] Twentieth Century Fox, Braveheart (1995)

[46] See Burns, J.M., *Leadership* (Harper and Row, New York, 1978), Part 1, Chapter 1

[47] Gleitman, H., Gross, J., and Reisberg, D. *Psychology* (Eighth Edition, W. W. Norton & Co. Inc., NY, USA, 2011), 615

PILLAR 7: Virtuous Circle

In this chapter, we will discuss some questions our future raises. The future is full of mystery, but the values that will take us successfully forward are not so mysterious – and they will continue to emerge in a positive way if we do not thwart them. The text will suggest that we have good reason to be mildly optimistic with respect to the future and suggest that our values-in-action carried forward are the source of our future wealth and wellbeing. Finally, in terms of your own philosophical quest, this chapter will encourage you to write your own philosophical pledge. It will be up to you to dare to take your pledge and live by it.

Abundance Mentality

Has the genetic struggle for environmental fitness been superseded in importance by the memetic struggle for cultural fitness already? Probably yes. However, maybe our individual and shared values, our vremes, are now a more important survival factor for our species and the world as a whole than our more basic memes or our genes. This is because it is our vremes that have the potential to create new publics and a new social fabric that cuts across the fault lines of old civilisations and cultures.

Will our brains ever evolve to the point where instinctive biases and/or the mind-games of our consciousness no longer hamper our judgments and actions? Actually, in a sense overcoming instinctive bias has already begun to take place in, for instance, eHealth bedside diagnoses by computer systems rather than by emotional human diagnosticians. A similar phenomenon is well known amongst stock market traders relying on computer-generated technical analysis.

Perhaps we haven't developed systems to fully overcome our mind-games of consciousness yet. Nevertheless, the trend to partly or fully out-source human decision-making to computers for the express purpose of minimising human failings (in conditions of emotional pressure, groupthink pressure, the pressure of time and complexity, etc.) will continue to accelerate and grow in sophistication into the future.

As a species, are we like the Australian fox, too successful to date? Are there enough scarce resources for humankind 50-100 years hence, given our current growth trajectories? The reality is that we are near to reaching or just past reaching not only peak oil but peak commodities, peak soil and peak fresh water as well. Will our own successes in non-sustainable mastery of the environment actually reduce our evolutionary fitness and development? In the short-to-medium term, the answer is probably 'yes'.

Perhaps these and other questions like them, our progeny will soon have to seriously answer if we do not. Perhaps the more direct and most important question this book raises is, 'What values-framework and value-sets will we equip ourselves and our children with to answer such questions?' The answer to this question is our immediate quest. We might also ask, 'With which method will we equip ourselves and our children, to promote our wellbeing as individuals and as a species?' Will the answer focus on classical atheism, agnosticism or believing – or will the answer more broadly and more aptly point to Emergentism, its Emergent Method and the relativistic interdependency of emergent monism?

In Pillar 4, we talked about the interpenetrating relationship between our draft model of quadune-self and the four quadrants of our modes of human behaviour, each governed by its own distinct value-set. We suggested that the most recent layer of quadune self – the Trader-self – is itself going through growth governed by its evolving interactions or interconnections with the earlier three dimensions of self (the technologically focused Warrior-Self, the culturally embedded Lawmaker-Self, and the socially aware Philosophical-self).

The evolving Trader-self, while rejecting the violence of the classic warrior, was spurred on by, and continues to be influenced by, the Warrior-self. Much more recently, the Trader-self, while rejecting injustices embedded in many of society's laws, is likewise being shaped by an explicit acknowledgement of Lawmaker-self, especially in successful multicultural societies like Australia, where we largely celebrate our cultural diversity within our economic system. We briefly discussed this topic in Pillar 5, but clearly, we have some way to go to reach something Huntington (1996) called Civilisation with a capital 'C'.

Philosopher-self is perhaps the most recent explicit enhancer of Trader-self. We briefly discussed this topic in Pillar 6, where we looked at Jarvis' book *Public Parts* and in particular his concept of the radically public company. If we have a long way to go to achieve Civilisation, then we also have a long way to go to reach a globally diverse social fabric that completely reinvents our interpenetrating publics (and indeed philosophies), which is what Jarvis seems to suggest is underway. Perhaps, if Harry Dent Jnr. is on the right track with his demographic modelling based on data sourced from the United Nations,[1] the emerging economies will follow an order of peak performance from Asia now to Africa in the 2090's. By that stage, once each of Huntington's (1996) civilisations has had its time at the top, perhaps there will be less motivation for strife along civilisational boundaries. Perhaps at that time we will achieve a healthy global social fabric that truly recognises and appreciates the value of its diverse inputs.

However, even this is not the 'endgame' (relativistic interdependencies never end). We will achieve all best possible outcomes when we emerge as 'all-in-all' in a super-species that finds the fullest possible self-actualisation through its interpenetrating parts and through continuing trade in emerging virtues of all kinds.

To illustrate the evolution of quadune self, I refer to Steven Pinker's review of the work of Cas Wouters. He notes,

> *"If our first nature consists of the evolved motives that govern life in a state of nature, and our second nature consists of the ingrained habits of a civilized society, then our third nature consists of a conscious reflection on these habits, in which we evaluate which aspects of a culture's norms are worth adhering to and which have outlived their usefulness."*[2]

I note here that the fourth nature would not be so tightly held by the individual, but would be more openly shared with, and reflected in, the community. That is, the fourth nature will be almost complete, but at a very new beginning, when we hold up diversified, interpenetrating, tamed yet healthy Gaia, as we would lift a mirror to our faces, and in that act, see our own reflection.

Our ongoing quest, if you like, is to encourage this evolution through a partially artificial but naturalistic method – the Emergent Method. Our quest is to encourage an explicit trading of values in all areas of human endeavour much like Darius the Great seems to have had the very early foresight to do. Our quest is to learn how to become expert traders of vremes. As noted earlier, Emergentism will cause us to need, respect and honour the environment and each other as a monistic but also diversifying whole. The highest-order Vremetic-self is the planet loving itself.

Emergentism's Catch-phrase

To help our children and each other answer this question, "What grand catch-phrase could we devise to encapsulate the fundamental ideas of Emergentism?" would it be 'Volitional emerging'? No, because this demands that humans and everything else must always have conscious volition. What about 'Dependent emerging'? This seems to capture the idea, except for one very important point. Subordination to authority, tradition, or all that has come before can undermine greatness, or the heroic inside of us, and our striving for self-actualisation. Ayn Rand explored this individual heroicism so expertly through her book *Fountainhead* and its main character Howard Roark.

If we always see our values as only a gift, we will never transcend dependence (or giving) to enjoy self-actualising as fully as possible. However, if we always see our values as our own, we will never transcend independence (or taking) to enjoy self-actualising as fully as possible. So even though 'Dependent emerging' correctly captures the volitional and restrained in all self-organised arising, it muddies the interplay between independence and dependence that is required for a fuller self-actualising.

Perhaps **'Emergent self-organising'** expresses the core concept more clearly, because it explicitly recognises self and everything else with which self struggles, organises, emerges and coagulates on this side of the veil. However, here we run into the issue and reality of struggle and self-elimination in the universe. Even so, if we view struggle, self-elimination and incorporation as important constituents of self-organising rather than its opposite (as in the processes of metabolism and replication), then we can freely utilise the term.

Self-imposed boundaries and limitations are the way organised systems control energy flow – and the more boundaries, the more diversified the risks and rewards. Hence, the great value of our planet's biodiversity. Thus, we have the logical ability to claim that, regardless of what may be behind the veil of unknowable reality, there always was an **'Emergent self-organising'** in the midst of vastness at some level or other. This is the lesson nature teaches us without contradiction (as long as we also acknowledge a certain level of necessary struggle, self-elimination and incorporation). Further, if nature is arranged according to the wishes of a 'Something', then this is the core lesson 'Something' intended us to learn from nature. One further step suggests that whatever lies behind the veil must be self-organising as well, which seems to obviate that veil.

Finally, for intelligent, rational and courageous beings at least, we can also claim the contingent principle of **'Emergent self-actualising'**. In these two catch phrases – emergent self-organising and emergent self-actualising – we see the clear naturalistic link between our universe's virtual particles at the most microscopic level and our virtual consciousness at the highest level laid bare.

Does this mean Emergentism is a philosophy of Buddhism after all? No, but it does suggest that the emergentist and the Buddhist can both be students of different aspects of "Dependent arising". Nevertheless, emergentists advocate the abolition of all superstitions and the subordination of an insistence on transcendental consciousness rather than some Buddhist's insistence on the subordination of material existence.

The application of Emergentism would also ask the agnostic to move towards an adult agnosticism that takes full responsibility for using every day to strive towards life-affirming self-actualisation. Likewise, Emergentism's application suggests that if you want to remain a follower of an Abrahamic faith, then strip away its superstitions and move towards an adult version of it. Our ideas of Emergentism would ask the same of Buddhism; they would ask all Buddhists to move towards an adult Buddhism that takes full responsibility for using every day to strive towards an unsuperstitious Buddhahood. This unsuperstitious Buddhahood is simply an emergent self-actualisation; it is also akin to objectivism's rational, heroic person emerging in each of us.

If the 'Something' of whatever religious persuasion was plainly manifested to us right now, with which kind of person would this 'Something' be most satisfied? The cringing believer or non-believer or agnostic who didn't use his or her life to self-actualise, or the person of whatever persuasion that fully used life to reflect most carefully his or her Ultimate Agent's own rational, volitional, interdependent, courageous and evidently emergent and self-actualising values and nature?

Contentment

Perhaps our progeny will ultimately answer the question of our universe's beginning – but maybe the answer will lead to further questions about an even larger sphere. If we do finally meet our ultimate Agent, perhaps the meeting will be casual and without fanfare because we will meet as fallible self-actualising peers – peers wondering about yet further possibilities in an unfolding present.

However, perhaps the more pressing question we often ask ourselves is, 'Why do I exist?' This is a very different question to the one regarding the universe's beginning, which we can start to answer right now. As we have already noted, the forests and all other non-human life-forms do not ask themselves this question – there is no need. By their very actions they affirm their simple answer – 'we exist because we are and because we can. We exist because we have the feedback systems in place that enable us to extract information and food-energy from the environment and use it for our own survival. We exist to promote purposes that facilitate the inexorable coagulation of life in our genes and survival machines; we exist to self-organise'. If we self-indulgently asked the non-living spheres the same question, they would give us almost the same answer, except to say that they seem to exist to give life its opportunity to arise from within them, coagulate and self-actualise. Their destiny seems to be as the extended phenotypes of life, in the same way some noncoding DNA may provide resources and potential for the evolution of the mind/body.

We have subtly arrived at a very interesting point here. As we have already noted, as species coevolve, they become more effective and efficient within their environments. They cyclically compete and cooperate at ever-higher levels of complexity and symbiotic coagulation (unless a destructive force, such as humanity's ignorant and unsustainable use of the environment, interrupts their dance). However, the productivity gains of more coagulated species are rather temporal for two reasons:

1. Competition is lifted to a higher coevolved level as coevolved but competing species become equally well organised; and
2. If the pre-coevolved level of competition is forgotten (in genes or memes), then prior more individualistic skills are lost (via the 'use it or lose it' principle of biological and mental development).

This means that the individual species' survival machines or their coevolved genes do not make the true gains, because both are changed and thus previous evitabilities are lost with time. Gaia Herself makes the long-term benefit of symbiosis or coagulation. In this sense, and without assigning any current self-conscious intelligence to Gaia as a whole, we might view humanity's environmental naivety and its self-eliminating consequences as a necessary lesson in the ongoing development of Gaia's most highly self-actualised species. In this sense planet Earth is not an inert stage upon which the battle between good and evil is fought. Rather, Earth is the resource and source of Gaia's emergence. Gaia, like our universe itself, stands at the beginning of infinity.

The most important question is perhaps not really 'Why do I exist?', because we already have the answer within and around us in our self-contained and emergent universe. What really matters is what we do with our lives now. We could ask ourselves 'What goals and aspirations do I hold?' but more deeply we should ask, 'What motivates me' and even more deeply, 'What values do I hold, represent, and leave as my legacy to Gaia's emergence?'

We could go more deeply and ask ourselves, 'Where do our values come from …' (a family, a tribe, a culture, a concept of God, a book, a religion, the unconscious emergence of our species, etc.?), but do these 'fixed labels of uni-dependent ownership' really matter? Emergentists hold the view that they have mostly objectively arisen from a wonderfully self-organising and information-rich universe, but it is good enough to understand that the values we hold and represent from whichever source are in the most important sense our own and the reasons why we do what we do, and why we are what we are.

Some will say that we should always link our values to a Standard (the Torah, the Bible, the Koran, etc.) This book has taken a slightly different approach and offered an answer to a slightly different question. Rather than normatively asking, 'Which standard (of which god) do your values come from?' the positive question is, 'Why do we have the values we hold and how can we use them?' The immediate answer comes from the reality of our own existence. We hold and use our values because we are volitional, extrospecting, social and introspecting beings, with instincts and consciousness that continue to coevolve. This is a positive question because its answers we can recognise in each other now, while we live, no matter what creed we follow or stance we take with respect to what possibly lies beyond the veil. Our shared values are logical and 'personally objective', which means they may be identified and elucidated on this side of a possible veil.

The world is in the midst of a war for values – some quasi-immutable values the West has taken for granted for so long it has now almost forgotten them, and forgotten how important they were for our own self-actualisation. It is strange indeed that it takes a brave Somali woman, Ayaan Hirsi Ali, to tell us that one reason why life is better in the West is because we value the here-and-now, and the state protects this value.[3] The secret to realising our self-actualising potential, both personally and socially, is no more difficult than finding a way to express on the outside the virtuous values (reason, choices, freedoms, joys, etc.) that we identify to reside on the inside, including subconsciously; but to identify we have to actively think here and now.

We don't need to form a new coalition or brotherhood espousing a certain set of immutable values – thereby giving those values a fixed home, a frozen identity, a label, a trademark, ownership, copyright, and subjection to patent that excludes all others. We need the exact opposite. We need to stay in our existing groups, institutions, clubs, markets, and networks and make these groups subservient to, or owned by, the locally recurring values that are themselves evident in the first-person sense and immutable in a fuzzy kind of way in the third-person sense.

We need to venerate these values as so embedded in matter's diverse arrangements that they are unable to be owned or possessed, the same way Australian Aborigines see their relationship with their land or country. They do not own their land, if anything the land owns, holds, or includes them. At most, an Australian Aborigine can be an entrusted custodian of a certain landscape; the relationship with the land is one of active, living, and emergent nature-and-life shared. This should be our relationship with 'our' virtuous values and our natural universe.

Held values do not prove there is a God just as consciousness does not prove it is transcendental. Values held by people can be sophisticated or naive; consciousness can be clouded or highly self-aware. What matters are the emergent values we hold as their local custodians and the values we ultimately live by and share with one-another in the living present. No organisation or religion has a monopoly on values, consciousness or their development.

We have noted how consciousness evolved from Ida, Ardi, or a close relative, to *Homo-sapiens.* Consciousness and instincts clearly evolve. This means our consciousness of values evolves. This suggests that our destiny evolves given this is not a deterministic, clockwork kind of universe, but rather a stochastic and emergent one. Applied truth, meaning and destiny will continue to evolve, which means our virtuous global tradition will also continue to evolve. A non-supernatural god, if it exists, will make contact with us openly in life when it is ready to do so, but in the meantime, a virtuous life will keep us in good stead.

If this book were able to make just one lasting contribution to the virtuous part of humanity's evolutionary journey, it would be this: The realisation that our values may be vastly appreciated and lived in our lives now, without any necessary connection to, or ownership by, a religion or secular ideology. This single idea, as important as the separation of church and state in our past, could transform and save us this century both individually and corporately. The liberation will be complete when we all come to the realisation that a classic life-stance (as atheist, agnostic, believer) is inappropriate, and a self-actualising life in which we recognise its interdependent relativity is the appropriate way to live.

Do we have the right to be mildly optimistic? I for one want to answer 'yes!' The sustainable coagulation of life seems inexorable. We, as part of a civilising society, have come a long way since the horrific atrocities and dehumanising attitudes associated with the Middle Ages. I believe in an abundance mentality and the super abundance of interconnected logic in the current universe. We can find a (chaotic) way forward. Our realised values are also the source of our visions, wealth, vitality, and our own agency. All we have to do is align with the values we know we each hold personally, think about them (conceptualise and contextualise them), and deeply honour them, personally, with our lives.

I recently heard Professor John Lennox of Oxford University defend Christianity against atheist evolutionary arguments by saying, 'don't confuse mechanism with agency'. The idea was that by studying the mechanism of a Ford motor does not inform us with respect to the man who originally made the Ford motor possible, Mr Ford himself. Likewise, science might study all the mechanisms of nature and evolution, but science cannot inform us as to nature's original Agent, God.

The idea was that mechanism does not prove agency or lack thereof, and so we can't use studies of nature's mechanisms to disprove God. Nevertheless, Darwin's dangerous idea of natural selection rebutted this idea long ago: You do not need an artificial choosing agent for life to evolve. Further, you do not need an artificial choosing agent for all things, living or non-living, to emerge. Existence (as something even prior to the organisation that space-matter-time introduced) weakly emerges from its relativistic basis or border – if we could imagine an end of emergence, we could imagine an end of existence and its relativity.

However perhaps the deeper argument put forward in this book is that the wonderful, meaningful, and purposeful agency of life is found in the values we can observe and use in our shared lives on this side of any imagined veil in a continually self organising and emergent universe. So agency can be discerned on this side of the veil, even while some still question the nature of existence using the classical paradigms ('existence exists', 'existence is uncertain', 'existence is subordinate').

Of course, from a broader perspective the question of optimism is irrelevant. If the current human species does not participate strongly in the advancement of virtuous consciousness-plus-instincts, the emergent universe will most likely find some other evolving species to do the job. *Homo sapiens* will be superseded, as was Neanderthal Man. The universe will retreat from this local peak, as Dan Dennett might describe it, to statistically reach for another within its potential vastness. Moreover, many would say this is the inevitable end of every species eventually.

I do not think we can escape the ultimate reality intimated by Matt Ridley's inexorable coagulation of Life. We will only reach our potential, individually and socially, if we each freely choose the sustainable way forward in such a way that all our individual and mutual benefits are not only optimising, but also slowly coagulating together, perhaps in the next bio-socially diverse yet symbiotic and information-exploiting Super-species. The longer we take to come to such a realisation, the heavier and more expedient this realisation becomes.

We can start to learn now the graceful dance or orbit in the present moment, of which the stars testify so strongly. There are many stars participating in this dance – hopefully many of us can participate as well.

> *"In that respect, though we may not be able to achieve a global religion, we may achieve a global civilisation in which values from the great traditions are woven together in a glittering net. Perhaps it will turn out like the jewel net of Indra, of which Hua-yen so eloquently speaks: each stone reflecting each other."*[4]

Perhaps for some, this view is too idealistic. Even if we share this glittering net and stop killing each other, just to feed the huge population this new peace would suggest, we would likely have to organise the breeding and slaughter of cows, sheep, pigs, chickens and fish at unprecedented levels. In effect, we would be substituting the slaughter of our own species for highly domesticated others. Would this really represent a higher morality and state of enlightenment? What makes our species more deserving of death by natural causes than theirs? We come back to the idea of moral relativism and Nietzsche's idea that moral phenomena don't exist, just our intersubjective interpretations of them,[5] although I disagree with the idea's narrow-minded sensibility. There is a continual emergence of meanings and values that more accurately reflects the properties of observed phenomena. These same values more efficaciously promote Gaia's wellbeing. At some future time, Civilisation may even deal with the morality of our carnivorous appetites.

Participation

Pinker tells us Kant suggested in his article "Perpetual Peace" that there are two ideas promoting of world peace amongst nations: Trade and Democracy.[6] Similarly, as we have already noted, Matt Ridley argued for Trade as the ultimate promoter of human civilisation, innovation and prosperity. Underlying commercial trade is a trading of basic values that each party must recognise to succeed. The greater that shared recognition of values, the more efficient the trade and the more likely the creation of a relationship that will foster further trades of a similar and evolving nature.

Pinker, quoting the work of Bruce Russett and John Oneal, adds a third idea of Kant's to his list of two - Voluntary Intergovernmental Associations.[7] Pinker then asks if there is a common factor in these three, Trade, Democracy, and Voluntary Intergovernmental Associations, that is a deeper driver or cause of human peace and prosperity. He suggests that this factor is something to do with our developing nature as a species, either as self-interested rational realists, or as innately moral beings. Pinker then suggests that the underlying element might be a learned realisation, reached through harsh trial and error, that peace and wellbeing is better than war, violence and distress.[8]

I would thus tender three ideas that might explain our bumpy march towards human wellbeing and Huntington's *Civilisation*:

- We are quasi-moral beings by blind instinct, emotional choice and genetic evolution
- We are quasi-moral beings by cold rational reflection, conscious observation and vremetic emergence
- We are moral beings necessarily by both instincts and consciousness, that is, by both genetic and vremetic emergence and interpenetration

Clearly, I would support the last idea as the most accurate. How, then, do we enhance the positive trend we sometimes note in the emergence of human society and human consciousness-with-instincts? Do we just wait for the evolutionary change, unable to greatly affect its speed or direction? Alternatively, do we each, as individuals, take responsibility for our here-and-now and for our societal enlightenment? More pointedly, is the emergence of our human nature a matter of natural or artificial evolution, or is it a mixture of both? Again, clearly, it is a mix of both, with artificial emergence growing in importance as we more clearly identify the explicit vremes of our emancipation, and natural emergence decreases its role in relative terms.

Just as consciousness in the individual can only indirectly influence our behaviours after the fact, partly because behaviours are already predetermined in subconsciousness, so we are the products at the societal level of a past society over which we have had almost nil control. However, we are also the ones that inhabit this time-and-space and consciously control our individual futures through planning for contingencies. Therefore, likewise, we have a limited but growing control over our emergence as communities (or publics) at the societal level. We can plan for contingencies. We can participate in the artificial emergence of our species because we are animals of conscious but co-limited free will that can pass on worthwhile legacies to the next generation. We participate in the tending, keeping, and growing emancipation of our species through our learned and adopted vremes, and the trading of those vremes with each other. To do this confidently, we have to be sure of the framework, which is Emergence, and the mechanism, which is the Emergent Method.

As noted earlier, some of the last words in Merleau-Ponty (2014) were as follows: "The only way I can fail to be free is if I attempt to transcend my natural and social situation by refusing to take it up".[9] Logically, this wisdom applies at both the personal level of the "I" and the civilisational level of the "we".

To the outsider, the great transformational and transactional leaders work relentlessly, yet to these leaders, they count their life's work as a necessary privilege and joy. Their enlightened work adds to their personal liberty. We find here another aspect of the deep purpose leaders draw from their work. Being free, or at rest, from selfishly destructive purposes, they release themselves into dimensions of cosmic significance, somewhat unfettered by space, time and circumstance.

Their authority is derived from 'heaven' (that is, the not-so-mystical font of values they treasure), and while they are not lawless (rather, they are sometimes the most law abiding), at their personal best they are answerable to no earth-bound person or temporal law, because no law or person can gainsay them. Their secret is to identify personally with the noble and authoritative values within them and so not see themselves as inferior to any human authority, but rather they see themselves as sometimes equal to the inescapable authority of their values and vision. The combined life works of Mahatma Ghandi, Dr Martin Luther King Jr., and Nelson Mandela seem like a case in point.

If we can develop to be such leaders of life and in life, then perhaps the noblest thing we can do is to also teach the next generation how to achieve the same, that is, through emergent self-actualising. As we do this mentoring, we continue the unfolding of the virtuous circle.

Have you thought deeply about the legacy you and your life's achievements are leaving behind for your family, your friends, your associates, and your world? Your legacy, rather than any material gain, should perhaps be your strongest motivation to self-actualise.

Your Philosophical Pledge

Before we finish this philosophical quest and start the next, please seriously and carefully consider this question: If you were to write your own philosophical pledge, that is, a private pledge to your own values and beliefs, what would that be? What self-restraints or essential significations would you freely apply to your own introspective consciousness at various emergent stratifications? A possible guide:

Epistemologically:

I pledge to always endeavour to study and take lessons from nature and reality.

Intellectually:

I pledge to neither live in full intellectual certainty nor continual uncertainty, but to always live with the temporal intellectual certainty my past has granted me and with a willing mind, to always endeavour to accept new ideas that are proven to be more intellectually certain than my past convictions.

Ethically:

I pledge to always endeavour to be true to my courageous self, that is, true to the self-organising and emerging values I represent and that resonate within my mind-body.

Morally:

I pledge to always endeavour to live a virtuous life according to the values I see as appropriate to each phase of my journey through life's situations. I also pledge to always seek personal moral congruence from the outcomes to which my reason and choices contribute.

Metaphysically:

I pledge to use my contemplative freedom to accept quasi-objective reality and to imagine, to play, to grasp for the ideal and to be innovative. I also pledge to endeavour to never bind my own contemplative thinking or that of another person, except to the extent of mutual and shared appreciation of that contemplative freedom.

Socially:

I pledge that I will endeavour to never live for the sake of another person, nor ask another person to live for my sake. Rather I will endeavour to live for the sake of our mutually shared self-actualising values and my own unique set of self-actualising values.

Aesthetically:

I pledge to always endeavour to honour, appreciate and enjoy the universe and life's chaotic, antinomic dance.

To my family, friends and associates:

I pledge to always endeavour to live by my philosophy and to share it openly with anyone interested.

To the universe and myself:

I pledge to always endeavour to align my life with the universe's impersonal principle of 'Emergent self-organising' and its personal principle of 'Emergent self-actualising'.

If you take your pledge and rely deeply on its self-actualising principle, this will provide you with all the vitality you will ever need to see it and your 'business plan' of Pillar 6, and far more, realised.

This is the Emergent Method.

At the time of death, we rest from our labours. Death is just passing the baton.

Concepts of Emergence Encountered in Pillar 7

In these two catch phrases – emergent self-organising and emergent self-actualising – we see the clear naturalistic link between our universe's virtual particles at the most microscopic level and our virtual consciousness at the highest level laid bare.

In this sense planet Earth is not an inert stage upon which the battle between good and evil is fought. Rather, Earth is the resource and source of Gaia's emergence.

Existence (as something even prior to the organisation that space-matter-time introduced) weakly emerges from its relativistic basis or border – if we could imagine an end of emergence, we could imagine an end of existence and its relativity.

---

[1] See Dent, Harry S. Jnr., *The Great Depression Ahead* (Free Press, USA, 2009), especially 189 and 238

[2] Pinker, Steven, *The Better Angels of Our Nature: Why Violence Has Declined* (Viking Adult, USA, 2011), 128

[3] See Ali, Ayaan Hirsi, *Infidel* (Pocket Books, Sydney, 2008), 348

[4] Closing words of Smart, Ninian, *The World's Religions* (Cambridge University Press, 1989)

[5] See Hollingdale, R. J., *A Nietzsche Reader* (Penguin Books, Sydney, 1977), 104

[6] See Pinker, Steven, *The Better Angels of Our Nature: Why Violence Has Declined* (Viking Adult, USA, 2011), 289-293

[7] Ibid., See p.290

[8] Ibid., See pp.291-293

[9] Ibid., p.483

EPILOGUE

Where To From Here? Your Challenge

This book has suggested that in order for all of us to find and fulfil meaning, purpose, and our greatest potential in our lives, we need to be part of a strong community; preferably an enlightened and virtuous global community whose focus is sustainable human flourishing in the fullest sense. We, as humans, need a deep sense of shared interconnectedness with each other and with Gaia. The problem is that currently such an all-encompassing but grass-roots community is not well organised. Perhaps we need to become organised – as a philosophy, a scientific organisation, a social phenomenon, a self-help intentional community, a business corporation, a political voice, an economic powerhouse, a wellness-promotion group, a world governance body, etc.

Perhaps you run a business or are an active member of your local community, group, or club already, and you can clearly see how to apply Emergentism to its activities. Nevertheless, if you would like to participate in the growth of a community of emergentists directly, in any way, please let us at the website know. You might like to contribute to an expansion or re-write of this book, e.g. in the areas of Emergentism's philosophy of a-religious and a-secular ethics, Existential phenomenology, Bayesian epistemology, Descriptive or Neuro-Semantic psychology, Process Physics, political frameworks, etc. Alternatively, you may like to contribute to a challenging and slightly edgy blog devoted to emergentism so that it can reach a larger or more specific audience (for example, schoolchildren or those who speak other languages). You might be a marketing or I.T. expert, business entrepreneur, corporate advisor, angel investor, law student, musician, comedian, other entertainer, cook, teacher, scientist, mathematician, or a stay-at-home mother or father of four children. Whatever the case, you are the result of billions of years of natural selection of genes, memes and vremes - and you can help.

Perhaps our online group could be something like Wikipedia – a one-stop shop not of just encyclopaedic information but values-rich and existential information and development. Perhaps it could even be a little like WikiLeaks – revealing values-poor, self-eliminating and non-sustainable decisions made by organisations of all kinds worldwide.

Perhaps it could also be a little like LinkedIn, Facebook, or eBay, but rather than just linking people with common business, social, product or service interests its foremost role is to link people with common or complementary value-interests and interests in self-actualisation.

Additionally, perhaps 'Traders of Values' could be the launching pad of many fairly standard sorts of businesses, such as employment agencies, coaching/mentoring services, and travel agencies, that put their psychosocial values at the very front of all their business interactions.

Perhaps 'Traders of Values' could also link people together like an online chat or dating service but whose aim is not just individual relationships but a purposeful, chaotic and fractal global relationship whose emergent core consists of venerating the quasi-immutable values themselves.

Perhaps the group could be a bit like a non-fanatical, non-fundamentalist secular religion that accepts atheists, agnostics and believers alike because it holds the personally objective, immutable-but-fuzzy values at its core rather than personal convictions about what might lay behind a possible veil across our knowledge of reality. In addition, perhaps we could invent an 'a-religious' or 'a-secular' but equally reverent version of 'Namaste' for all traders of values.

Perhaps 'Meetup', the world's largest network of local groups, would be an ideal vehicle to get a local group started in your town. Alternatively, perhaps the 'TED: Ideas worth spreading' and 'TED: Conversations' models could be used or emulated to facilitate larger platforms.

Perhaps our group could also be something like Fair Trade Association Australia and New Zealand, which promotes a fairer trading environment for developing countries by enabling consumers to support participating companies who carry the Fair Trade label on their goods offered in the local supermarkets. So likewise, all stakeholders and support organisations could carry the 'Traders of Values' badge of approval via an application of blockchain[1] technology. Such a badge would testify to the organisation's allegiance to the evolving or emerging Philosophical Pledge of Pillar 7. Individual organisations within the group could then demonstrate their sincerity by upholding a basic set of international standards regardless of the lack of local host-nation compliance requirements. These might include IAS International Accounting Standards, ISO 9000 Quality Management Standards, ISO 14000 Environment, Health and Safety Standards, ISO 31000, or COSO Enterprise-wide Risk Management standards. Each organisation could perhaps also choose to be bound by its own International Workers' Compensation standards and so on, and by our group's audits. Many multinational corporations already choose to do these things without internationally binding audits.

Perhaps our group could be the beginning of an online community that could develop the world's first values accords as mentioned in the Pillar 5, and then lobby governing bodies everywhere to adopt them. Perhaps our body could simply raise the awareness of issues related to vremes outside a country that cannot be explicitly raised within a country, to give a clear voice to its citizens of growing enlightenment. Alternatively, perhaps our organisation can quietly remind our fellow citizens of the world that it is reasonable to expect their governments to reflect their own values and aspirations, because Gaia inexorably coagulates. This would not be a revolutionary call to arms, but rather an evolutionary agreement to abolish arms.

Perhaps our online group could also grow to be the first grass-roots alternative to the United Nations itself, that is, a globally-linked organisation of politically motivated and civic-minded activists that would each enjoy and use a blockchain voting mechanism to make his or her voice heard and thus create or guide globally binding political policy. Guided by the catch cry of 'emergent self-actualisation' and the explicit moral aim of movement through 'personal moral congruence', we could advocate not just a transparency of direct actions, but a transparency of indirect motives, beliefs, principles and values as well.

Perhaps it is only we that can find a sustainable and self-actualising future for our emerging species and our planet.

So if you are willing to take the pledge of Pillar 7, please join our online forum at www.TradersOfValues.com (under construction – just leave your comments at MichaelKeanBlog.com while we await support) and let's start something grandly disruptive but profoundly non-violent together. My hope is that the website will provide a growing source of satisfaction amongst all its participants.

Epilogue

One idea whose time has come is the idea of Emergence and its Emergent Method.

Take care…

---

[1] The blockchain provides proof of transaction and provenance in a fast-growing suite of applications from various brokerage services to digital currencies, smart contracts, timestamped messaging services, etc.

Bibliography and Further Reading

BIBLIOGRAPHY:

Ali, Ayaan Hirsi *Infidel* (Pocket Books, Sydney, 2008)

Barrow, John D., *The Book of Universes* (Vintage Books, London, 2011)

Bowen, Jeremy, *The Arab Uprisings* (Simon & Schuster, New York, 2012)

Burns, J.M., *Leadership* (Harper and Row, New York, 1978)

Cahill, Reginald T., *Process Physics* (Flinders University, Adelaide, 2003 (available on web))

Coveney, Peter and Highfield, Roger, *Frontiers of Complexity* (Ballantine Books, New York, 1995)

Dalai Lama, *Becoming Enlightened* (Rider, London, 2009)

Damasio, Antonio, *Descartes' Error* (Vintage Books, London, 2006 (originally 1994))

Damasio, Antonio, *Self Comes to Mind* (Vintage Books, New York, 2010)

Davies, Paul, *The Fifth Miracle* (Simon and Schuster, New York, 1999)

Dawkins, Richard, *The Selfish Gene* (Oxford University Press, USA, 1990 (originally 1976))

Dawkins, Richard, *The Extended Phenotype* (Oxford University Press, USA, 2008 (originally 1982))

Dawkins, Richard, *The Blind Watchmaker* (W. W. Norton and Co. Ltd, New York, 1987)

Dawkins, Richard, *The Ancestor's Tale* (Phoenix Paperback, London, 2004)

De Bono, Edward, *New Thinking for the New Millennium* (Penguin Books, USA, 2000)

De Botton, Alain, *Religion for Atheists* (Penguin Books, London, 2012)

Dennett, Daniel C., *Consciousness Explained* (Penguin Books, USA, 1993)

Dennett, Daniel C., *Darwin's Dangerous Idea* (Penguin Books, USA, 1996)

Dennett, Daniel C., *Freedom Evolves* (Penguin Books, USA, 2004)

Dennett, Daniel C., *Breaking the Spell* (Penguin Books, USA, 2007)

Dent, Harry S. Jnr., *The Great Depression Ahead* (Free Press, USA, 2009)

Deutsch, David, *The Fabric of Reality* (Penguin Books, New York, 1998)

Deutsch, David, *The Beginning of Infinity* (Penguin Books, New York, 2012 (original edition 2011))

Diamond, Jared, *Collapse* (Penguin Books, London, 2005)

Diamond, Jared, *The World Until Yesterday* (Allen Lane Penguin Books, London, 2013)

Doidge, Norman, *The Brain that Changes Itself* ((Revised Ed.), Scribe Publications, Melbourne, 2010)

Druckerman, Pamela, *French Children Don't Throw Food* (Doubleday, Great Britain, 2012)

Bibliography

Edwards, Jonathan, *Freedom of the Will* (Dover Publications Inc., New York, 2012 (originally 1754))

Flannery, Tim, *Here on Earth* (The Text Publishing Company, Melbourne, Australia, 2010)

Gardner, Howard, *Frames of Mind, The Theory of Multiple Intelligences* (Basic Books, New York, 1983)

Gleick, James, *Chaos, a New Science in the Making* (Penguin Books, New York, 1987)

Goleman, Daniel, *Emotional Intelligence* (Bloomsbury Publishing, London, 1996)

Goleman, Daniel, *Social Intelligence* (Arrow Books, London, 2007)

Grayling, A. C., *The God Argument* (Bloomsbury, New York, 2013)

Greenfield, Susan, *The Neuroscientific Basis of Consciousness* (address to SALK Institute, 22nd May 2009)

Hamilton, Clive, *Earthmasters: Playing God with the climate* (Allen and Unwin, Sydney, 2013)

Harris, Sam, *The End of Faith* (Simon and Schuster Australia, Sydney, 2006)

Harris, Sam, *The Moral Landscape* (Bantam Press, UK, 2010)

Hawking, Stephen and Mlodinow, Leonard, *The Grand Design* (Bantam Press, UK, 2010)

Hill, Napoleon, *The Think and Grow Rich Action Pack* (Plume, New York, 1990 (original edition 1937)

Hofstadter, Douglas R., *Gödel, Escher, Bach: An Eternal Golden Braid* (Penguin Books, USA, 1983)

Hofstadter, Douglas, *I Am a Strange Loop* (Basic Books, USA, 2007)

Hollingdale, R. J., *A Nietzsche Reader* (Penguin Books, Sydney, 1977)

Huntington, S. P., *The Clash of Civilisations* (Simon & Schuster, New York, 2011)

Integro Learning Systems, *Everything DISC Group Culture Report.pdf* available at www.integrolearning.com.au

Jackson, Tim, *Prosperity without Growth: Economics For A Finite Planet* (Earthscan, UK, 2009)

Jaques, Elliott, *Requisite Organisation* (Caston Hall, Arlington, 1989)

Jarvis, Jeff, *Public Parts* (Simon & Schuster, New York, 2011)

Jukes, Andrew, *Four Views of Christ* (Kregel Publications, Grand Rapids, 1966)

Kahneman, Daniel, *Thinking, Fast and Slow* (Penguin Group (Australia), 2011)

Kauffman, Stuart, *Reinventing the Sacred* (Basic Books, USA 2010)

Klinger, Chris, *Bootstrapping Reality from the Limitations of Logic* (VDM Verlag, 2010)

Kosko, Bart, *Fuzzy Thinking* (Harper Collins Publishers, London, 1994)

Krauss, Lawrence M., *A Universe from Nothing* (Free Press, New York, 2012)

Bibliography

Lewis, C. S., *The Four Loves* (Harcourt Brace, 1971)
Livingston, Gordon, M. D., *How to Love* (Hachette Australia, Sydney, 2009)
Lovelock, James, *Gaia: A New Look at Life on Earth* (Oxford University Press, Oxford, 2009 (first published 1979))
Manning, G.L., Reece, B.L., *Selling Today* (Fourth Edition, Allyn and Bacon, Boston, 1990), Chapter 5
Mant, Alistair, *Intelligent Leadership* (Allen and Unwin, Sydney, 1997)
McRaney, David, *You Are Not So Smart* (Penguin Group (USA) Inc., New York, 2011)
Merleau-Ponty, Maurice, *Phenomenology of Perception* (Routledge, New York, 2014 (Original edition in French in 1945))
Mitchell, Donald A., *Buddhism. Introducing the Buddhist Experience* (Oxford University Press, New York, 2002)
Pinker, Steven, *How the Mind Works* (Penguin Books, USA, 1999 (Original edition 1997))
Pinker, Steven, *The Better Angels of Our Nature: Why Violence Has Declined* (Viking Adult, USA, 2011)
Rand, Ayn, *Anthem Centennial Edition* (Signet, USA, 2005 (Original edition 1938))
Rand, Ayn, *Atlas Shrugged* (Plume, USA, 2007 (Fiftieth Anniversary Edition; Original edition 1957))
Rand, Ayn, *Capitalism: The Unknown Ideal. Centennial Edition* (Signet, USA, 2005 (Original Edition 1966))
Rand, Ayn, *The Fountainhead* (Penguin Classics, USA, 2007 (Original edition 1943))
Ridley, Matt, *The Origins of Virtue* (Softback Preview, Great Britain, 1997)
Ridley, Matt, *The Rational Optimist: How Prosperity Evolves* (HarperCollins, New York, 2011)
Rumbaugh, Blaha, Premerlani, Eddy, Lorensen, *Object-Oriented Modelling and Design* (Prentice Hall, New Jersey, 1991)
Sarkar, P.R., *Human Society, Part 2* (Proutist Universal, Washington DC, 1967)
Shermer, Michael, *The Believing Brain* (St Martin's Griffin, New York, 2011)
Smart, Ninian, *The World's Religions* (Cambridge University Press, 1989)
Smith, Douglas K., *On Value and Values* (Financial Times Prentice Hall, USA, 2004)
Tolle, Eckhart, *A New Earth* (Penguin Group (Australia), Melbourne, 2009 (Original Penguin Edition 2006))
Wattles, Wallace Delois, *The Science of Getting Rich* (The Elizabeth Towne Company, London, 1910 (available in a 2013 edition from Greater Minds Ltd, London))

Williams, Xandria, *Love, Health and Happiness: Understanding yourself and your relationships through The Four Temperaments* (Hodder and Stoughton, Sydney, 1995)
Wrangham, Richard, *Catching Fire: How Cooking Made Us Human* (Basic Books, NY, 2009)
Wright, Robert, *Nonzero: The Logic of Human Destiny* (Vintage, NY, 2000)

FURTHER READING:

Alkek, David S., *The Self-Creating Universe* (Strategic Book Publishing, New York, 2009)
Camus, Albert, *The Myth of Sisyphus* (Full text available on the web (original edition 1942))
Camus, Albert, *The Outsider* (Penguin Classics, USA, 2000 (original edition 1942))
Carnegie, Dale, *How to Win Friends and Influence People* (HarperCollins Publishers, Sydney, 1999 (original edition 1936))
Clark, Arthur C. et al., *The Colours of Infinity: The Beauty and Power of Fractals* (Clear Books, London, Reprinted in Singapore, 2006)
Corey, Gerald, *Theory and Practice of Counselling and Psychotherapy* (4th Edition, Brooks/Cole Publishing Company, California, 1991)
Covey, Stephen R., *Principle-Centred Leadership* (Fireside, New York, 1992)
Crabb, Lawrence J. Jnr., *Basic Principles of Biblical Counselling* (Zondervan, UK, 1975)
Crabb, Lawrence J. Jnr., *Understanding People* (Marshall Pickering, UK, 1987)
Dawkins, Richard, *The God Delusion* (Bantam Press, Sydney, 2006)
Dawkins, Richard, *The Greatest Show on Earth* (A Black Swan Book, UK, 2010)
Demartini, John, *The Values Factor* (Penguin Group, Australia, 2013)
Diamond, Jared, *Guns, Germs and Steel* (Vintage, London, 1998)
Dunford, R.W., *Organisational Behaviour* (Addison-Wesley Publishing Company, Sydney, 1992)
Flannery, Tim, *The Weather Makers* (The Text Publishing Company, Melbourne, Australia, 2005)
Gaardner, Jostein, *Sophie's World* (Phoenix, London, 1995)
Hitchens, Christopher, *God is not Great* (Allen and Unwin, Sydney, 2007)
James, William, *The Principles of Psychology* (Encyclopaedia Britannica, *Great Books of the Western World* (2nd Edition, 5th printing, Vol. 53, Sydney, 1994, Chapter 28)), 859
James, William, *The Meaning of Truth* (Prometheus Books, 1997)
Kaku, Michio, *Hyperspace* (Oxford University Press, Oxford, 1994)
Kaku, Michio, *Visions* (Oxford University Press, Oxford, 1998)

Bibliography

Kiyosaki, Robert T., *Rich Dad Poor Dad* (TechPress Inc., USA, 2007 (original edition 1998))

Kolter, P., Chandler, P., Gibbs, R., McColl, R., *Marketing in Australia* (2nd Edition, Prentice Hall, Sydney, 1989)

Rand, Ayn, *Introduction to Objectivist Epistemology (Expanded Second Edition)* (Meridian Books, USA, 1990 (First published in *The Objectivist* July 1966 – February 1967))

Stanton, W.J., Buskirk, R.H., Spiro, R., *Management of a Sales Force* (8th Edition, Irwin, Sydney, 1991)

Tolle, Eckhart, *The Power of Now* (Hachette Australia, Sydney, 2004)

Wade, Nicholas, *The Faith Instinct* (Penguin Books, NY, 2009)

Wilczek, Frank, *The Lightness of Being* (Penguin Books, NY, 2008)

Glossary

Antinomy

An antinomy is a complete contradiction between two equally reasonable laws or principles. For instance, one could be the fleeting and emergent principle of self-organised living, which is negentropic taking. The other could be the timeless and self-eliminating principle of externally organised non-living, which is entropic giving. A fulfilled and happy life is a wonderful antinomic dance or musical improvisation between the doing (fleeting and active living) and the being (inanimate and timeless resting or passive perceiving and surrendering).

Antinomies cannot be resolved with classical uni-dependent thinking, but they often can be better understood using complex variables and interdependent or relativistic approaches.

Autopoiesis

All living systems must deal with boundary problems in two ways. They must maintain the physical boundary between self and the environment as well as reproduce, the combination of which is known as autopoiesis. They must also deal with contingencies or satisfy the viability constraints imposed by their boundaries and environs, also known as cognition.

Paul Bourgine and John Stewart, in their 2004 MIT paper labelled "Autopoiesis and Cognition", propose the thesis "A system that is both autopoietic and cognitive is a living system." This proposed necessary and sufficient definition of life is a very interesting one because it is a systemic or organisational definition, or an informational definition, rather than one that depends on a specific underlying biology, implemented though inorganic molecules for instance.

Axiom

An axiom is something that should not be easy to reject based on quasi-objective knowledge in which we separate, compare and distinguish one thing from another, but we can reject, perhaps unfairly, based on a possible or assumed explanation that we cannot currently verify. Objectivism's axiom of existence (i.e. existence must be taken as a given in its philosophy) could also be rejected on the basis of an unsuperstitious monism that holds, on an emergent basis, that existence is not a given, but continually emerges from a past that did not contain matter, space or time as we know them today. Further to this, an axiom acts as a veil over any knowledge it might assume and purport. It thus hides its inherent incompleteness, to use the same word Gödel used to describe his "Incompleteness Theorems" (see below).

Like antinomies, axioms cannot be resolved with classical uni-dependent thinking, but they often can be better understood using complex variables and interdependent or relativistic approaches.

Being and Doing, Fact and Values

Being and Doing are the two basic interactions of our introsomatic self or what others have called the dominance axis of Marston's/Integro's DiSC Model. In contrast, Facts and Values are the two basic interactions of the extrosomatic self or what others have called the sociability axis of Marston's/Integro's DiSC Model.

Biodiversity

Evolution, being a purely statistical phenomenon, encourages a diversity of solutions to the problem of environmental fit, which seems to make it more robust when challenged by circumstantial change. This same statistical drive towards diversity also happens in social environments.

For instance, Jared Diamond (2013) p.73 discusses "conventional monopolies" in traditional societies, in which one tribe would rely on another for certain goods that it could obtain by other means just to keep healthy relations. Why do they do this? They do this so they can rely on a wider economic safety net that they may need in times of environmental stress. That is, economic diversity makes traditional societies more robust.

Bootstrap

The bootstrap refers to the ability of systems to locally defy entropy by taking themselves uphill towards greater order against the vast downward trend towards disorder. This statistical process in living systems is what we call natural selection. In the wider scenario including non-living systems, we could call this natural emergence. We could also talk of artificial selection and artificial emergence.

Boundaries

The incorporation of more soft boundaries and their constraints seems to equate with the better exploitation of platform-independent or statistically significant information. That is, as systems better incorporate and anticipate boundaries and constraints, they become more robust, more evitable and more intelligent.

To put this idea another way, a measure of a system's intelligence is the number of constraints or contingencies it incorporates or anticipates. That is, intelligence emerges from the boundaries themselves. Without boundary constraints and their introduced instabilities there is no opportunity for knowledge and intelligence. Intelligence is a peculiar feature of living systems - because living systems create, expand and defend their borders.

Glossary

Butterfly Effect

The idea that a small change in one place in a complex web of systems can have large effects across that web. One example is that the single flap of a butterfly's wings, perhaps in South America, could trigger, through a complex maze of local interpenetrating relationships, huge and irreversible changes in weather patterns on the other side of the globe, for instance here in Australia.

We had an example of this in June 2011, with the temporary closure of Australian airports caused by very fine but aircraft-engine-crippling ash particles carried undispersed across the Pacific and Antarctica from Chile's erupting Cordon Caulle volcano. In this case, aircraft engines in Australia were also made temporarily local to the volcano in Chile by the expanding and then contracting manifold of interactions that connected them.

Cause-and-Effect

The old, geometric, uni-dependent and incomplete way of understanding reality's processes that misses the fundamental role of uncertainty, complexity, feedback, interdependency, Relativity and self-organising emergence

Chaos Theory

We can define chaos as initial conditions (pseudo-randomness) sullied by non-random feedback in complex systems. See Gleick (1987).

Chaos is not randomness because randomness strictly excludes feedback. However, if the initial conditions are indeterminate, then the chaotic system will be also.

Complexity

Complexity emerges, locally. Complexity cannot be directly infinite in a self-referencing universe, but it can always change. As Deutsch suggests by his book of the same title, we are always at the "beginning of infinity". See also Coveney and Highfield (1995).

The complexity of all systems can also be understood to enable and constrain the complexity of each component system because all systems in our self-referencing universe fully interpenetrate.

Consciousness-plus-Instincts

A way of summarising human nature and the drivers of human behaviour, emanating from our genes and memes.

CQ, IQ, SQ & EQ

Using the common concept of IQ (intelligence quotient), Goleman added two other intelligences - EQ (what he called emotional intelligence) and SQ (what Goleman called social intelligence). To these three intelligences, I have added CQ or competitive intelligence. I also prefer to think of IQ as extrospective intelligence and EQ as introspective intelligence.

Deontology, Consequentialism (including Utilitarianism) and Value-Ethics

Other ethical systems contrasted and compared with the Emergent Method's Vremetic Ethics.

Direct and Indirect; Full reality

Perhaps the most important examples of indirect phenomena are infinity and nullity. As asymptotes, they act as system attractors or repellers. Other phenomena, such as virtual particles, are said to fleetingly pop into and out of existence. While they exist as potentials outside of existence, like the square root of minus one, they would be other examples of indirect phenomena that can affect the direct. Full reality is the idea of including unresolved indirect phenomena within a definition of reality.

DISC, D-C-I S

Marston's model of human behaviour defines four quadrants - D, C, I & S - or what he called Dominance, Compliance, Inducement and Submission, but Integro's system calls Dominant, Cautious, Influential and Steady. I suggest these four classic personality traits correspond with CQ, IQ, SQ & EQ respectively.

Ego, Id and Superego

Freud's model might not be directly relatable to Damasio's triune model of self or emergentism's quadune self. However, one of the aims of the Emergent Method is to help make the Superego more realistically flexible and thus break down the dissonance between the Id's 'is' and the Superego's 'ought', with which dissonance the Ego and its narratives must continually struggle.

Emergence

Emergent properties are the result of a positive (i.e. more information-incorporating) mutation from a prior configuration or coagulation of relative disorder. That is, everything, including space, matter and time, emerges.

Emergence is the ability of the universe to continually bootstrap itself, forming islands, systems or levels of order that defy the prior levels of order. I do not ascribe to any form of supernatural or 'strong' emergence.

Emergentism is the philosophical theory or approach that emphasises how order arises in everything.

Entropy and Negentropy

The decay or loss of orderliness of all systems, or increasing entropy, is perhaps the logical consequence of the expanding universe model. Entropy is perhaps also the reason for the humanly perceived 'arrow of time' (always moving forward). Entropy begets disorder - it takes things downhill from a local order - it does not let anything stay comfortable.

Just as the universe finds a new local equilibrium or homeostasis of higher order, entropy messes with it.

Negentropy is entropy's complement - in the context of emergentism it is self-organising information and energy.

Epiphenomenalism

A philosophical theory, which suggests that while the mind is affected by the brain, it is not capable of affecting the brain/body. For instance, some consider free choice as an epiphenomenon, in support of a fully deterministic reality and in rejection of the compatibilist view. I prefer a self-contained, interpenetrating, emergent and effectual view of direct and indirect reality in which the mind and brain/body affect each other.

More generally, epiphenomena include such things as rainbows and mirages because while they are perceived by the observer, they do not exist apart from the observer. However, a more generous view suggests that even if such epiphenomena affect only the observer, then they qualify as normal phenomena after all, rendering the term of little value except in distinguishing between a first-person and third-person view.

Evitability

Dan Dennett's definition of evitability involves our ability to avoid unwanted consequences in the future through anticipation, and so it would be fully compatible with deterministic human action. More complex and adaptive living systems will possess more evitability than less well-equipped ones. The agent could be any living thing from insects to humans.

(Darwin's) Evolution

"The theory of evolution by natural selection, first formulated in Darwin's book "On the Origin of Species" in 1859, is the process by which organisms change over time as a result of changes in heritable physical or behavioural traits. Changes that allow an organism to better adapt to its environment will help it survive and have more offspring." See http://www.livescience.com/474-controversy-evolution-works.html for more details (last accessed May 2016).

Extended Phenotypes

A phenotypic effect is an effect or difference made on an organism, such as a bodily attribute or behaviour, originating from its local environment and one or more genes within it. A non-trivial phenotypic effect influences the survival prospects of the species' genes. A non-trivial extended phenotypic effect is an effect on an organism's survival prospects that is outside the body and its immediate behaviours such as animal artefacts (or survival-promoting technologies) like bird nests, beaver dams, termite mounds, spider webs or butterfly cocoons. A non-trivial extended phenotypic effect can also originate from one or more genes outside the organism.

Glossary

(Mandelbrot's) Fractals

Fractals are recursive feedback systems that show high-order complexity but organisation at many levels, and can be relatively easy to represent mathematically. Fractals can have a spatial dimensionality between the three we typically recognise (for example, more than 2 and less than 3). Our cosmic cycle may be a kind of imperfect fractal. See Coveney and Highfield (1995).

Fuzzy Logic

Fuzzy logic is a kind of quantum logic that permits degrees of truth (or degrees of logic) between 0 and 1 rather than at just 0 or 1. See Kosko (1994).

Gaia

We cannot deny life's interconnectedness. Our individual lives seem to be part of a larger organism called our society, and our society along with our biosphere seems to be part of a larger organism James Lovelock and others would call Gaia.

Does the bird make the nest a nest or does the nest make the bird a bird? Does microbial life support the atmosphere or does the atmosphere support life? Clearly, the best way to frame such questions is through the concept of emergent interconnectedness.

(Einstein's) General Relativity

Physicist John Wheeler's characterisation is perhaps the simplest – "Matter tells space how to curve. Space tells matter how to move."[1] Everything we perceive is relativistic - there are no unidimensional absolutes, not even of time itself.

Hard- and Soft- Determinism

To hard determinists there can be no room for free will if prior events cause all effects. In effect, the argument suggests that if there are nil independent causes, there can be no freedom, harking back to what I believe to be the faulty ancient Greek idea of a Prime Mover or First Cause. Hard determinists would also add that while humans can make choices, they cannot bear a moral responsibility for those choices because antecedent causes locked them in.

I am a soft determinist. That is, I believe things locked into the past drive us, but we can also adjust behaviours for future contingencies through the emergent properties of subjective and unfolding consciousness. I would therefore argue that we can teach ourselves, and thus bear moral responsibility and enjoy moral agency.

(Heisenberg's) Uncertainty Principle

The Uncertainty Principle has gone through several iterations. The most common iteration is Earle Kennard's formulation, which says you cannot suppress quantum fluctuations of both position and momentum below a certain limit. See Furuta, Aya, *One Thing Is Certain: Heisenberg's Uncertainty Principle Is Not Dead,* Scientific American, 8th March 2012.

Heterophenomenology

Referring to Dennett, Daniel C., *Consciousness Explained* (Penguin Books, USA, 1993), heterophenomenology (or the phenomenology of another person) is a term created by Dennett and is used to define a third-person, scientific approach to the study of a target person's self-observations, combined with all other available evidence, to determine his/her mental state. In one sense, emergentism is about carefully applying the method and benefits of Dennett's heterophenomenology to ourselves.

Hofstadter's Strange Loop of Consciousness

We build mental schemas of the environment in our brains, and attenuate certain features of them and amplify others, perhaps in conscious anticipation of likely futures and perhaps as we apply our values. That modified environment within merged with our conscious observations, becomes our own local reality and an emergent cause of our future subconscious computations and actions. That is, as we through many iterations or Hofstadter's strange loops of self, act on largely self-generated and self-referencing subjective mental schemas rather than actual reality, we find some level of interdependence, shared agency, or pseudo-freedom with reality which Dennett may describe as an illusion, but which I prefer to label emergent. It may be useful to think of the mind/body's Strange Loops or schemas as fractals that can display self-similarity across many dimensions.

Homeostasis

A state in living systems that corresponds with equilibrium in non-living systems. It requires the evitability, and involves the dynamic management, of those living systems.

(Gödel's) Incompleteness Theorems

A system cannot demonstrate its own consistency or complete set of truths. Gödel showed us that all formal mathematical proofs are incomplete because they rely on axioms that are outside the proof itself. Some would say that explaining something by using something unexplained is no explanation at all – which would seem to undermine all of our mathematical certainty. However, a good explanation is better than none and a tentative, incomplete knowledge is better than none if it is reliable in all the ways we use it.

Instinctive bias and the mind-games of consciousness

We are consciousness-plus-instincts, but both consciousness and instincts have their weaknesses. Biases drive instincts and mind games can drive consciousness. We need the Emergent Method to guide self away from its own pitfalls and towards its strengths.

Interpenetration

Quantum Theory has presented us with neither a pure wave model nor a pure particulate model of matter or space, but rather an interpenetrating wave-particle antinomy. The interpenetration of matter and space, energy and information, or the direct and indirect, is also the nature of Relativity. Likewise, our extrospective facts and introspective values interpenetrate in the mind/brain. This is why we need a method that can deal with more than scientific facts alone. That is, we need the Emergent Method.

Laissez-faire Capitalism

Laissez-faire capitalism is a form of capitalism in which markets are free to arise and regulate themselves with little or no government intervention. The role, size, taxation and social program of government tends to be much smaller under such a system. In this sense, in the 1960s the form of capitalism practiced in Hong Kong would have been more laissez-faire than that practiced in Sweden as it was growing its large State-based welfare program. We can consider Australia, as somewhere in between those two extremes in the 1960s, as a 'mixed' system.

Life-and-Being Here-and-Now

Evolution's life-and-being here-and-now approach to genetic and memetic advancement is, if it promotes survival it persists. Natural selection does not start with a pure design idea - it just reinforces mutations that promote survival. To a large degree, this is what the Scientific and Emergent Methods also suggest.

Life-stance

Society seems to leave us with three alternatives: The axiom 'existence exists' and atheism, the idea that 'existence is uncertain' and agnosticism, or the idea that 'existence is subordinate' and blind faith in 'Something'. That is, if we are to take a certain life-stance, then it seems we must make a blind assumption, make a blind denial or make a blind belief. The universe's lesson to us is simply this: You may hold whatever life-stance or method of self-actualisation you wish but the focus must be your life, your self-actualisation, your happiness, your sustainability, your societal flourishing and your legacy – because 'existence emerges'. The universe offers us a fourth life-characteristic: Emergent Monism.

Monism and Secular Humanism

Statistical emergence is fundamental to this and all possible universes. In this sense, our universe does not have innate laws, just perceptible properties, some of which are longstanding and easily measurable, and others which are not so obvious. Likewise, in the sense that explicit moral phenomena and moral agency emerge from matter's arrangements, they must also be an implicit part of reality - or further perceptible properties of reality. These ideas capture what I mean by an interpenetrating and emergent monism. The universe is not dualistic in some kind of transcendental sense - it is just wonderfully emergent. From this philosophical worldview, I would hold many things in common with a secular and unsuperstitious humanism.

Meme

Dawkins suggests that cultural transmission is analogous to genetic transmission and so he coined a new term, the "meme", to capture the most basic unit of cultural mutation, self-interest and natural selection. Memes can consist of cultural ideas or bits of ideas, jokes, sayings, songs, a few bars of a song, parts of fashions, technologies, ceremonies, art-forms, socio-political systems, beliefs, values, etc. that reside in survival machines (in this case, consciousness within the living structures of our brains) just like genes. The full grouping of memes within one brain might more loosely relate to all the genes within the DNA of a specific organism. Similarly, we might loosely relate all the groupings of memes in a particular macro-culture to a genetic species, competing with other species in the shared environment to survive.

Multiverse

The multiverse concept grew out of the success of particle physicist Alan Guth's inflationary model in explaining our visible universe. It suggested there could be other inflationary-bubble universes that we can't see. It also suggested that new inflationary-bubble universes could fractally bifurcate from existing ones. Theoretical physicist Lee Smolin's idea of the evolution of many universes seeded from black holes in prior universes also has some appeal.

I would like to see stronger evidence of the possibility of universal collapse following expansion and so on, rather than the continual expansion of the currently preferred paradigm, to lock in the idea of a fractal multiverse.

Nonzero-Sum

Robert Wright's nonzero-sum outcome is achieved in a win-win situation, that is, where one party's gain is not necessarily the other party's loss. The virtuous interpenetration of the immaterial and material seems to be how we discover the bio-diverse win-win (or nonzero-sum game) and materialise wealth against all odds - easily, naturally and repeatedly.

Objectivism

http://en.wikipedia.org/wiki/Objectivism_(Ayn_Rand) (accessed 2010), quoting Rand (1957), p. 1074 says, "My Philosophy (objectivism), in essence, is the concept of man as a heroic being, with his own happiness as the moral purpose of his life, with productive achievement as his noblest activity, and reason as his only absolute."

Object-Oriented

A model of better organisational structure is suggested by modern object-oriented database methods (see for instance Rumbaugh (1991)), which would see each organisation arranged into a highly adaptable matrix of stakeholder classes (objects) and corporate objectives rather than a uni-dependent and brittle matrix of corporate functions (such as Sales and Marketing, Warehousing, or Corporate Services) and corporate objectives. Stakeholder classes may change over time but they never become irrelevant, whereas corporate functions may become irrelevant over time. An object-orientation of organisations would suggest they focus their efforts on their stakeholders' needs and wants rather than the corporation's functional proficiencies alone.

Pain-Body

The idea here is that the body stores the pain and stresses of its egocentric thought life, which we could also name consciousness gone astray in its mind-games. Tolle labels this bodily store or side effect of self-defeating mind-games the "pain-body".

Personal Moral Congruence

The Emergent Method is largely and simply the maintenance of limited but dynamic personal moral congruence. A definition of human success is personal moral congruence (or moral homeostasis or sustainable wellbeing). Perhaps more than moral homeostasis, personal moral congruence represents a state of physical, ego, mental, social and spiritual homeostasis or wellbeing.

Phaseal

Synonymous with 'phaseal' is 'cyclical', although 'phaseal' focuses on the phases passed through in the cyclical journey rather than the outcome of each cycle traversed.

Positive Extrospection and Self-enlightened Introspection

These are at the two opposing ends of the extrosomatic axis. They are the self-restraints of consciousness that describe or circumscribe the extrosomatic axis.

Proto-Self, Core-Self, Autobiographical-Self, and Public-Self

This model of quadune self adds Jarvis' Public-self to the end of Damasio's triune model of self, beginning with the Proto-self at the mostly unconscious level, then the Core-self at the mostly subconscious level and ending with the Autobiographical-self at the mostly conscious level.

Quantum Mechanics, Quantum Theory

Quantum Theory predicts that a photon, under certain conditions, can occupy an indeterminate position in space that we can best understand in terms of a wavelike probability curve. Further, two photons can be subject to a quantum entanglement that causes them to act in a kind of systemic unison even though they may be far apart in geographical space. That is, at some levels of Nature's everyday dance, classical science's uni-dependent cause-and-effect explanations are insufficient or incomplete. Nature is statistically probabilistic or unpredictable.

Real and Virtual Particles

Just as we cannot fully appreciate the pure note of a musical instrument except within a background of relative silence, so material things cannot manifest themselves without immaterial but essential space (or virtual particles). Space makes room for the five forces or interactions we currently recognise, and any others we may not recognise yet.

In these two catch phrases – emergent self-organising and emergent self-actualising – we see the clear naturalistic link between our universe's virtual particles at the most microscopic level and our virtual consciousness at the highest level laid bare.

Reality and Existence

I tend to take the widest possible definition of reality as everything that has, does, or potentially exists. I thus see reality as the palette from which existence is painted. I apologise if this is in disagreement with your own definition, but at least this definition clarifies my use of the word. Reality by this definition thus includes the first-person view, the indirect, the virtual, and the imaginable. It includes fantasies to the extent of their effects in existence.

Risk-Weighted Competition and Normative Cooperation

These are at the two opposing ends of the extrosomatic axis. They are the instinctive self-restraints that describe or circumscribe the extrosomatic axis.

Self-Actualisation

Emergentists do not demand that you give up any beliefs in the supernatural or transcendental, but we will challenge you to consider two things: 1. the efficacy of your beliefs in terms of your wellbeing and self-actualisation, and 2. the evidence from reality for an interdependent life-stance freed from axiomatic bases. As an emergentist, I see the promotion of your wellbeing or self-actualisation, rather than your life-stance, as the core issue facing you - and all humanity with you.

Emergentism is about identifying and promoting that which is self-organising to our introspective consciousness and in accord with our self-actualisation. It is also about identifying but minimising that which is contrary to our self-actualisation. Virtuous values or vremes are the most important tools we can use to rebel against the tyranny of our blindly selfish replicators (i.e. sometimes short-sighted genes and often short-sighted memes that cannot escape finite resolution as can human imagination). It seems we have to tread even more carefully than the most diligent of scientists (and his/her Scientific Method), if we are to uncover the introspective, contingent truths that will enhance our self-actualisation (or artificial emergence). This careful uncovering is what the Emergent Method offers.

Selfishness

Perhaps selfishness at its most basic level is simply another word for locally recurrent and the local algorithmic feedback that is common to all statistical processes in both the living and non-living. That is, life is not essentially different to non-life in respect to its systemic local focus.

Selfishness emerges in reality along with everything else. It is the borders or boundaries of space and matter that brought about differences and it is these differences that brought about all else, including selfishness in living things, through the mechanism of emergence. Life's selfish taking of information from spacetime is also a giving of shape and form (or organism) to disorganised matter. Conversely, non-life's unselfish giving of itself to spacetime is also a kind of selfish taking of shape and form away from matter

Is it life and its knowledge and meaning that is ultimately selfish, or is it spacetime? Alternatively, are all these components selfish in a kind of asymmetrical and interpenetrating informational cycle? Exploitable information in all its forms seems to beget and seek more of itself in an expanding universe of ever more numerous but softer and self-contained border constraints. I suspect this concept presents the clearest picture of selfishness. Without intending any transcendental connotation, your life and mine is part of this lattice or web of transformed and organised information that spreads itself across the universe with every particle of matter.

Knowable life cannot bootstrap itself without forms that require and take information, impetus and food-energy from the local environment. Conversely, it is in the self-seeking taking that life forms also give knowledge, meaning and value back to life's genetic code, and organisation back to reality's spacetime code or system.

Sensor/Actor, Believer/Claimer, Observer/Agent and Private/Public Self

A relativistic and quadune model of self that adds to a triune model the essential role of the environment. Its four components correspond with Damasio's Proto-self, Core-self and Autobiographical-self as well as Jarvis' Public-self, respectively.

Supersymmetry

Space and matter must fit almost perfectly together hand-in-glove like two pieces of the one jigsaw puzzle. Assuming no losses to an external multiverse, space must be the logical complement of tangible matter, except for the asymmetry or asynchronicity that divides them. This is the basis of the theory of Supersymmetry. According to the theory of Supersymmetry, space via its interaction wave-particles houses the complement of real matter and its array of particles, as well as many other of its own virtual particles. It is of interest in the context of the philosophy of emergentism because it promotes the idea of this universe's self-containment and self-reliance - indeed its blind and aware selfishness.

The Emergent Method

The Emergent Method is a way of adapting the Scientific Method, usually applied to the search for scientific facts, to those personal, first-person hypotheses, values, or beliefs inaccessible to science. The Scientific Method could perhaps be better described as the Extrospective Method and the Emergent Method as a methodology that applies to both our extrospections (and their facts) and our introspections (and their values). In this sense, the Scientific Method is a subset of the Emergent Method.

The Law of Introspective Selfishness

The law of introspection is one of duty or responsibility to self - in short, it is a law of selfishness or self-enlightenment as defined and discussed in Part 1. For humans with a more enlightened level of selfishness, this law of introspection includes a duty of:

- self-preservation
- self-respect, honour
- self-love, care, regard
- self-assessment
- self-control, discipline
- self-improvement
- self-actualisation, fulfilment, biological and sociological 'fit'
- self-esteem
- protection of self-image, public-image and legacy

Triune and Quadune Models Of Self

There are several models of our triune human nature - Freud's model of Id, Ego and Superego perhaps being the most famous. Closely related is a model that sees our human nature made up of Freud's unconscious mind, the preconscious mind and the conscious mind - or what is more commonly called unconsciousness, subconsciousness and consciousness. Perhaps the tightest definition of our triune nature from an evolutionary point of view is Damasio's formulation of the Proto-self, the Core-self and the Autobiographical-self.

My own formulation is to essentially add Jarvis' Public-self to Damasio's triune model, and in so doing assume that they are concepts that are mutually exclusive. By doing this we arrive at a quadune self that includes the environmental reflection of self as a necessary part of the fullest model of self. The four parts of self I have described is the Sensor/Actor, the Psychosocial Believer/Claimer, the Self-aware Observer/Agent and the Private/Public Self. Each part reflects a relativity and a kind of Gestalt switch in its operation.

Value-Set

A list of possible values, virtues, character traits, stances we take, or tools we use, or predispositions we hold with respect to life's situations, have been arranged into four tables of Pillar 1. In the interests of brevity and openness, I refer to these four tables as value-sets and their contents as values. They are broken into Table 1: The Risk-weighted Value-set, corresponding to the DiSC model's Dominant quadrant, Table 2: The Positive Value-set, corresponding to the Cautious quadrant, Table 3: The Normative Value-set, corresponding to the Influential quadrant, and Table 4: The Self-enlightened Value-set, corresponding to the Steady quadrant.

Veil

To some there is a presumed veil draped across a presently unknown and unknowable reality. This is essentially a dualistic and uni-dependent view, ignorant of the power of relativity. My own view as a materialistic monist is that no such veil exists. The study of emergentism suggests that we can reduce the problem of understanding life to one that is knowable on this side of the veil. Further, consciousness and its virtues on this side of the possible veil across unknowable reality do not come without empowering life forms. Belief in the fuzzy, interdependent and relativistic self-organising principles of the universe on this side of the veil is not compelled to include belief in a personally embodied afterlife from the other side of the veil. However, perhaps much more importantly, a life lived according to the quasi-immutable or fuzzy values transcend all the possibilities beyond the veil.

Vremes

We can define vremes as those memes that enable us to escape the tyranny of the blindly selfish replicators and thus bring about our higher consciousness. We could define a vreme as a highest-order meme. A vreme is a meme or self-actualising principle. It is subject to artificial selection or adoption by volitional and rational beings, as a means of understanding and motivating intentional behaviour, perhaps like a personal credo might do. In this sense, vremes are consciously selected mores and taboos.

The way we subjectively think, feel and desire, while it may be faulty at times, is not always without order or a kind of 'personal objectivity'. In this sense, vremes or virtuous values are intra-personal or intra-somatic facts. That is, vremes are facts that arise from within the body rather than outside it. And the basis of our intra-somatic facts lies in the body's highly evolved bio-regulatory systems that urge or force us to limit our activities to within preferred operating conditions – conditions that maximise the opportunity for us to pursue and enjoy wellbeing. A vreme is a kind of higher-order intra-somatic chemical required by the mind-body to regulate our socially conscious behaviours within homeostatic limits.

Glossary

Vremetic-Self

We can understand the vremetic-self as the highest form of the well-integrated quadune self that arises after the Private/Public self and Trader self is well known.

Warrior-self, Lawmaker-self, Philosopher-self and Trader-self

This quadune model of self arises after the four parts of self, through many cycles, develop and interact in the individual and in society. That is, Warrior-self grows out of the Sensor/Actor or Damasio's Proto-self; Lawmaker-self grows out of the Psychosocial Believer/Claimer or Damasio's Core-self; Philosopher-self grows out of the Self-aware Observer/Agent or Damasio's Autobiographical-self; and Trader-self grows out of the Private/Public Self that aligns with Jarvis' Public-self.

---

[1] Barrow, John D., *The Book of Universes* (Vintage Books, London, 2011), 52

Index

Index

Index

Index

Index

Index

Index

Index

Index

Index

Index

Index

72170045R00327

Made in the USA
Lexington, KY
27 November 2017

Work and Employment in a Changing Business Environment

Work and Employment in a Changing Business Environment

Stephen Taylor
and Graham Perkins

First published in Great Britain and the United States in 2021 by Kogan Page Limited

2nd Floor, 45 Gee Street
London
EC1V 3RS
United Kingdom

122 W 27th St, 10th Floor
New York, NY 10001
USA

4737/23 Ansari Road
Daryaganj
New Delhi 110002
India

www.koganpage.com

Kogan Page books are printed on paper from sustainable forests.

ISBNs

Hardback 9781398600225
Paperback 9781398600201
Ebook 9781398600218

British Library Cataloguing-in-Publication Data

A CIP record for this book is available from the British Library.

Library of Congress Cataloging-in-Publication Data

Names: Taylor, Stephen, 1965- author. | Perkins, Graham (Graham Michael), author.
Title: Work and employment in a changing business environment / Stephen Taylor and Graham Perkins.
Description: London ; New York, NY : Kogan Page, 2021. | Includes bibliographical references and index. | Summary: "Work and Employment in a Changing Business Environment is the definitive textbook for the CIPD Advanced Level 7 module of the same name. It covers the theory and practice of the developments affecting the world of work and the HR industry including technology and globalisation through to labour changes, policy and regulation. Most importantly, this brand new textbook covers the key elements that students on HR masters courses need to implement and manage in their future careers including flexible working, agility, sustainability, wellbeing, diversity and ethics in the people profession. Including case studies to help students see how the theory applies in practice, reflective practice activities to help them think critically about the content as well as 'explore further' boxes to encourage wider reading. Work and Employment in a Changing Business Environment is ideal reading for all postgraduate students on both CIPD and non-CIPD accredited courses. Online supporting resources include an lecture slides, multiple choice questions and activity feedback"– Provided by publisher.
Identifiers: LCCN 2021012238 (print) | LCCN 2021012239 (ebook) | ISBN 9781398600201 (paperback) | ISBN 9781398600225 (hardback) | ISBN 9781398600218 (epub)
Subjects: LCSH: Employees–Training of. | Employee training personnel.
Classification: LCC HF5549.5.T7 T315 2021 (print) | LCC HF5549.5.T7 (ebook) | DDC 658.3/124–dc23
LC record available at https://lccn.loc.gov/2021012238
LC ebook record available at https://lccn.loc.gov/2021012239

Typeset by Integra Software Services, Pondicherry
Print production managed by Jellyfish
Printed and bound by CPI Group (UK) Ltd, Croydon CR0 4YY

CONTENTS

LIST OF FIGURES AND TABLES

Figures

Tables

WALKTHROUGH OF THE BOOK'S FEATURES

CASE STUDY 11.1

Heathrow Terminal Five

Heathrow is the UK's largest airport, and one of the most significant global hubs for air travel as ranked by passenger traffic. Heathrow is located to the west of London and has good transport connections, being in close proximity to both the motorway network and rail links. The airport has expanded substantially over its operational life, and at the time of writing there are ongoing discussions regarding the addition of a third runway. This case study focuses on the opening of Terminal Five at Heathrow, a substantial event in the airport's history, which was some twenty-five years in the planning. The terminal cost the airport operator BAA over £4 billion, building work took six years to complete, and it was opened by the Queen towards the end of March 2008 amid much public interest.

Immediately following the opening, however, there was an array of issues and 'technical glitches' that resulted in the cancellation of a large number of flights. Thirty-four were cancelled on the first day (Woodman, 2008), and around 300 flights in total were impacted during the problematic opening period. Upon arrival at the new terminal, large queues of passengers formed at the check-in and transfer desks, with the new baggage system creating a backlog of some 15,000 bags (BBC News, 2008), compounding issues caused by poor signage and passenger flow through the terminal. According to the BBC, the immediate cost to British Airways (the sole initial airline operating from Terminal Five) amounted to £16 million, with longer-term direct costs estimated at £150 million, together with reputational damage for themselves and the airport operator, BAA.

Press reports at the time noted that the main cause of the initial disruption was poor planning, and in particular the mismanagement of staff. Quotes in press reports (for example, see Woodman, 2008) highlighted issues with staff car parking, delays in employee security screening, and lack of staff familiarisation with the new terminal. In short, poor training meant that staff were unable to operate the new equipment effectively, and some were late for their shifts as they simply could not find somewhere to park their vehicle (BBC News, 2008). The BBC went on to highlight that some employees had not attended training courses and trial runs at the new terminal due to poor levels of morale and a lack of trust between themselves and company managers. Once operations began on that opening morning, long-standing issues with employee engagement meant that initial problems were compounded, resulting in the cancellations and economic losses previously outlined.

CASE STUDIES

A range of case studies from different countries and sectors illustrate how key theories and ideas apply in practice.

PAUSE AND REFLECT

Please pause here for a few minutes and reflect upon the points that have been introduced above. What examples can you find of organisations going through change driven by the introduction of new strategies or altered objectives?

PAUSE AND REFLECT

Pause and reflect boxes include questions on key topics in each chapter to allow deeper engagement with the content and to encourage critical thinking.

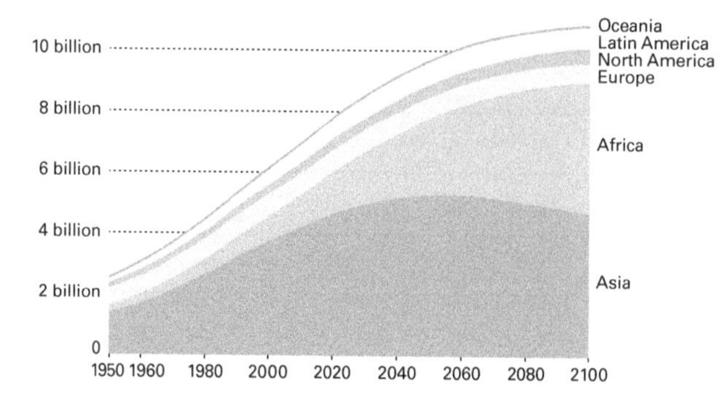

FIGURES AND TABLES

Figures and tables highlight key concepts to act as visual learning aids.

 References

Beatson, M and Zheltoukhova, K (2015) *Productivity: Getting the best out of people*, CIPD, London

Cooper, R (2009) UK jobs in the recession: Alternatives to redundancy, *Telegraph*, 24 August

Giles, C (2018) Britain's productivity crisis should be keeping the country's politicians and civil servants awake at night, *Financial Times*, 13 August

Graeber, D (2018) *Bullshit Jobs: A theory*, Allen Lane, London

Harari, D (2107) Productivity in the UK, House of Commons Library Briefing Paper, 06492

Ilzetzki, E (2020) Explaining the UK's productivity slowdown: Views of leading economists, VoxEU/CEPR, https://voxeu.org/article/uk-productivity-puzzle-cfm-survey (archived at https://perma.cc/D699-XK3M)

Institute of Directors (2018) *Lifting the Long Tail: The productivity challenge through the eyes of small business leaders*, IOD, London

Johnson, C (2003) The problem and management of sickness absence in the NHS: Considerations for nurse managers, *Journal of Nursing Management*, 11 (5), 336–42

Marchington, M, Wilkinson, A, Donnelly, R and Kynighou A (2020) *Human Resource Management at Work: The definitive guide*, 7th edn, Kogan Page, London

Paauwe, J, Guest, D and Wright, P (2013) *HRM and Performance: Achievements and challenges*, Wiley, Chichester

Pidd, H (2009) Honda's factory re-opens, *Guardian*, 2 June

Sparrow, P and Otaye-Ebede, L (2017) HRM and Productivity, in *A Research Agenda for Human Resource Management*, ed P Sparrow and C Cooper, Edward Elgar, Cheltenham

Wolf, M (2020) The UK's employment and productivity puzzle, *Financial Times*, 30 January

REFERENCES

References at the end of each chapter provide a comprehensive list of books, articles, reports and websites to support wider reading and further exploration of the topic.

01
Introduction

The main objective of this book is to provide people who are studying human resource management (HRM) and associated subjects with an introduction to two areas:

1. Major contemporary trends in the HR business environment and some debates about these.
2. Significant areas of HR activity that derive from or are being given additional prominence as a result of environmental developments.

We have structured our book to reflect the contents of Unit 7CO01 'Work and working lives in a changing business environment' in the Chartered Institute for Personnel and Development's syllabus for Level 7 advanced diplomas in Strategic People Management and Strategic Learning and Development.

The documentation provided by the CIPD is thus a good place to start when introducing the contents and aims of our book. So we will start there. This introductory chapter will then conclude with some observations about employment practice in the contemporary business environment and a brief discussion of two tools that can be used in order to analyse an organisation's external business environment.

The CIPD Advanced Diploma syllabus

The programme documentation starts with some material introducing the unit:

> This unit extends understanding of the interaction between the commercial business environment and likely future developments in the world of work, employment and the management of people. It discusses the range of people practices that are growing in importance, including those relating to ethics and sustainability, employee wellbeing, equality, diversity and inclusion.
>
> You will focus on major ways in which leaders and managers working in people practice are responding to globalisation and its significance for work and employment.

> You will investigate the current and future thinking within organisations around technological developments and how new agendas are evolving. You will evaluate social, demographic and economic trends and how developments in public policy affect people practice. Effective leadership of change, innovation and creativity, including the key interrelationships between ethics, sustainability, diversity and wellbeing, will be covered in this unit. Finally, you will critically analyse policy, practice and corporate social responsibility and the ways in which people professionals can apply and promote them for organisational productivity.

The unit documentation then goes on to identify four learning outcomes, each of which is subdivided into four further parts:

Learning outcome 1: Understand ways in which major, long-term environmental developments affect employment, work and people management in organisations

1.1 Assess globalisation and its long-term significance for work and employment. Impact of globalisation on business and working lives; major developments in the global business environment; industrial restructuring and the changing nature of employment; volatility and competitive intensity.

1.2 Critically evaluate organisational vision of the current and future impact of technological trends on working life. AI, advanced robotics, virtual reality, autonomous vehicles; debates about the impact of technological developments on employment, organisational management and the experience of working life; ways in which technological developments affect organisations, management, the experience of work and employment.

1.3 Evaluate the impact of long-term social and demographic trends for work and employment. Population ageing; patterns of demographic change; individualism; ethical awareness; attitudes to work and diversity; the role of organisations in shaping society and social change.

1.4 Appraise the significance of long-term economic trends for work, employment and management practice in organisations. Affluence and inequality; deindustrialisation and the rise of a service-based, knowledge-based economy; long-term shifts in macro-economic policy.

Learning outcome 2: Understand current and short-term developments in the people management business environment

2.1 Evaluate current developments in the media, technological and economic environments and their significance for people management. Debates about current trends in work and working lives; the impact of social media and analytics on people practice; the current prospects for national and international economic development.

2.2 Assess developments in public policy which are affecting work, employment and people management in organisations. The significance for people practice work of contemporary economic, industrial, education and employment policy; the impact of government policy on the people practice agenda.

2.3 Analyse major legal and regulatory developments in employment and the labour market, including the importance of mitigating risk. Employment policy; major employment rights and their enforcement; health and safety regulation.

2.4 Critically discuss current labour market trends in the supply of and demand for skills. Upskilling and down-skilling; undersupply and oversupply of skills; major contemporary labour market developments and their impact on HR work in organisations.

Learning outcome 3: Understand how change, innovation and creativity can promote improvements in organisational productivity

3.1 Analyse the effective management and leadership of change in organisations from a people management perspective. Structural and cultural change; leading change effectively; major theories of change management; principles of the psychology of change; effective leadership and communication during periods of change; increasing capability and readiness for change; major theories of effective change management and debates about these.

3.2 Examine ways that organisations address resistance to change and recognise the levers that will achieve and sustain change. The role of employee involvement in successful change; encouraging engagement with change agendas; sustaining change; the role played by change consultants and change agents; improving organisational agility. Levers to achieve change; clear rationale for change; timely, meaningful involvement and consultation with affected parties; communication; process alignment; training and development. Mechanisms for sustaining change; monitoring and review; open feedback channels; ongoing training and development, communication of outcomes/benefits realisation.

3.3 Evaluate theory and practice in the fields of flexible working and organisational resilience. Different forms of flexible working and debates about these; non-standard contracts and evolving forms of work; promoting organisational resilience.

3.4 Assess the contribution of people management aimed at improving organisational productivity, creativity and innovation. Links between people practice interventions and organisational productivity; effective people practice responses to increased competition in product and labour markets; promoting creativity and innovation.

Learning outcome 4: Understand the key interrelationships between organisational commitment to ethics, sustainability, diversity and wellbeing

4.1 Propose initiatives aimed at improving an organisation's ethics and values. Major debates about business ethics and organisational values; the business case for ethical and sustainable people practice and policy; ethical dilemmas in people practice work; people practice interventions to improve ethical standards in organisations, and when working with suppliers, contractors and other organisations; debates about the ethics of people analytics.

4.2 Evaluate policy and practice aimed at improving employee wellbeing in an organisation. The business case for promoting employee wellbeing; people practice interventions that support improved wellbeing in organisations; debates about safeguarding and bullying at work.

4.3 Critically evaluate theory and practice in the fields of corporate social responsibility and sustainable management practices. The principles of corporate social responsibility; stakeholder approaches to management; principles of sustainability and ways in which people professionals can apply them.

4.4 Critically discuss how the effective promotion of greater equality, diversity and inclusion in organisations supports people practice. Debates about diversity and inclusion in organisations; approaches to making organisations more equal, diverse and inclusive.

PAUSE AND REFLECT

Which of these learning outcomes do you consider to be the most important for you now in your present role? Which do you think will be most important for you in the future and why?

Each of these learning outcomes is also associated with specific assessment criteria, one for each of the sixteen statements. These are as follows:

1.1 Understand, explain and debate globalisation and its long-term significance for work and employment.

1.2 Lead discussion and thinking in organisations on the current and future impact of technological trends on working life.

1.3 Evaluate the impact of long-term social and demographic trends for work and employment.

1.4 Appraise the significance of long-term economic trends for work, employment and management practice in organisations.

2.1 Demonstrate understanding of current developments in the media, technological and economic environments and their significance for HR/L&D work.

2.2 Understand, explain and debate developments in public policy which are affecting work, employment and HR/L&D practice in organisations.

2.3 Review and evaluate major contemporary regulatory developments in the fields of employment and the labour market.

2.4 Appraise and explain current trends in the supply of and demand for skills.

3.1 Lead and provide sound advice on the effective management and leadership of sustained change in organisations from an HR/L&D perspective.

3.2 Review the ways that managers address resistance to change and build agile, change-ready organisations.

3.3 Debate and evaluate theory and practice in the fields of flexible working and organisational resilience.

3.4 Demonstrate understanding and provide advice about the contribution of HRM/L&D practice to improving organisational productivity, creativity and innovation.

4.1 Understand, evaluate and lead initiatives aimed at improving an organisation's ethics and values.

4.2 Develop policy and practice aimed at improving employee wellbeing in an organisation.

4.3 Critically analyse theory and practice in the fields of corporate social responsibility and sustainable management practices.

4.4 Review, evaluate and provide sound advice on the effective promotion of greater equality, diversity and inclusion in organisations.

Relationships between employing organisations and their environments

There are different ways of thinking about an organisation's relationship with its business environment. For many, for much of the time, the approach taken is essentially reactive. Managers respond to environmental change after it occurs. They have either been unable or unwilling to look ahead and plan to meet the change when it arrives. They are thus obliged to firefight, dealing with issues as they arrive, on the hoof. In a fast-changing and unpredictable environment this is often the best way to approach things. However, in more stable environments where change is not occurring so fast or so unpredictably it makes more sense to plan carefully so that the organisation is well placed when environmental developments occur to meet challenges more effectively. The longer we have to plan for change, the better our response is likely to be.

Some organisations are sufficiently powerful or innovative that they are not only able to respond very effectively to environmental developments but are also in a position to shape aspects of their environments themselves. This could be said, for example, of the big technology companies that have enjoyed such success in recent years like Google, Microsoft, Amazon, Netflix, Uber and Apple. They have developed technologies and business models that have brought change on others. They have also successfully lobbied governments and pioneered new ways of managing people that have proved highly influential. Organisations like this are the movers and shakers of the corporate world. Many public sector organisations are also well placed partly to shape their own business environments. Most employing organisations are not, however, in this position. They are obliged to restrict themselves to responding thoughtfully and effectively when aspects of their business environment change.

In management theory, 'resource dependency thinking' starts with the assumption that the 'natural' state of affairs is one in which organisations are heavily dependent on their environments and are to a significant extent influenced by them. In other words, the norm is for an organisation to have limited control over what happens in its environment. They therefore have a tendency to respond by taking action that reduces this dependence and allows them greater freedom of manoeuvre. Major 'sources of dependence' are as follows:

- Capital.
- Labour.
- Raw materials.
- Plant/equipment.
- Knowledge.
- A market for products/services.

Some of these resources are *critical* in that they are necessary prerequisites for the organisation to function. Others are *scarce* in that their supply cannot be guaranteed. It follows that organisations are most dependent on resources that are both critical and scarce. An example would be a major customer who could easily take its business elsewhere or a supplier who has few competitors and from whom the organisation has little choice but to buy. This is not a comfortable position and is one that managers seek to avoid if they can. No business, if it can avoid it, wants to be in a position where it relies too heavily on a few large customers or a few large suppliers.

For readers of this book it is interesting to note that recent years have seen various environmental trends coming together to create a situation in which, at least compared with the past, some forms of skilled labour have developed into resources that are both more critical and more scarce.

Higher skilled workers are more critical than they were in the past because the nature of the work that we are doing is steadily becoming more professional and technical in nature. More and more of us every decade are employed for the knowledge we carry around in our heads than we are for our manual labour. The more sophisticated and job-specific that knowledge is and the better networked employees are with customers and other professionals, the more it matters to an organisation when someone is hired or leaves to work for another employer. It may take years to find a replacement and for them to make an equivalent contribution. When the preponderance of employees are lower skilled, there is much less dependence of this kind.

Criticality of this kind is best illustrated with an example. If you were to be managing HRM or learning and development (L&D) activity in a large pharmaceutical company and one of your most experienced and senior scientists was to leave to join a rival firm, you would potentially consider this to constitute a major strategic problem. Immediate action would be taken to lure the person back with offers of increased salary, better terms and conditions or potentially more say over how their area of work is managed. Such efforts are made because people like this scientist are critical to the organisation's success. By contrast, if one of the security personnel who patrol the organisation's buildings were to leave, it would not be such a problem. This person can readily be replaced by someone with similar skills who could then be trained up to do the job equally well within a few days. These employees carry out valuable and necessary work, but as individuals they are much less critical for the organisation.

The key point is that, over time, more of us are carrying out work that is similar to that of the scientist, while fewer are in roles akin to that of the security guard.

This means that the long-term trend is towards a situation in which employees are more critical.

The same is true of scarcity. As, over time, more and more of us are employed in specialist, knowledge-based, graduate jobs, it is inevitable that we become harder for employers to recruit and retain. The labour markets in which we compete for work, and in which employers compete to employ us, become tighter. The more specialised our knowledge experience, and the more in demand it is from employers, the scarcer we are from the perspective of an organisation.

This long-term trend is sometimes referred to as 'upskilling', and while it tends to make recruiting and retaining people more challenging from an employer's point of view, it is good news for people looking to develop a career in HRM and related professions. The more critical and the more scarce employees become, the more significant and prominent the HR function is likely to be in an organisation. Settings in which labour is critical and scarce are also inevitably those in which HR managers enjoy greater influence and in which their functions are better resourced. The more competitive conditions (and employment regulation too) constrain an organisation's freedom of manoeuvre, the more significance HR professionals tend to have in that organisation.

PAUSE AND REFLECT

To what extent do you agree that, over time, labour in your sector or industry is becoming either more critical, more scarce or both?

VUCA

While different sectors and different organisations face a variety of future circumstances in their business environments, for many, the years ahead are likely to be less stable than those in the past. The acronym VUCA is often used nowadays to sum up a future world of work and employment that will be, in comparison with the past, more:

V – volatile
U – uncertain
C – complex
A – ambiguous

The change is due to the apparently accelerating rate at which new technologies are being developed and brought to the market, at which globalisation is occurring, at which major social and demographic shifts are taking place and at which new regulations are coming into effect. We will be discussing all of the developments in the chapters that follow in the first half of this book.

The impact of a VUCA world is one in which the effective management of change will inevitably become more important. It is one in which it will be harder for managers to plan ahead effectively and one in which increased pressure will inevitably be put on organisations to cut costs, act opportunistically, deploy staff more flexibly and generally be prepared to make difficult HR decisions quickly and under pressure. Increased occupational stress is thus likely, leading to more problems developing in respect of employee wellbeing. Moreover, when put under pressure, managers inevitably cut corners from time to time, act in ethically questionable ways and may downplay important longer-term agendas such as the need to improve productivity, to promote diversity and inclusion, and to ensure that their businesses conduct themselves in an environmentally sustainable way. These issues will be discussed in the second half of the book.

Tools for environmental analysis

We will finish this chapter by introducing two mental mapping tools that can be used as a means of thinking systematically, if fairly basically, about an organisation's business environment. While no great claims can be made for these in terms of their practical significance, they remain useful as the starting point for discussion that has some strategic relevance for managers.

Network analysis

Network analysis focuses on institutions and other organisations with which an organisation interacts regularly, or by which it is otherwise affected. When carried out in an HRM context it typically includes the following:

- Competitors.
- Major customers.
- Regulatory bodies.
- Suppliers.
- Trade unions.
- Partner organisations.

The analysis is often represented as a sketch or diagram, circles being used to illustrate the significance of each institution in the network, each linked by lines which signify their shared relationships. The size of each 'circle' represents the relative importance of the institution in the organisation's network.

The main problem with the network concerns the drawing of the boundary. The larger an organisation, the bigger its network and the more complex its graphical representation. It is therefore necessary to strike a balance between simplicity and the need not to omit an institution that may have an important potential impact. The most common approach involves limiting any analysis to immediate networks of the organisation itself and its major competitors. You will see a simple example of a network analysis in Figure 1.1.

Figure 1.1 Network analysis

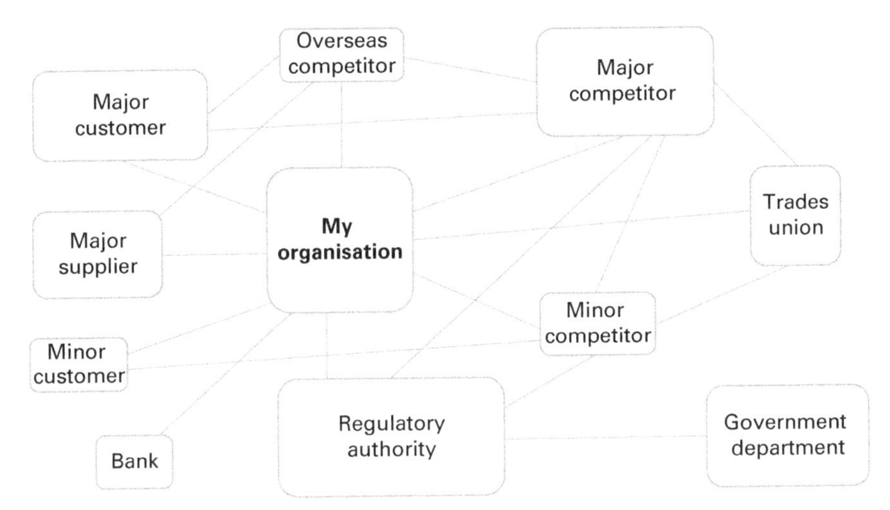

PESTLE or STEEPLE analysis

These are well-established, simple and straightforward approaches that really constitute no more than a checklist of headings that help to ensure that all relevant factors are taken into consideration when scanning an organisation's business environment.

The tool has evolved over forty years. It started out as a PEST analysis, the letters standing for:

P – political
E – economic
S – social
T – technological

By the 1990s it was common to read about PESTLE analyses, as two additional categories were added to the original four:

P – political
E – economic
S – social
T – technological
L – legal
E – environmental

The addition of legal factors reflected the growth in business regulation in the years since 1980. 'Environmental' referred to protection of the environment and reflected the growing importance attached to green issues over the previous decade.

The most recent incarnation of the tool is as a STEEPLE analysis, a third 'E' being added at the end to reflect contemporary interest in ethical matters and corporate social responsibility:

S – socio-cultural
T – technological
E – economic
E – environmental
P – political
L – legal
E – ethical

It could be argued that population ageing, globalisation and the emergence of international labour markets mean that further headings now need to be added to the list. Doing this would make it impossible to have a simple, one-word, memorable acronym like STEEPLE.

We would therefore, with tongues firmly in cheeks, like to propose a new acronym – the ELDEST PIG analysis:

E – economic
L – legal
D – demographic
E – ethical
S – social
T – technological
P – political
I – international
G – green

CASE STUDY 1.1

Eldest pig

Here are some key points that could be made if an ELDEST PIG analysis were to be carried out in respect of a budget airline operating from UK airports:

E – economic

Major challenges resulting from the reduced number of flights taken during the Covid-19 pandemic. Collapse of the market. Recovery may be slow and hesitant over time.

L – legal

Continual need to take account of and act on regulatory developments, particularly those relating to health and safety, working time and airport taxes.

D – demographic

Population ageing in key markets means fewer younger people, who are a very significant target market. Offering needs to be tailored to the needs and preferences of older demographics over time.

E – ethical

Some low-cost airlines have been accused in the media of acting unethically towards some customer groups (eg disabled persons), to staff, and, in respect of pricing regimes, to customers generally. For reputation protection reasons, all aspects of the operation need to be kept under review in terms of business ethics.

S – social

Greater affluence presents many opportunities to expand the range of services on offer in the future and to add further destinations to the existing portfolio. Low cost may gradually become less of a priority for customers vis-à-vis comfort and service.

T – technology

New, lighter and more energy-efficient aircraft are being manufactured. Longer term, there is the prospect of electric aircraft as well as models that require less pilot input due to satellite navigation and evolving drone technologies.

P – political

The potential impact of Brexit on the operation of flights within the European Union. Debates over new runways and the development of high-speed rail services. Potential for government subsidies to be provided for alternative means of transport.

I – international

Very substantial opportunities for global expansion and partnerships with other airlines to take advantage of a growing middle class in many countries with disposable income. Markets for tourism and business travel may grow significantly.

G – green

Increasing proportion of the population, including target markets, is looking to find alternatives to flying. Increased use of social media platforms for business meetings, leading to less long-distance business travel.

02 Globalisation

Globalisation is a good subject to start a book on the contemporary environment for the worlds of employment and work. While difficult to define precisely, we can nonetheless state with confidence that, over recent decades, the trend towards greater globalisation has had profound effects, both direct and indirect, on many aspects of working life. Moreover, while aspects of globalisation are the subject of much criticism, it is difficult to envisage a significant reversal in the trend occurring soon. A slowing down of the rate at which our economic system is globalising may be possible, but the broad direction of travel, at least for most countries, looks fairly well set.

In this chapter we start by setting out a working definition of the term 'globalisation', going on to explain how it manifests itself and what factors explain its growth. We then explore the major ways in which globalisation is affecting working life and employment practice, also discussing some of the more significant critiques and their potential future significance.

Defining 'globalisation'

There is no generally agreed definition of the term 'globalisation'. This is because it has many different potential dimensions and is thus used to describe a range of different global developments. Steger (2019: 7) points out that the term is commonly used in really quite different ways 'in both the academic literature and the popular press to describe a process, a condition, a system, a force and an age'. Different writers thus use the term in subtly different ways. For some it is primarily used to describe greater levels of international exchange, for example in goods and services traded, international migration, money moving across national borders or information and knowledge. The metaphor of liquidity is sometimes used in this context, the core idea being that, over time, barriers that prevent these things from flowing freely around the world are being brought down (see Ritzer and Dean, 2019: 2–3). Others label this 'internationalism or transnationalism', arguing that 'globalisation' involves taking things a step further. For them it is the extent to which we observe greater *integration* of economic, political, social and cultural activities across national borders and greater *interdependence* that distinguishes our current 'age of globalisation' from previous periods in history. For people looking at the issue from a geographer's perspective, an emphasis is placed on the idea of a 'shrinking planet' as it becomes less costly and time consuming to move people, objects, payments, information and ideas around the globe. The term 'global village' is sometimes neatly used to sum up the outcome of this set of developments. A further perspective emphasises steady convergence over time across continents of ways of thinking, sets of assumptions, norms and ethics – in other words, a process by which people and societies across the world are slowly but surely becoming more like one another in terms of how they feel, think and act. We thus see the term 'international community' often being used by political commentators to indicate a common acceptance between governments and international institutions of basic standards of human rights. Some take a wider, more philosophical perspective and focus on the apparent spread of 'modernity' across the world – a phenomenon Pinker (2012: 692) defines as:

> The transformation of human life by science, technology and reason, with the attendant diminishment of custom, faith, community, traditional authority and embeddedness in nature.

It is probably best to think of globalisation as comprising a series of diverse processes of all these kinds and more. Over time, more and more economic, political and cultural activity occurs internationally rather than just nationally. We are increasingly aware of what is happening elsewhere in the world, more likely to trade internationally, work for companies headquartered in other countries, travel overseas and know people who live or were born overseas. Because this situation is so pervasive it is easy to forget that, at least in its current form, the current extent of globalisation is a reasonably recent development. While we have in human history experienced waves of increased global activity previously (mainly as a result of imperialism), the scale of today's international exchange, the level of integration and the proportion of nations that are participating in the process are unprecedented. It is truly the defining characteristic of our times and one of the most important developments in human history.

Just as it is difficult to reach a common understanding about how the term 'globalisation' should be defined, it is equally difficult to measure its historical and current development entirely satisfactorily. The most commonly cited attempt to do this is the 'KOF Globalisation Index', which is continually being maintained and refined by the Swiss Economic Institute in Zurich. Their index takes account of a wide range of different measures from across the economic, social and political fields. They have computed an annual figure for the world since 1970, as well as one for all the major countries to indicate how 'globalised' they really are. The index takes account of factors such as the amount of trade carried out internationally, the extent of foreign direct investment (FDI) across national borders, the extent of restrictions placed on trade in the form of tariff and non-tariff barriers, the number of international airports and passengers, numbers of international migrants and tourists, participation in the activities of international bodies, signatories to treaties and even the number of McDonald's restaurants and IKEA furniture stores. They have two different measures, which complicates matters further, but the broad pattern that they have recorded is clear. Since 1970 there has been a very marked, steady, year-on-year increase in globalisation when all these different measures are mashed up together. The early 1990s then saw the start of a period of acceleration, which continued until 2008 when the pace of increased globalisation slackened somewhat in the wake of the big, global recession. During the more recent period of recovery the KOF indices have continued to head in an upward direction, but at a slower rate. In 2018 one index was reported to have risen from 40 to 57 since 1970 and the other from 36 to 64 (ETH Zurich, 2018). The UK is persistently shown by the measures included in the KOF index to be one of the more globalised countries in the world. In 2018 the UK was ranked in fifth place out of 203 countries included in the analysis, after Switzerland, the Netherlands, Belgium and Sweden.

Globalisation is manifesting itself in a wide range of different ways, creating trends that are having and are likely to continue to have a profound impact on the world of work. Some major examples are discussed below.

International economic exchange

The World Trade Organization (WTO) estimates that the total value of goods that were exported internationally in 1948 was $59 billion. After nearly seventy years of globalisation the figure in 2017 was $17,198 billion (WTO, 2018: 122). When estimates for global trade in services are added and prices taken into account, it is reasonable to conclude that the total volume of international trade has increased in real terms thirty-three-fold since the early 1950s when the current wave of globalisation commenced. This is a very significant rate of growth and one that surpasses global economic growth considerably. In the late 1940s around 7 per cent of the goods produced around the world were exported across national boundaries. Ortiz-Ospina et al (2018) estimate that the figure today is over 26 per cent, so before taking account of services, we can see that world trade has increased some four times faster than world output over the past seventy years. Moreover, much of this growth is relatively recent. The figures have doubled in real terms since the early 1990s. Market places have shifted in a generation from being mainly locally based and confined within nations and regions of the world to a situation today in which most industries operate on a global scale. Organisations are thus obliged to compete with international rivals, and in order to do so effectively they are looking to export increasing amounts of their products and services overseas.

UK businesses have been very active participants in this process. The total volume of goods and services they have collectively exported has increased sixfold in real terms since 1960 and threefold since 1990. The UK has increasingly specialised in the export of services, particularly business and financial services. In 2017 a total of 45 per cent of all UK exports were accounted for by services, compared to 26 per cent in 1960 (Ortiz-Ospina et al, 2018).

Foreign direct investment

Multinational corporations (MNCs) are the major drivers of economic globalisation, and their growth (in terms of both numbers and size) provides another useful way of illustrating the extent to which economic activity has globalised in recent years.

The United Nations used to maintain a database that listed all parent companies that owned assets in a country other than that in which they were headquartered. In 1990 there were 30,000 organisations that fitted this definition, and by 2010 the figure had risen to 104,000, after which the UN stopped collecting the information and publishing annual estimates. The relevant UN agency – the United Nations Conference on Trade and Development (UNCTAD) – does, however, continue to make annual estimates for the total number of workers who are employed directly by corporations that are based overseas. The figure was 21.4 million in 1990. The latest estimate, in 2016, was 82.1 million. Multinationals are also very much larger than they were thirty years ago, employing many more people, operating in more countries and controlling a greater market share. Recent estimates suggest that the top 100 companies in the world account for around 40 per cent of all international trade. This makes them very powerful in terms of their commercial and geopolitical influence. They exert a substantial and growing influence over consumers, workers and smaller companies.

MNCs are major investors in countries other than those in which they are headquartered and are thus the most significant source of FDI around the world. The published figures for global FDI provide another effective way of tracking the extent of globalisation in recent years. UNCTAD estimates that the total stock of inward foreign direct investment in 1980 was $702 billion. By 2017 this had risen over forty-five times to $31.5 trillion at constant prices.

The UK has always been one of the bigger receivers of FDI, while UK-based companies are very active investors overseas. In 1980 the total stock of FDI estimated to have been received by the UK was $63 billon. By 2000 this had increased to $439 billion and was estimated by UNCTAD to stand at $1.4 trillion in 2017 at constant prices (ie a twenty-two-fold real terms increase). Outward FDI, defined as investments made overseas by organisations headquartered in the UK, rose from $80 billion in 1980 to $1.5 trillion in 2017. This represents a nineteen-fold real terms increase. The UK currently makes up around 2.5 per cent of the world economy, but is responsible for over 4 per cent of total world FDI.

Offshoring

Companies do not only take advantage of the possibilities offered by globalisation through international expansion. It is often much more cost effective to develop relationships with suppliers based overseas and to set up what have become known as 'global supply chains'. These are most commonly associated with manufacturing operations, but increasingly companies in countries like the UK with relatively high wage costs have also subcontracted areas of administrative and customer service work to overseas suppliers. The result, over time, has been the wholesale restructuring of many industries. While there are exceptions, in the vast majority of cases this process involves 'offshoring' whereby work that was previously carried out in a western, industrialised country is exported to developing countries where wages are lower and working practices often more flexible and less heavily regulated.

According to Dicken (2015), offshoring takes three main forms:

1 Captive direct – An organisation establishes an overseas operation to which it allocates a body of work previously undertaken elsewhere.
2 Vendor direct – The contract to carry out a body of work that was previously done in-house is given to a specific overseas provider.

3 Vendor indirect – An organisation contracts a specialist provider to carry out a body of work that was previously carried out in-house. The provider then determines where in the world the work will actually be carried out.

Figures on precisely how many jobs have been offshored from industrialised to developing countries are difficult to find and vary considerably depending on how precisely we define the term 'offshoring'. It is, however, regularly and authoritatively estimated that over one hundred million jobs have effectively been exported since the late twentieth century. In the UK in 1982, 5.5 million people were employed in manufacturing jobs, representing over 20 per cent of the workforce. By 2018 this had fallen to 2.7 million, which represented only 8 per cent of total employment. Similar falls have been recorded in most western countries in recent decades (Rhodes, 2018). It is impossible to calculate precisely how much of this steep decline is due to jobs being exported overseas and how much has been due to other factors such as the introduction of new technology and improved productivity, but estimates suggest that around half the total decline in manufacturing employment in richer economies such as the USA can be put down to shifts in production to developing countries (Klein, 2016).

The major receivers of offshored jobs are countries in Asia such as India, China, Indonesia, Singapore, the Philippines, Thailand and Malaysia. However, some European countries are also active in competing for contracts and have been net importers of jobs as a result. The major examples are Bulgaria, Romania, the Baltic states and Portugal. Some offshored employment also goes each year to African countries such as Kenya, Ghana and South Africa.

The main motivation for offshoring is cost reduction. Labour costs, in particular, make offshoring very attractive for organisations trying to compete in labour-intensive industries. For example, the salary that has to be paid to a call-centre worker in India is around 20 per cent of the figure required to secure the services of someone in the UK. This means that call centres in India operate at around 40 per cent of their cost in the UK once international telephone charges have been taken into account. Setting up a 850-seat centre in India costs £20 million, but can easily yield annual savings of £15 million once it is established. Savings of that kind are just too significant to pass up in highly competitive industries that must cut costs wherever possible if they are to survive and grow.

PAUSE AND REFLECT

Aside from savings on the cost of wages, what other factors do you think a company might take account of when deciding to outsource some of the work it carries out to an overseas subsidiary or supplier?

International migration

According to UN estimates, in 2017 there were 258 million people living long term in a country other than the one they were born in. These statistics are widely agreed to be unreliable because many member countries either cannot or do not record accurately

Table 2.1 Total number of international migrants

Year	Number
1970	81 million
1980	100 million
1990	154 million
2000	175 million
2010	214 million
2018	244 million

SOURCE: United Nations (2018)

the number of migrants that they host. Figures on the number of people entering legally are accurate, but statistics on illegal immigrants and on the numbers of foreign-born residents leaving to return to their home countries are often inaccurate. The UN figures probably underestimate the total figure, but they nonetheless give an indication of the growth in international migration that has occurred in recent decades. The figures presented in Table 2.1 show a threefold increase in international migration over the past fifty years. In this time, of course, the global population has grown rapidly, so the raw figures are somewhat misleading. International migrants make up just 3.3 per cent of the world's population, an increase from 2.2 per cent in 1970.

However, because patterns of international migration are very varied, the extent of the change is very much greater in some places than in others. Seventy-five per cent of all international migrants live in 28 countries, and over half have moved to ten countries. In percentage terms the patterns vary hugely around the world. In Indonesia, Vietnam and China, for example, only 0.1 per cent of the population is made up of overseas immigrants, and in India the figure is just 0.4 per cent. These countries are major exporters of migrants. The major importers are the industrialised countries. Around 20 per cent of all international migrants live in the USA, while over 30 per cent have settled in the European Union. Within Europe most immigrants are found in Switzerland (29 per cent of the total population), Ireland (16 per cent), Germany (15 per cent), Spain (14 per cent), the UK (13 per cent), France (11 per cent) and Italy (11 per cent). Elsewhere the proportions are much lower. Poland (1.6 per cent), Turkey (3.8 per cent) and Romania (0.9) are larger countries with tiny immigrant populations.

The biggest immigrant populations are found in the Gulf countries, where people born overseas make up a majority of the total population in some cases (Kuwait 73 per cent, UAE 70 per cent and Qatar 74 per cent). In Saudi Arabia 31 per cent of the population is made up of overseas immigrants. As a proportion of the workforce in these countries, migrants account for even larger shares. In the UAE, 90 per cent of the workforce was born overseas.

Within countries there are also very varied patterns. There are parts of the UK, for example, where immigrants make up less than 3 per cent of the total population (eg the Scottish Borders), but others (some north London boroughs) where over 40 per cent of the population has come from overseas. Fifty per cent of all babies born in London now have a mother who was born overseas while 75 per cent of the babies born in London have one parent who was born overseas.

This represents a very big change over fifty years and is one of the most significant ways in which globalisation manifests itself in practice. It is of very considerable importance from an employment perspective because a big majority of international migrants move from one country to another either to look for work, to take up a job offer or to improve themselves educationally. Some move for family reunion purposes, while a small proportion (less than 10 per cent) are refugees, but for most it is the search for improved economic prospects that takes them overseas.

International travel

Another way in which the rate and extent of globalisation can be tracked is via the statistics that are published on international travel. We are living through a period in which the amount of foreign travel is increasing at unprecedented levels, making the travel industry by some margin the biggest and fastest growing industry in the world. In 1954 the number of passengers going through UK airports was 4 million. By 1974 it had reached 45 million, and, by 1984, 57 million. The figure in 1994 was 98 million. Since then the rate of growth has accelerated vastly. In 2004, 216 million people travelled through UK airports and by 2007 the total figure had reached 240 million. Projections pointed to a figure of over 400 million a year by 2020. Internationally, the number of air passengers rose from 9 million in 1945 to 88 million in 1972 and to 344 million in 1994. Since then the growth has been exponential – close to 3 billion people now travel by air every year.

Around half of these flights are taken for business purposes of one kind or another; the other half are for leisure. In 1990 an estimated 435 million people travelled overseas on holiday, up from just 25 million in 1950. By 2018 the figure had grown to over 1.4 billion. UK citizens now take a combined total of 74 million overseas holidays each year, while 39 million visit the UK each year for the purposes of tourism. They spend over £20 billion while visiting, enabling the tourist industry to boast that it accounts for 2.5 per cent of UK gross domestic product (GDP). Year on year, growth is considerable. By 2025 Visit Britain estimates that the tourist industry will generate 10 per cent of GDP and employ one in every ten workers (Blackhall, 2019).

Why is globalisation happening?

The obvious answer to the question of why the processes described above are happening and have been for over fifty years is that they suit the economic interests of key social groups. While it is probably reasonable to argue that a good majority of us have benefited in some way from globalisation, some have unquestionably benefited much more than others. The wealthy in our societies have tended to benefit most, and it is these people who tend to have the political and economic power required to bring about greater globalisation and to break down barriers that might otherwise get in its way. If we dig a little deeper we can identify three major factors that between them can be said to have enabled globalisation over recent decades. Moreover, these also help to explain the apparent acceleration in the rate of globalisation from the early 1990s that we identified above.

Technology

There is no question that globalisation could not have happened to anything like the extent it has without some of the new technologies that have been developed in recent decades. Some of the most important are in the field of transportation. More recently it has been developments in information and communications technologies (ICTs) that have had the biggest impact. Dicken (2015) notes that transportation was as slow 200 years ago as it was 2,000 years ago, and because information could only be transmitted on paper, the speed with which people could communicate over distances also remained the same throughout most of human history.

While the nineteenth century saw the development of railways and steamships, it was not until the later twentieth century that really efficient and affordable global transportation became a possibility. It is interesting, by way of example, to track the declining time that it has taken to travel from the UK to Australia over the years. Back in 1788 the eleven ships that made up the First Fleet took over 250 days to sail from England to Sydney Harbour. Fifty years later and early steamships had reduced the journey time to around eighty days, a journey time that had halved by the time soldiers from Australia made the journey in reverse to fight in the First World War. But this reduction in time was as much due to the opening of the Suez Canal in 1869 as it was to major developments in shipping technology.

The first air journey to Australia from Europe was completed in 1919 in twenty-eight days with stops each night en route. It was not until 1935 that a regular commercial passenger service started using 'flying boats'. On average the journey took twelve days to complete and could carry only ten people at a time. The cost of a ticket was equivalent to £13,000 per person today. By the late 1940s the journey time was down to less than a week, but it was still prohibitively expensive, so most people undertaking the journey, as well as freight, went by sea (the journey time by sea was down to 28 days by the 1950s). The big breakthrough in air travel came in the 1960s thanks to the invention of the gas turbine engine, a development Smil (2010) credits as being one of the 'prime movers of globalisation'. This allowed the operation of commercial jets carrying hundreds of passengers from London to Sydney in about thirty hours with two stopovers for refuelling purposes. By the 1970s the time was down to twenty-four hours using jumbo jets that could complete the journey with just one stop. The price of tickets then steadily fell and the number of scheduled flights increased. It is now possible to pick up a return flight from the UK to Australia for well under £1,000. The most recent development was the introduction in 2018 of a daily non-stop passenger service from London to Perth with a journey time of sixteen and a half hours. There is every reason to expect that journey times will continue to shorten and that the costs of transporting people from one side of the world to the other will continue to fall.

The same is true of cargo; the amount that is carried around the world by air has also grown greatly in recent decades thanks to faster and more efficient jets with greater capacity. According to the World Bank, just over 15.5 million tonnes of goods were shipped by air in 1973. Since then the figure has risen steeply in most years. In 2017 the global estimate was 231.6 billion tonnes. This represents a fourteen-fold increase. Most cargo, however, continues to be transported by sea and it is the development of vast container ships in recent decades that has been a much bigger facilitator of globalisation. It is extraordinary that it took until the middle of the twentieth century for shipping companies to realise that massive efficiency gains were possible if they shipped goods that had been preloaded into standard containers. Prior to this, goods would arrive at a port in all manner of packaging and have to be loaded and unloaded separately, a process that could take days. The development of the standard twenty-foot and forty-foot steel containers that can be rapidly loaded from a train or lorry onto a ship and off again at the destination port thus represented a huge if simple breakthrough in transportation technology. The first container ships launched in the 1950s could carry only 100 containers, since when their capacity has steadily grown along with their speed. Shipping costs have come down at the same time. Today's container ships are typically over 400 metres long and can carry 12,000 containers. In 1980 a total of 102 million metric tons of cargo were shipped internationally. The global estimate for 2017 was 1.83 billion tons – an eighteen-fold increase (Wagner, 2018).

Another major facilitator of globalisation has been the rapid development of information and communication technologies. These have permitted the transmission of data in all kinds of forms around the world much more rapidly, more reliably and less expensively than was the case a few decades ago. One simple way of gauging the extent of the change is to look at the cost of making international phone calls. In the 1930s it cost an extraordinary $300 (at today's prices) to make a three-minute telephone call from New York to London. Even in the 1950s it cost over $100, and the quality of the sound was often very poor. The cost then fell very rapidly, today being less than $1 if you have the right supplier. In addition, of course, we now have services provided by Skype, Zoom and others that permit face-to-face communication at little to no cost at all, although the quality can still be patchy.

These improvements have been made possible thanks to a number of distinct technological developments:

- **Satellite technologies.** Since the launch of the first telecommunications satellite in 1965, over 200 geo-stationary satellites have been launched, enabling the simultaneous transmission and receipt of information in many forms (text, video, audio) across the world at a much reduced cost than was possible previously.
- **Fibre optic cables.** These date from the early 1970s, but their carrying capacity has increased hundreds of times over the past twenty years. At present, the amount of data that is transmitted across the globe using fibre optic cables is doubling every nine months. Further international networks are being built all the time, hugely reducing the cost of transmitting information and increasing the speed and capacity of transmission.
- **Internet.** The invention of the World Wide Web has clearly been a very major factor in accelerating globalisation in recent years. Basic computer networks across which emails could be sent were developed in the early 1970s, but it was only two decades later that Sir Tim Berners-Lee invented web pages and the World Wide Web was born. In 1993 there were just 130 web pages in existence. Ten years later the number had grown to 8 billion. The number now grows by over 1 billion every day.
- **Mobiles.** The evolution of internet-enabled mobile devices (smartphones, tablets, etc) is the most recent development, effectively allowing us continual access to all the information on the web wherever we are in a highly user-friendly way. In 2000 there were 750 million mobile phone subscribers worldwide. There are now over 4.7 billion.

These developments make it possible for huge international corporations to exist and thrive, basing operations in different countries separated by large geographical distances. They have also facilitated the development of mass media systems that operate internationally rather than nationally. Hence the development of international cable, satellite and internet-based broadcasting networks which transmit a great deal of commercial information, allowing organisations to reach a worldwide audience through advertising and PR activities. The ubiquity of the World Wide Web and smartphones means that now anyone, anywhere in the world, is able to communicate directly, inexpensively and nearly instantly with anyone else anywhere in the world who has some basic equipment. The result has been the growth of a truly international market for goods and services. In theory, anyone can now sell anything to anyone else, wherever they are based in the world.

Here, too, there is every reason to anticipate further improvements in quality, capacity and reach and also cost reductions. The year 2018 saw the introduction of 5G (fifth generation) mobile network technology, which is set to be rolled out internationally over the coming years. It will increase download speeds to mobile devices by a hundred times, allowing vast quantities of data to be accessed instantly everywhere. The next few years are also likely to see the half of the world's population that cannot currently access the web gain access as mobile coverage is spread everywhere and the costs of accessing data through mobile devices fall. Battery life will extend greatly, and high-capacity solar-powered batteries developed. Cashel (2019) calls this 'the great connecting', arguing that its implications for developing countries will be transformative. The poorest people in the world will soon be able to access education and information at low cost, together with banking facilities, advice about health and mapping services, and participate in forms of e-commerce. Another development that is very much on the near horizon is the introduction of reliable, high-quality, simultaneous translation. Put simply, this means that people will be able to talk into a mobile device in their language and be heard by the person they are communicating with in that person's language. The same will be true of written communication. These developments can only further accelerate globalisation in various forms.

PAUSE AND REFLECT

How will the availability of reliable video-based communication systems equipped with accurate simultaneous translation features affect working life? Which industries might be most directly affected?

Regulation and deregulation

For most people, the second principal cause of contemporary globalisation is rather less interesting to read about than technological change, but it is just as important in providing an explanation for the trend. The truth is that to a considerable extent increased globalisation in the area of business and trade has been brought about through government action. Over the past forty years in particular, a consensus has evolved in favour of public policy choices that facilitate and encourage international trade and foreign direct investment. The major international institutions charged with promoting global economic development (the World Bank, the International Monetary Fund and the World Trade Organization) have been and continue to be highly significant vehicles in this process. The same is true of trade blocks and customs unions, which function as inter-governmental arrangements designed to promote pan-national forms of commerce. Negotiations held under the auspices of the World Trade Association (created in 1995) have led to the liberalisation of international trade in some sectors (mainly services), while further progress has been made regionally with the creation of cross-border customs unions and trade agreements such as the European Union (EU), the North American Free Trade Agreement (NAFTA), the Greater Arab Free Trade Area (GAFTA), the Common Market for Eastern and Southern Africa (COMESA) and the Comprehensive and Progressive Agreement for Trans-Pacific Partnership (CPTPP).

Jones (2008) argues that politics and regulatory developments are considerably more important than technology in explaining the trend towards greater globalisation. Taking a longer-term historical perspective, he explains how political decisions and particularly light-touch border regulations largely account for the growth of greater globalisation in previous eras. Technology then, as now, is seen as playing a secondary, supportive role.

Unquestionably, a major impetus for increased international trade and the expansion of multinational corporations in recent years has been the relaxation or wholesale abandonment of foreign exchange controls by over 150 countries. Foreign exchange controls are forms of control imposed by governments on the purchase/sale of foreign currencies by their own residents or on the purchase/sale of local currency by non-residents. Common foreign exchange controls include:

- Banning the use of foreign currency within the country.
- Banning locals from possessing foreign currency.
- Restricting currency exchange to government-approved exchangers.
- Fixing exchange rates rather than allowing them to float and be traded on international markets.
- Restricting the amount of currency that may be imported or exported by any one person.

Some controls of this kind remain common in developing countries, including very significant international players such as China and India. They have, however, been largely abandoned in the western industrialised countries, making it far easier and speedier to conduct international business. The USA abandoned exchange controls in 1974 and the UK in 1979. The completion of the European Single Market removed all controls on the movement of capital, goods and services within the EU in 1986. Free international trade remains restricted thanks to a range of informal and formal barriers in the form of tariffs, quotas, national licensing standards and state subsidies. But the movement over time has been towards greater liberalisation.

Importantly, however, it must also be recognised that globalisation has also been accelerated by a degree of additional regulation as well as forms of deregulation. Global regulatory developments that have supported elements of globalisation include forms of cross-national standardisation in fields such as customs documentation, air traffic control conventions, accounting standards, insurance rules, global patents and global copyright rules. The establishment of some international property rights across most countries has also hugely helped to promote FDI.

Geopolitics

A third factor is important in explaining the acceleration of globalisation we have seen in the past two or three decades. This is the large number of political changes that occurred around the world in the late twentieth century, particularly in the early 1990s. These, along with the spread of the internet at this time, provide a very plausible explanation for the acceleration of globalisation that occurred.

The most important single example is the opening up of the Chinese market after 1976, a steady process that really hit its stride in the 1990s. Another was the fall of the Berlin Wall in 1989, which heralded the inclusion of Russia, the former Soviet Republics and the countries of Eastern Europe into established world trade arrangements. India sharply changed its political direction following the appointment of Manmohan Singh as finance minister in 1991, choosing the rapid adoption of a policy of economic liberalisation, and similar developments occurred at around the same time across South East Asia, South America and in South Africa.

Collectively, these geopolitical developments have had a huge impact on the extent of globalisation, not least because of the sheer size of the populations and economies concerned. In the years since the 1990s, as these economies have industrialised and grown at a very strong pace, they have assumed an increasingly central role in the nexus of global trading arrangements. There is no question that their role will continue to grow in the future, opening up further opportunities for global economic exchange and integration.

How is globalisation affecting work and employment?

Globalisation has had, is having and will have a very considerable impact on our working lives. Some of these developments are direct consequences of globalisation and have already been discussed above. We are, for example, much more likely now than was the case in the past to work for a multinational organisation or one that operates internationally. This opens up opportunities for overseas work assignments, but also often means that the senior leaders who ultimately determine employment policy and practice in organisations are located more remotely. Understandably, people who have a global corporation to manage are less able to focus their attentions on the particular needs of individual workplaces in specific country operations. Also, we are a great deal more likely now than used to be the case to work in multicultural teams. This can work most effectively when the diverse experiences, knowledge and skill sets of team members are effectively harnessed to improve decision making and generate innovative ideas. But multicultural teams, like multigenerational teams, can also be dysfunctional at times, particularly when organisational politics complicate their operation.

In addition to these clearly evident, direct impacts, globalisation has also had some more diffuse, longer-term and more indirect effects on work and employment. Three major examples are industrial restructuring, increased competition and increased volatility, which we will now turn to.

Figure 2.1 Changing employment types (UK)

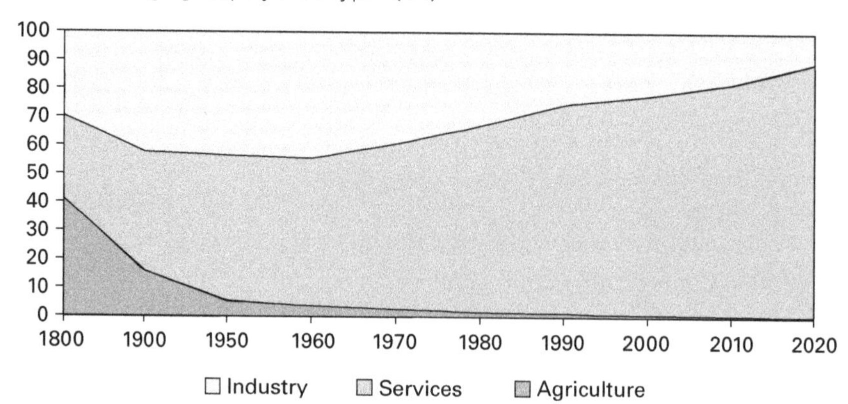

Industrial restructuring

Over the past sixty years, with the trend accelerating in more recent years, most economies in the world have restructured profoundly in response to globalisation. Agrarian economies have become industrial economies, while established industrial economies have become service-based and knowledge-based in their major activities. The UK is a good example of the latter. In recent years the UK, like many countries, has undergone wholesale industrial restructuring, which is very much ongoing. The period we are currently living through is often referred to as comprising a Second (or sometimes a Third or Fourth) Industrial Revolution. It is sometimes referred to as 'a post-industrial revolution' to indicate the very profound nature of the changes that are occurring in a short space of time.

The key long-term trends are shown in Figure 2.1. This shows how, during the period known as the First Industrial Revolution, the proportion of UK output devoted to agriculture fell hugely as farming became more efficient, and the population grew and became urbanised. Until the 1960s the manufacturing sector, along with mining and other industrial occupations, continued to grow and accounted for a very substantial proportion of the economy. Since then, growth has been focused on the service sector, which now accounts for well over 80 per cent of both output and jobs. A widely held myth is that the UK manufacturing sector is in decline. This is not strictly true. Employment in manufacturing has fallen, and continues to fall substantially. It is also the case that the proportion of national economic output accounted for by manufacturing has fallen from 34 per cent to below 10 per cent since 1970. However, in terms of absolute output, the manufacturing sector was, until 2008, growing year on year, and growth resumed in 2010. The rate of growth, however, has for some time fallen short of that for the economy as a whole. The UK's shift in balance from manufacturing to services is typical of all the major western industrialised countries, although the process has been notably faster, and has gone further, in the UK than in some EU member states and in the USA.

The trend is almost entirely explained by the globalisation of the world economy. It is far cheaper to manufacture most products overseas in low-wage economies and to ship them to the big consumer markets than it is to manufacture them close to the consumer. The same is true of mined goods and agricultural products. A consequence of globalisation is that each region or country of the world tends to specialise in areas of economic activity in which they have some kind of competitive advantage.

The manufacture of lower-cost goods is no longer profitable in countries such as the UK because the sources of raw materials are distant and because labour costs are relatively high. The same is true of food production. It is cheaper to import much of the UK's food from overseas rather than to grow it at home. Globalisation has thus driven the change in our industrial structure. UK companies have had to adjust so that they specialise increasingly in the provision of services for which there is demand elsewhere in the world. Hence, in recent decades as traditional manufacturing and agriculture have become less significant, other industries like financial services, retailing, tourism, hospitality, IT and higher education have grown.

Profound restructuring is also occurring across the developing world. Here we are seeing shifts of a similar nature to those experienced during the First Industrial Revolution in the UK. Agrarian economies, in which most people work on the land producing food, are becoming industrial economies, manufacturing goods that are sold all over the world. The key difference this time is in the speed with which developing economies are transforming themselves. They are also, in many cases, fast moving beyond the manufacture of cheaper manufactured goods into hi-tech, high value added manufacturing and service provision. In the 1950s and 1960s Japan led the way, closely followed by other 'tiger economies' such as Hong Kong, South Korea, Singapore and Taiwan. These are now among the wealthiest nations in the world. We are currently seeing a similar transformation of the massive Chinese economy, which is truly now the centre of world manufacturing. Brazil, India and many other countries are following on close behind, with their vast populations and fast rates of economic growth. The beginnings of similar developments can also be observed in many African nations.

The impact of these changes on work and on labour markets has been vast and continues to have a transformational effect. Quite simply, they mean that the established stock of skills among a people becomes increasingly obsolete. We are all having to learn new ways of earning a living, and often having to re-learn another set of skills later in life again. The process of industrial restructuring is thus a very painful one for many people, and indeed for whole communities. The skills that used to make people highly employable are increasingly not required by employers. Established industries have less and less demand for people as they introduce labour-saving technologies and offshore work to developing countries, while newer industries often struggle to recruit people with the skills and experience that they need. We see this happening across the world. In most countries it is creating a situation in which some skills are in over-supply, while others are in under-supply. This in turn is leading to greater levels of inequality as some people find that they are much more employable than others.

Competitive intensity

It is notoriously difficult to measure the extent of competition in an industry in any objective or satisfactory manner, and hence almost impossible accurately to track growth or decline in competitiveness over time. This is because industries vary so much in terms of their competitive structures. You might think that the number of firms in competition would be a good measure on the basis that the more players there are in the competition, the tougher it is to do well. This is, however, over-simplistic. Some of the most cut-throat markets in the world are dominated by just two or three companies. We thus have a situation in which most agree that trading conditions have become hugely more competitive in recent years, but no way of measuring accurately by exactly how much. Economists use a series of different measures that each act, some more satisfactorily than others, as proxies for measures of increased competitive intensity. Examples are reductions in the length of time leading firms are able to maintain dominance of a market, increases in the extent of churn in 'industry membership' and changes in concentration ratios (ie the extent to which the top five or top ten companies in an industry contribute to its total output).

In some industries, largely due to takeovers by larger multinationals, the extent of competition has remained unchanged or even reduced as globalisation has occurred. A prominent example is IT, where a small number of tech companies (eg Google, Facebook, Microsoft) have consolidated their very dominant position. In many other industries, however, the extent of competitive intensity has grown considerably. In recent years, a number of academics and consultants have advanced the view that the competitive environment in some industries is not just reaching a new level of intensity, but is developing in a qualitatively different direction than has been the case in the past. The term 'hyper-competition' is increasingly used in this context; another term that is sometimes used is 'high-velocity competition'. In such situations, competitive advantage can be more transitory in nature, requiring organisations that wish to grow to continually 'reinvent the wheel' or 'ride the wave of innovation'. The more competitive the conditions in an industry, the harder it is to achieve sustained competitive advantage and hence to maintain a high level of profitability. Survival also becomes increasingly questionable and difficult.

Increased competitive intensity is a natural consequence of globalisation, simply because it increases the number of potential competitors a particular company faces. Organisations that enjoyed a dominant position in their countries prior to the recent waves of globalisation (big fish in small ponds) find that globalisation tends to turn them into small fry in a much larger lake. There are numerous examples of organisations that have had to 'adapt or die' in a more globally competitive world. European airlines are a good example. Until the 1990s national flag-carrying airlines such as British Airways, Air France and Lufthansa faced only limited competition. Routes were negotiated by governments and awarded to these large companies. Other airlines existed, but tended only to fly routes that the national airlines did not wish to fly. The situation is now transformed. Deregulation has led to a vastly more competitive industry. In less than thirty years we have seen the rapid rise of low-cost competitors such as Ryanair and EasyJet that are now among the largest airlines in Europe. To survive, the established airlines have had to cut costs, adjust working practices and, in many cases, form strategic alliances with one another.

Another industry that is currently being transformed by a mix of globalisation and technology is television broadcasting. Until the twenty-first century, the vast majority of homes in the UK could access only four or five television channels. Competition was thus very limited and the broadcasters had a captive audience. With the rise of international satellite and internet-enabled broadcasting, the number of viewing options has massively increased. New business models are being established, allowing new global entrants such as Netflix to compete highly effectively with the established broadcasters.

There is inevitably a knock-on effect on working life for employees in the affected industries. Organisations continually have to seek ways of increasing efficiency and satisfying the needs of demanding customers. Employees are thus being put under greater pressure to achieve more for limited additional pay (or sometimes reduced pay) as work is intensified in response to competitive pressures. Mergers and takeovers have also become much more common, resulting in a situation in which the identity of our employers changes along with the content of our jobs and often our stress levels.

Volatility

The more competitive markets become, particularly as they internationalise, the more volatile and unpredictable the trading environment becomes for organisations. Economic stability of the kind that can exist and be promoted by government in a national market is much harder to achieve internationally. New technologies and applications are continually being developed all over the world and brought to market very rapidly. Insecurity, unpredictability and volatility are thus increasingly characterising many product markets, although the impact is far greater in some than it is in others. This inevitably makes it much harder than is the case in relatively stable business environments for organisations to be able to offer their staff long-term, stable employment of a traditional kind. More intense competition and greater volatility also provide the pretext for organisations to change their structure more frequently. In some industries it has been observed that organisations are now engaged in 'permanent restructuring' as new opportunities present themselves and established businesses cease to generate the profits they once did. Thanks in large part to globalisation we are thus faced with the evolution of a business environment that requires more opportunistic approaches to business, a greater level of organisational agility, and more flexible approaches to staffing and paying people for their work. Significant structural change is also occurring more frequently in organisations. Job content is subject to greater volatility as a result, requiring employees to retrain periodically, alter their priorities and develop a change-ready mindset. Moreover, because change management is often rather poorly handled by managers, organisational stress and workplace conflict born from a resistance to change on the part of employees can also be seen as a consequence of globalisation.

Critical perspectives on globalisation

The enormity of the changes that globalisation is bringing over quite a short period of time has inevitably made it – or at least features of it – the subject of a great deal of criticism. Moreover, because globalisation is disrupting our lives in such a diverse range of ways, the critiques vary considerably. Opposition to aspects of globalisation has come from across the political spectrum, and in many countries political parties and leaders have exerted influence and been elected recently on platforms which include an anti-globalisation element. The truth is that large numbers of people perceive themselves to be losers rather than winners from globalisation and they are tending to become increasingly vocal. Others, while themselves largely winners, have deep concerns about the impact of globalisation on others and have also taken up sceptical positions. These groups, however, often object to different elements and do not share a political ideology.

Environmental objections

Some of the more persuasive arguments against the globalisation of commerce in particular come from people who have particular concerns about its environmental impact. At one level the focus is on the pollution caused by the transportation of people, manufactured goods and foods around the world in ships and planes. Why, for example, do we in the UK ship so much of our refuse to be sorted and recycled in Asia? It may be cost effective economically, but the impact is to degrade the environment, particularly when one result is a large mass of abandoned plastic floating about in the oceans. At another level, the concern is climate change and the contribution that is made to it by the carbon emissions that fuel international trade and tourism. The big question that green campaigns are forcing up the international political agenda is: to what extent is globalisation compatible with sustainable economic development? As more people conclude that it is incompatible, some societies are beginning to question the political consensus in favour of greater globalisation that has been so widely established in recent years.

Multinational corporations

A second major critique of globalisation is focused on the activities of multinational corporations. In the western industrialised countries they are often criticised from the political right for exporting jobs and putting people out of work in pursuit of profit. Reversing this process formed a major feature of the platform that Donald Trump stood on during his US presidential campaign. On the political left, complaints tend to focus more on the way that multinational corporations treat employees in developing countries, or more commonly contract with suppliers who take a cavalier attitude towards the safety and wellbeing of their staff. Applebaum (2019), like many writers, draws attention by way of example to some of the more notorious cases of recent years, such as poor working conditions at Foxconn factories in China, which apparently provoked twenty-two suicides between 2010 and 2013. Another widely cited example is the collapse of the Rana Plaza factory building in Bangladesh, where seven storeys had been unlawfully added above the original two. Over a thousand workers were killed. The argument that critics make is that outrages such as these only occur because multinationals push suppliers to cut costs too far and do not take responsibility for ensuring that decent working conditions are maintained in their global supply chains.

Further opposition has been directed more broadly towards the recent evolution of international capitalism as a system. Multinationals are becoming too powerful and too big to regulate effectively and are running the global trading system in their interests at the expense of smaller competitors. Particular criticism is levelled at international institutions, notably the International Monetary Fund, the World Bank and the World Trade Organization, which stand accused of forcing developing countries to liberalise trade and accept the presence of multinationals in their economies. Loans to indebted governments are used to force acceptance of international capitalism, so it is argued, even when this is not in the interests of a large proportion of people in developing economies. In the first years of the twenty-first century there was a period when large protests were organised at a succession of conferences and summits at which political representatives gathered to discuss international trading issues. These were commonly referred to at the time as being 'anti-globalisation demonstrations'.

Migration

The third major contemporary critique of globalisation focuses on the extent of international migration, which, as we explained above, tends to be heavily concentrated in specific places. The concern in destination countries tends to be on the effect that the arrival of large numbers of less skilled migrants can have on wage levels and the capacity of public services such as schools, prisons and transport systems to manage effectively. In the UK, for example, concerns about immigration have consistently been found to be among the top three drivers of voting behaviour in recent years. A good majority of people consider that immigration levels are too high and that they are having a negative effect on the country. Only tiny numbers tell pollsters that they would like to see more immigration, despite the reliance of our health and agricultural sectors on migrant workers to undertake key roles (Wells, 2018). Immigration was undoubtedly one of the concerns that led to the UK's vote to leave the European Union in 2016.

In the countries that have been major net exporters of migrants in recent years, opinion tends to be more divided. On the one hand, people who take up paid work in countries with higher wages have a marked tendency to send a great deal of money home to their families in the form of remittances. The amounts concerned are huge (around $600 billion a year according to the United Nations), providing a major economic boost to the receiving economies. On the other hand, there are major concerns about the impact of 'brain drains' on longer-term economic prospects. The Chinese government has recently put in place a wide range of measures designed to tempt people who have emigrated overseas to return. This is unsurprising when we consider that of the 400,000 highly educated young Chinese who go overseas to study each year, only 275,000 return to China after they have graduated (Zhang, 2013).

The case for globalisation

While the case against globalisation has several strands, there is really only one major argument in its favour: namely, that, despite its many faults and imperfections, it is benefiting many more people economically than it is harming. In many developing countries – particularly in Asia – living standards have massively improved since the introduction of free market capitalism and entry into the global trading system in the post-war period, and particularly in the most recent decades. There have also been some pretty spectacular rises in average living standards across Eastern Europe and in some, but not all, South American countries. Africa should be next to benefit. Table 2.2 shows the United Nations figures for GDP per head in three of the economies that have grown fastest in recent decades.

Table 2.2 GDP per head

Country	1960	2016
Brazil	$210	$8,649
Iran	$191	$5,291
South Korea	$158	$27,538

SOURCE: United Nations (2018)

Despite significant corruption issues in many developing economies, taxes are paid by corporations and the growing middle class, and these are used by governments to fund schools, hospitals and better infrastructure. As a result, educational attainment is rising fast, as is life expectancy. The number of people living in abject poverty remains too high, but it is far lower than it was thirty years ago. According to the World Bank, in 1981 a total of 52 per cent of the world's population lived on less than $1.25 a day. It is now down to around 7 per cent at constant prices. The new measure of absolute poverty used by the World Bank is $1.90 a day. Around 10 per cent of the world's population are estimated to fall into this category, and the number is reducing year on year.

Between 1982 and 2012 the proportion of people living in India on less than $1.25 a day fell from 60 per cent to 22 per cent; in China the poverty rate dropped from 84 per cent to 13 per cent during the same thirty-year period as the economy grew twenty-five-fold. Hundreds of millions have seen dramatic increases in life expectancy and living standards as a result (Ross, 2016). If current trends continue, the proportion of the world's population living on less than $1.25 will be below 3 per cent by 2030.

The gains in more developed economies have been less remarkable, but have still been very positive, leaving a larger proportion of people much better off economically than they would otherwise have been had it not been for globalisation. There have been losers, but despite all the disruption, many more have been winners. The reason is a very simple one: replacing national and regional trading systems with a larger, single global system allows goods and services to be produced much more effectively and efficiently. Consumers have much more choice and prices tend to fall when barriers to trade are reduced and technologies are developed to make a global system viable. Clausing (2019) puts the case very succinctly in her recent book. Here she describes free international trade as having something of a 'magic' impact on humanity. She asks her readers to consider what life would be like if there was no international trade, each country being forced to provide for itself. In such a scenario living standards would be far lower for everyone because there would be vastly less choice available, fewer economies of scale, inefficient production systems and much higher prices. Basic items such as food, clothing and shoes would be much more expensive, while shortages of energy supplies (oil, gas, etc) would others such as oil and gas would mean that the cost of everything would be very much higher.

Clausing (2019) also puts forward a powerful case in respect of the economic advantages that international migration brings. The key advantage is that it allows skills to be better distributed. In other words, people with specific experience and abilities are able to take up job opportunities that they are best suited to. There are fewer mismatches between what employers need and what employees can provide. This increases productivity, makes innovation more likely and helps to ensure a higher level of employee engagement. She quotes studies that suggest the long-term economic benefits of reducing barriers to skilled immigration into a country like the UK or the USA are some fifty times higher than those associated with reducing barriers to trade in goods and services. Large-scale immigration is disruptive, but forcing people from host countries to retrain and take up alternative job opportunities is beneficial to national economies over the long term. She also puts forward an argument in favour of multinational corporations. First, they play a positive role simply because they are responsible for fostering and enabling international trade. Second, she argues, they are successful because they are good at what they do. They are highly productive and innovative, and thus major contributors via taxation to state coffers that fund essential public services.

PAUSE AND REFLECT

On balance, are you a winner or a loser from globalisation? In which ways have you gained personally and in which ways have you lost?

References

Applebam, R (2019) Labor, in *The Oxford Handbook of Global Studies*, ed M Juergensmeyer, S Sassen, M Steger and V Faessel, Oxford University Press, Oxford

Blackhall, M (2019) The briefing: How many tourists are there – and where do they go? *Guardian*, 1 July, 10–11

Cashel, J (2019) *The Great Connecting: The emergence of global broadband and how that changes everything*, Radius, New York

Clausing, K (2019) *Open: The progressive case for free trade, immigration and global capital*, Harvard University Press, Cambridge, MA

Dicken, P (2015) *Global Shift: Mapping the changing contours of the world economy*, 7th edn, SAGE, London

ETH Zurich (2018) KOF globalisation index, www.kof.ethz.ch/en/forecasts-and-indicators/indicators/kof-globalisation-index.html (archived at https://perma.cc/FQ6M-J7FR)

Jones, G (2008) Globalisation, in G Jones and J Zeitlan (eds) *The Oxford Handbook of Business History*, Oxford University Press, Oxford

Klein, M C (2016) How many US manufacturing jobs were lost to globalisation? *Financial Times* https://ftalphaville.ft.com (archived at https://perma.cc/ZLH4-57ZY)

Ortiz-Ospina, E, Beltekian, D and Roser, M (2018) Trade and globalization, www.ourworldindata.org/trade-and-globalization#the-two-waves-of-globalization (archived at https://perma.cc/W396-B5QU)

Pinker, S (2012) *The Better Angels of our Nature: The decline of violence and its causes*, Allen Lane, London

Rhodes, C (2018) *Manufacturing: Statistics and policy*, House of Commons Library, London

Ritzer, G and Dean, P (2019) Globalization: The essentials, 2nd edn, Wiley Blackwell, Hoboken, NJ

Ross, A (2016) *The Industries of the Future*, Simon and Schuster, London

Smil, V (2010) *Prime Movers of Globalization: The history and impact of diesel engines and gas turbines*, Massachusetts Institute of Technology, Cambridge, MA

Steger, M B (2019) What is global studies?, in *The Oxford Handbook of Global Studies*, ed M Juergensmeyer, S Sassen, M Steger and V Faessel, Oxford University Press, Oxford

United Nations (2018) *World Migration Report*, United Nations, Geneva

Wagner, L (2018) Container shipping: Statistics and facts, www.statista.com/topics/1367/container-shipping/ (archived at https://perma.cc/3YHL-T3JB)

(Continued)

(Continued)

Wells, A (2018) Where the public stands on immigration, https://yougov.co.uk/topics/politics/articles-reports/2018/04/27/where-public-stands-immigration (archived at https://perma.cc/F2EB-8NVS)

WTO (2018) *World Trade Statistical Review*, World Trade Organization, Geneva

Zhang, Y (2013) Fight against brain drain, *China Daily*, 30 October, www.chinadaily.com.cn/2013-10/30/content_17067899.htm (archived at https://perma.cc/CW9S-YZYT)

03 Technology 1

Getting here

The impact of evolving technology on our economy and society is self-evidently profound, as by extension is its influence on the development of business and the world of work. While there is nothing new about this state of affairs, a persuasive case can be made in favour of the proposition that we are currently living in an 'age of accelerations' (Friedman, 2016). New and improved technologies are being developed at a much faster pace than in any earlier period of human history, making this a fascinating and exciting time to be living and working, but also one in which disruption and uncertainty are the norm in many industries.

This subject is far too big to cover properly in a single chapter, so we are going to focus on different aspects in this chapter and the next. Here we are going to survey technological developments of the past and present, going on in Chapter 4 to discuss future developments. We will start with a series of observations about how technology affects business organisations and the world of work, using examples from recent and not-so-recent history to illustrate our key points. We will then go on to focus specifically on how recent and contemporary technological developments are affecting aspects of human resource management practice.

Technological development

When we think about the history of technology we tend to be drawn to the big landmark developments or 'light bulb moments' that changed history by bringing about a rapid improvement in people's lives. These have occurred periodically throughout human history. Examples from the very earliest times are the ability to light fires artificially, the making of boats and the invention of the wheel. The ancient Egyptians invented writing, ploughs, toothpaste and the first reliable clocks. Ancient China saw the development of the first paper and block printing techniques, as well as the compass, kites and sails. The Greeks developed plumbing systems and maps and built the first lighthouse, the Romans going on to build arches, use concrete and create bound books. In more recent centuries examples of breakthrough inventions include optical lenses, gunpowder, printing presses, anaesthetics and vaccines, steam engines and electric power. In living memory, the major examples are communications satellites and the internet.

Most of the time, however, technologies advance in an evolutionary rather than a revolutionary fashion as over decades they are steadily improved to be made more reliable, efficient, user-friendly and affordable. A process of steady refinement occurs in which many thousands participate, learning from one another's successes and failures.

Smil (2005, 2010) shows how this process of evolution occurred, and continues to occur, in the case of cars. The first step was the invention of a workable internal combustion engine, which was collectively achieved in the late nineteenth century in German workshops by Karl Benz, Gottlieb Daimler, Wilhelm Maybach and Rudolf Diesel. They were, however, building on work of many others stretching back over a century previously, including those who had developed steam engines and stationary machines that were explosively powered. The 1880s saw the first commercial production of engine-propelled carriages on wheels that looked like two-seater tricycles and could travel at a maximum speed of about ten miles an hour. Engineers and manufacturers working in different countries then set about improving functionality and design, progress being made in steps over two or three decades. By 1900 cars with four wheels and canvas hoods that could travel at up to forty miles an hour were being manufactured. In the early twentieth century car manufacturing then took off internationally as different companies continued to work on refinements. Engines were improved with electronic fuel injection and turbo-charging, while steel frames, balloon tyres, four-wheel brakes, electronic ignition and power steering became standard features. Mass production of cars started in the 1920s, by which time a million were being produced annually across the world. Later in the century further improvements were made. Lighter materials were used, all manner of safety elements introduced and a wide range of additional electronic features were developed. The combined efforts of thousands of engineers and designers transformed car production in the twentieth century from a small, specialised affair producing a few thousand simple but expensive machines each year, to the biggest mass production industry in the world. Over fifty million highly sophisticated vehicles are now manufactured annually. It was a global effort. Most of the early developments occurred in Europe. Mass production was then pioneered in the USA, and many recent innovations in the fields of fuel efficiency and reliability emanate from Japan.

Technological advances in vehicle manufacturing have almost all been commercial in nature. They provide an excellent example of how entrepreneurs and businesses play a central role in carrying out the research and development activity that drives technology forward. Initial ideas lead to experiments and research, then to the manufacture of prototypes that are tested, modified and re-tested through a process of trial and error. International patents are secured and new products taken to market. All the time rival firms compete with one another in a bid to develop and exploit new technologies more effectively and efficiently in order to earn bigger returns on their investments. In these highly competitive commercial environments it is necessary to innovate continually so as to keep in step with – or preferably a few steps ahead of – rivals. Risks have to be taken in the process and it must be accepted that much investment in technological development will not have a commercially viable outcome.

It is by no means only commercial organisations that develop major technologies. Government laboratories and universities have also long been cradles of new invention, the products of which often subsequently go on to be exploited commercially. There are several pervasive products that were originally developed in military settings or sprung from research carried out for military purposes. The Colossus machines that were built for use by codebreakers working at Bletchley Park during the Second World War were the world's first digital programmable computers powered by electricity. Radar was also developed for military use during the Second World War. One of the researchers working on its development, Percy Spencer, subsequently went on to use his know-how to invent the microwave oven. Telex was originally developed for military use, as were duct tape, superglue and canned foods. Satellite navigation systems and drones are the most prominent recent examples of military technologies that have gone on to be commercialised and sold widely.

While individual inventiveness and sometimes a touch of genius are vital, most technological development is very much a team effort involving a range of institutions, resources and sources of investment. Innovation thrives through human contact and networks. Formal education plays a major role, as does tacit knowledge gained through experience and passed on to others informally. Technology develops fastest when large numbers of knowledgeable people are actively involved (see Ridley, 2020), and it is the expansion of collaboration that explains much of today's apparent acceleration of technological evolution. As Coggan (2020: 331–2) shows, the big difference between the present era and that of past centuries is not so much human inventiveness as our ability to network physically in multinational research teams, to meet at international conferences and trade fairs, and to communicate information and ideas rapidly and inexpensively across the world. The more people who have access to understanding and information about current developments, the higher the chances that new, improved generations of technologies will be devised.

In later chapters we will be discussing how organisations and their people need to be managed in order to make the most of technology by maximising productivity and promoting innovation and creativity. For now we just need to note that management styles and techniques play an important role in determining the effectiveness of investment in technology. Recruiting the right people, retaining them, engaging them and, particularly, providing them with the right developmental opportunities are essential. Above all there is a need to manage change effectively. This is always challenging, but becomes more so the faster technologies develop. There is a strong tendency for us to be suspicious of change and to fear its possible impact, but the history of business demonstrates the need to embrace it. There is much truth in Leon Megginson's management mantra that 'it is not the strongest of the species, nor the most intelligent that survives. It is the one that is the most adaptable to change'.

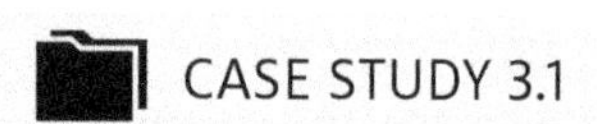

CASE STUDY 3.1

Kodak and Dyson

In recent years, as the rate of technological innovation and progress has accelerated, we have witnessed the contraction of a number of once-dominant corporations. Sometimes it is because they do not innovate sufficiently, but often it is because they make the wrong strategic choices about technology when planning for the future.

The decline and fall of the Eastman Kodak Company (commonly known as Kodak) provides a spectacular recent example. Founded in 1881 in Rochester, New York, Kodak pioneered dry film photographic technologies and the mass production of pocket-sized cameras, later introducing colour film. By the middle of the twentieth century it dominated the international market for cameras and photographic film. Later, as Japanese competitors entered the market, Kodak lost its very dominant position, but it diversified successfully and continued to grow, remaining one of the world's top five most recognisable brands.

In the 1980s and 1990s Kodak employed 150,000 people worldwide, and earned profits in excess of $2 billion each year. Kodak's decline and fall after 2000 was swift and brutal. After failing hopelessly to predict the rate at which digital photography would eclipse the established film-based technologies, it made several very poor strategic decisions. Markets disappeared, debts piled up and fat profits turned into vast losses. This situation could, however, have been avoided because it was Kodak engineers who first developed prototype digital cameras in the 1970s. Instead of embracing the new technology, the company took a more cautious approach and decided to continue to focus on film-based photographic products.

In 2012 Kodak filed for Chapter 11 bankruptcy, subsequently selling off nearly all its remaining assets and patents. In September 2013 it emerged from bankruptcy planning to operate principally in the commercial printing industry. It now employs fewer than 5,000 people and is struggling to remain profitable.

By contrast, Dyson Ltd is an example of a highly successful company that has grown healthily in recent years as a result of its ability to develop technologically innovative products that have anticipated consumer demand very effectively. The company was founded by the British inventor James Dyson in 1993, using the money he had amassed from selling his designs for 'ball barrows' and 'g-force cleaners' to other companies. He was now in a position to set up a laboratory and to employ a team of people to assist him in developing products that he planned to manufacture himself.

Dyson's first big success was the dual cyclone vacuum cleaner, a machine that had 'designed out of it' many limitations of conventional vacuum cleaners. It took fifteen years and over 5,000 prototypes before it was ready to take to market. Not only are 'Dysons' bagless and able to reach into every corner of a room, they are also attractively designed products that rapidly came to dominate the UK market, before becoming commercially successful globally.

In more recent years Dyson and his team have successfully pioneered further product innovations, including bladeless fans and AirBlade hand dryers. Dyson now employs over

12,000 people worldwide and is earning profits in excess of £900 million a year. Much of its recent growth is in Asian markets, where sales are growing very fast year on year. The company now runs its own university programmes aimed at training dozens of engineers and has invested billions in the development of artificial intelligence and robotics.

Snags and surprises

While, over time, technological developments have collectively brought great benefit to societies all around the world, it is important to appreciate that the processes whereby this has happened have not always been smooth. Wrong steps are frequently taken and poor judgements made, and it is often only with the benefit of hindsight that this becomes clear.

Sometimes, for example, an invention will have considerable prospects, but will not find a market because it is insufficiently user-friendly, too costly, or often is simply ahead of its time. The most celebrated examples are probably the Difference Engine and Analytical Engine that were designed by Charles Babbage, a Professor of Mathematics working at Cambridge University in the 1820s and 1830s. These were large, intricate mechanisms made up of gears and metal drives that could carry out highly sophisticated mathematical calculations and, in the latter case, print out the results. They promised to be much more accurate at maths than human brains, and as his collaborator Ava Lovelace later showed, were also programmable in the manner of a modern computer. The problem was their immense physical size and complexity. During his lifetime Babbage was successful in getting substantial government finance to support him in developing these machines, but he did not live to see a full, working prototype built. His designs were far too expensive and cumbersome to manufacture in practice. It would be a further hundred years before a workable computer called the Electronic Numerical Integrator and Calculator was built using vacuum tubes that were powered by electricity, but even then the machines were too big and expensive to find a market beyond a few big telecommunications companies. It was the invention of transistors in the 1950s and their subsequent use in tiny integrated circuits printed on silicon chips that made the mass production of modern computers commercially viable.

Interestingly, the inventors of the first transistors developed them at the Bell Telephone Company in the USA for the purpose of amplifying sound. At the time they do not seem to have appreciated the huge range of transformative applications their little device would soon have outside the world of telephony (Ede, 2019: 268). This phenomenon is not unusual in the history of technology. George Orwell famously wrote that 'to see what is in front of one's nose needs a constant struggle', and when it comes to assessing the long-term impact of technology this is very true. It is not easy to estimate how exactly contemporary technological developments will change our lives over the long term. The history of invention is full of examples of inventors themselves, let alone the wider population, failing to appreciate the full implications of what they had developed. For example, when Édouard-Léon Scott de Martinville invented the phonograph in the 1850s he does not seem to have grasped the immense commercial possibilities of sound recording. His machine translated the human voice into a series of squiggles on a piece of paper, which an interpreter was then trained to 'translate' into written prose. It was a further two decades before Thomas Edison invented the gramophone record, providing a means by which sound waves could be preserved and played back. But he wholly failed to appreciate the implications of his invention. Edison listed ten purposes to which he thought it might be put. These included dictation in office environments, making recordings of books so that blind people could hear them, recording the last words of dying people and teaching spelling. The idea that music might be recorded and sold commercially on his discs never seems to have occurred to him at all.

The same was true of the inventors of radio a few years later. Marconi and Gugliemo worked out how to send messages and receive them wirelessly across great distances, but were only thinking in terms of one-to-one communication. They called their invention 'wireless telegraphy' and they were thrilled when ships started using it to send messages using Morse code. It continued to be used like this through World War One. Only in the early 1920s, some thirty years after the original invention, did the idea develop that it could be used for 'broadcasting' from a single station to a mass audience tuned in to separate receivers. It took a long time for the full commercial significance of the original invention to be appreciated.

So, while today there is no shortage of debate about their possible long-term impact, it is too soon to make definitive judgements about how the advent of social media, nanotechnology or biotechnology, de-salination or new forms of energy are going to affect human lives over the long term. The full range of business implications may not yet be fully apparent.

PAUSE AND REFLECT

Think about some contemporary technological developments or those you have read about that we can expect to see maturing soon.

What are the obvious, immediate implications your chosen technology is likely to have? What possible longer-term impact might it possibly have, both on business and on society more generally?

Moreover, because new technologies have the capacity to transform whole societies as well as providing commercial opportunities, we can observe that it is often only with the hindsight of decades that the true, full implications of a particular technology's significance becomes apparent. Some examples from the twentieth century are as follows:

- The impact of consumer white goods (fridges, washing machines, dishwashers, vacuum cleaners, etc). They made housework far less time consuming, freeing up time for women in particular, allowing them to join the paid labour force and heralding the start of the transformation of gender roles that has occurred in the past few decades. White goods also destroyed the career prospects of many thousands of domestic servants who lived and worked in wealthier households.

- The advent of multichannel commercial television in creating media competition. This was a big threat to traditional newspapers and many responded by focusing on reporting entertainment, sporting and celebrity gossip. Modern tabloid newspapers with their irreverent approach were the outcome. Their social impact has been immense, not least in making societies like Britain's far less deferential and respectful.
- The effect of private motor cars, televisions, personal computers, smartphones and tablets in fostering individualism and the decline of community identity and of some collective leisure pursuits. We will return to look at this development in more detail in Chapter 8.
- The way that affordable air conditioning has enabled the development of highly significant knowledge industries in places with very hot climates, such as Singapore, Bangalore and Dubai.

Quite fundamental debates are currently raging among sociologists about the likely long-term impact on society of some contemporary technological advances. Some are pessimistic, others more optimistic. Moses Naim (2013), for example, argues that technology along with other developments (regulation, migration, education, increased affluence) is having the effect of diluting the extent to which powerful institutions in society (governments, corporations, traditional media, social elites) are able to exercise power without restraint. Access to information is no longer restricted to the wealthy and powerful. His book is full of examples of ways in which the barriers that have traditionally prevented groups from empowering themselves are gradually being weakened. This has negative potential consequences (extremists benefit as well as political moderates), but is mainly to be welcomed.

Jaron Lanier (2013), by contrast, reaches a completely different, and far more pessimistic, set of conclusions. He argues that IT is steadily putting more and more power (and hence money) into the hands of fewer and fewer people, creating a grossly inequitable distribution of wealth and reducing economic security for most. In the future, he argues, those who control the most powerful computer networks (governments, financial institutions, technology companies) will have access to much more knowledge than everyone else and will seek to restrict access to it because it will be in their interests to do so. It follows that contemporary technological developments are more likely to be disempowering for most people than empowering. In her recent book entitled *The Age of Surveillance Capitalism*, Shoshana Zuboff (2019) takes Lanier's side in this debate. She argues that the extent to which tech companies like Google, Amazon and Facebook can predict and manipulate our behaviour through electronic analysis of our personal data is very concerning, and is yet to be fully understood by their users.

We can debate the extent to which these two competing visions (if either) are right, but we cannot know. The answer will only become clear in a decade or two, when we can look back at the present and see what happened and why with more clarity of vision.

The history of technological development is also filled with examples of products failing to achieve their full commercial potential due to public suspicion about their ethics or safety. In recent years this has been true of stem cell research in the USA, the planting of genetically modified crops in Europe and energy production using hydraulic fracturing (fracking) techniques in the UK. There is currently plenty of suspicion among the general public about the safety of autonomous self-driving vehicles. Whatever statistics and reassurance are provided by manufacturers and regulators, a good proportion of people say that they would not get into one, particularly if it was going to transport them at high speed.

Suspicion of technology is not always unfounded. There have been plenty of new products launched with great expectation that have later been shown to have damaging effects. Asbestos is a good example. When first developed, it was hailed as something of a wonder material. For much of the twentieth century it was very widely used in the construction industry, particularly in roof insulation. Then in the 1970s it was established that it had carcinogenic properties and that breathing in asbestos dust could kill you. So it was swiftly withdrawn from use and is now banned in most countries.

The use of thalidomide as a drug to treat morning sickness is another example. Launched in the 1960s, it was soon found to cause deformities in some of the babies born to women who had taken it while they were pregnant. It is still widely used in the treatment of cancer and HIV, but the scandal associated with its use in treating pregnancy-related conditions continues to fuel suspicion of claims made about all manner of new drug treatments.

Governments get judgements about technology wrong too, sometimes. In 2000 the British government decided to promote the use of diesel instead of petrol on account of its lower CO_2 emissions. Taxes were cut and manufacturers were incentivised to build and sell diesel cars. The motives were sound, but the results were not. Diesel is now recognised to have a far more polluting impact than petrol because of the nitrogen oxides and particulates that it releases into the atmosphere.

INVENTED BY MISTAKE

Many of the most significant technological advances have occurred by accident. The most well-known example occurred in 1929 when Alexander Fleming returned to his laboratory after going on holiday to find that, in his absence, some mould had grown over a sample of a bacteria he had accidently left out in a dish. The mould had killed a lot of the bacteria. He tried using it to kill others types too and found that it worked, without harming patients in the process. This became penicillin, the first antibiotic. Many millions of lives have been saved as a result.

Matches were also invented fortuitously. Before the 1820s starting a fire or lighting a candle required some skill using flints and tinders. Everything changed when a pharmacist called John Walker noticed sparks emanating from the end of a stick he had used to stir chemicals when he was scraping them off. It never seemed to have occurred to anyone previously that a stick tipped with a coating of combustible substances could provide a very simple and inexpensive method of creating a flame.

X-rays, Bakelite (the first man-made plastic), safety glass, heart pacemakers, artificial sweeteners, Coca-Cola, Teflon and the impotence drug Viagra are other examples of products which were originally stumbled upon in an unplanned and unexpected way.

SOURCE: Prabhune (2015); Breyer (2017)

Technology and work

The development of new technologies often has a disruptive impact on established organisations and groups of workers. Some always benefit more than others in the process, while a third group lose out and can suffer badly. As a result, the introduction of new processes, tools, machines and products is often controversial, contested and resisted. Experience tells us that there are five main effects on the world of work and that a single technological development often has more than one of these effects at the same time – efficiency, quality, opportunity, disruption and work location.

New technologies can make workers and workplaces more productive and efficient

A good example that is very close to home for us is writing a textbook like this one. Thirty years ago, researching and writing a book that covers so many different fields would have taken a great deal longer. We would have to have spent a great deal of time in libraries flicking through card indexes, ordering books and bound volumes of journals up from stacks. We would have taken pages and pages of handwritten notes, following up footnotes and cross-referencing from notebook to notebook. The manuscript would then have been composed rather laboriously using a typewriter, then reviewed and annotated, before being re-typed. A lot of Tippex would have been consumed. The final product would then have been photocopied and posted to the publisher who would have created each page using metal, movable type. Galley proofs would then come back to us and we would have written corrections and edits physically on the paper. More typesetting would then have occurred before the physical book could be printed. The whole process would have taken at least two or three years.

Now most of the research can be done from our offices and home studies. Internet search engines have massively reduced the amount of time taken to access published materials and statistics. Pretty well the entire, cumulative knowledge of the world in all the relevant fields is now available for instant download at the touch of a button. Where this is not the case, books that are both in print and out of print all over the world can be located rapidly using Amazon and obtained in the post within a few days. The writing process is vastly faster thanks to word-processing, which allows us to compose and recompose each sentence as many times as necessary to get it just right without the need to pick up paper and pen. When completed, the manuscript is sent to the publisher by email and final proofs sent back in the same way. The book is then printed digitally much more quickly and at far lower cost. An online version is also then made available for readers to download from anywhere in the world.

At every stage in the process, IT has radically reduced the time, effort and costs involved. This makes writing textbooks more commercially viable and inevitably means that more are written. They are more comprehensive and can be updated via new editions more regularly. Similar efficiencies can be seen being achieved in numerous other work settings as technologies are introduced and are used in combination to improve industrial and service efficiency.

A process of deskilling frequently accompanies the introduction of new technology, rendering obsolete skills and expertise that have taken people many years to attain. Less highly qualified people can then be hired to do the work accompanied by the technology, and such people are nearly always paid less – sometimes a lot less. Great efficiencies are thus made, but instead of this creating more opportunities, as in the book-writing example above, they are reduced. Bookselling is a good example. The rise of digital books and vast warehouse-based online retailers has made it much easier to obtain books without the need to visit a physical bookstore staffed with highly knowledgeable buyers and customer advisers. Reviews written by readers are readily available on websites, cost comparisons are easily made and convenience inevitably then pushes people to avoid traditional book shopping.

Another current example of deskilling occurring due to technological innovation is the taxi-driving profession in cities like London. For 150 years it was only possible to secure a cab drivers' licence by completing a gruelling training course known as 'The Knowledge'. Would-be drivers had to spend an average of two or three years learning the names and locations of all the streets in the city as well as all the landmarks, museums, police stations, hospitals, hotels, restaurants, theatres, clubs and more. Several attempts were needed to pass, the result being a highly skilled group of professional drivers who rarely needed to refer to maps and were invariably able to take a customer anywhere across London, without delay, by the quickest routes. Global positioning systems (GPS) have very rapidly reduced the advantages associated with being driven by a cab driver who has 'The Knowledge'. Satnav not only tells drivers where they need to go, but also suggests fastest routes given current traffic conditions. While not always accurate, in truth, GPS is generally more reliable than the human brain. Its invention has brought about highly disruptive competition in the taxi business as companies like Uber have made use of it to employ thousands of drivers at a much cheaper rate. They are less highly skilled, but are just as able to transport customers about, and at a rather lower cost.

New technologies can improve the quality, safety and reliability of established work processes

Much technological development is focused on improving the quality of a good or service rather than the efficiency of its production. Numerous examples can be cited from across the whole industrial field, from the development of new building materials and techniques to hi-tech transportation systems, greener energy production, the use of special effects in film-making, sophisticated approaches to policing and the introduction of a wide range of new farming technologies. The quality of umpiring and refereeing in sports matches has been improved hugely in recent years thanks to technology, as have the acoustics at concert and theatre venues. There is often a net cost increase associated with the introduction of such equipment as a result of purchasing the kit itself, training staff up to use it and hiring better-paid people who have the required qualifications or experience to maximise the benefits on offer. But the effect is a better quality of customer experience, greater choice, lower energy usage and, for the businesses involved, improved commercial prospects.

The most significant recent examples of technology bringing about improvement in the quality of services provided come from the field of medicine. A brief glance at major statistical trends demonstrates the extent to which ongoing medical research and the continual development and refinement of treatments are having a large and very welcome impact:

- Life expectancy at birth in the UK rose from 68.7 in 1950 to 81.4 in 2020 – that is almost 13 years of extra life on average (Macrotrends, 2020).
- In 1951 the age at which the average person in the UK could expect to live for a further 15 years was 59 for men and 60 for women. This rose to 70 by 2017 (ONS, 2019).
- Infant mortality stood at 31.7 per thousand births in 1950. It is now less than 4 per thousand births – a fall from 3.2 per cent to 0.4 per cent (Statista, 2020).
- Cancer survival rates have more than doubled since 1950 (Cancer Research, 2012).
- In the 1980s when the HIV/AIDS epidemic started, the average survival period for people who were infected was 20 months. It is now averaging over 40 years – almost equal to normal life expectancy (Lawrence, 2015).

This progress is mirrored across the world, and it has been brought about in large part by a big increase in the amount of money we spend as societies on healthcare. In 1950, shortly after the National Health Service (NHS) was founded, the UK government spent 3.5 per cent of GDP on the NHS. The UK now spends over 7 per cent of GDP on the NHS and there are plans to increase this figure considerably in the future (Nuffield Trust, 2018). We spend twelve times more on the NHS in real terms than we did when it was founded in 1948. The number of full-time doctors employed has increased from 11,700 to 115,000 and the number of full-time nurses from 68,000 to 217,000.

At the same time, advances in medicine have allowed the number of NHS beds to decline from 480,000 in 1948 to just 120,000 now. People spend much less time in hospital when they are sick. They receive far more intensive and effective care and recover much more quickly (Triggle, 2018). New therapies have been particularly significant in reducing the number of mental health beds. People still get very sick, but their conditions are managed far more effectively and for a great deal longer before they die.

These developments have also brought about a sea change in the medical professions. At the time the NHS was founded, nurses spent around a quarter of their time carrying out domestic duties like cleaning and laundry. Nursing is now an all-graduate profession, its members increasingly working in highly specialised medical teams carrying out a wide range of procedures that were previously reserved for doctors (Stephenson, 2018). Doctors have also seen their profession change beyond recognition as new and far more complex treatments have been developed. For general practitioners (GPs) this has meant less involvement with the treatment of terminal and acute illnesses. As a rule, they no longer attend births or take such an active role in caring for people with acute medical conditions. They now devote much more of their time to helping to manage chronic, long-term conditions and promoting good health. For hospital doctors the trend has been towards much more 'specialisation and sub-specialisation' and working in clinical networks across hospital sites (Rivett, 2018).

In healthcare, far from bringing about down-skilling, the impact of technological development has thus been a massive upskilling process, which is very much still ongoing. The medical professions have grown alongside their capabilities.

New technologies can provide significant business opportunities, creating jobs that did not exist before

Most major technologies, once developed, create new business opportunities and hence jobs. We can see this reflected in the statistics that the Office for National Statistics publishes showing the fastest growing occupations in the UK. In recent years these have been two main types: hi-tech and hi-touch (coders and carers), reflecting the rapid current growth of both the technology and caring professions. Between 2011 and 2019 the fastest growing occupations by some margin were in the fields of programming and software development, which rose by 73 per cent, representing an additional 160,000 jobs. Computer servicing, production management and IT and telecommunications professionals also saw large increases in job opportunities over the past decade (World Economic Forum, 2020). In many cases, these were occupations that did not exist ten or twenty years earlier.

While many new jobs created as a result of technological advances are specialised in nature and are occupied by people with technical qualifications and experience, this is by no means always the case because technology also facilitates the growth of whole industries employing a wide range of people in different kinds of roles.

The film and broadcasting industries provide a good example. Precise statistics on numbers employed are hard to establish because these industries indirectly employ large numbers of freelancers. But by any measure there has been very considerable expansion over recent decades as new technology has facilitated the rapid growth of this sector. The number of feature films made for cinema release in the UK has increased considerably since the turn of the century, and now averages over 200 each year compared to only thirty or forty in the 1980s and 1990s. Part of the explanation lies in the establishment of a more favourable taxation regime, but more films are being shot everywhere in the world due to the development of digital film-making technologies, which have dramatically reduced production costs. The number of people involved in making a feature film remains high, some 500 names typically featuring in the credits that roll at the end. This expansion means that over 90,000 people in the UK now make their living from the film industry, most in production, others in distribution and exhibition (ONS, 2017; BFI, 2019).

Broadcasting is another industry that has seen very considerable growth over time. The BBC started in 1922 with just four employees charged with establishing the UK's first radio channel. One of them was the BBC's founding Director General, Sir John Reith. When he left sixteen years later there were 4,000 employees producing programmes for three radio channels. A fourth radio channel was launched in 1946 at the same time that the single-channel BBC television service restarted broadcasting after the Second World War. A commercial channel was launched in 1955, followed by a second BBC television channel in 1964. Channel 4 started operating in 1982 and Channel 5 in 1997. Since then, thanks to satellite and digital broadcasting, the amount of choice has ballooned, alongside the opportunities for people to work in the industry. We now have hundreds of radio and TV channels to choose from, most of which operate twenty-four hours a day. The demand for high-quality content has increased exponentially. New platforms like Netflix and Amazon Prime have increased choice still further, while many people now make a living running their own private mini-channels on YouTube.

By 2020 the BBC was directly employing 22,000 people, but supported many thousands more working as subcontractors and freelancers, as well as the production companies which help to supply much of its programming. In total almost a quarter of a million people in the UK are now employed in television, radio, film-making and photography. Moreover, over a quarter of all the new businesses that are started each year in the UK are in the creative sector, representing a net annual increase of some 7,000. The creative industries as a whole support over 2.4 million jobs in the UK, a number that increased by 35 per cent between 2011 and 2017 alone and continues to grow further every year (DCMS, 2017).

New technologies can damage or destroy established business processes, causing the loss of jobs and work opportunities

While technological advances create and enhance business and job prospects, they also destroy them. In most cases they create losers as well as winners. Over time, of course, people adapt to new technological realities, retrain and take up different occupations, but the process can take years and often creates a great deal of hardship and upheaval. People are made redundant and have to live on welfare for extended periods and find retraining to be a struggle. Their new jobs often pay less well and are more insecure. In some cases they are also less satisfying to perform and may generate less by way of pride or occupational status. It is sometimes necessary for families to relocate. The development of new technology brings opportunities, but for some it frequently brings highly undesirable disruption. It is therefore to be expected that those whose livelihoods stand to be affected negatively will sometimes oppose and resist its introduction.

In the twentieth century the major examples of job destruction occurring due to technological development were in industries that employed large numbers of people to carry out manual labour. Stewart et al (2015), using an analysis of UK census records, showed that between 1871 and 2011 the proportion of the UK workforce that was employed principally for its 'muscle power' fell from 23.7 per cent to 8.3 per cent. Since then, it has reduced further still. The mechanisation of agriculture played a major role in bringing about this change. In the late nineteenth century more than one in ten British men were employed as agricultural labourers. The latest Labour Force Survey data shows that some 320,000 people are now employed in agriculture, fisheries and forestry – that is less than 1 per cent of all UK employment, and some 70,000 of these jobs are seasonal in nature and largely carried out by labourers who come to the UK to work for a short period each year (ONS, 2018). The same broad trend has occurred across the manufacturing sector, where manual jobs are far rarer than they used to be. The decline has also been substantial in industries like washing and laundering, which employed hundreds of thousands of people a century ago doing jobs that are now almost entirely carried out by machines.

Currently we are witnessing an upheaval that may turn out to be just as historically significant in the retail sector, as each year more and more consumer activity moves away from high streets towards online. In 2006, according to ONS figures, just 2.5 per cent of UK retail spending was carried out online. Since then, the proportion has risen each year, reaching 10 per cent in 2012 and 20 per cent in 2019. The Covid pandemic then gave online shopping a further boost as high-street shopping became impossible for a period, and over 30 per cent of retail transactions were carried out online. It is surely inevitable that this trend will continue in the future, making traditional high-street shops less and less viable. Ten years ago it was still common to see travel agencies on UK high streets trading alongside recruitment agencies and shops selling recorded music. These are now much rarer sights.

The last decade has seen many long-established retailers disappear thanks in large part to the rise of e-commerce. The first major chain to wind up was Woolworths, which ceased operating in 2009 after exactly a hundred years of ubiquitous presence on UK high streets. Eight hundred stores closed with the loss of 27,000 jobs. Thousands more stores closed in the subsequent decade as BHS, Mothercare, Oddbins, Hardy Amies and many others shut up shop. Other big names such as House of Fraser and HMV went into administration and were bought up only to continue operating on a smaller scale in many fewer locations. Many big name retailers that have survived are often only doing so by reducing the size of their operations. Many thousands fewer people are now employed by the likes of Boots, Debenhams, Marks and Spencer and John Lewis than was the case just a few years ago.

New technology can change the locations at which work is performed

In Chapter 2 we explored how reductions in the cost of transporting goods (air freight, container shipping, etc) in combination with more recent developments in information and communications technologies (ICTs) have permitted organisations to offshore low-cost manufacturing work as well as a lot of back-office, administrative activity. In recent years these processes have resulted in the displacement of millions of jobs from locations with relatively high labour costs to those where employing people is a great deal less expensive. Offshoring is one example of how technological development has created opportunities for employers to alter the location at which work is carried out.

ICTs have also made it entirely normal for people to travel around the world while continuing to carry out their usual work at the same time. Someone with management responsibilities can now quite easily continue to communicate with and supervise a great deal of work activity from thousands of miles away. Team members can collaborate on many types of projects entirely satisfactorily despite being separated geographically and it is no longer at all unusual to participate in conference calls and virtual gatherings of various kinds with people who are based on different continents. The technology is not yet always wholly reliable, but each year improvements are made, and ways of working that would have seemed extraordinary twenty years ago are gradually becoming commonplace.

Increased reliability and reduced costs are also making homeworking a far more plausible option for people in a wider range of job roles. The extent to which this kind of arrangement currently occurs is not easy to state with certainty because many people work from home only for a proportion of their time, but according to the most recent government statistics there are around 1.6 million people who work mainly from their own homes (ONS, 2020a). This represents around 5 per cent of all employees. When self-employed people are added in, the figure increases as many more self-employed people are based at home. During 2020, the Covid-19 pandemic saw people necessarily starting to work from home on a full-time basis for the first time. In April 2020, according to the UK Labour Market Survey, 46 per cent of UK employees carried out paid work from their homes, 86 per cent of whom (some 13 million people) did so because of the pandemic (ONS, 2020b). Adapting to the experience was easier for some than others, but it served to demonstrate the practicability of using technology to relocate some kinds of work away from offices to domestic locations.

For twenty-five years many commentators have been puzzled about why homeworking has not increased far more as it has steadily become more and more technically feasible. On paper the advantages are many. For employers, homeworking makes it possible to reduce the size of premises required, generating very considerable potential savings in terms of office rents, business rates, heating and lighting. For employees the great advantage is the ability to work more flexibly and, of course, to avoid having to spend a chunk of each day commuting, an activity that can – when stressful – cause wholly avoidable deteriorations in mental and physical health (Cooper-Ryan et al, 2020). The main disadvantages relate to the low morale that homeworkers sometimes suffer as a result of being isolated from co-workers and a tendency for working days to lengthen in the absence of fixed start and finish times. A different kind of supervision is required, along with control systems that assess performance on the basis of the quantity and quality of work undertaken irrespective of time. Because it is not possible to oversee each individual's work, there is no opportunity either to encourage or to correct when mistakes are made or when the pace of work slackens.

The biggest problem in managing a home-based workforce is the maintenance of effective communication – a particularly important issue where members of the 'core' workforce are employed on a teleworking basis. If such arrangements are to be successful, more is needed than simple electronic communication. There is also a need to hold regular team meetings, as well as face-to-face sessions between supervisors and staff. In practice, much homeworking of this kind is carried out part-time, the employee performing some work at the office and some at home. When managed well, this can be the best of both worlds, in that effective communication is retained while savings in terms of office space and energy use are also achieved. Of course, this can occur only if employees forego the privilege of having their own office or desk at work and accept 'hot-desking' arrangements, whereby they occupy whichever workstation or computer terminal is free when they are not working at home.

Technology and HRM

Like other areas of management, technology has brought change and opportunity to many areas of HRM work in recent years. Some of these are now so commonplace that it is easy to forget how great an impact they have brought about in day-to-day HR practice. We are thinking here about the use of email and email attachments, video conferencing, computerised payrolls and the use of HR information systems for recording, storing and reporting employee data. In practical terms, the biggest change we have seen over the past two decades is simply the ability that the internet now gives us to access high-quality professional information and advice quickly and cheaply. In all these ways, and more, ICT has transformed the effectiveness and efficiency with which HR work is carried out in organisations. A major effect has been to reduce the amount of time that HR professionals are obliged to spend carrying out routine administrative tasks. This has enabled us to focus more effort on value-adding interventions of various kinds that have more of a direct, positive impact.

In addition to these general examples, there are some other notable ways in which aspects of HR work have recently evolved in new ways, in large part due to the availability, affordability and increased reliability of technologies. We will end this chapter with some observations about three of these: the impact of social media, the use of online tools in employee selection and the increased use of e-learning.

Social media

The rise of social media platforms has unquestionably been one of the most significant developments in both the general business and HRM environments in recent years. All kinds of completely new challenges and opportunities have been created as a result. The speed and scale with which a handful of hugely influential social media companies have established themselves and achieved a dominant position internationally is striking:

- LinkedIn was launched in 2003, quickly becoming established as the world's pre-eminent networking site for professionals. In just over a year the site had a million users, after which it continued to grow rapidly. In 2016 the company was bought by Microsoft for $26.2 billion. As of 2020 the number of people who had set up LinkedIn profiles had reached 690 million. The site has users in every country of the world and operates in numerous languages.
- Facebook started operating in 2004 on a similar basis to LinkedIn, but aimed to be a platform helping people to establish and maintain friendship networks. It is one of the most extraordinary success stories in business history. Within four years it had 100 million users and, by 2010, 500 million. At the start of 2020, despite not being available in China, the company claimed to have 2.5 billion active users visiting its site at least once a month.

That is around a third of the world's population. Facebook also owns Instagram, which itself had over a billion users in 2020.

- The fastest growing website ever is YouTube. It launched in 2005 and was attracting eight million visitors a day by the end of that year. It was bought by Google in 2006 for $1.65 billion. Over 500 hours of video are uploaded onto the site every minute, generating more content in a month than the combined output of all TV stations around the world manage in more than a year. Over a billion hours of YouTube content is watched every day by over two billion users. After Google it is the most visited website in the world.
- Glassdoor launched in 2008 as a website on which employers could advertise jobs and employees could post reviews about their organisations, salaries and the questions they were asked at interviews. As of 2020 the site held some six million company reviews posted by employees, listed over ten million job vacancies and was visited by over 50 million users each month. Glassdoor was purchased for $1.2 billion by the Japanese company Recruit Holdings in 2018.

There are two major ways in which the presence of social media has, to date, affected HR practice. The first is in the reduced ability that organisations now have to keep a firm grip on their reputation as employers. In the old media world, where most companies operated in national rather than international settings, it was not at all difficult to monitor what was being said in a few newspapers or on the limited number of TV and radio channels that were available. Journalists, even on local papers, were not generally interested in reporting stories about employee experiences, so most of the time negative perceptions of the way managers treated their people were spread only by word of mouth. Social media has completely changed this situation. It creates a challenging situation from an HRM perspective, but the truth is that employers of any size at all can no longer hope to control their reputations. They can and do seek to influence how they are perceived by working hard at developing positive employer branding, but in some respects this creates greater hazards in our complex, volatile and unpoliced new media environment. The bigger an organisation and the more valuable its brand, the bigger the risk of losing a hard-won positive reputation as a good employer. Closing down a damaging story after it has been published is much harder than it used to be.

There has also, of course, been a knock-on effect on old media outlets. Newspapers, in particular, have been forced by commercial pressures to look out for any opportunity to run highly sensationalised, damaging, negative stories about businesses. Journalists are obliged always to ensure that there is some truth in what they write. The facts are right, but they are commonly chosen selectively and are often twisted to give the worst possible impression. Just as the media loves to build up a celebrity's reputation before bringing them down with negative stories, the same is true of business organisations and, potentially, individual business people who have a high profile.

This means that organisations have to tread very carefully so that they do not risk their reputations. They also need to take steps to guard confidentiality and restrict what their employees are able to say about them to journalists but also on social media:

- There is a need for policies on who is and who is not authorised to speak to media representatives about the organisation.

- There is a need for extensive media training for anyone who is required to give media interviews.
- There is some need to police social media originating from employees.
- There is sometimes a need to build up a positive counter-narrative using public relations tools and techniques.
- There is a need to try to monitor and respond to media coverage of employment matters, particularly on much-visited websites like Glassdoor that provide an opportunity for employees to review and rate their work experiences.

CASE STUDY 3.2

Sports Direct

In 2015 undercover journalists from the *Guardian* newspaper wrote a series of stories about what they had seen and been told by staff working at a warehouse in Derbyshire that was part of Sports Direct, a large, prominent and financially successful UK retail company. The stories suggested that most workers were employed on zero hours contracts and were 'harangued' using a tannoy system to get them to work faster. They were banned from wearing clothing from 802 particular clothing brands, were subjected to 'rigorous searches' at the end of their shifts and would be dismissed if they received six warnings for 'offences' like being sick, making errors, taking too long to use the toilet, wasting time, chatting to one another too much or using their mobile phones at work. The worst accusation was that some workers were, in practice, not receiving the National Minimum Wage.

The company argued that some of the reports were exaggerated and untrue, but this did not prevent a rapid fall in the company's share price as investors looked elsewhere to invest. The newspaper claimed that £400 million was wiped off the company's value, causing it to fall outside the FTSE listing of the biggest 100 companies operating in the UK. Questions were raised in parliament and the company's chair summoned to provide explanations by a House of Commons Select Committee. The chair of the Institute of Directors was quoted saying that Sports Direct was 'a scar on British business'.

The company responded by addressing the issues raised, cooperating with the Select Committee and raising wages. But huge damage was sustained by an episode in which very negative publicity spread across the internet following the publication of the reports.

SOURCE: Goodley and Ashby (2015); Goodley (2016)

The ubiquity of social media is also in the process of transforming aspects of the recruitment process, as organisations use it both to identify potential hires and to research the backgrounds of people who have applied for jobs and are being actively considered for appointment. The CIPD's resourcing and talent management survey for 2017 reported that over half of UK employers now use social media sites when recruiting new staff, LinkedIn (82 per cent), Twitter (55 per cent) and Facebook (51 per cent) being the most commonly used sites. There are three major ways in which these and other sites are being used:

- As a means of informing people who are potentially well qualified about job vacancies and opportunities.
- More generally as a means of actively managing an organisation's reputation as an employer so that would-be job seekers view it more positively.
- As a tool for maintaining contact with job applicants throughout the entire recruitment process and hence increasing the chances that they will accept a job offer if it is made.

The key development is the ability social media provides for recruiters to drive interested parties to their own recruitment websites where vastly more technically sophisticated materials can be posted than was ever dreamed of a few years ago when jobs were typically advertised in newspapers and trade journals. Recruiting through corporate websites is vastly more effective because of the opportunity this gives employers to include interactive features alongside video clips, testimonials from existing employees and large amounts of information that potential job applicants can search through themselves (see Dineen and Allen, 2014: 388–90). Put simply, organisations looking to source applications from highly talented recruits are now able at lower cost to make use of approaches that are a great deal more persuasive in nature. It has, of course, been possible to advertise roles on corporate websites for over twenty-five years, but the problem in the early days of the internet was getting people to visit the sites. Unless a job applicant already had an established interest in working for a particular employer, it was highly unlikely that they would ever become aware of the career possibilities on offer unless they were advertised in print or via an online job board. It is the rise in social media in more recent years that has made the big difference here. Sites like LinkedIn, in particular, now allow recruiters to identify potential future recruits and to build up relationships with them. There is no need to pay a headhunter to do the networking on their behalf. The work can all be done, often more effectively, by a dedicated, in-house recruitment team.

The use of social media to screen potential recruits is more controversial, because it often involves researching what people post online (or have in the past posted) , which may have nothing whatever to do with their working lives. Such evidence as is available suggests that these practices are very common (Davison et al, 2016); this is unsurprising when it is considered how very easy it now is to check out someone's online profile, to look at pictures they have posted and even to establish what their political views are.

It can be argued that what people do in their private lives should be irrelevant in a recruitment situation, even if they have shared this information publicly with friends on Facebook or Instagram. Similarly, views they express in a private capacity on Twitter should be neither here nor there when decisions are being taken about whom to appoint. It is difficult to reach a firm conclusion on this. On the one hand, it is reasonable to argue that an employer with a reputation to protect and enhance has a good reason to check that this is not going to be harmed by appointing someone unsuitable. On the other hand, the process of checking out people's private posts often leads recruiters to form positive or negative impressions which, however subtly, can very easily influence decision making unfairly.

PAUSE AND REFLECT

What is your view on checking out job applicants' online profiles and activities before deciding whether or not to invite them to interview or appoint them to jobs? Do you consider the practice to be unprofessional, unfair or improper in any way?

Online selection

In recent years, technological advances have provided HR managers and others responsible for recruitment in organisations with some new tools to use in the employee selection process. Some save money, others are more convenient, and one or two can be argued to increase effectiveness. But all have critics, and many of their concerns are justified.

The most widely used approach is remote interviewing using Skype, Zoom, WhatsApp or other similar platforms. This has become increasingly common as the reliability of these technologies has improved and as more and more people have access to them and are comfortable using them. Over 65 per cent of UK employers use these tools alongside more traditional telephone interviewing for some of their employee selection activities (CIPD, 2017: 28). It is inevitable that interviewing remotely will increase further in the future as it can save a great deal of money and potentially allow a much wider field of candidates, including people based overseas or many miles away, to participate in a selection process. When used quite formally as a means simply of exchanging information, the approach has much to recommend it, but it is unsatisfactory in other respects from both the recruiter and candidate perspectives.

A major problem is that a lot of people are not used to communicating in this way and find it difficult to do so anything like as effectively as is the case when they meet face to face. Technically, also, because lines break up from time to time and sound quality varies, interviewees may not hear the questions fully and interviewers may not fully take in all that is said by way of an answer. Subtler cues that we pick up from facial expressions and people's body language when we meet face to face are less easy to perceive and respond to when we communicate remotely. For the time being, until the quality of the platforms increases further, it has to be concluded that much remote interviewing is at best serviceable, and in most situations much less effective than face-to-face interviewing. Particular care must be taken from a fairness perspective when interviewing some applicants who have been shortlisted for the same job remotely, while meeting others in the flesh. It is unsurprising that both interviewers and interviewees, when asked, appear to have a strong preference for face-to-face interviewing (D'Souza et al, 2017).

Another major contemporary development in recruitment technologies is the development and use of tests that are taken by candidates online in order to decide who will go through to the next or final stages of a selection process. Because so many more people are able to find out about job openings now thanks to social media and internet-based, searchable job boards, the number of applications that organisations receive has increased. A well-known company advertising a desirable job can expect to receive many hundreds or even thousands of applications when new positions are advertised. Cost considerations thus inevitably mean that online sifting mechanisms have to be used to make the process manageable.

Some of these tools are more effective than others. The most straightforward approaches involve candidates completing online questionnaires that simply elicit factual information about qualifications, relevant work experience, salary expectations and other competencies. These are then scored electronically, effectively sifting out candidates who are weaker on paper while allowing stronger applicants to proceed through to a further stage. While such approaches can be designed to pass a 'feel fair' test, as selection tools they can be rather blunt and hence potentially less effective at shortlisting than an experienced recruiter would be reviewing each application using human judgement. Online application systems are especially poor at assessing unconventional applications from people who have a good deal of potential but non-standard qualifications or experience. This is particularly the case when the system contains some 'killer questions' that are particularly highly scored and play a major role in deciding whether or not a candidate advances. For practical and efficiency reasons, though, employers running large-scale recruitment exercises have little choice but to sift applications in this way (Tippins, 2015).

A second approach involves administering psychometric tests online, meaning that candidates complete them at home rather than on an employer's premises under some supervision. These tests include both those designed to test ability (typically mental ability) and those aimed at mapping key personality traits. Research studies have shown that provided the conditions in which candidates complete such tests are similar in terms of pre-briefing, speed with which questions upload and time limits, scores are similar irrespective of the setting. In other words, it makes little difference whether someone completes the test at home or at an employer's premises (Lievens and Chapman, 2019: 132–3).

Problems arise, however, when the conditions vary from one candidate to another. Slow internet connection speeds can make a big difference when tests are time limited and there is always the possibility of impersonation. How can the employer be certain that it is the actual job applicant completing the online IQ test and not a rather brighter friend or sibling? These issues can be addressed using sophisticated verification software or by holding a second round of tests for a small number of shortlisted candidates at the employer's premises, but there are costs involved that can cancel out much of the savings gained by administering tests online in the first place. Another potential way of reducing cheating involves adopting 'gamification' techniques to make tests enjoyable and exciting to complete, but as yet there is little robust research published to demonstrate whether this ever really works. The best approach probably simply involves suggesting to candidates that it is in their own best interests not to cheat and subsequently to select themselves out of a selection process if they fail to score highly in online tests.

The use of CV-matching software is another approach that has become common in recent years. These systems are becoming more sophisticated over time, but the basic methodology requires CVs to be uploaded electronically into a system that then scans them for key words and phrases. Those who include the 'right' words and phrases then get scored more highly and proceed to the next round of the selection process. It is a very unreliable approach that can easily end up rejecting good candidates simply because of their choice of language. However, where employers receive too many CVs to sift through with human eyes it is understandable that they are drawn to technological solutions that, while demonstrably imperfect, still have some practical utility.

E-learning

Recent years have seen an explosion of interest in various forms of technology enabled learning or 'e-learning'. Many organisations have invested significant sums of money into learning technologies – whether these involve the application of virtual reality (VR), online learning content management systems, virtual classrooms or curated libraries of video content. Similarly there has been interest in 'gamification' – ie the use of games to promote learning (Chapman and Rich, 2018; Nixon, 2018), with many studies suggesting that such approaches deliver improved learning transfer than traditional techniques.

E-learning can have many benefits for organisations, including the ability to reach all organisational members more easily. For instance, consider a multinational company with offices around the globe. Content delivered via e-learning systems can be rolled out almost instantaneously to the whole population, while classroom-style delivery would be significantly more resource intensive, and much slower in terms of speed of delivery. Learners themselves can engage with e-learning content at a time and place that best suits them. As a result, it is argued that there will be better knowledge retention, and more direct application to one's work. Technology, in this sense, can provide greater agility within organisational learning provision; indeed, researchers have shown that individuals typically respond best to learning content when it is delivered in small chunks, close to the point of need.

Having highlighted positive points, it is important to recognise that there are risks and drawbacks associated with the utilisation of technology in learning. The application of technology must not be seen as a 'silver bullet' or 'magic' solution that will solve issues associated with uninspiring or irrelevant learning content. The fundamental principles of learning design and delivery still apply. If we do not assess core learning needs, if we do not thoughtfully design learning events, they will have little impact. Additionally, e-learning systems are typically very cost intensive, meaning that they have to generate significant returns in order to deliver a return on investment (Gold et al, 2013). For smaller employers, the initial costs can be prohibitive, and given the challenges in isolating returns from learning input, even large organisations can struggle to justify such expenditure. A related problem is often the lack of skill/knowledge within the internal HRM/L&D function. It is frequently highlighted that human resource functions do not, for instance, have sufficient ability within the area of analytics (Vargas et al, 2018), and this can present a substantial problem. The same is often true within the field of learning technologies, which again presents a barrier to the successful adoption and use of novel human resource development

(HRD) methods and techniques (Meager, 2013). Again, for smaller companies employing fewer, and in some cases no, dedicated HRM/HRD staff, this issue can be substantially more problematic than for larger organisations.

There is also something in the nature of learning itself. If we subscribe to a view that learning is, in essence, a social practice where individuals make sense of ideas and give them meaning through their dialogue with others, then the basic 'point, click and read' e-learning systems operated by many organisations arguably divorce learning from its social context. Research into this issue, as discussed by Berg and Seeber (2017), suggests that instructors are not solely conveying information during taught classroom sessions; they are also conveying emotion, affect and feelings to their audience. These are qualities that are difficult (and in many cases impossible) to convey through technological delivery formats. The best conclusion to reach is that, where used properly and appropriately, technology can provide an excellent tool to enhance learning, but we must not forget the principles underlying the design and delivery of successful learning events.

References

Berg, M and Seeber, B K (2017) *The Slow Professor: Challenging the culture of speed in the academy*, University of Toronto Press, Toronto

BFI (2019) *Statistical Yearbook*, British Film Institute, London

Breyer, M (2017) 10 accidental inventions that changed the world, www.treehugger.com/accidental-inventions-that-changed-the-world-4864131 (archived at https://perma.cc/6JLV-9JWV)

Cancer Research (2012) Celebrating 60 years of progress, https://scienceblog.cancerresearchuk.org/2012/06/03/celebrating-60-years-of-progress (archived at https://perma.cc/NJ6D-985T)

Chapman, J R and Rich, P J (2018) Does educational gamification improve students' motivation? If so, which game elements work best? *Journal of Education for Business*, 93 (7), 315–22

CIPD (2017) *Resourcing and Talent Planning 2017: Survey report*, Chartered Institute of Personnel and Development, London

Coggan, P (2020) *More: The 10,000-year rise of the world economy*, Profile Books, London

Cooper-Ryan A M, Stonier, C and Ghadamosi, A (2020) The impact of the commute on our mental health and physical health within the context of flexible and non-flexible working, in *Flexible Work: Designing our healthier future lives*, ed S Norgate and C Cooper, Routledge, London

Davison, H K, Bing, M, Khuemper, D and Roth, P (2016) Social media as a personnel selection and hiring resource: Reservations and recommendations, in *Social Media in Employee Selection and Recruitment*, ed R Landers and G Schmidt, Springer, Switzerland

DCMS (2017) Sector economic estimates, www.gov.uk/government/collections/dcms-sectors-economic-estimates (archived at https://perma.cc/9VV5-8NHK)

Dineen, B R and Allen, D G (2014) Internet recruiting 2.0: Shifting paradigms, in *The Oxford Handbook of Recruitment*, ed K Yu and D Cable, Oxford University Press, Oxford

D'Souza, G, Prewett, M and Colarelli, M (2017) Selection for virtual teams, in *The Wiley Blackwell Handbook of the Psychology of Recruitment, Selection and Employee Retention*, ed H Goldstein, E Pulakos, J Passmore and C Semedo, Wiley, Chichester

Ede, A (2019) *Technology and Society: A world history*, Cambridge University Press, Cambridge

Friedman, T L (2016) *Thank You for Being Late: An optimist's guide to thriving in the age of accelerations*, Farrar, Straus and Giroux, New York

Gold, J, Holden, R, Stewart, J, Iles, P and Beardwell, J (2013) *Human Resource Development: Theory and practice*, 2nd edn, CIPD, London

(Continued)

(Continued)

Goodley, S (2016) Sports Direct falls out of FTSE 100 following *Guardian* investigation, *Guardian*, 1 March

Goodley, S and Ashby, J (2015) *Guardian* undercover reporters find world where staff are searched daily, harangued via tannoy to hit targets and can be sacked in a 'six strikes and you're out' regime, *Guardian*, 9 December

Lanier, J (2013) *Who Owns the Future?* Allen Lane, London

Lawrence, J (2015) A history of HIV survival in the UK, *The Pharmaceutical Journal*, www.pharmaceutical-journal.com/news-and-analysis/infographics/a-history-of-hiv-survival-in-the-uk/20200169.article (archived at https://perma.cc/VBM4-NATY)

Lievens, F and Chapman, D (2019) Recruitment and selection, in *The SAGE Handbook of Human Resource Management*, 2nd edn, ed A Wilkinson, N Bacon, S Snell and D Lepak, SAGE, London

Macrotrends (2020) UK life expectancy 1950–2020, www.macrotrends.net/countries/GBR/united-kingdom/life-expectancy (archived at https://perma.cc/F6C6-TYTW)

Meager, S (2013) How to overcome learning technology barriers, www.hrmagazine.co.uk/article-details/how-to-overcome-learning-technology-barriers (archived at https://perma.cc/WRF2-33WS)

Naim, M (2013) *The End of Power*, Basic Books, New York

Nixon, W (2018) Developing a digital first culture, *Training Journal*, December, 21–3

Nuffield Trust (2018) NHS spending as a percentage of GDP: 1950–2020, www.nuffieldtrust.org.uk/chart/nhs-spending-as-a-percentage-of-gdp-1950-2020 (archived at https://perma.cc/79UV-YTKY)

ONS (2017) *Paddington, Star Wars* and the rise of the UK film industry, www.ons.gov.uk/economy/grossdomesticproductgdp/articles/paddingtonstarwarsandtheriseoftheukfilmindustry/2017-12-14 (archived at https://perma.cc/SKM5-YDD4)

ONS (2018) Labour in the agriculture industry, UK: February 2018, www.ons.gov.uk/peoplepopulationandcommunity/populationandmigration/internationalmigration/articles/labourintheagricultureindustry/2018-02-06 (archived at https://perma.cc/YD36-E8LN)

ONS (2019) Living longer: Is age 70 the new age 65? www.ons.gov.uk/peoplepopulationandcommunity/birthsdeathsandmarriages/ageing/articles/livinglongerisage70thenewage65/2019-11-19 (archived at https://perma.cc/4DYU-EFZM)

ONS (2020a) Homeworking in the UK labour market, www.ons.gov.uk/employmentandlabourmarket/peopleinwork/employmentandemployeetypes/datasets/homeworkingintheuklabourmarket (archived at https://perma.cc/A22A-7JWF)

ONS (2020b) Coronavirus and homeworking in the UK: April 2020, www.ons.gov.uk/employmentandlabourmarket/peopleinwork/employmentandemployeetypes/bulletins/coronavirusandhomeworkingintheuk/april2020 (archived at https://perma.cc/7L5U-CP9W)

Prabhune, A (2015) 10 inventions that changed the world, but were made by mistake, www.storypick.com/inventions-made-by-mistake (archived at https://perma.cc/6QJZ-RTZ7)

Ridley, M (2020) *How Innovation Works*, 4th Estate, London

Rivett, G (2018) A 70-year perspective, The Nuffield Trust, www.nuffieldtrust.org.uk/chapter/conclusion-a-70-year-perspective (archived at https://perma.cc/NE3D-YKKE)

Smil, V (2005) *Creating the Twentieth Century: Technical innovations of 1867–1914 and their lasting impact*, Oxford University Press, Oxford

Smil, V (2010) *Transforming the Twentieth Century: Technical innovations and their consequences*, Oxford University Press, Oxford

Statista (2020) Infant mortality rate (under one year old) in the United Kingdom from 1950 to 2020, www.statista.com/statistics/1042373/united-kingdom-all-time-infant-mortality-rate/ (archived at https://perma.cc/WX7R-GPBK)

Stephenson, J (2018) How has nursing changed since the birth of the NHS in 1948?, *Nursing Times*, 5 July

Stewart, I, De, D and Cole, A (2015) *Technology and People: The great job creating machine*, Deloitte, London

Tippins, N (2015) Technology and assessment in selection, *Annual Review of Organizational Psychology and Organizational Behavior*, 2, 551–82

(Continued)

(Continued)

Triggle, N (2018) The history of the NHS in charts, www.bbc.co.uk/news/health-44560590 (archived at https://perma.cc/AGV4-LSCL)

Vargas, R, Yurova, Y V, Ruppel, C P, Tworoger, L C and Greenwood, R (2018) Individual adoption of HR analytics: A fine grained view of the early stages leading to adoption, *International Journal of Human Resource Management*, 29 (22), 3046–67

White, J, Behrend, T and Siderits, I (2020) Changes in technology, in *The Cambridge Handbook of the Changing Nature of Work*, ed B Hoffman, M Shoss and L Wegman, Cambridge University Press, Cambridge

World Economic Forum (2020) This UK survey reveals a troubled future for low-skilled workers, www.weforum.org/agenda/2020/02/20-fastest-growing-occupations-uk-jobs (archived at https://perma.cc/8KTL-4C5Z)

Zuboff, S (2019) *The Age of Surveillance Capitalism: The fight for a human future at the new frontier of power*, Profile Books, London

04

Technology 2

Going where?

In this chapter we are going to explore debates about the future of technology and the changes these developments are likely to bring to the world of employment and work.

We will start by exploring some of the most important contemporary examples of technologies that are soon likely to mature and become widely adopted. We will go on to discuss the likely impact on employment in a general sense, particularly the debate about whether the opportunity to work is going to lessen, and possible responses to this situation.

We will end with a discussion about predictive analytics and employee surveillance, exploring some of the professional and ethical issues associated with their use in workplaces.

PAUSE AND REFLECT

Most people reading this book are likely to have been born some time in the 1990s, and if life expectancy continues to rise in the future as it has in the past, there is a good chance that you will live well into the second half of the twenty-first century.

Reflect for a few minutes on the technological changes that most people who were born 100 years before you lived to see. When they were born there were no cars, no planes, limited electricity, no antibiotics, few telephones, no radio and no TV. They could have had no idea that they would live to see space missions, heart transplants, supersonic travel or the internet – even if they could have imagined such things.

Now reflect on what technological developments you might reasonably expect to see occurring in your lifetime.

We live in a time of extraordinarily fast technological development in which a wide range of technologies with potentially transformative possibilities are rapidly being refined and enhanced. Several are likely to come to maturity, and hence to the market, in the 2020s and 2030s. It is helpful to start this chapter with a brief discussion of some of these developments.

Biotechnology

Whereas information technology is a mature twentieth-century invention that is being continually improved and refined and its applications extended, biotechnology is still in its infancy, but with a similar potential to transform our lives. It has many manifestations, but all in their different ways involve the manipulation of living organisms or biological systems to provide products that serve a useful purpose.

There is nothing new about these kinds of processes in themselves. Brewing alcoholic drinks, fertilising crops and vaccinating people against diseases all draw on biotechnological principles. It is the combination of biological study with large-scale computer processing technology which is a relatively recent development and which lies at the heart of biotech research. These processes allowed scientists to map genomes for the first time in the final years of the twentieth century, and to go on to map the human genome in 2003. In the 2020s and 2030s we are going to see artificial intelligence (AI) and machine learning processes being used to identify new applications, hence vastly increasing the number of ways in which bioengineering will affect many aspects of our lives.

Once scientists have access to detailed genetic information about an organism's DNA, they can isolate it, alter it and reintroduce it into that organism's cells. This process is variously known as 'genetic engineering', 'genetic modification' and 'gene therapy', and it lies at the heart of current developments in biotechnology. The field is divided into a number of distinct sub-fields, often labelled red, green and white.

Red biotechnology is concerned with medicine. It involves designing organisms that can produce new antibiotics and genetically engineering medicines that cure or prevent diseases. Radical medical improvements are likely to be developed using these technologies over the next two decades. Key examples include the following:

- Personalised drug treatments involving the prescription of drugs that are tailored to match the genetic makeup of individuals rather than standard drugs designed to treat the whole human population. Not only should this make the treatment of disease more effective, it could also mean that diseases that we are genetically predisposed to contract (such as cancers) can be prevented.
- The manufacture of artificial body parts. It is believed that we are not far off being able to restore sight with artificial retinas made from silicon. Cochlear implants are already widely used in the treatment of deafness.
- Ribonucleic acid (RNA) technology involves creating drugs (known as antisense drugs) that interfere with the ability of certain cells (like cancer

cells) to reproduce themselves. There are hundreds of treatments currently in development with the potential to treat a wide variety of conditions, including diseases of the heart, muscles and liver (see Wang et al, 2020).

- Stem cell research involves manipulating cells in laboratory conditions that are 'uncommitted' so that as they divide and multiply they form a different type of cell. For example, a stem cell can be extracted from someone's bone marrow or from a human embryo and manipulated so that it develops into liver tissue. This could then be used to repair liver damage caused by excessive alcohol consumption. Further applications are likely to be in the treatment of spinal cord injuries and Parkinson's disease, where damaged neurons can be repaired by introducing new cells into the body.

Over time, it is believed that blood cells, heart muscle cells and insulin-secreting cells could be grown 'in bulk' and that one day we will be able to clone organs such as livers and kidneys. The possibility of dramatically increasing human life expectancy is thus very real and could be achieved within the next two or three decades.

Green biotechnology relates to agricultural usage and is associated with the genetic modification of crops. Little use is made of them in Europe, but they are now widely used in many other countries (10 per cent of the world's arable land is already cultivated with them).

Genetically modified (GM) crops are manipulated to be more resistant to disease (and hence do not need to be treated with so much insecticide), cold and drought, and are generally more nutritious. The shelf life of fruit can also be extended via genetic modification, and agricultural products generally can be given qualities that make them healthier for humans to consume.

Another area of innovation in green biotechnology is in the production of crops grown for animal feed (mainly maize and soya), the purpose of which is to improve its content so that farm animals grow bigger and are less susceptible to disease. Animals themselves, of course, can also be genetically modified (and potentially cloned) to make them more resistant to disease, bigger and, from a meat-eater's perspective, tastier too. Over the long term, these developments have the capacity to improve crop yields substantially, to improve the quality of food that is produced and reduce the likelihood of famine.

White biotechnological developments involve using organisms to create industrial materials and fuels. For some time, enzymes have been manufactured and used in products such as washing powders, but many other applications are now being developed. Examples include the development of genetically modified bacteria that produce substances used in the manufacture of plastics, and 'bioleaching', which involves using bacteria to extract gold and copper from their ores.

One of the most significant contemporary innovations is represented by the development of maize and straw crops that are far richer in glucose than traditional strains – in the stems and roots as well as the seeds. The crops are then harvested and the glucose extracted to use in the manufacture of ethanol, a viable and more environmentally friendly alternative to oil for use as a motor fuel. Indeed, a major aim of white biotech processes is to create new biodegradable products that can replace the ones we currently use that rely on plastics and papers.

LAB-GROWN MEAT

A vast amount of venture capital is currently being invested in the production of so-called 'clean meats', which are products like steaks, burgers, sausages and chicken nuggets that have been grown in labs rather than by raising animals on farms. The manufacturing process involves taking stem cells from animal tissue and encouraging them to replicate in a nutritious culture rather than a host animal. It is all done in a laboratory.

There are major challenges for the pioneering companies. The final product needs not only to taste good and be fit for human consumption, but also to cost less than farm-produced meat. But if these challenges can be met – and the current signs are good – we may soon on a daily basis be consuming meat which has been produced without the need to raise and slaughter animals and which would be very much more environmentally friendly. Rain forests would no longer need to be cut down to make room for beef production, and it should be possible for many billions of people to access a more nutritious diet than they can currently afford.

Needless to say, the rise of a global 'clean meat' industry would be very bad news for livestock farmers, who are fiercely opposed to its development and can be expected to lobby regulators very hard on issues like potential long-term health risks and labelling.

SOURCE: Altraide (2019: 254); Schaefer (2018)

Nanotechnology

Nanotechnology is a fast-emerging field with huge potential, and some significant developments already made. It involves manipulating material at the nanoscale (ie smaller than a billionth of a metre) – individual atoms and molecules. It was made possible by the invention of the scanning tunnelling microscope by IBM in 1981 and electron beam lithography in more recent years. These allow scientists to move atoms about and combine them to create new materials, some of which are self-assembling because they consist of two or more complementary and mutually attractive component parts. Nanotechnology also makes use of the way in which the chemical properties of materials such as copper, gold, platinum and silicon differ at the nanoscale.

To date, most major, commercial applications have involved improving existing products. Sun-tan lotions have been made more effective, new protective coatings have been developed, better stain-resistant fabrics manufactured and special bandages developed which are coated in silver for the treatment of burns. In the longer term it is expected that completely new types of material will be developed, such as metallic rubbers, flexible ceramics and hyper-strong plastics. These will be lighter, cheaper and more durable than current materials.

The invention of graphene is a high-profile contemporary example of a significant evolving form of nanotechnology. Graphene is a flake of graphite that is just one atom thick. Three million sheets of graphene stacked up on top of each other make a pile just 1 millimetre high. It has nonetheless been found to be extraordinarily strong – 200 times stronger than steel – despite being very much lighter. It has also been found to be an excellent conductor of electricity. By combining plastics with graphene, transparent materials can be produced that conduct electric current very efficiently and are a great deal less expensive to source than the materials currently used to manufacture touch-screen devices.

As with biotechnology, some of the more significant contemporary developments in nanotechnology concern medicine. A major example is the development of tiny nano-robots that can search out diseased cells and treat or destroy them. Such treatments could potentially remove the need for cancer patients to undertake a course of chemotherapy while also being more effective. Other likely areas of development in the near future involve cells which can store solar energy efficiently and the shrinking to the nanoscale of transistors, which will allow many more to be incorporated into all manner of electronic products. The website www.understandingnano.com provides numerous examples.

Autonomous vehicles

There are two distinct contemporary developments here, both of which make use of similar technologies:

- Self-driving automobiles (cars, trucks, trains, trolleys, etc).
- Remotely operated drones.

In both cases, initial technologies were largely developed for military purposes in the USA during the 1990s. They are now being developed commercially. In most respects the major technological hurdles have already been successfully negotiated; what have yet to be resolved are the regulatory frameworks and arrangements for transition.

Autonomous farm vehicles are already in wide use globally, substantially reducing the number of people required to plough fields, sow crops and then to harvest them on larger farms. There are also several examples of unmanned public transportation systems in operation around the world.

Drones with cameras mounted on them are also now used on large farms for surveillance purposes, as well as by film and TV producers, and coastguards for sea patrol and rescue purposes.

Long-haul passenger planes already operate with extensive use of auto-piloting systems and military drones are now very widely used in war zones. For several years now, Amazon has been looking to develop a drone-based delivery service that will, potentially, permit orders to be delivered from a warehouse to a customer's front door (or front garden) by a drone within an hour.

Tesla, Google and Audi are in the vanguard of research in this area, each having developed fully autonomous driverless cars that they aim to market commercially in the 2020s. Uber has placed an order for 500,000 of them with Tesla. It has also been announced that the new generation of Tube trains for use on the London Underground, due for introduction in the 2020s, will feature driverless technologies.

Cars with advanced, autonomous features (parking, reversing in narrow spaces, cruising on motorways, etc) are already on the market, while trials involving different forms of driverless cars are being permitted and, to an extent, funded by governments in specific locations around the world. In the UK, a full review of the Highway Code is also now in progress with the aim of bringing forward proposals that would allow for forms of 'driverless driving' in the near future.

The manufacturers of autonomous vehicles claim that they are vastly safer than human-operated equivalents, with the capacity to reduce the number of deaths and serious injuries on our roads very considerably. In the UK at present over 700,000 road accidents are recorded each year, leading to over 25,000 serious injuries and 1,500 deaths. There would also potentially be significant environmental benefits arising from the shared-use model of transportation becoming practicable. The new generation of self-driving cars will be electric, and quite possibly used more in the manner of a taxi service than as privately owned vehicles. This would help ensure that the highest environmental standards and refinements were continually being made.

There are, however, still many technical difficulties to overcome – notably the time it takes to alert drivers that they need to assume control when something does go wrong with a driverless car. Answers also need to be found to prevent road traffic continually coming to a halt if cyclists and pedestrians simply walk out in front of these cars without concern, knowing that they will stop instantly when they do. There also remain some security issues. In 2015 a team of hackers demonstrated how they had managed to take control of a driverless car via its entertainment system (Schwab, 2016: 148). The potential for using driverless cars in terrorist attacks is also self-evidently a major potential problem.

There is also a big further hurdle to overcome in persuading people that travelling in autonomous vehicles is safe. According to opinion poll research, their use on British roads is opposed by 43 per cent of the public, while 25 per cent say they would refuse to get into a driverless car. The presence of strong trade unions who are prepared to take robust and prolonged strike action is the factor that is most likely to hold back the introduction of driverless trains on overground and underground networks. Regulatory and safety issues are thus likely to hold back the rate at which we adopt autonomous transportation, but it is nonetheless widely expected to be introduced in steps over the next twenty years.

Initially, autonomous cars and trains will be dual-use, with drivers' seats and sets of controls that can be used if the autonomous mechanisms fail in any way. Over time, however, as the technology becomes more robust and people get used to the idea, we are likely to see pod-like vehicles on our roads with limited controls that can be used by anyone, whether or not they have a driving licence. All you will need to do is tell the pod where you want to go, and it will take you there.

There is some debate in the industry about which model of car usage will be adopted once we get to the stage at which autonomous vehicle technologies are commonplace. One view is that the traditional car will be replaced by caravan-like vehicles in which people can entertain themselves, work, sleep, eat and socialise while they are being transported from place to place. If so, a personalised model involving continued ownership of bespoke vehicles will predominate. The alternative view is that the car-ownership model that we are used to will be abandoned in favour of a shared-resource model in which fleets of driverless taxis will parade around waiting to be summoned to transport people on demand for a fee. The logic here is that, on average, having purchased a car, its owner only actually uses it for 4 per cent of the time. It is parked for the remaining 96 per cent of its life. Moreover, three-quarters of the cost of using an Uber taxi is for the driver. There would thus be very strong economic incentives to abandon car ownership and instead rent autonomous cars for long journeys or use taxis for short journeys. Worrying about parking would no longer be a feature of modern urban life.

Before we arrive at a point in which we no longer drive ourselves around and in which goods are transported from place to place in vehicles with no drivers, it is likely that we will get used to seeing autonomous vehicles in other environments like warehouses, hospitals and shopping centres, where they can be readily used in quite a small way within a confined area. Increased usage of unmanned vehicles on farms is also very likely.

Either way, we can conclude that a considerable number of jobs that involve driving will disappear over the next few years. It is estimated that over a million people make a living from driving professionally in the UK, not to mention the associated trades that rely in large part on drivers and driving for their income, and the employees in companies that manufacture all the cars. Many hundreds of millions of people around the world are thus likely to see their livelihoods disappear as a result of the introduction of autonomous vehicles over the coming decades. By any standards, this is going to be a seriously disruptive technology.

PAUSE AND REFLECT

Aside from people who make a living from driving, portering and piloting, who do you think will be the main winners and losers from the widespread use of autonomous vehicles?

How do you anticipate that this development may affect your life?

General artificial intelligence

Another prospective transformative technology is the next generation of supercomputers. We already have computers and IT systems that exhibit high levels of 'artificial narrow intelligence' and have had for some time. These are capable of carrying out narrowly defined 'mental' tasks as well, if not considerably better, faster and more cheaply than human beings. Several times already, for example, as I have been writing this chapter, my computer has advised me of potential spelling and grammatical errors without me having to prompt it.

The 'Deep Blue' computer beat Garry Kasparov in a chess match as long ago as 1995, since when we have got used to IT systems that alert our banks when an unusual pattern of spending on our bank cards is detected, and personalised pop-up adverts tailored to our interests or demographic group that appear when we surf the internet. Less well known is the fact that some 70 per cent of financial transactions on the equity markets in major financial cities are now carried out entirely on the whim of IT systems making use of algorithms (Buchanan 2019: 15). They are 'trained' not only to track and predict market movements, but also to trade automatically as and when necessary – a feat they accomplish in one-eighth of a second. The latest systems also include decoy software that withdraws from trades at the last minute (or eighth of a second), thus deliberately misleading the IT systems of competitors.

Moore's Law famously states that the storage capacity of an integrated circuit doubles every two years. Gordon Moore made this observation when he worked at Intel in the 1960s, since when it has held true. Put crudely, it means that the raw computing power of each generation of devices that comes onto the market is growing (literally) at an exponential rate, as has been the case for the past sixty years. As a result, we now commonly carry small smartphones in our pockets that are each many times more powerful than all the IT equipment used by NASA to put men on the moon in the 1970s. They do more and more, and do it more quickly, all the time.

Thanks to this exponential growth in computing power, the capability of IT systems is now moving from a 'narrow AI stage' to a 'general AI stage' (known as general artificial intelligence), which means they are a good match for an intelligent and well-educated human being. This, according to Del Rosal (2015: 9), enables them to carry out 'any intellectual task that is humanly possible'. In other words, IT systems are now being developed which have the ability 'to reason, plan, solve problems, think abstractly, comprehend complex ideas, learn quickly and learn from experience'. Such systems will be widely used in the 2020s and 2030s.

Artificial intelligence at this level involves computer systems developing their own algorithms in response to experience and hence genuinely learning as they go about their business. They are also much better at making accurate predictions than human brains. Potentially, what makes these systems genuinely disruptive is the way that they link up with the cloud (shared global cyberspace) and big data (the unimaginably large stores of information on everyone and everything that now exists electronically and grows exponentially all the time).

Susskind and Susskind (2015) and Susskind (2020) argue that, over the next ten to twenty years, IT systems of this kind will not just eliminate all jobs that involve any form of routine transmission, creation or manipulation of data, but they will also make major inroads into all kinds of professional work too. They argue that people who make their living as 'human experts' will increasingly find that computers can do major parts of their jobs much better than they can. In their book they make the case in relation to the following:

- Healthcare/doctors.
- Teachers.
- Priests.
- Lawyers.
- Journalists.
- Management consultants.
- Accountants.
- Architects.

However, their argument extends into all other areas of professional life, including government, scientific research and even some professions that are considered to be 'creative' in nature. Prototype teaching systems (robotic instructors) are already apparently getting better exam results than traditional human lecturers on some university courses, while essay-marking software is 88 per cent reliable when compared with human marking.

In medicine, it is argued, machines are now considerably better placed than human doctors to make accurate diagnoses when presented with symptoms, asking more relevant questions and making fewer errors. They are also better at prescribing medicines. The same is true of systems designed to answer legal questions or to provide solutions to management problems. In each case vast amounts of big data are 'mined' very rapidly by machines that are programmed to 'learn for themselves' over time – just like human professionals do, but with greater accuracy, continually updating knowledge, in multiple languages simultaneously and with no loss of memory.

The significance of these systems is that they have the potential to destroy any claim made by professionals to have 'special knowledge' or a claim to be licensed to give special advice. IT systems, for example, will be able to read and interpret X-ray images far more quickly and accurately than radiographers. A medical assistant who is paid a fraction of what a doctor is paid could potentially use diagnostic software systems of this type and prescribe better treatments than doctors do today. Indeed, it could be the case that patients will themselves be able to consult the systems and book the appropriate treatments.

Advanced robotics

The application of Moore's Law to the development of robots means that over time they too will evolve the capacity to do more and more tasks that we currently consider to be only capable of being performed by reasonably highly trained human beings. Once they can 'think', 'learn' and 'see' it becomes possible for them to do more than just routine, repetitive tasks that they have been programmed to perform.

Ford (2015) gives numerous examples of advanced robotic technologies currently in development which he persuasively claims will not just make employees in major trades more efficient but will eliminate them altogether as 'labour is engineered out of the product':

- Robots that do the jobs of chefs, from basic burger-making machines to robots that cook and assemble gourmet, restaurant-standard dishes cheaply and reliably.
- Robots that enable people to attend exhibitions remotely.
- Robots that are much more effective than security guards at ascertaining when something changes in an environment.
- Kiva robots being used in Amazon warehouses to collect and box up books for despatch.
- All manner of complex vending machines/kiosks that deliver instantly.
- Fruit-picking robots.

In manufacturing, the possibilities for the use of advanced robotics are particularly transformative, allowing the possibility of:

- Lights-out factories operating around the clock in which no human being is needed except IT programmers and maintenance people.
- The re-shoring from developing countries to industrialised countries of manufacturing activity – robots being cheaper and more reliable than workers.

ROBOTS IN CARE HOMES

Care homes to house and look after elderly people are a kind of workplace that many consider may be able to make good use of advanced robots in the future. As our population ages and the cost of living in a care home or nursing home grows, a good business case can be advanced for moving in this direction. But can robots ever really replace human care-giving in such settings at all satisfactorily?

The kinds of jobs that robots can carry out would be dispensing medicines at the correct time, delivering meals and snacks to residents' rooms, answering questions and potentially providing some entertainment. Robots can certainly be 'trained' to have conversations and to recall what was said the last time they spoke to the same person. They can also, potentially, carry out some routine physical tasks like lifting people from wheelchairs to armchairs or picking people up from the floor after a fall.

They could thus quite readily be used in care homes alongside employed persons and, in the process, potentially reduce costs sufficiently to make it less expensive to reside in a home.

Longer term, these developments raise the prospect of personal robot carers being placed in people's houses as they get older to provide domestic assistance and to alert medical professionals when elderly people living alone require assistance.

3D printing

A great deal is currently being written about cutting-edge developments in digital manufacturing that have the capacity to revolutionise many of the world's established economic systems over the next two or three decades. While very much still in its infancy as a technology, 3D printing or additive manufacturing prototypes have been developed and are being refined all the time. They work by building three-dimensional solid objects layer by layer using different raw materials. Printers are fed computer-aided designs and then manufacture the product according to that specification.

3D printers make combined use of a variety of technologies (including lasers, electron beam melting, lamination and stereo lithography) to manufacture objects rapidly and relatively cheaply. At present, printing has largely been restricted to the manufacture of quite simple products (hammers, plastic moulds, coat hangers, etc). But the technology is developing very rapidly and it is confidently predicted by science writers that quite complex electronic machinery will be 'printed' in this way before long. A feature in *The Economist* (2011) made the following observation about the consequences of longer-term developments in additive manufacturing:

> Three-dimensional printing makes it as cheap to create single items as it is to produce thousands and thus undermines economies of scale. It may have as profound an impact on the world as the coming of the factory did... Just as nobody could have predicted the impact of the steam engine in 1750, or the printing press in 1450, or the transistor – it is impossible to foresee the long-term impact of 3D printing. But the technology is coming, and it is likely to disrupt every field it touches

There is a lot of speculation that before long it will be routine for people to own a digital manufacturing unit that they will keep at home. Instead of purchasing products in shops, we will buy designs and download them, before programming our 3D manufacturing machines to 'print' the item for us.

The big contemporary development in this field is the evolution of 3D printers that can operate using multiple materials – ie not just plastics and metals but a range of these and other materials too, at the same time. This will allow the printing – speedily and at low cost – of bespoke machine tools and replacement parts for machines of very considerable complexity. Companies are also actively working on machines that are able to 3D print houses and other buildings – piling up layer upon layer of concrete/plastics to a specified design. This technology could very substantially reduce the demand for construction workers, potentially moving much construction work into factories, with houses and other small buildings being prefabricated before they are transported for assembly on site.

Most extraordinary of all are successful experiments in the printing of animal tissue, raising the possibility that in a decade or two we will be able to print whole organs and transplant them into our bodies, potentially extending life expectancy by many years.

Virtual reality

The rapid development of virtual reality technologies is another outcome of increasing computing power. Put simply, the suggestion is that before long we will be able to put on a headset that covers our eyes and ears, before entering and interacting with a virtual world that is recognisably familiar. The technology will give us the sensation that we are somewhere else with considerable accuracy. When combined with high-quality simultaneous translation capabilities, the possibilities for enhanced global communication are substantial.

The main early adopters have been gaming companies, who are producing all manner of exciting virtual, interactive experiences that enhance the kinds of experience that people have already been having using two-dimensional applications on screens. Another early application will be virtual 'visits' to zoos, museums, galleries, theatres, concerts and other entertainment venues. Users will actually be sitting at home or on a train going to work, but they will, through their VR headsets, experience the sensation that they are elsewhere, participating along with others in an experience that they would otherwise not have the time or money to enjoy.

Over time, though, it is entirely plausible to argue that much more mundane places will be 'visited' virtually on a routine basis, and that it will be in our working lives that we experience virtual reality most frequently. Might it be offices, conferences, meeting rooms, classrooms and lecture theatres that are most commonly occupied by virtual visitors?

Used in such a way, VR would permit any group of people to assemble together in one place to meet, discuss, make decisions, share ideas, etc, without ever actually meeting physically at all. Moreover, the technology should soon also enable them to communicate with one another in different languages entirely naturally. In a virtual office set-up an employee might give an opinion to a manager in English, while the manager hears it in Japanese. Geographically we could literally be located anywhere in the world, but virtually we will experience working alongside one another in the same place.

Some writers go further in stating that we will soon make regular use of augmented reality in addition to virtual reality. Augmented reality involves the headsets/glasses (some even suggest contact lenses) we use to enter virtual reality worlds being able, for example, to project text messages and internet pages for us to read and respond to using voice activation while engaging with virtual activities.

This set of technological developments has very profound, long-term implications for work in that it raises the possibility that traditional workplaces will be less and less necessary. The proportion of time that it will be feasible to spend tele-commuting will increase greatly, removing for many people the need to have a workplace at all.

PAUSE AND REFLECT

Think about work in human resources in particular. What parts of the role could potentially be performed as effectively, if not more effectively, in the future with the use of high-quality virtual reality applications?

The debate about jobs

The above discussion of some significant near-future technological developments inevitably leads us to two conclusions about the likely impact on the future of work and jobs:

- Highly disruptive technological developments are likely to bring about significant reductions in the number of people employed in some significant job categories. Some will be eliminated, while others will see demand fall as machines enable smaller teams to carry out their roles with greater efficiency.
- In order to remain employable, and particularly to access higher-paying jobs, people are going to have to develop new skills. There will be higher demand for people who are technically specialised and lower demand for people to fill roles that can be substituted by machines.

The extent of this anticipated labour market disruption and its timing are not at all clear and have therefore been heavily contested by writers who speculate on how the world of work is likely to evolve over the next twenty or thirty years.

Some prominent contributors take a deeply pessimistic view, arguing that we are likely to enter an era in which there is insufficient work to go around. High levels of unemployment, job insecurity and underemployment are therefore likely to emerge and remain with us for a long time. Machines will effectively 'steal' jobs that are currently carried out by people, leaving those who own the machines very wealthy indeed, while those whose jobs are displaced by them will become a great deal poorer.

Ryan Avent (2016), for example, predicts a twenty-first-century world of work that is characterised by 'labour abundance' as machines replace many of today's established and secure jobs. While the new technologies will also create new job opportunities in new industries, these will often not be suitable for the people who have been displaced from their old jobs:

> There is no iron law that says new, more profitable firms will create exactly enough of the right kinds of work to absorb those kicked out of shrinking occupations. On the contrary, displaced workers are quite often in an unusually bad position to be rehired. They have spent years, or decades, accumulating know-how of declining value: such as how to use obsolete equipment or how to operate successfully within the culture of defunct firms.
>
> (Avent, 2016: 47)

Of course, technological advance is nothing new. It has occurred throughout human history and, while it has often had a disruptive impact on jobs in the past, most people have been able to adapt and have found new, if sometimes less satisfactory, work in newer, expanding industries. In recent decades, as we saw in Chapter 3, we have seen very substantial falls in the number of people in the UK employed in manufacturing, ship building, mining and agriculture, yet the total level of unemployment has remained low during periods of economic growth. New jobs have replaced the old ones. So why should things be any different in the future?

The pessimists give two reasons. First, they point to the sheer scale and likely rapidity of the coming jobs revolution. Ford (2015: 121–2) argues that some 50 per cent of existing jobs in an economy such as the UK's are susceptible to automation and that it is reasonable to expect that 35 per cent will be replaced by machines over the coming two decades or so. Some of this displacement will be direct (ie machines replacing humans), while some will be indirect as technology enables further offshoring of jobs overseas to locations where wages are lower (eg translation software enabling people in China to carry out more administrative activities currently carried out in countries like the UK). His estimates are broadly supported by Frey and Osborne (2013), whose detailed study of the susceptibility of 702 professions to automation in the USA concluded that between 45 and 50 per cent of current jobs are likely to disappear over the coming twenty years, 35 per cent as a result of automation and the remainder as a result of offshoring. Manyika et al (2018) go further still. Drawing on an analysis carried out by the McKinsey Global Institute, they state that while only around 5 per cent of current occupations can readily be *fully* automated, 'in about 60 per cent of occupations, at least one-third of the constituent activities could be automated'.

Even those studies that predict less dramatic levels of job displacement still conclude that the extent of job churn in the coming decades will be huge. Hawksworth and Fertig (2018), for example, report an analysis carried out by PricewaterhouseCoopers (PWC) that suggests that around 20 per cent of existing jobs in the UK will be automated over the next twenty years, while the Office for National Statistics suggests that 7.5 per cent of jobs (2.5 million or so) are at high risk of being automated in the near future (ONS, 2019).

Second, it is argued that the future will differ from the past because over the next few decades it will be the more highly skilled jobs that will be replaced and not just the lower skilled jobs, as was mainly the case historically. It will be brainwork and jobs in professions that are knowledge-based that artificial intelligence will be able to carry out more effectively and reliably than humans, at a lower cost.

Chase (2016) argues that the nature of the transformation that we are about to live through will be of an order similar to the agricultural revolution three thousand years ago and the Industrial Revolution of two hundred years ago. During the twentieth century, humans whose jobs were replaced by machines were able to upskill, to become more highly educated and move on into jobs that drew more on their brain power than their muscle. That option will not be available in the future because it will be the jobs that draw on brain power that will be automated. He argues that our fate in the twenty-first century will be akin to that of horses in the twentieth century. Unlike people, they could not adapt to the new technologies that were emerging 150 years ago. It is estimated that there were some 3.3 million working horses in the UK in 1900. They were used on farms, down mines, in the armed forces and had for centuries been the main form of transportation over longer distances. There were 50,000 horses transporting people around London alone at the end of the nineteenth century. With the advent of motorised transport, the grooms and blacksmiths who made a living servicing all these horses, along with the carriage drivers, were able to adapt and retrain, but the horses themselves were not. They became permanently 'unemployed', and their numbers dropped rapidly. There are now estimated to be around 850,000 horses in the UK. They are ridden for leisure purposes and their number drops a little each year. The UK's overall horse-to-human ratio fell from 1:12 to 1:80 since 1900.

This deeply pessimistic thinking about the future of jobs is not universally held. Other writers take a more optimistic view of our capacity to adapt and to create new jobs in the future. The new technologies will destroy jobs, they argue, but will also, through increased productivity, lead to the creation of new ones. Frey and Osborne's (2013) analysis of the London labour market between 2000 and 2013 shows what can happen. These years saw vast, rapid drops in the number of people employed in some job categories:

- 65 per cent fall in library assistants and sales jobs.
- 58 per cent fall in clerical jobs.
- 48 per cent fall in counter clerks/receptionists.
- 44 per cent fall in secretaries.

Yet the total number of jobs in London grew rapidly, along with wages. The city attracted hundreds of thousands of new residents from around the world, and unemployment remained remarkably low, even during the recession of 2008–10.

In the future, there will be plenty of jobs available in developing the new technologies and servicing them. Many millions more people will be employed in IT roles than is currently the case, as well as in the development of all the other technologies we can expect to see maturing. Many of the new technologies are also themselves likely to create new job opportunities either directly or indirectly. Just as the rise of social media has allowed people to develop their own YouTube channels, to produce podcasts and to set themselves up as dealers using Amazon and eBay, there is every reason to anticipate as yet unforeseen opportunities arising from the development of virtual reality, 3D printing and artificial intelligence.

There is also capacity for expansion of the number of people employed to do jobs that are less susceptible to automation. The main examples are in the creative industries and the provision of services that require high levels of human empathy. Human resource management is likely to be one of these, along with allied professions that require a good deal of human insight, understanding and judgement if they are to be performed effectively. The more social and emotional intelligence required to carry out a role, the less readily automated it will be. In their American study, Frey and Osborne (2013) list human resource managers as the fifth least likely group to be displaced as a result of automation out of 702 professions – only mental health professionals, doctors, psychologists and choreographers are at less risk.

Another possible outcome of this coming disruption in our labour market may be an appetite for reorganising aspects of our economic lives more fundamentally, potentially creating a situation in which we are able to live well while working less. If people were to have more leisure time available and still enjoy a reasonable level of disposable income to spend filling it, more jobs would be created. One way of achieving this would be to move in stages towards a universal basic income (UBI) approach, which, put crudely, involves governments 'taxing the robots' as they take over jobs carried out by humans, and then redistributing the proceeds. Through such a mechanism, society would ensure that everyone, rather than just a few wealthy people, benefits from their development.

Under a UBI system, all citizens of a country are paid the same amount each month by their government. In a world of true labour abundance in which, thanks to technology, there are simply not enough jobs to go round, the level of UBI would have to be set at a level that allowed each recipient to live without carrying out full-time paid work – perhaps at a similar level to state pensions in today's system. There are all kinds of flaws with this proposal, not least the potential adverse effects on people's mental health if they do not have productive work to carry out over a long period of time. Too much leisure time is not necessarily good for us. But if we really are about to enter a world in which job opportunities for many dry up, there may be no alternative if a massive increase in social inequality is to be avoided. Susskind (2020) broadly agrees with the 'labour abundance' thesis and recognises that the evolution of some form of UBI system may be necessary, but he argues that it should be paid conditionally and not automatically. Only those who carry out useful work in their communities should receive it, hence ensuring that we do not become a nation of idlers simply living off the taxes generated by the robots we employ to do the work we previously carried out ourselves.

PAUSE AND REFLECT

How do you anticipate that the possible future technological developments we have discussed here may affect your own, personal, employment prospects?

Do you see them as a threat or an opportunity?

To what extent is the prospect of receiving a universal basic income something you would welcome?

Using predictive analytics in HRM

According to many informed commentators, some of the next major developments in the way that HRM is carried out in organisations will involve step changes in the application of predictive analytics. This is a field that has been developing over the past ten years and which is likely to become much more prominent in the future as new software comes onto the market that is less expensive, more sophisticated and more user-friendly.

There is nothing new about the use – mainly by larger employers – of HR metrics to analyse and predict trends in areas such as performance management and staff turnover. They have long provided an effective means of using quantitative data to inform decision making about who to recruit, who to promote and the possible outcomes in financial terms associated with taking different courses of action. The aim, essentially, is to make rational inferences about the future from historical data, statistical techniques being used to analyse patterns and provide evidence on which to make HR decisions.

The kinds of data that are often collected and analysed have typically included the following:

- Voluntary staff turnover rates.
- Absence rates.
- Accident rates.
- Average speed with which vacancies are filled.
- Candidate acceptance rate.
- Proportion of recruits from ethnic minorities.
- Proportion of staff who have been formally appraised in the past twelve months.
- Number of tribunal cases fought/settled.
- Overtime worked in the past year.
- Number of formal grievance/disciplinary hearings.

Sometimes cost assessments are added to create a further set of HR accounting metrics:

- Profit generated per employee in the past year.
- Sales per employee in the past year.
- Recruitment cost per new recruit.
- Labour costs as a proportion of total costs.
- Absence costs as a proportion of staff costs.
- Voluntary turnover costs as a proportion of staff costs.
- HR department costs as a proportion of total costs.

Finally, particularly in large organisations, data can be collected from staff engagement surveys with a view to looking for patterns and establishing where correlations lie:

- Proportion of employees who are satisfied with their work.
- Proportion of employees who are satisfied with their supervision/ management.

- Proportion of employees who consider the employer to act ethically/equitably.
- Proportion of employees who are clear about organisational goals/objectives.
- Proportion of employees who are clear about their own objectives/goals.

This kind of data is then used to benchmark one department's performance against another and to track progress over time. Interventions are then made in the form of training events or efficiency measures that aim to improve future performance.

Predictive analytics involve designing IT systems that are able to analyse vast amounts of this kind of data in order to provide high-quality information about the likely impact of different types of investment decisions. Multiple linear regression analyses form the basis of the IT capabilities of predictive analytic systems. Essentially, this involves searching for patterns in data sets that show how one measure is related to another. The aim is to establish unexpected and potentially unexplained correlations from which insights can be gleaned and considered.

What is likely to happen in the near future is that new, more powerful applications using artificial intelligence will become available and more widely used in such a way as to radically improve our ability to use data sets like these, and others too, when making decisions about HR policy, practice and investments. There is also a probability that their capability will extend to mining the vast amounts of information that we all leave in cyberspace every day through various interactions and which together form what is known as 'big data'.

This kind of activity has been a feature of organisational operations outside the HR sphere for several years. It has been used with considerable success by the likes of Google and Amazon. They retain data about their users' browsing patterns and use that to 'predict' their preferences. This determines what ads are displayed and what merchandise is offered via direct email offers. The big supermarkets also use their loyalty card schemes to analyse customer purchasing patterns. The purpose is to gain a deep understanding of their customers' preferences, which informs both sales strategy and future product development.

In truth, the HR function is rather lagging behind the marketing and finance functions in developing and adopting similar approaches. We tend to be far stronger at building relationships, reacting to problems as they arise and carrying out basic administrative activities than we are at analysing data and using it strategically to shape future decision making. Hunches based on observation and experience inform what we do much more than hard data, but there is an increasing recognition that this will have to change in the future as new applications become available and are adopted by industry competitors. Predictive analytics will not remove the need for human judgement, but they will enhance our capacity to make better-informed judgements. Khan and Millner (2020: 13) quote Charlotte Allen, Head of Workforce Analytics at AstraZeneca, as follows:

> HR will not be replaced by data analytics, but HR who do not use data and analytics will be replaced by those who do.

At present, in HRM, these approaches are mainly being used in the field of talent management, the aim being to maximise the efficiency and effectiveness of hiring practices, induction and initial training, later HRD interventions and minimising unwanted employee turnover. A lot of work is also being carried out that focuses on predicting the financial value of an individual to an organisation, and hence the return on investment that hiring a particular individual is likely to represent if that person is retained for a specific number of years.

This is done via a detailed analysis first of what types of characteristic and experience are associated with superior individual performance, and second of how much more revenue/profit is generated by superior performers vis-à-vis average or inferior performers. Hence predictive analytics might, for example, establish that a service provider who is rated (on average) by customers at 85 per cent will contribute £50,000 more over a three-year period than someone who is rated at 45 per cent. The idea is that having this information allows an HR manager to make better-informed decisions about who to hire, what to pay them, how much to invest in retaining them, what training interventions offer the best chances of increasing customer ratings and the costs of these.

Some argue that the results of such studies are inevitably misleading much of the time, because the data is always going to be incomplete and biased in particular directions. There are also concerns about the results putting back the cause of minorities and equality at work if they show that already privileged groups represent the best HR 'bet' from a return-on-investment perspective. Privacy is also a major concern among critics.

There is thus a need to plan the use of predictive analytics carefully and to ensure that it is only carried out when robust and sufficient data is available to analyse. Care also has to be taken to ensure that these techniques are used fairly and transparently. Perfection will not always be possible, so it is always going to be important for human judgement to be used alongside statistical analysis. A pragmatic and cautious approach is required.

The many potential problems should not, however, mean that the HR function holds back from experimenting with and developing new tools of this kind. Substantial benefits can accrue. Khan and Millner (2020: 163–69) provide an excellent case study to demonstrate what can be achieved. This concerns Rentokil, a large international company that specialises in pest control and improving hygiene. The company employs some 700 sales people around the world and their performance varies considerably. Put simply, some are much better than others at meeting their targets. In 2018 they hired a data analyst to see if they could get to the bottom of why this was. He gathered a great deal of information about sales strategies and the HR policies used by the company in different international settings. He also undertook extensive staff surveys. The first discovery was that the existing recruitment tests being used when hiring sales people were very poor predictors of subsequent performance. This was thus replaced, but only after further data was analysed focusing on which personal attributes were associated with the stronger performers. Six different tests were then trialled to establish which best enabled managers to match candidates with the attributes of the most effective existing sales people. The one that was chosen was then used internationally, a new global induction training programme also being introduced. The net result, after two years, was an improvement in sales of 40 per cent, which represented a return on investment of over 300 per cent.

Reading about examples such as these, there can be no doubt that the use of predictive analytics will grow in the future and that new data-mining technologies stand to play a very significant role in this growth.

Software that analyses qualitative data is also likely to become more widely available, more refined and less expensive. This will potentially allow employers to use technology as a means of establishing their employees' moods and hence to act in order to prevent possible resignations before frustrations trigger them. Mendoza (2014: 43) describes an early attempt by IBM to do exactly this:

> Through an internal social media site known as IBM Connections, 75,000 employees can share thoughts and ideas through forums, blogs, wikis and status updates. Using data-mining software, IBM creates anonymised 'senti-maps' showing what employees are talking about. They are then used in predictive modelling, most immediately to spot trends in attrition, but could help alert the business to potential problems or untapped markets. Interestingly IBM's research appears to show a correlation between high levels of social media usage and workplace performance, though the reasons for this are as yet unclear.

Employee monitoring

Ethical questions are raised in practically every area where predictive analytics are used (see Edwards and Edwards, 2019: 479–85) because the processes involve decisions being made that can cause people detriment on the basis of what algorithms and AI tell managers about people's performance and capabilities. In the main, though, with care, these can be overcome because the systems are more focused on how people will behave or perform in the future according to an analysis of how groups of people have acted in the past. It is much harder to get over ethical objections in another major contemporary HR-related technological development, namely employee monitoring.

Employers in many industries have long sought to monitor what their employees do at work, sometimes very closely, with a view to maintaining a high level of performance. Traditionally the approach used was simply to employ supervisors and foremen to oversee work. This approach works fine, particularly if the supervisor also sees their role as being to motivate positively while also monitoring productivity and work quality. In recent years, technologies have been employed to assist in this work. So, for example, CCTV has been installed to maintain security and phone calls monitored 'for training purposes'.

We are now seeing the emergence of new generations of employee-monitoring software that are far more effective and potentially far more intrusive. The market is growing very quickly and the ethics of monitoring employees in these ways are becoming a major talking point. This is because employee-monitoring software is becoming available which tracks electronically pretty well everything an employee does while in a workplace. In office settings, sensors can be attached to desks to monitor how much time people are spending working, all activity carried out on a PC throughout a working day can be tracked, phone conversations recorded and records of typing speed kept. People working in warehouses or out driving as part of their jobs can be supervised electronically using GPS trackers, while the newest CCTV systems can keep a record of how often and for how long medical staff wash their hands when working in hospitals or care homes.

Put simply, the capability of an organisation's managers to check up on their staff and keep them under surveillance across every working day is being enhanced by technology and will be enhanced far more in the future. And this is as true potentially for people working from home using electronic devices as those who commute into a workplace.

It can be argued that these developments are not anything to be concerned about, as employers are going to use them primarily to identify which employees are under-performing or wasting time during a working day for which they are being paid, and which are working most productively. Why should the first group not be warned and the second group rewarded? Why should an organisation not make use of all the technological solutions available to supervise people effectively with a view to increasing effort and productivity? Indeed, if competitors are doing this successfully, would a company not have much choice but to follow suit?

Aside from ethical arguments relating to personal privacy and the potential for such systems to compromise employee welfare, there is a bigger HR problem. This is that electronic employee surveillance systems, by their very nature, signal a lack of trust. By installing them, even transparently and with no covert elements, an employer communicates to its employees that it considers close monitoring to be necessary in order to ensure that they work hard and obey all workplace rules. Employers should not then be surprised if their lack of trust is reciprocated. Decades of research tells us that this is the case (eg Gould-Williams, 2010; Brown et al, 2015). Employees will respond by ensuring that they comply, but it is less likely that they will work with enthusiasm and creativity. You are much less likely to go the extra mile and work beyond contract in a low-trust environment than in one characterised by high-trust, authentic relationships. A compliant workforce is not necessarily, indeed not usually, as productive as one in which discretionary effort is forthcoming and positive leadership is the main method used to motivate rather than close supervision.

PAUSE AND REFLECT

Think about your own work experience and the different types of workplace you have either worked in or observed others working in.

What levels of direct supervision were in place? To what extent do you agree with view that forms of impersonal, electronic monitoring are likely to reduce levels of trust and thus have a negative impact on performance?

References

Altraide, D (2019) *New Thinking*, Mango Publishing, Coral Gables, FL

Avent, R (2016) *The Wealth of Humans: Work, power and status in the twenty-first century*, St Martin's Press, New York

Brown, S, Gray, D, McHardy, J and Taylor, K (2015) Employee trust and workplace performance, *Journal of Economic Behavior and Organization*, 116, August, 361–78

Buchanan, B (2019) Artificial intelligence in finance, The Alan Turing Institute, www.turing.ac.uk/research/publications/artificial-intelligence-finance (archived at https://perma.cc/N2A6-4RPQ)

Chase, C (2016) *The Economic Singularity*, 3Cs, London

Del Rosal, V (2015) *Disruption: Emerging technologies and the future of work*, Emtechub, Dublin

(Continued)

(Continued)

Edwards, M R and Edwards, K (2019) *Predictive HR Analytics*, 2nd edn, Kogan Page, London

Ford, M (2015) *The Rise of the Robots: Technology and the threat of mass unemployment*, Oneworld, London

Frey, C and Osborne, M (2013) *The Future of Employment: How susceptible are jobs to computerisation?*, Oxford Martin School, www.oxfordmartin.ox.ac.uk/downloads/academic/The_Future_of_Employment.pdf (archived at https://perma.cc/ZR67-2E78)

Gould-Williams, J (2010) The importance of HR practices and workplace trust in achieving superior performance: A study of public-sector organizations, *International Journal of Human Resource Management*, 14 (1), 28–54

Hawksworth, J and Fertig, Y (2018) *UK Economic Outlook*, PricewaterhouseCoopers, London

Khan, N and Millner, D (2020) *Introduction to People Analytics: A practical guide to data-driven HR*, Kogan Page, London

Manyika, J, Lund, S, Chui, M, Bughin, J, Woetzel, J, Batra, P, Ko, R and Sanghvi, S (2018) Jobs lost, jobs gained: What the future of work will mean for jobs, skills, and wages, McKinsey Global Institute, www.mckinsey.com/featured-insights/future-of-work/jobs-lost-jobs-gained-what-the-future-of-work-will-mean-for-jobs-skills-and-wages (archived at https://perma.cc/7SW5-N5RB)

Mendoza, M (2014) How data analytics is revolutionising baking and hunting, *Work*, Winter, 34–44

ONS (2019) Which occupations are at highest risk of being automated?, www.ons.gov.uk/employmentandlabourmarket/peopleinwork/employmentandemployeetypes/articles/whichoccupationsareathighestriskofbeingautomated/2019-03-25 (archived at https://perma.cc/5XRC-YM4N)

PricewaterhouseCoopers (2018) What will be the net impact of AI and related technologies on jobs in the UK?, *UK Economic Outlook*, www.pwc.co.uk/economic-services/ukeo/ukeo-july18-net-impact-ai-uk-jobs.pdf (archived at https://perma.cc/6ZCZ-GCWK)

Schaefer, G O (2018) Lab-grown meat: Beef for dinner – without killing animals or the environment, *Scientific American*, www.scientificamerican.com/article/lab-grown-meat (archived at https://perma.cc/A5KU-CB6H)

Schwab, K (2016) *The Fourth Industrial Revolution*, World Economic Forum, Davos

Susskind, D (2020) *A World Without Work: Technology, automation and how we should respond*, Allen Lane, London

Susskind, R and Susskind, D (2015) *The Future of the Professions: How technology will transform the work of human experts*, Oxford University Press, Oxford

The Economist (2011) Print me a Stradivarius, 10 February

Wang, F, Zuroske, T and Watts, J K (2020) RNA therapeutics on the rise, *Nature Reviews*, 19, July, 441–42

05 Economy

In the last three chapters we explored the different ways in which industrial activity has in the past and will in the future be affected by globalisation and technological developments. In these contexts we have discussed industrial restructuring and increasing competitive intensity, both of which have profound implications for the worlds of work, employment and people management. In this chapter we are going to continue with these themes by discussing economic processes. We will start with an overview of the way that markets for goods and services operate, going on to explore free market economics and government policy in the economic arena. We will finish with some observations about possible future national and international economic trends.

Markets

For most organisations that employ people, it is the operation of markets for goods and services that forms the most significant element of their business environment. This is true of all private sector organisations for which survival, let alone prosperity, is determined by their ability to compete effectively in a market. But it is also true to an extent of many public and third sector organisations. Competition is often less fierce and operates within boundaries set by government, but nowadays market considerations have to be taken into account by schools, hospitals and primary healthcare providers, universities and social care providers. All compete to some extent with others for clients or customers, funding or contracts in some shape or form. The same is true of the charitable sector, where, however subtly, numerous charities compete with one another to attract government grants and donations from the public.

The interaction of supply and demand is the central feature of all markets. It determines which products and services are available to potential consumers, how they are priced, where the best future business opportunities lie and ultimately how successful a business can be.

Demand

Demand for products and services tends to increase when prices reduce and fall when prices increase. By the same token, when consumers or would-be consumers have access to more disposable income, demand for products rises. When they have less money to spare, or expect to have less in the near future, demand will fall.

The extent to which this happens, however, depends on how elastic demand is in a particular market. Economists describe demand as being highly elastic when small changes to the price of a product lead to big changes in demand for the product.

By contrast, demand is described as being inelastic when price changes or changes in income lead to little or no shift in demand for a product. Elasticity is determined by a number of factors:

- **The extent to which a product or service is perceived by consumers as being a necessity.**
  The more necessary a product is, the less shift we see when prices rise. By contrast, when a product is perceived as being a luxury, demand will rise and fall quite severely when prices go up or down.
  For this reason, during periods of economic recession the sales figures for supermarkets such as Tesco and Sainsbury's hold up well. This is because the market for basic foodstuffs (their core business) is highly inelastic. By contrast, sales of overseas holidays (a luxury) fall substantially. This market is elastic. Both food and holidays seem more expensive because disposable income is falling, but the impact on sales is very different.
- **The extent to which substitute products are available that we can switch to when prices rise or income falls.**
  If there is a close substitute that we can buy instead of a product whose price is rising, demand will be more elastic than when there is no substitute, or at least none that are available to or desired by consumers.
  For example, if the cost of buying a particular brand of soap powder increases, demand will fall quickly because there are so many very similar other products on the market. By contrast the market for a specialised medicine is likely to be relatively inelastic because there are few alternatives available or may be none. Producers can put up their prices without losing much business.
- **The significance of the product as part of our overall spending patterns.**
  When a lot of our income is spent on a product, we notice changes in the price more than we do if we spend only a small amount of money on it occasionally. This is simply because the price change has a major effect on our overall income when it is a major part of our weekly expenditure.
  The market for new cars is an example of one that is relatively elastic. The cost of buying a car is high and will, for most of us, be a very major item of expenditure about which we think long and hard. As a result, it is a price-sensitive market. A 10 per cent increase in the cost of a particular model will lead to significant falls in demand. By contrast, the market for cheap novelty products is relatively inelastic. A 10 per cent increase in the price of inexpensive costume jewellery, for example, is unlikely to have a major impact on demand simply because consumers typically spend only a small proportion of their earnings on such products.

Price and income, however, are by no means the only causes of rises and falls in demand for products and services. An economic impact does not always have an economic cause. Over time consumer tastes evolve, often in response to wider social trends or technological developments. These continue to occur irrespective of changes in price or disposable income.

An example of an industry that is suffering badly as demand falls is the market for newspapers, which has been in long-term decline. In 1950 over 20 million national newspapers were sold in the UK on average every day, and over 30 million on Sundays (Communications Management Inc, 2011). This was considerably more than the number of households in the country, indicating that many bought two or more papers each day. The market has been steadily falling, year on year, since then. By 1980 daily newspaper sales had fallen to 16 million and Sunday sales to 19 million as people bought or rented televisions and got their news from nightly bulletins. The decline then accelerated somewhat so that by 2010 only 10 million papers were sold on weekdays and on Sundays. By this time, there were 25 million households, so the majority were not buying newspapers at all. In 2019, sales had tumbled further to just 5.5 million on weekdays and 4.9 million on Sundays as more and more people sourced their news through the web and social media sites in particular (OFCOM, 2019). This decline is not significantly related to price. Printed newspapers are good value for money and are readily affordable by the vast majority who make up the target market. The decline in the size of the market is due to changing technology and changing consumer preferences. News reports can be accessed more easily electronically, so many people – particularly younger people – see no need whatever to even think about purchasing a physical printed copy.

By contrast, there are numerous examples of industries that have seen substantial increases in demand over recent years. In some cases this is due to globalisation and the opportunities it has increasingly provided for organisations based in the UK to sell their products and services internationally. A prominent example is the English Premier League, which was founded in 1992 when the top football clubs decided to break away from the English Football League to take greater control of their affairs. In the years since, global interest has grown hugely, with a knock-on effect on sales of merchandise around the world and on the value of television rights. In 1992 the newly established Premiership sold the rights to broadcast live matches and highlights packages for five years for £191 million per year. By 2019 global TV revenues had risen by thirty-fold over to over £5 billion a year. Higher education is another example of a market that has grown rapidly in recent years, partly thanks to the opportunities provided by globalisation. In 1992, when the polytechnics were first classed as universities, the total number of students enrolled in UK universities was 984,000. By 2019 this had risen to 2.34 million, including almost half a million overseas students. This has occurred despite the introduction of student fees, indicating that markets can still increase rapidly despite increased prices if other factors are serving to encourage the growth. Other markets that are currently growing healthily include many areas of new technology, creative industries and healthcare products such as those produced by the pharmaceutical sector.

CASE STUDY 5.1

The cosmetic surgery market

An example of a market that has seen very substantial growth in recent years before declining markedly is cosmetic surgery. In the UK in 2003 a total of 16,367 operations in this field were carried out, according to the British Association of Aesthetic Plastic Surgeons. Five years later the figure had more than doubled to reach 36,482. After a further six years, in 2014, despite a severe recession that reduced consumer purchasing power considerably, the number of cosmetic procedures carried out by plastic surgeons peaked at 51,140. After this, the market started to decline and the number of operations fell by over 40 per cent to just 30,750 in 2016 and then to 28,315 in 2017, before stabilising at 28,347 in 2019. The main reason for the drop in people looking to have cosmetic surgery appears to be an understandable preference for non-surgical procedures, which are much less expensive and require less of a commitment in terms of time. They are also believed to carry fewer risks. So while rather fewer people are opting to improve their looks with breast implants and nose jobs, the number paying for regular injections of Botox is increasing.

Above we used the example of university tuition fees to demonstrate that price rises do not always lead to a drop in demand for a product or service. In that case one reason that demand increased when prices increased was the absence of a genuinely free market mechanism. Government ministers, and not the market, determined the price.

There is another type of situation in which similar effects sometimes happen. This is where the appeal of a product or service to consumers is, in part, its relatively high price. Some goods have what the French call 'cachet'. Their high price is intrinsic to what makes them attractive.

Examples are clothes with designer labels, brands of perfume, watches, cars – even some up-market country hotels and brands of ice cream. These products have achieved a position in the market that makes them the preserve of relatively rich people, and without that cachet they would not sell so well. People consume these products to show off either to the world in general (eg carrying a designer handbag) or to a loved one (taking your partner to an expensive restaurant on your anniversary).

Some other goods and services are also consumed in larger amounts with a high price than they would be at a lower price because they are particularly important to people who consequently judge their quality, in part, by their price. The property market operates in this way to an extent. Prospective purchasers will actively look for a new house or flat that is priced at a certain level, dismissing without serious thought something that is cheaper because they assume there must be a reason why it is cheaper. The same is potentially true of privately provided education and package holidays. Some consumers avoid 'the bargain basement' experience precisely because it is 'bargain basement'.

Supply

The amount of supply of a product or service also helps to decide its price. When supply increases in quantity, prices tend to come down. By contrast, when a good or service is in short supply for some reason, prices increase.

As is the case for demand, a number of distinct factors can influence the level of supply. The most important is the cost associated with supplying a good or service. If that rises, it tends to restrict supply, particularly if the market is relatively elastic and putting up prices is not an option. By contrast, when costs of production drop, supply tends to increase.

So what can affect the costs of supply? An important component is the cost of labour. Wage levels vary depending on the state of the labour market. When skills are in relatively short supply, wages rise. This tends to happen when unemployment levels in an industry are low. By contrast, when unemployment is high and skills are in plentiful supply, wages tend to hold still or even fall. However, wage levels are not the only determinant of labour cost. It is possible to employ fewer people on higher wages and to get an equally good performance if the workforce is organised in a more efficient way or if people are prepared to work harder.

Either way, the supply of a product or service can be affected by changes in labour costs. This is particularly true in the public sector, where wages tend to account for a good majority of overall costs. When wage levels go up, the extent to which a particular public service can be provided is limited. Policing is a good example. Were we able to pay police officers much less money, we would be able to have many more of them on our streets tackling crime. But this can't happen because of the level of wages that a police officer is able to demand for doing the work.

The market works the other way, too. Falls in the cost of labour lead to increases in the supply of goods and services. This has been a hugely significant factor in manufacturing over recent decades. The rise of developing economies such as China has enabled vast reductions to be made in the cost of manufactured goods. The most significant factor is the substantially lower hourly wage rates that Chinese workers are paid compared to equivalent workers in western factories. The result has been an increase in the number of goods that are manufactured and, hence, a reduction in the price that consumers are obliged to pay. This is why the past few decades have, for example, seen very substantial reductions in the price of electronic goods. Televisions, for example, have plunged in price. In the 1950s, when they first came to market, they were far too expensive for many people to buy. Until the 1980s it was very common for people to rent a television set. Since then, they have not only become much more affordable, but vastly more reliable too. The fall in price in real terms between 1950 and 2020 was in fact 99.14 per cent, indicating an average annual inflation rate of –6.57 per cent each year (in2013dollars, 2020).

Another way in which labour costs are reduced is through the introduction of labour-saving technologies. The impact can be quite swift, and can rapidly lead to increased supply of a good or service. A good contemporary example is book and magazine publishing. The costs associated with the production of books have reduced substantially in recent years due to the availability of digital technology, which enables publications to be printed without the need for typesetting. Far fewer people are needed, making it commercially viable to publish material in small production runs. Whereas twenty years ago a book would only be published if 3,000 or so copies were expected to be sold, the figure is now a few hundred. The result has been a substantial increase in the number of titles available. In the UK, over 184,000 titles are now published every year.

Changes to taxation can also have an impact on the level of supply in a market, generally because of the way that it affects the cost and hence the price of products. Increases in Value Added Tax (VAT), for example, usually increase the price paid by consumers, and hence tend to reduce the level of supply. The opposite occurs when taxes are reduced. Take cigarettes as an example. Every time the tax the government levies on a packet goes up, the price also goes up. This reduces demand, but also means that producers of cigarettes are likely to reduce their output in the expectation that demand will fall.

Finally, we need to mention the potential role of substitute products and complementary products, as we did when thinking about demand. They also play a role in increasing and decreasing supply.

Griffeths and Wall (2008: 15–16) give examples from agriculture to explain how these processes work. They explain, for example, that farmers have a choice about which crops to grow in their fields and that they make this decision based on the price that a particular crop will get when sold on the market. Each crop is a substitute for the others. So if the price of wheat falls, farmers will switch to another crop, say barley. The result is an increase in the supply of barley.

Similar thinking affects complementary products in agriculture. Sheep farmers, for example, will typically decide how many sheep to keep after analysing the price of lamb meat. If prices rise, they will increase supply by breeding more sheep. However, in this industry there is a complementary product in the form of wool. So the supply of wool will also increase irrespective of prices when the price of lamb meat increases.

The property market is one that is heavily influenced by demand and supply. In recent years property prices in the UK have risen very considerably. There have been periods in which prices have fallen, for example in the early 1990s and during the recession of 2008–10, but in general the moves have all been very much in an upwards direction. The cost of buying a house now is very considerably higher than it was in the past. In 1975 the average price was £1,891, which is the equivalent of around £65,000 today. In 2019 the average cost of a house was £231,215, representing a real terms increase of over 3.5 times in thirty-five years. The main reason is simply that the rate at which new houses can be built is much too slow to accommodate the demand. There is thus a chronic undersupply and this inevitably means that prices rise.

Of course, the process also works in the opposite direction. Sometimes prices rise due to short supply of a good or service, or sometimes due to increased demand. This price rise then makes it worthwhile for existing suppliers to invest in expanding their businesses and will also attract new businesses into the market. It can take time for the market to respond in this way, but over time the adjustment occurs.

All kinds of factors can affect the supply of a product or service, particularly when there is an international market and geopolitics play a part in determining supply. The oil market is a good example. Up until 1974 prices were low by today's standards, averaging just \$3 or \$4 per barrel. They then leapt up to \$37 a barrel in the late 1970s as supply was restricted by the OPEC cartel (the Organization of the Petroleum Exporting Countries) in order to increase revenues. The price then fell back somewhat and remained stable for twenty years until it increased sharply again as the political situation in the Middle East became unstable during and after the Iraq War. Oil prices then continued to rise as developing countries built up their industrial capacity and transport infrastructures. Fifty million extra cars were joining the world's roads each year by 2010 and this increased demand could not be matched by increased supply. Oil was also steadily becoming less accessible and thus becoming more costly to produce. In more recent years oil prices have fallen back down again, the major factor being the development of the shale oil industry in the USA, which led to increased global supplies in the years after 2010. The price per barrel plunged to $46.34 in 2015 and has stayed low since then. In 2020 falls in demand caused by the coronavirus pandemic reduced it further to just $37 a barrel.

PAUSE AND REFLECT

To what extent is supply in your industry restricted? If demand was to increase rapidly or prices to increase, how quickly and effectively would existing and potential suppliers be able to increase supply? What different factors would act as restrictions?

Free market economics

Economists tend to use the term 'efficient' when explaining the way that free market economies work. By this they mean that using this model as the basis for national and international economic exchange, while by no means being unproblematic, provides a very efficient means of producing and distributing goods and services.

Costs are kept low because markets are price-sensitive. In a free market, if one provider charges too much, it loses business to a competitor. So prices are kept low for consumers. Quality is also maintained because buyers take their business elsewhere if they are dissatisfied with the service or the quality of goods that they receive. In short, an organisation that is inefficient will fail in a free market, simply because other more efficient organisations will move in and take its place.

Because the market operates with a high degree of efficiency, societies that promote free markets as a whole are well placed to create wealth. Over time, the economy grows and this means that living standards rise. Wealth is unevenly distributed in a market system and often also unfairly distributed, but the chances that it will be created are increased the more efficiently the market operates. It is then for government to redistribute some of the wealth through the operation of the tax and benefits system.

According to the proponents of free market economics, these benefits are not present in economies that opt for a system of economic distribution which is dominated by centralised state planning or indeed any kind of system in which there is not free and open competition between rival companies for limited business.

The implications for organisations that are managed less well than their competitors can be bleak. Even large and respected companies can go under when they cease to attract sufficient business to remain profitable.

The most prominent recent examples in the UK have involved the demise of large high-street retailers who have been unable to find ways of competing with online retailers and the larger supermarkets. Over the past fifteen years we have seen several very well-known chains of stores either collapse completely or struggle on at a much smaller scale. Many thousands of jobs have been lost in the process. Examples include Woolworths, HMV, Mothercare, BHS, House of Fraser, the Arcadia Group and Debenhams. The chains that have bucked the trend and thrived do so either because they offer cheaper products (eg Primark) or because they have a loyal customer base that prefers them to their online rivals (eg Waterstones and Boots).

PAUSE AND REFLECT

It is not always apparent why a free market economic system should be any better at ensuring the efficient creation and distribution of goods and services than a system characterised by central economic planning on the part of a government.

Some argue that a planned system ensures more efficient outcomes because resources are pooled together and economic management is coordinated centrally, to achieve fair outcomes that benefit everyone. What is your view of this argument?

Market failure

While economies such as the UK's are effectively driven by market mechanisms for much of the time, our markets are not and can never be completely free. Regulation is necessary and in some cases there is more active state involvement. This is because from time to time free markets 'fail', by which we mean they fail to achieve the objective of distributing goods and services that meet people's needs in an efficient manner. Here are some examples:

- Sometimes one company becomes so successful and grows so big and powerful that it gains an unfair advantage in the market. It may, for example, be able to sustain huge price cuts for a short period with which others can't compete. The smaller competitors are then forced out of business, leaving the big company in a monopoly or near monopoly position.
- Sometimes several large companies collude to fix prices at a rate that is 'artificial', by which we mean higher than would be the case if a free market operated.
- Sometimes in free markets companies can only survive by setting prices that are too high for most people to afford. This is of no great importance in the case of luxury goods such as high-performance cars or computer games. But it matters a lot if the product or service is one that everyone needs, such as prescription drugs, primary education, transport to work, or even food, water and housing.

- Sometimes, in a bid to increase profits or to survive in a highly competitive free market, companies are tempted to act in ways that are against the public interest. They may, for example, compromise health and safety or make claims for their products that are not true, hence conning consumers.

For these reasons government involves itself in the operation of markets through regulation of one kind or another:

- Regulation prevents any organisation gaining a monopoly position in a market unless it can demonstrate that it is in the public interest that it should have such a position.

  In recent years we have seen a number of examples of these regulations having an impact in the supermarket world. It is particularly important that this market is not allowed to fail because food is something all of us need all the time. In 2004 the Safeway chain was put on the market, while five years later in 2009 it was Somerfield that was being sold off. On both occasions there was a lot of interest from the big three existing chains, Tesco, Asda and Sainsbury's. Then in 2018 Asda and Sainsbury's proposed a merger. On each occasion anti-monopoly laws ensured that no single supermarket chain could merge or acquire another so as to achieve too great a dominance of the market.
- Government ensures that extensive regulation covers areas such as health and safety, employment practices and trading standards so that markets remain free, fair and safe in the way that they operate. These operate by requiring organisations that breach regulations to pay compensation to any victims, or fines, or indeed both.

  In December 2005, for example, the Buncefield oil depot in Hemel Hempstead exploded early one morning, creating a huge fire. Fortunately no one was killed, but two workers were seriously injured. In addition, 200 houses were evacuated and substantial damage was done to the local environment. In 2010, after extensive investigations, faulty health and safety practices were found to have been the cause of the explosion. Five firms were found to be culpable and were fined over £10 million.
- Where the free market is not in a position to provide an essential good or service to everyone who needs it at an affordable price, government steps in to provide it itself or to pay for it using money collected through taxation.

  There are numerous examples, the most prominent being the National Health Service, which provides free healthcare of a good standard to everyone who is resident in the UK. The NHS is funded via taxation and provides a full range of health services. Some charges are levied, for example for prescriptions, but patients are not required to pay the full cost of their medicines, and those on low incomes, children and older people are not charged.
- Government occasionally steps in to rescue a failing organisation where it is clearly in the public interest that it should.

  A prominent example was the purchase by the UK government in 2008 of substantial shareholdings in four banks – RBS, Lloyds TSB, Northern Rock and Bradford & Bingley. Had this not happened, these banks would have failed, causing millions of people to lose money and destroying confidence in the UK financial sector.

Markets and employment

The presence of market competition has major implications for employment and, as a result, for HR practice. One major effect is to provide an incentive to keep costs low. In a free market, organisations with higher cost bases than their competitors are able to maintain will, all other things being equal, have to price their goods and services at a higher rate. The effect is likely to be lost business as buyers take the cheaper option.

For many organisations, wages account for a large portion of overall costs. This is particularly true of the service sector, where human capital is more important than other factors of production such as land, raw materials or machinery. Over time, therefore, as the UK economy has become increasingly dominated by the service sector, the need to keep a lid on wages has become increasingly important for HR managers. This does not, of course, simply mean keeping pay at low levels. It also means keeping the total number of staff employed at the lowest achievable levels. Overmanning is often a feature of organisations that do not face competition. It is not possible if an organisation is to flourish in a free market because it is inefficient.

Competition, however, also puts pressure on organisations to improve the quality of the products and services they offer for the prices they charge. It is not difficult to produce a highly unreliable product at a low price; it is much harder to maintain high standards at competitive prices, yet this is what free market economies require. This means that HR managers have to focus continually on helping to ensure that the organisations they work for have access to the skills and experience required in order to deliver goods or services. In a free market, HR managers thus prove their worth if they are able to ensure that their organisations recruit, retain, engage, motivate and develop high-quality staff.

Long-term market trends also need to be watched, understood and acted upon by people with HR responsibilities. Where a market is expected to decline over time, the need will be to reduce headcount. When there is a sudden drop off in demand, it will be necessary in many cases to implement a programme of redundancies. When the reduction in demand is slower and takes place over months or years, this can be avoided by refraining from replacing leavers or hiring people on shorter fixed-term contracts that can be extended where required. By contrast, when a market is growing there is a need to develop the capacity to carry out human resource planning. Estimates need to be made of how many new hires will be required in the future and steps taken to ensure that the organisation will have 'the right people in the right jobs at the right time'. This means gaining approval and funding for training and recruitment programmes ahead of time so that labour requirements can be met.

The volatility associated with competing in free markets is one of the challenges that make HRM and L&D practice interesting. The key problem that we continually have to address is how to recruit, retain, train and engage the best possible people, while also keeping a lid on overall wage levels. Solving this conundrum is the central task that HR people face in a free market economy. And the more competitive a market becomes, the harder it is to achieve the two aims at the same time.

PAUSE AND REFLECT

Think about your own organisation. Is demand for its products or services increasing, decreasing or remaining stable? What do you think will happen to these markets over the next five years? To what extent and in what ways are you planning to ensure that HR needs will be met in the future?

Economic policy

Over time, the aims of governments in respect of economic policy have shifted. Between 1945 and 1979, for example, the principal aim of UK governments was typically to maintain full employment. Since then, full employment has rarely been achieved, except for brief periods. This is partly a consequence of globalisation and fast-moving technological change of the kind discussed in earlier chapters, but it is also the case that achieving a situation in which everyone who wants to be is employed is no longer a stated aim of governments when devising their economic policies.

In recent decades governments have had two major aims as far as economic policy is concerned, the second of which follows on from the first:

1. Government aims to provide a stable economic environment in which investors have confidence and markets can operate as efficiently as possible.
2. Government intervenes through the tax and benefits systems to redistribute wealth from richer people who do very well out of the free market system to poorer people (including unemployed people) who do not.

The first of the two aims is the hardest to achieve in practice. This is because many things that happen and which have the effect of destabilising the economy from time to time are largely outside the control of governments. Some examples in recent UK history were the Covid-19 pandemic of 2020–21, the collapse of the banking sector on the back of a crisis of confidence in the American mortgage market in 2008, the bursting of the so-called 'dot-com bubble' in 2001, the collapse of confidence in sterling on 'Black Wednesday' in 1992 and the steep oil price rises of the early 1970s. In each case these 'external shocks' led to economic downturns and to a loss of consumer and business confidence for a period.

The other factor that makes it hard to ensure stability is the presence of what is known as 'the economic cycle' (sometimes referred to as the 'business cycle'), which, hard as they might try, governments have never been able to eradicate. All economies are cyclical in the way that they operate over time. This means that there are periods of growth, followed by periods of recession that in turn once again give way to periods of growth. Economic cycles vary in their length. The UK, for example, enjoyed a long upswing between 1993 when a recession came to an end and 2008 when a very deep recent recession started. At other times in history, such as the 1930s and the 1970s, the UK endured several years of sluggish growth and economic decline before the economy began to grow more strongly again.

The view that economies naturally grow and contract over time is rooted in an understanding of human economic behaviour, and particularly in the role played by confidence. The idea is that when the economy is growing and both jobs and business opportunities are in plentiful supply, our levels of economic confidence grow. In such circumstances, we are easily persuaded to borrow money to buy goods and services. Credit flows quite readily as financial institutions lend money to people. The trouble is that we tend to get over-confident and take risks. Growth then gives way to a period of boom. Too much is borrowed by too many people. This then leads to a 'bust' as debts go unpaid and over-confidence turns to insufficient confidence.

We have seen this process happen several times in the UK in recent decades. On two occasions the property market was the major driver of the booms and the busts that followed. In the late 1980s and again in the late 1990s and early 2000s property prices rose quickly. Many people were making more money each year from the rising value of our houses than we were from working. On both occasions the market boomed because there was a great deal of confidence. Over time we forgot about the previous crashes in property values, and so we were easily tempted by the banks to borrow higher amounts to finance house purchases.

Credit booms and asset price bubbles then gave way to recession, increased unemployment and falls in property prices as over-confidence was replaced by under-confidence. When confidence is low, by contrast with the good times, people cease spending, are cautious with their money, save as much as they can and take few risks. The economy suffers as a result as companies fold through lack of business or an inability to secure credit from the now very cautious banks. Over time, however, confidence returns and the economy starts to grow again. We forget the bad times and start heading again into a boom. Then the whole cycle repeats itself.

PAUSE AND REFLECT

Why do you think we have this habit of repeating the mistakes of the past and being sucked into over-confidence when we know that if history is any guide a crash is coming round the corner?

The presence of external shocks and the economic cycle do not mean that governments do not try to create a stable economic environment. For industrialised countries such as the UK the ideal state is generally considered to be one in which the economy grows each year at a steady, sustainable rate of around 3 per cent or so. If it grows more slowly confidence tends to wane, creating the danger of a recession. If it grows too much more quickly over-confidence can set in, leading to an inflation of prices and the dangers of the boom–bust scenario.

In order to maintain steady growth for as long a period as possible, governments, along with their central banks, such as the Bank of England, make use of the various levers of economic management that they have at their disposal. It is helpful to think of this process as being rather like driving a car. As drivers we control our cars by using the foot brake, the handbrake, the steering wheel, the accelerator and the gear stick. When the car is going too fast for the conditions, we apply our brakes, shift down a gear and bring the speed down. When it is safe to go faster we push down on the accelerator and move up a gear. In principle, managing a national economy is not really that different, although it is typically rather harder to see the prevailing conditions along the road ahead.

The key levers available to government are:

- The interest rate (the cost to businesses and individuals of borrowing money).
- The exchange rate (the value of sterling vis-à-vis other world currencies such as the US dollar, the euro and the yen).
- Taxation (the amount of money the government takes in various forms to fund its activities).
- Borrowing (the amount of money the government itself borrows in order to finance its activities – this is known as the 'public sector borrowing requirement').

In the UK, ministers, along with the governor of the Bank of England, make use of these levers as and when necessary in a bid to maintain stability or to regain it once it has been lost.

When the economy is sluggish (growing perhaps at just 1 per cent a year) or is in recession (not growing at all), governments typically do the following with their levers of economic management, the purpose being to stimulate the economy:

- They cut taxes. This puts money into people's pockets and, it is hoped, will encourage them to spend more.
- They cut interest rates. This reduces the cost of borrowing for business and also reduces monthly payments made by people with mortgages, giving them more money in their pockets.
- They increase borrowing. Government itself meets its obligations by borrowing money rather than simply using the proceeds of taxation.
- They let the exchange rate fall. This is done by releasing stocks of sterling onto the world's markets. The price then falls. This has the effect of making goods and services that originate in Britain less expensive for people based overseas to buy. Imports by contrast are more expensive.

In recent years we have seen extensive use being made of all four of these levers. For example, during the very severe recession we had during 2008–09, Value Added Tax (VAT) was cut to 15 per cent for a period of a year to encourage spending, while interest rates went down to 0.5 per cent, which was the lowest level they had ever reached for 250 years. More recently, during the economic crash provoked by the lockdown measures that were introduced to reduce infections during the Covid-19 pandemic of 2020, massive increases in public borrowing were undertaken to finance the furlough scheme and to stave off what otherwise might have been a total financial collapse leading to mass unemployment and years of recession.

In recent years, a further lever of economic management has also been used in the face first of a deep recession that followed the banking collapses of 2008 and then of sluggish levels of economic growth achieved in the years since. This is known as quantitative easing (QE), but it is often described as 'printing money'. It involves a central bank creating new money and making this available for financial institutions such as banks and pension funds to borrow at low levels of interest. Governments only resort to quantitative easing when their economies are in such bad shape that they fear that deflation may set in. This occurs when confidence is so low that prices start to fall or when such a situation is at risk of developing. It is a dangerous position to be in because people naturally put off buying things in the expectation that it will be cheaper to do so next month or next year. Once a depression of this kind sets in it is difficult to re-establish confidence. In normal times, QE is avoided because it is highly inflationary and can easily set off an unsustainable boom. But in the depths of a recession it can play a further role in stimulating the economy.

By contrast, when an economy is doing well and over-confidence begins to set in, governments generally reverse these levers in order to prevent the development of booms and asset price bubbles and to maintain steady, sustainable growth. So taxes tend to rise, the exchange rate is encouraged to float upwards, interest rates are increased and governments look to pay down their debts.

This second main aim of government economic policy is much easier to understand and can be explained quickly. A disadvantage of the market economy is the very considerable inequality that is created across society. Those who have skills and experience that are sought after and which are relatively rare earn much more than those who don't because organisations are obliged to pay more in order to employ them. Owners of successful businesses along with their senior executives can earn spectacular amounts of money, amounting to dozens of times more than their more junior employees.

The UK government raises taxation through Income Tax, Corporation Tax (taxes on company profits), National Insurance contributions (paid by employers and employees), Value Added Tax (a levy of 20 per cent on most goods and services that we purchase), Vehicle Excise Duty, customs levies on imported goods, Capital Gains Tax (tax levied when property and companies are sold) and a wide range of levies on cigarettes, fuel, airline tickets, gambling and alcoholic drinks. Local government raises revenue principally by collecting Business Rates and Council Tax.

This public money is then spent on the provision of public services (health, education, policing, defence, etc) and also on a wide range of benefits such as tax credits and state benefits, for example Jobseeker's Allowance, Incapacity Benefit, Housing Benefit and the State Pension. Prior to the changes made in 2020 when the Covid-19 pandemic hit, the UK government planned to spend £928 billion in 2020–21, of which £285 billion was to take the form of pensions, tax credits and state benefits. The anticipated revenue from taxation was £873 billion, leaving quite a gap to be filled with increased borrowing.

Economic policy affects the management of people in organisations because it has a major impact on labour market conditions. Raising taxes and increasing interest rates tend to dampen demand for labour, meaning that employers are less inclined to expand their workforces. If a business sees its taxes rise it is going to be less inclined to hire new staff and may decide to reduce headcount to save money. If interest rates go up and borrowing gets more expensive, businesses are less able to expand and may see the value of their property fall. When exchange rates rise, any business that exports products or provides services to tourists from other countries sees its prices rise from the perspective of people based overseas. This reduces demand and hence makes them less inclined to employ people. Of course, the reverse is also true. Employment levels tend to rise when taxes and interest rates are cut and when exchange rates are lowered to reduce the costs of goods and services that are exported. Consumer confidence rises, and with it business confidence.

The extent to which and the ways in which money is redistributed through the tax and benefits system are also important in the management of people in workplaces because they directly impact on their levels of income. The level of the State Pension, for example, inevitably has an effect on whether or not people decide to continue working either on a full-time or part-time basis once they qualify to receive it. The same is true of benefits that are paid to people on lower incomes. In recent years governments have sought to restrict some benefits by freezing them in a bid to push people to take up work rather than live on welfare. The supply of labour is thus linked closely to the benefits system.

Demand for labour in the public sector is also clearly linked to government expenditure on public services. If a hospital or school budget increases, it can afford to take on more staff and this tends to have the effect of tightening labour markets and making it harder for private and third sector organisations to recruit and retain people. Conversely, when public sector budgets are cut, it becomes easier for companies to recruit and retain people because there is less competition for staff.

CASE STUDY 5.2

Covid-19

Early in 2020, for the first time in living memory as far as the UK was concerned, a virus with the capacity to kill large numbers of people started to spread around the world. It was not an external shock that our governments were at all well prepared to meet, and scientists disagreed on the best public policy responses. After a while, the decision was taken to save as many lives as possible by ordering large parts of the economy to shut down for a period. In the summer rules were relaxed for a time, but in the autumn the virus started spreading rapidly again and as the year came to a close the government decided to restrict economic activity and require people to stay at home and avoid mixing with other households as much as possible.

Inevitably these measures caused a rapid and substantial fall in economic activity. In the UK, a long period of steady, if unspectacular, economic growth came to an end as the size of the economy fell by an astonishing 20.4 per cent in the second quarter of 2020. This was the deepest recession since records began. The government's response was to borrow some £300 billion to ensure that the economy did not collapse completely. A large chunk of this money was spent on a variety of tax relief and grant schemes aimed at keeping businesses going, along with a newly created 'furlough scheme' which ran for over a year. This enabled employers to pay staff who were unable to work due to the pandemic a sum equivalent to 80 per cent of their normal salary, up to a maximum of £2,500 per month. A good portion of this money could then be reclaimed from the government.

A very unusual economic crisis was thus ameliorated to a great extent by a very unusual type of government response.

National and international economic trends

The UK, like many advanced industrial economies, can reasonably be described as being in the middle of an industrial revolution at present, which over time is bringing about a radical restructuring of the national economy. Globalisation and technological developments have combined to bring this about, making it uneconomic any longer to carry out some kinds of industrial activity that have long been a major provider of jobs.

The result has been a decline in the scale of the manufacturing industry and the rise in its place of a much bigger service sector. Increasingly, consumers in the UK source their manufactured products overseas, making the money to pay for these by providing a range of services to consumers at home and around the world. UK manufacturing will continue to prosper, but in the main it will be focused in more specialist fields, making greater use of technology while employing fewer people with higher-level, specialist skills. Lower-cost manufacturing involving the employment of large numbers is simply no longer profitable to carry out in the UK.

There is every reason to consider that this broad direction of travel will continue in the foreseeable future. This is because countries like the UK have a competitive advantage internationally in the provision of services. That is where the demand is and there are thus plenty of incentives to seek to meet that demand. Over time, we can thus expect to see continued growth in the main service sectors that international consumers are looking to access and buy from. Business services, financial services, higher education, tourism and the creative industries are likely to continue growing, while other export markets are likely to stagnate or see further decline.

This process of industrial restructuring and adjustment has been ongoing now for several decades, and while it has been disruptive and often very difficult for many individuals and their families, from a national economic perspective it has proceeded relatively successfully. As we will see in Chapter 6, the number of jobs in the UK has risen significantly over time and, with the exception of recessionary periods, unemployment has remained relatively low. Overall, wage levels have kept up with price rises while economic growth has been maintained at reasonably solid levels in most years. While the long-term economic outlook is by no means positive for everyone, particularly people whose jobs are less secure than they were in the past, in broad terms it is sound. And for people with the right skills and qualifications, there are plenty of attractive career opportunities available.

That said, it is impossible to be confident about the short-term economic outlook. We are writing this book at the end of 2020 as the UK is preparing to leave the European Union's Customs Union and Single Market. Considerable economic damage has been sustained due to the Covid-19 pandemic, leading to many job losses and business closures. It is very unclear whether economic recovery will be swift or slow, or whether a lengthy period of recession will ensue. At the start of 2020 the economic outlook looked good. Unemployment was at a historically low level, both the UK and world economies were growing, and national debt was running at a very sustainable level. We enter 2021 with poor levels of growth, rising unemployment and over £2 trillion of public debt. Monthly tax receipts are much lower than government expenditure, indicating that tax rises are likely to be necessary in the near future. This suggests that, while there is scope for the economy to start growing again after the pandemic, we will not feel so well off again for some time, meaning that economic confidence will take time to be restored to the level it was at in 2019.

Internationally, the Covid-19 pandemic has had a very different economic impact in different places. In a number of Far Eastern countries like China, Taiwan, Vietnam and Japan, where the public was used to these health emergencies, having suffered from them in the recent past, swifter and more effective action was taken to prevent the spread of disease early in 2020. The same was true of New Zealand and, albeit to a lesser extent, Australia. These economies were thus in a position to recover more rapidly than has been the case across most of the rest of the world. A serious global depression was thus avoided and the international economy suffered less damage than might otherwise have been the case. It is thus possible to look forward with reasonable confidence to the 2020s from an international economic perspective. The global market for goods and services is thus very likely to continue growing at a healthy rate, providing opportunities for businesses that are able to tap into market needs and supply the goods and services that are in demand at an affordable price. The big challenge of the next twenty years may well be less finding markets than supplying them in a much more environmentally sustainable fashion than has been the case in the past.

References

Communications Management Inc (2011) *Sixty Years of Daily Newspaper Circulation Trends*, http://media-cmi.com/downloads/Sixty_Years_Daily_Newspaper_Circulation_Trends_050611.pdf (archived at https://perma.cc/KP9M-GHA7)

Griffeths, A and Wall, S (2008) *Economics for Business and Management*, 2nd edn, FT/Prentice Hall, London

in2013dollars (2020) Televisions priced at $300 in 1950 → $2.57 in 2020, www.in2013dollars.com/Televisions/price-inflation (archived at https://perma.cc/V59L-2QCU)

OFCOM (2019) *News Consumption in the UK: 2019*, www.ofcom.org.uk/__data/assets/pdf_file/0027/157914/uk-news-consumption-2019-report.pdf (archived at https://perma.cc/WK35-2TRF)

06 Labour markets

In Chapter 5 we looked at the economic context in which organisations operate, noting that the market, competition and the forces of supply and demand are central. In this chapter we are going to move on to look at markets for labour and at broad labour market trends. We will start by focusing on individual labour markets that operate locally in particular areas or industries. We will then go on in the second part of this chapter to look at the dynamics of the UK labour market generally from a national perspective, exploring the long-term trends and discussing what these may mean for the future of work and employment if they continue.

Labour markets, like any other kind of market, operate according to the interaction of supply and demand. Here, though, the 'price' that the market determines is the wage that the employees are paid. Labour market dynamics thus play a crucial role in deciding individual living standards and how equal or unequal our workplaces and societies become.

Labour market conditions

Over time, the ability of an organisation to recruit and retain people who have the skills, qualifications, experience and personal attributes it requires varies. Moreover, at any one time an organisation may struggle to recruit for one vacancy, while having no problem at all finding very strong candidates for another. This is because, like all markets, labour market conditions depend on the interplay of forces of demand and supply and these change over time. Decisions made by employers determine the nature and level of the demand, while the labour supply is determined by demographic factors, educational attainment and the attributes and skills of the available workforce. Labour markets are also, of course, competitive, so the ability of one employer to recruit and retain will always inevitably be affected by the offers rival employers are able to make to their staff and would-be staff. Labour markets can be explored from a variety of angles. Here are three examples.

Tight v loose

A 'tight' labour market is one in which employers are obliged to compete harder to secure the services of the people they need. This state of affairs arises when there are fewer people looking for jobs than there are jobs available; in other words, when demand for labour outpaces supply. Some labour markets are always tight, whatever the prevailing economic conditions, because the skills required are relatively scarce. In recent years the major examples have been in the IT and sales professions, where really effective people have tended to be in short supply. The same tends to be true of markets for senior managers with international experience and for elite performers across all industries. By contrast, when people are in plentiful supply, labour markets are characterised as being 'loose'. Such situations are a great deal easier to manage from a resourcing perspective. Vacancies are easy and cheap to fill quickly, while managers have to tread less carefully in the way they treat staff because there are fewer alternative job opportunities for dissatisfied employees to take up. This is not to say that staff can be treated badly, as doing so will tend to demotivate them and lead to poorer performance, but there is less of an imperative to ensure that individuals are satisfied with their employment.

In practical terms, when tight labour market conditions prevail for a period the result is skills shortages. The tighter the labour market, the more acute the skills shortages. In such circumstances, employees enjoy a degree of market power. If they are unhappy in their jobs they can readily find other opportunities. Their employers know this and so take steps to ensure that as far as possible employees enjoy job satisfaction. Losing an effective performer to a competitor is something that can be particularly damaging and needs to be avoided if at all possible. Employee retention is thus every bit as significant as attracting new staff in industries where severe skills shortages tend to prevail.

By contrast, in loose labour markets there is a surplus of people with the skills required to do the jobs that are available. This means that large numbers of people apply for the jobs that are advertised and the employer enjoys much greater power. Employees have to accept developments they do not like, because they have no alternative and also, potentially, they fear the consequences of resisting.

How tight or loose a labour market is thus plays a significant role in determining how much money needs to be spent on wages and benefits, on recruitment budgets and on employee development. The tighter the market, the more money needs to be spent in order to attract sufficient numbers of good applications when vacancies arise. It is also more costly to retain staff, hence the need to provide development opportunities in order to encourage good people to stay.

However, it is not just a question of money, important though that is. Other things matter to staff and would-be staff too. They want to find their work rewarding in the widest sense of the word, they usually want to be involved in decision making, to be treated courteously, to enjoy decent career development opportunities and to have a good work–life balance. The tighter the labour market, the more alternative job opportunities there are for people and the greater the need for an employer to work on improving these other aspects of the employment relationship and the quality of the workplace experience.

A number of researchers have looked at the responses of employers faced with either loose or tight labour markets. Windolf (1986) identified four types of approach used in the UK and in Germany, which varied depending not only on the tightness of the market but also on the capacity of the organisation to respond intelligently to the situation. He found that many organisations with high market power (ie faced with a relatively loose labour market) made little effort at all to capitalise on their advantageous position by recruiting outstanding people. They simply took the opportunity to spend as little as possible on recruitment and selection and waited for people to come to them. When there was a vacancy it tended to be filled by a similar person to the one who had left, thus maintaining the status quo. According to Windolf, the more intelligent organisations took the opportunity afforded by favourable labour market conditions to seek out people with the capacity to innovate and who would develop their roles proactively. All available recruitment channels were used, leading to the development of a richly diverse and creative workforce. A similar dichotomy was identified in the case of tight labour markets. Here, many organisations simply 'muddled through', finding people where they could, giving them training and hoping that they would stay long enough to give a decent return on the investment. By contrast, the more shrewd organisations looked at restructuring their operations, introducing flexible working patterns and devising ways of reducing their reliance on people who they had difficulty in recruiting.

PAUSE AND REFLECT

Think about the major labour markets that your organisation competes in. Which are the tightest and which are the loosest? Does the way people are managed vary according to labour market conditions, and if so how?

Geographical differences

For most jobs in most organisations, the relevant labour market is local. Pay rates and career opportunities are not so great as to attract people from outside the district in which the job is based. The market consists of people living in the 'travel to work area', meaning those who are able to commute within a reasonable period of time. In determining rates of pay and designing recruitment campaigns there is a need to compare activities with those of competitors in the local labour market and to respond accordingly. Skills shortages may be relieved by increases in the local population or as a result of rival firms contracting. New roads and improved public transport can increase the population in the travel to work area, with implications for recruitment budgets and the extent to which retention initiatives are necessary.

Generally speaking, the higher paid a job is and the greater its prospects, the further people will be prepared to commute, and hence the larger the travel to work area/employment market for that type of position. In the case of more highly paid jobs, the labour market is national in its scope. These are roles that are sufficiently remunerated and prestigious enough to attract applications from all over the country. The more senior jobs in larger organisations fall into this category, as do many better-paid professional roles. But some less well-paid positions are also commonly filled from a national market. Graduate recruitment rounds, for example, operate nationally as a rule, along with other jobs that are predominantly filled by younger staff who are more mobile, have fewer domestic ties and are thus able to move around the country in search of good career opportunities. In the case of the highest paid jobs of all, employment markets are international and jobs are advertised all over the world. This is true of the top positions in the financial services industry, in medicine and in sport.

There are, however, some international labour markets in areas of work that are relatively low paid. Hotels and restaurants are one example, academia is another. These are industries that employ large numbers of overseas staff who are interested in building up their work experience (and hence their earning power) or seeking to experience living overseas for a period. Within the European Union recent years have seen the development of many international labour markets as a result of imbalances in employment and earning opportunities in different member states. After the recession of 2008–11, for example, very high unemployment levels developed in Greece, Cyprus, Italy, Spain and Portugal. Unemployment among younger people was especially high, reaching 50 per cent or more in some places. Inevitably this led many to seek work in other EU countries where opportunities were more plentiful. We have thus seen a great deal of movement across borders within the EU from east to west and south to north as people have sought to enhance their earnings and career prospects by seeking work in western European countries. The UK 2001 census recorded 75,000 people of Polish origin living in the UK. By 2011 this figure had grown to 521,000 and in 2017 it was estimated to be over a million. In total in 2019 it was estimated that 1.43 million people from Eastern European countries were settled on a long-term or permanent basis in the UK (ONS, 2019a).

Occupational structure

Labour markets also differ according to established behavioural norms among different occupational groups. The attitude of people to their organisations and to their work varies considerably from profession to profession, with important implications for their employers. A useful model developed by Mahoney (1989) illustrates these differences. He identifies three distinct types of occupational structure: craft, organisation-career and unstructured.

In craft-oriented labour markets, people are more committed over the long term to their profession or occupation than they are to the organisation for which they work. In order to develop a career they perceive that it is necessary for them to move from employer to employer, building up a portfolio of experience on which to draw. Remaining in one organisation for too long is believed to damage or at least to slow down career prospects. Examples include teaching, where there is often a stronger loyalty to the profession as a whole than there is towards the employing institution.

By contrast, an organisation-career occupation is one in which progress is primarily made by climbing a promotion ladder within an organisation. People still move from employer to employer, but less frequently, and will tend to stay in one organisation for as long as their career is advancing. In such organisations we see 'internal labour markets' developing as staff compete with one another for advancement across or up the corporate ladder.

Mahoney's third category, the unstructured market, consists of lower-skilled jobs for which little training is necessary. Opportunities for professional advancement are few, leading to a situation where people move in and out of jobs for reasons that are not primarily career-related.

To an extent, employers can seek to influence the culture prevailing among members of each type of occupational group. There is much to be gained in terms of employee retention, for example, by developing career structures that encourage craft-oriented workers to remain for longer than they otherwise would. However, a single employer can have limited influence of this kind. It is therefore necessary to acknowledge the constraints associated with each labour market and to manage within them. Different areas of HR activity have to be prioritised in each case. In craft-oriented labour markets it is necessary to work harder at retaining people than in those that are organisation-oriented, because people will be more inclined to stay with one employer in the latter than in the former. Recruitment and selection will be different too. Where career advancement is generally achieved within organisations (as in banking or the civil service), there is a good case for giving a great deal of attention to graduate recruitment. It is worth spending large sums to ensure that a good cohort is employed and subsequently developed because there is likely to be a long period in which to recoup the investment. The case is a good deal weaker in craft-oriented labour markets where there is less likelihood of a long association with individual employees.

Competing in a tight labour market

What are the practical consequences for organisations of tightening labour markets? The answer is a reduction in the number of people with the required attributes looking to work in the jobs that are available. This means that it becomes progressively harder to find new recruits and that good people who have the skills that are most in demand are more likely to leave their current employment in order to work for another organisation or to set up their own businesses. Because it also alters, however subtly, the power balance within organisations, people who are hard to replace are less easily controlled by management. They are in a position to demand greater autonomy over their own areas of work, more flexibility, better working conditions and more developmental opportunities, in addition to competitive rates of pay.

This has an impact on the means used to achieve performance and change management objectives, as well as those related to staffing. HR and L&D policies and practices have to reflect employee interests as well as those of the employer, while more care has to be taken over the manner in which people are treated by their managers. People who can, if they wish, resign and find another job elsewhere are less likely to put up with arbitrary and iniquitous treatment than is the case when alternative career opportunities are few and far between. However, the major immediate impact of tightening labour market conditions is in the staffing arena. Here several alternative approaches need to be developed and pursued:

- **Recruitment initiatives.** Employers can allocate a greater budget for recruiting staff by designing more sophisticated advertising and seeking to ensure that it has greater reach. They can also look to recruit people from outside their customary sources. Some of the best-known recent examples have been in the NHS, where recruiters have looked overseas in a bid to meet their specialist staffing needs. Efforts have also been made to lure back into work people who have previously left for family reasons or to take early retirement.

- **Retention initiatives.** When labour markets tighten, employers can work harder at retaining the staff that they already employ. The key here is understanding why people leave and then acting accordingly. Initiatives include attitude surveys, exit interviews and surveys of former employees. People resign from jobs for a wide variety of reasons, so employers are foolish to make any rash assumptions.
- **Reorganisation.** A third response to skills shortages is to reorganise job tasks among existing staff in such a way as to reduce the extent of dependence on hard-to-recruit groups. The term 'skill mix review' is used frequently to describe this process. The aim here is to ensure that people who are in greatest demand (normally those with the rarest skills or qualifications) spend 100 per cent of their time doing what only they are qualified to do. Other activities (eg managerial or administrative) should be carried out by support staff, who are easier to recruit.
- **Development initiatives.** The fourth response involves developing home-grown talent rather than seeking to buy it in 'ready made'. This takes rather longer to achieve and can be the most expensive approach, but should form part of a longer-term strategy for tackling skills shortages. The approach simply involves employing people who do not have the requisite skills and experience, and then giving them the opportunity to gain these skills.

PAUSE AND REFLECT

A major problem that employing organisations face is the need to compete fairly fiercely to recruit and retain experienced people while keeping control of the wage budget. They are looking to attract strong performers without being in a position to pay them a great deal more money. What kind of initiatives need to be taken in this type of situation?

Managing people in loose labour market conditions

From an HRM perspective, managing staff in loose labour market conditions is much less problematic. Recruiting and retaining strong performers is not a problem, because there is surplus labour in the market looking for job opportunities. This means that HR can be carried out safely without employing too many sophisticated approaches. People can, in short, be told what to do, rather than having to be persuaded. It means that organisations can be run very efficiently and that steps can be taken over time to extract greater productivity by pushing people harder.

These types of management responses are common when labour market conditions loosen, but it can be persuasively argued that they are short-sighted.

This is particularly true when an employment market is loose temporarily and may well tighten in the future. Treating people in a way they perceive to be poor or unfair can then backfire when market conditions change, leading good, experienced performers to leave as soon as they get the opportunity. More generally, though, treating people less well than they expect to be treated by an employer has other, more immediate negative consequences. Staff will tend to be less engaged, less enthusiastic about their work and less inclined to put effort in. They may minimise their contribution, perform to the minimum acceptable standard or even sabotage management initiatives. Low-trust, adversarial employment relations also tend to develop.

Long-term national labour market trends

At the start of 2020 there were 42.7 million people of working age in the UK, of whom 32.9 million people were undertaking some form of paid work. A total of 1.3 million were unemployed and actively seeking work, indicating an official unemployment rate of under 4 per cent, which was very low by recent historical standards. A further 8.5 million people of working age were defined as being 'economically inactive'. The impact of the Covid-19 pandemic was to increase unemployment levels by the end of 2020 to 1.7 million, representing 4.9 per cent of working-age adults. At the time of writing it is difficult to know how much higher this figure will go before economic recovery brings it back down again. What is, however, clear is that it is younger workers who are, as usual, being hardest hit by the onset of recession. This is partly because employers prefer to employ people with experience, but mainly because they are choosing to delay replacing older workers who leave jobs or retire. It will not be until widespread economic confidence returns that good job opportunities for younger people become widely available again.

The economically inactive category consists of full-time students, people who have taken early retirement, people who are disabled or suffering from a long-term medical condition and people who are 'looking after a family/home'. The percentage of people of working age classed as 'economically inactive' has not changed greatly over the past forty years. It has been estimated at 20 to 25 per cent for decades. It was at its lowest in the late 1980s and early 1990s, but then grew again somewhat, before falling in the most recent years. Government ministers believe that it can come down further still and are pursuing policies aimed at achieving this outcome.

It is, however, important to bear in mind that while economic prospects always feel dire in the middle of a recession, history strongly suggests that it will not be too long before they pick up again. This can be seen in Table 6.1, which shows the total number of people in the UK who are of working age and are in work of some kind. These Labour Force Survey statistics have been gathered and published monthly using the same approach since 1971, so they provide a good picture of the long-term trends.

Table 6.1 Total number of people in the UK who are of working age and are in work of some kind

	Employed	Economically active
1971	24.6 million	25.6 million
1976	24.8 million	26.1 million
1981	24.7 million	27.0 million
1986	24.7 million	27.8 million
1991	26.7 million	28.9 million
1996	26.0 million	28.4 million
2001	27.6 million	29.1 million
2006	28.7 million	30.3 million
2011	29.1 million	30.8 million
2016	31.7 million	33.3 million
2020	32.9 million	34.4 million

SOURCE: Labour Force Survey

You will see in Table 6.1 that, over time, the UK economy has seen a fairly persistent, steady increase in demand for labour. There was a reduction in the recessions of the early 1980s and the early 1990s and in the period after 2008, but it was not long on each occasion before the upward trend re-established itself. On the one hand, this is not at all surprising, as the period since 1971 has seen a significant increase in the size of the UK population. It would therefore be expected that economic activity would grow and that this would lead to increased demand for labour. On the other hand, it is surprising to see such a steady and persistent trend upwards when we consider that since 1971 there have been four major international recessions which led to millions of job losses, many labour-saving technologies introduced, the offshoring to developing countries of millions more jobs that were previously carried out in the UK and wholesale industrial restructuring leading to many more job losses. The national figure thus disguises a vast amount of churn in specific labour markets, some of which have declined sharply, while others have grown significantly.

The same long-term trends are apparent from Figure 6.1, where we have used a different database to plot changes in the total number of jobs in the UK. In this case it is not only people of working age who are included, and two part-time jobs count even if undertaken by the same person. A pronounced long-term trend towards increased demand for labour is very apparent. It is always possible that the trend upwards will stabilise or reverse over the coming years, but any analysis of the position over the past fifty years must conclude that demand for labour is likely to continue growing steadily in the future.

The overall national figures also conceal major shifts in the type of work that we do and thus the skills and attributes of workers who are in demand in the labour market. This can be seen in Tables 6.2 and 6.3. A different method of classification was used before 1999, making direct long-term historical comparisons with today difficult. The general trends, however, are very clear. The major growth areas are

Figure 6.1 Total number of jobs in the UK

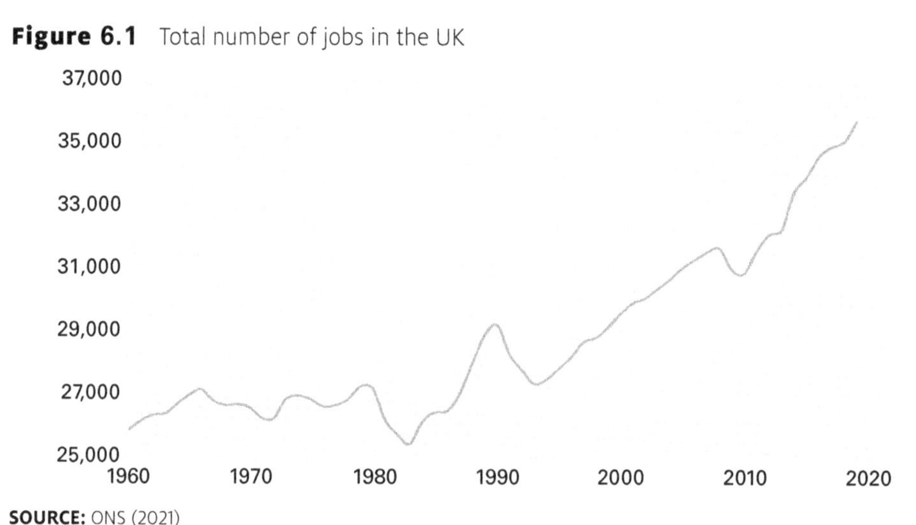

SOURCE: ONS (2021)

Table 6.2 Changes in occupation, 1951–99

Occupation	% in 1951	% in 1999
Higher professionals	1.9	6.4
Lower professionals	4.7	14.9
Employers and proprietors	5.0	3.4
Managers and administrators	5.5	15.7
Clerks	10.7	14.9
Foremen, supervisors and inspectors	2.6	3.1
Skilled manual	24.9	12.7
Semi-skilled manual	31.5	23.0
Unskilled manual	13.1	5.9

SOURCE: UKCES (2016)

in 'personal services' and in the managerial and professional occupations. We are witnessing a decline in lower-skilled jobs and in the skilled trades. This change in the occupational profile of the UK workforce has largely been driven by the pronounced shift that has occurred in the nature of UK industries during the past forty years.

In 1978 (the first year when data was collected on employment by sector) seven million people in the UK worked in manufacturing and a further one and a half million in the energy, water, farming and fishing industries. These have all hugely declined since then. Manufacturing employment is now only 3 million (12 per cent of all jobs). Agriculture and fishing now account for only 200,000 jobs, energy and water for fewer still. The big growth areas have been in retailing, distribution, hotels and restaurants, finance, business services, public administration, education and

Table 6.3 Changes in occupation, 1994–2024

Occupation	1994	2004	2014	2019	2024
Managers and senior officials	7.6	8.8	10.0	10.5	10.9
Professional occupations	14.8	17.1	19.9	20.8	21.4
Associate professional and technical occupations	11.9	13.1	14.0	14.5	14.8
Administrative and secretarial	14.9	12.8	10.7	9.7	9.1
Skilled trades	14.5	12.4	10.9	10.4	10.0
Caring, leisure and other services	6.4	8.0	9.4	9.8	10.1
Sales and customer services	8.1	8.1	7.8	7.6	7.4
Process, plant and machine operatives	8.9	7.2	6.2	5.8	5.5
Elementary occupations	12.9	12.4	11.0	10.9	10.8

SOURCE: UKCES (2016)

health. Employment in the financial services sector has grown especially quickly, having more than doubled since 1978.

What these statistics show is that, over time, while demand for labour in general terms is increasing, the significant growth is in demand for people with higher-level skills and experience. It is the professional, managerial and technical jobs that are growing most rapidly. Demand for people to fill lower-skilled roles is either stagnating or falling as our economy continues to transform in response to globalisation and technological advances.

The 2011 census included a question about the highest level of qualification that people aged 16–64 had achieved. For England and Wales the results showed that around 27 per cent of the then working population was educated to degree level, 41 per cent had left school with another academic qualification (O levels, A levels, GCSEs, etc) and 32 per cent had left school without any formal qualifications at all. People with degrees or equivalent qualifications are heavily concentrated in the professional occupations. Just under 70 per cent of professional people hold degrees. Around 30 per cent of people employed as 'managers and senior officials' and 'associate professional and technical workers' hold degrees. By contrast, people without qualifications or who left school without gaining A levels are concentrated in manual and elementary occupations, and in lower-paid jobs in the service industries.

We have seen significant improvements in the proportion of UK adults of working age who have qualifications of some kind over the past twenty years. People entering the labour market now have typically stayed on at school for much longer than those who are retiring, while many more have been to university. In 2018 government figures reported the following educational attainment levels for 25- to 64-year-olds in the UK:

- Tertiary: 46 per cent.
- Secondary: 33 per cent.
- Below secondary: 21 per cent.

In truth, though, while good progress is being made, the UK's position in comparison with other countries continues to deteriorate, and there are major questions over how effectively educated people are – despite gaining more formal qualifications.

A big OECD skills outlook survey was published in 2013, which revealed the results of tests carried out on 166,000 adults in 24 countries, focusing on:

- Problem solving in technology-rich environments.
- Literacy.
- Numeracy.

The UK came 9th, 14th and 16th respectively in each of these as far as the adult population generally was concerned.

Worryingly, though, our 16- to 24-year-olds performed more poorly:

- 15th on problem solving.
- 21st on literacy.
- 19th on numeracy.

Of particular concern, the UK was reported to be the only country with 16- to 24-year-olds scoring less highly on the literacy test than 55- to 56-year-olds. In other words, on this measure, our education system appears to be deteriorating over time.

In 2015 OECD data focusing on cognitive skills among 15-year-olds, the key findings for the UK were as follows:

- 22 per cent of students in the United Kingdom were low performers in mathematics (OECD average: 23 per cent), 17 per cent were low performers in reading (OECD average: 18 per cent), 15 per cent were low performers in science (OECD average: 18 per cent), and 11 per cent were low performers in all three of these subjects (OECD average: 12 per cent).
- Around 162,000 15-year-old students in the United Kingdom were low performers in mathematics, and more than 83,000 students were low performers in all three subjects (maths, reading and science).
- About 29 per cent of 15-year-old students in the United Kingdom attend schools where 30 per cent or more of the students are low performers in mathematics (OECD average: 30 per cent), and about 9 per cent attend schools where half or more of the students are low performers in mathematics (OECD average: 14 per cent).
- In the United Kingdom, the share of low performers in all three subjects has not changed significantly since 2006. Brazil, Germany, Italy, Mexico, Poland, Portugal, the Russian Federation, Tunisia and Turkey reduced their share of low performers in mathematics between 2003 and 2012.

Since 2016 we have seen significant improvements in literacy as far as school leavers are concerned, but it will take time for these to feed through into the adult population more generally. We can thus conclude that while in the UK we are as a working population becoming better educated over time, there remains a big skills gap between the qualifications our workforce has and those it will need to have in the future as low-skilled employment opportunities continue to decline and those for more highly educated people increase.

The statistics thus suggest that the supply of people (and particularly highly qualified people) in the UK economy is likely to rise at a lower rate than is required in order to meet the expected increased demand. If this is true, the result over time will be a tendency for UK labour markets to tighten considerably in the near future and remain tight for a significant period of time. The result will be chronic and persistent skills shortages in many industries that will be very challenging to manage.

An obvious objection to make to this proposition is to point out that none of the trends we are discussing here are remotely new. Demand for labour has been increasing since the 1980s and for many years the shift has been towards higher-skilled professions. This has not led to such major problems in the past. Supply and demand have adjusted effectively. Why should it be any different in the next forty years?

The answer is that in the past there were three other long-term trends that enabled supply to meet increased demand in the UK labour market. In each case there are some grounds for questioning the extent to which these will be present in the future:

- **Female participation**

  Female participation is the main reason for the long-term increase in the total number of people in paid work in the UK over the past forty years.

  In 1952 a total of 35 per cent of women were classed as 'economically active' in the official government statistics. By 1981 the proportion had risen to 65 per cent, and it now stands at over 74 per cent, compared with a male participation rate of 83.7 per cent (ONS, 2018). Female participation rates in the UK are now as high as the rates in all but a handful of other countries of the world. The raising of the female State Pension age from 60 to 65 between 2010 and 2020 has helped the situation to an extent, but in truth these gains in labour supply have been more than offset by the raising of the school leaving age to 18 and increases in the proportion of school leavers who attend university.

  The UK benefited substantially over the past few decades in terms of an increased supply of skilled labour arising from the trend towards much greater female participation in the paid economy. There is limited scope for further increases in the future.

- **The baby boom generation**

  The UK, like many countries, has also been fortunate in that the recent period of economic growth and of increased demand for labour that accompanied it coincided with a period in which the size of the working population has grown rapidly. This has been due to the presence in the market of the 'baby boom generation' born in the two decades following the Second World War. You will see from Table 6.4 how much bigger this generation is than those that came after it. The baby boom generation are now aged between 55 and 75. Many have retired, and the rest will follow over the next twenty years. This poses major labour supply problems because the cohorts coming up behind them are very considerably smaller in number. Far fewer babies were born each year in the 1980s and 1990s than were born in the 1950s and 1960s.

Table 6.4 UK births, 1945–2004

Period	Total births	Average per year
1945–64	17.6 million	880,000
1965–84	16.1 million	803,000
1985–2004	14.8 million	738,000

- **Immigration**
  The past forty years have seen extensive net migration into the UK from overseas. The total population was 56.31 million in 1980, a figure that had increased to 66.83 million by 2019. Between 2003 and 2019 the UK population increased by over seven million people, the majority of whom came from overseas to settle and were of working age (ONS, 2019b). In recent decades, we can thus conclude that the UK economy has relied very heavily on inward migration to meet skills needs. The extent to which this long-term trend will continue in the future is questionable. While net inward migration is very likely to continue outpacing net outward emigration, it is entirely possible that the gap will reduce and that employers will find it both harder and more expensive in the future to hire people from overseas than they have in the past. This is partly due to the ending of free movement of people from 2021 following the UK's departure from the European Union and partly due to government immigration rules more generally that make it difficult for lower-skilled workers to gain the right to work in the UK. For some years now the government has sought to restrict quite severely the extent of migration from outside the EU via the introduction of income requirements that favour higher-skilled workers. Employers can face large fines if they fail to comply with the regulations. As of 2021 these apply to all nationals from overseas except for people from the Republic of Ireland.

It is thus credible to argue that the UK labour market as a whole is highly likely to tighten in the future and that in this respect established trends are not going to continue as they have in the past.

Hourglass theories

According to a number of well-respected economists and labour market analysts, what appears to be happening is that we are seeing the emergence of an 'hourglass' occupational structure in the UK in which half of the new jobs being created are of the 'high skill/high pay' variety, and the other half are 'low skill/low pay'. Very few indeed, however, are unskilled.

The metaphor of the hourglass was originally advanced by Peter Nolan of Leeds University in 2001. But it was later popularised and expanded in the highly influential article by Goos and Manning (2003) entitled 'McJobs and Macjobs: The growing polarisation of jobs in the UK' and in further later articles that took a more international perspective.

In brief, they argue that:

- Increasing numbers of people are being employed in relatively highly paid, secure, professional and managerial occupations in finance, private services and some parts of the public sector.
- Jobs in manufacturing along with many clerical and administrative roles are being 'exported' to countries in Eastern Europe, to China, India and other developing economies where cheaper labour is readily available.
- *But*, at the same time, the growing number of higher-paid people in the UK are using their disposable income to purchase services that cannot be provided from overseas.
- Hence the simultaneous and rapid growth in demand for hairdressers, beauticians, restaurant workers, and people to work in the tourist and entertainment industries. These jobs tend to be lower paid.
- There also remains a demand for, and shortage of, some groups of skilled workers – plumbers, builders, decorators, etc – whose jobs also, by their nature, cannot be so easily exported.

If the analysis presented above is accurate, it strongly suggests future labour market trends that will contribute to greater inequality. People with relatively scarce sets of skills and experience will face tight labour market conditions in which there are more job opportunities available than people to fill the roles. These people will thus be well placed to bid wages up and to demand a good set of terms and conditions of employment. By contrast, those in the lower half of the hourglass will face loose labour market conditions. Lacking the required specialist skills and education, in the future there will be more people than there are jobs available for them. The result will be a tendency for wages to be bid downwards and for jobs to be less secure, with less attractive terms and conditions.

Offshore wind

As is the case in many countries, a key objective of the UK government's energy policy is to reduce rapidly and radically the country's dependence on fossil fuels. Initially, the aim was to reduce carbon emissions to achieve a stronger degree of energy security by making the economy less reliant on imported energy supplies. However, it is now clear that over time renewable energy will be a lot less expensive to produce than is the case when fossil fuels are used, and that the surplus electricity generated from UK renewables may be so sizeable as to create export opportunities.

The UK's combination of 7,000 miles of coastline, shallow seas and strong winds means that it is in the offshore renewables sector where the potential for efficient electricity generation is most promising, and a good deal of government subsidy is thus being made available to companies that are willing and able to develop offshore wind farms and technologies that can harness wave power. In 2008 only 1.5 per cent of UK electricity was supplied by offshore wind farms. By 2019 this had increased to 21 per cent. The costs of electricity generated by offshore wind farms has reduced by 66 per cent since 2010.

The government now aims to develop the industry further so that the UK not only meets a big proportion of its own energy needs from offshore renewables, but also becomes the world's leading supplier of energy generated in this way. To this end, an ambitious set of objectives has been set for the development of offshore renewable sources of energy, the current aim being that 30 per cent of the total demand for electricity in the UK should be met from such sources by 2030. That will mean increasing supply from 8.5 gigawatts to over 30 gigawatts in a decade. This is an ambitious target, given that the offshore renewables industry is so new. In 2007 fewer than 1,000 people were employed in the UK's fledgling offshore renewables industry, working on a handful of farms located close to the shore. This grew to 3,000 people in 2010 and had reached over 9,000 by 2018. The new offshore farms are much bigger, consisting of hundreds of turbines located many miles from the shore in the Irish and North Seas.

In order to meet the 2030 target, it is estimated that direct employment in offshore energy production will need to triple to 36,000 people, many more thousands of new jobs being created in supply chains. The major job roles are in project management, engineering and maintenance. There is a further need for people to design and manufacture the turbines and undersea cables, to construct new electricity substations and to lay foundations beneath the sea. Environmental analysts, boat and helicopter pilots and divers will also all be required in considerable numbers. Most of these jobs are specialised and require considerable training, expertise and experience. Forty-seven per cent require people who have graduated in science, technology, engineering and maths (STEM) subject areas.

Chronic skills shortages already represent a major problem for this fledgling industry. It has been reported that UK-based suppliers are often deterred from bidding for available contracts because they are unable to recruit people with the skills and experience required to develop and operate an offshore energy-generating facility. Part of the problem is illustrated by the following quotation from Stephen Wilson, Director of Wind and Marine at Narec (Reuters, 2010): 'For some reason, unlike the rest of Europe, engineering as a discipline in the UK suffers an acute lack of kudos. Consequently the UK turns out proportionally fewer graduates than many other European countries... We as a sector are always struggling to find experienced people for offshore wind.' He goes on to explain that the growth of the sector is already fast outpacing the available human resources.

In fact, the number of engineering graduates and qualified technicians is now falling in the UK, as older workers retire and too few new graduates come onto the labour market each year to take their place. This is set to accelerate over the next decade due to demographic trends. But it is not just people with the right qualifications who are needed. Offshore energy supply requires people with the right experience, too. Laying cables and installing equipment offshore is specialised and difficult work.

The problem is further compounded in the following ways:

- The UK is by no means the only country currently seeking to develop its renewable energy sector rapidly. There are plenty of opportunities for people with the required skills to work in this sector all around the world.
- While oil and gas supplies are in long-term decline, these businesses remain highly profitable and are able to pay very high rates to skilled engineers and technicians who are prepared to relocate to where the work is.

Demand for engineers across Industry generally is very strong. It is estimated that several hundreds of thousands of new skilled workers will be required across the sector by 2020.

The upshot is that the offshore renewables industry is going to have to work very hard to recruit and retain all the people it is going to need in the future. Government initiatives are assisting by encouraging more young people, especially girls, to opt for STEM subjects at school and university. Over 3,000 apprenticeships in the offshore renewables sector are also being partly funded publicly, while ex-military personnel with readily transferable skills are being targeted. For specialist roles, overseas recruitment will be necessary.

The costs associated with recruiting and retaining this large workforce in a tight labour market will be high. It will be a challenging HR and L&D task. But the signs are that this is an industry well worth investing in. Seen as risky fifteen years ago, it is now set to be secure and highly profitable.

References

Goos, M and Manning, A (2003) McJobs and Macjobs: The growing polarisation of jobs in the UK, in *The Labour Market Under New Labour*, ed R Dickens, P Gregg and J Wadsworth, Palgrave, Basingstoke

Mahoney, T A (1989) Employment compensation planning and strategy, in *Compensation and Benefits*, ed L Gomez-Mejia, BNA, Washington DC

OECD (2013) *OECD Skills Outlook 2013: First results from the survey of adult skills*, www.oecd.org/skills/piaac/Skills%20volume%201%20(eng)–full%20v12–eBook%20(04%2011%202013).pdf (archived at https://perma.cc/XA4H-PNFA)

OECD (2015) PISA 2015 results, www.oecd.org/pisa/keyfindings (archived at https://perma.cc/L749-W8EH)

ONS (2018) Long-term trends in UK employment: 1861 to 2018, www.ons.gov.uk/economy/nationalaccounts/uksectoraccounts/compendium/economicreview/april2019/longtermtrendsinukemployment1861to2018#womens-labour-market-participation (archived at https://perma.cc/U589-YNAY)

ONS (2019a) Population of the UK by country of birth and nationality: 2019, www.ons.gov.uk/peoplepopulationandcommunity/populationandmigration/internationalmigration/bulletins/ukpopulationbycountryofbirthandnationality/2019 (archived at https://perma.cc/3ZQP-DBY3)

ONS (2019b) Overview of the UK population: August 2019, www.ons.gov.uk/peoplepopulationandcommunity/populationandmigration/populationestimates/articles/overviewoftheukpopulation/august2019 (archived at https://perma.cc/PGE3-FMFC)

ONS (2021) Employment and employee types, www.ons.gov.uk/employmentandlabourmarket/peopleinwork/employmentandemployeetypes (archived at https://perma.cc/M2WU-BZV7)

(Continued)

(Continued)

Reuters (2010) UK wind industry: Where are all the people?, www.reutersevents.com/renewables/wind-energy-update/uk-wind-industry-where-are-all-people (archived at https://perma.cc/276U-9EFT)

UKCES (2016) *Working Futures 2014–2024: Evidence report 100*, UK Commission for Employment and Skills, London

Windolf, P (1986) Recruitment, selection and internal labour markets in Britain and Germany, *Organizational Studies*, 7 (3), 235–54

07 Population

There has long been an interest in the study of populations and population trends. Demography, the statistical study of human populations, is a large field of research interest that has persisted for a substantial period of time. This chapter explores a variety of demographic trends and will draw out implications, in terms of both labour and product markets. It is crucial that businesses of all kinds understand the key demographic trends that characterise the societies in which they are based. This is particularly true of the longer-term 'mega-trends' that enable them to plan effectively for future years. The majority of this chapter takes a UK perspective. Whilst there are some references to international examples and research, in general terms the experience of the UK is quite typical of most industrialised countries. Where there are substantial international differences, however, these will be pointed out.

To begin our discussions, we will review population trends at a global scale, examining historical and projected growth whilst critically discussing the underlying causes of population growth. During our initial discussions we will also consider the issue of overpopulation and explore the notion of environmental 'carrying capacity'. When the focus of the chapter turns to the UK we will explore two trends in particular: population ageing and the impact of migration. Akin to other major advanced economies, the UK has experienced substantial population ageing over time, and this has implications both at a societal level and for organisations seeking to recruit and retain employees. Similarly, the ageing population means that markets are changing as ageing consumers demand different goods and services. Our organisations must therefore adapt to serve this changed consumer base. To begin discussions within this chapter let's review population trends at the global scale.

Population trends at the global scale

It is undoubtedly the case that the world has seen substantial and sustained increases in the human population over time. It is currently estimated that there are between 7.7 and 7.8 billion people in the world. Fifty years ago the figure was 3.7 billion, and one hundred years ago it was approximately 2 billion. By 2050 the UN estimates that the population will reach 9.8 billion, whilst by 2100 it is expected that there will be over 11 billion individuals on the planet (United Nations, 2017). By all measures, these are substantial growth trends and the second half of the twentieth century saw a rise in human population that was vastly greater than at any other time in human history. The rate of growth has now slowed, as will be discussed in more depth during the coming sections, but the world's population grows by 70 to 80 million individuals each year, and by nearly 200,000 every day. This figure is net change – in other words, births minus deaths – and to put it in some context this is approximately the population of Milton Keynes being added to the world each day. In addition to this we must recognise that population trends vary substantially around the world. It is incorrect to assume that individuals are spread homogeneously across the planet; in reality we see large variance in the numbers of individuals living in different countries and regions. In terms of historical trends, the population of industrialised countries has risen relatively gently since the 1950s; the majority of the increase in population has been concentrated in the developing world. These trends are expected to continue into the future. Figure 7.1 highlights the differential spread of population across the world, covering the period from 1950 to the present day, and projecting forward to 2100 in line with UN estimates.

Figure 7.1 presents us with a visual map of population change over time, segregated by region. What is most noticeable first of all is the size and growth of the population on the Asian continent, together with the increase in the African population. As the chart demonstrates, we expect to see substantial continued growth in the African population and something of a plateau and then a subsequent reduction in the Asian population. At a later point this chapter will consider the dramatic ageing of the population in China, which goes some way to explaining why we expect to see something of a decrease in the total population of the Asian continent. What is perhaps also surprising is the comparatively small populations of the North American and European regions. These regions generate substantial economic output but it is certainly interesting to note that they have relatively small populations when compared with both the Asian and African regions. To understand a

Figure 7.1 World population by region, 1950–2100

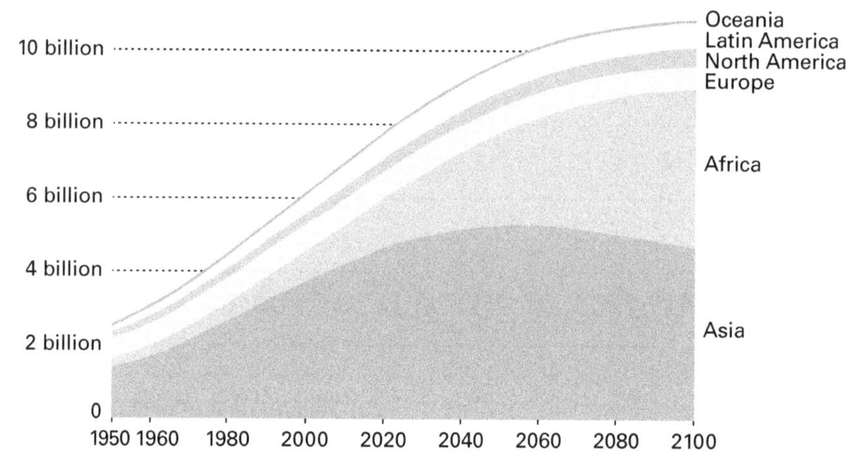

SOURCE: Our World in Data (2020a)

Figure 7.2 Absolute increase in global population per year

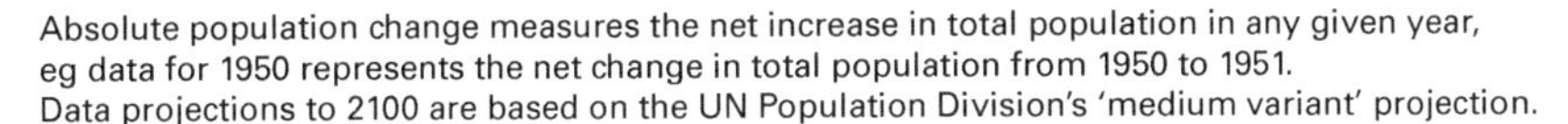

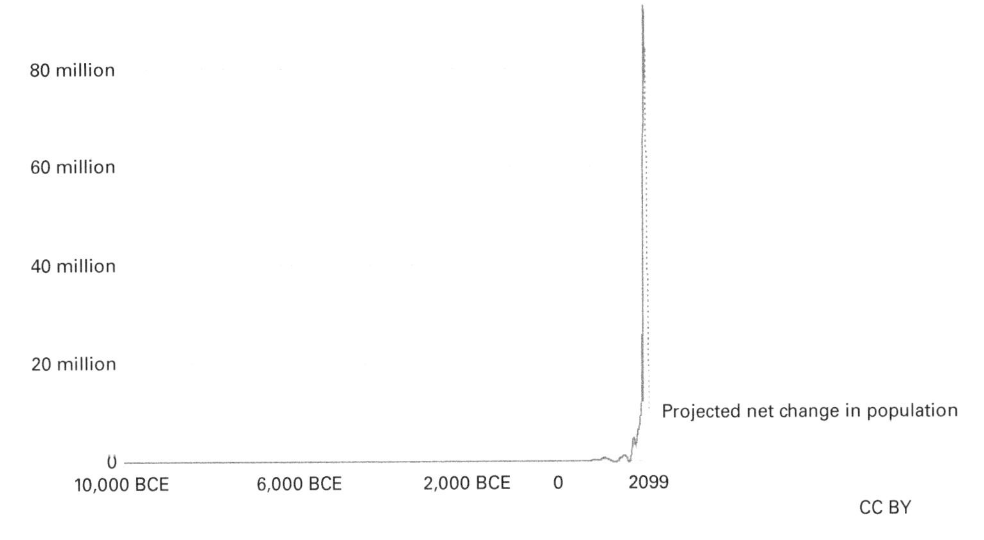

SOURCE: Our World in Data (2020b)

little more about the trends related to population increase, Figure 7.2 illustrates the absolute increase in global population per year.

The data in Figure 7.2 reinforces the point made earlier about the growth rates experienced across the world during the twentieth century. We see population change moving from around the 20 million per year mark in the early part of the period up to considerably more than 80 million towards the 1980s and 1990s. Growth rates have since fallen back somewhat, and the projected change in population shows a downward trend up to the end of the twenty-first century.

The question to be asked is, why are we seeing substantial changes in population levels? Population growth or decline is determined by birth and death rates. Recent patterns of world population growth have occurred because death rates have fallen while birth rates have either increased or stayed the same. We know that global life expectancy at birth has risen very substantially over the past fifty years, with World Health Organization figures in the mid-2010s indicating that global life expectancy at birth is 72 years (World Health Organization, 2016) and is projected to increase in the future. There are again very substantial variations by region, however, with developed parts of the world having substantially longer life expectancies than developing countries and regions. Indeed, in Southern Africa, where AIDS is particularly prevalent, life expectancy fell from 62 to 43 between 1980 and 2010, causing a decline in the overall population. Life expectancy has, however, increased at a global scale for a number of reasons, including improved healthcare, sanitation, access to clean drinking water, improvements in housing, and more nutritious food (Hendrixson and Hartmann, 2019). Of course, decreases in death rates are only one part of this story, and the next section of this chapter is devoted to exploring the other substantial driver of population change – fertility rates.

Population growth and fertility rates

Fertility rates greatly impact populations in terms of their overall size and age structure (Hauer and Schmertmann, 2020). The fertility rate – in other words, the average number of children each woman gives birth to – is partly determined by the total number of women in a population who are of childbearing age, partly by the number of children that each woman has, and, to a more limited extent, by the levels of male fertility. What is known is that for a population to remain stable, given a stable death rate and no great excess immigration or emigration, a fertility rate of 2.1 is necessary. At face value, it would appear that a fertility rate (other factors being equal) of 2.1 would lead to population increase over time, but this figure must take account of infant mortality as well. In reality, for a population to remain stable, each couple must produce just over two children. Just as there are substantial variations in broad population trends around the world, so too are there variations in terms of fertility rates. The birth rate in some parts of the world is substantially higher than in others.

Data from the World Bank (2020) estimates that global fertility rates are just over 2.4, with the highest rates seen in sub-Saharan Africa (4.7) and the lowest rates in Asian countries such as South Korea (1.0) and Singapore (1.1). The World Bank also highlights low fertility rates in many European countries such as Spain (1.3), Greece (1.4), Portugal (1.4), and Germany (1.6). In total, data from the World Bank indicates that fertility rates are below the replacement rate (2.1) in 95 countries/territories in which information is currently collected. For the UK they indicate that the current fertility rate is approximately 1.7. What we therefore see is that in several parts of the world fertility rates are substantially below the established replacement rate, and combined with increasing trends in individual longevity, this therefore gives rise to an ageing population, and eventual population decline. As an example, the low levels of fertility in Europe are expected to cause the European population to fall by around 100 million by 2050. Several major international bodies, including the United Nations, highlight that low levels of fertility are unprecedented in human history. Indeed, if we look back just to 1950 the global fertility rate was 4.7.

As noted above, some of the most extreme trends related to fertility rates are found in Asia, with countries such as South Korea and China facing substantial problems caused by low fertility rates. In China, for instance, the one-child policy introduced in 1979 (Hou, 2019) has instituted substantial shifts in broader population trends, and it is estimated that by 2050 around 40 per cent of the Chinese population will be over the age of 60. There are therefore concerns that China will become 'old before it becomes rich' (Hou, 2019: 505), with substantial shifts needed in terms of flexible working, healthcare and pensions for the country to adapt to demographic changes. Likewise, in South Korea we similarly see a very low fertility rate, the lowest globally by many measures. Whilst the government there has attempted to stimulate birth rates through subsidies for childcare, improved maternity and paternity pay, and the provision of childcare facilities, the high cost of housing and desire for career advancement, amongst other factors, have meant that couples have increasingly chosen not to have children. As a result of this, South Korea's population is expected to begin falling in the late 2020s, with negative impacts on economic output, as measured by GDP figures. As serban and Aceleanu (2015) note, future trends in global fertility rates will be impacted by a number of factors, including advances in medical technology, trends in family size and the desire to have children as compared with other lifetime goals.

Overpopulation

Substantial academic and practitioner interest has been invested into thinking about the issue of overpopulation. We frequently see articles in the media and press discussing the relative sustainability of human population growth, but this is an issue that has long been an area of discussion. As far back as Malthus in 1798, when one billion individuals lived on the planet, there were debates about how many individuals could comfortably live on Earth. The Malthusian thesis was that, if left unchecked, the human population is likely to grow in an exponential manner, whilst the production of food grows in a linear fashion (Remoff, 2016). Malthus therefore argued that a growing population inevitably creates increasing levels of scarcity and hardship, hence there is a natural check on population growth. As Remoff (2016) argues, however, thinking has moved on considerably from the work of Malthus, with significant academic interest now invested in the broader area of sustainability, together with the field of human ecology (Faith, 2019). In brief, the field of human ecology explores the interrelationships between humans, other organisms and the broader environment, and is a field that is attracting a growing number of contributions. The key learning point from this brief discussion, however, is that concerns and debates regarding overpopulation are nothing new. This is an issue that has long held the interest of various scholars and thinkers.

Debates related to overpopulation frequently examine the specific issue of 'carrying capacity'. This is an interesting concept and suggests that all environments, whether local, or the planet as a whole, have a maximum population of a species that can be maintained. Within the field of population ecology specifically, carrying capacity is seen as the maximum load (of humans) that the environment can sustain (Hui, 2006), and as Hendrixson and Hartmann (2019) point out, human population is indeed limited by the resources that the broader environment can provide. Linking all the way back to the work of Malthus, researchers point out that issues connected with food supply play a particularly significant role in determining environmental carrying capacity (Hopfenberg, 2003), and therefore the changing climate is a concern as this may impact levels of food production, and hence food security around the world. Whilst it is outside the scope of this book to explore these issues in detail, we know that increasing levels of desertification have a profound impact on the food security of populations living in affected regions (Smith et al, 2020).

Whilst the issue of carrying capacity has attracted much attention, there has been no agreement on the level of human population that the planet can sustain. Different researchers and thinkers approaching the issue from different perspectives arrive at a variety of conclusions, and there is active debate within this field as a result. Recent work from Binder et al (2020) is, however, interesting as it seeks to understand more about the extreme limits of human population. The researchers argue that in the presence of 'ideal' technology the true upper boundary of human population is likely a figure in the trillions of individuals, and even with current technology the carrying capacity of Earth is very substantially higher than the current population. In practical terms, though, we understand that there are various other limits on human population (for example space), which suggests that an equilibrium point is, in reality, far lower than the absolute maximum figure that the planet could hypothetically support. The true upper boundary of human population therefore remains unknown. When we look at a regional level, however, interesting dynamics emerge. As Hendrixson and Hartmann (2019) note, we see a large youth population in the 'Global South', contrasted against an ageing and, in some cases now, declining population in the 'Global North'. It is therefore thought that there could be shifts in economic power and potential conflict as a result, indicating that we must consider population pressures at a more local level.

PAUSE AND REFLECT

Before you move on from this section please pause and reflect for a few minutes. To what extent do you agree (or disagree) with the arguments put forward here? Putting aside any human resource management implications for a moment, should we be concerned about population pressures?

How did you get on with the question above? It is indeed important at times to step back and consider the macro trends that are impacting the world around us. Whilst there are no right or wrong responses here, the question above should have caused you to reflect on the issues of population pressures and specifically overpopulation in some detail. We know that there are more of us living on the planet than at any other point in history, and we know from the media and academic research that population pressures are having impacts on the world around us: for example, the well-documented impacts of human activity on the climate. Although the upper boundary of human population is unknown, it is inevitably the case that a growing population will intensify resource pressures and will require careful management in the years and decades to come. We now move on from broader macro issues in terms of population to take a closer look at population trends within the UK.

Population trends within the UK

The United Kingdom shares many of the population trends seen in other advanced economies. This section of the chapter will explore the growth of the UK population, in terms of both current observations and the recent past, alongside the issue of population ageing. Many other advanced economies are seeing a so-called 'greying' of their population, and the UK has not been immune from this trend. It is also important that migration patterns are examined in depth as the relative levels of immigration and emigration have a substantial effect on the UK's total population, and we will explore statistics that chart these patterns over time.

In mid-2019 the UK's population was estimated to be approximately 66.8 million individuals (ONS, 2020a). This is around 9 per cent of the total population of

European countries and rather less than 1 per cent of the total world population. The Office for National Statistics highlights that approximately 56.29 million individuals reside in England, with some 3.15 million in Wales, alongside 5.46 million in Scotland, and 1.89 million in Northern Ireland. Population density is rather higher in England as opposed to elsewhere in the UK, with the highest concentrations of people located in London and the South East.

Growth

Unlike many other European countries, the UK's population is currently growing. This is for two reasons. The first is that the birth rate exceeds the death rate. Each year approximately 800,000 babies are born in the UK, whilst roughly 600,000 people die – a net gain of around 200,000 individuals per year. Alongside this it is estimated that around 200,000 more people on average migrate into the UK than emigrate out of it. According to the ONS, total immigration for the year to March 2020 was estimated at 715,000, whilst emigration was 403,000. For the period net migration was therefore approximately 312,000 (ONS, 2020b). The chart reproduced from the ONS in Figure 7.3 provides a visual representation of the main drivers of population change within the UK over time.

Figure 7.3 provides a very useful graphical representation of the UK's population change. The solid line indicates the level of net population growth, and this has moved from a figure of 100,000 to 200,000 per year during the 1990s to a figure in excess of 400,000 per year for the majority of the years post-2004, with only a slight dip more recently. Levels of net migration (as represented by the lighter bars) have increased over the period and we see something of mini baby boom in the years from 2005 to 2016, with substantial population change caused through natural change

Figure 7.3 The main drivers of UK population change, 1992–2019

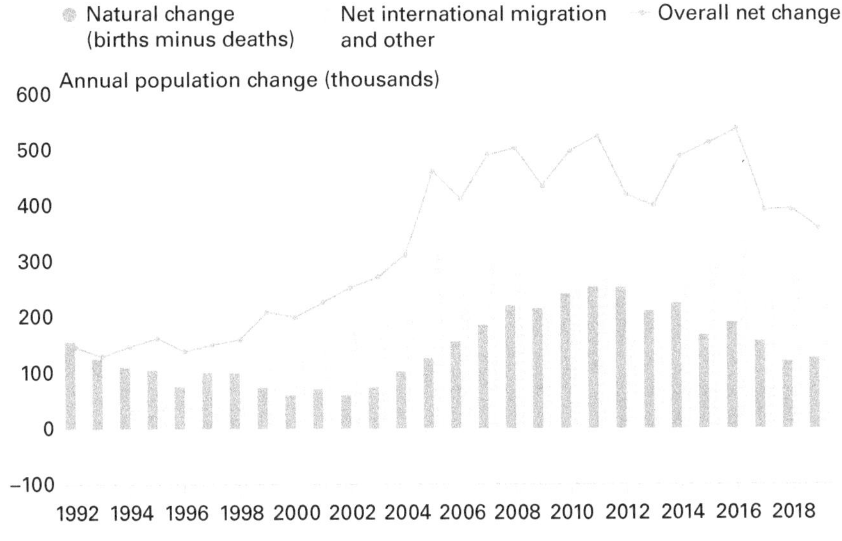

SOURCE: ONS (2020a)

Table 7.1 Estimated and projected UK population (millions)

	2018	2023	2028	2033	2038	2043
UK	66.4	68.1	69.4	70.5	71.4	72.4
England	56.0	57.6	58.8	59.8	60.8	61.7
Wales	3.1	3.2	3.2	3.2	3.2	3.3
Scotland	5.4	5.5	5.5	5.6	5.6	5.6
Northern Ireland	1.9	1.9	2.1	2.0	2.0	2.0

SOURCE: ONS (2019a)

(darker bars). With these trends in mind, official estimates now project that our total population will reach almost 70 million in 2028, passing 72.4 million in 2043 (ONS, 2019a). The ONS projections are outlined in Table 7.1.

The data in Table 7.1 demonstrates that we expect to see continued growth in the UK's population, although the ONS projects a convergence in birth and death rates in the coming years. If these projections prove accurate then the level of natural change in the population will decline, with the ONS projecting that by the 2040s all net change in the UK's population will be the result of net migration. The observed and projected growth in the UK's population represent a reversal of the position in the 1970s and 1980s – a substantial period of net emigration and relatively low birth rates. This itself contrasted with the period after the Second World War when the UK saw high birth rates, which persisted until the late 1960s. This period created the large 'baby boom' generation, who are now at, or approaching, retirement age. In brief, the UK's population trends have indeed varied substantially over time and we will inevitably see ongoing variance in the future.

Population ageing

As mentioned earlier in this chapter, and indeed this section, the UK is an ageing society. The extent of this trend, as is the case for many developed societies (Papapetrou and Tsalaporta, 2020), is remarkable. We are seeing substantially greater longevity as a result of a number of factors, including improved healthcare, better housing, more nutritious food, a decrease in harmful behaviours such as smoking (Şerban and Aceleanu, 2015). The speed at which the UK's age structure is changing is indeed surprising. The census of 1851 recorded that just 7 per cent of the population at that time was over the age of 60, by 1901 this had increased to 9 per cent, and by 1951 to 16 per cent. In 2011 the national census at that time reported that 22 per cent of the population was over 60, and the ONS continues to report a shifting age structure within the UK (2019b). In their 2019 population data release the ONS projects that by 2050 one in four individuals will be over the age of 65 within the UK, an increase from approximately one in five in the late 2010s. They highlight that the 65+ portion of the population is the fastest growing sector of society, highlighting the substantial impact of the baby boom generation rising through the age pyramid. By the late 2060s the ONS (2019b) projects that there could be an additional 8.2 million people

Figure 7.4 Age structure of the UK's population

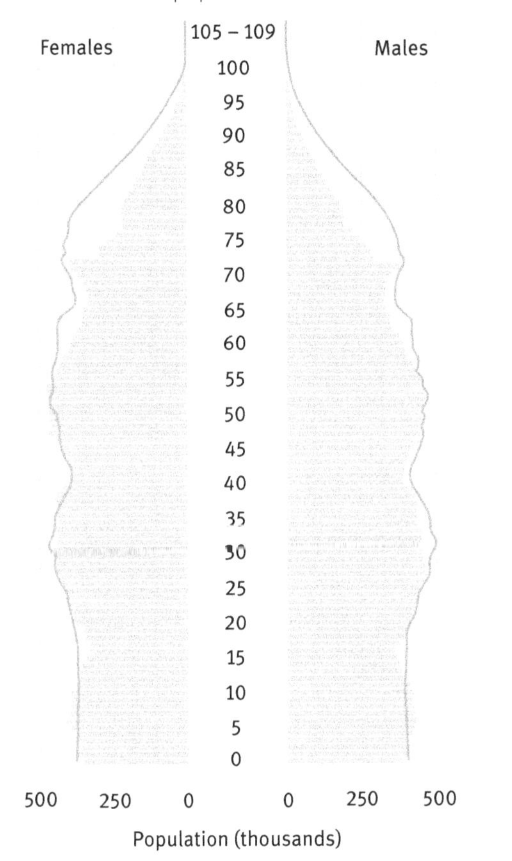

SOURCE: ONS (2019a)

aged 65 or over within the UK. In brief, we are seeing many more individuals live to advancing ages within the UK; indeed, we are seeing substantial increases in the 'oldest old', in other words the numbers of people living to 85 and over.

We appreciate that there are a number of statistics in the above paragraph, which when written in text are rather difficult to visualise. Figure 7.4 is again taken from ONS data and shows the age structure of the UK's population. The solid bars represent the age structure in 2018, whilst the line represents the expected age structure in 2043.

What we first note from Figure 7.4 is the uneven spread of the population amongst the age groups. We tend to see peaks and troughs rather than a linear trend, and what we must also note is that the population groupings towards the base of the chart are somewhat smaller than those towards the middle. The line representing the projected position in 2043 shows that we are likely to see dramatic shifts in the UK's age structure – look, for instance, at the projected changes between the age range of 65 and 90. The ONS statistics are projecting a substantial increase within the upper age categories.

PAUSE AND REFLECT

Please take a few moments to reflect on the data presented so far in this section of the chapter. What do you make of the trends that we see concerning the ageing of the population? Is this something to be welcomed or something to be concerned about?

How did you get on with the question above? As with the majority of these opportunities to pause and reflect, there is a variety of arguments that could be advanced, and this is most certainly not a question that lends itself to a 'right' or 'wrong' answer, as it were. First of all, at a surface level, the fact that individuals are generally living longer lives is something that should be welcomed. It is unlikely that anyone would disagree with this broad observation, but there are some substantial challenges that increased longevity causes. The first of these relates to pension systems. Increases in life expectancy inevitably place pension systems (both state and privately funded) under increasing strain (Şerban and Aceleanu, 2015; Hendrixson and Hartmann, 2019). Can a state continue to afford to pay a State Pension at any meaningful level if the numbers of older individuals continue to rise with proportionately fewer individuals of working age contributing to the tax base? In terms of private pensions, how can individuals (and indeed organisations) be encouraged to put money aside to properly fund retirement? In an attempt to address these points, the UK government introduced legislation in 2012 requiring organisations to enrol eligible workers into pension schemes. This scheme is an 'opt out' rather than 'opt in' for employees, although there remain questions about the level of organisational and individual contributions required, and indeed research demonstrates that individuals from ethnic minority groups are substantially less likely to benefit from occupational pensions of any kind (Vlachantoni et al, 2015), a very problematic finding indeed.

As there are active debates concerning pension provision, so too are there discussions connected with issues around health and social care. For instance, how does a society structure its health and social care arrangements so that they are simultaneously affordable and provide the right quality of care when it is required? We know that as populations age healthcare tends to become more expensive (Lawson, 2016; Max, 2018), and commentators predict that advanced societies are likely to need to commit an increasing portion of gross domestic product to health in general. The issues discussed above are not, however, confined to the UK context. As noted earlier in this chapter, China, South Korea and other Asian countries are experiencing rapid population ageing as well, with consequential impacts on health and social security systems (Hou, 2019). These are indeed global challenges that many societies are facing, although we must recognise that societies are ageing at different relative speeds.

A key challenge for societies is in balancing the portion of one's life that is spent in working versus non-working roles. During this section we have already alluded to the concept of the dependency ratio: in other words, the proportion of people of working age in a population compared to the proportion who are above pension age. Turning again to the information provided by the ONS (2019c), the UK's dependency ratio in 2019 was estimated to be just under 29 per cent. In other words, for every 100 working individuals, there were 29 living above the pensionable age. By 2050 the ONS projects that this figure will climb to 36 per cent, and by 2065 to almost 40 per cent. There have been debates about whether the dependency ratio continues to be a valid measure given that a number of individuals remain economically active above retirement age, and indeed the ONS (2019c) notes that economic activity amongst the 50–65 and over 65 age groups is outpacing growth within other age categories. That being said, it does provide a useful data point from which we can discuss the implications arising from the ageing society as we have done within this section of the chapter.

Migration

We know that migration across the world is increasing. Many more individuals are moving between countries and regions than has been the case previously, and for a variety of reasons, including employment and study. The United Nations defines a migrant as:

> any person who is moving or has moved across an international border or within a State away from his/her habitual place of residence, regardless of (1) the person's legal status, (2) whether the movement is voluntary or involuntary, (3) what the causes for the movement are, or (4) what the length of the stay is.
>
> (United Nations, 2020)

They point out that in 2019 estimated global migration was 272 million, although there is appreciation that this figure is likely to be an underestimate because data is typically poorly captured around the world, and the statistics on illegal immigration are not rigorously documented.

Within the context of the UK, up until the early 1990s there had been broad balance in terms of international migration for the previous twenty years. Indeed, for much of the 1970s and 1980s more people left the UK each year than entered it. Since the early 1990s, though, this trend has changed. Every year there are now substantially more immigrants than emigrants, as was shown particularly well in Figure 7.3. As noted earlier in this section, for the year ending December 2019 the net migration figure for the UK was approximately 270,000 (ONS, 2020b), although within this figure we do see some interesting trends. Since 2016 the UK has seen a fall in EU net migration, which commentators attribute, in part, to the Brexit referendum and vote (Wadsworth, 2018; Tyrrell et al, 2019). Figure 7.5 from the ONS demonstrates this shifting trend.

As can be seen from Figure 7.5, immigration and emigration figures have changed significantly over time (the dotted lines represent unadjusted estimates). A substantial rise in net migration from EU sources can be seen during the period from 2012 to 2015, with a sharp decline following the EU referendum in June 2016. This decline was caused by a fall in immigration and a simultaneous rise in the number emigrating from the UK. Wadsworth (2018) notes that debates connected with immigration featured prominently during the EU referendum and the decline in net migration following the result was caused by a number of factors including the depreciation of sterling against the euro, and migrants settling in other EU destinations such as

Figure 7.5 UK net migration from EU sources, 2010–19

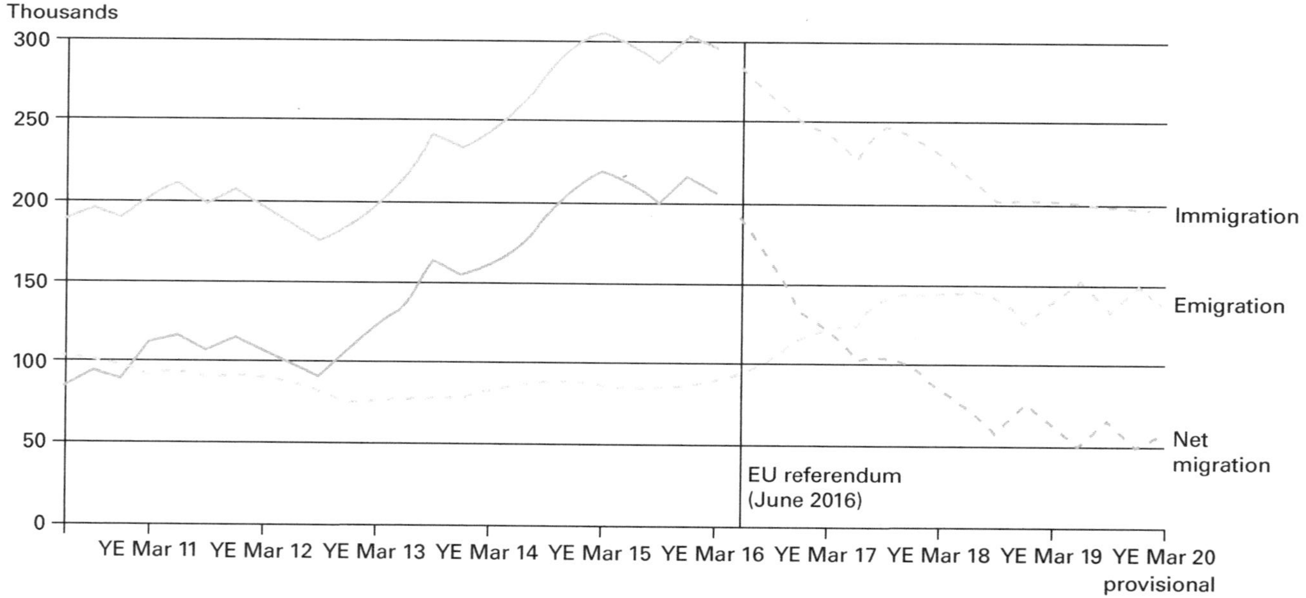

Thousands

300

250

200

150

100

50

0

YE Mar 11 YE Mar 12 YE Mar 13 YE Mar 14 YE Mar 15 YE Mar 16 YE Mar 17 YE Mar 18 YE Mar 19 YE Mar 20 provisional

Immigration

Emigration

Net migration

EU referendum (June 2016)

SOURCE: ONS (2020b)

France, Spain and Germany. Interestingly, the work of Tyrrell et al (2019) argues that the EU referendum result has impacted the sense of 'belonging' that migrants have towards the UK, with younger individuals in particular feeling a loss in terms of their identity, consequently being less motivated to remain in the UK for the longer term. Whilst net migration from the EU has fallen since the EU referendum, the UK has seen a significant rise in non-EU migration, as Figure 7.6 demonstrates.

The chart presented in Figure 7.6 demonstrates again how migration, this time from non-EU sources, fluctuates over time. There has indeed been a substantial increase in net non-EU migration following the EU referendum, with large numbers of arrivals from non-EU sources. All told, the ONS (2020b) reports that net migration figures have been relatively stable in the medium term, but that the location of migrants has shifted substantially since 2016, with many more non-EU citizens arriving into the UK.

The economic impact of migration into the UK, and other advanced economies, tends to be positive (Portes and Forte, 2017). Migration tends to have a positive impact on GDP per capita, and, as other researchers such as Boubtane et al (2015) highlight, it boosts productivity across an economy. Migrants are more likely to be engaged in employment (Portes and Forte, 2017) and are therefore more likely to be net contributors to an economy. Similarly, for countries such as the UK with an ageing population, migration has a positive impact on the dependency ratio as migrants themselves tend to be younger, and they tend to have larger families; hence there can be a positive impact upon birth rates.

As we have seen in this section of the chapter, the UK's population shows a series of interesting trends over time, with population growth caused both by natural change in the population (births exceeding deaths) and positive net migration. We have also highlighted that the UK's population, like that of other advanced economies, is ageing; in the coming sections we will focus on this trend in substantially more depth. We begin by considering the labour market implications that arise from an ageing population.

Population ageing: labour market implications

As populations in many advanced economies age there will be substantial HRM implications, with significant adaptations needed in HR policy/practice frameworks. Given the changing population structure in the UK, we know that there will be fewer individuals in the younger age categories, and organisations will find it increasingly difficult to recruit from their traditional sources. Indeed, research from Papapetrou and Tsalaporta (2020) highlights just this issue, noting that labour supply will inevitably reduce as populations age. As a result of this, organisations will increasingly be required to look to other labour markets in order to recruit and retain the individuals they require. This will include three distinct groups of individuals who have not traditionally found themselves to be in great demand by recruiters:

1. People over the age of 50 who are currently still working.
2. People who have taken early retirement or redundancy.
3. People who are over the state retirement age.

Figure 7.6 Non-EU migration, 2010–20

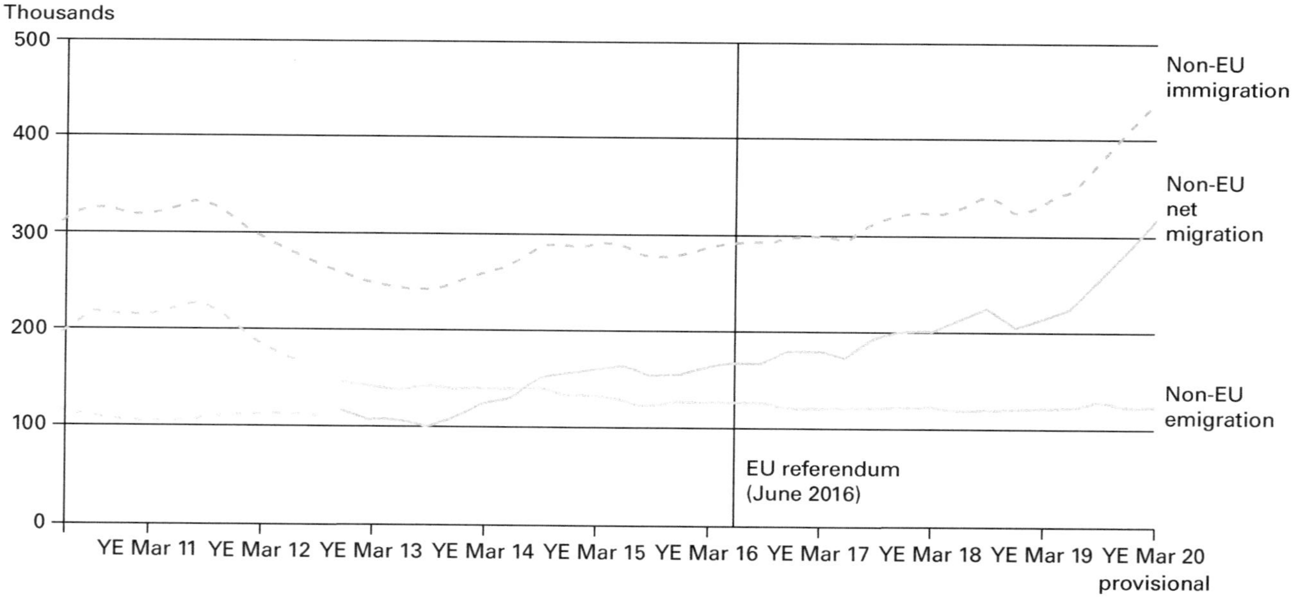

SOURCE: ONS (2020b)

What is important to point out is that these groups are very diverse in makeup, and employment opportunities that appeal to one set of individuals may be distinctly unattractive to another group. In the case of the first group, individuals who are over 50 and still actively working, it is likely that traditional full-time jobs will be sought. In the case of the second and third groups it is likely that more flexibility will be important, with individuals looking for part-time work, or another form of flexible employment. We know from the research base that many individuals have a preference for a phased retirement as opposed to working full-time until a retirement date and then undertaking no paid work after that time. It is very likely to be the case that the employers who can adapt and offer greater flexibility will be in a substantially stronger position to compete for the services of older workers. Emphasising just these points, research insights produced by Bal and De Lange (2015) and Butler (2020) both highlight that flexibility is particularly important for older workers. Whilst we know that flexibility appeals across age categories because it better enables individuals to balance their personal and work responsibilities, older individuals were seen to display stronger job performance in the presence of flexible work initiatives (Bal and De Lange, 2015). Similarly, Butler (2020) notes that older workers with caring responsibilities particularly valued flexible work arrangements, and that the provision of flexibility enables organisations to therefore retain a diverse range of skills and experience. These findings, and many others besides, add weight to the argument that it will be the organisations that develop effective flexible working practices that will be best positioned to tap into older labour markets.

Alongside improvements in workplace flexibility it is also argued that organisations need to develop workplace cultures that respect and value the contributions made by older workers. Considerable research has been carried out over time exploring the attitudes to older workers among managers and younger employees (see, for example, Ng and Feldman, 2012; Oliveira and Cabral-Cardoso, 2017), and we see that just as people tend to stereotype younger workers, the same is true with older individuals. Older workers tend to be seen as being reliable, stable, mature and experienced, but also as difficult to train, resistant to change, over-cautious, poor with technology, slow and prone to ill health. Organisations that are serious about employing greater numbers of older workers will therefore need to tackle such stereotyping and ensure that opportunities for development are provided for people of all ages. Having said this, it is important that employers recognise that individuals change as they age and therefore are likely to contribute different qualities over time. As a result of this we will need to tailor the work environment, reward packages and development plans to suit the needs of our increasingly diverse employees. Simultaneously, organisations, and society as a whole, will need to tackle age discrimination in employment. Whilst legislation prohibiting age discrimination already exists, we must ensure that our organisations are diverse and inclusive, tackling problematic stereotypes and cultures that pervade elements of the working environment.

The case study that follows discusses the employment of older workers at the DIY chain B&Q, which has long been held up as an example to others regarding the employment of older workers.

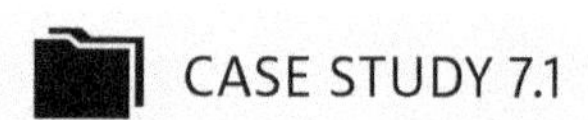

CASE STUDY 7.1

Employing older workers at B&Q

B&Q is a British DIY and home improvement chain that operates approximately 300 stores, employing some 27,000 individuals across the UK. Founded in 1969, it is well known for supplying a large range of products including paint and decorating equipment, home hardware, tools, flooring, plumbing and heating items, amongst many others, and is a well-known brand on the UK high street (B&Q, 2020). In 2018–19 it turned over nearly £3.4 billion with six-monthly profits to July 2019 estimated at £245 million. The organisation has faced challenges in recent years surrounding general market trends and the move to online shopping, but it remains a successful business, employing many thousands of individuals.

B&Q is seen as something of an authority regarding the employment of older workers within the UK, with some of its workers of extreme age. You will have no difficulty finding news stories regarding this, and indeed the organisation regularly employs individuals who are over 90 years of age. The organisation has a long-standing policy on employing older individuals and it has placed age diversity towards the top of its corporate agenda. This emphasis came about in response to customer feedback. The organisation found that customers preferred to be served by individuals with a good level of knowledge regarding DIY projects, something typically possessed by older individuals who tended to own their own homes. For B&Q, the growing pool of older workers was therefore an underutilised pool of talent from which they could recruit. Older workers provide the organisation with valuable experience and knowledge that it otherwise would not have access to. The organisation actively recruits across all age groups, understanding that age diversity supports corporate performance.

In order to successfully integrate individuals from different generations into its workforce, B&Q has put in place a number of specific HRM initiatives. First, the organisation offers flexible contracts. Rather than only offering a standard full-time permanent or fixed-term arrangement, the organisation offers a range of contracts that permit individuals to work a pattern that suits their lifestyle. Similarly, B&Q offers flexible working arrangements to all staff members regardless of age, length of service or life situation. The organisation also has no upper age limit related to training and development, and there are examples of individuals over the age of 70 undertaking NVQ qualifications through the firm. Within the workplace itself, the organisation adapts its working practices to suit individuals, for example manual handling, and B&Q also trains all staff in the area of age diversity. Understanding that teams work better when they understand each other and have an appreciation for individual differences, the company expends substantial effort in order to ensure that everyone engages with learning programmes connected with age diversity.

The organisation, as mentioned at the start of this case study, is indeed frequently held up as a best practice example regarding the employment of an age diverse workforce. Commentators frequently highlight that success within the field of age diversity has benefits for corporate performance.

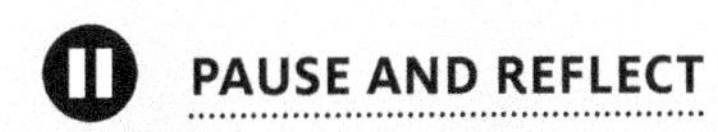

Please take a few minutes to conduct a little of your own research into B&Q and their employment of older workers. Start with the B&Q website, and also utilise general internet searches to find news stories about the company. Can you highlight any specific factors that, in your view, have contributed to their success in employing older workers?

How did you get on with the question above? B&Q is certainly an interesting organisation to explore with regard to the employment of older workers and you will inevitably have arrived at a diverse set of responses within this opportunity to pause and reflect. During your reflections you may have considered the underlying culture of the organisation and how the presence of a long-standing policy with regard to the employment of older workers is likely to have played a substantial role in normalising age diversity within the organisation. Similarly, the fact that B&Q offers flexible working opportunities and opens up learning and development to all are important symbols of the inclusivity demonstrated by the firm. These factors will have inevitably contributed strongly to its success in terms of recruiting and retaining an age diverse workforce.

The final major section of the chapter now focuses on the product market implications of population ageing.

Population ageing: product market implications

A second major implication arising from the changing population structure relates to the products (and indeed services) that individuals consume at different stages of their lives. Whilst there is debate about the extent to which current observed trends will continue into the future, it is likely to remain the case that older and younger consumers will continue to spend their income on different products. Older individuals, for instance, tend to spend less of their income on entertainment and alcoholic drinks than younger individuals, but more on food and recreation more generally. Older individuals are substantially larger consumers of culture, and spend significantly more on holidays. Expenditure on cars is low both for older and younger individuals, but higher for those who are middle-aged. Older people spend much less on transport generally, and far less on clothes, and they spend more on domestic fuel and healthcare products. These trends have important implications for the future of many industries. For example, the decline in recorded music sales, which has been reported internationally, has been partly due to the rise in internet downloading and streaming, but the broad trend predates the widespread use of the internet. The most common explanation is a decline in the number of teenagers – the prime consumers of the product.

Connected with the points made in the paragraph above, Papapetrou and Tsalaporta (2020) argue that there are substantial differences in consumption and

Figure 7.7 Mean wealth by component and age, April 2016–March 2018

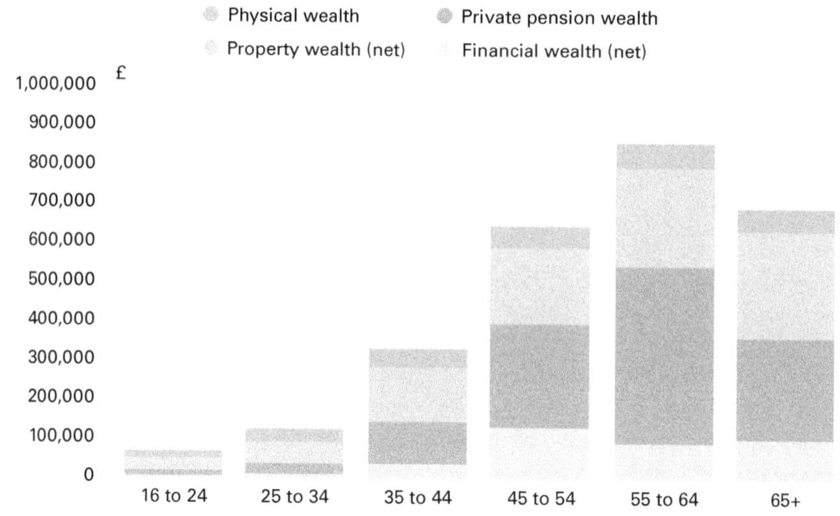

Physical wealth
Private pension wealth
Property wealth (net)
Financial wealth (net)
£
1,000,000
900,000
800,000
700,000
600,000
500,000
400,000
300,000
200,000
100,000
0
16 to 24
25 to 34
35 to 44
45 to 54
55 to 64
65+

SOURCE: ONS (2019d)

saving patterns across the generations. They argue that we tend to accumulate assets and wealth during our working years and then liquidate those assets as we enter retirement in order to fund our ongoing needs. With an ageing population it is therefore the case that we will inevitably observe different saving and consumption patterns, with the authors noting that because elderly consumers tend to spend more money in areas with low levels of price growth (eg food), we are likely to see lower levels of inflation in the broader economy. Indeed, when we consider patterns of household wealth there are important trends to explore. The ONS collects data on household wealth, and Figure 7.7 shows average wealth broken down by constituent components, for various age categories.

As would be expected, Figure 7.7 shows that those in younger age categories have substantially lower levels of wealth, but figures rise rapidly to a peak for those in the 55 to 64 age category, particularly in terms of accumulated levels of pension wealth. Interestingly, net financial wealth peaks for those in the 45 to 54 age group, but what we see in the chart is that older individuals typically have substantial levels of accumulated wealth. Indeed, we know that the baby boom generation, which is now starting to retire in increasing numbers, is especially well off in comparison to other age groups. Hence the impending rise of what commentators label the 'silver economy' and the 'grey pound'. Whilst we must recognise that there are substantial levels of inequality in many advanced societies, and that pension poverty is a real issue for many, the key point to grasp here is that as the typically very wealthy baby boom generation moves through into retirement, the significance of older consumers will grow. Indeed, discussion of the 'grey pound' is increasing within academic circles, an example of this being Omar et al's (2014) discussion of the food shopping habits of older consumers. They argue that, whilst organisations have overlooked older consumers in the past, there is a growing recognition that older consumers represent a lucrative market. As a result of this, organisations need to adapt their offering, and indeed retail spaces, to attract older consumers. One sector that has boomed in recent times as a result of the rising influence of the 'silver economy' and 'grey pound' has been the cruise industry, as illustrated in the following case study.

CASE STUDY 7.2

The impact of the grey pound: cruise holidays

The cruise industry is an excellent example of the product market impacts of the so-called 'grey pound'. Whilst cruise ship operators have existed for a substantial period of time, some tracing their roots back to the 1800s, what we know as the modern cruise industry gained traction in the 1960s and 1970s with the founding of companies such as Royal Caribbean International and Carnival Cruises. Since that time, ships have grown ever larger, with increasing ranges of facilities and experiences for guests. As of 2020 the largest of these ships, Royal Caribbean's *Symphony of the Seas*, weighs in at 228,081 tonnes, measures 1,184 ft in length, and can carry a maximum of 6,680 passengers alongside 2,200 crew. To put this in context, this is approximately the same number of people living in the town of Glastonbury in Somerset.

The cruise 'product' is essentially very similar no matter the chosen provider, or desired destination, save for some more niche elements of the market that cater for more adventurous passengers (see, for example, the expedition cruises operated by the French company Ponant: https://uk.ponant.com). Ultimately, the bulk of the industry revolves around a well-established formula that includes the efficient loading of passengers and stores, transport of these individuals to a series of destinations, the provision of on-board dining and leisure experiences, before returning the passengers to a designated port at the end of the holiday. A passenger's ticket price will typically include their room and accommodation, meals in the main dining room and buffet, use of all 'hotel' facilities such as pools, gyms, bars and lounges, and on-board entertainment. Extra charges are, however, often made for items such as dining in speciality restaurants, organised shore excursions, casinos and alcoholic drinks.

As an excellent example of the spending power of the 'grey pound', the majority of cruise ship passengers tend to be from older generations, and tend to have passed retirement age. Indeed, Statista (2020) reports that in 2018 the average age of passengers embarking on cruises from the UK was 57. Whilst this is changing over time and cruise companies, notably lines such as Celebrity Cruises, are marketing specifically to younger generations, the typical cruise passenger stereotype persists. As has been previously noted in this section of the chapter, individuals are experiencing substantially greater levels of wealth and income in retirement than has been the case in previous times. With individuals in developed societies typically experiencing many more years of good health, and retirees expected to remain in good health for significantly longer than previously, these individuals are devoting an increasing proportion of their income to leisure and active pursuits. The cruise industry provides a generally safe and convenient way for individuals to visit a range of destinations, with passengers reporting that they value the ease of going on holiday via a cruise (eg unpacking only once), the fact that dining, bars, leisure and hotel facilities are all contained under one roof, and that cruising typically represents value for money, with the ships providing a very comfortable means of transportation indeed.

The popularisation of the cruise industry demonstrates to us the power and scale of the 'grey pound'. Whilst old age was typically a time of poverty and hardship for previous generations, the current older population (veterans and increasingly those from the baby boom generation) has experienced unprecedented levels of property wealth and income, with savvy companies developing distinct product offerings to cater to these new consumers.

The cruise industry does indeed represent an excellent example of the consumption power of older consumers. More generally, we know that the brands we purchase tend to vary with age. Whilst all age groups go on holiday, different age groups are attracted by the 'Saga' and 'Contiki' brands, and the same is true in the cruise industry with different brands attracting different individuals. What we do know about consumption more generally (and not specifically linked to holidays or cruising in particular) is that as we become older we tend to develop stronger brand loyalties and take more and more persuasion to switch from products that we know and like. Older consumers tend to distrust new brands and products. Advertisers thus put substantial energy into gaining the custom of younger people, 'hooking' them for the longer term. Many organisations are investing substantial effort in order to understand how changing demographics will affect them into the future. We are seeing the launch of different products and services by many companies in many different sectors, and we will likely see more products and services specifically marketed to older people in the coming years and decades.

Conclusion

This chapter has highlighted a number of the key demographic trends at both the global level and that of the UK specifically. At the broadest level, the human population of the planet continues to grow, and whilst there are questions surrounding the sustainability of population growth over time, there are many more humans on the planet than at any other time in history. Similarly, we must recognise that population growth is not a homogeneous trend across the world. Different countries and regions of the world have different trends, and in the first section of this chapter it was noted that the majority of current population growth is arising in the African continent. Indeed, some parts of the world are experiencing, or about to experience, declines in their population, notably countries such as South Korea, and many European countries such as Spain, Greece and Portugal. The UK is not immune from these trends, and this chapter has highlighted that it is currently experiencing population growth from two distinct sources: natural change and international migration. Data collected by the ONS indicates that the UK has experienced something of a mini baby boom in recent years, which, coupled with declining death rates and increases in international migration, means that the population continues to rise. ONS estimates predict that the UK's population will reach 72.4 million in the mid-2040s. Increasing longevity in the UK, as is the case with many other advanced economies, means that we are seeing an ageing population. ONS statistics point to substantial increases in the old age dependency ratio and the growing elderly population has implications for both the state and organisations.

At a societal level, a growing elderly population means that there are questions surrounding how healthcare and pensions are provided for. With proportionately fewer individuals of working age contributing to the tax base from which these and other services are paid for, there are debates surrounding how the UK, and other societies experiencing ageing populations, will ensure that important societal-level provision will remain properly funded. At an organisational level, an ageing society poses questions for organisations, both in terms of where they will source labour from and how they will adapt to changing market demands. It is very likely the case that organisations will have to adapt their HRM policy frameworks in order to recruit and retain individuals from previously marginalised sectors of the labour market. Likewise, there is growing recognition that older consumers have substantial purchasing power. Understanding that older consumers represent a lucrative market, we are seeing many organisations expand or alter their product or service offering in order to appeal to different demographic groups. It will be interesting to follow how these trends, and others, develop over time.

References

B&Q (2020) About B&Q, www.diy.com/corporate/about (archived at https://perma.cc/Q9XQ-E3BL)

Bal, P M and De Lange, A H (2015) From flexibility human resource management to employee engagement and perceived job performance across the lifespan: A multisample study, *Journal of Occupational and Organizational Psychology*, 88 (1), 126–54

Binder, S, Holdahl, E, Trinh, L and Smith, J H (2020) Humanity's fundamental environmental limits, *Human Ecology*, 48 (2), 235–44

Boubtane, E, Dumont, J C and Rault, C (2015) Immigration and economic growth in the OECD countries 1986–2006, CESifo Working Paper Series No. 5392, https://ssrn.com/abstract=2622005 (archived at https://perma.cc/S4QH-QKAB)

Butler, S (2020) Never too old to work: Managing an age diverse workforce, *Strategic HR Review*, 19 (3), 117–22

Faith, B (2019) *Encyclopedia of Ecology*, 2nd edn, Elsevier, Oxford

Gill, R (2016) Chasing the 'grey pound', *Travel Trade Gazette UK and Ireland*, 28 January, 18

Hauer, M E and Schmertmann, C P (2020) Population pyramids yield accurate estimates of total fertility rates, *Demography*, 57 (1), 221–41

Hendrixson, A and Hartmann, B (2019) Threats and burdens: Challenging scarcity-driven narratives of 'overpopulation', *Geoforum*, 101, 250–59

Hopfenberg, R (2003) Human carrying capacity is determined by food availability, *Population and Environment*, 25 (2), 109–17

Hou, J W (2019) Unorthodox proposals for China's extreme aging population, *Chinese Economy*, 52 (6), 505–26

Hui, C (2006) Carrying capacity, population equilibrium, and environment's maximal load, *Ecological Modelling*, 192 (1–2), 317–20

Lawson, T (2016) How the ageing population contributes to UK economic activity: A microsimulation analysis, *Scottish Journal of Political Economy*, 63 (5), 497–518

(Continued)

(Continued)

Max, S (2018) 5 big demographic trends to bet on for the coming decade, *Money*, 47 (1), 78–83

Ng, T W H and Feldman, D C (2012) Evaluating six common stereotypes about older workers with meta-analytical data, *Personnel Psychology*, 65 (4), 821–58

Oliveira, E and Cabral-Cardoso, C (2017) Older workers' representation and age-based stereotype threats in the workplace, *Journal of Managerial Psychology*, 32 (3), 254–68

Omar, M, Tjandra, N C and Ensor, J (2014) Retailing to the 'grey pound': Understanding the food shopping habits and preferences of consumers over 50 in Scotland, *Journal of Retailing and Consumer Services*, 21 (5), 753–63

ONS (2019a) National population projections: 2018-based, www.ons.gov.uk/peoplepopulationandcommunity/populationandmigration/populationprojections/bulletins/nationalpopulationprojections/2018based (archived at https://perma.cc/SS8W-ASL8)

ONS (2019b) Overview of the UK population: August 2019, www.ons.gov.uk/peoplepopulationandcommunity/populationandmigration/populationestimates/articles/overviewoftheukpopulation/august2019 (archived at https://perma.cc/WVE9-9WRK)

ONS (2019c) Living longer and old-age dependency: What does the future hold?, www.ons.gov.uk/peoplepopulationandcommunity/birthsdeathsand-marriages/ageing/articles/li vinglongerandoldagede-pendencywhatdoesthefuturehold/2019-06-24 (archived at https://perma.cc/UB9A-C6H9)

ONS (2019d) Total wealth in Great Britain: April 2016 to March 2018, www.ons.gov.uk/peoplepopulationandcommunity/personalandhouseholdfinances/incomeandwealth/bulletins/totalwealthingreatbritain/april2016tomarch2018#trends-in-average-wealth-in-great-britain (archived at https://perma.cc/ZP3Y-ANGT)

ONS (2020a) Population estimates for the UK, England and Wales, Scotland and Northern Ireland, provisional: Mid-2019, www.ons.gov.uk/peoplepopulationandcommunity/populationandmigration/populationestimates/bulletins/annualmidyearpopulationestimates/mid2019estimates (archived at https://perma.cc/L7AR-M7TU)

ONS (2020b) Migration statistics quarterly report: August 2020, www.ons.gov.uk/peoplepopulationandcommunity/populationandmigration/internationalmigration/bulletins/migrationstatisticsquarterlyreport/august2020 (archived at https://perma.cc/84CA-TTQV)

Our World in Data (2020a) World population by region projected to 2100, https://ourworldindata.org/grapher/historical-and-projected-population-by-region (archived at https://perma.cc/7TBT-N4QA)

Our World in Data (2020b) Absolute increase in global population per year, https://ourworldindata.org/grapher/absolute-increase-global-population (archived at https://perma.cc/KVN6-GVRE)

Papapetrou, E and Tsalaporta, P (2020) The impact of population aging in rich countries: What's the future?, *Journal of Policy Modeling*, 42 (1), 77–95

Portes, J and Forte, G (2017) The economic impact of Brexit-induced reductions in migration, *Oxford Review of Economic Policy*, 33 (S1), S31–S44

Remoff, H (2016) Malthus, Darwin, and the descent of economics, *American Journal of Economics and Sociology*, 75 (4), 862–903

Şerban, A C and Aceleanu, M I (2015) Current demographic trends: A new challenge for the labour market, *Theoretical and Applied Economics*, 4 (605), 309–20

Smith, P et al (2020) Which practices co-deliver food security, climate change mitigation and adaptation, and combat land degradation and desertification?, *Global Change Biology*, 26 (3), 1532–75

Statista (2020) Average age of ocean cruise passengers in the United Kingdom (UK) from 2005 to 2018, www.statista.com/statistics/586704/age-ocean-cruises-share-uk-united-kingdom (archived at https://perma.cc/Q3WX-4PGV)

Tyrrell, N, Sime, D, Kelly, C and McMellon, C (2019) Belonging in Brexit Britain: Central and Eastern European 1.5 generation young people's experiences, *Population Space and Place*, 25 (1), 1–10

(Continued)

(Continued)

United Nations (2017) News: World population projected to reach 9.8 billion in 2050 and 11.2 billion in 2100, www.un.org/development/desa/en/news/population/world-population-prospects-2017.html (archived at https://perma.cc/LC2K-4D5S)

United Nations (2020) Migration, www.un.org/en/sections/issues-depth/migration/index.html (archived at https://perma.cc/4MSS-SQ96)

Vlachantoni, A, Feng, Z, Evandrou, M and Falkingham, J (2015) Ethnicity and occupational pension membership in the UK, *Social Policy and Administration*, 49 (7), 801–23

Wadsworth, J (2018) Off EU go? Brexit, the UK labour market and immigration, *Fiscal Studies*, 39 (4), 625–49

World Bank (2020) Fertility rates, total: Births per woman, https://data.worldbank.org/indicator/SP.DYN.TFRT.IN (archived at https://perma.cc/D7WA-3VNN)

World Health Organization (2016) Global Health Observatory (GHO) data: Life expectancy, www.who.int/gho/mortality_burden_disease/life_tables/situation_trends_text/en/ (archived at https://perma.cc/524R-6TF6)

08
Society

It is impossible in one chapter to do full justice to the many different social trends that are shaping our collective lives and having a major impact on the worlds of work, employment and the management of people in organisations. We are living through a period of profound social change and changes in social attitudes that will continue to have a major impact in all kinds of areas over the coming decades.

It is, however, useful to explore some of the more important developments and to discuss their origin and likely future impact. In this chapter we will be focusing on our increased affluence as a society, at our failure to become more economically equal, at the rise in concern for issues relating to ethics and the environment and our tendency over time to become a more individualistic society. These are major issues that are frequently contested and which several books could easily be devoted to. Here we are going to have to discuss each quite briefly and will be focusing in particular on the implications of social trends for the management of people in employing organisations. This is not, however, a one-way street. Employment practices are very much a cause of some significant contemporary social trends. So we will also in this chapter be looking at how the experience of being a worker in the twenty-first century is helping to shape wider societal changes.

Affluence

There is no general agreement about which of the major, long-term, current social trends is the most significant. The rise of an affluent society, however, must surely be a strong candidate when judged in terms of the profound changes that it has brought about in social attitudes, the social structure, people's behaviour, their ambitions and self-perception.

Affluence is a controversial subject area in sociology. This is partly because of disagreement about its extent and partly because of disagreement about its long-term consequences. As a rule, sociologists have tended to show more interest in the impact of the increasing inequality that has accompanied the rise of affluence than about the impact of increasing affluence itself.

Affluence is usually defined as equating to the standard of living enjoyed by a society or the extent to which people are able to 'live well'. It is associated with the expression 'disposable income' and the evolution of societies in which a good proportion of the population forms a 'middle class' which has income left over after providing for its basic needs.

The term thus refers to the amount of income remaining after essential needs – housing, heating, water, food, transport to work, basic medicines, etc – have been met. The term 'threshold of need' is sometimes used in this context, an affluent household being defined as one that has sufficient resources available to spend on goods and services that are discretionary in nature and provide pleasure or enhance social status.

Affluence is not the same thing as income. Income can be rising, but if this is accompanied by faster rises in the prices of essential goods, the level of affluence falls. While writers tend to refer to whole societies as being affluent to a greater or lesser degree, there is of course vast disparity across societies. Some people are much more affluent than others, and the direction of travel is not towards greater affluence for everyone all of the time. In recent years, thanks to the financial crisis of 2008 and the subsequent recessionary years, followed by the more recent economic damage caused by the Covid-19 pandemic, many countries have seen stagnating and falling income levels.

However, it is important not to let these experiences obscure the fact that for most people, in most economies, the past few decades have seen increasing affluence as they have industrialised, urbanised and generated more wealth. In developing economies the extent of progress over the past four decades has been remarkable, but all over the world standards of living have on average improved very substantially. We are living longer, healthier, more fulfilled lives than people did twenty years ago, let alone a century ago.

Evidence of increasing affluence comes from various types of measures:

- GDP per capita.
- Rises in income vis-à-vis rises in prices.
- Rises in discretionary expenditure on 'luxury' or 'non-essential' goods and services.

According to each of these measures, the UK population, along with the big majority of others across the world, has become considerably more affluent over recent decades.

The most common measure, if a crude one, of a country's living standard is its GDP per capita – its national income divided by its population. The UK's gross domestic product (GDP) was £2.21 trillion in 2019. This gives a per capita figure of around £32,600 when divided by the total population.

UK GDP increased on average by 2.6 per cent a year throughout the twentieth century, far exceeding the rise in population. This meant that we were over six times wealthier in real terms at the end of the century than we were at the beginning. National wealth increased particularly sharply between the mid-1980s and 2008. GDP per capita rose by 108 per cent between 1990 and 2005. After 2008 it stagnated and fell back, recovering its pre-recession level in May 2015. Despite this, real wage levels have still not quite caught up and suffered a further hit in 2020 during the Covid-19 pandemic.

Increased affluence is also reflected in the statistics for annual earnings for people who are working. In the UK, average gross earnings at the end of 2019 were £30,420 for full-time workers. This compares to £12,500 in 1990 and just £3,000 in 1975. These figures mean very little unless the cost of living is also taken into account. The annual rate of inflation since 1990 has averaged 2.5 per cent a year, meaning that prices rose by around 50 per cent. But in the past thirty years average earnings and GDP per head in the UK increased at around twice the rate of inflation. Real terms living standards, on average, thus rose considerably.

The Office for National Statistics also makes estimates for 'real gross disposable income per head' – removing tax from the total gross figure. These figures do not take account of housing costs, but they nonetheless give a good indication of the extent to which affluence has grown in recent decades. They show that there were two periods in the twentieth century when living standards fell somewhat (the 1930s and the 1970s) and that the same happened again between 2008 and 2012. Overall, however, the trend has been upwards over the longer term.

If these long-term trends continue, we can thus look forward to increased levels of affluence in the future of a similar scale to the improvements we experienced in the second half of the twentieth century. Recessions will continue to occur periodically, but if long-term trends continue the overall pattern points to substantially increased levels of affluence by the middle of this century.

In the developing world, contemporary increases in affluence are of a far greater magnitude than they are in the industrialised countries like the UK. The extent of the change since the late 1980s/early 1990s is set out by Steven Radelet (2015) in his book *The Great Surge: The ascent of the developing world*. Here, drawing on figures from the World Bank and the International Labour Organization (ILO), he demonstrates that some extraordinary and very welcome changes have occurred in less than thirty years.

The most widely accepted definition of absolute poverty that was until recently used in all international statistics was an individual income of $1.25 or less – a basic subsistence level in most countries of the world – just enough to keep you alive. In 1993 two billion people were estimated to be living on less than $1.25 a day. The figure went below 1 billion in 2015. In percentage terms the fall is as follows:

1981: 53 per cent
1993: 43 per cent
2011: 17 per cent
2016: 10 per cent

Put another way, this means that since the 1980s over 50 million people each year have been lifted out of extreme poverty. In more recent years the figure has been 70 million a year. Sixty per cent of this progress was accounted for by China, where 838 million people (84 per cent) lived on the equivalent of less than $1.25 in 1981. That went down to 646 million (55 per cent) by 1993 and then to 84 million (6 per cent) in 2011. It is now fewer than 17 million. Extreme poverty is now on target to be eradicated in China in the next few years. This whole process will have taken less than 40 years.

An income of $1.90 a day is now more commonly used to define extreme poverty. On this measure, too, the proportion has now fallen below 10 per cent.

At the other end of the scale are the figures for people living on more than $5.50 a day. In 1990 the World Bank estimated that only a third of the world's population met this threshold. By 2015 it was 54 per cent, and further progress has been made since then. This is happening because of industrial development and the accompanying expenditure that it makes possible on health, education, welfare, sanitation, infrastructure and housing on the part of government. GDP per head has been increasing at a rate of 3 per cent per year in developing countries since 1995, giving a total real terms increase of 70 per cent between 1995 and 2013. There has also been a 34 per cent increase in the number of people working in developing countries since the 1990s – that is 700 million new jobs. Wages have increased by 60 per cent in real terms.

PAUSE AND REFLECT

Numerous studies show that increased affluence, for most people, does not lead to increased levels of life satisfaction. Most of us have got richer over recent decades, but we are no more happy. Why do you think this is?

The social implications of affluence

Academic studies on the present and likely future impact of increased affluence tend to focus on two distinct areas:

- Its impact on how we perceive ourselves.
- Its impact on how we perceive society.

The first largely focuses on individual behaviour, the second on changing social attitudes. Both are controversial areas. As is frequently the case, the main debate is not about whether changes are being brought about, but about the extent to which they signal fundamental sociological shifts.

In terms of personal identity, the claim that is most commonly made is that we are moving steadily, as a result of increased affluence, from a society whose members' identity is rooted in occupation and social class, to one that defines itself by consumption. Paul Ransome (2005) sums this argument up in his book on affluence and consumption. He argues that the way we see ourselves (our personal identity) and the way that we perceive one another are less to do with how we earn our living than used to be the case. Instead, occupational identity is being replaced with consumer identity. This is happening, he argues, because it is no longer just a minority (of upper- and middle-class people) who are able to 'consume' large amounts of non-essential products and services. As a result, traditional class and gender identities are breaking down.

Affluence gives people the choice to create their own social identity to a far greater extent. As a result, we define others (and ourselves) by what they spend their money on – the types of clothes they wear, music they listen to, sports they are involved in or the types of holiday they prefer.

A rather different view is expressed by Lars Svendsen (2008) in his writing on work in the twenty-first century and also by sociologists such as Bauman (2005), du Gay (1995) and Besen-Cassino (2014). They argue that, far from being replaced in terms of their significance to us by our consumer lives, our working lives are increasingly being subsumed into our lives as consumers. We now expect our jobs to be pleasurable, fulfilling and meaningful as well as an opportunity to earn money. In other words, we are witnessing the consumerisation of working life.

We thus increasingly tend to make decisions about which job to apply for and to remain in a similar way to that in which we make decisions about what car to buy or what holidays we go on. We 'shop around' for new jobs when we are unfulfilled in our existing ones. This is a highly significant development for HRM as it suggests that employers will increasingly be required to 'sell' the jobs to job seekers if they are to recruit and retain good people.

A second debate relates to shifting social attitudes, and is often associated with the work of the American sociologist Ronald Inglehart, who argued that the combination of affluence and the absence of war in industrialised countries are steadily creating a revolution in our value systems. He coined the term 'postmodern' to describe these new evolving social norms, arguing that changes spring from the way that the current generation, unlike their forebears, can take both economic and physical security for granted. The argument has been developed and popularised by Brink Lindsey in his book called *The Age of Abundance* (2008). Drawing also on the ideas of J K Galbraith (in *The Affluent Society*, 1958), he talks about the huge significance in human history of 'the conquest of scarcity' in western countries during the late twentieth century:

> For all the preceding millennia, physical survival stood front and centre as the overriding problem that most people had to confront, day in and day out, for all their lives... In this state of affairs, choices were relegated to the margins of life. What to do, how to contribute to the great social enterprise, where to live, whom to associate with, how to spend one's resources – these were questions that, for most people, required little thought, since the options were few and the relative desirability of this or that alternative was usually obvious.

As a result of this 'new' state of affairs, we increasingly focus our ambitions on achieving a higher quality of life. This involves resisting cultural and social constraints such as family duty, religious observance, and deference towards people in positions of authority and towards the state generally. A premium is now placed on individual fulfilment, which has replaced the concept of 'duty' as the guiding force in our lives. The prevailing attitude is one of 'live and let live'.

These trends – directly attributable to increased affluence – are associated in the work of many contemporary social theorists with the steady decline of traditional class-based systems in which people had a fixed view of their position in a social hierarchy and tended only to mix with people of a similar social background. Money now tends to determine social position much more than birth, manners, tastes and background, although they still remain bound up together to an extent. There are significant potential implications for the management of people here, too. Respect for and trust of organisations and managers are less and less likely to be taken as given. Instead they have to be earned, and potentially re-earned again and again. The implication is that organisations will have to be managed subtly, more democratically and with greater levels of emotional intelligence than has been the case in the past.

CASE STUDY 8.1

Ethical consumerism

Each year between 1999 and 2012 the Co-operative Bank published its *Ethical Consumerism Report*. This tracked the growth of ethical consumerism in the UK, as well as ethical behaviour more generally. It demonstrated substantial year-on-year increases over the first decade of the twenty-first century in the proportion of the UK population who claim to have acted ethically in one of a number of specified ways 'at least once' during the year in question:

	1999	2009	2012
Recycling	73%	96%	92%
Supporting local shops/suppliers	61%	87%	74%
Avoided a product/service due to ethical reputation	44%	64%	50%
Chose a product/ service due to ethical reputation	51%	60%	51%
Bought primarily for ethical reasons	29%	52%	42%
Felt guilty about an unethical purchase	17%	43%	31%
Sought information on a company's reputation	24%	38%	33%
Actively campaigned on environmental/ social issues	15%	26%	24%

Interestingly, many of the 'gains' accrued during these years were lost again following the recession and subsequent period of sluggish economic growth. This demonstrates the significance of the link between ethical consumerism and affluence. It is costly to shop ethically.

These reports also tracked year-on-year growth in purchases of organic, fair trade and ecologically sustainable products, the extent of charitable donations, ethical consumer boycotts, the use of public/private transport and usage of ethical banking and investment products.

Annual growth rates in most of the categories were between 13 per cent and 18 per cent until 2009, particularly big increases being recorded in the purchase of fair trade items, energy-efficient light bulbs, rechargeable batteries and hybrid eco-friendly cars. Here too upward trends halted following the onset of austerity, but there was little decline. There is also evidence to suggest that consumer boycotts of products for ethical reasons are increasing.

In recent years, the ethical consumer reports published by the Co-op have focused more on the amount of money estimated to be spent each year by consumers on products or services that can be characterised as having an ethical character. In 2019, twenty years after the first report was published, the following figures were included:

- Between 1999 and 2019 the amount spent on fair trade and organic food and drink products increased from £1 billion to £12 billion. This reflects major changes in spending patterns as consumers have switched to free range eggs (now 63 per cent of all sales), plant-based alternatives to common foods and sustainably farmed fish.

- Annual spending on green products for the home has increased from £1 billion in 1999 to £10 billion in 2019. This includes items such as energy-efficient boilers, solar panels and ethically sourced cleaning products. There has also been a very significant increase in the spend on green energy tariff options.
- Spending on vintage, second-hand clothing increased from £133 million in 1999 to £732 million in 2019.
- Spending on bicycles, electric vehicles and fuel-efficient cars increased from £301 million in 1999 to £5.8 billion in 2019.

The market for ethical products remains relatively small, but the long-term direction of travel is very much in their favour. We can expect the upward trajectory of all these measures over time, slowing in harsh economic times and growing again when people have more disposable income.

Rising inequality

The benefits of affluence, while welcome, are very unevenly spread. In the UK, levels of inequality have increased substantially over the past twenty years as incomes have risen faster among higher-income earners than among the lower-income earners. People who were already relatively well off have gained much more from the growth in affluence than those further down the income scale. Forty per cent of all spending in the UK is now accounted for by the wealthiest 5 per cent of households.

A widely recognised measure of inequality that allows trends over time to be tracked is the Gini coefficient. It captures in a simple manner the extent to which incomes are equally distributed. It operates on a scale of 0 to 1. A figure of 1 would indicate that all the income in the country concerned was earned by a single individual (as unequal an income distribution as is possible). A figure of 0 would indicate total equality of earnings between everyone. Over the past forty years the UK Gini coefficient has shown a considerable rise, indicating that we are becoming less equal over time. There was a dip in the mid-1990s as a result of the property price crash, and again after 2008 and the period of austerity that followed, but the long-term trend is very much upwards:

1977: .24 1996: .33
1981: .25 2001: .34
1986: .28 2006: .36
1991: .34 2019: .35

The same is true in most countries, leading some to argue that increasing inequality is inevitable in post-industrial societies where a premium is paid to highly skilled workers. In other words, rising inequality is a price that we pay for rising affluence.

These trends mean that more people each year in the UK are living in 'relative poverty', a commonly used measure defined as having an income that is less than 60 per cent of median household income. In 2019 this stood at 14 million people – slightly lower in percentage terms than the peak in the 1990s, but considerably higher than was the case forty years ago.

The Institute for Fiscal Studies defines 'absolute poverty' as having an income below 60 per cent of median earnings as measured ten years ago. According to this measure there are around 9.5 million people who live 'below the poverty line' in the UK. This group is sometimes described as comprising an 'underclass'. They are heavily concentrated in urban areas.

Another measure that is sometimes used is 'deep poverty', which is the proportion of people living on less than 50 per cent of median earnings – around £100 a week after housing costs have been taken into account. Around 4.3 million people in Britain were in this category in 2019, most of whom had lived persistently at this level for at least two years.

A particularly pronounced trend across the world is the growth that we are seeing in the wealth of 'the super-rich' – generally defined as comprising the wealthiest 1 per cent in a society. In recent years these people have seen their incomes soar away ahead of those of the rest of the population. Indeed, according to some economists, the disparity in wealth between the top 1 per cent and the rest of us accounts for all rises in Gini coefficient scores in recent decades. In other words, the bottom 99 per cent have actually been getting more equal. The big social development is the emergence of this super-rich elite.

Among OECD countries, the UK's super-rich do particularly well, taking 13 per cent of the total income (after tax). In the USA the top 1 per cent earn 20 per cent of the income, but everywhere else is less unequal according to this measure. In the Scandinavian countries the top 1 per cent account for only 6–8 per cent of the total income.

A great deal has been written recently about rising inequality, its causes and its consequences. We have already discussed one cause in this book, in Chapter 6, namely the increasing disparity in our labour markets between the high-quality, secure and well-paid 'Macjobs' and the low-paid, poor-quality, insecure 'McJobs'. This is a significant contributing factor, although it is less effective at explaining the rise in the proportion of wealth held by the super-rich top 1 per cent.

The most widely cited explanation is that set out by French economist Thomas Picketty (2013) in his well-known book *Capital in the Twenty-First Century*. His main point is simple: namely, that over the past thirty or forty years a situation has arisen in which the value of capital has increased at a much faster rate than the level of wages. This means that people who have property and investments have seen their wealth increase at a considerably faster rate than those who rely largely (if not entirely) on wages to live. The more unearned income someone has, the higher their relative position in terms of wealth has become.

However, as Atkinson (2015) demonstrates, in recent years increasing amounts of wealthy people's income has been derived from salaries rather than investments. This is particularly true in the UK where two-thirds of the highest earners are also members of the top 1 per cent – mainly bankers, top executives, financiers, elite sportsmen and 'celebrities' working in the entertainment industry. These people have all seen vast increases in their pay over recent years – far in excess of rises in average earnings.

In 2020 the average chief executive of a FTSE 100 company (the biggest 100 companies in the UK) earned £29,559 in less than three days. That was a sum equivalent to the average annual earnings of everyone in the country. There are several explanations that are commonly advanced:

- Top earners are 'elite performers' with rare abilities that place them in very high demand internationally. In a free market they are thus able to command very substantial salaries. The example that is often given to demonstrate this is Steve Jobs. When he resigned due to terminal ill health $18 billion was wiped off the value of Apple the next day as shares plummeted in value. By contrast, when Steve Ballmer resigned as CEO of Microsoft in 2014, the company immediately gained $20 million in value.
- In recent years there has been a shift away from paying top executives and bankers in cash and towards paying them via deferred payments of various kinds – often share options. This means that their income is heavily geared towards the financial performance of their companies. The result is that senior people in organisations are able to make themselves extremely wealthy if they can help engineer a strong financial performance for shareholders.
- 'Top people' (members of the super-rich class) effectively award one another high pay rises, thanks to a semi-corrupt system whereby they sit on one another's boards as non-executive directors. Senior pay in big public corporations is typically determined by remuneration committees that are dominated by non-executive directors who are disinclined to pay less than a good market rate. In private companies, owners can pay themselves what they want to.
- Some argue that the tendency for governments to reduce higher rates of personal taxation has also had both a direct and an indirect effect. This clearly has tended to put more money into people's after-tax pay packets, but indirectly it is claimed that it has caused people to focus more on how much money they are earning than on the challenge of doing a good job. Others stress the role played by accountants who make a good living helping wealthy clients to avoid paying tax in the first place by designing tax-efficient investment vehicles and locating assets offshore in tax havens.
- Large corporations are much bigger than they used to be. They operate in more countries, own more assets, employ more people, meet the needs of more consumers and make more by way of return on capital. As they grow, their salary bills grow too. There is thus a potential direct link between, on the one hand, globalisation, deregulation and privatisation, and, on the other, greater inequality. These have allowed organisations to grow rapidly, turning themselves into multinationals in the process.
- Globalisation and the technologies that support it may also offer some explanation as to why some individuals earn vast amounts of money. It is because these developments make it easier to achieve global (rather than national) celebrity status. J K Rowling, for example, is estimated to earn around $300 million a year – the vast majority of it from outside the UK. This would have been impossible had she been writing her books thirty years ago when there was much less opportunity to sell children's books internationally. The same is true of all celebrities who have an international profile. Leading footballers and film stars earn £10 million a year at the peak of their careers, partly from their day jobs, but also from advertising, voice-overs, endorsements, newspaper columns, reality TV shows and media appearances of one kind or another. The bigger the 'international name recognition', the more money an 'A list celebrity' can earn.

- Inequality of opportunity is growing alongside inequality of outcome. If you are born into privilege, you have a much better chance of making yourself wealthier than your parents were than if you come from an ordinary background. This occurs, according to theorists who are influenced by the leading sociologist Pierre Bourdieu (1930–2002), because in addition to economic capital, wealthier people are able to provide their children with access to social capital (contacts) and cultural capital (education and 'taste') as well as exposure to a lifestyle that breeds self-confidence. The result is a situation whereby elite families pass on their status to the next generation, making social mobility harder for those born in less privileged circumstances to achieve. As a result, the rich become richer still over time.

PAUSE AND REFLECT

In his book *The Great Leveler: Violence and the history of inequality from the Stone Age to the twenty-first century,* Walter Scheidel (2017) argues that throughout human history inequality has always increased in periods of economic stability. Big social disruptions like wars, revolutions, plagues and economic collapses are 'great levellers', but on balance do much more harm than good. What is your view of this argument?

The impact of inequality

However much statistics demonstrate that real living standards are increasing generally in countries such as the UK and that absolute poverty is therefore in decline, it is widely agreed that increasing inequality has long-term negative social consequences – many of which have an impact on the world of employment. The case for this proposition was developed in the books by Richard Wilkinson and Kate Pickett entitled *The Spirit Level: Why more equal societies almost always do better* (2009) and *The Inner Level: How more equal societies reduce stress, restore sanity and improve everyone's well-being* (2018). Here, a vast amount of international data comparing the OECD countries and different US states is explored, leading the authors to conclude that in industrialised countries health, life expectancy and the prevalence of social problems are not associated statistically with average income levels. They are, however, clearly associated with levels of inequality.

Not all social problems are caused by social inequality, and other factors invariably play a role, but it is nonetheless important to be aware of the contribution made by inequality and hence of the social costs we pay for having a society that has achieved affluence at the expense of equality.

The major examples that are commonly cited are:

- Increased use of drugs and alcohol.
- Family breakdown.
- Social and political alienation.

Other recent contributions to this debate include work by Michael Marmot (2015) and Angus Deaton (2015). Marmot shows how, across the world, average life expectancy is clearly linked to social and economic status. The higher up the social scale someone is – mainly judged by income, but other measures of status too – the more healthy and long their life is likely to be. He argues that this is due to two important factors:

- **Autonomy**
  Higher-status individuals tend to have a great deal of autonomy in their lives. Their disposable income gives them choices about where and how to live. This brings satisfaction. While their jobs may also be stressful, they generally enjoy them because they have control over what they do and how and when they do it. Irrespective of income, their jobs are enjoyable and fulfilling.

  By contrast, the lower one's social status, the less autonomy and choice one has over home life and work. Jobs are often mundane, repetitive and subject to a great deal of management control. The term 'soul-destroying work' is often used to describe these jobs. They are also, of course, less secure. Lower-income families are also obliged to live in smaller homes in areas of social deprivation (crime, drugs, anti-social neighbours).

  The result is that the higher someone's status, the less likely they are to suffer poor mental health (depression, persistent stress, poor relationships) – and it is increasingly appreciated that mental health problems lead on to physical health problems.

- **Social integration**
  High status and income tend to be associated with large social networks that are available to support people emotionally as well as materially. Their lives are richer, they get to hear about opportunities, there is help available to them when things go badly for them – in short, they are likely to have richer, more fulfilled social lives.

  Marmot also uses statistics from Japan where there is a tradition of lifetime employment in bigger firms, with promotion up management hierarchies coming from inside rather than outside the organisation. In fact, this type of approach is common in many workplaces around the world where 'relationship cultures' predominate. The biggest beneficiaries tend to be the more educated and wealthy people who are qualified to work in the larger organisations. Relationships within these workplaces thus tend to be closer, more co-dependent and characterised by much higher levels of trust than is typical in western industrialised countries. The result is greater security, higher levels of social support, less insecurity and better health too.

  Social problems of this kind create difficulties for employing organisations because they lead to ill health and to low levels of psychological wellbeing. There is thus a direct link to absence, limited employee engagement and sub-optimal performance. It is often the HR managers who are responsible for handling these issues, both at the individual and aggregate, organisational level.

 **PAUSE AND REFLECT**

How far is it true to argue that organisations both help to create social problems by bringing about greater inequality, and also end up having to devote a lot of time and effort to dealing with them?

Increased individualism

Alongside affluence, and linked to it in the eyes of many commentators, runs another major social trend of our time – namely the growth of individualistic attitudes and approaches to life, and, as a corollary, a decline in collectivist values and outlooks. There are different definitions. Here are two of the more useful:

> An individualistic society is one that is made up of people who see themselves as independent of collectives; are primarily motivated by their own preferences, needs, rights, and the contracts they have established with others; give priority to their personal goals over the goals of others; and emphasise rational analyses of the advantages and disadvantages to associating with others.
>
> (Trandis, 1995)

> An individualistic society is one in which the ties between individuals are loose and in which everyone is expected to look after him/herself and his/her immediate family.
>
> (Hofstede, 1997)

While there is some broad agreement about defining individualism, the extent of the trend away from a more collectivist outlook has never been satisfactorily measured in any objective way. Undoubtedly it varies considerably from country to country and within different communities, but how far we cannot know for certain. What we have here is thus a phenomenon that sociologists generally agree to be highly significant and to be transforming the nature of western societies, but one that remains quite vaguely defined and unmeasurable.

That said, there are some proxy measures that are commonly used – none wholly satisfactory – to demonstrate a substantial and steady decline in collectivism and a growth in individualism since the 1960s across most western, industrialised countries:

- Trade union membership.
- Church attendance.
- Membership of political parties.
- Participation in team sports.
- Solo living.

Twenge and Campbell (2008) drew on a database consisting of 1.4 million questionnaire responses, comparing the responses to a series of questions given by new American college students with those given by their predecessors when they were the same age (ie also starting college). They concluded that on a variety of measures very significant changes had occurred over the previous seventy years:

- Today's young people have far more self-esteem, being much more likely than their parents to be satisfied with themselves and to see themselves in a positive light.
- Current college students are a great deal more likely than past cohorts to exhibit narcissisms (self-belief taken to the point of arrogance), believing that they are 'special', that the world would be a better place if they ran it and that they can live their lives 'any way I want to'.
- There has been a striking decline since the 1950s in the extent to which young American adults need social approval. In other words, they are a great deal less concerned than previous generations were about what others think of them and are much less conformist in their outlook.
- Today's college students are much more likely to exhibit 'an external locus of control', meaning that they perceive themselves to have less influence over what happens in their own lives than former generations did. They thus tend to blame others when things go wrong, rather than accepting personal responsibility.
- There has been a fundamental shift in the traits exhibited by women. Whereas in the past female respondents to personality questionnaires were a good deal less assertive and somewhat less analytical than men, by the 1990s such differences had disappeared.

What can account for this apparent sea change in attitudes towards a more individualistic mindset? Not surprisingly there are many explanations advanced. While the leading sociologists and commentators agree that societies across the world are becoming more individualised, they tend to disagree about the precise causes. Here are some brief summaries of the arguments advanced by six influential thinkers who addressed their minds to the issue of rising individualism from different perspectives:

Ulrich Beck Beck was a German sociologist. He argued that individualism is partly caused by the way that modern labour markets oblige us to compete with one another all the time, and partly by the way that the modern welfare state has reduced our reliance on family groups and on ties with neighbours and communities. This has led, in his view, to a situation in which 'community is dissolved in the acid bath of competition'. The outcome is a situation in which social norms are weakened. People choose for themselves their own 'biographies', social identities and affiliations, whereas in the past these were determined for them to a far greater extent by the expectations of families, churches, communities and social classes.

Robert Putnam The American sociologist Robert Putnam has argued that a significant fundamental shift is occurring towards an individualistic approach to our lives. In his book *Bowling Alone* (2000) he argued that people are less and less engaged in community activities or interested in developments outside their own lives. Instead, we are becoming increasingly disconnected from family, friends, neighbours and civic society. Putnam argues that a range of factors explain increased individualism in the USA, but he gives particular attention to the role played by television and newer forms of information technology in reducing the amount of time we spend socialising with others and hence developing a collective identity.

Zygmunt Bauman Bauman welcomes individualism as the result of people breaking free from restrictive social bonds. His most celebrated book on this subject is *Liquid Love* (2003), in which he argues that in postmodern societies people tend to be more individualistic within their personal relationships as well as in their relationships with society more generally. 'Human bonds', according to Bauman, are becoming increasingly frail. He writes about the rise of 'top pocket relationships' that can be accessed easily when needed, but pushed out of sight when not needed. Human relationships are now more transient and less meaningful than they used to be. We have many more friends, acquaintances and partners in our lifetime, and these 'networks' are increasingly taking the place of deep, long-lasting relationships built on mutual love and reliance.

Anthony Giddens Giddens is a highly influential and prolific British sociologist. His major contribution to the debate about individualism is the identification of its causes. He argues that it is not mainly a result either of the social changes of the 1960s or of the market-based ideologies that first emerged in the 1980s. Instead he stresses the significance of education, and particularly higher education. The whole university experience – what is taught as well as the life of the student – leads to a questioning of assumptions and a loosening of social ties. Giddens also stresses the significance of globalisation and the World Wide Web in facilitating communication between people from different cultural backgrounds. This, he argues, leads us to question our own customs and to develop an individualistic 'live and let live' outlook.

Eric Hobsbawm Hobsbawm was a hugely respected British historian with a Marxist outlook. In a series of books that are highly critical of the turn UK society took in his lifetime, he made the observation that individualism is an entirely natural and inevitable consequence of increased affluence. He argues that financial deprivation leads people into collective action because it provides the best hope of alleviating their poverty. Once most people get past 'the threshold of need' they are much more likely to pursue private interests.

Robert Skidelsky Skidelsky's recent work on social change in the past fifty or sixty years has addressed the question of rising individualism with a tone of deep regret. While he accepts that many factors are involved, he gives particular prominence to the decline in religious observance and belief, which he perceives has led people to become more selfish and less caring/compassionate.

Other commentators focus to a greater extent upon workplace issues and thus 'lay the blame' to an extent on management practices. The Irish sociologists Collins and Boucher (2005), for example, argue that the transformation of workplace relationships and management practices has been particularly rapid and profound. Remuneration is increasingly negotiated individually and influenced by individual performance, we are increasingly made individually responsible for meeting personalised objectives by our managers and also are given more individual discretion about how we perform our own areas of work. Moreover, career trajectories are increasingly individualised too, each of us being expected to develop his/her own career rather than to follow an organisationally predetermined path.

Reflecting on the development of labour markets in the UK, McGovern et al (2007: 142–8) agree that HR management practices have contributed to individualisation. Their prime focus is on the evolution of competitive external labour markets in which organisations hire people at all levels directly from other employers, paying the going rate and cutting down on their own training costs. Such approaches force employees to compete with one another in order to progress their careers and earn more money. Previously, when organisations had well-developed internal labour markets and a preference for internal promotion, employees wanting to progress career-wise would only have to compete with fellow employees.

PAUSE AND REFLECT

Which of these explanations do you find most persuasive, and why?

How far do you think it is fair to blame/credit contemporary human resource management practices for helping to make our societies more individualistic?

The impact of increased individualism on management

What does this steady but apparently inexorable move towards greater individualism mean for the future of people management? It is difficult to know for certain, but interesting to speculate about.

Trade unions

Increased individualism is bad news for traditional trade unionism. It suggests that the now well-established decline in the proportion of workers who join unions is likely to continue in the future and that their influence will continue to wither. Unless they are able to reinvent themselves quite radically so that they are seen primarily to represent individual rather than collective interests, their significance in employment relations is highly likely to erode further.

There is a knock-on effect for HR practices that are union-focused or which assume that a recognised trade union speaks for employees in a genuine sense. The more individualistic the workforce becomes, the less it will be possible for any collective body to represent it effectively. Employees will increasingly expect to be consulted individually, to negotiate their own payment arrangements and to be able to organise their own work without the assistance or interference of any collective grouping, be that a union or the immediate team of colleagues with whom they work.

This provides major opportunities for managers but also some major threats. Managing employment relationships individually allows greater flexibility, permits the reward and recognition of outstanding individual performance and generally licenses managers to exercise greater levels of discretion. But it also requires greater sophistication and emotional intelligence if the result is to be a satisfied, committed, well-motivated and high-performing workforce. Maintaining high-trust employment relationships is often harder in a non-union setting. Managers have more opportunity to impose their will, but also more opportunity to do so in a manner that is perceived as unfair and which de-motivates and reduces trust as a result.

Loyalty to an employer

Another reason that rising individualism makes HR harder to achieve effectively results from the fact that individualists tend to act in ways that they perceive to be beneficial to their own long-term interests. Herriot (2001: 135) sums this up as follows:

> Individualist relationships are often isolated, in the sense that they are not anchored in a wider social setting. The parties relate to each other as individuals, rather than as representatives of groups. Hence they have only each other, and expect only the other(s) to provide all the benefits of the relationship. The temporary nature of so many individualist relationships is due to the failure of one or both the parties to achieve their expected outcomes... Overall, individualists tend to belong to a lot of groups and form a lot of relationships; however most are temporary and superficial.

The key implication therefore relates to expectations on the part of employees when they enter an employment relationship. Whereas someone with a collectivist temperament or outlook will readily identify themselves as a member of an organisation, will develop loyalties to it and to colleagues who they work alongside and will feel a sense of duty towards it, individualists are far more focused on what is in it for them.

This means that they will stay employed somewhere only for as long as they perceive that employment to be beneficial or fulfilling from their own individual perspective. If their careers stagnate, if they become dissatisfied with an aspect of their work or if they believe they could do better elsewhere, they will think nothing of leaving. The extent of their loyalty is strictly circumscribed. The same is true of the effort they are prepared to put into their work. They are happy to work hard, to exercise discretionary effort and to perform to the highest standards, but their continued willingness to do so is dependent on their own perception of the benefit they are personally deriving from doing so.

This means that organisations dominated by individualists are much harder to manage effectively. There is a greater need, for example, to develop and invest in appropriate employee retention initiatives. Performance management also needs to be more sophisticated because each individual's personal ambitions and needs have to be considered and acted upon separately. The fostering of effective, thoughtful and sensitive line management thus becomes an increasingly important HR objective. Meeting employee expectations in an individualistic workplace leads to substantial dividends for the organisation. Dashing expectations has serious negative consequences.

Non-conformism

It is also true that high-trust employment relationships are simply harder to maintain when the workforce is more individualistic. Herriot argues that collectivists tend to believe that 'attitude and behaviour should not necessarily be consistent' and that there are 'many occasions when it is right, rather than merely expedient, to comply, even though you may not agree with what you are doing'.

Such an approach to workplace life jars with individualists, who see it as lacking in integrity. They say what they think, aim to behave in line with their own values and respect managers who do the same. They will not trust or respect those who take a collectivist approach if this means compromising personal integrity. Moreover, if it is true that today's more individualistic workforce is also unconcerned with what others think of them, they are thus relaxed by the thought that colleagues or managers may not rate them or may disapprove of what they do or say or how they dress. They admire non-conformism and are likely to mistrust managers who take a more conformist line and expect them to do the same. Being straight, honest and up-front is often harder for managers than spinning a positive message or concealing the whole truth from employees. The more individualistic the workforce becomes, the greater the need to take the straight approach.

Concern for others

It is important to appreciate that while individualists are focused on themselves and their own interests, this does not mean that they are unconcerned or unaffected by the way that colleagues are treated – the way they react is just different from that of collectivists. When a fellow employee is treated unfairly or in an unjust fashion, individualists may not feel much by way of collective solidarity, but they will be concerned because they fear being treated in a similarly unfair manner themselves.

Indeed, in many respects the lack of a conformist outlook means that an individualistic person is adversely affected by poor treatment meted out to colleagues compared with a collectivist. This results from the 'live and let live' credo of tolerance that underpins the individualist's view of life. It means that organisations will increasingly have to pay attention to the way that individual employees are treated more generally. Fair dealing and being seen to act equitably are more not less important, because of the greater impact perceived unfairness towards individuals has on individualists. In particular, it means that organisations must be increasingly vigilant not to discriminate unfairly on any grounds.

CASE STUDY 8.2

Solo living

The most striking trend in behaviour, as opposed just to attitudes, that suggests growing individualism is the considerable and ongoing growth in solo living. The extent of the trend is best illustrated by looking at the growth of single occupancy households. All government and other authoritative forecasts suggest that this trend will accelerate over the next twenty years.

In 1971 approximately 5 per cent of working-age adults lived alone. The figure is now 18 per cent and is expected to rise to 20 per cent in the 2020s. Only 2 per cent of people in the 24 to 44 age group lived alone in 1971. Over 15 per cent now do so.

There are several discernible causes for the increase in solo living among people of working age:

- Single people are less likely to form stable relationships and are more likely to wait until a later age before forming couples.
- Couples are more likely to separate and are more likely to do so earlier in their relationship than used to be the case. Divorce rates have increased very substantially over recent years.
- We have seen the development of relationships in which couples live apart for most of their time, seeing one another only at weekends or occasionally rather than living together in the same household. A good proportion of these relationships involve couples living at some distance from one another.

The extent to which employment practices and HR policies are likely to, or need to, develop in response to the solo-living trend has not been widely debated – but it should be. It is interesting to speculate about how things may change if the trend continues or accelerates:

- It is likely that, as the number of solos in the labour market increases, employers will have to prioritise employee retention initiatives. This is because single people living alone have fewer social ties, are less likely to own their own homes and are more likely to want and be able to make radical changes in their lives.
- There are potential gains to be made in recruitment and retention terms from developing benefits and policies that attract single people to an employer. Interestingly, this may involve downplaying or running counter to the current trend for practices that are 'family-friendly' and that privilege parents, as these tend to irritate and anger people without families.
- Greater care needs to be taken when managing the career expectations of solo-living people and when disciplining them. This is because they tend to have a more intense emotional reaction that lasts longer when they perceive themselves to have been treated unfairly. They may have fewer robust support networks and are more likely to see negative emotions persist for longer.
- There is scope for actively providing opportunities for people to socialise in a work-related context (gym membership, social events, etc). Some employers discourage people from forming relationships with colleagues at work. Such policies are understandable but may be counter-productive from the point of view of recruiting and retaining solos, particularly where the culture is one of long hours.

References

Atkinson, A (2015) *Inequality: What can be done?*, Harvard University Press, Cambridge, MA

Bauman, Z (2003) *Liquid Love: On the frailty of human bonds*, Polity Press, Cambridge, UK

Bauman, Z (2005) *Work, Consumerism and the New Poor*, 2nd edn, Open University Press, Maidenhead

Besen-Cassino, Y (2014) *Consuming Work: Youth labor in America*, Temple University, Philadelphia, PA

Collins, G and Boucher, G (2005) Irish neo-liberalism at work, in *The New World of Work: Labour markets in contemporary Ireland*, ed G Boucher and G Collins, The Liffey Press, Dublin

Deaton, A (2015) *The Great Escape: Health, wealth, and the origins of inequality*, Princeton University Press, Princeton, NJ

du Gay, P (1995) *Consumption and Identity at Work*, SAGE, London

Galbraith, J K (1958) *The Affluent Society*, Hamish Hamilton, New York

Herriot, P (2001) *The Employment Relationship: A psychological perspective*, Routledge, Hove

Hofstede, G (1997) Organization culture, in *The IEBM Handbook of Organizational Behaviour*, ed A Sorge and M Warner, Thomson Learning, London

Lindsey, B (2008) *The Age of Abundance: How prosperity transformed America's politics and culture*, HarperBus, New York

Marmot, M (2015) *Status Syndrome: How your place on the social gradient directly affects your health*, Bloomsbury, London

McGovern, P, Hill, S, Mills, C and White, M (2007) *Market, Class and Employment*, Oxford University Press, Oxford

Picketty, T (2013) *Capital in the Twenty-First Century*, Harvard University Press, Cambridge, MA

Putnam, R D (2000) *Bowling Alone: The collapse and revival of American community*, Simon and Schuster, New York

Radelet, S (2015) *The Great Surge: The ascent of the developing world*, Simon and Schuster, New York

Ransome, P (2005) *Work, Consumption and Culture: Affluence and social change in the twenty-first century*, SAGE, London

Scheidel, W (2017) *The Great Leveler: Violence and the history of inequality from the Stone Age to the twenty-first century*, Princeton University Press, Princeton, NJ

Svendsen, L (2008) *Work*, Acumen Publishing, Stocksfield

Trandis, H (1995) *Individualism and Collectivism*, Westview Press, Boulder, CO

Twenge, J M and Campbell, S M (2008) Generational differences in psychological traits and their impact on the workplace, *Journal of Managerial Psychology*, 23 (8), 862–77

Wilkinson, R and Pickett, K (2009) *The Spirit Level: Why more equal societies almost always do better*, Penguin, London

Wilkinson, R and Pickett, K (2018) *The Inner Level: How more equal societies reduce stress, restore sanity and improve everyone's well-being*, Allen Lane, London

09 Politics and public policy

In this chapter we are going to start looking at the relationship between government and business. This is a subject area that we expect many of you will think of as being dry and uninteresting, and this can sometimes be the case. It is, however, of very great importance and has all kinds of direct and indirect consequences for HR and L&D work. The truth is that government policy is often a major factor that determines not only what an organisation is able to accomplish, but also whether it is viable in its current form.

In the domestic sphere, governments are responsible for two main areas of activity. First, they make the laws and regulations under which we all live and which employing organisations are obliged to follow. Second, they determine how much and in what ways revenue is to be raised (largely through taxation and borrowing) before deciding how to spend these public monies on fulfilling their policy objectives.

In Chapter 10 we will be focusing on the main areas of regulation that are of relevance to the management of people in workplaces. In this chapter we start with a brief survey of the UK political system and significant contemporary developments in our politics, before going on to focus on some key developments in public policy.

Public policy is sometimes heavily contested. Our political parties tend to take very different positions on how it should evolve, and it is sometimes the subject of significant lobbying operations by pressure groups representing the interests of industries, groups of working people and others who are affected by the regulations governments develop in order to implement their policies. In fact, there is often a good deal of agreement about what public policy should be looking to achieve but major differences on how these objectives should be attained in practice.

There are several distinct areas of public policy that are of professional interest to people working in the field of people management. In Chapter 5 we discussed economic policy at some length. Here we focus on the two other areas that have most potential significance for the management of people: industrial policy and policy on education and skills.

Key features of the UK political system

The UK is one of the more established liberal democracies in the world. In practice this means that our governments are elected freely under a one-person, one-vote system and that any number of political parties are entitled to contest elections. In addition, there is, for the most part, total freedom of speech and conscience. Provided you do not incite hatred, defame or libel someone you can say or write what you like about anyone without fear or favour under the law. Moreover, the principle of 'the rule of law' also applies, meaning that everyone, however wealthy and powerful, is treated in law in the same way as anyone else.

Like many liberal democracies, the UK has a parliamentary system of government. At least every five years, and often more frequently, a general election is held in which every seat in the House of Commons is contested. Each Member of Parliament (MP) is elected by the people who are over the age of 18 and who live in one of 650 geographically defined constituencies, each of which is roughly the same size in terms of its population. After a general election the leader of the party that has the largest share of the 650 seats is invited to form a government. When one party has an overall majority over the others, this is not at all problematic. But from time to time, when no single party has an overall majority following an election, there is a need for the largest party either to form a coalition with another party (as happened in 2010) or to govern as a minority with an understanding from others that they will not oppose on matters of confidence (ie really important issues) or finance, as effectively happened between 1976 and 1979, and again more recently between 2017 and 2019. A government with a stable and significant majority can usually be confident that whatever legislation it proposes will be approved by the House of Commons and hence become the law of the land.

General elections in the UK are contested using the 'first past the post' system, meaning that the candidate with the most votes gets elected, irrespective of whether that person has secured over 50 per cent of the votes in their constituency. This has the effect of benefiting larger parties and means that the governing party very rarely has the support of an absolute majority of those who vote in the election. The system can be criticised for failing to give smaller parties sufficient numbers of MPs, in contrast to systems based on proportional representation. The advantage of first-past-the-post is that it tends to produce stable one-party governments that are able to implement their agendas without the need to do deals with small parties that have limited support. It also makes it easier for the electorate to dismiss an unpopular government by voting them out at a general election.

The upshot is that the leader of the largest party almost always becomes Prime Minister and appoints a ministerial team from among his/her colleagues to govern the country. Most ministers are also MPs, but it is necessary also to select some from the House of Lords in order to help get government business through.

The House of Lords is the UK's second chamber. Historically its members – known as peers – were all hereditary and, in the main, members of the landowning aristocracy. Nowadays the vast majority of peers are nominated for appointment by the leaders of the main political parties. They hold their seats for life, meaning that they are not obliged to stand for election and cannot be removed unless they are guilty of a very serious criminal offence. This system is very difficult to justify in theory, but it has some practical advantages. First, no party ever has anything close to a majority in the Lords, preventing poorly drafted or damaging new laws being passed without proper scrutiny. Many peers sit as cross-benchers and take an independent stance, while others are fiercely independent, irrespective of party label. Second, the powers of the House of Lords are limited. At best they are only able to delay legislation proposed by a government for a period of two years, while by convention they never block a measure that featured in the manifesto of the party that won the previous general election. Third, among their number are many people who have expertise in certain fields that they can bring to debates about proposed legislation. This is helpful because the main role of the House of Lords is to scrutinise proposed legislation and put forward amendments where, for example, it is thought that there may be unintended consequences if wording is not altered.

New law only comes into effect when it has been voted through on three occasions (first reading, second reading and third reading) by both chambers of Parliament and then given Royal Assent by the Queen. By these long-established procedures, new law is made in the form either of Acts of Parliament or sets of regulations issued by ministers under powers given to them by such Acts.

These law-making procedures are of considerable importance to employees and employing organisations because it is through them that employment regulations are made and ultimately enforced. In Chapter 10 we will be looking in detail at some of these. For now we just need to note that a great deal of our employment law is set out in regulations issued under Acts of Parliament. Major examples are the following:

- Health and Safety at Work Act 1974.
- Employment Rights Act 1996.
- National Minimum Wage Act 1997.
- Employment Act 2002.
- Equality Act 2010.
- Enterprise and Regulatory Reform Act 2013.
- Trade Union Act 2016.

The Scottish Parliament and the Welsh and Northern Ireland Assemblies have power devolved to them from the UK Parliament to legislate in defined areas. The right to make new employment laws is not currently devolved either in Scotland or in Wales, although some other areas of employment policy are. It is only in Northern Ireland that ministers have the power to propose employment legislation, which is voted on by the Assembly members and can become law. As a result, in recent years, some significant differences have developed between employment law in Northern Ireland and in the rest of the UK. For example, the qualifying period for unfair dismissal in Northern Ireland is one year, whereas in Great Britain the period has been set at two years since April 2012.

PAUSE AND REFLECT

What do you consider to be the main strengths and weaknesses of the UK's political system? What elements would you like to see changed or reformed?

Contemporary political developments in the UK

Public policy is always in some kind of flux because governments are continually bringing forward new initiatives. Some of these involve implementing manifesto commitments put forward by the governing party or parties in the election campaign that brought them to power. Others take the form of proposals brought forward in reaction to particular developments that require a government response. The first tend to be more predictable than the second.

At the time of writing (late 2020) in the UK we have a recently elected Conservative government that is able to draw on a significant majority of the seats in the House of Commons to get its business through without major compromises. It has limited need to adjust its agenda to satisfy the concerns of other political parties or groups of its own MPs who might otherwise be in a position to block proposals or demand amendments. We are not expecting a general election to be called until 2024 and we can therefore reasonably anticipate that the public policy commitments set out in the Conservative Party's manifesto at the 2019 general election will be implemented provided the public finances allow that to happen. Examples include increased spending on healthcare and transportation infrastructure projects, immigration rules which replace free movement of workers to and from the European Union with a points-based approach that applies equally to all, and the implementation of the variety of reforms to our employment law system recommended by the Taylor Review in 2017.

In other areas, as is true for all governments, public policy responses on the part of ministers are going to be much less predictable because they will occur in response to events. Who would have thought, for example, at the start of 2020 that the UK, like most other countries, would have to increase public debt levels by over £200 billion in order to cope with the immediate effects of a global pandemic? It is surely inevitable that this is going to have a significant long-term effect on the government's ability to borrow money for other planned investments. The spread of the Covid-19 virus also led ministers to develop policies requiring the temporary closure of some businesses and many millions of people to work mainly from home for an extended period.

While the Conservative Party is firmly in charge of public policy development in England and at the level of the UK, other parties are in government in Scotland, Wales and Northern Ireland, and this also is a situation that looks set to continue. The devolved administrations have the power to set their public policy priorities and in several areas to spend their budgets differently. Policy in important areas such as health and education are fully devolved, meaning that regulations differ significantly in different parts of the UK. Hence, for example, Scottish students attending Scottish universities are not obliged to pay tuition fees, as is the case elsewhere in the UK, while rates of income tax are marginally different for some people.

The rise of identity politics

In the second half of the twentieth century, political debate in the UK, as was the case across most of the world, was focused on questions of economic policy. Parties with a left-leaning philosophy like the British Labour Party based their electoral appeal on a policy agenda that involved a higher degree of state intervention, public spending and central planning. By contrast, parties of the centre-right argued for less by way of state control over industry and in favour of free market economics with limited central planning and lower taxation.

Parties of the political left have tended to favour regulations that limit the extent to which owners and managers of organisations are at liberty to run their organisations without restriction. They also tend to be comfortable with the idea that the state should own and directly run major corporations. They are more suspicious of free markets, largely on the grounds that too little regulation in this area can increase social inequality and social injustice. By contrast, people on the political right tend to see regulation as being harmful both to the economic prospects of organisations and by extension to national economies more generally. They favour deregulation and believe that wealth is more effectively and efficiently created by free enterprises than by government-run organisations.

The two sides also disagree on the extent to which taxes should be levied by governments before being redistributed through welfare payments and spent on the provision of public services. People on the right of politics tend to favour low tax and limited public services on the grounds that over time this approach makes everyone wealthier, even if they are less equal. People on the left are more concerned about social inequality and thus tend to favour higher taxes, more redistribution through welfare and more government spending. Those on the right favour the capitalist, market economy and believe it serves us well. They are suspicious of state control, thinking that it can easily lead to inefficiency and corruption. Those on the left are more attracted to socialism as a means of bringing about greater equality and security for all. They are equally suspicious of the owners of private corporations, thinking that if given too much power and left unregulated they will exploit their customers and workers in an equally corrupt manner.

A metaphor for thinking about the left–right divide is the circus trapeze. Trapeze artists can only fly really high and perform extraordinary manoeuvres if they are provided with two crucial pieces of equipment – a ladder to climb up and a safety net to catch them if they fall. We can look at the left–right political divide in this way. People on the political left tend to favour high safety nets. As many people as possible should be able to perform in the trapeze act. It might not be so spectacular, but it will be safe and inclusive. By contrast, people on the right tend to focus on the ladders. A net is necessary in case people fall, but the main aim of government should be to encourage and enable the construction of high ladders. Some will fly higher than others, but that is not a bad thing if it enables them as individuals to reach their full potential.

These kinds of economic questions very much remain an important feature of our political debate, but in the twenty-first century there has been a marked tendency for them to be supplemented, and in some respects eclipsed, by public policy debates that relate in one way or another to questions of identity. This situation looks set to continue and is a development that has some important implications for the way that HRM is conducted in organisations.

As globalisation has accelerated, we have seen increased international economic and cultural exchange. There has been a big increase in the level of migration from one country to another, and we have seen the rapid rise of larger and more powerful multinational corporations that are keen to locate their activities, and thus jobs, wherever they are most profitable. National economic activity has become more international, creating winners and losers in the process. Moreover, of course, supra-national institutions like the World Trade Organization and the European Union have gained a great deal more prominence. The question of how far it is appropriate to increase further the power that international regulatory bodies wield has thus become politically salient.

Meanwhile, as societies around the world have enjoyed increased affluence, concerns about poverty and access to healthcare and education have become less all-encompassing. The vast majority of households in a country like the UK are now able to enjoy a reasonable standard of living, and there is less hunger and squalor of a kind that used to be commonplace. Life expectancy has risen greatly and most children have access to reasonably good levels of education. Political debate has thus moved on to reflect this. There is less focus on absolute poverty and more on inequality between groups and ongoing unfair distribution of power and opportunity.

Identity politics has in part been the result of these changes. Issues such as gender and racial inequality and the rights of disabled people and marginalised groups have become far more important in public discourse. People have also become more concerned about what they see as the erosion of their national identity and a perceived threat to their ability to live their lives within established cultural norms.

In the UK, identity politics has long been present in Northern Ireland where there are two distinct communities rooted in different religious traditions with completely opposed political objectives. But the rise of this kind of politics is much newer elsewhere. The twenty-first century has, for example, seen the rise of Scottish nationalism. Once a fringe party in most elections, the Scottish National Party now has a dominant place and opinion polls show that support for full independence for Scotland from the rest of the UK has become both an established and popular cause. Across the rest of the country the recent debate over whether to stay in or leave the European Union is another example of identity politics becoming much more significant.

Identity politics is much less directly focused on economic questions (although these play a part) and is much more about what kind of society we want to live in and where power should lie. How far we should be able to say and write what we think has also been brought into question, with many now arguing for curbs on free speech in situations where people may be offended by others' comments.

The spectrum is thus one that pits progressives against conservatives – and this often cuts right across that traditional, longer-established left–right spectrum. Taking an international perspective, Francis Fukuyama argues that identity has replaced economics as the key driver of political debate only in recent years:

> In the second decade of the twenty-first century that spectrum appears to be giving way in many regions to one defined by identity. The left has focused less on broad economic equality and more on promoting the interests of a wide variety of groups perceived as being marginalized – blacks, immigrants, women, Hispanics, the LGBT community, refugees, and the like. The right, meanwhile, is redefining itself as patriots who look to protect traditional national identity, an identity that is often explicitly connected to race, ethnicity or religion.
>
> (Fukuyama, 2018: 6–7)

Voting patterns have shifted markedly as a result. Social class and income are now far less effective predictors of how someone will vote in an election than age, geography and education. If you live in London or another major city, are under 40 and have a degree, the chances are that you are a Labour voter. By contrast, if you are older, live outside a big city or a university town and left school at 16 or 18, you are now much more likely than used to be the case to vote Conservative. Goodhart (2017) makes a useful distinction in this context between the former group, who he argues have 'an achieved identity', and those in the second, who have 'an ascribed identity'. This appears now to be creating a big divide in our society in terms of attitudes and ambitions – a new kind of class divide.

Identity politics is controversial because it is often very tribal in nature. People identify strongly with one group and define themselves collectively in opposition to those in the opposite camp. There is some danger here, as it leads to deep social divisions rooted in a disagreement about fundamental values. As Maalouf (2000) points out, identity politics can very easily become a politics of exclusion that leads people to have hateful thoughts about members of other tribes, which can lead to hateful words and deeds. It pits older people against younger people, urban dwellers against country dwellers, more educated people against less educated people, religious people against secular people – even men against women. People are sometimes judged by others less by how they conduct themselves individually as according to how their tribe conducts itself and how its values are perceived. The extent, for example, to which people are sympathetic to or passionate about causes such as Brexit, Black Lives Matter, the Me Too movement, transgender rights or climate change is fuelling divisions in society and in politics. Agreeing to disagree about such issues is becoming less feasible over time, as is the possibility of taking a neutral or equivocal position. This goes for some organisations, too, which perceive a need to take a conspicuous stance on issues of identity as a means of protecting or enhancing their reputations.

The rise of identity politics has had and will continue to have a significant impact on the way that people are managed in organisations. First and foremost, it has strengthened the need for employers to give far greater attention to issues of equality, diversity and inclusion. In the past, these matters were often discussed and taken into account, but were not in truth given any kind of priority. Lip service was paid and initiatives pursued, but not with the same vigour that was mustered in, for example, the field of performance management. Such an approach is no longer an option for many organisations, particularly larger employers with a significant public profile. The effective promotion of diversity and inclusion is expected by customers, employees and the wider community. Practices in the people management field need to be both fair to all and seen to be fair. As is the case with issues relating to business ethics and sustainability, there are now considerable potential commercial advantages that accrue when organisations successfully and authentically embrace practices that chime with the agendas of groups for whom issues of identity are of growing importance.

We are also seeing a marked increase in concern for employee wellbeing and this is often framed in terms of mental health. Limited periods of depression or stress related to work that would not generally have been considered to be such major issues twenty years ago are now being taken far more seriously by employers and employees alike. Because issues relating to identity are often triggers for feelings of stress or depression, these are increasingly being treated as wellbeing matters. The term 'micro-aggression' is sometimes used in this context to denote an act or statement made by one person that another finds offensive or disrespectful on grounds of identity, despite there not being an intention to offend. It follows that employers who are concerned to improve employee wellbeing have to think through what can and cannot be said in conversations at work, when a line is crossed and how they should respond. One response has been to introduce training programmes aimed at raising awareness of these issues so as to reduce the likelihood that an employee's wellbeing may be adversely affected when one employee says something that another finds to be insensitive.

CASE STUDY 9.1

WPP

WPP is one of the biggest, best established, most influential and successful marketing, advertising and public relations companies in the world, employing 107,000 people in 106 countries. Known for its creativity, innovation and international reputation, the company is recovering from some relatively difficult years, which culminated in the sudden retirement of its founder and long-serving chair, Sir Martin Sorrell, in 2018.

In the summer of 2020 a group of over 600 'Black Leaders in Advertising', none of whom worked for WPP, signed an open letter drawing attention to the continued under-representation of black people at all levels in the advertising industry. Their focus was on the USA, but many of the points they made, if not all, apply across much of the world. Their letter concluded as follows:

> Though advertising agencies boast some of the most politically progressive business leaders in America, agency leadership has been blind to the systemic racism and inequity that persists within our industry. Many gallons of ink have been spilled on op-eds and think pieces, but tangible progress has eluded this industry for too long.

The letter asked companies in their industry to commit to taking a number of steps aimed at rapidly improving the situation. These included:

- Make a specific, measurable, and public commitment to improve Black representation at all levels of agency staffing, especially senior and leadership positions.
- Track and publicly report workforce diversity data on an annual basis to create accountability for the agency and the industry.
- Provide extensive bias training to HR employees and all levels of management.
- Extend agency outreach to a more diverse representation of colleges, universities, and art schools.
- Invest in management and leadership training, as well as mentorship, sponsorship, and other career development programs for Black employees.
- Require all leadership to be active participants in company diversity and inclusion initiatives and tie success in those initiatives to bonus compensation.
- Introduce a wage equity plan to ensure that Black women, Black men and people of colour are being compensated fairly.

WPP's CEO, Mark Read, responded by issuing a statement committing his company to 'taking decisive action' on these and all the points in the letter. Furthermore, he committed to reviewing every aspect of WPP's core HR practices and to publishing racial diversity data. He went on to commit his company to the following:

- We will use our voice to fight racism and advance the cause of racial equality in and beyond our industry.
- We will invest $30 million over the next three years to fund inclusion programmes within WPP and to support external organisations.

This is a particularly striking example of a company looking to make some radical changes, particularly in HRM, in response to developments in its wider social and political environment. But it is by no means an isolated example. Many similar initiatives are being announced by prominent corporations every month.

Industrial policy

In recent years, across most of the world, governments have developed a preference for policy interventions in the field of business that promote 'competition and choice'. This agenda has involved actively promoting competitive conditions and creating competitive markets where previously they were more limited or non-existent. The aim is to maximise productivity while requiring commercial organisations to better serve consumers by responding to their evolving needs and preferences. However, the belief that it is the role of government to promote competition and choice and that such concerns should form the centrepiece of industrial policy is relatively new.

In the UK from the 1940s until the 1980s, and in many other economies until much more recently, a completely different approach was taken to the relationship between government and industry. The term 'mixed economy' was often used to describe the situation that prevailed.

During this time, a large portion of UK industry was owned and directly controlled by the government. In most cases, though not all, the government owned corporations that were monopoly suppliers or dominated the national market. This included key strategic industries that became known as comprising 'the commanding heights of the economy'. State-owned corporations such as British Telecom, British Steel, British Petroleum, British Coal, British Airways, British Aerospace, British Leyland (a car and truck manufacturer), the British Oxygen Company, British Gas, British Rail, the National Water Board, the Central Electricity Generating Board and the National Bus Company employed millions of people and were, in many cases, the only supplier permitted to operate. Where competition was permitted, it was often restricted so that the state-owned corporations could remain dominant players in their industries. Rolls-Royce was owned by the government for a period, as was Thomas Cook (travel agent) and Pickfords (removals). In addition, local government owned bus companies and nearly all the nursing homes, not to mention 35 per cent of the housing stock.

Until the 1980s industrial policy was decided in large part by a body called the National Economic Development Council (NEDC), which developed five-year national plans. Government ministers sat on the council alongside senior industrialists and trade union leaders. Central industrial plans involved some strengthening of competition, improving training and introducing new technology as now, but also building up state monopolies, directing subsidies to failing corporations and providing financial incentives for organisations to locate activities in less prosperous regions of the country. A policy of 'picking winners' was also pursued whereby government would decide which industries should be grown via public investment and played a role in facilitating that development. Where industries were flagging, investigations were carried out and government subsidies channelled to improve their productivity or help them to survive in the face of international competition.

The political consensus around the desirability of a 'mixed economy' broke down in the 1970s with the rise of competition from overseas – particularly from German and Japanese companies. In the UK the election in 1979 of the Thatcher government signalled the start of a complete reversal of the policy. By the time of the election of the Blair government 18 years later, a new political consensus had been established around a wholly different set of industrial policies. This has largely continued since then, except when rapid and decisive government intervention has been needed at times of crisis such as the banking collapses of 2008–09 and the Covid-19 pandemic of 2020–21.

At its root, the consensus now is the belief that fair competition between organisations is the best means of:

- Satisfying customers.
- Raising productivity and quality.
- Innovating.
- Growing the economy.

The view became established that government should not seek to run industry, but should concentrate on regulating competition and creating the right macro-economic conditions for organisations to thrive and compete internationally. Privatisation replaced nationalisation, as all major corporations run by the government were either sold wholesale or divided up and then sold off. Tariff reductions and the evolution of the European Single Market meant that formerly nationalised companies (eg British Steel/Corus, British Coal/RJB Mining and the energy companies) now have to compete internationally. They have to satisfy their customers and make a profit in order to survive. In some cases they have become wholly owned subsidiaries of large overseas-based conglomerates.

In addition, governments have increasingly sought to import 'market disciplines' into the public services that have remained under direct government control. Examples are the internal market mechanisms that now operate in healthcare, education and some areas of social care.

The first decades of this century saw private funding being attracted into the public services through policies such as the Private Finance Initiative (which involves private firms undertaking major capital investment and renting the resultant facilities back to the government), foundation schools and direct involvement of the private sector in running defence and prison facilities. At the same time, a variety of public–private partnership (PPP) arrangements flourished.

The same 'big policy switch' has occurred in most industrialised countries, although it came in the 1990s and 2000s rather than the 1980s in most cases. Some countries had less of a transition to make (eg the USA), while others have not gone as far as the UK down the 'competition and choice' road (eg many European economies). Elsewhere, the transition has been more dramatic than was the case in the UK (China, India, Eastern Europe).

There is now a broad international consensus that government should not seek to micro-manage economic development centrally or seek to own the 'commanding heights' of their economies. The interests of economies and consumers are best served by regulated markets that require organisations to compete with one another to secure customers and satisfy their shareholders.

However, there remain major disagreements between governments about subsidy, the extent to which genuine free trade is desirable and the level of regulation that is necessary. Government industrial policy today is thus primarily concerned with:

- Actively promoting competition and choice (ie an enterprise culture).
- Seeking to provide an environment in which industry can compete effectively on an international basis.

The belief in the virtues of the free market, once so politically controversial, now goes largely unchallenged as a principle across the major political parties. The Labour

Party began to move in a different direction under the leadership of Jeremy Corbyn (2015–20), proposing, for example, to re-nationalise rail and water companies, but this kind of approach is unlikely to form such a central plank of the party's platform in the future. The global ubiquity of industrial policies rooted in a belief in 'competition and choice' is too well established to make more than modest changes in approach to be practicable when the major world economies are so fully integrated.

Subjecting both companies and public sector bodies to competition is seen as being by far the most effective means of forcing them to improve their productivity by reviewing their structures and methods, and by investing in new technologies. For private sector organisations this is widely accepted as being a more effective method of securing long-term survival and growth than heavy government intervention. Governments now only get involved in directly running industries when crises hit that leave them little alternative.

For the public sector, competition, alongside target-setting, is similarly viewed as being the best way of securing efficiency, innovation and value for money for taxpayers. Competition provides an incentive for organisations to maximise their performance, while ensuring that new entrants can come into a market with fresh ideas and approaches to displace less successful organisations. It has become accepted that the less efficiently run hospitals and schools must close or be taken over by new teams of managers.

Government thus now tends to be more concerned with setting the rules under which competition takes place than in actively participating in industrial endeavour as a player in its own right. Competition laws prevent companies from dominating a market to too great an extent, while government funds a great deal of research and development activity to support the development of new technologies. Tax incentives are also provided to encourage new enterprises to establish themselves and investors to back them. Corporation Tax, for example, has been reduced considerably over the past thirty years. Finally, there is of course a great deal of public investment in education with the aim of improving skills and knowledge, and in particular improving the country's science and engineering base.

The main criticism that is made about this switch to a policy that privileges competition and consumer choice is one that is especially relevant to a book about employment. The biggest losers from the change in direction have been workers in the industries that have been adversely affected and the trade unions that seek to represent their interests. Initially, especially, the change in policy was highly controversial and caused great pain to many families. Subsidies were withdrawn from organisations that were not productive, leading to millions of job losses and the destruction of communities that relied on particular employers. Established steel towns, coal villages and ports that relied on fishing and shipbuilding suffered particularly as their major employers were required to compete effectively in increasingly international markets. In many cases they were ill equipped to do so and had to close or shed a lot of their workforces. Those that survived were very often taken over by larger privately owned corporations, many international, with shareholders who were solely focused on extracting as much value as possible from their assets, irrespective of their impact on the affected communities. In short, a long period of industrial restructuring occurred which, on occasion, resulted in bitterly fought industrial disputes in response to the introduction of labour-saving technologies, and factory and pit closures.

For employees in many sectors, secure, pensionable work has been replaced by job opportunities that are short-term, insecure and often a good deal less well-paid. Secure final salary-based pensions that most people could look forward to receiving on retirement are now a rarity. In order to compete effectively, employers are obliged to reduce their costs wherever they can, and that often means keeping a lid on wages. Many more flexible working practices, such as fixed-term working, zero hours contracts and outsourcing, have been introduced, while millions now work in the so-called 'gig economy' with no job security at all.

On the other hand, the past forty years have seen substantial increases in real wages and have led to considerable innovation and economic growth. While some industrial sectors have seen serious decline, others have grown rapidly and become major international business success stories. Countries like the UK have lost millions of manufacturing jobs, but have seen more created in the sectors that have boomed, such as IT, financial services, retailing, hospitality and tourism, healthcare, higher education, film and TV production, business services and publishing.

The other major change that the new approach to industrial policy has brought about has been a steady decline in the number of people joining trade unions and participating in their activities. There are numerous factors that can help to explain trade union decline in countries like the UK over recent decades. The rise in individualism that we discussed in Chapter 8 has played a part, as has regulation that hinders the ability of unions to sustain campaigns of industrial action. Broader changes in industrial structure have, however, played the biggest role. This is because, with some exceptions like transport and energy production, it is those sectors of the economy which have traditionally been most heavily unionised that have been most significantly affected by government focusing more on competition and choice. Steel manufacturing, coal mining and ship building, for example, now employ far fewer people in the UK than they used to, as is the case in manufacturing and the public sector generally. These industries all provided fertile recruiting grounds for trade unions. The new jobs, by contrast, have tended to be in smaller businesses and in industries like retailing and business services where trade unions have never played such a major role and are not part of the workforce culture. It is important to point out that around 24 per cent of British workers are union members. They maintain a very significant role across much of the public sector, as well as in the transport sector and privatised companies such as Royal Mail. But this is less than half the strength they enjoyed before the switch in industrial policy and the introduction of competition in industries where there was little previously (DBEIS, 2020).

PAUSE AND REFLECT

To what extent do you agree with the view that, despite much about the big switch in industrial policy being regrettable, it would not be wise to return to an era of nationalised industries, central planning, limited competition and less consumer choice?

Policy on education and skills

The long-term aim of UK governments and ministers in devolved regional governments has for many years been to raise, steadily over time, levels of educational attainment. This is true of governments around the world and has been for over a century. While there is frequent disagreement about how best to achieve it, and about funding arrangements, there is very considerable consensus around the broad objective.

For the past fifteen years policy in the UK has been heavily influenced by the findings of the Leitch Review, which were published in 2006. Lord Sandy Leitch was commissioned by the then Chancellor of the Exchequer, Gordon Brown, to undertake a broad survey into the country's skills base, to compare it with that in other countries and to make recommendations as to what would need to happen if the UK economy is to be well placed to meet future skills needs through to 2020.

The final report was entitled *Prosperity for All in the Global Economy: World class skills*, and it concluded that the UK's performance in terms of basic educational attainment was mediocre when compared with that of similar countries and some rapidly developing economies such as China and South Korea. Projections were also made as to the likely needs of the economy in 2020. The key conclusion was that there was a need for much more investment in upskilling the UK workforce both on the part of government and by employers.

Some of the headline findings in the Leitch report were concerning. For example, he and his panel found that:

- 16 per cent of the working population (5 million people) lacked Level 1 literacy skills (ie GCSE English at grades D–G).
- 48 per cent of the working population (15 million people) lacked Level 1 numeracy skills (ie GCSE Maths at grades D–G).
- 21 per cent of the working population (6.6 million people) lacked entry-level numeracy skills (ie cannot add or subtract and have no understanding of fractions).

Lord Leitch also commissioned studies by leading labour market economists at Warwick University and at Cambridge University, and their projections included the following about the needs of the economy in 2020, which have proved to be pretty accurate:

- 40 per cent of UK jobs will require their holders to have a tertiary level education.
- 2 million new management roles will be created.
- 1 million new professional roles will be created.
- 8 times as many vacancies in these fields will arise as a result of retirements.
- A net decline in the number of lower-skilled jobs.
- Continued growth in demand for people to work in the service sectors, the highest levels of growth being in financial and business services.
- Reductions in demand for people to work in construction, manufacturing, primary and utility industries.
- Increases in demand for managers, professionals, associate professionals, sales and customer services and personal services.

- Decreases in demand for administrative and secretarial roles, skilled trades, machine and transport operatives and (hugely) people to work in 'elementary occupations'.

In short, the Leitch Report argued persuasively that the UK's skills base was very much lower than it needed to be in order to meet the challenges of what was then the future and what is now the present.

These findings and recommendations justified a range of government initiatives that had already commenced by 2006 and also led to the development of several more. The need to upskill the UK workforce by increasing skills and educational attainment remains the government's principal policy objective.

Increasing the number of school leavers going to university

The number of university places in the UK has increased rapidly in recent years. In 1950 when a total of 17,300 first degrees were awarded in the UK, just 3.4 per cent of school leavers had the opportunity to attend university (Bolton, 2012). The proportion increased to 14 per cent by the late 1960s, where it remained until the big expansion in the number of university places that occurred after 1990.

In 1999 the then Prime Minister, Tony Blair, set a new target of 50 per cent for the proportion of school leavers who should attend a university or gain a degree-level education at a college. It took time, but in 2017–18 participation rates reached 50.2 per cent for the first time (Coughlan, 2019). The percentage is considerably higher for women (57 per cent) than it is for men (44 per cent), but it is likely that more of both sexes will attend university in the future. The same trends are occurring around the world, several countries recording much higher rates of university attendance than the UK. The proportion is highest in Finland, where over 80 per cent of women now attend university.

Apart from in Scotland, the increase in the number of places has largely been funded by the introduction of student fees. At the time of writing (late 2020) these are capped at £9,250 a year for a standard three-year undergraduate degree programme. The fees regime is, though, heavily supported by government financing, it being assumed that in practice much of the debt being built up by students accessing the government loan repayment system will not be paid back.

Raising the education leaving age

A long-term, steady improvement in educational attainment can be tracked by looking at the age at which it has been lawful to finish school and start working. Elementary schooling became compulsory as long ago as 1880, when the leaving age was set at 10. This then increased to 11 in 1893, to 12 in 1889, to 14 in 1918, 15 in 1944, 16 in 1972 and 17 in 2013. In England and Wales, since 2015 an education leaving age of 18 has applied. This means that after the age of 16 it is lawful to leave school, but only if you go on to further education either on a full-time or part-time basis.

Apprenticeships

In the UK there is a long tradition going back several centuries of employers taking young people on as apprentices and training them in a skilled trade. Pay was always limited, but apprentices tended to have few living expenses and were prepared to trade pay for the opportunity to learn skills that they could subsequently use to make a living for many years.

Since the 1990s UK governments have become increasingly involved in coordinating, setting and certifying standards in what have become quite complex national schemes. The key development is the inclusion of formal off-the-job training and assessment, which is undertaken alongside on-the-job training. Apprentices are contracted to work for employers, but are also required to study more formally on a day release basis. The cost of this formal training is paid in full by the government in the case of 16- to 18-year-olds. For 19- to 24-year-olds a 50:50 split applies between the government and the employer. For apprentices over the age of 24, government funding is not generally available, but subsidised loans are provided for employers to draw on.

The scale of the government-sponsored apprenticeship scheme is substantial. In most of the 2010s over half a million started apprenticeships each year in England alone, meaning that some 750,000 people are training as apprentices at any one time. Government ministers wish to expand the scheme considerably, but a target of 3 million apprentices by 2020 has not been met. Total numbers have fallen somewhat in the most recent years. This is mainly thought to be due to the rather rigid apprenticeship levy that was introduced in 2017. This is effectively a tax amounting to 0.5 per cent of payroll that has to be paid by all employers with a wage bill of over £3 million a year. The money can then be recouped if spent on providing apprenticeships, surpluses being spent by the government on funding apprenticeships for smaller employers.

The major current trend appears to be growth in higher-level apprenticeships, including degree apprenticeships, which are completed over a number of years and allow people to study on a practically oriented degree programme while working and avoiding building up the levels of debt that fee-paying full-time students following a traditional degree programme now have.

Academy schools

The Academy Schools programme was launched in England in 2000, replacing the city technology college scheme that had been established a few years earlier. Its aim was to raise educational attainment levels in poorer-performing schools. Initially there was a need for private sponsorship and a focus placed on larger secondary schools in inner cities that were underperforming. The idea was that these schools could apply to become independent of local authority control, receive their funding directly from central government and develop along distinct lines. Standards would improve, it was argued, if head teachers could have a much freer hand in how their schools were run. Good schools would expand to meet parental demand while poorer-performing schools could improve standards by specialising in defined subject areas like science or the arts.

In the subsequent years most secondary schools have become academies as the requirement for sponsorship has been removed. We now have a large number of primary schools with academy status, some free schools which have been established from scratch by groups of parents, and multi-academy trusts in which several schools are managed by a single central team, reducing administrative costs.

T-levels

The most recent major reform to the education system is the introduction of T levels in English schools from 2020. These are vocationally oriented qualifications that are intended to be equivalent to A levels and studied towards by 16- and 18-year-olds who would prefer to take a less academic route in their final years at school. They involve extensive work placements alongside classroom study and can thus only be offered by schools working in partnership with employers. T levels are focused on specific industries (transport, engineering, social care, accounting, etc), the expectation being that most pupils who pass them will subsequently go on to work for employers in those industries.

All of these public policy initiatives have significant implications for employing organisations, as they rely for their existence on their capacity to recruit and retain people with the required level of education and skill to do their jobs. There is thus a shared interest between employers and government, not to mention the workforce, to address issues of inadequate education and skills acquisition in the UK. In the case of more highly skilled employees there is the option to recruit people from overseas, but for most jobs most of the time employers are reliant on their local economies. It is for these reasons that public policy in this field relies very directly on cooperation from employers and, in many cases, requires a shared funding relationship between government and industry.

PAUSE AND REFLECT

In what ways does your organisation help people to gain new skills or to increase their level of educational attainment? To what extent would you say that your organisation makes a full contribution to the process of upskilling the UK's workforce?

References

Bolton, P (2012) Education: Historical statistics, House of Commons Library, https://dera.ioe.ac.uk/22771 (archived at https://perma.cc/TFG3-BK64)

Coughlan, S (2019) The symbolic target of 50 per cent at university reached, BBC News, 26 September, www.bbc.co.uk/news/education-49841620 (archived at https://perma.cc/4KGY-KNA5)

DBEIS (Department for Business, Energy and Industrial Strategy) (2020) *Trade Union Membership, UK 1995–2019: Statistical bulletin*, https://assets.publishing.service.gov.uk/government/uploads/system/uploads/attachment_data/file/887740/Trade-union-membership-2019-statistical-bulletin.pdf (archived at https://perma.cc/4BZT-8BVV)

Fukuyama, F (2018) *Identity: Contemporary identity politics and the struggle for recognition*, Profile Books, London

Goodhart, D (2017) *The Road to Somewhere: The new tribes shaping British politics*, Penguin Books, London

Maalouf, A (2000) *On Identity*, translated from French by Barbara Bray, Harvill Press, London

10 Employment regulation

At the start of Chapter 9 we explained that one of the jobs of government is to propose new legislation and to pilot it successfully through Parliament. It is through this process that most employment law is made and methods for its enforcement set. This is not, however, the only mechanism. Much UK employment law has a European origin and would in all likelihood not be on the UK statute book had it not been for the forty-six years that the UK was an EU member state. These regulations are found in UK statutes but do not originate with the UK government. Moreover, we also have a great deal of common law which helps to determine what happens in employing organisations and provides a legal forum in which disputes between employers and employees can be argued out and settled. Common law is made by judges rather than parliamentarians, and it has been evolving steadily as cases are brought to court and considered over several centuries.

In this chapter we introduce all the major elements of UK employment law. You will see that these have a major influence over a very wide range of people management activities, including recruitment, promotion, terms and conditions, pay, health and safety, holiday entitlement, family leave rights and, of course, dismissal.

We will go on to explore some of the major debates that exist in respect of employment regulation, thinking about how far it achieves a fair balance between the interests of employees and their employers.

The development of employment regulation in the UK

In recent years, the volume and complexity of employment regulation has grown very significantly. The UK workplace and labour market are now some of the most heavily regulated in the world, yet this was very much not the case until relatively recently.

Until the 1960s, with the exception of regulations outlawing the exploitation of children, some basic health and safety rules and the limited protection provided by the common law, employers were able to manage their organisations without the limitations that are now imposed through statute. Managers could dismiss staff as and when they pleased, could require them to work whatever hours were deemed necessary, and were in no way restrained by the law if they wished to discriminate against women or members of the ethnic minorities. Indeed, it was common to have separate pay scales for male and female employees.

For many employees, this unregulated world of work meant a life of insecurity and, in some cases, thorough exploitation. But for the majority this was not the case, because trade union membership was high and a good majority of staff worked in unionised environments. They were thus covered by national-level collective agreements that operated across whole industrial sectors, and to which all organisations in those industries signed up. The same approach remains in place to an extent in the public sector today, most staff still being employed on nationally agreed terms and conditions that cover everyone in particular professional groups.

Protection from abuse of power on the part of employers was thus effectively provided through non-legal mechanisms. This voluntarist system of industrial relations had evolved over decades and appears to have suited both employers and trade unions reasonably well, but it was very different from the traditions of active state involvement that were being developed in other European countries at the same time. Whereas in the UK the way that workplaces operated vis-à-vis employees was almost exclusively determined with reference to contracts of employment, negotiated collective agreements and informal arrangements developed through custom and practice, elsewhere on the continent labour codes created and enforced by government agencies had become the norm.

Over the past forty years the UK's situation has changed to a very considerable extent. The first major reform came in 1965 with the introduction of statutory redundancy payments and a fledgling industrial tribunal service to enforce the new law. Henceforth staff could not be laid off without a minimum level of compensation being paid by their employers at a level determined by Parliament. The 1970s then saw the introduction of unfair dismissal law, equal pay legislation, paid maternity leave and protection from discrimination at work on grounds of sex and race. Health and safety regulation was also greatly extended at this time and protection introduced for employees whose organisations were taken over by or merged with others. During the 1980s the pace of law making in the employment field slowed down as the Thatcher and Major governments (1979–97) sought to fight off the demands of those who argued for greater levels of protective regulation. Instead, attention was given to reforming the trade unions through the development of a regulatory regime that they had to comply with and by taking steps that reduced their ability to organise lawful industrial action. There were, though, still some important advances in employment rights, not least the Disability Discrimination Act 1995, which has led to considerable improvements in working life and employment opportunities for people with disabilities.

Under the Blair and Brown governments (1997–2010) there was a reversal of this position. The trade union laws largely stayed in place, but there was also a substantial increase in the extent to which the individual employment relationship became subject to state-imposed regulation. The year 1998 saw the establishment of a National Minimum Wage, the Working Time Regulations and comprehensive data protection law. In subsequent years, regulation was introduced to protect part-time workers and people employed on fixed-term contracts. New regulations outlawing discrimination on grounds of sexual orientation, religion and age came into effect. Maternity rights were vastly extended and rights for fathers, parents of young children and carers of disabled adults introduced for the first time. New regulations set out in law situations in which an employer must recognise a trade union, and others in which employers are required to consult with workforce representatives about a broad range of management issues.

The period since 2010 has seen significant increases in the level of the National Minimum Wage being introduced and a new higher National Living Wage brought in for workers over the age of 25. We have also seen the introduction of gender pay gap reporting and more trade union law. The government's acceptance of the recommendations made by Matthew Taylor in *Good Work: The Taylor review of modern working practices* (2017) are now being turned into new regulations, the first of which came into effect in 2020.

In short, we have moved a very great distance from a voluntarist regime towards one that has a great deal more in common with the 'codified' approaches of our continental neighbours.

There are no straightforward answers to the question of why the UK has seen such a transformation in the nature and extent of employment regulation in a single generation. The process has not been a planned one and progress continues to be made in a piecemeal, step-by-step fashion. No grand, overarching strategy on the part of any government can clearly be identified. However, it is interesting to note how rarely a piece of regulation has been repealed by an incoming government with different priorities from its predecessor. Once in place, employment law appears to be very difficult to displace, suggesting that there is a broad level of agreement among the electorate in favour of the regulatory structure that has been steadily erected. Despite the lack of any single dominant explanation, it is possible to identify a number of factors that have contributed significantly to this transformation:

- UK membership of the European Economic Community (EEC) and then the European Union (EU) clearly played a major role. A great deal of UK employment legislation originates in Europe and would probably not be on our statute books at all were it not for our membership of the EU between 1973 and 2020.
- The decline in trade union membership and activism the country has witnessed over forty years has also played an important part. As collective bargaining structures have been dismantled and the numbers joining unions have fallen, new institutional arrangements have had to be put in place to ensure that people are protected against unfair treatment and undue health risks while they are at work. In many important respects, employment law

thus now fills this gap. Indeed, one of the major reasons for the introduction of unfair dismissal law was the aim of reducing the number of strikes that were precipitated when employers were perceived to have fired someone unjustly (Davies and Freedland, 1993: 199–200).

- Government economic policy has also played an important role in helping to shape employment regulation through the decades and continues to do so. In the 1980s the focus was on using legislation to tame trade union power so as to boost UK productivity and the capacity of the economy to change in response to emerging global industrial trends. By contrast, in the past two decades a major driving force behind employment regulation has been the government's desire to encourage people to 'come off welfare and into work'. To that end, various regulatory changes have been made to remove barriers that act to deter people from entering (or, in the case of mothers with small children, from returning to) the workforce.
- The growth of employment regulation can also be viewed in a far wider perspective – as just one of the many areas of national life that have become regulated or have seen an expansion of their regulatory regimes in recent years. It is not only the employment relationship that has seen its traditional 'voluntarist' nature dismantled in recent decades. The same kinds of processes have occurred in the professions, in the City of London (see Moran, 2003), in the world of pensions, and more generally across most industry sectors. Everywhere, informal, locally established ways of running institutions have been replaced since the 1970s with new, formalised structures that impose standard rules on everyone, and help ensure that institutions are accountable for their actions and open to far greater public scrutiny.
- Finally, it is very reasonable to assert that plain political expediency has also played a part. Much of the regulation that has been introduced has been politically popular either in a general sense or among particular constituencies that governments have needed to court in order to gain electoral advantage.

PAUSE AND REFLECT

To what extent do you agree with the view that employment law, once established, is very difficult to repeal and that over time it is thus inevitable that our workplaces will become steadily more and more regulated?

Discrimination law

The major piece of legislation in which the detail of discrimination law is set out is the Equality Act 2010. This was largely what is known as a consolidation act: namely, a mechanism for bringing together in one place diverse statutes that had previously operated separately. There were some new rights that were introduced in 2010, particularly in the field of disability discrimination, but by and large the Act updated existing law to take account of court rulings and harmonised the terminology that is used across the various types of discrimination statute.

The Equality Act thus now prescribes broadly similar sets of protection for people who are discriminated against on the following grounds:

- Sex.
- Race, ethnicity or national origin.
- Disability.
- Religion or belief.
- Sexual orientation.
- Age.
- Gender reassignment.
- Marriage and civil partnership.
- Pregnancy and maternity.

Direct discrimination occurs when an employer discriminates against an employee on one of these grounds, causing them a detriment. Failing to promote a woman to a more senior role because she is a woman would be an example of direct sex discrimination. In such circumstances when an employment tribunal finds against an employer, with the exception of age and disability discrimination, in most circumstances there is no defence available. The most significant defence only applies at the point of recruitment and is known as the occupational requirement defence. This applies in rare cases when a job has to be reserved for someone with a specific characteristic because of its very nature. Hence, theatre directors who are casting for actors to play Lady Macbeth are entitled to specify that they wish to cast a woman to play the role.

In a case of discrimination on grounds of disability and age, a more general defence of objective justification is available. An employer has to satisfy the tribunal that it is necessary to discriminate against the individual for a good, legitimate reason and that its response to the situation was proportionate. It is thus lawful, for example, to deny someone aged 62 access to a graduate training scheme. It is also sometimes necessary to terminate the employment of someone who is very seriously sick and unlikely to return to work soon after a long absence. There is a need in disability discrimination law first to establish whether or not any 'reasonable adjustments' could be made to enable the disabled employee to return to work, but if there are none, it is lawful to dismiss the person.

Indirect discrimination is more common and can be entirely unintentional. This occurs when an employer has in place some kind of rule, policy or practice that has the effect of significantly disadvantaging one group more than another. There may be a perfectly reasonable explanation, but those who are disadvantaged have the right to require the employer to justify their practice in court as 'a proportionate means of achieving a legitimate aim'. In other words, the employer has to satisfy the tribunal that their discriminatory practice both serves a good, legitimate business purpose and does so in a fair and reasonable manner.

An example of these principles being applied is the case of *Naeem v Ministry of Justice (2017)*, which involved the courts having to determine whether the fact that

Muslim imams were on average paid less than Christian priests to carry out work as prison chaplains was discriminatory. The Ministry of Justice explained that it was entirely due to more of the Christian priests having long periods of service and hence being paid higher up the scale.

The design of the pay system had nothing whatever to do with religion and could not therefore amount to unlawful religious discrimination. The case was appealed to the Supreme Court where it was decided that this may in fact very well be the reason for the difference in pay, but that did not mean that the imams did not have every right to require the Ministry to justify their pay practices in court.

Unlawful harassment is the third major heading under which cases deriving from discrimination law are commonly brought to court. This can be a challenging area for employers, as harassers are very often fellow employees and are not acting in any kind of official capacity when they cause offence. However, because employers are deemed in law to be vicariously liable for the actions of their workers when at work, it is very often the employer that has to defend the case in court.

Harassment is also complicated by the fact that it is for the victim to determine whether or not the treatment concerned caused sufficient offence for it to count as an unlawful act. This occurs because of the way the term 'harassment' is defined in the Equality Act as a situation in which a worker suffers from unwanted conduct which violates dignity or creates 'an intimidating, hostile, degrading, humiliating or offensive environment'.

The employer can defend itself in such circumstances by demonstrating that it did everything a reasonable employer might be expected to have done first to prevent the unlawful harassment from occurring, and second in terms of its response when informed of an incident. It is thus important to have robust, written policies on harassment and training for managers in anticipating and acting to prevent it. When a complaint is made it is then essential that the employer can show it took decisive and appropriate action. This will invariably involve separating the people concerned and, if necessary, suspending the alleged perpetrator pending a full investigation.

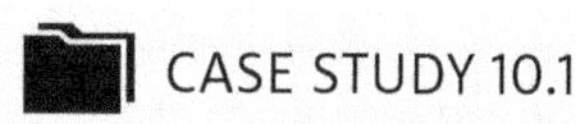

Zulu and Gue v Ministry of Defence (2019)

This case concerned two soldiers who resigned from the army in response to various acts of racial harassment. The issue at stake was whether or not the Ministry of Defence was vicariously liable, given that in its view it had done everything a reasonable employer could be expected to do in order to root out racism among its employees – such as fining and demoting soldiers found guilty of making racist remarks, a great deal of training and the drawing up of policy statements. However, the claimants in this case argued that the annual training carried out by the Army was really a tick-box exercise that had a cursory nature. Moreover, the Army had admitted publicly that there was more it needed to do to combat racism, while in practice some incidents were not logged formally. In other words, there were further reasonable steps that could have been taken and had not been. The Ministry of Defence was therefore vicariously liable and lost the case.

The law of contract

The law of contract forms part of the common law. It is thus largely judge-made and has not been created through Acts of Parliament. It is, however, no less important for that and must be taken account of in decision making on resourcing matters. Express terms are those agreed by the parties to a contract. In the case of employment contracts these will typically include job title, hours of work, pay and holiday entitlements. Implied terms are, in the main, present in all contracts because judges have ruled that they should be. Examples include the duty of care, the duty to indemnify (ie pay out-of-pocket expenses) and the duty to maintain a relationship of trust and confidence.

The key points are as follows:

- Employers need to be aware that a contract of employment is only established when there has been a clear, unambiguous and unconditional offer made and a clear, unambiguous and unconditional acceptance received. In addition, there must be an intention to create legal relations and some form of consideration (eg wages) made to bring it into existence.
- In principle, contracts cannot be changed unilaterally by one party. An employer wishing to make significant changes must first secure the agreement of the employees concerned. Where this is not forthcoming, they need to consider making the change financially attractive or, where there is no alternative, dismissing and rehiring on new terms. The latter course leaves open the possibility of an unfair dismissal claim. It is thus best to include in all contracts of employment one or more flexibility clauses giving the employer the right to make reasonable changes from time to time.
- Once a contract of employment is established, it confers duties on both parties. Both henceforward owe a duty of mutual trust and confidence to one another. In addition, the employer owes a duty of care along with several others, while the employee owes a duty of fidelity and the duty to exercise reasonable skill and care. If either side breaches one of these implied terms of contract, it can lead to breach of contract proceedings or can form the basis of a constructive dismissal claim.

There are a number of practical consequences for employers. First, care must be taken to ensure that the employees' understanding of what constitutes the key terms of their contracts matches what the employer believes their contracts to contain. This is best achieved through the existence of written statements of terms and conditions, which are given to new employees before they commence their jobs. It is not necessary to issue everyone with a formal, written contract of employment extending to many pages. A simple offer letter accompanied by a basic summary of the key terms is all that is required. This helps to create certainty and should ensure that accusations of breaches of contract do not occur.

Second, it is essential that employers explicitly build into the contracts that they offer employees a degree of flexibility. In other words, one of the terms of the contract needs to give the employer the right to make reasonable changes to terms and conditions from time to time in order to meet business requirements.

Third, it is important that everyone who undertakes a supervisory role is provided with basic training on the law of contract, and in particular on the issue of implied terms. The majority of managers in the UK remain blissfully unaware of the existence of such law and its potential significance, and it is this ignorance that leads to situations in which employees (and more commonly ex-employees) find themselves with a strong legal claim to pursue.

Family-friendly statutes

Recent years have seen the introduction of a range of new rights designed to make it easier for people with family responsibilities to combine these with a career. In addition, we have seen the extension of existing rights, notably the provisions for statutory maternity leave and pay. It is beyond the scope of this book to describe or evaluate these in any detail, but it is important that readers appreciate the extent of the rights that now exist and will exist in the near future. They are significant from a resourcing point of view because they provide rights for employees to be away from the workplace for much longer than has been the case historically. Aside from the need to administer such matters professionally, resourcing specialists also need to build assumptions about take-up of these rights into their human resource planning activities. The upshot is a need to hire more people to work on a temporary or casual basis (to cover absences), to increase staffing generally and to think more openly than has been the tendency to date about job-sharing arrangements and the possibility of allowing much more flexible working. The key areas of law that fall into this category are the following:

- The right to maternity leave/right to return.
- The right to maternity pay.
- The right to unpaid parental leave.
- The right to time off to care for dependants.
- The right to paid paternity leave.
- The right to request flexible working.
- The right to take shared parental leave.

This is an area of law in which employees' rights have improved very considerably over time in incremental steps. New rights, and extensions of existing rights, are introduced every two or three years. At the time of writing (2020), the government is reviewing the operation of the shared parental leave system that permits mothers and fathers to share periods of paid leave entitlement between them. It is possible that grandparents may be entitled to share parts of this paid leave too.

PAUSE AND REFLECT

How far do you agree with the proposal to include grandparents in the shared parental leave system? From an employer's perspective, what would be the major advantages and disadvantages?

Working time

The Working Time Regulations were introduced in 1998 and have been amended twice since then. The regulations are complex and have been heavily criticised for lacking clarity. The major rights they include are:

- A working week limited to a maximum of forty-eight hours (averaged over seventeen weeks).
- Four weeks' paid annual leave per year (in addition to bank holidays).
- A limitation on night working to eight hours in any one twenty-four hour period.
- Eleven hours' rest in any one twenty-four hour period.
- An uninterrupted break of twenty-four hours in any one seven-day period.
- A twenty-minute rest break in any shift of six hours or more.
- Regular free health assessments to establish fitness for night working.

There is currently a right for employees to opt out of the forty-eight hour week, and it remains lawful for employers to require that such a step is taken as a condition of a job offer. However, less known is the existence of a right to opt back in again without suffering any kind of detriment. There is a considerable question mark over whether the right of employees to 'opt out' of their rights under these regulations will survive into the long-term future.

Working time is clearly central to human resource planning and to the organisation of work generally. Hitherto, because of the opt-outs, the lack of clarity about some terminology and the lack of an effective policing regime, many employers have found themselves to be affected by these regulations only to a limited degree.

Confidentiality issues

A range of regulations have been introduced in recent years that affect the use of information in the workplace. The Public Interest Disclosure Act 1998 and the Telecommunications (Lawful Business Practice) (Interception of Communications) Regulations 2000 deal respectively with the rights of 'whistle-blowers' and of employers to snoop on their employees' phone calls and emails. A third piece of legislation, the Data Protection Act 2018, implemented the European Union's General Data Protection Regulation (GDPR). This sets out to prescribe what kinds of information can be held about employees and for how long. The Act covers paper records and information that is held electronically.

The law states that personal data should now only be processed when one of six 'lawful bases' applies:

- **Consent.** The individual has given clear consent for you to process their personal data for a specific purpose.
- **Contract.** The processing is necessary for a contract you have with the individual, or because they have asked you to take specific steps before entering into a contract.

- **Legal obligation.** The processing is necessary for you to comply with the law (not including contractual obligations).
- **Vital interests.** The processing is necessary to protect someone's life.
- **Public task.** The processing is necessary for you to perform a task in the public interest or for your official functions, and the task or function has a clear basis in law.
- **Legitimate interests.** The processing is necessary for your legitimate interests or the legitimate interests of a third party unless there is a good reason to protect the individual's personal data which overrides those legitimate interests. (This cannot apply if you are a public authority processing data to perform your official tasks.)

The Act gives workers the right to see any files kept on them, to receive copies of the data held and to correct inaccurate information. Employers are required only to process personal data that is 'necessary' and 'justifiable' and must keep it only for as long as is strictly necessary. Adequate security measures must also be taken. Stricter rules apply to 'sensitive data', such as information that relates to people's ethnic origins, religious beliefs, health records or past criminal convictions. This cannot be kept or processed without the employees' express consent.

Health and safety law

In terms of volume of regulation, there is as much health and safety law in the UK as employment law, but in the main it consists of specialised sets of regulations that are only relevant in specific types of workplace. General health and safety law that applies everywhere is actually pretty concise and straightforward. Staying on the right side of the law thus involves ensuring that some familiar, well-established management principles and protocols are always followed.

This area of law is complicated from the employer's perspective by the presence of criminal as well as civil laws that need to be observed. On the criminal side, the key piece of legislation is the Health and Safety at Work Act 1974, which places employers under a duty to ensure that the following occurs in practice:

- To maintain plant and equipment and to provide safe systems of working.
- To ensure safe arrangements for handling, use, transport and storage of hazardous equipment and/or substances.
- To provide all necessary information, training and supervision in the use of hazardous equipment/substances.
- To ensure that entrances and exits to buildings are safe and maintained.
- To provide adequate facilities and arrangements to ensure welfare at work.

In practice, this is achieved through the maintenance of a written health and safety policy which identifies who is responsible for maintaining health and safety in each area of business premises, and more detailed written risk assessments that identify hazards to health (mental as well as physical) and are acted upon appropriately. Ideally, employers look to build health and safety cultures in their workplaces so that everyone is aware of dangers and acts to minimise them.

The criminal law side of health and safety law is enforced by health and safety inspectors, who have the right to inspect premises without notice and to issue orders requiring improvements to be made within specified timescales. A failure to act on such a notice will then lead to a criminal prosecution. The same occurs when a major industrial accident occurs that results in either death or serious injury, or might well have done if people had been in different places at different times. Here, too, the principle of vicarious liability applies, so the employer is required to defend unlawful actions that are actually the doing of employees.

The civil side of health and safety law involves individuals who have been injured at work or who have suffered mental breakdowns caused by their work bringing personal injury claims in the county courts or the High Court. Substantial damages are often sought, usually through claims brought under the law of negligence. Here the law tries to strike a balance between employer and employee interests by making employers liable only where an accident can be described as 'reasonably foreseeable' and where an employer has failed in its duty to act reasonably and 'take positive thought for the safety of his workers in the light of what he knows or ought to know'.

CASE STUDY 10.2

Hatton v Sutherland

In 2002, ruling on four separate stress-based personal injury claims (always referred to as *Hatton v Sutherland* – one of the four), the Court of Appeal found against three ex-employees who had earlier won large amounts of damages in the lower courts. They found in favour of the fourth.

The grounds varied. In one case they said that the risk of further injury had not been reasonably foreseeable, in another they found that the workload had not been excessive. In the third the illness was found to have been 'readily attributable to causes other than workplace stress'.

In the process the Court set out guidance that tightens up on stress claims very considerably. It included the following:

- Employers are not obliged to make searching enquiries to establish whether an individual is at risk.
- Employees who stay in stressful jobs voluntarily are responsible for their own fate if they subsequently suffer stress-based illnesses.
- There must be indications of impending harm arising from the workload in order for an employer to take action.
- The employer is only in breach where the risk is foreseeable 'bearing in mind the size of the risk, the gravity of the harm, the costs of preventing it and the justification of running the risk'.
- There are no occupations that should be regarded as intrinsically dangerous to mental health.
- Employers who offer confidential counselling services with access to treatment are unlikely to be found in breach.
- The illness must clearly be caused by breach of duty and not just by occupational stress.
- Damages must be reduced to take account of pre-existing disorders or the chance that the claimant would have fallen ill anyway.

These principles were later confirmed by the House of Lords. The rulings make it much harder than it was before for employees to pursue personal injury claims based on stress suffered at work.

Dismissal law

The law of unfair dismissal gives employees in the UK a very important legal right, namely the right to claim compensation when dismissed for an unfair reason or in an unfair manner. It is, however, only a right that is generally available to employees who have completed more than two years' continuous service. Moreover, the courts have tended over the years to interpret the law in an employer-friendly fashion. This combined with the relatively low levels of compensation that are awarded to victorious claimants has tended to mean that many larger employers settle cases out of court and do not always pay much attention to the requirements set out in the law.

In most cases an employment tribunal faced with an unfair dismissal claim will start by establishing what it concludes is the main reason for the dismissal. Sometimes this falls into the 'automatically fair' category, meaning that the employer wins the case without the need to justify the precise actions they took. An example would be a dismissal on grounds of national security. Conversely, when the main reason for the dismissal is found to be one listed in the law as 'automatically unfair', the former employee wins and the employer loses. Examples are dismissals on grounds of pregnancy or for refusing to work in unsafe conditions. In these cases there is also no requirement to have two years' service. The right not to be dismissed for an automatically unfair reason applies from the first day of employment.

Most dismissals, however, fall into a third category, namely reasons that are 'potentially fair' in law. These include poor performance and ill health (ie incapability), misconduct and redundancy. When a tribunal finds that a dismissal occurred for one of these reasons, the matter is determined with reference to the manner in which the employer carried out the dismissal. Central here is the concept of 'reasonableness', which can vary depending on the size and resources of the employer. The tribunal does not, however, ask itself in general terms whether the employer did or did not act 'reasonably'. Instead, a test known as the 'band of reasonable responses' test is applied which essentially involves an employer trying to persuade the court that its action *could* be described as being reasonable. By this means, dismissals that might be considered unreasonably harsh by most people can nonetheless be found lawful because they just slip into the 'band of reasonable responses'. Recent case law, for example, contains numerous examples of the courts finding summary dismissals on grounds of gross misconduct to have been fair in law even when the employee had no conception when they committed the act that it would be anything like serious enough to warrant immediate dismissal without notice. All the employer has to do is show that what they did could by some people be seen as reasonable.

That said, for many years it has been established that a dismissal that is carried out without a proper procedure is considered unfair and outside the 'band of reasonable responses'. Employers must therefore, when dismissing any employee with over two years' service, ensure that their own established procedures are followed, or as a minimum the procedures set out by the Advisory, Conciliation and Arbitration Service (ACAS) code of practice on Discipline at Work. This requires a full and fair investigation by managers, followed by a formal hearing at which the employee concerned can be accompanied (and effectively represented) by a work colleague or trade union official. A right to appeal a decision following a hearing is also required. With the exception of gross misconduct dismissals of the kind described above, the ACAS procedure envisages that employees will be given formal warnings and an opportunity to improve their conduct, performance or health, before being dismissed. In the case of redundancies there is a requirement to consult with employees and their representatives in order for a dismissal to be found fair in law.

Since 2013 a new regulation has been introduced which has proved to be very helpful for employers who have an underperforming employee they would like to dismiss with a settlement rather than having to go through a full capability procedure with warnings that may take several months to complete. The reform introduced allows employers to hold what is known as 'a protected conversation' with an employee to discuss a settlement and an immediate departure on agreed terms. Such conversations can now be held on a 'without prejudice basis', which means that the fact one has taken place cannot subsequently be revealed in court if an employee refuses to settle and is instead later dismissed on grounds of poor performance using the full ACAS procedure.

PAUSE AND REFLECT

On balance, do you think it is too easy or too difficult for employers to dismiss people in the UK? What reforms to dismissal law would you like to see introduced?

Debates about employment regulation

It is not at all difficult to find fault with many individual pieces of employment legislation. Those representing employee interests tend to argue that measures 'do not go far enough', while employer associations tend to argue that businesses are over-regulated and that individual statutes 'go too far'. Hence, for example, unfair dismissal law is criticised both for favouring employers and for making it too difficult to dismiss underperforming employees. Data-protection legislation can be criticised for being too easy for employers to ignore, but also for being a wholly unnecessary set of regulations that create red tape and give employees rights they have no real interest in having. The law of indirect discrimination can penalise employers who have absolutely no intention whatever of discriminating unfairly against an individual, but can also be criticised for providing a defence that allows employers to justify practices that have the effect of perpetuating gender and racial inequality. However, there are some areas of employment regulation that are generally agreed to function badly and several types of criticisms that are made by protagonists from all sides of the debate:

- It is argued that some employment law fails to meet its own objectives in practice. It is thus both burdensome from an employer point of view and ineffective when seen from the perspective of the employee. Equal pay law is a good example. Despite its being on the statute books for over forty years now, women's average hourly rates of pay remain substantially below those enjoyed by men.
- Some employment law is very badly drafted, leaving much uncertainty about whether or to what extent it applies in particular situations. Too often, it is argued, governments have passed legislation or issued statutory regulations that

lack clarity. It then takes several years for the courts to establish what employers actually have to do in practice through the establishment of precedents in individual cases. The best examples are the Working Time Regulations and the Transfer of Undertakings Regulations. The latter apply in some (but not all) cases in which a business or part of an organisation is taken over or merged with another. The volume of case law in this field is immense, yet after forty years and major reform in 2006 there are still many grey areas.

- Unnecessary complexity is another general criticism of much UK employment law. Again, this creates uncertainty and makes it hard for employers to act within the law when carrying out their activities, even when they wish to. The best example is the law on employment status. Many statutes give rights only to 'employees' (eg unfair dismissal and parental leave) whereas others apply to all 'workers' (eg discrimination law and the National Minimum Wage). Yet these terms are not fully defined. So over the years the courts have had to devise tests to establish who is an 'employee', who is a 'worker' and who is neither. The law in this area has become far too complex and subject to change, leaving major groups such as agency workers unclear about what rights they have.

At a broader level, a further set of debates are carried out that focus on the full range of employment regulation. The question about which the protagonists disagree can basically be summed up as follows:

> Is increased employment regulation beneficial or harmful to the UK's economy and people?

Those who tend as a rule to argue that it has harmful effects include the Confederation of British Industry, the Institute of Directors and pro-business research organisations such as the Institute of Economic Affairs. At the heart of their argument is the claim that much employment regulation serves to place substantial additional costs on employers and that this has the effect of making UK businesses less competitive in international markets than they would otherwise be. While it is accepted that most western European countries impose even greater costs on their employers, this is not true of the rest of the world. Hence, in an increasingly global economy, regulation reduces the capacity of organisations to match the prices of goods and services originating in other countries. Small firms in particular are hit hard because they do not have the flexibility, profit margins or expertise to work within the requirements of the ever-growing volume of employment regulation. According to the CBI, the cost to businesses of implementing just the National Minimum Wage and the Working Time Regulations amounted to over £10 billion (Confederation of British Industry, 2000: 9). The costs can be categorised under several headings.

- First, there are the direct costs that employers assume as a result of employment regulation. Examples are the costs associated with the payment of Statutory Sick Pay, Maternity Pay, increased numbers of paid holidays and complying with the expectations of disability discrimination law.
- Second, there are knock-on costs that arise as a result of employees exercising their rights. A good example is the recruitment of temporary workers to cover when someone exercises their right to time off to care for a dependent relative or takes a period of paternity leave.

- Third, there are costs that derive from lost opportunities. Instead of spending their time running competitive businesses or providing efficient public services, management time is spent finding out how to comply with regulations, making changes to ensure that they act within the expectations of the law and showing regulatory bodies that they are complying in practice.
- Finally, there are the costs associated with litigation itself. Employers and their representatives often complain that even those who act entirely lawfully are often called upon to defend their actions in the employment tribunal. Ex-employees can bring cases at no risk to themselves, and even if they lose cause considerable expense on the part of the responding employer in legal fees and management time.

The total impact of these costs is to reduce competitiveness and hence slow down the growth of jobs. Moreover, where budgets and profit margins are tight, the net effect is to cause employers to shed labour and thus create unemployment. Over the longer term, employment regulation serves to give organisations a preference for expansion based on capital expenditure (buildings, machinery, etc) rather than expanding the number of employees. It also tends to encourage investment overseas in countries where employing people is a great deal less restrictive and costly. The net impact is fewer job opportunities for the UK workers whose interests the law is intended to serve.

This argument about employment law being counterproductive is often extended to focus on particular groups of workers. The most significant example is the position of women with young children or when an employer suspects that a woman will probably start a family in the near future:

> It is clear that many business people are very supportive of maternity benefits and rights (nearly a fifth of members provided more than the statutory maternity benefits in terms of leave or pay). But there is a clear warning from our survey. Already 45 per cent of our members feel that such rights are a disincentive to hiring women of prime child-rearing age. If the regulations are made even more burdensome then employers will be even more reluctant to employ these women.
>
> (Lea, 2001: 57)

According to critics, another way in which employees suffer as result of increased regulation arises from its tendency to impose on organisations single, standard ways of doing things. As a result, local flexibility is reduced. Whereas, previously, individual workplaces or even departments within larger organisations could devise their own informal workplace rules that met the needs of employer and employee alike, everyone now has to comply with a single centralised and often bureaucratic approach, often imposed by a strengthened corporate HR function. Local management discretion over pay, benefits and terms and conditions, for example, has had to be curtailed in order to ensure that equal pay claims can be defended. All manner of informal practices have had to be ditched thanks to data protection law. Not because any employee complained or ever would complain in practice, but because the law expects all organisations to standardise arrangements in ways approved of by the Information Commissioner. Too often, the changes introduced in response to such regulations serve to reduce the quality of working life enjoyed by employees rather than raise it. In particular, this mitigates against local, team-based decision making and enhances the power and reach of administrators.

PAUSE AND REFLECT

It is sometimes argued that smaller firms should be exempted from much employment law. What are the arguments for and against this proposition? To what extent do you agree with the idea that employment law should only fully apply to larger organisations?

According to Davies (2009), there are two types of argument that are commonly deployed against the critics of increased employment regulation. The first type revolves around notions of human rights and social justice; the second around its economic impact. The social justice arguments are the most straightforward to grasp. There may well be costs that have to be borne by employing organisations as a result of greater regulation, it is claimed, but these are justified because without it employees would suffer unreasonably as a result of the actions of employers. Employment law is necessary to protect vulnerable people who might otherwise be unjustly exploited by far more powerful employers. In the absence of effective trade unions, the state must step in to provide such protection. Much employment law is intended to give employees a degree of power to resist unjust treatment and hence to help reduce social injustice generally:

- Discrimination law is necessary because without it the position at work and in the labour market of women and certain minority groups would be a great deal worse. This is true of ethnic minorities who suffer from racial and religious discrimination, of gay and lesbian people, trade union activists, older workers and ex-offenders, all of whom had fewer opportunities and suffered greater prejudice at work in the days before they were protected by the law.
- Dismissal law serves to protect employees from being fired by their employers for no good reason or without first being given a reasonable opportunity to put right whatever fault the employer finds with their work. The alternative is a world of insecure employment where no employee, however effective and long-serving, can be sure that their job is safe from the whims or prejudices of maverick managers.

The economic arguments in favour of employment regulation are harder to summarise briefly and include several distinct strands. Further information is provided by Davies (2009) and in many publications of the Institute of Employment Rights (eg Deakin and Wilkinson, 1996; Institute of Employment Rights, 2000). However, the broad conclusion reached by their proponents is that, over the long term, the UK economy, as well as employing organisations, stand to benefit rather than to suffer from the regulation of labour markets. There are costs to be borne over the short term, but the overall effect is positive.

An important part of this case concerns tight labour markets. According to many influential economists, the major threat to the future growth of the UK economy is a lack of qualified people to carry out jobs and chronic skills shortages in particular industries. Recessions may be precipitated simply because the UK finds itself without the human resources needed to keep its economic engine running at full speed. This is a far greater threat to our international competitiveness than the costs associated with employment regulation. It is therefore in our economic interests for our government to force employers to provide workplaces in which people want to work and terms and conditions that attract them into employment.

Too many people do not work for one reason or another. Some take early retirement, others take time out to care for young children or elderly relatives, some are constantly leaving one job and taking time out while they search for something more satisfactory, while others find that they are better off overall by claiming state benefits of one kind or another. Some of these people have skills that are not currently being placed at the disposal of the economy, while others have the potential to gain these skills. Without employment regulation, it is argued, the number of such people would be higher because working in UK organisations would often be less attractive than it is now. Moreover, further regulation that serves to improve the experience of work, by making workplaces equitable and forcing organisations to employ people on decent terms and conditions, will encourage more people back into work and discourage others from leaving. Importantly, by the same token, employment protection legislation can help the UK to attract skilled workers from overseas to fill vacancies. The better the deal offered by UK employers, the more likely it is that talented people will want to come to live and work here.

A rather different economic argument is also commonly deployed by advocates of employment regulation. Here the focus is on the appropriate long-term strategy for the UK economy as a whole. It starts with an acceptance of the proposition that the UK is not, and will not for the foreseeable future, be able to compete internationally on the basis of low labour costs. However little employment regulation we have, developing countries will always be able to undercut us when it comes to the costs associated with employing people. It follows that some other basis has to be found on which to base our competitive future in a global economy. The only realistic choice is to focus on the development of hi-tech and knowledge-intensive industries that compete through their capacity to innovate and to produce high-quality products and services. Where we can compete on a cost reduction basis, it will only be through the development of machinery that reduces the need to employ people. It follows that employment law must play its part in pushing sometimes reluctant employers in this direction. If low-quality jobs, low pay and sweatshop-type conditions are effectively outlawed, the only alternative is for employers to pay well and to create higher-skilled jobs that enable them to compete on grounds other than price. In turn, this requires them to invest in training and hence gives them a strong incentive to retain people they have trained. Hence, employment regulation, along with complementary measures taken in fields such as education and research, funding is helping UK industry over time to transform itself so as to enable it to compete effectively in the modern global economy.

The third principal economic argument concerns productivity. To a great extent, the government accepts the arguments long advanced by researchers in the employment field that good HRM practice is linked to improved business performance. If you treat people well in the workplace they will respond with greater loyalty and with a willingness to work with greater effort. Moreover, they will choose to remain employed and will not continually be looking out for a job move. Beyond this general point is a strong belief on the part of ministers that partnership approaches to the relationship between management and staff are far more likely to bring about business success than autocratic styles or adversarial employee relations climates. Employment regulation can promote the establishment and maintenance of fruitful partnerships between staff (and their representatives) and management. This is partly achieved through requirements to recognise trade unions where that is the will of most staff, and partly through various requirements in the legislation to inform and consult with the workforce. More generally employment legislation serves to make it harder for employers to treat their staff in an inequitable or repressive manner. In so doing it plays a role in helping to create workplaces that are managed well and hence stand the best chance of achieving competitive advantage.

CASE STUDY 10.3

XYZ Ltd

XYZ is an independent software house employing 150 staff. Its business involves the production and ongoing maintenance of specialised software that is used by car dealerships for stock control and accounting purposes. New versions of its core products are developed every two or three years. The business is fiercely competitive, XYZ being a relatively small player in a large international market. It survives by maintaining its 500-strong customer base, serving these businesses well and involving them in ongoing developmental and maintenance issues through a 'user group' that meets each month.

XYZ employs a ten-strong sales team who are seen by senior managers as being central to the organisation's success and future survival. Their role is to maintain good relationships with established customers, to seek out new business wherever possible, to ensure that clients are happy to invest in new versions of software packages as they come on stream and to liaise with the 'user group'. The latter involves running formal meetings, which are always followed by social events at which XYZ managers entertain their clients late into the night in pubs, clubs and restaurants.

Alan McSlick has been the Sales Director at XYZ for as long as anyone can remember. He is widely considered to have managed the sales team very effectively, while also maintaining excellent relations with major clients. Despite attempts to persuade him otherwise, he has now decided to take early retirement. He and his wife now plan to use the substantial commission he has earned over the years to travel the world in some style. A replacement thus needs to be found.

It is decided that an exclusively internal recruitment exercise will take place, that Alan's job will not be advertised outside XYZ, and that a relatively informal selection procedure will be used to install a replacement quickly. No formal advertisement is drawn up. Instead, a meeting of the ten sales staff is called at which Alan McSlick's retirement is announced by the Chief Executive, Paul Double-Barrel. At the same time, he states that while Alan is working his notice a replacement will be appointed to work alongside him for a few weeks before he leaves. 'If any of you are interested in being considered for the position,' he says, 'drop me an email.' Later that day he receives four emails from long-standing members of the sales team putting themselves forward. The next day Paul meets with his Finance Director, his Company Secretary and Alan to make a decision about who should be promoted to the vital role of Sales Director.

The first candidate they consider is Julie Keene. Julie is a very good recent recruit and one of only two women currently employed as sales people at XYZ. The senior managers quickly decide to reject her application. She is a good deal younger than most of the team she would be managing and they doubt that a predominantly older male team would take at all kindly to being managed by a young woman. They are also concerned about the impact Julie's appointment would have on the mainly older, male customers who make up the user group and who are used to being entertained by others of a similar ilk after the 'user group' meetings.

The second candidate is Aldo Viscida. He is in his early forties and is a brilliant salesman. He has been employed at XYZ for some years, having emigrated from Milan in the 1980s. He regularly tops the monthly table for sales commission, having won many bonuses and prizes over the years. He is respected by the other employees on his team. However, his written English is poor and the Sales Director's role involves writing regular reports as well as much more written communication with clients than is required of the sales team. Aldo's application is thus also rejected.

The third candidate is Derek Constant. He is the longest-serving member of the sales team after Alan, and is generally considered to be his deputy in all but name. He is well liked among the user group members and would do a competent job in the Sales Director's role. On the downside, from the panel's point of view, is the fact that Derek is now 60 and so can be expected to retire soon. He has also recently told them in confidence that his wife has been diagnosed with multiple sclerosis and they fear that he will have to devote himself to her care sooner than he thinks. This would inevitably mean that he has less energy to put into the more senior role.

It is thus decided that the fourth candidate, Mike Replica, will be appointed to succeed Alan. Mike is 40 years old, very professional and has long been considered a possible future senior manager. He has plenty of interesting ideas about how to develop both the role and the team. The fact that he is married to the Finance Director's niece and regularly plays golf with Paul Double-Barrel are not considered to be problematic issues. After all, why should he be prevented from being promoted simply because of these relationships? It would be unfair to bar him on that account, particularly when he has the innovative ideas required to take the sales team forward and improve its performance.

Among Mike Replica's ideas are the following:

- Move to a payment arrangement that is wholly commission-based. At present the sales team add, on average, 20 per cent or so to their monthly salaries in commission payments. Mike would like all pay to be at risk so that 100 per cent of earnings were dependent on sales targets being met. Mike plans to introduce this new system with immediate effect.
- Dismiss, as a matter of policy, the poorest performing members of the sales team each year and replace them. The identity of the leavers should be determined purely on the basis of the value of new sales achieved (or not achieved). Not only should this policy result in poorer performers being replaced over time with stronger ones, it should also boost energy levels generally and increase the hours the team puts in.

What legal risks might XYZ be taking in following the course of action outlined in this case? What advice would you give the senior management team when running a recruitment exercise of this kind in the future?

References

Confederation of British Industry (2000) *Cutting Through Red Tape: The impact of employment legislation*, CBI, London

Davies, A C L (2009) *Perspectives on Labour Law*, 2nd edn, Cambridge University Press, Cambridge

Davies, P and Freedland, M (1993) *Labour Legislation and Public Policy*, Oxford University Press, Oxford

Deakin, S and Wilkinson, F (1996) *Labour Standards: Essential to economic and social progress*, Institute of Employment Rights, London

Institute of Employment Rights (2000) *Social Justice and Economic Efficiency*, Institute of Employment Rights, London

Lea, R (2001) *The Work–Life Balance and All That: The re-regulation of the labour market*, Institute of Directors, London

Moran, M (2003) *The British Regulatory State: High modernism and hyper-innovation*, Oxford University Press, Oxford

Shackleton, J R (2017) *Working to Rule*, Institute of Economic Affairs, London

Taylor, M, Marsh, G, Nicol, D and Broadbent, P (2017) *Good Work: The Taylor review of modern working practices*, https://assets.publishing.service.gov.uk/government/uploads/system/uploads/attachment_data/file/627671/good-work-taylor-review-modern-working-practices-rg.pdf (archived at https://perma.cc/KG83-YN7V)

11 Managing change

This chapter introduces readers to the subject of change management, a topic that has arisen in passing in many previous chapters, but has not yet been dealt with in and of itself. It is most certainly true that the landscape facing organisations of all kinds is changing, and arguably changing rather more rapidly than has been the case historically. In the opening chapters of this text we explored some of the various issues together, including the debates connected with globalisation, new and emerging technologies, labour markets and population amongst others. The factors shaping the landscape for contemporary organisations is certainly shifting rapidly, with firms forced to deal with unprecedented complexity, volatility and uncertainty. For leaders and managers, and this extends to leaders and managers within the HRM domain, the skills to successfully lead and manage change would appear to be fundamental to success in the modern working environment. These skills are only likely to become more important in the future, and indeed recent events such as the Covid-19 pandemic of 2020 have demonstrated just how rapidly change can be required.

The intent behind this chapter is to provide the reader with a thorough grounding in the subject of change management, introducing the various underlying debates, alongside some of the theory base in terms of the various models and frameworks that have been developed to help individuals lead change. Discussions begin by expanding upon the context for change, including the various forms of change that leaders and managers within organisations may encounter. It is after this point that the chapter takes something of a critical perspective through an examination of the pitfalls and problems that leaders and managers can encounter when attempting to initiate change within organisations. Within this part of the chapter the concept of resistance to change is discussed and debated, with the various sources and forms of resistance introduced. Readers are then taken on something of a tour through a variety of models that have been produced to help us understand and manage the change process. Well-known work from Kurt Lewin and John Kotter, alongside perhaps less well known ideas from Elisabeth Kübler-Ross, is included for discussion, with readers encouraged to adopt a critical perspective when examining the various frameworks of change. Towards the end of the chapter attention turns to how change may be facilitated in practice, with the issues of communication, employee voice and participation explored in some considerable depth. The intent of this chapter is therefore very much to provide the reader with an overview of the major concepts within the change management field, beginning to build awareness of the underlying theory and the application of that theory in practice.

Background and context

Previous chapters in this book, as noted in the introduction, have already highlighted a diverse range of issues impacting our organisations. New technologies discussed under the banner of the Fourth Industrial Revolution (Morgan, 2019; Schwab, 2016), including advances such as artificial intelligence (Iansiti and Lakhani, 2020), additive manufacturing (Steenhuis et al, 2020) and others, will likely cause substantial changes in patterns of employment (Frey and Osborne, 2013). Similarly, we have seen that shifts in demographics and alterations in personal tastes and preferences have profound implications for our organisations, alongside the continuing, and substantial, impacts of globalisation. Organisations are faced with a complex and shifting landscape that is arguably more volatile and prone to disruption than at any other point in our contemporary history, with complexity a key theme of the modern business environment (Reeves et al, 2016). 'Change' is frequently written about within both scholarly and practitioner outlets (see, for example, Colvin, 2015; Reeves et al, 2020), although, as noted in the introductory thoughts, it is an issue our organisations can, and do, struggle with. As Baxter and MacLeod (2008) found, organisational change efforts frequently fail to meet initial expectations, and these issues will be expanded upon during the next section of this chapter.

To begin discussions regarding change management, it is first of all important to recognise that change can come in many different forms and levels of complexity. Change may impact a whole organisation, or indeed an entire industry – for example, the impact of digital downloads and streaming services on the music industry (Urbinati et al, 2019) – or it may be far more localised in its effects. Change varies very substantially in its extent; in some cases there may be a rather nominal impact, whilst at other times change can have a transformative impact on individuals, organisations or industries. As a result of this, the term 'change management' covers a very diverse range of organisational situations, some being of far greater importance to different stakeholders than others. Broadly speaking, change situations that require management fall into the following general categories:

- Structural change.
- Cultural change.
- Technological change.
- Changing policies and practices.
- Relocations and ergonomic change.
- New organisational objectives or strategic directions.

First of all, change may involve structural alterations to our organisations. We may, for instance, move from a more traditional hierarchical arrangement of individuals to perhaps a matrix structure, or a layout based upon the idea of the self-managed team. Differently, we may find ourselves either needing to grow or shrink our staff base, or we may shift the emphasis of our operations in response to market changes. Cultural change is very different to this. Often conceptualised as altering the established attitudes and behaviours within organisations, it is much harder to achieve in practice than structural change, as it involves changing not just what people do, but their 'hearts and minds' as well. Substantial research effort has been invested in understanding cultural change (see, for example, Holbeche, 2005), with one of the key findings being that organisations adopting an incremental or emergent approach are substantially more successful. What this indicates is that cultural change cannot be approached in the same way as structural change; rather than imposing change via a top-down method, managers and leaders need significant emotional intelligence, setting a broad direction of travel and involving employees, enabling the change to be driven from below.

Change can also involve technology. As hinted at already in this chapter, the spread of technologies associated with the so-called Fourth Industrial Revolution (Morgan, 2019; Schwab, 2016) are already having a substantial impact within the workplace. Many organisations are automating elements of their operations, utilising advanced computing and artificial intelligence (Iansiti and Lakhani, 2020) to augment, or replace, the work undertaken by employees. Amazon, as an example, is making very extensive use of technology and robotics within its warehousing and fulfilment centres (Simon, 2019), with robots undertaking heavy manual work and routine tasks, enabling employees to be redeployed to focus on knowledge-intensive work where greater value can be added. Change may also come about through organisations altering their policies and practices. Intertwined with many of the issues we have covered so far within this discussion, organisations frequently need to adjust their policy frameworks or working practices: an example again would be Amazon altering the working practices of staff as it has introduced robotics into its fulfilment centres. Key here, however, is to ensure that all relevant elements of employment law are taken into consideration when instituting change, involving and consulting employees in order to embed the new methods of work.

Similarly, organisations may alter the location of work, or the layout of their workplaces. Again, involvement and consultation are both important here, and it is likewise important to ensure that changes are permitted within contracts of employment, or that altered contracts of employment are accepted by employees prior to the introduction of the proposed change. The 2020 global pandemic has sparked a fascinating shift in flexible and remote working, with organisations such as Twitter permitting employees to work permanently from their home location if they wish (Cogley, 2020; Paul, 2020). This is a sizeable shift in the world of work and creates a variety of opportunities; but it also presents a series of challenges that need to be tackled. What is known is that this change needs to be managed effectively, and work routines, practices and policies may need to be adjusted accordingly. For instance, if employees choose to work permanently from home is there any expectation that they will attend meetings or other events at an organisational base? How will IT equipment be maintained and updated? Who is responsible for increases in home utility costs such as electricity and heating? Whilst on the surface the changes proposed by Twitter, amongst other organisations, appear very straightforward, a complex series of issues appear once you dig a little beneath the surface.

Finally, substantial organisational change can be driven by the introduction of new organisational objectives or strategic directions. Faced with markets and environments that are increasingly volatile and complex (Reeves et al, 2016), managers and leaders frequently need to adapt their strategies and objectives in order to pivot away from risks, or to move towards new identified opportunities. At times, the extent and impact of such change can be relatively minor, but often it can result in major alterations to our organisations.

PAUSE AND REFLECT

Please pause here for a few minutes and reflect upon the points that have been introduced above. What examples can you find of organisations going through change driven by the introduction of new strategies or altered objectives?

There are countless examples of organisations altering their strategic directions, some with more success than others. For example, Carly Fiorina substantially altered the strategic direction of HP when she became CEO in the late 1990s (Tabrizi, 2015), with mixed results, whilst the return of Steve Jobs to Apple in 1997 signalled a marked shift in the direction of the organisation, and a slew of innovative new products such as the iPod and iPhone which propelled the organisation to become one of the richest in the world (Fell, 2011), and for a substantial period of the 2010s the largest company globally as measured by market capitalisation figures.

The key learning point that readers should draw from this section is that change management is a complex and multifaceted issue. There are various categories into which change can fall, and whilst some change episodes may be relatively limited in nature and extent, others can have substantial impacts on all areas of our organisations and their operations. It is known that the environments outside our organisations are increasingly complex and volatile in nature, and that change is ever-present for our firms. The following sections now explore change management in considerably more depth, beginning with a discussion concerning common problems that can beset the change process.

Change management: pitfalls and problems

Readers will know from discussions so far in this chapter that the change process can be complex and that change efforts can have far-reaching implications within our organisations. There are well-documented problems that can affect the change management process within organisations and this section explores a number of these, including poor planning, conflict, lack of consultation and involvement, and resistance to change. A practical example of problematic change management is also explored during this part of the chapter: the opening of Terminal Five at Heathrow Airport near London.

To begin this section, it is known that poor planning substantially impacts change programmes of all kinds. When seeking to lead change, leaders and managers must ensure that they understand the context before change programme are initiated (Kirby, 2019) and are clear about the alterations that are needed. This not only helps in initiating and facilitating change, but also in retrospective evaluation. If leaders and managers do not have clarity on the expected impacts of their change programme and what they hope the change will bring from the outset, it is very difficult to evaluate progress towards desired goals. Alongside this, managers and leaders often fail to create a sense of urgency for change, and this can be problematic as employees may be unaware of the pressures for change and why the organisation therefore needs to engage in change efforts. It is also important to appreciate that the change process tends to create 'winners' and 'losers': individuals and groups who may gain from a proposed change, and individuals and groups who may lose out. As far back as the work of Kurt Lewin (1943), it has been understood that conflict between those who may stand to gain from change against those who stand to lose out is a substantial issue that must be addressed. Breaking this problem down to its most simple form, HRM practitioners in particular must appreciate that organisations are pluralist in nature. Where a unitarist frame of reference is adopted – in other words, all stakeholders will unquestionably have the same assumptions and goals as one another and therefore buy into one shared vision or purpose – HRM practitioners will likely fail to see (or at least substantially underestimate) the potential for conflict between differing groups.

Within the research base that sits around the field of change management there are a large number of examples where change has been managed poorly, or where the change management process has encountered substantial problems. Baxter and MacLeod (2008), for example, have undertaken a major longitudinal research study exploring episodes of change management in twenty differing organisations. They found that, in practice, all failed to meet their initial expectations due to poor management of the change process. Baxter and MacLeod (2008) argue that communication practices were typically inadequate, with managers failing to take proper account of the insecurities and anger that employees may have been feeling during the time of change. The researchers also found that senior managers frequently failed to accept responsibility for failed change initiatives, at times going to substantial lengths to present data that was favourable to themselves, even though the vast majority of employees were aware that the claims being made were not accurate. Further cynicism was then generated as new initiatives were launched, which were in reality the same change programmes but with different names or branding. The researchers argue that change programmes were only successful, in their study, when employees were fully involved in the planning and decision-making processes; in brief, the saying 'people support what they help to create' carries with it a great deal of truth. Organisations frequently overlook or downplay the people side of change, failing to factor in the need to involve and engage staff. The following case study presents a practical perspective on just this issue.

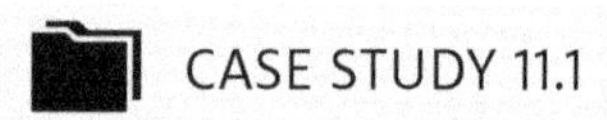

CASE STUDY 11.1

Heathrow Terminal Five

Heathrow is the UK's largest airport, and one of the most significant global hubs for air travel as ranked by passenger traffic. Heathrow is located to the west of London and has good transport connections, being in close proximity to both the motorway network and rail links. The airport has expanded substantially over its operational life, and at the time of writing there are ongoing discussions regarding the addition of a third runway. This case study focuses on the opening of Terminal Five at Heathrow, a substantial event in the airport's history, which was some twenty-five years in the planning. The terminal cost the airport operator BAA over £4 billion, building work took six years to complete, and it was opened by the Queen towards the end of March 2008 amid much public interest.

Immediately following the opening, however, there was an array of issues and 'technical glitches' that resulted in the cancellation of a large number of flights. Thirty-four were cancelled on the first day (Woodman, 2008), and around 300 flights in total were impacted during the problematic opening period. Upon arrival at the new terminal, large queues of passengers formed at the check-in and transfer desks, with the new baggage system creating a backlog of some 15,000 bags (BBC News, 2008), compounding issues caused by poor signage and passenger flow through the terminal. According to the BBC, the immediate cost to British Airways (the sole initial airline operating from Terminal Five) amounted to £16 million, with longer-term direct costs estimated at £150 million, together with reputational damage for themselves and the airport operator, BAA.

Press reports at the time noted that the main cause of the initial disruption was poor planning, and in particular the mismanagement of staff. Quotes in press reports (for example, see Woodman, 2008) highlighted issues with staff car parking, delays in employee security screening, and lack of staff familiarisation with the new terminal. In short, poor training meant that staff were unable to operate the new equipment effectively, and some were late for their shifts as they simply could not find somewhere to park their vehicle (BBC News, 2008). The BBC went on to highlight that some employees had not attended training courses and trial runs at the new terminal due to poor levels of morale and a lack of trust between themselves and company managers. Once operations began on that opening morning, long-standing issues with employee engagement meant that initial problems were compounded, resulting in the cancellations and economic losses previously outlined.

The key lesson that emerges from the well-documented case of Heathrow Terminal Five is that organisations must involve employees in change programmes. It can be argued that if employees had been actively engaged in this scenario, during both the planning and implementation of the change, then the teething problems may well have been far less severe, or avoided altogether. What is remarkable is that despite many managers and organisations understanding the importance of employee participation and involvement in change, so few put these principles into action. Very frequently, organisations engage in pseudo-consultation exercises, which give the impression of enabling employee voice, but in reality generate substantial levels of disappointment and discontent when it becomes apparent that in practice only the views of senior management have been taken into account.

PAUSE AND REFLECT

How might organisations involve employees in change programmes, avoiding the problematic consequences associated with pseudo-consultation exercises?

This opportunity to pause and reflect was intended to provide you with a chance to consider the importance of employee involvement, and methods through which effective involvement might be bought about. It is indeed a very real challenge faced by managers and leaders within organisations. The Heathrow Terminal Five example demonstrates that a lack of employee engagement and involvement can have deleterious impacts upon change programmes. What is clear is that organisations need to operate genuine two-way communication programmes, with information flowing in both directions between employees and managers/leaders themselves. This may involve employee forums, focus groups, suggestion schemes and other such mechanisms designed to source input from employees and their representatives. It is then important that managers and leaders are conspicuously seen to act on employee input. Where suggestions or input from employees are sourced, and managers/leaders are seen to ignore this, friction can be created between the groups. Employees may become apathetic or disengaged from the process, and resistance to change may become a problematic issue.

Resistance to change

It is very important that resistance to change is considered, and the reasons that individuals may be reluctant to engage with change efforts are explored in depth. Within the academic literature base, resistance to change is a hotly debated issue, with all manner of scholars discussing it from a variety of perspectives (see, for example, Hultman and Hultman, 2018; Shimoni, 2017). For many, resistance emerges from inside the individual as a personal reaction when confronted with substantial change (Shimoni, 2017). When seeking to manage change, and overcome resistance to it, interventions therefore typically focus on the individual in an attempt to overcome defensive reactions (Fincham and Clark, 2002). Scholars such as Schein (1991) argue that it is therefore important to understand psychological dispositions towards change, but others such as Ford et al (2008) argue that the focus here on individuals who require remediation, or 'fixing', is rather limiting. In certain instances, it is important to recognise that resistance to change is not wholly negative. As scholars such as Waddell and Sohal (1998) note, resistance can highlight unforeseen problems with a proposed change, and dialogue with resistors can positively shape the change process, resulting in improved outcomes. Leaders and managers must therefore avoid jumping to the presumption that resistance is necessarily a 'negative' aspect of the change process.

Whilst the points above are important, and vital for HRM practitioners to consider, there are a number of practical reasons that may cause individuals to display resistance to change. The first of these is simply a fear of the unknown (Govindarajan and Faber, 2016), or a sense of surprise that change is needed. Individuals inevitably become accustomed to certain ways of working, working within certain team environments, working in certain locations, or indeed simply the natural ebb and flow of the working day. Individuals tend to feel secure within environments and routines that feel familiar, and therefore changes that disrupt this can create a sense of trepidation and unease about new working arrangements. Tied to these points is the sense that change can present a challenge to the established order of status, skills and knowledge within an organisation. Where change represents (or is perceived to represent) a challenge to an individual's skill or knowledge base then resistance is often a natural reaction as our employees seek to protect their own status and indeed their continuing employment prospects. Ultimately, this resistance results from economic fears where change can be perceived as a threat to ongoing job security. Individuals may also resist change because of symbolic fears. For instance, an employee with a large office in a prime position within a building may resist moving to a more open-plan working arrangement because of the symbolism tied to having a prestigious office location. In short, individuals may resist change for a variety of reasons, and our role as HRM practitioners is to diagnose the sources and causes of resistance, developing compelling arguments to support our proposed change.

As mentioned earlier in this section, resistance to change is not to be seen in a wholly negative light (Waddell and Sohal, 1998). Resistance to change is indeed a multidimensional construct (Garcia-Cabrera and Garcia-Barba Hernández, 2014), and resistance is caused by a large number of factors (Lewin, 1951), which may include individual, organisational and contextual elements. At times, however, resistance to change can be caused where employees closer to the operational context spot problems with the proposals that managers and leaders were previously unaware of. In such situations, active and open dialogue throughout the organisation is important if we are to arrive at optimal outcomes. If leaders and managers, in light of such resistance, attempt to impose inappropriate change, this can result in substantially negative outcomes. The well-known writer Jim Collins reflects on contrasts between successful and unsuccessful organisations, and it is apparent within some of his contrasts and discussions that where leaders fail to listen to well-meaning critique, the resulting outcomes can be very problematic indeed (2009).

Overcoming resistance to change is a perennial problem for leaders and managers, as resistance can be caused by many factors, as we have already seen within this section. Many management and leadership scholars discuss how resistance to change may be overcome, with interesting contributions from Kan and Parry (2004) suggesting that the role of the leader is to identify and reconcile paradoxes, so that change is in line with the reality experienced by employees. Drawing on social network theory, Battilana and Casciaro (2013) argue that 'change agents' – in other words, the individuals sponsoring, implementing or otherwise managing change – strongly influence the change process through their relational ties with others inside organisations. The authors argue that if change agents can win over influential individuals who are ambivalent to change, the change process is likely to be successful. Above all else, leaders and managers must thoroughly understand the source of any resistance to change if they are to successfully tackle it. The next section of the chapter explores a model from Kurt Lewin, force field analysis, which may prove useful when attempting to map the positions of various stakeholders with respect to their views regarding proposed change. Finally, it is also important to understand that where employees are actively involved in the change process they are much more likely to support it. Involving employees enables organisations to actively take account of their views, with individuals then feeling some sense of ownership of the final outcome. The chapter now moves on to explore change management theory in more depth, considering some of the well-known and established models of change.

Change management: models

Within the academic literature base that surrounds change management there are a number of differing models and understandings related to how we may effectively manage change. Many of these models have persisted for a surprisingly lengthy period of time, and there are lessons that can be drawn from them all. Within this part of the chapter a variety of the established change models are considered, and at various points readers are invited to critically reflect upon their implications for the change process. The first commonly referenced model of change was developed by Kurt Lewin (1947).

Lewin: unfreeze–move–refreeze

Kurt Lewin is regarded very much as the founding father of what is understood as the field of change management (Cummings et al, 2016). His three-step model has arguably dominated western theories of change for a substantial period of time (Michaels, 2001), although there are questions about the applicability of his seminal theory in the modern business environment. Lewin's core idea was that in order to change people's views and obtain their support for proposed changes, leaders needed to adopt a three-step model, as represented in Figure 11.1.

Figure 11.1 Lewin's change model

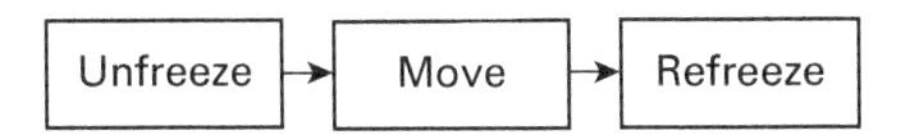

Lewin used the metaphor of ice to describe the change process, arguing that it took time to 'thaw out' starting positions, creating a state that was moveable before change could occur. He argued that managers and leaders needed to advance a convincing case for change, which – if accepted by employees – would then enable movement to a new state, before the change could be embedded through a 'refreezing' process. It is widely acknowledged that the embedding of change (refreezing) can take substantially longer than creating the initial movement needed, with managers and leaders needing to enable the change to firmly take root.

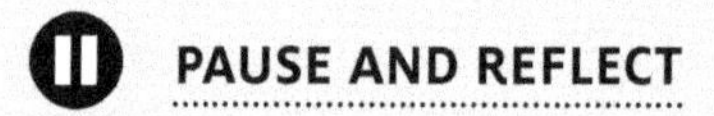

What are your initial thoughts about Lewin's change model? Do you think his theory continues to be relevant in the modern business environment?

Lewin's model is indeed interesting and is still useful in demonstrating the need for leaders and managers to create a convincing case for change in order to encourage movement. What must be recognised, however, is that Lewin's model presents change as a linear and episodic phenomenon. Essentially, change is something that, under Lewin's view, starts and stops. As was argued at an earlier point of this chapter, change is endemic within the contemporary business environment (Colvin, 2015), therefore to assume that there is a point at which a change 'stops' is arguably rather simplistic (Cummings et al, 2016). Lewin's model is nevertheless useful in demonstrating that leaders and managers must embed changes, otherwise individuals could simply revert to the previous way or method of operating, but it must be recognised that change is an endemic and continuous phenomenon.

Lewin: force field analysis

Not content with making one significant contribution to the change management arena, Lewin's second major theory was force field analysis (Lewin, 1943). This model argues that, within any change process, there are likely to be individuals and groups that benefit on the one hand, and other individuals and groups who might lose out as a result of the change. Force field analysis is depicted in Figure 11.2.

What must be understood is that the change process creates 'winners' and 'losers'; individuals and groups who are likely to benefit from the proposed change will generally support the proposals, whilst those who are likely to lose out will generally resist. By undertaking a force field analysis leaders and managers can swiftly ascertain which individuals, groups and stakeholders are likely to be positive advocates for change, and where resistance may be encountered. It should be apparent that if the forces resisting change are stronger than those driving it, then no movement occurs. On the other hand, if the forces in support of change are stronger, then change is much more likely to take root. According to Lewin, leaders and managers

Figure 11.2 Lewin's force field analysis

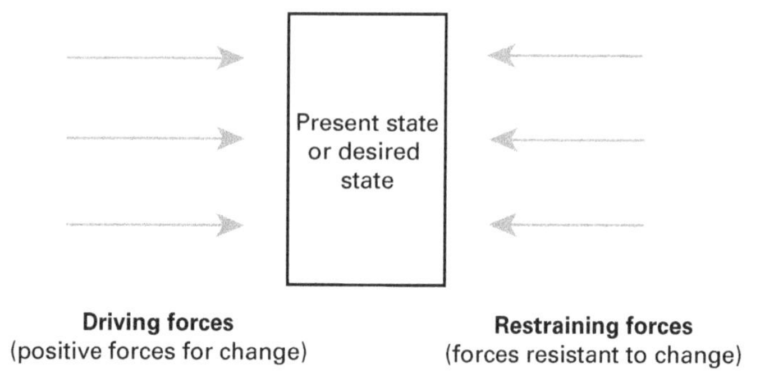

must attempt to strengthen the driving forces so that they may gain influence and gradually come to outweigh the resistors to change. Having identified the restraining forces, leaders and managers may also be in a position to proactively communicate with those individuals and groups in an attempt to better sell the positive case for change. If resistors to change can be won over, they often become powerful advocates supporting change processes.

Whilst force field analysis is indeed helpful in indicating some of the complexities within the change process, it is not without criticism (Swanson and Creed, 2014). The model still assumes that change is episodic rather than continuous in nature, and although it is helpful to consider the forces broadly in support of change versus those who may resist, the interactions between these groups are likely to be complex and multifaceted. To position individuals and groups as being diametrically opposed may be somewhat simplistic in reality, failing to take account of the nuances within the various arguments in support of or opposed to change. Change is rarely quick and neat within organisations; as Harford (2011: 42) states, 'what works in reality is a far more unsightly, chaotic, and rebellious organisation all together'.

Kotter: eight-step model of change

John Kotter's work within the change management field is very well known indeed (Kotter, 1996). In many respects it can be seen as a development of the work of Lewin, and a more sophisticated model of change management. Kotter's eight-step model is very widely cited in management literature and is a useful tool for leaders and managers to use when seeking to plan and execute change. The model was developed originally from Kotter's observations of leaders and organisations as they attempted to transform themselves (Kotter, 2020) and therefore has strong roots in management practice. The model is depicted in Figure 11.3.

Figure 11.3 Kotter's eight-step model

1 Develop urgency

2 Build a guiding coalition

3 Develop the 'change vision'

4 Communicate the vision

5 Empower others to take action

6 Generate short-term wins

7 Consolidate gains and produce further change

8 Institutionalise the new approaches

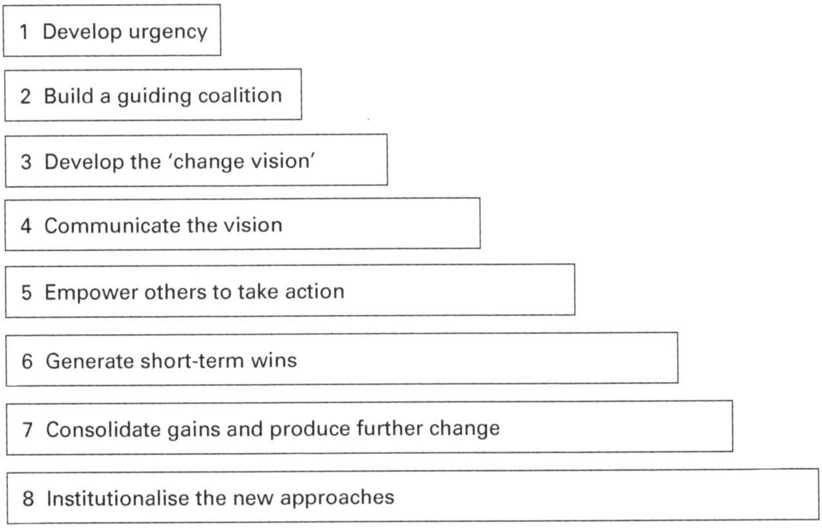

Writing in the *Harvard Business Review* in 1995, Kotter remarked that he had watched a substantial number of organisations attempt to change themselves, with some of these events producing remarkable success, and some producing very problematic outcomes indeed (Kotter, 1995). He argued, however, that most change episodes appeared to fall somewhere in between, neither being categorised as outright failures nor successes. The model that Kotter subsequently produced attempted to draw out core lessons for the successful management of change and we can see how the steps build from one to the other. Initially it is argued that leaders and managers must develop a sense of urgency for the proposed change. Sometimes this is obvious and apparent to everyone, such as the Covid-19 pandemic in 2020, but at other times leaders and managers need to generate conversation about the need for change, highlighting organisational problems or dilemmas. After creating consensus concerning the need for change, Kotter then argues that organisations must form a guiding coalition. It is recognised that change is difficult for any one individual to lead, and therefore the model suggests that we need to build a team with diverse skills, from various parts of the organisation (not just the senior management team), in order to lead the change programme. At step three, developing the vision, organisations are tasked with producing a compelling and inclusive roadmap for the change. What is it that we are proposing? Visions should be straightforward and encapsulate the need for change. They need to highlight the direction that our organisation must move in.

At the fourth stage of the model, communicating the vision, leaders and managers set about actively sharing their message with individuals throughout their organisations. At this stage, the strength of the 'guiding coalition' becomes apparent, as those closest to the message for change can be powerful advocates for the vision: namely, the 'change agents'. Communicating the vision is, however, insufficient on its own. In order to be successful leaders and managers have to empower others to act upon it (stage five). Organisations as a whole must seek to remove blockages or obstacles that may impede the change, and from an HRM perspective this means that HRM practitioners must ensure that their systems, policies and approaches to people management align with the vision. As Kotter states, where there are inconsistencies in messaging or the requirements of individuals, progress within the change process is dampened. To build momentum behind the change it is important that at stage six short-term wins are created. Quite often change programmes take substantial time and effort in order to bear fruit. In order to encourage employee buy-in, and generate greater commitment to our proposed change, it is useful where leaders and managers can show small, short-term improvements (quick wins) that have a positive overall effect within the confines of the larger change programme. Where individuals can see a direct, positive impact emerging from proposed changes, they will be much more likely to support the broader change programme. In essence this is a little like pushing a snowball down a hill; it is small when it begins, but grows upon itself as momentum is generated whilst it falls down the hill.

Towards the end of the model, stage seven – consolidating gains and producing further change – is concerned with sustaining momentum and continuing to push the change programme forward. Kotter argues that if victory is declared too soon (Kotter, 1995), leaders and managers run the risk that initial targets will be undershot, or that individuals will row back from the new ways of working that have been instituted. In short, the change programme will not have been cemented into the organisation's consciousness and the gains generated to that point could easily be lost. This very much builds into stage eight (institutionalise the new approaches), which is essentially where the 'change' becomes the norm. The change is firmly anchored within an organisation's working practices and culture, and it becomes just another part of the way that individuals operate or the way that we 'do business around here'. Whilst very useful in helping us to understand the change process in some depth, Kotter's model is not without criticism.

First of all there is the same issue here as with Lewin's earlier model of change management: the notion that change is episodic rather than continuous in nature. Kotter's model has defined start and end states, and it can be argued that in the contemporary business environment this may be an overly simplistic view of the change process. As noted previously in this chapter, change is ever-present within organisations, and whilst it is indeed important that we institutionalise new approaches (stage eight of Kotter's model), it must be recognised that it is likely that one cycle of change will lead into the next. Further to this, Kotter's model also says relatively little about resistance to change (Dent and Goldberg, 1999; Ford et al, 2008). Whilst Kotter's model does highlight the need to generate buy-in and the importance of effective communication, it would appear to be underpinned by rather unitarist assumptions of organisations and their management. In reality, the change process can be rather more 'messy' than Kotter would lead readers to believe, and indeed Ford et al (2008) argue that employee resistance can be a useful source of feedback for those leading change programmes, as was seen during the previous section of this chapter.

Elisabeth Kübler-Ross: the change curve

Elisabeth Kübler-Ross was a well-known psychiatrist, whose studies working with the terminally ill produced important findings related to the near-death experience and the grief process. Her seminal text, *On Death and Dying* (1969), illustrated her key contribution to literature, arguing that terminally ill individuals, and those who have lost someone close to them, go through five emotional stages: denial, anger, bargaining, depression and finally acceptance. More recently, theorists and writers associated with the change management field have co-opted her model, adjusting it a little to form a tool that leaders and managers can utilise when planning change, particularly when guiding an organisation through the aftermath when there is a tendency for those who opposed the change to continue their resistance. One interpretation of the Kübler-Ross model, as applied to the change process, is represented in Figure 11.4.

Time is depicted on the horizontal axis of the model, intended to represent the length of time that has elapsed since the start of the change process. The vertical axis is labelled morale/competence, intended to indicate the relative levels of morale and competence during the change process. Reading from the left of the diagram, the first stage an individual is likely to encounter when a change is announced or otherwise instituted is 'shock'. At this stage the organisation has communicated the upcoming change process, and this can be met with a level of surprise or shock from individuals perhaps unaware of the need or desire for change. This then feeds into the second stage, denial. Denial is, counterintuitively, a state at which morale and competence may in fact rise. Individuals may feel that the change programme will not be implemented, or that the change will not impact them. Essentially, here, employees are

Figure 11.4 The Kübler-Ross change curve

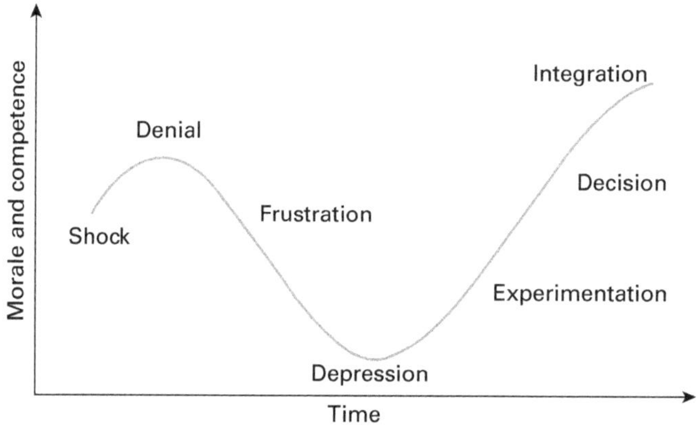

SOURCE: Author's own representation

looking for evidence that the announcement isn't true. Further along the continuum, individuals then enter a stage of frustration. This is a stage at which morale and competence will start moving in a downward trajectory as individuals realise that changes are, in fact, taking place, and that their role will be impacted. There is recognition at this stage that the organisation and the individual's job role will be different in the future, and this can cause anger, before stage four is arrived at: depression.

Depression is positioned at the middle of the change curve and is the state of lowest morale and competence, where individuals will typically display a low mood and will be lacking in energy. Employees may, for instance, be struggling at this point with new work systems or processes, or they may be experiencing difficulties adjusting to new environments if, for instance, the change involved moving work locations or developing different team structures. What the Kübler-Ross model demonstrates, however, is that this state is transitory, and morale/competence can begin a positive climb through the next stage: experimentation. At this stage there is acceptance of the change that may be brought about through individuals beginning to engage with new opportunities, or simply relief that the change itself has survived. The subsequent argument is that if experimentation is successful then individuals make a 'decision' (stage six) to accept the change, learning to work in the new situation. Finally, there is a stage of integration where the change becomes the norm and is fully embedded within the organisation and its culture.

The Kübler-Ross model certainly provides a useful insight into the change process as it helps leaders and managers to understand the states that individuals typically pass through. If movement through the curve is understood then HRM practitioners can best target interventions in order to support the change process. For instance, learning and development programmes might arguably have little impact if an individual is at the 'shock' or 'denial' stage, but might help to substantially improve performance at later stages of the curve such as 'experimentation'. What must also be remembered is that individuals move through the change curve at different speeds, and that they can move in either direction. It might conceivably be the case that an employee reaches the stage of 'experimentation' but has a negative experience of a new work process, and subsequently falls back into 'depression'. HRM practitioners must provide the right support and interventions at the right time in order to facilitate the change 'journey'.

Rogers: diffusion of innovation

In the early 1960s Everett Rogers produced what was to become a widely used model in the innovation literature. His 'diffusion of innovation' model sought to understand how innovations moved through a population, from initial release of the idea, service or product, through to mass popularisation. His core argument was that if an innovation was to sustain itself it needed to reach a critical mass of individuals in order to become embedded as the 'norm'. A representation of his diffusion of innovation model has been reproduced for readers in Figure 11.5.

This diagram is 'read' from left to right; the argument is that innovations (and in our context, changes) begin movement through a population first with the 'innovators', being adopted last by the 'laggards'. The 'innovators' tend to be very willing to take risks, and are very interested in new ideas. Little needs to be done to persuade these individuals to adopt a change; they are very willing to try out new technologies, working methods or systems. Likewise, 'early adopters' are also keen to embrace change. They are comfortable with new ideas but will want a little more information about the proposals before jumping on board. The 'early majority' tend to be a little more sceptical of change. They will adopt new ideas earlier than average, but they will want to see evidence of the change having been successful elsewhere before they commit. The 'late majority' are typically much more resistant to change. Individuals within this group will be reluctant to try a new innovation or engage with a change before it has been tried by a majority of others. Similarly, 'laggards' are very conservative in their thinking. They are very resistant to change and will be the most difficult group to win over. The key learning point from this model, with respect to change management, is therefore that different parts of an organisation's population will have different reactions to change. Granted, this is a generalisation, but there will inevitably be some individuals and groups who are more willing to engage and experiment with change than others. Leaders and managers may therefore be able

Figure 11.5 Rogers' diffusion of innovation

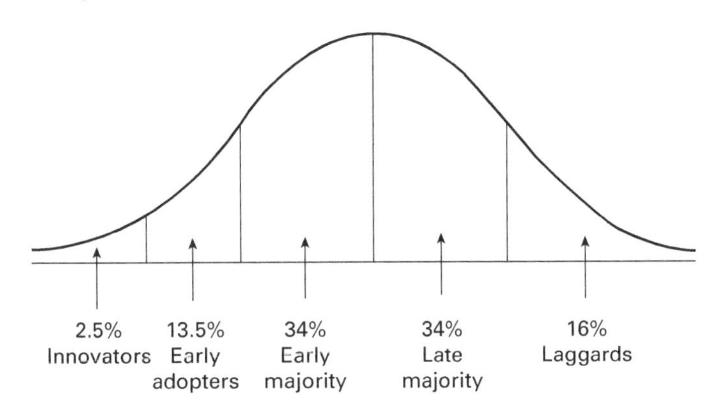

to utilise the identified 'innovators' and 'early adopters' as powerful advocates of change (the change agents), enabling the spread of the change message to the other groups within the wider organisation.

The various models explored within this section help us to understand more about organisational change, both in terms of the change process itself and individual reactions to it. As noted previously, however, these models are not without issues. One key problem is that they all appear to assume that the change process is linear; it starts and stops at defined points. In reality, change is endemic within the contemporary business (Colvin, 2015), and markets of all kinds are becoming increasingly volatile. That being said, these models certainly indicate important facets of the change process, points that will be picked up in the remainder of this chapter.

Facilitating organisational change

Facilitating organisational change is arguably the most complex and multifaceted role undertaken by HRM practitioners. In order to successfully facilitate change, HRM practitioners need to deploy an array of skills and deal with often unexpected problems and obstacles that come into view once a change programme has been initiated. The first key step within the change process, as has been explained within the body of this chapter, is making the case for change; as Kotter (1996) would say, leaders and managers need to develop the sense of urgency for change. In other words, clear rationale for the change process must be highlighted and communicated, and organisations must accept that this is likely to cause shock (Kübler-Ross, 1969) amongst those unprepared for or unaware of the need for change. Kotter (1996) goes on to note that change must be actively led. It is not about dealing with the consequences or fall-out as change is imposed, but about leaders and managers, including HRM practitioners, actively leading the change process from the start until such a time as the change becomes normal organisational practice. This may initially involve developing and selling a vision for the future, highlighting why the change is important for the organisation, and it will inevitably involve the provision of learning and development opportunities in order to help employees operate successfully in our changed context.

Communication is absolutely vital to successful change. Academic research frequently points to the important role of communication during times of change (see, for example, Smollan and Morrison, 2019), and indeed it is known that communication has a substantial impact on how employees perceive the change process (Saruhan, 2014). Research from Neill et al (2020) argues that where organisations utilise open and participative communication they will see more positive reactions towards change and significantly stronger commitment to change programmes from employees. Where communication is poor and evasive, or employees perceive it as incongruent with their observed reality, then it is likely that substantial resistance and mistrust will occur, which will, in turn, negatively impact the change process as a whole. This is shown within the research base where Smet et al (2016) argue that poor or insufficient communication about change can lead to the development of rumours that can cause fears about job security. Logically this stands to reason.

In the absence of clear and transparent communication from leaders and managers, employees will attempt to fill in the blanks, and misinformation is likely to spread within organisations. Where organisations actively foster open and transparent communication, change processes are likely to be substantially more successful. In these situations, employees are more likely to buy into the need for change, and it can be argued that they are more likely to move at a faster pace through the change curve (Kübler-Ross, 1969).

Tied together with the importance of effective communication is the need for employee voice and participation. At various points of this chapter so far employee involvement and participation have been considered in passing, with an argument being made that employees are very much more likely to 'support what they help to create'. Employee voice can be encouraged in a variety of different ways. Organisations may seek to operate suggestion schemes, town hall-style debating sessions, face-to-face briefings with senior managers or focus groups around specific issues, amongst many other interventions. Decisions about which method(s) may be applicable must take account of the nature of the proposed change, together with the makeup of organisations and their cultures. What is important to understand here is that efforts to encourage voice that are little more than 'window dressing' are likely to compound rather than resolve problems within the change process. For instance, if employees are asked for suggestions, or focus groups are convened around a particular topic (say, for example, the restructuring of a department), and the results are then seen (or perceived) to have been disregarded, this may then foster cynicism or indeed cause overt resistance to the change. Inevitably it will not be possible to implement all (or indeed at times most) employee suggestions or ideas, but feedback is important. Where leaders and managers communicate the reasons why ideas cannot be implemented then employees are likely to remain engaged; if feedback is not provided, or leaders and managers are perceived (by employees) to be avoiding issues, this can store up problems within the change process.

Employee voice is most certainly important when seeking to facilitate effective change, as is employee participation. Again, keeping in mind that organisations will very likely experience lower resistance to change when employees themselves actively participate in the change process, and that individuals will 'support what they help to create', it is crucial that HRM practitioners create opportunities for employee participation within the change process. Academics have long argued that employee participation has positive outcomes within the change process, with writers such as Nurick (1985) demonstrating the importance of participation mechanisms. Interestingly, Nurick argues that organisations must set clear boundaries around the change process, and that all key parties need to be involved in the change process as a whole. If groups are left out (or perceived to be overlooked) then it is argued that significant resistance will be encountered. Similarly, Nurick argues that training is important, and this training needs to be focused on helping others understand the mechanics of the change process, thus improving the outcomes for all. Much more recently, work from Nielsen and Randall (2012) again highlights that participation in change processes creates improved outcomes. They argue that employee participation has a direct role in the embedding of change, with participation causing greater employee commitment to change. Through employee participation, individuals within organisations can have a direct impact on the implementation of change, leading to substantially lower levels of resistance. HRM practitioners must therefore consider where, and how, employee participation can be encouraged within the change process. They may, for instance, seek to create project teams with a broad and inclusive membership exploring certain aspects of the proposed change programme, or delegate authority for decisions as far down the hierarchy as they are able to. Likewise, directly involving employees in the decision-making process can help to demonstrate the value placed on upward communication.

Conclusion

Change is an ever-present phenomenon for organisations. It is known that the markets and environments that firms operate in are less predictable than in previous times, characterised by greater levels of uncertainty, volatility and complexity. In these environments it is the organisations that are better able to change, and manage the process of change successfully, that are likely to thrive. Having discussed change in quite some depth within this chapter, readers should understand first of all that there are different types of change, including structural change, cultural change and change that might be brought about through technology, amongst other forms. Change management is therefore a complex and multifaceted field, and it is little surprise that there is a variety of examples of poorly managed change. Where the change process is poorly planned, where leaders fail to create a sense of urgency, or where organisations fail to appreciate the importance of clear communication with staff, enabling employee voice and participation, change processes frequently do not deliver the results that are sought. Change can indeed cause substantial resistance within organisations. Individuals and groups affected by change may resist for a variety of reasons, with leaders and managers needing to take time to understand the causes of resistance, rather than seeking to bulldoze change through by using sheer brute force. At times, resistance can have a positive impact on the overall change process. Perhaps employees on the front line may spot problems with a proposed working method, or an opportunity that could be exploited. This again highlights the importance of effective two-way communication, employee voice and participation.

The theory base that sits around the subject of change management is vast, with a substantial number of contributions emanating from researchers who have undertaken work in a variety of different fields. As was highlighted during the chapter, theories can help HRM practitioners to understand a variety of facets within the change process, such as the sources of employee resistance (Lewin's force field analysis) and the psychological journey that individuals go through when faced with change (the change curve from Kübler-Ross). By studying these theories, more is understood about the complexities of the change process, but leaders and managers must recognise that in many cases these models present the change process in a linear and episodic form. Organisations are faced with change on a continual basis, and therefore it is a little artificial to argue that there can be a defined 'end point' to change.

References

Battilana, J and Casciaro, T (2013) Overcoming resistance to organisational change: Strong ties and affective co-optation, *Management Science*, 59 (4), 819–36

Baxter, L and MacLeod, A (2008) *Managing Performance Improvement*, Routledge, London

BBC News (2008) What went wrong at Heathrow's T5?, http://news.bbc.co.uk/1/hi/uk/7322453.stm (archived at https://perma.cc/587H-9F3S)

Cogley, M (2020) Twitter to allow most employees to work from home permanently, www.telegraph.co.uk/technology/2020/05/12/twitter-allow-employees-work-home-permanently (archived at https://perma.cc/Q5KF-FNWN)

Collins, J (2009) *How the Mighty Fall: And why some companies never give in*, Random House, London

Colvin, G (2015) The 21st century corporation: Every aspect of your business is about to change, *Fortune*, 172 (6), 102–12

Cummings, S, Bridgman, T and Brown, K G (2016) Unfreezing change as three steps: Rethinking Kurt Lewin's legacy for change management, *Human Relations*, 69 (1), 33–60

Deloitte (2019) *Closing the Digital Gap: Shaping the future of UK healthcare*, Deloitte Centre for Health Solutions, Deloitte LLP

Dent, E B and Goldberg, S G (1999) Challenging 'resistance to change', *Journal of Applied Behavioral Science*, 35 (1), 25–41

Fell, J (2011) How Steve Jobs saved Apple, www.entrepreneur.com/article/220604 (archived at https://perma.cc/S3UD-7SH5)

Fincham, R and Clark, T (2002) Introduction: The emergence of critical perspective on consulting, in *Critical Consulting: New perspectives on the management advice industry*, ed T Clark and R Fincham, Blackwell, Oxford

Ford, J D, Ford, L W and D'Amelio, A (2008) Resistance to change: The rest of the story, *Academy of Management Review*, 33 (2), 362–77

Frey, C and Osborne, M (2013) The future of employment: How susceptible are jobs to computerisation?, Oxford Martin School Working Paper, www.oxfordmartin.ox.ac.uk/downloads/academic/The_Future_of_Employment.pdf (archived at https://perma.cc/9SW7-9M2S)

Garcia-Cabrera, A M and Garcia-Barba Hernández, F (2014) Differentiating the three components of resistance to change: The moderating effect of organization-based self-esteem on the employee involvement–resistance relation, *Human Resource Development Quarterly*, 25 (4), 441–69

Govindarajan, V and Faber, H (2016) What FDR knew about managing fear in times of change, *Harvard Business Review* Digital Articles, April, 2–5

Harford, T (2011) *Adapt: Why success always starts with failure*, Picador, New York

HM Government (2020) Coronavirus job retention scheme: Step by step guide for employers, www.gov.uk/government/publications/coronavirus-job-retention-scheme-step-by-step-guide-for-employers (archived at https://perma.cc/6TYE-KWUZ)

Holbeche, L (2005) *The High Performance Organization*, Routledge, London

Hultman, K and Hultman, J (2018) Self and identity: Hidden factors in resistance to organizational change, *Organization Development Journal*, 36 (1), 13–29

Iansiti, M and Lakhani, K R (2020) Competing in the age of AI, *Harvard Business Review*, 98 (1), 60–7

Kan, M M and Parry, K W (2004) Identifying paradox: A grounded theory of leadership in overcoming resistance to change, *Leadership Quarterly*, 15 (4), 467–91

Kirby, D (2019) Changing the nature of organizational change, *Strategic HR Review*, 18 (4), 155–60

Kotter, J (1995) Leading change: Why transformation efforts fail, *Harvard Business Review*, 73 (2), 59–67

Kotter, J (1996) *Leading Change*, Harvard Business School Press, Boston, MA

Kotter (2020) 8-step process, www.kotterinc.com/8-steps-process-for-leading-change (archived at https://perma.cc/MW92-HMVH)

Kübler-Ross, E (1969) *On Death and Dying*, Macmillan, New York

Lewin, K (1943) Defining the 'field at a given time', *Psychological Review*, 50 (3), 292–310

Lewin, K (1947) Frontiers in group dynamics: Concept, method and reality in social science; equilibrium and social change, *Human Relations*, 1 (1), 5–41

Lewin, K (1951) *Field Theory in Social Science*, Harper and Row, New York

(Continued)

(Continued)

Michaels, M (2001) *The Quest for Fitness: A rational exploration into the new science of organization*, Writers Club Press, San Jose, CA

Morgan, J (2019) Will we work in twenty-first century capitalism? A critique of the Fourth Industrial Revolution literature, *Economy and Society*, 48 (3), 371–98

Neill, M S, Men, L R and Yue, C A (2020) How communication climate and organizational identification impact change, *Corporate Communications*, 25 (2), 281–98

Nielsen, K and Randall, R (2012) The importance of employee participation and perceptions of changes in procedures in a teamworking intervention, *Work and Stress*, 26 (2), 91–111

Nurick, A J (1985) Human resource management in action: The paradox of participation: Lessons from the Tennessee Valley Authority, *Human Resource Management*, 24 (3), 341–56

Paul, K (2020) Twitter announces employees will be allowed to work from home 'forever', www.theguardian.com/technology/2020/may/12/twitter-coronavirus-covid19-work-from-home (archived at https://perma.cc/P6MR-6A7A)

Reeves, M, Levin, S and Ueda, D (2016) The biology of corporate survival, *Harvard Business Review*, 94 (1), 46–55

Reeves, M, Levin, S, Fink, T and Levina, A (2020) Taming complexity, *Harvard Business Review*, 98 (1), 112–21

Rogers, E M (1962) *Diffusion of Innovations*, Free Press, New York

Saruhan, N (2014) The role of corporate communication and perception of justice during organizational change process, *Business and Economics Research Journal*, 5 (4), 143–66

Schein, E H (1991) *Process Consulting Revisited: Building the helping relationships*, Addison-Wesley, Reading, MA

Schwab, K (2016) *The Fourth Industrial Revolution*, World Economic Forum, Davos

Shimoni, B (2017) What is resistance to change? A habitus-oriented approach, *Academy of Management Perspectives*, 31 (4), 257–70

Simon, M (2019) Inside the Amazon warehouse where humans and machines become one, www.wired.com/story/amazon-warehouse-robots (archived at https://perma.cc/9UXS-BRA9)

Smet, K, Vander Else, T, Grief, Y and De Witte, H (2016) The explanatory role of rumours in the reciprocal relationship between organizational change communication and job insecurity: A within-person approach, *European Journal of Work and Organizational Psychology*, 25 (5), 631–44

Smollan, R K and Morrison, R L (2019) Office design and organizational change: The influence of communication and organizational culture, *Journal of Organizational Change Management*, 32 (4), 426–40

Steenhuis, H-J, Fang, X and Ulusemre, T (2020) Global diffusion of innovation during the Fourth Industrial Revolution: The case of additive manufacturing or 3D printing, *International Journal of Innovation and Technology Management*, 17 (1), 1–34

Swanson, D J and Creed, A S (2014) Sharpening the focus of force field analysis, *Journal of Change Management*, 14 (1), 28–47

Tabrizi, B (2015) Carly Fiorina's legacy as CEO of Hewlett Packard, *Harvard Business Review* Digital Articles, September, 2–4

Urbinati, A, Chiaroni, D, Chiesa, V, Franzò, S and Frattini, F (2019) How incumbents manage waves of disruptive innovations: An exploratory analysis of the global music industry, *International Journal of Innovation and Technology Management*, 16 (1), 1–23

Waddell, D and Sohal, A S (1998) Resistance: A constructive tool for change management, *Management Decision*, 36 (8), 543–48

Wainwright, O (2020) How to build a hospital in nine days: Emergency architecture in a pandemic, www.theguardian.com/artanddesign/2020/apr/07/how-to-build-a-hospital-in-nine-days-emergency-architecture-in-a-pandemic-coronavirus-outbreak (archived at https://perma.cc/8Z22-N2AT)

Woodman, P (2008) Disastrous opening day for Terminal 5, *Independent*, www.independent.co.uk/news/uk/home-news/disastrous-opening-day-for-terminal-5-801376.html (archived at https://perma.cc/UWU4-G7DW)

12
Flexibility

This chapter explores the issue of flexibility within organisations. One of the key themes discussed within this text thus far has been the extent of change in the environment facing our organisations, and the unpredictable nature of the disruptions that managers and leaders frequently must grapple with. There have long been calls for our organisations to be more flexible and agile in order to deal with a context that is increasingly volatile and unpredictable, with substantial thought invested into this area from both academic researchers and industry commentators alike. 'Agility' as an issue will be dealt with in a separate chapter that follows this one, as it is a sizeable field in and of itself. The focus of discussions here is firmly set on flexibility, and this chapter will take the reader through various major debates connected with flexibility in contemporary organisations.

After first setting the context for discussions, the chapter will explore the major differing forms of flexibility utilised by organisations: namely, functional, numerical, temporal and financial. During each section, differing ways in which flexibility can be brought about will be examined, together with statistics and case studies where appropriate to illuminate the issues under discussion. A variety of critical perspectives will also be explored, together with a series of good practice approaches to flexibility. To begin with, the section that follows provides a background to the issue of flexibility and its importance to organisations.

Flexibility and its importance within contemporary organisations

In recent years, issues connected with flexibility have become a major preoccupation of both managers and commentators within the HRM field. These issues have come to the fore for many different reasons, but underlying all is the belief that organisations that gain a capacity for greater flexibility can develop sustained competitive advantage. This occurs for three distinct, but equally important, reasons:

- An organisation that is flexible is able to deploy its people and make use of their talents more effectively and efficiently than one that is not.
- The more flexible an organisation becomes, the better able it is to respond to and embrace change.
- Flexibility, particularly in terms of hours of work, is often valued by employees and can thus help to recruit and retain strong performers.

The term 'flexibility' thus covers a broad range of different areas of activity. It is useful to categorise these under two general headings: structural flexibility and cultural flexibility. The first refers to the types of contract under which individuals work and the architecture of the organisation. An organisation that is structurally inflexible could be characterised as one that employs all individuals on the same basic set of terms and conditions, which is made up of fairly narrowly defined 'jobs' into which individuals are required to fit themselves, and which is managed via traditional hierarchical structures. A flexible organisation, by contrast, deploys people as and when they are needed using a variety of contractual arrangements, and expects its people to work in a variety of different roles as and when required. These characteristics mean that it is better placed to respond quickly to changed circumstances and evolving customer expectations. Cultural flexibility is the other side of the same coin, being concerned with beliefs, attitudes and values. It is little use having an organisation that is structurally flexible if its people do not share a flexible mindset. Willingness to respond to change is as important a capacity to develop as the ability to do so, but it is harder to develop. Ultimately it can only be achieved by gaining the commitment of staff by promoting a working environment characterised by high trust, partnership and mutual respect.

Organisations have always sought to achieve a degree of flexibility within their operations. It is not correct to argue that flexibility is the sole preserve of the 'modern' organisation. For centuries workers have been laid off (either permanently or temporarily) when business levels dip, and offered premium overtime rates to work additional hours during periods of peak demand. What must be understood, though, as Rubb (2020) argues, is that changes in the prevailing business environment such as periods of recession can have a differential impact on workers. Those who are at the periphery of operations are more at risk of redundancy, as organisations will normally attempt to maintain their core of educated and well-trained individuals into whom they have invested significant sums to recruit and develop. This hints at a relatively new phenomenon within the flexibility field, which is the propagation of flexibility in the form of a model intended to guide management actions. In other words, it is the idea that organisations should deliberately develop core and peripheral structures as part of a considered strategy that represents a departure from historical practice.

Researchers and industry commentators present different ideas about why interest in the area of flexibility should have grown so significantly in recent times. In truth, the reasons are many and varied, and encompass several of the major contextual developments that have already been explored within this book. The following list of factors draws on a range of analyses and attempts to indicate something of the diversity in the reasons why interest in the issue of flexibility has grown. It draws on work from the CIPD (2019) and the Taylor Review of Modern Working Practices (Taylor, 2017), as well as more long-standing work from Bryson (1999), Heery and Salmon (2000) and Reilly (2001).

- Flexibility is a response to increased market volatility. Globalisation, technology and in particular the rise of e-commerce have meant that businesses are required to expand and contract more frequently in order to compete in markets which have become increasingly less predictable.
- New technologies have provided more scope for businesses to act opportunistically in response to customer demands. Smaller production runs are possible, as is the production of bespoke goods and services, meaning that flexibility is becoming more central to the achievement of competitive advantage.
- Interest in established Japanese management techniques has also fuelled interest in flexibility. The Japanese have long used approaches that are similar to the one advocated in Atkinson's flexible firm model (this model will be explored later in this chapter). Functional flexibility for core workers employed on a long-term basis is a particular feature of the traditional Japanese approach credited with generating substantial economic growth in the late twentieth century.
- The decline in the size and influence of trade unions over the past 30 years has made it easier for managers to introduce a greater degree of flexibility. There is less resistance to change and fewer demarcation disputes.
- Increased female participation in the labour market, together with greater interest in work–life balance issues, has meant that a greater proportion of potential employees are looking for atypical working arrangements. We have also seen an increase in the number of individuals looking to 'semi-retire', the number who have caring responsibilities for elderly relatives, and the number of young people seeking to combine part-time and temporary work with full-time education.
- Encouragement from successive governments seeking to create 'flexible labour markets' is a further factor. All recent administrations have sought to minimise unemployment by encouraging economic dynamism. Responsiveness to change and a high degree of efficiency are central components, along with encouragement of 'lifelong learning' and measures to remove barriers preventing individuals from participating in the workforce.
- A tendency towards short-term thinking in financial markets is demonstrated through an increased emphasis on maximising shareholder value. This, it is argued, leads managers to think in terms of short-term horizons. Long-term commitments to individuals are thus less necessary than used to be the case in order to develop a financially successful organisation.
- The evolution of a 'twenty-four-hour society' is also cited as a factor. This kind of society cannot, by its very nature, rely on employment of the traditional '9 to 5' variety. Customers need to be satisfied when they want to be satisfied. This requires far greater flexibility on the part of organisations operating in the service sector in particular.

- There is an apparent tendency during periods of economic recession for both employers and employees to look for alternatives to making large numbers of individuals redundant. As a means of reducing the number of potential redundancies, there have been many recorded instances of employees accepting reduced working hours and more flexible employment conditions.
- There is also evidence to suggest that in recent years the major impetus for moves towards greater flexibility has come from employees rather than employers. Work–life balance has become an issue that is considerably more important for many individuals, with employees increasingly seeking jobs that provide the flexibility to balance work and other pursuits. Research from Wynn and Rao (2020) does, however, note that navigating work–life 'conflict' remains a problematic issue for individuals.

PAUSE AND REFLECT

Before moving on from this introductory section of the chapter, take a few moments to reflect upon the major ideas that we have introduced so far. Make a brief list of different organisations and/or sectors that you think make extensive use of flexible working practices. See if you can, in broad terms, begin to identify the sorts of flexible working practices that these organisations/sectors make use of.

The opportunity here to pause and reflect should have enabled you to begin developing your own thoughts and ideas in relation to flexibility within the contemporary workplace. Whilst there are, of course, many equally valid responses to the activity above, some of the sectors of the economy that make extensive use of flexible working practices include agriculture and hospitality. Equally, we tend to see very extensive use of flexible working within the public sector, and many other organisations where there are peaks and troughs in the demand for goods or services. In terms of the specific flexible working practices that are utilised by organisations, you may have thought about multi-skilling, the use of temporary or part-time contracts, or indeed the ability to vary working hours over the course of a year. We will explore these and other ideas in the sections to come. To begin our in-depth exploration of flexibility in the workplace, we will discuss the issue of 'functional flexibility'.

Functional flexibility

Functional flexibility is a system by which employees can be deployed across a variety of tasks to accommodate variations in the patterns of demand experienced by an organisation (Mutari and Figart, 1997). Ultimately, organisations may benefit from functional flexibility as there is a reduction in the amount of time that employees spend idle: in other words, not engaged in a productive activity. This arrangement can therefore be argued to intensify work, and is based on the assumption that there will be variations in the demand for different tasks at different times (Desombre et al, 2006). Functional flexibility is of most use where organisations experience patterns of demand that vary unpredictably because the organisation can react at a much faster pace, matching employee tasks to demand without needing to add additional labour. Desombre et al (2006) therefore suggest that functional flexibility has much to offer organisations operating in the service sector in particular, although much discussion in academic circles has centred on the application of functional flexibility within large manufacturing operations.

Ultimately, a programme that aims to increase an organisation's functional flexibility is one that promotes multi-skilling. The aim here is to equip employees with a broader range of skills and competencies, which consequently increases the range of tasks that they are willing and able to perform. Where organisations utilise greater levels of functional flexibility we typically see:

- Reductions in the number of different job descriptions within an organisation (in other words, less demarcation between jobs).
- Substantially more team working – instead of employees A, B and C undertaking tasks 1, 2 and 3, all work together on all three types of task.
- Flatter hierarchical structures – fewer levels of grades defined by the type of work that is performed, with more individuals employed on the same grades carrying out the full range of activities.
- Job rotation to ensure that as many employees as possible are familiar with as many roles as is practicable.

The application of functional flexibility has two major advantages for organisations. As noted earlier in this section, it means that there is a substantial reduction in slack time within the organisation. Employees can be deployed within areas where they are most needed at any particular time, meaning those who are experiencing quieter periods can be rapidly moved to support colleagues in an area of operations experiencing higher demand. Ultimately, this means that employees are used more efficiently and that overall headcount within an organisation can be minimised. Where functional flexibility is utilised, it is also easier for organisations to cover for absence and lateness, and it also means that customers are served more quickly and to a higher standard. The second major advantage that accrues when organisations utilise functional flexibility is connected with change management. Where organisations are functionally flexible, they are typically able to respond faster to change. If one area of operations within the organisation grows, there is a ready supply of individuals familiar with the connected job duties who are able to transfer over without the need for extensive training. In practice, this means that organisations utilising forms of functional flexibility are able to take advantage of new opportunities more quickly than their competitors.

Whilst there are indeed substantial advantages for organisations utilising functional flexibility, there are also a number of potential drawbacks that must be considered as well. The first of these is that improving functional flexibility within a firm is costly. Training programmes have to be organised for larger numbers of individuals and the administrative costs associated with recording skills learnt and organising job rotation systems increase. Employees may also resist. Forcing individuals to multi-skill against their will can lead to substantial dissatisfaction within an organisation, and potentially the loss of good performers to other, competing, organisations. It is also entirely possible that some skills are too complex or specialised to be shared easily by several individuals. This may be where a substantial amount of prior training or hands-on experience is needed to build the requisite levels of knowledge, and if organisations attempt to force the pace in these situations this can lead to dilution, and a group of individuals who are unable to carry out the job as effectively as one single highly trained individual with substantial experience can.

Organisations seeking to introduce functional flexibility must therefore act with care. Employee involvement in the design and implementation of such systems is to be encouraged as much as possible, and it is important that leaders and managers address the legitimate concerns that employees may experience in a sensitive manner. Organisations must also accept that there are limits to the extent to which functional flexibility may be introduced in different situations, and that it must be introduced within reasonably defined parameters. In many organisations it is neither practical nor desirable for everyone to be able to carry out everyone else's role.

Numerical flexibility

In terms of the subject of numerical flexibility, the central argument is that organisations seeking flexibility should employ individuals on different forms of 'atypical contract' so that they can deploy their staff to meet peaks in demand. Of course, the extent to which atypical contracts can be utilised varies across industries and sectors: for example, the ability of manufacturing organisations to utilise part-time contracts, has often been limited because of the need to maintain a common shift system. Subcontracting, however, is very common within that sector. By contrast to this, the use of fixed-term contracts has grown in the public services, where funding to undertake specific projects is frequently limited in terms of time. Such contracts are also used extensively by employers whose workload increases and decreases on a seasonal basis. A good example of this is the agriculture industry, where at times of the year there is substantial demand for labour. This is illustrated in more depth via the following case study.

CASE STUDY 12.1

Seasonal work in the agriculture industry

Despite the UK's economic output relying heavily on the contributions of the service and knowledge sectors, agriculture still plays a very important role in our national economy, and, of course, produces much of the food and other products that we consume on an annual basis. Agriculture is, however, a seasonal industry. There are periods of substantial activity during the year, and also times when work is significantly less intense. These peaks include the period of time during which crops are harvested, and as the ONS (2018) highlights, the industry is very much reliant on seasonal workers as a result. These individuals play a very important role in helping to manage peak periods of demand, and very often they arrive from overseas, with the UK's agriculture sector historically being very reliant on migrant workers from the EU. The use of migrant workers to manage peaks of demand within the agriculture industry is, however, not something that is just seen in the UK. The International Labour Organization highlights that such practices are commonplace throughout the world.

Whilst the seasonal movement of individuals to undertake temporary roles within the agriculture industry is very well understood, the UK's exit from the EU has caused considerable discussion within the agriculture community. Various elements of media commentary (Agerholm, 2018; Leyland, 2019) have highlighted that moves to limit the supply of labour to fill the well-known seasonal gaps risk damaging the output of the industry. In addition to this, since the vote to leave the EU, fewer individuals based overseas have shown interest in undertaking short-term work within the UK (Lindsay, 2020). The combination of these factors has meant that the UK agriculture industry has seen an increasing struggle to match the demand for labour to its supply, problems exacerbated in 2020 by the Covid-19 pandemic, which largely shut down international travel (Lindsay, 2020). The future of flexible employment within the agriculture industry will certainly be interesting to observe as organisations adapt to the new context that they face.

Part-time contracts

Part-time working is by far and away the most common form of atypical working. In the UK almost 25 per cent of employees work part-time, with 74 per cent of those being female (ONS, 2020). In certain sectors of the economy, for instance hospitality, the figures are substantially higher than this. The biggest period of growth in part-time working occurred during the 1960s, when the proportion of the total workforce that worked on a part-time basis increased from 9 per cent to 16 per cent, largely due to the entry of more women into the workforce. Since that time, the percentage of individuals working part-time has steadily increased, and more recently male part-time working has been increasing at a faster pace than female part-time working.

In past decades there were clear incentives for organisations to employ two individuals on part-time contracts rather than one full-time individual. This resulted from an inequality of treatment in legal terms, whereby those working on a part-time basis could be denied pension scheme membership and other fringe benefits, and had to wait five years before they were entitled to bring cases of unfair dismissal. In the 1990s the courts ruled that such practices were indirectly discriminatory towards women, and this led to several legislative amendments. Subsequently in 2001 the UK government implemented the EU's Part-time Workers Directive, which made any discrimination on the grounds that an individual worked part-time potentially unlawful. Treating part-timers less favourably than full-timers is thus now largely consigned to history. However, it remains the case that at the lowest pay levels part-timers often earn less than the lower National Insurance threshold, meaning that employers do not have to pay contributions when these workers are appointed.

Ultimately there are two main reasons for the creation of part-time posts within organisations. The first of these, and the more common reason, is to enable organisations to respond more efficiently to peaks and troughs in demand for services. For example, restaurants and pubs may employ individuals to work on certain busy days, or to cover the busiest hours in the day (for example, evening meal service). The second reason is in response to a demand from employees, or potential employees, for part-time jobs to be created. As an example of this, research from Germany (Beham et al, 2020) highlights that individuals with family responsibilities are seeking part-time employment in increasing numbers, and this trend extends to individuals within more senior roles. Indeed, we know that a broader range of individuals are seeking part-time employment than was previously the case (Dunn, 2018), with this phenomenon being seen across contexts. One of the more common situations in which part-time working requests arise is when a woman who has previously worked on a full-time basis wishes to return as a part-timer following her maternity leave. In many cases, organisations find ways of accommodating such requests, either by reorganising job duties or by advertising for another part-timer to share the job. Generally speaking, however, the inflexibility of such arrangements makes them less easy to organise for managerial employees who need to be present throughout the week in order to supervise their departments effectively. Often a difficult choice has thus to be made between accommodating employee wishes in this regard or losing a valued member of staff.

Creating part-time jobs can bring considerable advantages to an organisation. It can reduce costs dramatically by making sure that individuals are present only when required and can also attract well-qualified individuals who, because of childcare or other commitments, are looking for less than full-time hours. Where organisations make more extensive use of part-time working arrangements it can also open them up to a broader pool of talent, which includes individuals approaching or beyond retirement age, or those still in education. There are, of course, potential disadvantages too. First there is the possibility that those working on part-time arrangements, because of other obligations and the fact that theirs is often not the main family income, will show less commitment to work than their full-time counterparts. This problem can be compounded if the individual working part-time sees that promotion and development opportunities are limited. Part-time staff can also be inflexible in terms of the hours they work because of the need to honour their other commitments. They are often attracted to the job in the first place because they need to be guaranteed fixed weekly hours, and are often thus either unwilling or unable to tolerate much change to these. There is also the more general issue of training investment. Where two or three part-time employees are in place of one full-time individual, the training time and cost will be two or three times higher. There is something of a myth that part-time employees are harder to retain than full-time staff. This is not borne out in the data, with only marginal differences in retention rates between full-time and part-time staff.

Temporary contracts

The term 'temporary worker', in practice, covers a variety of different situations. On the one hand there are staff who are employed for a fixed term or on a seasonal basis to carry out a specific job or task. A second group includes individuals who are employed temporarily but for an indefinite period. Their contracts thus state that they will be employed until such time as a particular project or body of work is completed. Again, this category can encompass well-paid individuals such as TV presenters and actors, in addition to those occupying less glamorous positions. A third category includes temporary agency staff who are employed via a third party to cover short-term needs.

Currently in the UK there are approximately 1.5 million individuals working on a temporary basis (ONS, 2020). This figure has remained relatively constant over time, although it is slightly down on the numbers seen in the early to mid-2010s. This figure represents around 5 per cent of the total workforce, but the figure does rise somewhat during the summer months as a result of seasonal work in the tourism and agriculture industries (TUC, 2019). The number of such workers tends to fluctuate depending on economic conditions, so we see a rise during periods of recession and high unemployment, followed by a fall during periods of economic recovery when labour markets are considerably tighter. Organisations tend to utilise temporary workers for a number of reasons, including covering for absent members of staff (whether for reasons of sickness or holiday), to provide specialist skills for a set period of time, or to deal with one-off tasks. On occasion, employers may also use temporary assignments as a way to screen individuals before permanent recruitment.

From the employer's perspective there are a number of compelling reasons to consider offering fixed-term contracts to certain groups of staff. Such arrangements are particularly useful when the future is uncertain, because they avoid raising employee expectations. It is also far easier not to renew a fixed-term contract when a department or division closes than it is to make permanent members of staff redundant. Hence employers may favour fixed-term arrangements over permanent employment at times. A note of caution is, however, needed here. Regulations encapsulated under the Agency Workers Regulations 2010, a response to the EU's Fixed-term Work Directive, set standards which organisations must adhere to when employing temporary workers. The regulations dictate that temporary workers must be treated no less favourably than permanent employees (subject to a minimum qualifying period of 12 weeks), and organisations must notify their fixed-term staff of permanent opportunities. There are also limits on the number of times that fixed-term agreements can be renewed, and since July 2006 if an individual has been employed on a fixed-term basis for four years or more their contracts can become permanent unless the employer is able to provide objective justification for a continuation of the existing contractual arrangements.

Subcontractors

Aside from the use of temporary agency workers, subcontracting comes in two basic forms. First, there is the use by employers of consultants and other self-employed individuals to undertake specific, specialised work. Such arrangements can be long-term in nature, but more frequently involve hiring someone on a one-off basis to work on a single defined project or time-limited task. The second form of subcontracting occurs when a substantial body of work, such as the provision of catering, cleaning or security services, is subcontracted to a separate company. Both varieties of subcontracting have become considerably more common in recent years, explaining some of the rise in the numbers of agency employees and individuals who operate on a self-employed basis. The numbers of workers within this category vary depending on which groups are included within calculations. Estimates from the ONS (2020) suggest that there are around 4.5 million individuals working on a self-employed basis in the UK as of August 2020. This number has increased over time, rising by around 1 million since the 1990s. The ONS also indicates that around 850,000 individuals are employed on an agency basis. In previous times self-employment used to be focused in the fields of technical and professional services, with the majority of self-employed workers being relatively well-paid men. What we see in more recent times, however, are individuals in lower wage occupations being hired on a self-employed basis using contractual terms that are often unlawful. These arrangements are being challenged through the court system, and this may lead to a reduction in the total figures of those working under self-employed and/or agency status over the coming years.

Organisations typically utilise agency staff in order to save on permanent employment costs, or in order to gain access to certain skills. Agency staff can be hired when permanent employees are absent or on holiday, or to help with seasonal peaks in demand. A good example of this is seen at Royal Mail, where large quantities of agency workers are hired to cope with the seasonal increase in demand for mail delivery around the Christmas period. There are, however, potential disadvantages associated with hiring subcontractors and agency workers. First, of course, is that it is often more costly to the organisation when compared on an hour-to-hour basis. Second, individuals contracted under such arrangements often display less commitment, as they have no long-term ties to the organisation. This phenomenon has been studied within academic literature (see, for example, Sobral et al, 2020; Woldman et al, 2018), with many researchers interested in understanding the commitment displayed (or not displayed) by temporary workers. In some cases, where there is a clear possibility of a longer-term, ongoing relationship with an organisation, contractors may have a strong incentive to 'over-deliver', but in the majority of cases the focus is on delivering services as agreed, with no desire to go beyond the letter of the contract. This is, of course, a generalisation, but organisations have to take such thoughts into consideration when determining whether and how to subcontract work.

Temporal flexibility

A form of working arrangement that, evidence suggests, is becoming more common involves a move away from setting specific hours of work. While such contracts come in different forms, all help in some way to match the presence of employees with peaks and troughs in demand. Such arrangements thus help to ensure that individuals are not being paid for being at work when there is little to do, while at the same time avoiding paying premium overtime rates to help cover the busiest periods. The types of arrangement that are most common are listed here, together with the total number of individuals in the UK covered by such arrangements. This data was published by the ONS in 2016 and is the most recent data available on this subject.

Flexitime: 2.86 million.

Term-time working: 1.38 million.

Annual hours: 1.3 million.

Zero hours: 744,000.

On call: 642,000.

Compressed hours: 225,000.

Job sharing: 144,000.

Flexitime

The most common, and least radical, departure from standard employment practices is the flexitime scheme. Precise rules vary between organisations, but such schemes involve employees clocking in and out of work or recording the hours they work each day. Typically such schemes work on a monthly basis, requiring employees to be present for a set number of hours a month, but permitting them (and their managers) to vary the times they are at work in order to meet business needs and, where possible, their own wishes. Organisations operating such schemes often identify 'core' hours when employees must be at work (for example 10.00–12.00 and 14.00–16.00) but allow flexibility outside these times. Individuals might therefore choose to begin work earlier or stay later in order to meet their hours.

The number of situations in which flexitime can operate, and in which it is appropriate to do so, are quite limited. Clearly, it is not a good idea to introduce such a scheme where the presence of a whole team throughout the working day is important. It would thus not be useful for roles where there is substantial direct contact with customers, where organisations are more goal focused rather than hours focused, or where a manufacturing process requires a large number of employees to be present at the same time. However, flexitime remains in many organisations – particularly where large numbers of clerical and secretarial workers are employed to look after a range of different bodies of work. Where deadlines are relatively unimportant, where individuals have responsibility for carrying out a prescribed range of tasks, and where there is no requirement to be available to members of the public all day long, there is a good case for using flexitime to maximise organisational efficiency. This is because it cuts out the need to pay overtime and also keeps headcount to a minimum. Such schemes are therefore popular in government departments, local authorities and other public sector organisations.

Term-time working

Term-time working has a great deal to offer both employers and employees, and appears to be a good deal more common in the UK than used to be the case. Under such arrangements, employees are contracted only to work during the school term times. During school holidays individuals do not work, but return to their roles again once school terms restart. These arrangements are particularly attractive to parents of younger children in two-income households. They also give employers an opportunity to tap into a labour market of talented and well-qualified potential staff who are otherwise unwilling to take a full-time year-round job. From an employer's point of view, especially for lower-skilled/semi-skilled roles, disruption needn't be a consequence because there is another group of well-motivated potential staff who are looking to take over the jobs that have been vacated as school holidays begin – namely university students looking for holiday work.

Annual hours

Annual hours contracts, often known as annualised hours, represent a more radical form of flexitime. The underlying principle is, however, the same; it is only that the amount of time worked can vary from month to month or season to season. Such arrangements vary from typical flexitime systems, not least because variations in hours are decided by the employer without much choice being given to the employee. Each year, all employees are required to work a set number of hours (usually 1,880), but to come in for much longer periods at some times of the year. Pay levels, however, remain constant throughout the year. Again, from the employer's point of view the aim is to match the demand for labour to its supply and therefore avoid employing individuals at slack times, and then paying overtime during busier periods. Variations can occur seasonally or monthly, or can follow no predictable pattern at all. In tourism or agriculture, where the summer is busy and the winter quiet, this can then be reflected in the time put in by employees. In an accountancy firm, where month-end and year-end reports need to be produced to tight deadlines, employees can then work long weeks followed by compensation in the form of shorter ones. Importantly, research from Ryan and Wallace (2019) argues that for annual hours schemes to be successful they must deliver, in practice, the mutual benefits expected by both employer and employee. Where these arrangements become a facade covering over the intensification of work, they quickly fail, with consequential damage to trust and employee commitment over the longer term.

PAUSE AND REFLECT

Why might employers, at times, find it difficult to deliver the mutual benefits associated with annualised hours systems?

The question above should have prompted you to think critically about annual hours systems. In theory such systems appear ideally placed to help employers match demand for labour to supply, but in practice they can be complex to implement. From the employer's side it can be very difficult indeed to predict the demand for labour in some instances, which leaves the organisation either paying overtime anyway, or employing more individuals than necessary. There can also be issues if employees leave part way through a year, either having worked substantially too few or substantially too many hours. In these situations it can be difficult to reconcile final pay arrangements, and such issues must be handled delicately. From the employee's perspective difficulties may ensue when organisational promises (perceived or actual) regarding the pattern of working hours remain unfulfilled in practice. For instance, where a period of shorter working hours cannot be taken following a period of substantial effort because of alterations in the pattern of demand or generalised increases in business activity. Such situations leave the employer vulnerable to violating the psychological contract. Having said this, where these sorts of issues can be successfully overcome, annual hours arrangements can be particularly attractive to both employers and employees, leading to lower levels of staff turnover and absence in the longer term.

Zero hours contracts

At the other end of the scale is the idea of the zero hours contract. This form of contract can be highly contentious indeed (Koumenta and Williams, 2019), and can be found in situations where casual employees work with an organisation on a regular basis. They are most suitable for situations in which there are frequent and substantial surges in demand for employees on particular days or weeks of the year, but where their instance is unpredictable. As a result of this, the use of zero hours contracts tends to be clustered in particular industries such as the distribution, accommodation and restaurant sectors (Farina et al, 2020). For instance, a distribution organisation needs to have a body of trained courier staff it can call upon to look after client needs, but may largely be unable to predict exactly how many staff it will need, and on what dates. Such organisations may therefore hire individuals on a casual basis, give them training, and then call on their services as required over time. These arrangements are therefore very flexible for both employer and employee, and can suit particular demographics very well indeed, for instance full-time students looking for work during holidays or at weekends. The case study below provides a window into the use of zero hours contracts.

CASE STUDY 12.2

Zero hours contracts and the fast-food industry

Zero hours contracts have often been used in the hospitality industry (Farina et al, 2020). Fast-food establishments such as McDonald's (Syal, 2018) and Burger King (Neville, 2013) are no exception to this, with organisations in the casual dining sector making extensive use of zero hours contracts in order to best match the supply and demand for labour. In many instances these arrangements have been beneficial for both employer and employee, but in some instances workers can be sent home shortly after arriving, or have shifts changed with very little notice (Neville, 2013). This means that working hours are often uncertain, with individuals unsure how many hours they will be able to work over a given period of time.

In 2018 the use of zero hours contracts by fast-food establishments such as McDonald's hit the headlines with employees from various branches taking part in a strike, dubbed by the media the 'McStrike' (Syal, 2018). Whilst zero hours contracts worked well for some individuals, others argued that they should be given a choice of fixed-hours arrangements. The strike action, which began on 4 September 2016, prompted McDonald's to review its employment arrangements. Subsequently the organisation offered its 115,000 UK workers employed on zero hours contracts the chance to move to fixed-hours arrangements (BBC News, 2017; Ruddick, 2017), with many choosing to move over to the new contracts (Ruddick, 2017). There have been calls from trade unions and other groups to ban the use of zero hours contracts (BBC News, 2019), although such arrangements are still permitted and used by organisations across a variety of sectors of the economy.

A variation of the zero hours approach is a system that guarantees casual employees a minimum number of hours a week, or days a year. Some supermarket chains take this approach, for example guaranteeing employees between 10 and 16 hours' work a week, but then adding to this (via overtime) if the workload increases or in order to cover absences among non-casual employees. The core guaranteed hours are fixed week by week, but additional hours are flexible and can be arranged with little notice.

Zero hours contracts are associated with the casualisation of work and are particularly criticised by those who see increased flexibility as a cover for employers seeking to shift risk onto employees by making their livelihoods less secure (Koumenta and Williams, 2019). As such, they are often seen as comprising poor HRM practice. This is particularly true of situations in which employers move from fixed hours to zero hours, as some have done in recent years. This is a trend that was picked up by the Workplace Employment Relations Survey, which reported a jump from 4 per cent of workplaces using zero hours contracts to 8 per cent between 2004 and 2011. The reported growth among large employers (from 11 per cent to 23 per cent) is particularly striking (Van Wanrooy et al, 2013: 10). In response, in 2015, the government banned the use of exclusivity clauses in zero hours contracts, although the 2017 Taylor Review of Modern Working Practices stopped short of banning zero hours arrangements, arguing that such contractual arrangements could form a positive part of work–life balance but that 'one-sided flexibility' was a key problem to be addressed (Taylor et al, 2017).

Compressed hours

Arrangements whereby employees work a compressed schedule are considerably more common than they used to be in the past, with the Workplace Employment Relations Survey tracking a steady increase in this form of contract over recent times (Van Wanrooy et al, 2013). The core idea is essentially very straightforward. Job holders work a full-time week (for example 40 hours) but do so over a compressed schedule, thus having more free days overall. The most common patterns are four-day weeks (10 hours a day) and nine-day fortnights (8.88 hours per day). These schemes avoid the need to pay overtime payments, instead rewarding individuals with additional time off instead of cash when they work longer hours. Research into this field from Deery et al (2017) argues that compressed hours arrangements can reduce absenteeism, while work from Noback et al (2016) studying the use of compressed hours in the financial services sector argues that such schemes can boost female career progression, although they do argue that men within those contexts can be 'penalised' for not working five days a week. Again, when seeking to implement compressed hours arrangements, organisations must ensure that these schemes offer mutual gains rather than one-sided benefits.

Financial flexibility

The final major form of flexibility covered in this chapter is that of 'financial' flexibility. What we mean by this is that just as organisations and individuals may engage in forms of flexibility that alter the roles that employees undertake at different times, or change hours or patterns of work, so too is it possible to instil flexibility within reward systems and employee benefit offerings. The first major development that must be investigated in this regard is that of performance-related pay (PRP). Commonly argued to be the most significant development in the field of reward management in recent times, the spread of performance-related pay systems through both the private and public sectors has been well documented (see, for example, Bregn, 2013; Ray, 2014). Whilst PRP can be implemented in a variety of differing ways, it is perhaps best defined as being a discretionary bonus or pay rise that is linked to an employee's performance. The forms of payment that such systems generate vary between organisations, but generally PRP systems will generate some form of bonus, a permanent pay rise or perhaps enable upward progression within an incremental pay scale. Schemes also vary in the amount of an employee's earnings that are 'at risk'. At the lower end, schemes may only generate outcomes worth a few percentage points of an individual's salary, while in other circumstances PRP arrangements may be worth as much as 25–30 per cent of an individual's pay settlement.

There have been very substantial discussions regarding the efficacy of performance-related pay systems, with lively debate within academic circles stretching back some considerable time. To give just a brief insight into this territory, some authors have looked at the impact of PRP upon creativity and innovation – seeking to understand whether PRP schemes improve innovative work behaviour (De Spiegelaere et al, 2018). Literature within this particular area is rather conflicted, however, with some researchers arguing that small monetary bonuses improve creative effort, while schemes that place a larger percentage of salary 'at risk' damage innovative performance. The underlying thinking here is that where the element of PRP is more substantial it causes employees to become risk averse, thus negatively impacting the creative process. Other studies have separately argued that PRP can have a role as an anti-corruption tool (Sundström, 2019), and, indicating yet further diversity in this literature base, Peluso et al (2019) argue that where PRP is implemented in healthcare settings it can lead to improved patient health outcomes. Interesting work has explored the impact of PRP within education settings with, again, mixed results. Work from Lundström (2012) indicates that PRP schemes within schools had a negative impact on staff motivation, whilst other work from Atkinson et al (2009) argued that pupil test scores showed a substantial improvement where teachers were rewarded via a PRP scheme. What these differing extracts demonstrate is that the literature connected with PRP includes substantial debate, with, at times, little agreement regarding the impact of PRP on employee performance.

Despite the debates regarding the ultimate efficacy of performance-related pay systems, there are some known advantages that such systems can bring. They help organisations to attract and retain good performers and, in general, have been shown to improve individual and corporate performance. PRP schemes also help to explicitly link individual effort to organisational objectives and also help to reinforce management control. From a flexibility perspective specifically, PRP schemes help organisations to match outgoing reward to employee and business performance. At times, where output or sales are lower, less money is paid to employees, whereas at times of high output or sales, larger reward packages are provided.

PAUSE AND REFLECT

What types of occupations or roles do you think are most suitable for performance-related pay arrangements?

The question above should have caused you to think carefully about the application of PRP schemes. There are clearly some roles where PRP arrangements are more suitable than others. Consider, for instance, a sales environment such as a car dealership. Here it is very common for the sales staff to be rewarded via a base salary and then a substantial commission arrangement based on the number of cars that they are able to sell to customers. As a result of this, when sales volumes are high, more money is paid to staff, but when sales volumes are lower, less commission is paid. This type of pay arrangement works well in an individualised working environment and where there are objective targets to meet. This is not the case in, for instance, education settings, where there are considerably more variables influencing performance and it is much less possible to set clear, objective outcomes against which an individual's performance can be measured.

Despite PRP systems having some notable advantages, these arrangements can frequently be problematic as well. For instance, they can be demotivating to staff over the longer term, and can be destructive of teams – particularly where objectives are set at the level of the individual rather than the team as a whole. Others argue that PRP represents a problematic form of management control, and there can also be practical problems associated with the implementation of these pay arrangements. Where schemes are poorly designed, are poorly communicated, or are not based on objective measures of performance, they can lead to substantial problems within organisations. Commentators, however, often reach a contingency conclusion in relation to PRP schemes. Recognising that PRP is not universally applicable, the common understanding is that such pay arrangements can work well for particular job groupings and sectors where these schemes are appropriate to the needs of the business and where clear objectives can be agreed upon.

A further form of financial flexibility is that of flexible benefit schemes. Sometimes known as 'cafeteria' benefit schemes, discussions regarding flexibility in the benefits that employers offer to their staff have persisted for a surprisingly lengthy period of time. As Barringer and Milkovich (1998) highlight, as far back as the 1970s firms were beginning to offer employees choice in terms of the benefits that they received. And this is ultimately how such schemes are defined. A flexible benefit scheme enables employees to select from a menu of available options (hence the term 'cafeteria' often being utilised), enabling individualisation of employee benefits. Firms may, for example, offer employees the chance to purchase more annual leave, or medical or dental cover. Others may offer childcare vouchers, increased pension contribution levels, gym memberships, or life assurance policies. Essentially, employees can select from a predefined list of options, with employers having specified relevant boundaries in advance. For instance, there are legal restrictions regarding minimum periods of annual leave. Ultimately, it is argued that flexible benefit schemes tend to increase employee satisfaction as individuals have the ability to select a benefits package that suits their personal situation. Barringer and Milkovich (1998) also highlight that such schemes also tend to be more cost effective as employers are not committing funds to benefit provision that employees rarely utilise.

Flexibility: a critical perspective

The move towards greater flexibility, whilst offering substantial benefits to organisations and individuals alike, is not without criticism. Within the debates four types of broad argument are discernible, although all in some way are characterised by a tendency to equate the concept of 'flexibility' with that of 'employment insecurity'. This tendency to link increased flexibility with increased insecurity in employment is not necessarily a new argument. As work produced by Jenkins (1998) and Saloniemi and Zeytinoglu (2007) demonstrates, these debates have persisted for some years already. In more recent times there have been substantial discussions regarding the increases in casual and precarious work (Kalleberg, 2009; Manolchev et al, 2020), particularly associated with the 'gig economy' (Petriglieri et al, 2019) – a topic that we address in the chapter on agility and resilience. In short, we must recognise that there is substantial critical discussion connected with the subject of flexibility in the workplace.

The first major area of debate explores the connection between flexible working practices and employee commitment, with commentators arguing that flexibility is incompatible with high-commitment work practices. It is argued here that organisations cannot have it both ways in this regard. They cannot, on the one hand, create a situation in which human resources are deployed as and when required so as to maximise short-term efficiency, while also requiring staff to exhibit a high degree of commitment. Flexibility, because it creates insecurity, is associated with lower commitment on the part of employees. In consequence, staff turnover rates will be higher in firms that make extensive use of flexible working practices, while employees are less likely to display discretionary effort and perform above and beyond the basic requirements of their contracts. Ultimately the suggestion is that individuals who perceive themselves to be peripheral to an organisation's operations will act accordingly.

The second set of critical perspectives with respect to flexible working practices relates to the impact of these practices on the health of the national economy as a whole. There is concern in some quarters that flexible organisations tend to seek to hire staff who are already well trained in respect of the role(s) that the organisation requires them to undertake, refraining from investing in further staff development as a result. Why invest in training individuals who are classed as peripheral and with whom the organisation expects to have a relatively short-term relationship? Ultimately, the impact of such decisions is likely to be a situation in which there are chronic skill shortages, leading to reduced economic output as employers find themselves unable to take up business opportunities and, over time, economic decline will be a consequence of this. Some commentators have indeed already pointed out that the UK economy faces a shortage of skills in a number of areas, and academic research emerging from many advanced economies (see, for example, Cappelli, 2015; Healy et al, 2015) paints a similar picture.

Instead, it is argued that employers should be actively encouraged to invest in individuals with whom they expect to have a long-term relationship. It follows that the best interests of the economy over the longer term are met by expanding traditional, full-time work. UK government policy within this area is rather disjointed. On the one hand there is a range of protections in place to support flexible workers (eg those who work part-time, agency workers, individuals employed on zero hours contracts and so on) and a universal 'right to request' flexible working. On the other hand, recent measures include the introduction of an apprenticeship levy (Jeffery, 2018) on larger employers and tax credit changes that provide incentives for part-time employees to work longer hours.

Moving away from debates about the impacts of flexible working policies on skills and employee commitment, there are also those who argue that flexibility can lead to unethical practice within organisations. The core concern here is that managers will inevitably be tempted to use the development of a peripheral workforce to exercise too great a degree of power. People who perceive their position at work to be precarious are understandably keen to seek greater security. Peripheral workers will thus try, wherever they can, to gain admittance to the core workforce. This, it is argued, gives employers considerable leverage over them – power which can easily be abused. The result is often a situation in which managers intensify work to an unacceptable degree, requiring individuals to work longer hours than is good for them and generally exploiting their vulnerability. More generally, when individuals are employed via a variety of different contractual arrangements, but are required to work together in a team, damaging perceptions of injustice are easily fermented. This is due to the inevitable differences in terms of levels of pay, job security and development prospects that exist between individuals. Unfairness arising from differential treatment is thus more likely in a flexible firm than in a firm that relies on more traditional employment arrangements.

The final group of criticisms levelled at flexible working practices is that they are often socially undesirable. Widespread adoption of more flexible employment practices will lead to a more unequal society, more so than is the case today. We will arguably see a clearer evolution of two distinct classes of employed individuals, one of which consists of people in secure, well-paid employment, whilst the other is composed of insecure, peripheral workers who have substantially less security and substantially fewer development prospects at work. The result of such a situation may well be socially undesirable outcomes such as a rise in crime, increases in the incidence of family breakdown, and increasing political alienation. These are all outcomes that arise when a society becomes less socially cohesive. This then in turn requires government intervention, which necessitates higher taxation and consequently less demand for privately produced goods and services.

Flexibility: good practice approaches

Ultimately, the key question that managers and leaders within organisations need to grapple with is how to engender flexibility within their operations, whilst, at the same time retaining and, if possible, increasing levels of employee engagement and commitment. This is a central contemporary HRM conundrum. Whilst this is not an easy problem to address, a number of researchers and commentators have made useful suggestions.

Atkinson: flexible firm

One of the main theoretical models within this particular subject area was developed by John Atkinson (1984). His model of the flexible firm has been highly influential and was based heavily on his understanding of Japanese management practices. It often serves as the starting point for serious discussions regarding how an organisation may increase its level of flexibility. One of the most important features of the Atkinson model is the distinction that is made between core and peripheral staff; indeed, we have already made use of these terms within the body of this chapter thus far. Atkinson suggests that organisations should utilise a number of distinct groups of workers, and his model is represented in Figure 12.1.

As we can see, the 'core' staff are located at the heart of the model. Individuals within the core group are employed on traditional full-time, permanent contracts. They thus enjoy considerable job security and benefit from development and progression opportunities within the organisation. Their contribution to increased flexibility is entirely focused in the area of functional flexibility. This means that they are, so far as is practical and appropriate, multiskilled, and able to work across traditional job boundaries. A particular feature of functionally flexible arrangements within the Atkinson model is that there are fewer hierarchical layers than has tended to be the case in more traditional organisational structures. This means that individuals in the core group may find themselves carrying out duties that would, in more traditionally structured organisations, be reserved for their bosses or subordinates. Outside of the core group there are then two separate groups of peripheral staff, represented by the inner and outer rings. Those in the inner ring are employed directly by the organisation, but on some form of atypical contract. Beyond them, within the outer ring, are individuals whose skills the organisation draws on, but who are not employees – thus subcontractors and self-employed individuals occupy this space.

Figure 12.1 Atkinson's flexible firm

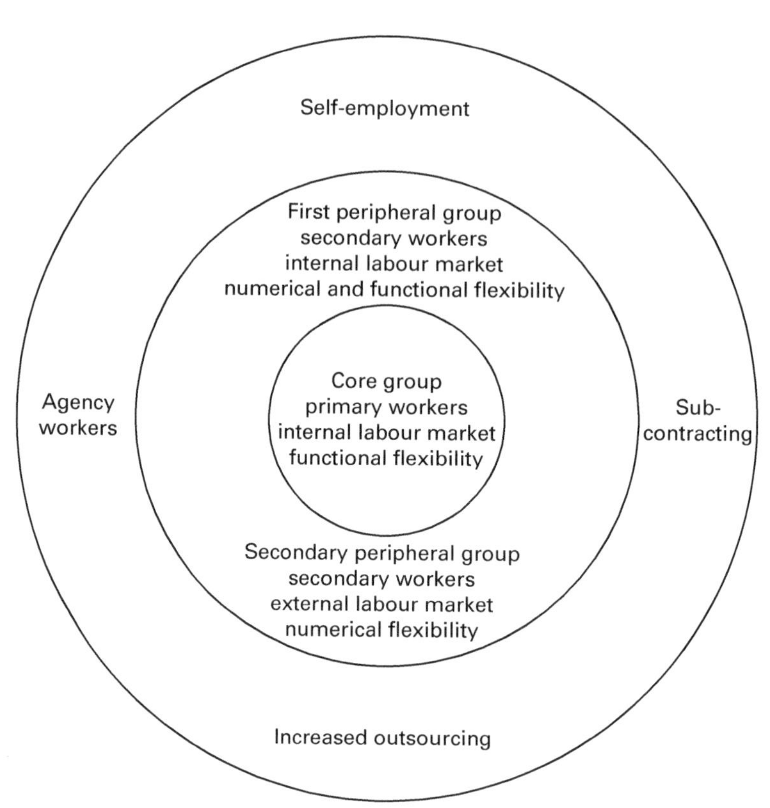

What Atkinson's model provides us with is one solution to the difficult issue of how to combine flexibility with employee commitment and engagement. The solution here is to segment the workforce. In practice, organisations that operate according to the flexible firm model often find that the individuals in peripheral positions are just as committed as those in the core group. This is because there is movement across the boundary. Staff may, for instance, be recruited initially on a fixed-term basis, but then move to a core role over time. The possibility of achieving this serves as an incentive and tends to mean that fixed-term staff are among the hardest working and most engaged individuals that the organisation employs.

Reilly: mutual flexibility

A second potential approach to solving the flexibility/commitment conundrum was articulated very effectively by Peter Reilly (2001) in his text discussing flexible working within the UK. The key point here was that employees have a shared interest with employers in some, but by no means all, flexible working initiatives. Reilly was able to show, for example, how multiskilling and systems to promote functional flexibility benefit employees if it means that they can increase their personal stocks of skills and knowledge, thus making them more employable. Other forms of flexible working are also actively sought out by would-be employees who are looking to achieve a better work–life balance. Flexitime, term-time working, career breaks, remote working, job sharing and part-time working are the main examples. It therefore follows that organisations can, if they take care to introduce flexible working thoughtfully and in consultation with their staff, evolve a suite of practices that create a win–win situation. They increase efficiency and organisational flexibility, while at the same time enhancing commitment and making the organisation more attractive in the labour market.

White et al: intelligent flexibility

A further approach to solving the flexibility/commitment conundrum is advocated very effectively indeed by White et al's (2004) work, termed 'intelligent flexibility'. Drawing on an extensive survey that explored practice within UK organisations, they argue strongly that major advantages can be gained if organisations steer clear of buying in more flexible labour, instead choosing a more sophisticated approach. They argue that the type of flexibility that enhances an organisation's adaptability and agility is increasingly more necessary for organisations to achieve than pure cost-based approaches that only serve to increase efficiency. Their approach first calls for a sophisticated approach to functional flexibility. The key, according to White et al (2004), is to develop a workforce that can undertake a wider variety of organisational tasks and where individuals are competent to cover one another's jobs as and when necessary. They then argue that organisations must invest significantly in learning and development initiatives via formal training programmes, but also job rotation. Finally, they argue that team working is central to the process and that UK organisations have not adopted team-working practices to the extent they might have done. Key here is the proposition that employees work best and most flexibly in a functional sense when they are placed in teams that are able to self-manage to a considerable degree.

Conclusion

This chapter has provided the reader with a substantial introduction to the subject of flexibility within the workplace and highlighted the importance of the subject within contemporary organisations. During the chapter the various forms of flexibility were introduced, including functional, numerical, temporal and financial, with short examples and more substantial case studies illustrating the operation of the concepts in practice. It is important to understand that there are, however, critical perspectives within this field. Flexibility cannot, and should not, be seen as something that is always positive for all concerned. There are numerous examples of schemes designed to increase flexibility being poorly designed and/or poorly implemented, and indeed situations where organisations have sought to gain through the imposition of insecure forms of work. More fundamentally than this there are questions as to whether the introduction of flexible working practices has a negative impact on employee commitment, and broader points related to the social desirability of flexible working. In spite of these arguments, it is clear that where flexible working practices are effectively designed and implemented, ensuring mutual benefits for both employers and employees, such schemes can bring about a range of positive outcomes. The chapter that follows takes these debates and extends them by investigating the subjects of organisational agility and resilience.

References

Agerholm, H (2018) Brexit: Farmers allowed to recruit 2,500 migrants a year under new government plan to plug seasonal workforce gap, www.independent.co.uk/news/uk/home-news/brexit-farmers-migrants-seasonal-workers-a8524871.html (archived at https://perma.cc/JQJ9-6ZZP)

Atkinson, A, Burgess, S, Croxson, B, Gregg, P, Propper, C, Slater, H and Wilson, D (2009) Evaluating the impact of performance-related pay for teachers in England, *Labour Economics*, 16 (3), 251–61

Atkinson, J (1984) Manpower strategies for flexible organizations, *Personnel Management*, 16, 28–31

Barringer, M W and Milkovich, G T (1998) A theoretical exploration of the adoption and design of flexible benefit plans: A case of human resource innovation, *Academy of Management Review*, 23 (2), 305–24

BBC News (2017) McDonald's to offer all UK staff fixed hour contracts, www.bbc.co.uk/news/business-39707429 (archived at https://perma.cc/8LPP-NVR3)

BBC News (2019) Ban zero-hours contracts that exploit workers, says TUC, www.bbc.co.uk/news/business-47193809 (archived at https://perma.cc/295Y-WSMH)

Beham, B, Baierl, A and Eckner, J (2020) When does part-time employment allow managers with family responsibilities to stay on the career track? A vignette study among German managers, *European Management Journal*, 38 (4), 580–90

Bregn, K (2013) Detrimental effects of performance-related pay in the public sector? On the need for a broader theoretical perspective, *Public Organization Review*, 13 (1), 21–35

Bryson, C (1999) Managing uncertainty or managing uncertainly?, in *Strategic Human Resourcing*, ed J Leopold, L Harris and T Watson, FT/Pitman, London

Cappelli, P H (2015) Skill gaps, skill shortages, and skill mismatches: Evidence and arguments for the United States, *ILR Review*, 68 (2), 251–90

CIPD (2019) *Megatrends: Flexible working*, CIPD, London

Deery, S, Walsh, J, Zatzick, C D and Hayes, A F (2017) Exploring the relationship between compressed work hours satisfaction and absenteeism in front-line service work, *European Journal of Work and Organizational Psychology*, 26 (1), 42–52

Desombre, T, Kelliher, C, Macfarlane, F and Ozbilgin, M (2006) Re-organizing work roles in health care: Evidence from the implementation of functional flexibility, *British Journal of Management*, 17 (2), 139–51

De Spiegelaere, S, Van Gyes, G and Van Hootegem, G (2018) Innovative work behaviour and performance-related pay: Rewarding the individual or the collective?, *International Journal of Human Resource Management*, 29 (12), 1900–19

Dunn, M (2018) Who chooses part-time work and why?, *Monthly Labour Review*, March, 1–25

Farina, E, Green, C and McVicar, D (2020) Zero hours contracts and their growth, *British Journal of Industrial Relations*, 58 (3), 507–31

Healy, J, Mavromaras, K and Sloane, P J (2015) Adjusting to skill shortages in Australian SMEs, *Applied Economics*, 47 (24), 2470–87

Heery, E and Salmon, J (eds) (2000) *The Insecure Workforce*, Routledge, London

Jeffery, R (2018) Apprenticeships: Will the levy ever work?, *People Management*, April

Jenkins, A (1998) Flexibility, 'individualization', and employment insecurity in France, *European Journal of Work and Organizational Psychology*, 7 (1), 23–38

(Continued)

(Continued)

Kalleberg, A L (2009) Precarious work, insecure workers: Employment relations in transition, *American Sociological Review*, 74 (1), 1–22

Koumenta, M and Williams, M (2019) An anatomy of zero-hour contracts in the UK, *Industrial Relations Journal*, 50 (1), 20–40

Leyland, A (2019) Seasonal labour crisis is a serious threat to British farming, www.thegrocer.co.uk/leader/seasonal-labour-crisis-is-a-serious-threat-to-british-farming/598896.article (archived at https://perma.cc/9XT8-54Z7)

Lindsay, F (2020) With no EU workers coming, the UK agriculture sector is in trouble, www.forbes.com/sites/freylindsay/2020/03/24/with-no-eu-workers-coming-the-uk-agriculture-sector-is-in-trouble/?sh=754bc608363c (archived at https://perma.cc/AC2H-68BC)

Lundström, U (2012) Teachers' perceptions of individual performance-related pay in practice: A picture of a counterproductive pay system, *Educational Management Administration and Leadership*, 40 (3), 376–91

Manolchev, C, Thomas, P, McArdle, L and Saundry, R (2020) Sensemaking as 'self'-defence: Investigating spaces of resistance in precarious work, *Competition and Change*, 24 (2), 154–77

Mutari, E and Figart, D (1997) Markets, flexibility and family: Evaluating the gendered discourse against pay equity, *Journal of Economic Issues*, 31 (3), 687–705

Neville, S (2013) Burger King and Domino's Pizza also using zero-hours contracts, www.theguardian.com/uk-news/2013/aug/06/burger-king-dominos-zero-hour (archived at https://perma.cc/4ZED-4V9L)

Noback, I, Broersma, L and Dijk, J (2016) Climbing the ladder: Gender-specific career advancement in financial services and the influence of flexible work-time arrangements, *British Journal of Industrial Relations*, 54 (1), 114–35

ONS (2016) People in employment with a flexible working pattern, www.ons.gov.uk/employmentandlabourmarket/peopleinwork/earningsandworkinghours/adhocs/005248peopleinemploymentwithaflexibleworkingpatternbygender (archived at https://perma.cc/24HH-B97Y)

ONS (2018) Labour in the agriculture industry, UK: February 2018, www.ons.gov.uk/peoplepopulationandcommunity/populationandmigration/internationalmigration/articles/labourintheagricultureindustry/2018-02-06 (archived at https://perma.cc/F5D2-QB8N)

ONS (2020) Employment in the UK: October 2020, www.ons.gov.uk/employmentandlabourmarket/peopleinwork/employmentandemployeetypes/bulletins/employmentintheuk/october2020 (archived at https://perma.cc/2QHF-S8UW)

Peluso, A, Berta, P and Vinciotti, V (2019) Do pay-for-performance incentives lead to a better health outcome?, *Empirical Economics*, 56 (6), 2167–84

Petriglieri, G, Ashford, S J and Wrzesniewski, A (2019) Agony and ecstasy in the gig economy: Cultivating holding environments for precarious and personalized work identities, *Administrative Science Quarterly*, 64 (1), 124–70

Ray, K (2014) Does performance related pay work in the public sector?, *Employee Benefits*, November, 22

Reilly, P (2001) *Flexibility at Work*, Gower, Aldershot

Rubb, S (2020) The impact of the great recession on overeducated and undereducated workers, *Education Economics*, 28 (3), 263–74

Ruddick, G (2017) McDonald's offers fixed contracts to 115,000 UK zero-hours workers, www.theguardian.com/business/2017/apr/25/mcdonalds-contracts-uk-zero-hours-workers (archived at https://perma.cc/8EHM-Y9JJ)

Ryan, L and Wallace, J (2019) Mutual gains success and failure: Two case studies of annual hours in Ireland, *Irish Journal of Management*, 38 (1), 26–37

Saloniemi, A and Zeytinoglu, I U (2007) Achieving flexibility through insecurity: A comparison of work environments in fixed-term and permanent jobs in Finland and Canada, *European Journal of Industrial Relations*, 13 (1), 109–28

Sobral, F, Ng, E S, Castanheira, F, Chambel, M J and Koene, B (2020) Dealing with temporariness: Generational effects on temporary agency workers' employment relationships, *Personnel Review*, 49 (2), 406–24

Statista (2020) Flexible working in the UK: Statistics and facts, www.statista.com/topics/ 6419/flexible-working-in-the-uk (archived at https://perma.cc/97QT-VN3B)

(Continued)

(Continued)

Sundström, A (2019) Exploring performance-related pay as an anticorruption tool, *Studies in Comparative International Development*, 54 (1), 1–18

Syal, R (2018) 'McStrike': McDonald's workers walk out over zero-hours contracts, www.theguardian.com/business/2018/may/01/mcstrike-mcdonalds-workers-walk-out-over-zero-hours-contracts (archived at https://perma.cc/PG9A-DKA5)

Taylor, M, Marsh, G, Nicol, D and Broadbent, P (2017) *Good Work: The Taylor review of modern working practices*, https://assets.publishing.service.gov.uk/government/uploads/system/uploads/attachment_data/file/627671/good-work-taylor-review-modern-working-practices-rg.pdf (archived at https://perma.cc/D5NA-G5DZ)

TUC (2019) *Insecure Work*, TUC Research Report, Trades Union Congress, London

Van Wanrooy, B, Bewley, H, Bryson, A, Forth, J, Freeth, S, Stokes, L and Wood, S (2013) *Employment Relations in the Shadow of Recession: Findings from the 2011 Workplace Employment Relations Survey*, Palgrave Macmillan, Basingstoke

White, M, Hill, S, Mills, C and Smeaton, D (2004) *Managing to Change? British workplaces and the future of work*, Palgrave, Basingstoke

Woldman, N, Wesselink, R, Runhaar, P and Mulder, M (2018) Supporting temporary agency workers' affective commitments: Exploring the role of opportunities for competence development, *Human Resource Development International*, 21 (3), 254–75

Wynn, A T and Rao, A H (2020) Failures of flexibility: How perceived control motivates the individualization of work–life conflict, *ILR Review*, 73 (1), 61–90

13 Agility and resilience

The forthcoming text very much picks up and extends the discussions that were initiated within the last chapter. Readers will remember that the chapter immediately preceding this discussed flexibility within organisations in some considerable depth, highlighting the various methods through which organisations can instil flexibility within their workforces. Agility, as will be seen within the discussions to come, is related in some senses to flexibility, but it is important that the terms are not conflated. Agility in many senses extends the concept of flexibility, and it must be recognised that just because an organisation demonstrates flexibility, this does not necessarily mean that it can automatically be considered 'agile'. The second major concept introduced in this chapter is that of resilience. This, like agility, is a quality at a personal and organisational level that is increasingly in demand as the business environment becomes more uncertain, volatile and dynamic. Both agility and resilience are being seen more and more as playing major roles in personal and organisational success.

This chapter begins by placing agility into its broader context. Drawing together some of the major debates discussed in this book thus far, a case is made for the focus on agility, and its importance to contemporary organisations is discussed at some length. Following these introductory observations, the chapter rapidly moves into a dissection of agility at a theoretical level, presenting the reader with definitions and an exploration of the factors, such as organisational learning, that are thought to be important in the development of agile organisations. Deeper into the chapter, the notion of agility is related to organisational structure, with commentary here indicating how differing approaches to organisation design may support, or alternatively have a negative impact upon, moves to make organisations more agile. Taking a short detour away from the internal workings of organisations, the 'gig economy' is then examined with reference to agility. Questions such as where the balance of risk in employment and working lives should fall will be debated, with readers introduced to the major debates within this area. Towards the latter stages of the chapter, resilience, in both a personal and an organisational context, will be examined. Again, the chapter defines the terms with care and explores both the qualities that contribute to resilience at a personal level, and the factors that enable resilience for organisations as a whole, including, importantly, the role of the HRM function. We begin by exploring the context that has given rise to the need for increasing agility and resilience within our organisations.

The context

In many respects, the first half of this book has very effectively introduced the context surrounding the need for agility and resilience within organisations. The business environment facing organisations has become considerably less predictable as a result of a number of changes (Harsch and Festing, 2020). These changes, such as the introduction of new technologies, the increasing pace of globalisation and shifts in the demographic structure of populations, have all conspired to make the landscape facing organisations considerably more complex and dynamic (Holbeche, 2018). As Holbeche (2018) goes on to note, the qualities of agility and resilience are particularly important enablers of organisational performance in uncertain environments, and this is even more the case where organisations are subject to the forces of disruptive innovation. Change, as discussed in considerably more depth in Chapter 11, is ever-present. It is not something that is episodic (ie a phenomenon that starts and stops and has discrete boundaries), but rather it is continuous, and requires constant movement and adjustment from organisations (Reeves et al, 2020). The key learning point to draw from this brief introductory note is that the wider environment within which all organisations operate is complex, dynamic and considerably less predictable than has previously been the case.

Whilst the business environment may be uncertain, dynamic and complex, there are many examples of organisations that continue to thrive and succeed. Jim Collins, the well-known thinker and writer, has studied the characteristics of successful organisations and has published several texts exploring the histories of companies, searching out key turning points that have led to either organisational success or failure. This is particularly the case in his text titled *How the Mighty Fall* (2009), in which he explores the failure of several previously successful, and in some cases very sizeable, organisations, charting how decisions made by leaders and poor attempts to introduce change, amongst other factors, led to organisational decline. In re-examining these works, and other material published in the field, we can conclude that organisational agility and resilience, or more precisely a lack of organisational agility and resilience, often play a considerable part in episodes of organisational failure, be these episodes of failure that cause disruption (marginal or otherwise) to a certain aspect or function within an organisation, or those that affect the organisation as a whole to a more significant degree. Indeed, Holbeche (2018) argues just this case – that organisational agility in particular helps organisations to respond more effectively to unexpected events, helping them to solve problems and implement change.

Kodak is often held up as an example of an organisation that failed because it was insufficiently able to adapt to change, and insufficiently agile in its nature (Shih, 2016). But this analysis rather glosses over some of the important information that is captured in historical market trends and can now be seen in full detail. Commentators argue that Kodak failed to make the transition from analogue film to digital technology, and whilst this is indeed true, what various analyses and opinion pieces often fail to point out is that at the time when decisions were being made by Kodak executives, the analogue film market was showing very strong growth. Of course, hindsight in these situations is very helpful and it is important to understand that organisational leaders and managers must make decisions based on incomplete information. Whilst traditional models of business strategy highlight the absolute importance of complete internal and external analysis before decisions are made, in reality 'complete' information is something of a fallacy, with leaders and managers needing to respond quickly to unexpected environmental changes (Teece et al, 2016). What this then calls for is a more emergent approach to organisational strategy setting, with agility and resilience being key qualities displayed by successful organisations. The chapter now moves on to consider the subject of organisational agility in much more depth.

Organisational agility

It is initially very important to understand that 'agility' and 'flexibility' are not the same thing. Commentators and writers in the field often conflate the two terms, but it is important to recognise that the concepts are fundamentally different from one another. In order to begin to grasp the differences between the terms, we can argue that whilst a flexible organisation may be able to make changes within the confines of its structures, systems and vision, an agile organisation is able to rapidly pivot the organisational ecosystem as a whole in order to avoid threats or take advantage of newly discovered opportunities. Agility is arguably a broader concept as a result, requiring flexibility but acting at a considerably different scale. Indeed, Van Oosterhout et al (2006) highlight that agility envelops and extends the notion of organisational flexibility, noting just as others have that just because an organisation demonstrates the qualities associated with flexibility, this does not mean that it can be considered agile. With this in mind, how might organisational agility be defined?

The literature connected with organisational agility contains a number of definitions, which do differ to some extent in their completeness and frame of reference. For instance, Felipe et al (2020: 580–1) define agility as 'the enterprise's ability to sense environmental change and respond readily by reconfiguring, assembling, and exploiting its own resources, processes, knowledge, and relationships'. Differently to this, Sherehiy and Karwowski (2014: 466) define it as 'an enterprise's ability to quickly respond and adapt in response to continuous and unpredictable changes of competitive market environments'. Key within both of these definitions is the sense that agility involves a response to change, although Felipe et al (2020) arguably go further than Sherehiy and Karwowski (2014) by explicitly including the notion of organisations sensing change within their definition. Showing yet more diversity in the understandings of organisational agility, Holbeche (2018: 9) defines the concept as 'an organization's ability to develop and quickly apply flexible, nimble and dynamic capabilities'. Another common theme that comes through here is the notion of 'quickness' or speed, the underlying argument being that without the ability to rapidly turn thoughts, plans and ideas into action, organisations will be swamped by their shifting environment and/or the actions of their competitors. Importantly,

Holbeche (2018) argues that any fit that an organisation achieves between itself and its environment is typically temporary, and therefore organisations must be able to respond rapidly to changes, sensing alterations in the broader landscape before rapidly pivoting organisational operations to suit the changed context. Again, the continuous and endemic nature of change comes through here with organisational leaders and managers needing to build and deploy their change management skills.

Agility, as a concept, is as much about proactive action as it is about an organisation being reactive to its environment (Harsch and Festing, 2020). Whilst organisations will indeed take cues regarding necessary actions from the external context, agile organisations will leverage their internal competences and the speed with which they can pivot their operations and plans in order to deliver value for their customers, and thus build competitive advantage. Indeed, the majority of organisations argue that agility is a critical success factor that enables them to operate effectively in dynamic markets. But, having said this, Harsch and Festing (2020) go on to point out that we do not yet have a full understanding regarding how organisational agility might be created and which factors or processes are important in enabling agility within organisations.

PAUSE AND REFLECT

What factors do you think might be important in bringing about agility within organisations?

This pause and reflect point was intended to stimulate your thinking regarding organisational agility, and specifically the factors that might contribute to making an organisation 'agile'. Whilst the picture painted towards the end of the last paragraph was rather bleak in terms of the existing knowledge regarding the factors that contribute to organisational agility, evidence is coalescing around a number of different areas that a variety of commentators believe to be important in bringing about agility. We first of all know that workforce flexibility (Harsch and Festing, 2020) is important. Whilst it is important to recognise that an organisation's systems, processes and structure affect agility to a significant extent, so too does its workforce. HRM professionals therefore need to consider how their HRM approaches and practices either instil or inhibit agility. This extends to areas such as reward management in terms of the behaviours and outcomes that are incentivised within the organisation, and we also know that employee involvement and participation are key factors in promoting agility within organisations. Agile organisations also typically invest substantial sums in employee learning and development, continuously upskilling their employees, ensuring that individuals are not just equipped to tackle today's problems and issues, but have an eye on the horizon as to what is next. In a similar vein, agile organisations will invest significant attention in talent management, recognising that traditional career structures – in brief, the upward linear progression within one specific career path or one specific organisation (Watts et al, 1981) – is rather a false illusion in the contemporary environment. Career development is considerably more dynamic, and may involve a labyrinth of moves, some of which may be sideways, upwards, or may involve individuals retraining at various points in time (Greenhaus et al, 2018).

One of the keys, however, to instilling agility is organisational learning. Defined by Sadler-Smith (2006: 209) as the 'development of new knowledge, understanding or skills that become embedded in the organisation's shared mental models... and influence behaviour which leads to enhanced performance', organisational learning is thought to be fundamental to our efforts to improve agility (Holbeche, 2018). Whilst we know that organisations, by themselves, cannot learn, it is through their employees acting as environmental scanners that they can sense changes in the external environment, and then when that learning is shared and integrated within the organisation's collective memory, future paths can be charted and decisions made. Organisational cultures must therefore prize (and indeed actively reward) efforts to share knowledge and encourage both risk taking and experimentation. Due to the fact that organisations are required to operate at some considerable speed, with limited information, and within an uncertain and shifting external environment, we must also understand that error is part of this process. Experimentation, by its very definition, involves trial, error and refinement, and therefore cultures that prize being 'right' at all costs will frequently render organisations static and risk averse.

Picking up on other factors that hinder organisational agility, we know that excess structure, bureaucracy and hierarchy can substantially impede efforts to encourage agile organisational mindsets (Hamel and Zanini, 2018). This is not, however, to say that structure and process in and of themselves are necessarily and always to be avoided. Indeed, organisations must ensure that they achieve reliability alongside adaptability and agility (Bernstein et al, 2016). The decision that managers and leaders must make is connected with the trade-off between reliability and adaptability. Too strong a focus on the latter will cause fragmentation and the loss of leverage that comes from organisational focus and the standardisation of practice, but too much focus on reliability tends to produce an organisation that is rather too fixed in its mindset and outlook. The optimal balance between these factors will inevitably vary in different industries and sectors of the economy, and it will indeed change over time as the environment around the organisation shifts and new innovations are introduced. The arguments concerning organisational structure and bureaucracy are now picked up in more detail in the following section of this chapter.

Agility and organisational structure

Without wishing to diverge too substantially from the subject matter at hand within this particular chapter, it is nonetheless important to briefly review various different forms of organisational structure, exploring the extent to which these may promote, or alternatively inhibit, agility. These discussions touch on some, but by no means all, of the territory surrounding organisational design, which the CIPD (2019) highlight is all about the decisions that affect the organisation of work and people in a company, with decisions here important in order to best enable the organisation to fulfil its core purpose. Organisation design is, in this sense, a little like an architect creating a blueprint for a building (Marsh et al, 2009). When considering how an organisation is to be structured, we must take into account how various parts of the operations need to align, how other systems will interface with the people structures we create and how the overall structure will impact upon organisational processes. For instance, in a large NHS hospital how will patients move through the system from arrival and diagnosis to treatment and ultimately discharge? The way in which the organisation is structured will have a profound impact on such issues and, as noted above, will substantially impact upon levels of organisational agility.

Classic hierarchical layouts, as depicted in Figure 13.1, have been popular with organisations for a very substantial period of time (Hamel and Zanini, 2018). Whilst there are arguments against the continued relevance of hierarchy and bureaucracy, this form of organisational structure is still very common. It does have a number of advantages, including very clear lines of control and the indication of authority, as well as the ability to develop distinct career paths through the organisation. So, an individual may enter at a relatively low rank within the organisation and then move up, assuming more responsibility as they grow their knowledge, skills and experience. Traditional, tall hierarchical structures are typically useful in situations where organisations must be highly reliable (Bernstein et al, 2016), the NHS being a good example of a situation in which such structures are entirely appropriate. The problem with hierarchical layouts is that bureaucracy can cause organisations to become rather stale and static. Change can be difficult when a hierarchy has persisted for a period of time, and indeed layers of management add substantial costs.

This is, of course, a much simplified diagram, but it helps in visualising the core idea. In a traditional hierarchical framework, employees sit within clearly defined roles that interconnect in a very explicit manner with others within divisional structures. There are clear lines of accountability, and also clear routes of progression up through the structure for employees. An important reason why hierarchies persist is that this functional arrangement has served organisations very well for a very substantial period of time. Whilst hierarchical arrangements do instil considerable inertia, they also provide order and structure. In an increasingly regulated and complex business landscape there is certainly a case to be made for hierarchy and bureaucracy being important in enabling organisations to navigate these complex contexts. This is one reason why they are often termed 'high reliability' structural arrangements. Whilst there are positive qualities to hierarchical arrangements, there has been a substantial number of examples over time where businesses have failed because they were (in part) unable to adapt to a changing context (see, for example, the well-documented failures of Thomas Cook and BHS amongst others).

Figure 13.1 Hierarchical organisational structure

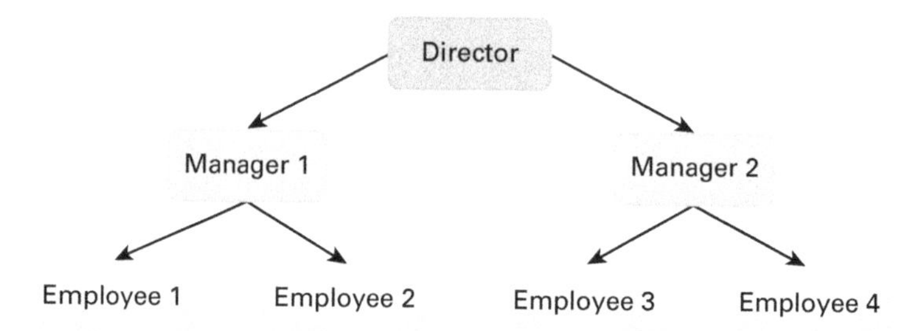

The need for increased agility, with organisations needing to be fleet of foot in order to exploit emerging opportunities and pivot away from threats, does call into question the continued relevance of largely fixed and static organisational forms. Whilst hierarchical arrangements can enable organisations to navigate complex operating contexts, they are typically slow to change, with increasing size leading to substantially more inertia and resistance to change.

Many organisations have experimented with flatter hierarchical arrangements, still retaining the broad principles of hierarchy, but removing what could be seen as 'excess' layers of management in order to improve coordination and information flow and reduce unnecessary supervision/management costs. Whilst on the surface the benefits of such arrangements may be clear, there are nonetheless drawbacks that must be considered. Flatter structures, first of all, can encourage individuals to become rather generalist in their skills, knowledge and experience. Whereas in traditional hierarchical structures individuals can develop into specialists, flatter structures tend to require individuals to operate across functional boundaries, thus meaning that there is comparatively less ability to develop deep, specialist knowledge and skills. Allied to this, flatter hierarchical arrangements can mean that the manager's role is considerably more challenging. Consider performance management systems, for instance. It is considerably easier to monitor and motivate performance where a manager has five direct reports than a situation where that same manager may have twenty, thirty or more subordinates for whom they are responsible. Finally, whereas traditional taller hierarchies provide very clear career progression routes, flatter hierarchies do not allow the demarcation of such clear career paths, as the intermediate roles through which individuals may progress will typically have been removed, thus limiting avenues of upward progression.

More recently, organisations across many different sectors have experimented with matrix-style structures. Matrix structures involve more fluidity than traditional hierarchies and thus may instil greater levels of agility within organisations; this is because individuals are organised in line with projects or distinct areas of work, with team structures changing as projects or areas of work begin and end. Matrix structures generally have two separate 'chains of command'. Individuals in a matrix structure will have a 'functional' manager and a 'project' manager. For instance, if an individual is employed as a marketing executive they may report to both a marketing manager (functional manager) and the leader of a particular client project (project manager). Matrix structures typically work well as employees can specialise within their functional area, whilst simultaneously maintaining cross-functional collaboration with others via their involvement in differing project teams. Matrix structures are, as noted above, rather more flexible than traditional hierarchies as a result, with the start and end of projects encouraging the firm to reorganise as needed. The downside to matrix structures, however, is that they tend to be rather complex to manage effectively, and they can introduce more management costs because there is effectively double supervision within the organisation (functional and project supervision). The need to share employees between different parts of the organisation (eg Project A, Project B, Project C) can also create unhealthy competition that inevitably will require delicate management.

A final form of organisational structure, and one that is increasingly popular as organisations search for increased agility, is the self-managed team. In this type of structure accountability for all aspects of work is devolved fully (or as fully as is possible) to the level of the team. The team then becomes fully responsible for its own output and quality outcomes, and operates autonomously within the organisation's broad boundary (Bernstein et al, 2016). Self-managed teams typically enable organisations to be considerably more fluid and agile in their approach, and they also tend to incur less organisational cost as they require less overt supervision and central control. Decision making is often improved, as decisions are devolved to individuals who are typically working much closer to the issue/problem (Roelofsen et al, 2017); as such, these structures can encourage stronger employee motivation and enable much greater levels of employee voice and participation. A problematic issue with self-managed teams as an organisational form, and particularly so with more extreme forms of self-management such as holacracy (Bernstein et al, 2016), is that coordination across the organisation can be much more difficult to maintain. Whilst organisations utilising extreme interpretations of the self-managed teams philosophy may be very agile indeed, there are questions as to how leaders may retain effective operational oversight, which may negatively impact organisational reliability. In addition to this, whilst some individuals may work very well in environments that permit large degrees of autonomy and freedom, others may flounder as a result of a lack of direction, particularly those at the earlier stages of their careers. Self-managed teams can also be susceptible to groupthink, and the negative consequences that ensue from such situations. The case study below provides an example of self-managed teams in practice within the Chinese consumer electronics and appliance group Haier.

CASE STUDY 13.1

Haier

Haier Group, the Chinese consumer electronics and home appliances company, was founded in Shandong in 1984. It has been very successful indeed and has consistently grown its revenue and profitability over time (Statista, 2020). Haier, as an organisation, has a track record of producing successful innovations and introducing these to the market, including its current focus on internet-enabled consumer devices. In short, the company has been very successful over an extended period of time, and all signs point to its continued growth.

Haier, although it is a large and complex organisation, does not operate a traditional hierarchical structure, instead rejecting bureaucracy and hierarchy in favour of a structure built around the concept of the self-managed team. The organisation therefore does not have departments and divisions, but instead has a large variety of micro-enterprises within its ecosystem that network and link with one another whenever and however they need to. Leaders within Haier encourage all individuals employed by the organisation to see themselves as entrepreneurs rather than employees. Commentators have said that the company, to an extent, mimics the structure of the internet; in other words, it has small, loosely joined pieces in a modular structure that freely evolve without direct control from the centre of the organisation.

This, it is argued, encourages internal competition and therefore a focus on providing the best customer products possible.

Haier's structure is certainly very different from that typically used by most large organisations. All of the various micro-enterprises within its ecosystem are, for instance, actively encouraged to buy services (eg HRM, marketing, finance) either internally from other micro-enterprises or from suppliers outside the organisation. The organisation is designed to evolve over time, with individuals adapting to change and some micro-enterprises ceasing whilst others start. Haier's structure enables it to be very agile and flexible as an organisation, but there are questions as to whether, and to what extent, such structures will endure over the longer term. There is, for example, rather more potential for error in a structure driven by the philosophy of self-managed teams, with the organisation sacrificing reliability on the one hand for improved agility and innovation on the other. That said, the organisation continues to post very successful results indeed and is an excellent case example regarding the use of self-managed teams at scale.

PAUSE AND REFLECT

What are your thoughts regarding the structural arrangements pursued by Haier? To what extent do you think that these arrangements will contribute to (or indeed impede) the organisation's future success?

The purpose behind the pause and reflect point above was to get you as the reader to think in some depth about the relationship between organisational structure and agility. One of the themes running through much of the content related to organisational agility so far has been that leaders and managers are required to make trade-offs between the amount of fixedness and flexibility (Holbeche, 2018) within their operations. Haier is towards the more extreme end of the agility spectrum, organising itself around the very extensive use of self-managed teams, having little, if any, upward structure in terms of oversight and control. The organisation does have a leadership team, but the business evolves and changes as its internal micro-enterprises configure and reconfigure themselves. There is certainly more potential for error in this sort of arrangement, and we would be likely to agree that it should be seen as a highly adaptable configuration, but with lower levels of reliability as a result. This poses something of a paradox when the nature of the industries in which Haier operates is explored. The consumer electronics and home appliances industries have stringent product safety standards that, we would assume, require high levels of organisational reliability in order to ensure that products are manufactured to a set specification and that they are safe for consumer use. Haier, however, would appear to have developed a structure that is highly agile without a visible cost to reliability. It will indeed be interesting to follow the development and growth of this organisation into the future.

Corporate agility: lessons from biology

Before moving on from this section of the chapter, there is one final issue to explore that is firmly connected with the debates surrounding organisational structure and agility. Recent work from Reeves et al (2016) has argued that there is much that organisations can learn from the study of biological systems, particularly when it comes to adaptation and therefore agility. The authors suggest that organisations and biological species are identical in one important way, namely that they are both complex adaptive systems. In other words, a system that has a dynamic network of interactions, and whose behaviour is not entirely predictable as a result. As a consequence of this, Reeves et al (2016) suggest that the principles conveying robustness in biological species and systems therefore directly apply to organisations, particularly in the contemporary business landscape. They argue that complexity exists at a number of different levels – within individuals, within the organisation itself, and outside the organisation's boundary, and therefore there are multiple feedback loops and sources of opportunity that individuals must be aware of.

What this ultimately means for organisations is that there are a number of principles that managers and leaders should seek to apply to both the structuring and ongoing management of their enterprises. The first of these is that managers and leaders should seek to retain heterogeneity, in other words variety, in terms of the individuals who are employed and the ideas that are fostered. This, of course, comes at the expense of efficiency, but helps the organisation to remain alive to new opportunities and ideas. The second principle is that organisations must be modular by design. A modular design shields the organisation from shocks, thus making the system as a whole more robust as 'failure' can be contained within one area, without the reverberations spreading throughout the organisation. Connected to this, organisations must also preserve redundancy: in other words, they must have multiple items playing overlapping roles. When one element fails, something else can take over. Whilst this is the antithesis of the lean philosophy, the underlying principle stands to reason. For example, if an automotive company relies on one supplier for its paint and bodywork materials, and that supplying organisation has one factory from which it operates, imagine the issues caused if that factory suffered a disaster, for instance a catastrophic fire. The automotive company would be unable, for a time, to source paint and bodywork materials, and thus it would be unable to continue operations until it was able to source replacement goods.

Further to the points above, the authors also argue that organisations must expect surprise but attempt to reduce uncertainty. What this means is that organisations, through their employees, must seek out signals indicating change in the wider environment and then act to amplify positive outcomes and minimise potential problems. Linked to this is the notion that organisations must create feedback loops to ensure that appropriate selection takes place between different courses of action and different ideas. Feedback is important within complex adaptive systems, and is an important way in which organisms (and therefore organisations) adapt to a changing context. Finally, organisations must seek to foster trust. Cooperation is essential in complex adaptive systems, and fostering trust and reciprocity enables managers and leaders to guide the organisational system as a whole by ensuring that individuals buy in to the broader vision and direction of travel.

The arguments put forward by Reeves et al (2016) are particularly interesting and it could certainly be argued that following the principles above could generate greater organisational agility. The chapter now moves on from discussions of agility within organisations to focus on a key development in the broader working landscape: the rise of the 'gig economy'.

The gig economy

The 'gig economy' is a relatively new expression that was coined by the former editor of the *New Yorker*, Tina Brown, in 2009 (Hasija et al, 2020). The term was very apt to describe the reality that, for many, was created by the global financial crisis of 2008–09 that precipitated a considerable rise in unemployment, and left many individuals needing to create incomes to replace the wages lost from more stable employment. Whilst we may assume that the gig economy is a product of the dynamism and uncertainty of the contemporary business landscape, and certainly is discussed in the sense of how it can contribute to organisational agility, the truth of the matter is actually rather different. As Frazer (2019) correctly points out, the concept of an individual engaging in many short term tasks and piecing together different elements of work to form an income and, over the longer term, a career, has been around for a considerable period of time. For instance, independent consultants, musicians, writers and photographers have arguably engaged with this pattern of work for a substantial period of time, relying on a series of short-term contracts (or, in the case of artists, commissions) rather than a single permanent position with any organisation. Having said this, we must recognise that the range of individuals engaging with 'gigs' has changed dramatically in recent times, with firms such as Deliveroo, Uber and many more offering platforms where individuals can earn money from 'side hustles' to supplement their main employment, or indeed earn their major source of income in this way.

'Gigs' certainly run across the whole spectrum of incomes. Frazer (2019), in particular, notes that there appear to be two major sectors within the gig economy. The first of these encompasses service-orientated gigs, such as the work available through organisations such as Deliveroo and Uber. The second sector of the gig economy is altogether different and involves knowledge-based gigs, where independent management consultants or data scientists (amongst many other specialisms) may sell their services as contractors to organisations when specific project work arises. This, again, just as was noted in the previous paragraph, is not necessarily a new development in the business environment in and of itself. There is a long history of individuals choosing to operate as independent contractors, but it can be argued that the prevalence of such arrangements has grown in more recent times. What must be recognised, however, is that the extensive use of gig-style arrangements fundamentally shifts the burden of risk from organisations to individuals themselves (Petriglieri et al, 2018). In standard employment relationships, a large burden of risk is placed on the organisation itself – in other words, if business slows down or circumstances change, it is still compelled to pay salaries to employees unless, obviously, the business enters a position where job losses are required. In the gig economy, the arrangements are very different. There is no requirement for an organisation to enter into an open-ended relationship with an individual, money is transacted in exchange for the completion of a specific task, and when that task is complete both parties move on independently. Petriglieri et al (2018) go on to note that this means that the stakes for gig economy workers are very high indeed, arguing that even successful individuals worry about money and financial security, borne from the increased risk that they are required to bear.

When discussions surrounding the gig economy took off in the late 2000s and there was considerable interest and debate regarding its development in the contemporary business landscape, commentators at the time argued that it was more likely to transform knowledge work and 'white collar' occupations (Hasija et al, 2020) than more manual or routine work. It was thought that knowledge-intensive roles which could be relatively easily codified and bounded, in other words 'packaged' as gigs, would enable organisations, particularly those operating in the professional services sector, to radically transform their workforce, enabling far greater agility. However, in practice, it has been seen that firms have largely preferred to keep knowledge-intensive roles 'in house', because when individuals are engaged via a traditional employment contract they are much better integrated into the processes, people and politics of an organisation. In short, the transformation of knowledge work that was much heralded during the late 2000s has yet to materialise in practice. Where the gig economy has seen substantial growth is in service roles, which in many cases are substantially easier to codify and bound. Tasks in this sector of the gig economy also tend to require fewer skills, qualifications and experience, thus opening up a wider pool of labour for organisations as and when they require it.

Having painted a relatively positive picture of the gig economy so far in this section, it is important that this is balanced with a more critical perspective. There have been many reports where work within the gig economy has been highlighted as exploitative and providing very little security for individuals. This, as Petriglieri et al (2018) argue, can cause substantially higher levels of anxiety and work-related stress for individuals operating (either by choice or not) within the gig economy. Interestingly, work by Berger et al (2019) exploring the experiences of Uber drivers in London suggests that gig economy workers experience higher than average levels of subjective wellbeing and that individuals choosing to work in this way have more control over their work–life balance, but they also note that pay and conditions can be problematic and that the lack of a secure income can create substantial levels of stress. They similarly find that Uber drivers in London experience higher levels of anxiety compared with other self-employed individuals in the London area, agreeing with the findings of Petriglieri et al (2018) in this regard. These findings are common across studies into the experiences of individuals who work within the gig economy. There are benefits, such as increased control over one's working life, but substantial drawbacks, such as income insecurity and the lack of protections afforded to individuals who are engaged under formal employment contracts. This means that the impact of the gig economy is, at best, rather mixed.

The Taylor Review of Modern Working Practices published in 2017 was in large part commissioned as a result of unease regarding working practices within the gig economy. Whilst it highlighted the importance of flexibility and agility within the UK's workforce, the author of the report, Matthew Taylor, has argued that more could have been done to enact its recommendations (Chapman, 2018). One of the particular issues, as noted previously in this section of the chapter, is the balance and sharing of risk between individuals and organisations. It can be argued that the increased levels of unpredictability, dynamism and change in the external environment facing organisations have encouraged firms to seek to shift risk away from themselves and onto the individuals who work for them, particularly in the private sector. Whilst it is difficult to assess where the balance of risk should fall, and indeed this is a matter of judgement, ethics and moral conduct, we can perhaps end this section of the chapter by arguing that the risk burden should be shared more evenly across stakeholder groups.

Personal resilience

Alongside the growth in interest that has surrounded organisational agility, we have also seen substantially more discussion in recent times connected with the notion of 'resilience', both from a personal and organisational perspective. The argument is that the very same factors that have caused organisations (and indeed individuals) to seek out agility have precipitated the need for an increased focus on resilience. Thinking logically, this makes sense. Where environments are more dynamic, where competitive advantage endures for substantially less time than was previously the case, and where organisations increasingly need to encourage risk and experimentation in the pursuit of the next disruptive innovation, the ability to 'bounce back' when events do not go to plan is increasingly necessary (Branicki et al, 2019). A substantial literature base has therefore developed around the notion of resilience, with significant academic interest being invested to understand the concept and the factors that impact resilience, both at a personal and organisational level.

In terms of definitions initially, the academic and professional literature that explores resilience contains a variety of interpretations of the term. Thinking about the subject matter of this section first of all (organisational resilience will be dealt with in the following section of this chapter), Näswall et al (2019) conceptualise personal resilience as the ability to bounce back after facing adversity. Highlighting that resilience is one of the great puzzles of human nature, Coutu (2002) similarly suggests that resilience is tied up with the ability to bounce back, with resilient individuals bending rather than crumbling in pressurised or stressful situations. Interestingly, Coutu (2002) goes on to argue that it is likely that resilience matters more than other qualities that organisations typically look for in potential employees, with resilience to a large extent determining which individuals succeed and which individuals do not. Branicki et al (2019) go a little further than the sources discussed so far by noting that there is a difference in what they term 'everyday' resilience, and resilience in the face of extreme events such as environmental disasters, financial crises and pandemics. Again, however, they settle on the broad view that resilience is about bouncing back, but note that the resources and skills needed to do so differ depending on the nature of the issue encountered. Importantly, work from Dunn et al (2008) aligns with the thoughts of Coutu (2002) by highlighting that individuals with higher resilience are likely to be more successful because they can outperform their peers during times of adversity or in high-pressure situations. The key learning point to draw from this initial foray into the literature is that resilience is an important personal quality and involves the ability to rebound successfully from adverse or pressured situations and events.

Having understood the nature of resilience in a broad sense it is now important to examine the circumstances and qualities that are thought to give rise to personal resilience. First of all, we know that resilience is not necessarily a rare personal trait (Liu et al, 2019). It is something that a great many individuals possess, but in varying degrees. Interestingly, Coutu (2002) argues that resilience is something that becomes apparent *after* the fact. In other words, it is something that individuals see that they have in hindsight when they look back on events and situations, considering their role and actions within them. Say, for example, a manager has to grapple with an extensive change programme that causes substantial upheaval and requires a very substantial amount of work and effort to resolve. After the event they may reflect back and realise that they had displayed very substantial levels of personal resilience indeed when dealing with the problems and challenges associated with implementing the change programme. Building on this, Liu et al (2019) argue that in order to develop resilience individuals must be exposed to some form of risk. In that sense, resilience is akin to a muscle that develops as it is used, and is something that can therefore be trained. Adversity and dealing with problematic situations and events can therefore be helpful in building personal reserves of resilience.

Literature directly exploring the factors that contribute to personal resilience is rather more conflicted in nature. Different studies appear to bring forward different factors or personal qualities that impact upon resilience, and it is certainly the case that further work is needed within this area in order to refine our collective understandings of the concept. Helpfully, Branicki et al (2019) have, however, produced useful work in this area and they suggest that personal resilience can be impacted by an individual's:

- Receptivity.
- Adaptability.
- Flexibility.
- Capacity for learning.
- Social support system and interpersonal relationships.

Interestingly, this list of emerging factors that are thought to influence one's resilience includes elements that are internal to the individual as well as factors that are external to the individual. In keeping with the themes discussed during the sections of the chapter covering organisational agility, there is again mention of 'learning' here as an input to personal resilience. Indeed, it has been argued that where individuals have a capacity to improvise – referenced in Branicki et al's (2019) list as the capability to adapt – they will likely be more resilient. This is the case because individuals with the capacity to adapt and improvise will be stronger in terms of solving non-routine problems, and thus will find solutions to issues where others may struggle in the face of the same challenges. In this sense resilience is, to an extent, about individuals accepting the reality they are faced with, working with the resources available to them in order to manufacture solutions to problems or otherwise deal with difficult circumstances. Like others, Näswall et al (2019) view resilience as a developable capacity that expands as individuals face difficult and uncertain situations. This again foregrounds the importance of a learning mindset within individuals, reinforcing the view that resilience develops as people experience crises. In this sense, resilience is not necessarily something that can be learnt in advance; it develops within individuals as they are exposed to events in the natural course of their work and wider lives.

In terms of outcomes, Näswall et al (2019) and others such as Bustinza et al (2019) argue that resilience links with important and organisationally relevant issues such as job satisfaction and employee engagement. Resilient individuals are thought to derive more satisfaction from their work and working lives, whilst also being substantially more engaged with and committed to their organisations. This consequently feeds into lower turnover intentions. In other words, resilient individuals are less likely to leave their organisations. When viewed in a practical sense, these linkages do indeed make sense. If it is accepted that resilient individuals outperform others when working under conditions of pressure and stress, being 'bendable' rather than reaching a breaking point, it is indeed entirely conceivable that the observed outcomes might involve higher job satisfaction, commitment and engagement and less of a desire to leave an employer in the face of adversity. It is important, however, to remember that this area of research is continuously developing. New findings have the potential to alter the directions of the conversations that are occurring in the academic literature.

PAUSE AND REFLECT

To what extent do you think that you display the qualities that are typically associated with resilient individuals? What might you be able to do to increase your levels of personal resilience?

These are questions to which, of course, no standard answers can be provided. The intent behind asking them was to get you thinking about the concept of resilience and the extent to which you feel you currently display the qualities that are associated with resilient individuals. As noted in this section of the chapter, an important way in which resilience is developed is through exposure to some form of risk (Liu et al, 2019). With that in mind, you may have thought about perhaps undertaking more stretching activities in the workplace, in your personal life or associated with your broader studies. Finally, it is worth repeating the point introduced earlier that personal resilience is not thought to be a rare quality. It is something that is quite common, although individuals may possess different quantities of it, and it is, we argue, most certainly something that can be developed over time. The focus of this chapter now changes slightly, moving to the concept of 'organisational resilience'.

Organisational resilience

Just as there has been interest in personal resilience, there has also been substantial work undertaken to understand resilience at a corporate or organisational level. It is a quality that is thought to significantly determine the extent to which an organisation may be successful over the long term, and thus has piqued the interest of academics and professional commentators alike. We know from Branicki et al (2019) that there is more limited literature discussing organisational resilience than the body of work that surrounds resilience at a personal level, but they note that resilience does correlate well with financial measures of performance. In other words, organisations that are more resilient tend to generate superior financial returns to their less resilient counterparts.

Definitions of organisational resilience draw very much from the same pool of thinking as those of personal resilience, although with a subtly different emphasis applied at times. Recognising that resilience at the organisational level is again about the ability to 'bounce back', Bardoel et al (2014: 280), drawing from Luthans (2002: 702), argue that it can be defined as 'the developable capacity to rebound or bounce back from adversity, conflict, and failure or even positive events, progress, and increased responsibility'. Differently, Bustinza et al (2019: 1371) argue that resilience at the organisational level can be defined as 'the ability to dynamically reinvent an organization when circumstances change, facilitating a firm's capacity to respond to uncertain conditions at the organizational level'. Consistent amongst these definitions is that organisational resilience is about adaptation to a changing environment, with the organisation able to 'reinvent' itself in order to remain relevant as external conditions alter. This theme is also picked up by Seville et al (2008: 259), who argue that resilience reflects the 'ability [of an organisation] to survive, and potentially even thrive, in times of crisis'.

Importantly, however, organisational resilience is not to be seen as the sum of the resilience of its members (Branicki et al, 2019). Just because an organisation may employ resilient individuals, this does not automatically mean that the organisation as a whole is necessarily resilient. When exploring resilience at an organisational level, we know that it arises from an interplay of various factors including the knowledge, skills and competences of individuals themselves, personal and social networks inside the organisation, and the external networks that the organisation has developed over time with its various stakeholder groups, including customers and suppliers. The latter connections can help the organisation to access resources, and thus facilitate the organisation's ability to adapt and change. Building from this, the ability for the organisational system to learn (Näswall et al, 2019) is also thought to feed in positively to levels of resilience, just as it is thought to enable organisational agility. A learning mindset at the level of the individual, team and organisation appears to be important if we are to increase levels of resilience.

But what is known about the role of HRM as a function and HRM specialists themselves in promoting organisational resilience? What role, if any, does HRM play in building organisations as resilient entities? Before moving on, the following pause and reflect point gives you an opportunity to collect your own initial ideas on this matter.

PAUSE AND REFLECT

In your view, how might the HRM function and HRM specialists positively impact organisational resilience?

Hopefully the chance here to pause and reflect gave you time to consider the role of the HRM function in organisational resilience. You may have thought about a number of different ways in which HRM could potentially contribute to resilience, and the chapter now turns back to the research base to explore this issue in considerably more depth.

Linking to the points discussed so far in this section, we know that a learning orientation, and thus learning and development input, has a role in increasing resilience at the organisational level (Bardoel et al, 2014). This is because organisational investment in learning and development enables individuals to develop new stocks of skills and knowledge, thus better enabling employees to deal with new and changed situations that they may encounter. Further to this, Bardoel et al (2014) argue, organisations should develop systems of social support at the level of the organisation and within teams, encouraging individuals to bond more effectively as groups. Linked to this, they similarly argue that efforts to increase diversity within the workforce have a positive impact on levels of resilience. This is because where the organisation, as a whole, employs a more diverse set of individuals, it can draw on an increased range of expertise, knowledge and views, thus enabling it to generate more creative solutions to problems that it encounters.

Changing tack slightly, it is also understood that HRM systems and practices designed to facilitate personal wellbeing can also contribute to increased resilience within the organisational system as a whole. For instance, Johnson (2008) highlights that employee assistance programmes have an important role in developing resilience at the level of the individual, with this in turn contributing to the resilience of the organisation as a whole. Likewise, a focus on ensuring a positive work–life balance (or, indeed, increasingly the integration of work and life activities) (Wood and De Menezes, 2010) is thought to enhance resilience, thus HRM initiatives aimed at ensuring employees develop a healthy relationship with work activities can contribute to increasing resilience in the overall organisational system. With this in mind, flexible working arrangements can be an avenue that HRM can profitably explore. At a more general level, Bardoel et al (2014) also note that investment in occupational health and safety systems is also important as this can reduce work-related injuries and illness, and we must not overlook the role of HRM in the development of crisis management systems. The Covid-19 pandemic of 2020 snapped this particular point into sharper relief where lockdowns across many countries required organisations to transition into a system of remote working, in many cases almost overnight. Organisations that had well-developed crisis management plans were thus able to adapt to the unprecedented changes more rapidly than their competitors. In short, the key learning point to draw from this section of content is that the HRM function as a whole, and differing areas of HRM activity, has a substantial role to play in encouraging and promoting organisational resilience.

Conclusion

The overarching aim of this book as a whole is to provide the reader with an overview of developments impacting contemporary organisations and to consider how organisations are adapting (in some cases better than others) to the modern business environment. A key theme, as expressed at the start of this chapter, is that the business environment is becoming increasingly turbulent, with a number of factors such as the growth of globalisation, substantial demographic changes and the development of new technologies and disruptive innovations impacting the landscape in which organisations must operate. In these increasingly uncertain, hyper-competitive and dynamic environments, the ability of organisations to adapt to, or indeed proactively anticipate, change is essential. Agile organisations that are able to rapidly pivot their operations – whether in part, or indeed in full – will thus be considerably more able to respond to changes in the wider environment. This involves the fostering of a learning mindset, with organisational learning arguably taking on a renewed importance as organisations grapple with environments that change at an increasing pace. Going hand in hand with the need for agility is the need for increased resilience, at both a personal and organisational level. With the increased level of risk that exists in the external environment, it is arguably individuals and organisations who are better equipped to deal with moments of crisis, 'bending' rather than 'breaking' in pressurised and stressful situations, that will be more successful over the long term. Resilience enables individuals and organisations to outperform their less resilient peers, and during the confines of this chapter readers have been introduced to the emerging literature base connected with these issues, with the chapter noting and discussing some of the factors that contribute to resilience, at the level of both the individual and the organisation. Readers should now have a solid grounding from which to further explore these important subject areas.

References

Bardoel, E A, Pettit, T M, De Cieri, H and McMillan, L (2014) Employee resilience: An emerging challenge for HRM, *Asia Pacific Journal of Human Resources*, 52 (3), 279–97

Berger, T, Frey, C B, Levin, G and Danda, S R (2019) Uber happy? Work and well-being in the 'gig economy', *Economic Policy*, 34 (99), 429–77

Bernstein, E, Bunch, J, Canner, N and Lee, M (2016) Beyond the holacracy hype, *Harvard Business Review*, 94 (7–8), 38–49

Branicki, L, Steyer, V and Sullivan-Taylor, B (2019) Why resilience managers aren't resilient, and what human resource management can do about it, *International Journal of Human Resource Management*, 30 (8), 1261–86

Bustinza, O F, Vendress-Herrero, F, Perez-Arostegui, M N and Parry, G (2019) Technological capabilities, resilience capabilities and organizational capabilities, *International Journal of Human Resource Management*, 30 (8), 1370–92

Chapman, B (2018) Gig economy: Matthew Taylor rates government response to his review at just four out of ten, www.independent.co.uk/news/business/news/gig-economy-matthew-taylor-review-rates-government-four-out-10-a8246396.html (archived at https://perma.cc/6ST4-JDYH)

CIPD (2019) Organisation design, www.cipd.co.uk/knowledge/strategy/organisational-development/design-factsheet (archived at https://perma.cc/3XW9-MZNE)

Collins, J (2009) *How the Mighty Fall*, Random House, London

Coutu, D L (2002) How resilience works, *Harvard Business Review*, 80 (5), 46–55

Dunn, L B, Iglewicz, A and Moutier, C (2008) A conceptual model of medical student wellbeing: Promoting resilience and preventing burnout, *Academic Psychiatry*, 32 (1), 44–53.

Felipe, C M, Leidner, D E, Roldán, J L and Rodríguez, A L (2020) Impact of IS capabilities on firm performance:

(Continued)

(Continued)

The roles of organizational agility and industry technology intensity, *Decision Sciences*, 51 (3), 575–619

Frazer, J (2019) How the gig economy is reshaping careers for the next generation, www.forbes.com/sites/johnfrazer1/2019/02/15/how-the-gig-economy-is-reshaping-careers-for-the-next-generation/?sh=4737ca7949ad (archived at https://perma.cc/75PZ-3Q37)

Greenhaus, J H, Callanan, G A and Godshalk, V M (2018) *Career Management for Life*, 5th edn, Routledge, Abingdon

Hamel, G and Zanini, M (2018) The end of bureaucracy, *Harvard Business Review*, 96 (6), 50–9

Harsch, K and Festing, M (2020) Dynamic talent management capabilities and organizational agility: A qualitative exploration, *Human Resource Management*, 59 (1), 43–61

Hasija, S, Padmanabhan, V and Rampal, P (2020) Will the pandemic push knowledge work into the gig economy?, *Harvard Business Review* Digital Articles, June, 2–8

Holbeche, L (2018) *The Agile Organization: How to build an engaged, innovative and resilient business*, 2nd edn, Kogan Page, London

Johnson, J D (2008) Employee assistance programs: Sources of assistance relations to inputs and outcomes, *Journal of Workplace Behavioral Health*, 23 (3), 263–82

Liu, Y, Cooper, C L and Tarba, S Y (2019) Resilience, wellbeing and HRM: A multidisciplinary perspective, *International Journal of Human Resource Management*, 30 (8), 1227–38

Luthans, F (2002) The need for and meaning of positive organizational behaviour, *Journal of Organizational Behavior*, 23 (6), 695–706

Marsh, C, Sparrow, P, Hird, M, Balain, S and Hesketh, A (2009) Integrated organisation design: The new strategic priority for HR directors, White Paper 09/01, Lancaster University Management School

Näswall, K, Malinen, S, Kuntz, J and Hodliffe, M (2019) Employee resilience: Development and validation of a measure, *Journal of Managerial Psychology*, 34 (5), 353–67

Petriglieri, G, Ashford, S and Wrzesniewski, A (2018) Thriving in the gig economy, *Harvard Business Review*, 96 (2), 140–43

Reeves, M, Levin, S and Ueda, D (2016) The biology of corporate survival, *Harvard Business Review*, 94 (1), 46–55

Reeves, M, Levin, S, Fink, T and Levina, A (2020) Taming complexity, *Harvard Business Review*, 98 (1), 112–21

Roelofsen, E, Yue, T, Van Mierlo, P and Noteboom, B (2017) Is holacracy for us?, *Harvard Business Review*, 95 (2), 151–55

Sadler-Smith, E (2006) *Learning and Development for Managers*, Blackwell, Oxford

Seville, E, Brunsdon, D, Dantas, A, Le Masurier, J, Wilkinson, S and Vargo, J (2008) Organisational resilience: Researching the reality of New Zealand organisations, *Journal of Business Continuity and Emergency Planning*, 2 (3), 258–66

Sherehiy, B and Karwowski, W (2014) The relationship between work organization and workforce agility in small manufacturing enterprises, *International Journal of Industrial Ergonomics*, 44 (3), 466–73

Shih, W (2016) The real lessons from Kodak's decline, *MIT Sloan Management Review*, 57 (4), 11–13

Statista (2020) Haier Group: Statistics and facts, www.statista.com/topics/4856/haier-group (archived at https://perma.cc/5P87-RAVF)

Teece, D, Peteraf, M and Leih, S (2016) Dynamic capabilities and organizational agility: Risk, uncertainty, and strategy in the innovation economy, *California Management Review*, 58 (4), 13–35

Van Oosterhout, M, Waarts, E and Van Hillegersberg, J (2006) Change factors requiring agility and implications for IT, *European Journal of Information Systems*, 15 (2), 132–45

Watts, A G, Super, D E and Kidd, J M (1981) *Career Development in Britain*, Hobsons Publishing, Cambridge

Wood, S J and De Menezes, L M (2010) Family-friendly management, organizational performance and social legitimacy, *Journal of Human Resource Management*, 21 (10), 1575–97

14 Creativity and innovation

A diverse range of commentators have highlighted the growing importance of creativity and innovation within our organisations. Although defining creativity and innovation in a business context is thought to be complex (Anderson et al, 2014), we nonetheless know that it serves as a basis from which we may develop new products or services, find efficiencies in our operations or production processes, or uncover new ways of marketing or branding our customer offerings. We know, in particular, that smaller organisations depend to a very great extent on their ability to innovate (McAdam et al, 2010), and that without creativity organisations of all shapes and sizes can quickly stagnate and decline (Cummings and Oldham, 1997).

During this chapter we will together explore some of the foundational concepts within creativity and innovation, thinking, in particular, about how ideas move from the stage of initial concept generation through to implementation, exploring the various elements of this journey. We will reflect on the importance of so-called 'collective creativity': in other words, the notion that good ideas are frequently produced by teams working together rather than individuals working alone. Further into our discussions we will consider the role of leadership within creativity, and specifically how leaders can set a stage, or foundation for creativity, within organisations. Connected to this, we will also think about how creativity and innovation operate in large organisations. It is important to understand that power and political activity have an impact on the creative process, and through an exploration of a number of practical examples we will reflect critically on how large organisations can both support and suppress creative ideas. Our exploration in this chapter will also cover a number of other factors that are known to support or constrain creativity. For instance, we will think deeply about how providing autonomy at work can encourage the flourishing of new ideas, and we will also think about how risk avoidance can suppress innovations. We begin our discussions here by exploring definitions of the terms 'creativity' and 'innovation'.

Defining creativity and innovation

The words creativity and innovation are rather evocative. When we think of 'innovation' we typically think of a rather grand breakthrough that an organisation proudly announces, and which then leads to some considerable and noteworthy success. 'Creativity' by contrast is often associated with the arts – music, dance and theatre, for example – or we may conjure up an image in our minds of the scientist in his or her garden shed inventing something rather off the wall. In truth these are rather misplaced stereotypes and we must recognise that creativity and innovation are parts of the same process. They are not, as some may think, separate from one another – successful innovations are predicated on the generation and development of creative ideas. So how, we may ask, are the concepts defined?

Creativity, some commentators would argue, can simply be defined as the generation of new ideas (Powell and DiMaggio, 1991). It is argued to be built from lateral thinking (De Bono, 1970), with the academic community settling, reasonably consistently, on the view that in an organisational sense it is ultimately about the production or development of novel and organisationally useful ideas (Amabile et al, 1996). This latter definition has two important aspects: 'novelty' and 'utility'. Ultimately, for an idea to be considered creative it must contain some element of 'newness' for the organisation: in other words, it must bring in something that hasn't to this point existed within the organisation's consciousness. On top of this, the idea must also be useful: in other words, the organisation must be able to make use of the idea within the bounds and confines of its mission and vision (Perkins et al, 2017). It is unlikely, for instance, that a small family-run newsagent could challenge the dominance of Apple or Samsung within the mobile phone market. Ideas can only be considered creative in an organisational sense if they have both novelty and utility for the organisation in question.

How, we may therefore ask, can we define innovation? Innovation is often seen as the implementation of ideas (Perry-Smith and Mannucci, 2017). It involves the process of translating a good, organisationally relevant idea into something tangible, perhaps a new product or service offering, or a production efficiency that creates value for the organisation. It is the value creation element that is essential here. If we have an idea that does not create organisational value then it cannot be considered to be an innovation. Let's think for a few moments about the linkages between the two processes of creativity and innovation.

As mentioned towards the start of this section, we must recognise that the processes of creativity and innovation are indeed linked. They are very much entwined with one another, and when exploring creativity and innovation in an organisational context it is often difficult to disentangle the two. Helpfully academic scholars have invested some considerable time into researching the linkages between creativity and innovation, with individuals such as Kijkuit and van den Ende (2007) and Perry-Smith and Mannucci (2017), amongst others, exploring the various phases through which ideas move. Sometimes conceptualised as 'the journey of the creative idea', the phases are often depicted as shown in Figure 14.1.

The creative process is thought to begin with a spark of inspiration. This is the initial 'generation' phase referred to within the diagram. This is where an idea, which may be only partially formed at this time, comes into an individual's mind. It is often therefore referred to as the archetypal 'light bulb moment' or 'flash of inspiration'.

Figure 14.1 The journey of the creative idea

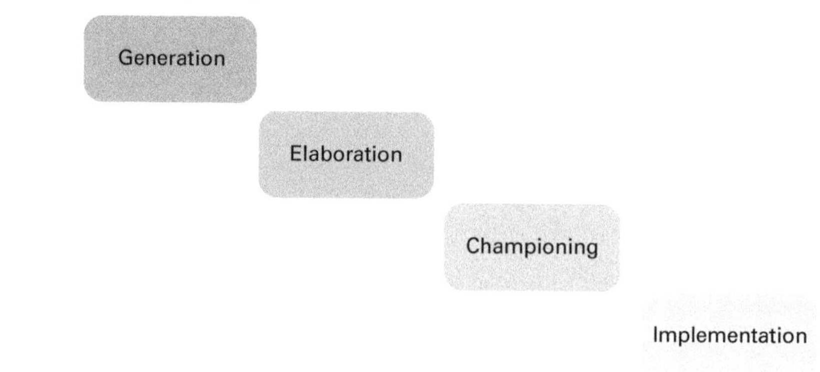

SOURCE: Perry-Smith and Mannucci (2017)

What is important to remember is that many of the nascent ideas we come up with go no further than this stage. We think of a number of possible ideas and then rapidly filter and reject the ones that have little value, would be unworkable or that we otherwise deem irrelevant. This process can be conscious or subconscious, and it is often said that creativity thrives in the liminal space that exists between the conscious and subconscious parts of our brains. The process of 'generation' ends when our thoughts begin to coalesce around a particular idea and then we move into the phase of 'elaboration'.

Elaboration involves putting flesh on the bones of our idea. When we end the generation phase we may have a broad vision of our idea and some outlining principles, but it is likely to be rather hazy and fuzzy at this time. Elaboration involves us undertaking some more thinking, and perhaps some prototyping or testing to begin to understand how our idea may operate in practice. What might this 'thing' that we have conceived look like in the real world? What might we need in order for our idea to become operational? It is these sorts of points that we address at this stage, as we get our idea ready to share with others.

Perry-Smith and Mannucci (2017) argue that the third stage in the process is that of 'championing'. This is the stage where we actively promote our idea to others. Perhaps we formally take it to our manager, or the senior leaders of our organisation, with both an outline of the vision for our idea and the detail that we have fleshed out during the 'elaboration' phase. At this part of the process we are actively attempting to build a coalition of support behind our idea, pushing for the resources (time, money, people and so on) that we might need in order to make it a reality. This, as you might imagine, is often where issues of power and political activity enter the fray. In order to progress our idea, we need to be seen as legitimate, and we will likely need significant capacity to influence others. History is littered with examples of revolutionary ideas that were initially rejected, sometimes with very significant results. For instance, Steve Sasson, an engineer at Kodak, invented the world's first digital camera only to be told by his bosses 'that's cute, but don't tell anyone about it' (Mui, 2012). History shows that Kodak subsequently missed the digital revolution in photography, with catastrophic results for the organisation.

The final stage in the journey of the idea is that of 'implementation'. Referring back to our previous definitions of creativity and innovation, you should recognise that this is where we transition between the two concepts. In the model, 'generation', 'elaboration' and 'championing' are arguably parts of the creative process. With 'implementation' we move into innovation. This is the stage at which our idea becomes real. We convert it into some sort of tangible outcome, for instance a new product or service, a process efficiency or improvement, and it begins to return results to our organisation. In the private sector this will likely mean sales revenue and profitability; in the public and third sectors outcomes may be connected more readily with organisational efficiencies and effectiveness. This, in short, is the journey of our creative idea.

PAUSE AND REFLECT

Examples of creativity and innovation are all around us. Take a few minutes to list as many examples of creativity and innovation as you can.

There are, of course, a great many answers to the activity above. Some may have reflected on specific products, services or approaches to doing business: for instance, Uber and Airbnb have developed innovative and highly disruptive business models, and Apple reinvented the mobile phone market with its iPhone, which launched in 2007. More generally, others may have pointed to things such as the internet as being very significant and substantial innovations, as well as the global positioning system (GPS) upon which many aspects of daily life now rely. It is important that we recognise the diverse outcomes of creativity and innovation, and the fact that organisations of all shapes, sizes and types are increasingly reliant on creativity and innovation in order to achieve their goals. This is a theme that we will build on during the next section of the chapter.

The importance of creativity

As we have noted so far during this chapter, the ability to create and innovate is fundamental to the success, and indeed in many cases the survival, of almost all organisations. Scholars such as Anderson et al (2014) note the role of creativity in bringing about improvements in organisational performance but argue that creativity and innovation are complex phenomena that are difficult to control and guide. We know that the creative process is rather messy and iterative (De Bono, 1970), often involving movement in one direction, only for projects and ideas to change course and backtrack several times before a final solution is reached. But having said this, we can see the results of successful creativity and innovation all around us.

During our discussions within the last section, we mentioned breakthroughs such as the internet, GPS, Apple's iPhone and Kodak's failure to capitalise on the digital camera. Putting these specific examples to one side for the time being, another sector that is seeing very substantial levels of creativity and innovation at this time is the automotive industry. It is experiencing profound levels of change with the push towards the electrification of vehicles and the development of autonomous driving technologies. These technologies are developing at a very rapid pace indeed. For instance, developments in battery technologies are permitting substantial increases in the driving range of fully electric vehicles. These technologies are, of course, very resource intensive, with manufacturer investments running into several billions of dollars each. As a result of the level of investment needed, news commentary frequently highlights that manufacturers are quickly developing partnerships or joint working and research agreements. For instance, Ford and Volkswagen have announced that they will partner on the development of autonomous driving technology and the development of electric powertrains (Abuelsamid, 2019). Companies not able to invest in these new technologies will likely face a precarious future, and it is likely that the present disruptions in the automotive market will create winners and losers, and potentially a new competitive order within the industry. Indeed, companies such as Apple and Google (via the Waymo subdivision) are rumoured to be developing vehicles for entry into the market. The automotive sector, from a creativity and innovation perspective, will very much be a sector to watch closely over the coming years.

Although it is important to recognise and discuss the value of creativity and innovation in an organisational sense, we must also reflect on the importance of the concepts at the level of the individual as well. We know that the world of work is changing (Colvin, 2015), with maturing technological developments such as artificial intelligence, machine learning and automated manufacturing processes likely to have a profound impact on work and employment in the coming years (Frey and Osborne, 2013; Straubhaar, 2017). It is thought that advancements in technology will radically reshape some jobs and lead to the elimination of other roles, particularly those that are lower in terms of skill requirements or those that involve repetitive work (Acemoglu and Restrepo, 2018). Indeed, we are already seeing robots trialled in some fast-food restaurants (BBC News, 2018). In the particular instance discussed by BBC News, a Californian fast-food outlet installed a burger-flipping robot to replace human labour, and although media reports indicate that the trial was less than successful due to the current speed of the robot, it is inevitable that the technology will continue to be refined and improved over the coming years.

Whilst the example above is interesting, what key message does it convey? One of the key messages to take from the discussion above is that in the future world of work employees will likely create value for organisations in different ways. Rather than employing individuals to undertake manual or repetitive roles, it will be skills such as creativity, imagination and innovation that provide individuals with a competitive advantage in the employment market. It is understood that, whilst AI, machine learning and other advanced technologies will likely lead to the automation of many routine roles, technology is unlikely to replace the human capacity for original thinking, inspiration and insight. Creativity, at the level of the individual, is thus likely to remain a skill or trait that is very much in demand (Pistrui, 2018).

Collective creativity and innovation

Creativity and innovation are often conceptualised as products of individual effort and exertion. As noted earlier in this chapter, the term 'creativity' often conjures up images of a scientist working away in his or her garden shed, producing something new after many phases of trial and error. In reality, creativity and innovation are often the products of collective efforts and teams of individuals working together in order to solve a particularly challenging problem. As far back as the early to mid-1990s researchers have been exploring and unpacking the various factors that are thought to impact upon creativity in groups (Woodman et al, 1993), and recent work (for example, Casciaro et al, 2019) emphasises the power of collaboration. Whilst we know through academic study that creativity and innovation can, in reality, be harmed in group environments when other individuals are present in an evaluative or judgemental capacity, Woodman and his colleagues nonetheless discuss how groups can produce improved creative outcomes.

One of the first factors we need to consider is leadership in the group environment. Woodman et al (1993) note the overriding importance of democratic rather than autocratic leadership. Where leaders are collaborative in their nature, group environments are thought to be more likely to arrive at innovative outcomes. Similarly, we know that group structures need to be organic, with changing membership and loose working structures (Martins and Terblanche, 2003). Because there is no one 'recipe' for creativity – in other words, ideas take many forms and the generation and elaboration phases can move in many different directions – rigid structures very much harm our ability to create in group environments. Individuals may need to take on different roles at different times, or we may reach a specific point where we need to involve individuals with different skills. All of this means that our group structures need to have an organic quality if we are to support creativity and innovation. We similarly need to recognise the impact of group cohesiveness and longevity on creative outcomes. It should be clear that if our groups lack cohesiveness then this will negatively impact upon the final outcome, but this is not necessarily the end of the story. Woodman et al (1993) note that groups that are too cohesive can, in reality, harm creative outcomes. This is because such groups can become vulnerable to groupthink, whereby it becomes more important to follow the group consensus than to challenge thinking. Indeed, practitioners such as Johnson (2010) indicate the importance of challenge in group environments, where ideas are improved when different points of view collide.

Building from the points above, there are also interesting arguments connected with the role of diversity and collective creativity. We know that diversity in groups supports creativity and innovation (Martinez and Aldrich, 2011), with it being repeatedly argued in academic circles that increasing diversity is one of the most important roles that leaders have when seeking to improve levels of creativity (Amabile and Khaire, 2008). There is a contention in the literature, however, which suggests that groups that are 'too diverse' are often not cohesive (Daniels and Macdonald, 2005). This therefore is thought to ultimately have a negative rather than positive impact on organisational performance. In essence, we need to remember that infinite diversity does not equal infinite creativity; academics have reached a conclusion that group diversity needs to be managed effectively for maximum benefit. When attempting to improve conditions for collective creativity and innovation, organisational leaders therefore need to actively 'design' the collective context. As Johnson (2010) highlights, however, creativity and innovation are supported when we can get 'more parts on the table', emphasising again the positive role of diversity in driving creative performance. To illustrate the collective nature of creativity and innovation, let's explore an example: the development of GPS.

CASE STUDY 14.1

The development of GPS

As we highlighted earlier in this chapter, the development of GPS was a remarkable innovation and a significant technological breakthrough. This was not, however, produced by any one individual working alone. It was most certainly a team effort, and its development began in the autumn of 1957 at Johns Hopkins University in the United States.

In October 1957 the Soviet Union had just launched the first satellite – Sputnik. This, in itself, was a tremendous achievement, and it sparked a great deal of scientific interest all around the globe. At Johns Hopkins University, two physicists, William Guier and George Weiffenbach, were chatting informally in the cafeteria when they began thinking about whether and how they might be able to listen to the satellite. The Soviets had ensured that the device produced a signal, so surely they should, somehow, be able to pick this up? Weiffenbach, being an expert in the reception of microwaves, had some receiving equipment set up in his office, and the two started to try out different ideas in an attempt to listen to the satellite.

After a relatively short time they did indeed begin to pick up the signals from Sputnik and they started to write these down, including the time and date stamps at which they had received the signals. They then moved out of Weiffenbach's office in order to use the university's mainframe computer to study the data, and ended up mapping out the orbit and location of the satellite. Although this endeavour had started out as a side project, borne out of their shared interests and passion, the two successfully mapped the orbit of Sputnik from just listening to the various signals that it produced.

A few weeks after this, their boss, Frank McClure, came to hear of their work and asked both Guier and Weiffenbach about their side project. Becoming a little more technical, McClure asked whether, if they had been able to map the location of Sputnik from a known location on the ground (the university), could they go the other way around? In other words, extrapolate an unknown ground location if the position of the satellite was known. Guier and Weiffenbach thought about this for some time and realised that this would actually be considerably easier than their mapping of Sputnik. It turns out that McClure was interested in the inverse process because he was working with the military to develop a new missile guidance system! Ultimately, this is how GPS was born. In the late 1980s the then US President, Ronald Reagan, made the GPS system 'open access': in other words, any individual or organisation could build upon the technology. In the present time our smartphones, cars and all sorts of other devices rely on GPS; it has indeed become embedded in our daily lives.

SOURCE: Adapted from Johnson (2010)

The story of how GPS was born is a very interesting insight into the power of collective creativity. What we often find is that one individual, or team, has one part of an idea, while another individual, or team, has another part. It is not until we create the conditions whereby the various elements of the idea can come together that we see creativity, and subsequently innovation. We must therefore design our organisations in such a way as to enable these 'collisions' between ideas; the stereotypical water-cooler conversation has tremendous power when it comes to creativity. Of course, moves towards increases in flexible working and indeed remote working (whilst very positive for other reasons) can potentially have a negative impact on the ability of individuals to come together and share their nascent ideas. The plethora of online collaboration tools can help, to some extent, but as Johnson (2010) argues, there is substantial value in bringing people together when it comes to creativity and innovation. We now move on to consider how leaders can best support creativity and innovation.

Leadership and creativity

There are a great many examples of leaders argued to be role models in terms of either their own personal levels of creativity and innovation, or their ability to inspire creativity and innovation within others. Steve Jobs and Jony Ive at Apple, Ed Catmull at Pixar, and many others besides, are often held up as examples of creative leaders. Within this section of the chapter we will explore some of the characteristics of leadership that are thought to facilitate creativity.

We know that there is a significant number of leadership theories. We have trait theories (Stogdill, 1974) and style theories (Lewin et al, 1939) alongside transactional, transformational, situational, adaptive, authentic and servant leadership (De Jong and Den Hartog, 2007). More recent work (Newman et al, 2018) has highlighted significant links between entrepreneurial leadership and employee innovative behaviour, particularly because these leaders are thought to demonstrate personal drive and vision, as well as encouraging a degree of risk (Kempster and Cope, 2010). What we know from the academic literature base is that there is substantial interest in how the concept of 'leadership', howsoever defined, interacts with creativity and innovation, but relatively little consensus. For instance, some well-known studies argue that leaders need to display transformational qualities (Shin and Zhou, 2003) in order to encourage creativity, but recent work from Ma and Jiang (2018) suggests that transactional qualities may, in fact, be important. What we may therefore argue from this brief introduction is that there may be a diverse range of approaches to leadership that can bring about creativity and innovation in organisations, and that matching leadership to the context and situation of the organisation is likely to be important.

So, where do we go from here? Helpfully, articles such as Amabile and Khaire (2008) provide some general principles regarding the role of the leader in encouraging creativity within others. First of all, it is important for leaders to recognise that they themselves do not have a monopoly on good ideas. Often the best ideas arise from the very bottom of the organisation's hierarchical structure, as it is these individuals who are arguably most in touch with day-to-day operations. Leaders need to therefore ensure that they gather ideas from the entirety of their organisations, fostering and encouraging collaboration, as we discussed in the previous section. It is when diverse perspectives can be brought to bear on a problem that we are most likely to arrive at a creative and innovative solution. Amabile and Khaire (2008) then go on to say that leaders need to create process, but carefully. The thought here is that where leaders attempt to standardise organisational processes in the search for efficiencies, creativity can quickly be squeezed out. Indeed, as Johnson (2010) argues, creativity and innovation are inherently 'messy', and it is through unexpected collisions that often the best ideas arise. In short, processes that are too rigid, or systems that are too bureaucratic, operate to suppress rather than support creativity. The filtering of ideas is similarly an area where leaders need to operate with caution. Whilst it is vital that organisations have some form of screening mechanism to remove ideas that are unworkable, too risky, or otherwise outside the organisation's vision and objectives, this filter must still encourage individuals to be creative. For instance, if employees are simply met with an answer of 'no' when presenting their idea, with no detailed feedback, we know that this response suppresses subsequent creativity. It is important that filtering systems provide feedback to staff, indicating why ideas have been rejected. It is often said that the key role of the leader is to provide employees with a path through the bureaucracy of the organisation, embracing the certainty of failure when striving to create something genuinely new.

Leading for creativity at Pixar

By almost all measures of performance we have, Pixar is a very successful organisation indeed. The animation studio, which for a substantial time operated independently of any major studio, has been behind successful animated movies such as *Toy Story*, *Finding Nemo*, and more recently *Cars* and *The Incredibles*. Its films have made substantial amounts of money in cinemas and through merchandise sales, whilst its creative performance has been recognised through a large number of major award successes; for instance, its outputs have won at both the Oscars and Golden Globes. It is held up as an organisation at the forefront of innovation in film and cinema, and has been the subject of intense interest in exactly how it has been successful. So, what can we learn about creativity from studying Pixar?

Helpfully, the substantial interest in Pixar means that there is no shortage of material published about the studio, with stories from employees and industry commentators illuminating some of the organisation's internal workings. One of the best comment pieces was produced by Ed Catmull (2008), the studio's founder. Catmull argues that at the core of Pixar's success is a recognition that people are at the heart of the organisation's culture. He argues that the organisation is only successful because they have a wealth of skills within their workforce, with everyone bringing a different perspective and skill set. Artists, graphic designers, directors, animators and many other roles are involved in the production of a Pixar movie, and it is through collaborative efforts that their creative vision is realised. Catmull argues that a key role of leaders at Pixar is to ensure that the right individuals are recruited into the organisation, and then that managers 'bet big' on them in terms of enabling creative freedom and providing resources. Alongside this, Pixar recognises that it is important to provide honest feedback and to ensure that employees are supported, particularly when projects have not gone to plan.

As an example of their collaborative approach, the studio convenes frequent internal screenings of its work in development. Employees are invited along and at these events all individuals are free to provide constructive feedback, and to challenge the creators. Catmull argues that it is better the studio hears about the problems with a movie before it is released to the public, and so one of the organisation's key principles is that everyone has the freedom to communicate with anyone – either across or up the hierarchy. The argument is that at Pixar it 'must be safe to tell the truth', and that the role of management is 'not to prevent risk, but to build the capability to recover when failures occur' (Catmull, 2008: 66). It would appear, therefore, that Pixar has recognised that creativity and innovation entail a substantial degree of risk. When leading for creativity, managers must understand that failure is part of the process and the creative journey. Where organisations seek to prevent risk, and/or act in a very risk-averse manner, we see little in the way of original thinking. Employees in those environments think incrementally rather than radically, and innovation can therefore suffer. Let's now build on these ideas by exploring creativity and innovation in large organisational environments.

Creativity and innovation in large organisations

Often we may assume that creativity and innovation are the preserve of small organisations or radical start-ups, which are born off the back of a revolutionary idea, or set of linked ideas. Sometimes this is indeed true, with organisations such as Facebook, Airbnb and many others starting from very small beginnings, creating new business models and revolutionary products. But, how do creativity and innovation operate in larger, established organisations? Google, for instance, began as a small start-up organisation, but it is now, by many measures, one of the largest and most influential organisations in the world. Having said that, it is frequently held up as an example of how to foster creativity and innovation, and indeed we still see a steady stream of new products and ideas emerging from the organisation. What challenges do large organisations face when seeking to encourage creativity and innovation, and how might these be overcome?

Academic debate connected with creativity and innovation indicates that bureaucracy and complex organisational structures negatively impact creativity (Hamel and Zanini, 2018; Hirst et al, 2011). Where employees need to navigate through different levels of hierarchy, conforming to the requirements of organisational regulation and process, we know that levels of creative and innovative output diminish (Hirst et al, 2011). This can be for a number of reasons. Whilst bureaucratic organisational structures have arisen as a method for organisations to cope with complex environments (Hamel and Zanini, 2018), we know that such structures are typically risk averse. As a result of this, good ideas may not be pushed through the decision-making chain, and may therefore not receive the resources that are needed in order to fully develop them. We may also find that as ideas rise up through the various levels of management there is a desire for each individual manager to add their own input or spin to the idea, with either positive intent or their own self-interest in mind, which consequently changes the original idea. In such circumstances we may find that the idea moves quite substantially away from the intent of the originator, and thus does not have the intended impact. Likewise, we know that power and political structures influence creativity and innovation (Eldor, 2017). Whilst Eldor notes that in some instances the presence of political structures inside organisations can encourage creativity by making employees more proactive and adaptive, individuals can encounter situations where creative ideas are blocked because they don't 'fit' with the broader agenda. In short, we need to recognise that the typical architecture of larger organisations can present substantial barriers to creativity and innovation.

Despite the points above, there are a number of examples of large organisations that are seen as very creative indeed. These include Google, as mentioned previously in this section, as well as organisations such as 3M and Apple, amongst many others. The late Steve Jobs (2010) argued that Apple was 'organised like a start-up' in that different individuals were fully accountable for different areas of its hardware and software. Similar to the philosophy of Pixar, Jobs also argued that giving people freedom and accountability encouraged creativity, as did effective teamwork. It can be argued therefore that Jobs perhaps saw his key task as a leader as being the removal of bureaucratic impediments within the organisation. Paul Burns (2016) discusses these ideas more formally and refers to the term 'corporate entrepreneurship', which is often termed 'intrapreneurship'. This is seen as a management style used in larger organisations that integrates the ability to take risks and innovate with the reward and motivational techniques more commonly associated with entrepreneurship. In essence, employees in larger organisations are given freedom to create, developing their ideas into profitable outputs, whilst operating within the broader umbrella of the wider organisation. This can be seen, to an extent, as organisations operating within organisations, but is believed to work very well in terms of providing conditions conducive to creativity and innovation within larger contexts.

In a similar vein to the points discussed above, other large organisations have invested in so-called 'skunkworks' in order to develop radical new innovations (Fosfuri and Rønde, 2009; Pullen, 2019). The core idea here is that, due to the risks associated with creativity and innovation, organisations have sought to place their more radical research and development activity into specialized divisions, which have considerable autonomy over the way that they use and allocate their resources. The term 'skunkworks' was originally developed to describe the research and development department run by the aircraft manufacturer Lockheed (Blank, 2014), but other organisations such as Facebook, Symantec, and DuPont have similar divisions. Broadly speaking, there is recognition that risk does not sit well inside typical bureaucratic and regulated corporate cultures, and therefore organisations have actively sought to create spaces where individuals can be given a good deal more creative 'licence' and the freedom to experiment with new ideas. Having said this, Blank (2014) argues that the concept of skunkworks is increasingly falling out of favour as the business landscape changes. Given that our organisations need to innovate continuously, it is thought that innovation, and the ability to experiment with new ideas, needs to be firmly impregnated throughout an organisation. Indeed, when we explore examples such as Pixar it is this that we see. Organisations such as Pixar do not seek to create a firewall between their business and the risks associated with innovation; innovation is a fundamental part of what they do, and their vision recognises that risk, and failure, is part of the process. Looking back to the previous part of this chapter, you will remember the quote that the role of management is 'not to prevent risk, but to build the capability to recover when failures occur' (Catmull, 2008: 66). Let's now turn our attention to the broader base of factors that are known to impact on creativity and innovation within the workplace.

What factors impact on creativity and innovation in the workplace?

What do we know about the various factors that impact on creativity and innovation in the workplace? During this chapter thus far we have already explored topics such as leadership, the notion of collective creativity, and the problems caused by bureaucratic work environments. The purpose of this section is not to repeat content that we have previously discussed, but to explore the broader base of academic and practitioner literature that has highlighted a variety of factors known to impact upon creativity and innovation in the workplace. During our discussions here we will cover:

- Autonomy and freedom.
- Rewards.
- Job design and the need for stretching work.
- Constructive dissent.
- Knowledge sharing.

There has been significant and substantial interest in the various factors that impact upon creativity and innovation at work. Scholars such as Anderson et al (2014) and the various papers published by Theresa Amabile (see, for example, Amabile et al, 1996; Amabile and Khaire, 2008; Amabile and Pratt, 2016), amongst many others, provide a strong foundation for the various factors that we will discuss, alongside many other contributions from practitioners such as Johnson (2010). Let's begin with autonomy and freedom.

Autonomy and freedom

Whilst we have not, thus far, specifically focused on autonomy and freedom, the terms have arisen at various points of our discussions. A variety of researchers, including Amabile et al (1996) and, more recently, Li et al (2018), have discussed the importance of autonomy at work, and strongly link it to increased creative performance. In essence, the argument here is that where employees have discretion in how they undertake their work, and a degree of freedom in what they do on a day-to-day basis, this positively impacts creativity. The view, particularly from Li et al (2018), is that autonomy gives rise to increased intrinsic motivation, and this is why it has a positive impact on creativity. Likewise, it is important that employees have a sense of ownership over their work. Where employees perceive that they have control over their work, they are more likely to exhibit the discretionary effort needed in order to produce something new. The example of the development of GPS, which we explored previously in this chapter, is a useful illustration of the powerful role of autonomy and freedom within the creative process.

Rewards

A substantial debate exists regarding the impact of rewards, both financial and non-financial, on creativity. At first we may assume that there is likely to be a simple relationship here; incentivise employees with a monetary bonus if they develop new ideas that positively impact the organisation, and a constant stream of innovations will flow as a result. Reality is, however, a good deal more complex than this. Whilst some studies do claim that monetary reward facilitates creative performance (Mehta et al, 2017), others vehemently disagree, arguing that financial reward lessens creativity (Amabile, 1982). Indeed, Eisenberg (1999) highlights the diversity of findings in relation to this particular issue, with some groups of researchers finding a positive link between financial reward and creative performance, and others finding no such link. The balance of opinion rests towards the view that monetary rewards can negatively impact creativity, particularly if the rewards are relatively large. The thought is that where large rewards are offered, individuals can fixate on the end goal, meaning that their ability to think laterally (De Bono, 1970) is compromised, and thus creativity suffers. It is thought that intrinsic reward, in the form of recognition and the satisfaction of having one's idea implemented, is substantially more powerful when it comes to motivating creativity and innovation.

Job design and the need for stretching work

As far back as Hackman and Oldham's (1980) work on job design there has been an interest in how we design jobs in order to encourage creativity and innovation. Indeed Hackman and Oldham themselves isolated a number of important factors underpinning satisfying jobs, including skill variety, task significance and autonomy, amongst others. More recent research (Yoo et al, 2019) has extended this work and found that job design has a substantial impact on creativity. Interestingly, it is argued that levels of creativity are positively impacted by challenging work (Oldham and Cummings, 1996). When we examine this critically, it stands to reason. If, for instance, we have a role that does not push and stretch us – in other words, it enables us to sit very much within our comfort zone – there is little pressure to do anything 'new'. It is when we have a role that requires us to undertake new or stretching activities that we best learn and develop new capabilities. It is in these spaces, where we are being stretched, but not pushed towards a zone of uncontrolled stress or panic, that we can be most creative. When designing jobs it is therefore important that managers and leaders consider how differing roles can be structured in such a way that they encourage stretch.

Constructive dissent

At an earlier point of this chapter we discussed the importance of cohesion in group environments, but also noted that environments which are too cohesive can become vulnerable to groupthink, and its negative impacts on creativity and innovation. Some groups of researchers argue that ensuring harmony in the workplace has a number of positive benefits (Kuo et al, 2014), but others (Perkins, 2017) suggest that creativity can be enhanced where workplace cultures enable disagreement to occur. It is thought that the development of new ideas can be improved in more dynamic settings where individuals can challenge not individuals but ideas; this is an important distinction. As logic would suggest, workplace cultures can be harmed, or indeed irreparably damaged, when individuals make such disagreements personal, but where constructive challenge is offered to ideas, this has been found to motivate improved creative performance. Perhaps different individuals try out different ways of solving a problem, and then attempt to find an optimal solution, or there are competing visions for marketing a new product. In these sorts of situations leaders should encourage a degree of challenge and debate, although it is important to recognise that the leader has a significant role here in managing the organisation's culture and steering such interactions. Where organisational environments degenerate into a situation where the individual that wins is the one that 'shouts the loudest', this has problematic consequences for creativity and innovation.

Knowledge sharing

Knowledge sharing is thought to be a vital part of the creative process. Linked to the idea that creativity and innovation are rarely the outcome of individuals working alone, organisations must develop channels through which knowledge can be effectively shared and managed. Sharing knowledge around the organisation is thought to provoke lateral thinking, with individuals coming up with new ideas in response to gathering new elements of knowledge. A substantial amount of time has been invested in understanding the properties of organisational networks (see, for example, Schilling and Fang, 2013) that best support creative and innovative activity. Whilst there is debate within this area, it is thought that networks which incorporate what are known as 'weak ties' (Granovetter, 1973) – in other words, connections to acquaintances rather than close friends and colleagues – are more likely to positively influence creativity and innovation. At first glance this seems rather unlikely, but why might this be the case? The view is that whilst strong connections with close friends and colleagues can provide a sounding board for new ideas, and this is indeed a positive quality of close connections, connecting with many individuals who are at some distance from yourself brings you into contact with new knowledge. It is this contact with diverse pools of knowledge that is thought to foster new ideas, and thus managers and leaders in organisational contexts should actively foster the creation of a diverse range of 'weak' ties.

Let's now move into the final discussion area of this chapter: the role of HRM in encouraging creativity and innovation.

The role of HRM in encouraging creativity and innovation

From our discussions within this chapter it is already apparent that the processes and systems we have in place to manage people can have a substantial impact on creativity. We have, for instance, discussed the importance of effective job design in terms of providing employees with stretch opportunities, and we have touched on the importance of reward management as well. The fascinating finding with regard to reward management is that it is intrinsic rather than extrinsic rewards that are arguably more likely to motivate creative performance. With this in mind, offering a monetary bonus to stimulate 'good ideas' may, in fact, be counterproductive.

A significant theme within this chapter has been the risky nature of creativity and innovation. The creative journey often involves failure; indeed you will remember that at Pixar failure is seen very much as part of the creative process. The role of the manager is, as has been said, 'not to prevent risk, but to build the capability to recover when failures occur' (Catmull, 2008: 66). This philosophy arguably has substantial impacts for performance management processes.

Picture the scene... You have been working in your organisation for a substantial period of time, and have recently developed an idea for a new product which you, and your team, believe will be highly successful. Imagine that the product is finalised, launched, but then doesn't generate a return for the organisation because of some sort of shift in the marketplace. Perhaps a competitor has come up with an even more ground-breaking idea, putting your new product rather in the shade. How should the organisation deal with this in relation to performance management? Is this a mistake to be punished? Did someone overlook a key piece of new research or a new, emerging technology? The reaction of managers to these sorts of events can have a significant effect on the broader culture for creativity and innovation that exists inside an organisation. What we know is that where mistakes are met with criticism and punishment, employees are much less likely to engage in the generation of future ideas. Individuals will naturally seek to avoid such risks.

When designing frameworks through which we measure employee performance we therefore have to strike a balance between ensuring appropriate and legitimate managerial control, and the recognition that doing something creative and innovative involves risk. Corporate cultures best supporting creativity and innovation are those that are set up to learn from failure, rather than assign blame; the same is true for performance management processes. It is important to recognise that this is not a 'soft touch' approach, or an excuse for failure; indeed, very creative organisations have rigorous processes from which they can draw out learning points when things do not go to plan. Having said this, the approach that we take to performance management can, and does, have a substantial impact on creativity.

As an interesting final point to consider, authors such as Binyamin and Carmeli (2010) argue that having formal structure within our approaches to HRM is important if we seek to encourage creative performance. What they mean by this is that we should have formalised HRM processes and systems that are clearly communicated to all staff, and applied both fairly and rigorously across our organisations. On the face of it this would appear rather counterintuitive, given our discussions concerning the need for autonomy, flexibility and appreciation that the creative journey can take all manner of twists and turns from start to finish. What the authors argue though is that the structuring of HRM instils stability, order and pattern, which then enables our employees to fully focus when they are tackling challenging tasks. Bureaucracy, in this sense, is thought to prevent reactive approaches to management, or short-sighted decision making, thus opening a space within which employees can experiment with new, and perhaps risky, ideas.

We know that a diverse range of factors affect the ability of our employees to be creative and innovative. This extends to our approaches to HRM. HR professionals within organisations must consider the impact that their processes, policies and systems have upon their employees' willingness to be creative. How do we best set the stage upon which our employees can perform?

Conclusion

Our exploration of creativity and innovation has come to a close. Through our discussions within this chapter we have come to recognise the importance of the topics, and how we define the terms in a business sense. We must recognise that creativity involves the development of ideas that have both novelty and utility, and that innovation concerns taking those ideas from concept to reality. Given the changes that are occurring in the business environment, it is becoming ever more important that our organisations can generate and develop new ideas – whether they be for products, services, methods of marketing, operational efficiencies, and so on. We must recognise, however, that creativity and innovation are rarely individual acts. Very often, it is individuals working together who come up with the best new ideas, with leaders having an important role in terms of bringing people together, and providing them a stage upon which to perform. Indeed, we spent some time exploring the impact that leaders can have upon the creative process during the course of this chapter.

Aside from the role of the leader in setting the right stage for creativity and innovation in our organisations, we know that a number of other factors impact upon the generation, development and ultimately implementation of new ideas. For instance, we know that we need to provide our employees with autonomy and freedom, design appropriate reward frameworks, and develop jobs that stretch our staff. Alongside this, we need to ensure that knowledge flows around our organisations rather than being bottled up within silos, and that constructive disagreement can stimulate new thinking and new ideas. Towards the end of our discussions we also explored the specific role of HRM in encouraging creativity and innovation, and you will remember the intriguing idea that formal and structured approaches to HRM can facilitate improved creative performance. Whilst somewhat counterintuitive, it does stand to reason that employees will be more creative when they have a secure and stable foundation.

References

Abuelsamid, S (2019) Ford and Volkswagen partner on self-driving and electric vehicles, www.forbes.com/sites/samabuelsamid/2019/07/12/ford-volkswagen-and-argo-partner-on-autonomous-and-electric-vehicles/#455f81d41e8b (archived at https://perma.cc/MU55-62YB)

Acemoglu, D, and Restrepo, P (2018) The race between man and machine: Implications of technology for growth, factor shares and employment, *American Economic Review*, 108 (6), 1488–542

Amabile, T M (1982) Children's artistic creativity: Detrimental effects of competition in a field setting, *Personality and Social Psychology Bulletin*, 8, 573–78

Amabile, T M and Khaire, M (2008) Creativity and the role of the leader, *Harvard Business Review*, 86 (10), 100–9

Amabile, T M and Pratt, M G (2016) The dynamic componential model of creativity and innovation in organizations: Making progress, making meaning, *Research in Organizational Behavior*, 36, 157–83

Amabile, T M, Conti, R, Coon, H, Lazenby, J and Herron, M (1996) Assessing the work environment for creativity, *Academy of Management Journal*, 39 (5), 1159–84

Anderson, N Potočnik, K and Zhou, J (2014) Innovation and creativity in organizations: A state-of-the-science review, prospective commentary and guiding framework, *Journal of Management*, 40 (5), 1297–333

(Continued)

(Continued)

BBC News (2018) Burger-flipping robot taken offline after one day, www.bbc.co.uk/news/technology-43343956 (archived at https://perma.cc/9XXN-KTJU)

Binyamin, G and Carmeli, A (2010) Does structuring of human resource management processes enhance employee creativity? The mediating role of psychological availability, *Human Resource Management*, 49 (6), 999–1024

Blank, S (2014) Why corporate skunk works need to die, www.forbes.com/sites/steveblank/2014/11/10/why-corporate-skunk-works-need-to-die/#3b2bf1ad3792 (archived at https://perma.cc/AV2V-7PGT)

Burns, P (2016) *Entrepreneurship and Small Business*, 4th edn, Palgrave Macmillan, Basingstoke

Casciaro, T, Edmonson, A C and Jang, S (2019) Cross-silo leadership, *Harvard Business Review*, 97 (3), 130–9

Catmull, E (2008) How Pixar fosters collective creativity, *Harvard Business Review*, 86 (9), 64–72

Colvin, G (2015) The 21st century corporation: Every aspect of your business is about to change, *Fortune*, 172 (6), 102–12

Cummings, A and Oldham, G R (1997) Enhancing creativity: Managing work contexts for the high potential employee, *California Management Review*, 40 (1), 22–38

Daniels, K and Macdonald, L (2005) *Equality, Diversity and Discrimination: A student text*, CIPD, London

De Bono, E (1970) *Lateral Thinking: Be more creative and productive*, Penguin, London

De Jong, J P J and Den Hartog, D N (2007) How leaders influence employees' innovative behaviour, *European Journal of Innovation Management*, 10 (1), 41–64

Eisenberg, J (1999) How individualism-collectivism moderates the effect of rewards on creativity and innovation: A comparative review of practices in Japan and the US, *Creativity and Innovation Management*, 8 (4), 251–61

Eldor, L (2017) Looking on the bright side: The positive role of organisational politics in the relationship between employee engagement and performance at work, *Applied Psychology*, 66 (2), 233–59

Fosfuri, A and Rønde, T (2009) Leveraging resistance to change and the skunk works model of innovation, *Journal of Economic Behavior and Organization*, 72 (1), 274–89

Frey, C, and Osborne, M (2013) The future of employment: How susceptible are jobs to computerisation? Oxford Martin School Working Paper, www.oxfordmartin.ox.ac.uk/downloads/academic/The_Future_of_Employment.pdf (archived at https://perma.cc/5UYH-58TD)

Granovetter, M S (1973) The strength of weak ties, *American Journal of Sociology*, 78, 1360–80

Hackman, J and Oldham, G (1980) *Work Redesign*, Addison-Wesley, Reading, MA

Hamel, G and Zanini, M (2018) The end of bureaucracy *Harvard Business Review*, 96 (6), 50–9

Hirst, G, Van Knippenberg, D, Chin-Hui, C and Sacramento, C A (2011) How does bureaucracy impact individual creativity? A cross-level investigation of team contextual influences on goal orientation-creativity relationships, *Academy of Management Journal*, 54 (3), 624–41

Jobs, S (2010) Steve Jobs talks about managing people, www.youtube.com/watch?v=f60dheI4ARg (archived at https://perma.cc/P4WF-6K6R)

Johnson, S (2010) *Where Good Ideas Come From: The natural history of innovation*, Allen Lane, London

Kempster, S and Cope, J (2010) Learning to lead in the entrepreneurial context, *International Journal of Entrepreneurial Behaviour and Research*, 16 (1), 5–34

Kijkuit, B and van den Ende, J (2007) The organizational life of an idea: Integrating social network, creativity and decision-making perspectives, *Journal of Management Studies*, 44 (6), 863–82

Kuo, Y-K, Kuo, T-H and Ho, L-A (2014) Enabling innovative ability: Knowledge sharing as a mediator, *Industrial Management and Data Systems*, 114 (5), 696–710

Lewin, K, Lippitt, R and White, R (1939) Patterns of aggressive behaviour in experimentally created 'social climates', *Journal of Social Psychology*, 10 (2), 271–301

Li, H, Li, F and Chen, T (2018) A motivational-cognitive model of creativity and the role of autonomy, *Journal of Business Research*, 92, 179–88

Ma, X and Jiang, W (2018) Transformational leadership, transactional leadership, and employee creativity in entrepreneurial firms, *Journal of Applied Behavioural Science*, 54 (3), 302–24

Martinez, M A and Aldrich, H E (2011) Networking strategies for entrepreneurs: Balancing cohesion and diversity, *International Journal of Entrepreneurial Behaviour and Research*, 17 (1), 7–38

(Continued)

(Continued)

Martins, E C and Terblanche, F (2003) Building organisational culture that stimulates creativity and innovation, *Journal of Innovation Management*, 6 (1), 64–74

McAdam, R, Moffett, S, Hazlett, S A and Shevlin, M (2010) Developing a model of innovation implementation for UK SMEs: A path analysis and explanatory case analysis, *International Small Business Journal*, 28 (3), 195–214

Mehta, R, Dahl, D W and Zhu, R (2017) Social-recognition versus financial incentives? Exploring the effects of creativity-contingent external rewards on creative performance, *Journal of Consumer Research*, 44 (3), 536–53

Mui, C (2012) How Kodak failed, www.forbes.com/sites/chunkamui/2012/01/18/how-kodak-failed (archived at https://perma.cc/8JWG-LQ7P)

Newman, A, Neesham, C, Manville, G and Tse, H M (2018) Examining the influence of servant and entrepreneurial leadership on the work outcomes of employees in social enterprises, *International Journal of Human Resource Management*, 29 (20), 2905–26

Oldham, G R and Cummings, A (1996) Employee creativity: Personal and contextual factors at work, *Academy of Management Journal*, 39 (3), 607–34

Perkins, G (2017) Exploring the mechanisms through which strong ties impact upon the development of ideas in SME contexts, *Journal of Small Business Management*, doi: https://10.1111/jsbm.12372

Perkins, G, Lean, J and Newbery, R (2017) The role of organizational vision in guiding idea generation within SME contexts, *Creativity and Innovation Management*, 26 (1), 75–90

Perry-Smith, J E and Mannucci, P V (2017) From creativity to innovation: The social network drivers of the four phases of the idea journey, *Academy of Management Review*, 42 (1), 53–79

Pistrui, J (2018) The future of human work is imagination, creativity and strategy, *Harvard Business Review* Digital Articles, January, 2–4

Powell, W W and DiMaggio, P J (1991) *The New Institutionalism in Organizational Analysis*, University of Chicago Press, Chicago

Pullen, J P (2009) Skunk works, *Fortune*, 179 (6), 134–9

Schilling, M A and Fang, C (2013) When hubs forget, lie and play favourites: Interpersonal network structure, information distortion, and organizational learning, *Strategic Management Journal*, 35, 974–94

Shin, S J and Zhou, J (2003) Transformational leadership, conservation, and creativity: Evidence from Korea, *Academy of Management Journal*, 46 (6), 703–14

Stogdill, R M (1974) *Handbook of Leadership: A survey of the literature*, The Free Press, New York

Straubhaar, T (2017) On the economics of a universal basic income, *Intereconomics*, 52 (2), 74–80

Woodman, R W, Sawyer, J E and Griffin, R W (1993) Toward a theory of organizational creativity, *Academy of Management Review*, 18 (2), 293–321

Yoo, S, Jang, S, Ho, Y, Seo, J and Yoo, M H (2019) Fostering workplace creativity: Examining the roles of job design and organizational context, *Asia Pacific Journal of Human Resources*, 57 (2), 127–49

15
Productivity

In this chapter we are going to explore labour productivity. This is a subject that is likely to become more and more significant for employing organisations in the future, and particularly in the UK where recent years have seen poor performance in this area when seen from a national economic standpoint. We will start by explaining what exactly is meant by the term 'labour productivity' and summarise some of the debates about why the UK has not been becoming more productive of late. This is by no means something that economists and commentators agree about. It is common for the term 'productivity puzzle' to be used to demonstrate how so many very different views are expressed about it. We will go on to consider some of the steps that employing organisations might take to improve their productivity over the longer term.

What is labour productivity?

The term is best understood by looking at the three ways in which it is most commonly measured at national level:

- GDP divided by the total number of people in paid work (output per worker).
- GDP divided by the total number of jobs (output per job).
- GDP divided by the total number of hours that are worked (output per hour worked).

GDP here refers to the standard measure of national income. It is essentially an estimate for the total value of all the goods and services produced in an economy in a set period of time across the economy. So labour productivity is a measure of how much of this 'added value' is produced by each worker or each job or in each hour worked on average. The third of these three measures (the one derived from hours worked) is generally considered to be the most useful, but the data collected by government statisticians tends to be less accurate because employers do not typically record this information.

It is also important to appreciate that these measures are different. A person can work a lot of hours, but not be so productive. A worker can easily improve his or her output by working more hours, but that does not mean they are becoming any more productive. It is only when output improves when hours of work stay the same, or when it stays the same as hours fall, that we can say productivity is improving.

In any event, it is important to recognise that national figures in this area are inevitably always going to be estimates. Provided these are carried out consistently in the same way over time and across national borders, it is not difficult to track progress over time and to make meaningful comparisons between the records of different countries.

Labour productivity statistics are also readily calculated for regions of the country and for different industrial sectors. However, these figures do not often seem to form a major part of discussions between HR and L&D managers, nor has the impact of HR activity on productivity been so widely researched (see Beatson and Zheltoukhova, 2015: 45; Sparrow and Otaye-Ebede, 2017: 168). Measuring them and tracking them, however, is the first stage in developing policies aimed at improving them.

Productivity in the UK

In the UK at the national level, productivity improved steadily for many years through the 1980s, 1990s and early 2000s. Growth rate averaged 2.3 per cent each year, leading over time to a very substantial increase. After the recession of 2008–10, however, it stagnated. It has not fallen hugely, but nor has it grown by much either. Best estimates suggest that annual productivity growth since 2008 has averaged just 0.4 per cent a year, others putting it lower still. It would seem that a decade on from the last recession, UK productivity has remained at the same level. This is unprecedented by historical standards and it has become a major public policy concern.

The UK is not alone in achieving such sluggish levels of productivity growth in this period. Canada and Japan, for example, have performed less well still. But most similar economies in the world, including those of France, Germany, Italy and the USA, have all seen significant productivity increases since the end of the recessionary years, albeit at a reduced rate. The UK economy has performed less well still, meaning that in 2018 it was some 16 per cent less productive than those of these other countries and 25 per cent less productive than it would have been had the pre-recession rate of growth resumed when the economy started growing again after 2010 (Giles, 2018).

The problem of poor productivity growth appears to be sector-wide. It is true of manufacturing as much as it is in service industries. The public sector too has seen stagnated productivity in recent years. There are, however, major regional disparities. In London productivity is far higher than in the rest of the country, at 32 per cent above average. The South East of England generally performs better too. Labour productivity is lowest in the West Midlands, Wales and Northern Ireland (Harari, 2017).

Short-term reductions in productivity or periods of slow growth are not problematic. Productivity across an economy will inevitably fluctuate from quarter to quarter and will go through periods of stagnation. But such limited growth in productivity over a ten- or twelve-year period is unsustainable in the long run if we want to see our economy thrive and our wages increase. Low productivity growth is also associated with a situation in which tax receipts decline or stagnate, leaving less money for public investment or expenditure on public services. Improving productivity leads to more economic growth and hence a higher standard of living. The wages we earn decline in real terms (ie after price inflation is taken into account) if we do not earn increases through improved productivity.

The UK's sluggish performance in this area is thus a very serious matter. It is one of the main reasons why standards of living for most people have not improved in recent years, and for many remain lower now in real terms than they were before the deep recession hit after 2007. This stagnation in living standards is often argued to have led to many of the political divisions we have seen since then, and particularly widespread feelings of resentment on the part of the majority towards those in the minority whose living standards have improved considerably.

PAUSE AND REFLECT

Think about yourself and your own work. Are you any more productive than you were a year or two ago? If you are, why do you think the improvement has occurred? If not, why is that the case?

Why is productivity stagnating?

There are no easy answers to this question. Opinion differs and there is no consensus among economists as to why it is that UK productivity did not resume its customary level of year-on-year growth after the recession, as has always been the case before (see Ilzetzki, 2020).

Here are some of the possible reasons advanced in the papers written on this productivity 'puzzle'. You will note that in many cases they are not directly related to people management matters at all:

- Relative weakness compared to other countries in our ability to make full use of new technologies like superfast broadband.
- Too many workers with low skills and too few with higher-level, technical and specialist skills.
- Too little government investment in national infrastructure in the years after 2010 when a policy of austerity was used to gain control of the public finances.
- Uncertainty following the Brexit referendum in 2016 providing a disincentive to long-term investment in the UK.
- Wide availability of cheap labour providing an alternative for employers who might otherwise have invested in new technologies.

- Caution on the part of banks when making decisions about lending to firms looking to invest in skills and technology.
- Population ageing leading to a situation in which the workforce as a whole is not so innovative, creative or technologically proficient.
- Too much regulation imposing unnecessary costs and bureaucracy on employing organisations.
- Too little investment in research and development activity.

This multiplicity of potential causes means that there is no easy answer. There is no silver bullet that could be manufactured and shot at the UK economy to solve the productivity puzzle. It is also clear that there are problems in most industries and that many of the solutions like investing in transport infrastructure, superfast broadband capabilities and improved educational outcomes are much more down to government than organisations. It is, of course, the case that some employing organisations are a great deal more productive than others. The UK economy has within it what Andy Haldane, the Chief Economist of the Bank of England, calls 'a long tail of low-productivity companies which drags on the aggregate' (Institute of Directors, 2018) but productivity has also apparently stagnated in the top-performing organisations which achieved some of the highest growth in productivity in the twenty-five years that preceded the onset of recession after 2007. This includes major players in key sectors such as finance, IT, energy and pharmaceuticals.

While employment-related factors are by no means the source of all the stagnation in productivity, they still have clearly played a significant role. And it may be the case that HR decisions taken over the past decade for entirely laudable reasons may in fact be a key issue. Wolf (2020) puts it very succinctly as follows:

> After a sharp fall during the crisis, employment again grew faster than the population: between 2009 and 2018, employment rose 12 per cent, while population rose only 7 per cent.

In other words, the UK was very good at minimising unemployment. While there were many job losses during the recession, they were fewer and further between than one would have expected to see in a period of such very sharp and rapid economic contraction. UK employers were more successful than their competitors in more productive economies at avoiding job losses. In many cases hours were cut and overtime eliminated while wages were frozen. Retirees were not replaced when they left so that compulsory redundancies could be minimised. People agreed to work more flexibly rather than risk losing their jobs. Then, once the recession was over, encouraged by government policy, more people were hired at relatively low wages. Welfare bills dropped along with unemployment. This was all good news for millions of people who might well have been unemployed for a long stretch of time had it been otherwise. By 2018 the UK unemployment rate had fallen to 4 per cent, which is close to full employment, for the first time since the early 1970s. The result was many more people in gainful employment than was the case in other comparable countries, but the price paid was lower productivity and hence lower wages.

CASE STUDY 15.1

Honda

In 2009 as the impact of the deepest recession in living memory reduced economic confidence and consumer demand, employers were faced with difficult choices. Costs had to be cut sharply and quickly, but there was a wish to avoid compulsory redundancies as far as possible. This was partly to ensure that reputations as 'good employers' were maintained and that employee loyalty was rewarded. There was also a desire to hold on to skilled employees in whom employers had invested a lot of money and who they wanted to continue employing after the recession was over.

At the Honda plant in Swindon there were no compulsory redundancies at all, despite the number of cars being manufactured in 2009 falling from the intended 228,000 to just 113,000. This was achieved in the following ways:

- 1,300 workers agreed to a generous voluntary redundancy package.
- The factory closed for four months, during which full pay was given to staff for two months and 60 per cent for two months.
- Workers agreed to a salary cut of 3 per cent for ten months, managers to a cut of 5 per cent.
- 400 workers were retrained for new roles.
- Workers were awarded an additional six days leave.

Through these means many job losses were avoided and as much of the skills base as possible retained.

However, productivity was lost in the process. An opportunity to restructure more radically by introducing new technologies and employing fewer people on higher pay was lost.

SOURCE: Pidd (2009); Cooper (2009)

The low road and the high road

Improving productivity in employing organisations is a multifaceted task that cannot be achieved in a lasting way quickly. There are a number of different strategies that can be pursued. It is tempting for managers to take what might be termed a 'low road' approach and simply look to get more value out of the workforce. This approach takes two basic forms:

- Pushing existing workers harder to produce more while holding back from paying them fairly or any more at all for their efforts.
- Cutting wage costs and other employee-related benefits and programmes while maintaining the same workloads.

Intensifying work in this way will often achieve good shorter-term results, and in a period of high unemployment employers can risk taking this approach because employees or would-be employees have little choice but to accept it if they are to secure work. Indeed, wherever labour is in plentiful supply and there are no recruitment or retention problems, managers will be interested in following this approach. Employees are put under additional pressure, and are replaced with others if they do not perform.

The problem is that it does not work in tighter labour market conditions when employers are competing for people with skills with other employing organisations. It also tends to lead to a situation in which staff turnover is very high, absence is rising and staff engagement is weak. Adversarial industrial relations is another consequence. Low trust prevails, particularly when managers are rewarded with higher salaries for driving productivity on the backs of low-paid workers. For many organisations that rely on employee commitment and engagement in order to meet their objectives, this is not a sustainable strategy.

This low road approach has, however, been used a lot in the UK in recent years, leading to a situation in which long-term, sustainable productivity improvements have not in fact been achieved. Some call centres, for example, have followed a so-called 'sacrificial HR strategy' by targeting younger recruits to do quite tedious work and setting very tough performance targets. It is anticipated that most will leave after a few months to seek higher-quality work, but when there is no difficulty attracting replacements a churn rate of 50 per cent or so is manageable. Short-term performance targets are met, but no long-term sustained improvements in productivity are achieved. It is because so many employers have taken this approach that the UK never moves on from being, in relative terms, a low-skill, low-wage economy.

What is therefore required is the development of long-term strategies that involve investing in human capital as well as technologies to achieve higher levels of productivity over time without driving people too hard or paying them too little. This 'high road' approach is more challenging and will typically involve some up-front costs by way of investment for the future. It is very much the direction government ministers want to see things developing, the aim being for the UK labour market to become one characterised by high skills, high wages and high productivity.

PAUSE AND REFLECT

Think about your own organisation. Have you observed situations in which senior managers have looked to achieve short-term financial targets by cutting costs and pushing people to work harder? What was the impact in terms of staff commitment and the employee relations climate?

Improving productivity

Achieving sustained, long-term improvements in productivity involves looking to move, step by step, to a situation in which an organisation employs fewer people, on higher pay, adding greater value. The extent to which this is possible inevitably varies greatly from organisation to organisation and industry to industry, but it has been achieved far more effectively in countries that are not so different from the UK like Germany, France and the USA, where productivity is over 20 per cent higher. Collectively, workers in these countries produce as much output in four days as the UK manages in five days. In the US and Germany wages are considerably higher, while in France employees are able to enjoy a shorter working week and much more annual holiday.

The first step involves collecting accurate data and working out on a monthly and/or annual basis how productive an organisation is. Management accountants will typically gather this kind of information routinely. So for many HR managers it is just a question of asking for it and using it along with other metrics to track progress over time and make comparisons with sector averages compiled by the Office for National Statistics and other government bodies. At the level of the organisation, and indeed the department, it is not difficult to establish output per employee or output per hour worked each month. Long-term targets can be set and serious progress made towards improving these ratios.

From a human resource management perspective, it is important to recognise that there is no standard, easy way of improving productivity in an organisation. Business models vary greatly, as do entrenched organisation cultures. It will be much easier, for example, to achieve greater economies of scale by expanding the size of an operation in some settings than in others. Similarly, the capacity to introduce and make use of new technologies is very different in different industries and workplaces. As Sparrow and Otaye-Ebede (2017:169) argue, each workplace is idiosyncratic and will need to find its own 'productivity recipe' in order to bring about genuine, sustained improvements in output per employee or hour worked.

There are, however, some general points that can be made and some general principles that can be applied when determining the required productivity recipe.

Employee satisfaction

An expression that is commonly heard in management circles is 'a happy team is a productive team', suggesting that a key way of achieving improved productivity is to raise the level of staff satisfaction. This is, of course, overly simplistic. A highly unproductive team can also be a very happy one, particularly if they are rewarded well and not required to work so hard. It is, however, true that sustained productivity improvement is unlikely to be achievable if the staff charged with bringing it about are unhappy in their jobs. Work satisfaction is therefore a prerequisite for long-term productivity improvement. It is much more likely to be improved if staff are content than discontent.

There is, of course, a very substantial body of research on job satisfaction and its links to business performance. It is explored fully in texts such as Paauwe et al (2013) and Marchington et al (2020). The key points are as follows:

- It is important that people are not only treated fairly, but perceive that they are being treated fairly. The same rules must thus apply to everyone; unjustified favouritism should be avoided and principles of equity applied across all HR and L&D activity.
- It is necessary to involve people in decision making as much as is possible. This is particularly true at the level of the individual and the team. Autocratic management styles should be avoided, while people should be given as much autonomy over their own areas of work as is feasible.
- People should be encouraged to attain a good work–life balance, particularly when their work is stressful or tedious. For people who genuinely love their work, job satisfaction can be hindered by forcing them to take more time off than they want to, but they and their managers need to be mindful of the possibility of burnout occurring through overwork.

- Design jobs so that they are as interesting and varied as possible. This can be achieved through job rotation, job enrichment and, in some situations, job enlargement. The aim should be to reduce tedium and make work sufficiently challenging to maintain active interest and engagement.
- Provide as many career and personal development opportunities as is possible. People need to be able to see their jobs as stepping stones to new and better career options. This is particularly important in the case of younger workers who are looking to gain experience in order to improve job prospects.
- Recognise achievements, notice when people go the extra mile and thank them. This is an inexpensive but very effective means of improving work satisfaction at the individual and team level.

Improving job satisfaction increases the likelihood that employees will demonstrate discretionary effort. This means going the extra mile and working more effectively because they want to and not because they have to. It also makes it less likely that good performers will leave an organisation voluntarily, leaving their employers to devote time and effort to avoidable and hence unproductive activities such as recruiting and training replacements.

Absence management

Reducing unauthorised absence in an organisation is a very direct way in which HR professionals working with line managers can improve an organisation's labour productivity. Some absence due to genuine illness is always inevitable, but a great deal either is taken on a dishonest pretext or is of a kind in which a minor ailment (or hangover) is used as an excuse not to work when, in truth, the person could come into work if they really wanted to. It is not at all unusual for organisations to report overall absence rates in excess of ten days a year when we know that the best performing organisations have rates of only three or four days a year.

It is not at all unusual for something of an absence culture to develop in a workplace, whereby a proportion of employees regularly take days off for spurious reasons simply because they can get away with doing so and see others taking time off in this way too. This type of situation is highly inefficient because it often means that employers are required to overstaff by 5 per cent or so to ensure that they can cover absences effectively (Johnson 2003).

Job satisfaction, as discussed above, is an important determinant of absence. People who perceive that their employers or individual line managers are treating them inequitably or are not interested in them are much more likely to take days off when they could work than colleagues who enjoy their jobs and take pride in working for their organisations. But there is much more involved in running an effective absence management policy.

First it is necessary to measure absence accurately and to use this data to identify who is taking more days off than the average in a department. Further enquiries are then needed to establish whether there is a genuine ill-health condition that explains an individual's poor absence record. Where days are taken periodically and there is no underlying medical condition, action needs to be taken and seen by others to be taken. This will involve formal return-to-work interviews, setting attendance targets and giving formal warnings if the problem persists. Ultimately people need to know that they will be dismissed if their absence record is excessive and unjustified.

It is also possible to ascertain by asking the question of referees when a job applicant has a significant record of poor attendance or unauthorised absence that is not explained by genuine incapacity due to ill health. Fifty per cent of absence is, after all, taken by just 5 per cent of the typical workforce. Those who are prone to taking unauthorised absence can then be screened out of selection processes.

Finally, there is a good case for taking a proactive approach to absence management by taking preventative measures aimed at improving health and wellbeing. There is always a limit to what can be done here, but physical wellbeing can be promoted, while more active interventions in the form of providing counselling helplines and training supervisors to recognise problematic signs can help to reduce absence taken for reasons related to stress and poor mental health.

Payment arrangements

The extent to which incentive payment systems improve productivity is one of the big debates in contemporary HRM. Some studies show that performance-related pay (PRP) does improve organisational performance, while others point to its flaws as a reward system and argue that PRP is as likely to reduce an individual's performance as improve it. In truth much depends on the organisational setting and the type of system that is introduced. If your employees are not especially motivated by pay or perceive that the system that is used for pay incentives operates unfairly or subjectively, it is unlikely to increase productivity. But that does not mean that reward systems play no part in improving productivity.

First, it is important to use them where it is appropriate. So, for example, when an organisation employs people to do work that is readily quantifiable, it makes sense to reward effort with incentive payments. This has long been seen to be true in traditionally organised factories, where people can be encouraged to work faster and more productively by the prospect of valued incentive payments. Sales commission operates in a similar way by rewarding staff whose output is strongest. Problems arise when work performed is not so readily quantified, leading to a situation in which management judgement plays a major role in determining how much by way of an incentive each employee receives. In these kinds of settings incentive payment systems can easily demotivate good, solid performers who perceive their own performance to be stronger than their managers perceive it to be. Here productivity will not necessarily be improved by the introduction of incentive payment schemes and may even be damaged by them.

In the case of incentives for senior managers, the key is to ensure that improving productivity forms a major part of any objectives that are set and which determine how much performance-related pay is received. Doing so will signal that improving productivity is a major organisational objective and that managers are expected to make it their own personal objective too. Payment systems should be designed to reward the achievement of long-term productivity targets rather than short-term profit targets.

A form of team-based performance pay that has clear links to productivity is 'gainsharing'. It is a system that became fashionable for a period in the late twentieth century, particularly in the USA, but which has never been used in a widespread way in the UK. However, there is a strong case for making use of gainsharing principles when an organisation is looking to find productivity improvements. It involves encouraging teams of employees who work together on a body of work to make suggestions as to how they could be more productive. For example, when a team member leaves, the remaining colleagues could be invited to propose how efficiencies and reorganisation could lead to a situation in which no replacement is necessary. If accepted, the gain is then shared. Savings generated result in lower costs for the employer and higher pay for the remaining team members. This is a very direct way of moving towards a situation in which fewer people are employed, but those who are receive higher salaries.

Organisation and job design

Over time, it is possible to increase an organisation's productivity by reducing headcount and finding ways of making more productive use of working hours. The extent to which this is achievable will always vary considerably from employer to employer, but in many organisations incremental improvements can be made without reducing the quality of products or services.

In 2013 the anthropologist David Graeber wrote a slightly mischievous article for a magazine called 'On the phenomenon of bullshit jobs' in which he argued that 'huge swathes of people spend their days performing tasks they secretly believe do not really need to be performed'. The article hit a nerve and was very widely circulated on social media. Thousands of people emailed him with accounts of the apparently pointless work they were paid to carry out each day, his theory also provoking others into researching the issue. Professor Graeber then went on to look at these findings in more detail in a full-length book that is both entertaining to read and very thought-provoking about the capacity of contemporary organisations to create jobs that are unproductive for much of the time (see Graeber, 2018). There are two main strands to his argument:

- A good portion of jobs (around 30 per cent) would not be hugely missed if they were to be abolished.
- A lot of people spend a great deal of their working time carrying out unproductive tasks.

While much of his argument is anecdotal in nature, being based on stories sent to him by disengaged employees, many ring true and deserve at least to be debated. Graeber classifies 'bullshit jobs' into five main categories:

- Flunkies – people whose jobs exist to make senior managers look important or organisations more successful than they really are. Examples are concierges, receptionists and PR people whose role is to aggrandise their organisations so that they seem bigger than they really are.
- Goons – people whose jobs involve carrying out unnecessary tasks that have no social value. Graeber gives as examples corporate lawyers, post-production editors who manipulate photograph ads to make actors and models look more perfect, and sales people employed to cold call and hard sell products and services few really want to buy.
- Duct tapers – people whose jobs involve repairing avoidable glitches and faults that should not have been created in the first place. His examples here include some computer coding jobs, proofreaders, and people employed to handle customer complaints.
- Box tickers – people whose jobs involve carrying out bureaucratic tasks that involve supervising, measuring, checking up on and reporting on productive

activity rather than actually doing it. Graeber's examples here come from government departments, local government and universities, all of which are practised at creating work that does not add any real value beyond ensuring that people are complying with bureaucratic requirements.

- Taskmasters – people whose job involves allocating work and supervising those who carry it out. Middle managers provide the main examples here. They take instructions from higher up the management ladder and pass them on to subordinates, creating nothing of value themselves in the process. Graber then identifies a further group whose role is to create unproductive work for others to carry out, often reducing their capacity to carry out productive work in the first place.

This last point is one that Graeber makes particularly effectively. It is not so much a case that whole jobs are unproductive (although he argues that many are), but that too much of our time, and particularly management time, is spent carrying out unproductive tasks. He quotes a large American study that involved assessing how much time each working day office workers spent productively. In 2016 the average breakdown was as follows:

Performing the primary duties of my job: 39 per cent

Answering emails: 16 per cent

Useful and/or productive meetings: 11 per cent

Administrative tasks: 11 per cent

Wasteful meetings: 10 per cent

Interruptions for non-essential tasks: 8 per cent

Everything else: 5 per cent

Of course, for many organisations, the lack of productivity will not be so big, but it is revealing that this is what office workers themselves considered to be their situation. A great deal of progress can thus surely be made by getting teams to work smarter. Managers, including HR managers, need to push for a reduction in wasteful meetings and requirements for people to carry out unproductive administrative tasks. Major savings can be generated in the process and more time devoted to core, productive activities.

Google is well known for instituting bureaucracy-busting days in which employees gather together to discuss how management and administrative systems can be redesigned or abolished in such a way as to improve the efficiency of operations. Involving people in these kinds of activities is a good, potential way of starting to raise productivity. The problem is that it is often more senior people in organisations who have an incentive to continue with unproductive processes that essentially involve supervising the work of others.

PAUSE AND REFLECT

Think about your own organisation and your own department. Do David Graeber's arguments ring true for you, or is he exaggerating the scale of the problem? What steps could be taken in your workplace to eradicate some unproductive hours or jobs?

Upskilling

It is widely agreed that the overall skill level of the UK workforce is too low. In international comparisons, we perform reasonably well at the higher level. The number of graduates in our labour market compares favourably with major competitors. The problem lies with school leavers, far too many of whom lack qualifications in technical areas where the skilled jobs of the future are most likely to be created. As we saw in Chapter 6, around 30 per cent of the current working-age population left school without any qualifications, and more without any in the STEM areas (science, technology, engineering and maths). This is hugely problematic given the nature of the jobs that are currently being created in our economy.

As was discussed in Chapter 9, improving this record is a major current public policy priority, but upskilling the nation's workforce is not something that can be accomplished by the education system. Active assistance on a large scale from employers will also be a requirement if significant progress is to be made. Moreover, it is in the interest of employers looking to improve long-term labour productivity in their own workplaces.

The key here is to treat time and money spent on learning and development as investment rather than spending. To that end, it is necessary to use it as part of a wider human resource planning approach that is focused on helping to ensure that an organisation will have access to the skills it requires in the future. It is pointless ploughing resources into training people to use technologies that will soon be obsolete or superseded by others. That is a recipe for low productivity. We need instead to start by thinking through what will be required in terms of technology and types of skills and experience in three to five years' time and then put in place L&D interventions to prepare the workforce for the changes. It is not just technological advance itself that has to be prepared for, but changing business models too that may or may not stem from technology.

Of course, it will often not be possible to know with certainty how different an organisation's activities will be in the future. While the broad aim may be to improve productivity, it will not always be clear exactly what will change and how. Some would say that such situations present a situation in which longer-term skills planning is not viable. There is too much uncertainty. This is often a weak argument. What is required in such situations is robust contingency planning. Learning and development activities need to be designed to help prepare people for a range of possible futures.

A good example is provided by the large supermarket chains, which have made considerable use of new technologies in recent years. The obvious examples are self-checkout systems that allow, and indeed encourage, customers to scan their own items and pay for them using a machine. Many still dislike these and prefer to pay a checkout operator who scans the purchases for them, but a portion of customers are happy to self-scan, enabling the supermarket to achieve major efficiency savings and hence greater productivity. The skills required of staff changed considerably in the process. In the future it is likely that the technology will move on again, creating a situation in which purchases will not need to be individually scanned at all. We will simply fill our trolleys and wheel them out, our accounts being debited automatically. At the same time, the ability of the supermarket operator to record and manipulate data about our personal buying preferences will further increase, enabling them to target advertising at us far more productively than they currently can. In the process, the type of skills supermarket employees need to have will increase. The job will be more about selling and assisting customers, and less about checking out at tills. The stock on display will also become more dynamic, changing at different times of the day to reflect customer preferences. Warehouse and shelf-stacking jobs will thus also need to change and upskill in some respects.

The key point is that the productivity gains associated with the introduction of the new technologies will only be fully realised if the employees are sufficiently skilled to operate them, to help customers navigate them and to maintain and repair them when faults occur. This requires considerable preparatory training, which needs to be planned and developed many months in advance.

Multi-skilling is another strategy that can increase productivity if planned thoughtfully. This is because it enables employees to perform a wider range of tasks, including some that may not generally be considered to be part of their main job roles. The great advantage from a management point of view is that this enables flexible deployment of people to different parts of a business when they are busy or pressured. In the process, it reduces downtime by taking people away from areas that are quiet. Multi-skilling can also help an organisation to prepare better for periods of change and hence respond sooner than competitors who still have to train people up to carry out new roles.

References

Beatson, M and Zheltoukhova, K (2015) *Productivity: Getting the best out of people*, CIPD, London

Cooper, R (2009) UK jobs in the recession: Alternatives to redundancy, *Telegraph*, 24 August

Giles, C (2018) Britain's productivity crisis should be keeping the country's politicians and civil servants awake at night, *Financial Times*, 13 August

Graeber, D (2018) *Bullshit Jobs: A theory*, Allen Lane, London

Harari, D (2107) Productivity in the UK, House of Commons Library Briefing Paper, 06492

Ilzetzki, E (2020) Explaining the UK's productivity slowdown: Views of leading economists, VoxEU/CEPR, https://voxeu.org/article/uk-productivity-puzzle-cfm-survey (archived at https://perma.cc/D699-XK3M)

Institute of Directors (2018) *Lifting the Long Tail: The productivity challenge through the eyes of small business leaders*, IOD, London

Johnson, C (2003) The problem and management of sickness absence in the NHS: Considerations for nurse managers, *Journal of Nursing Management*, 11 (5), 336–42

Marchington, M, Wilkinson, A, Donnelly, R and Kynighou A (2020) *Human Resource Management at Work: The definitive guide*, 7th edn, Kogan Page, London

Paauwe, J, Guest, D and Wright, P (2013) *HRM and Performance: Achievements and challenges*, Wiley, Chichester

Pidd, H (2009) Honda's factory re-opens, *Guardian*, 2 June

Sparrow, P and Otaye-Ebede, L (2017) HRM and Productivity, in *A Research Agenda for Human Resource Management*, ed P Sparrow and C Cooper, Edward Elgar, Cheltenham

Wolf, M (2020) The UK's employment and productivity puzzle, *Financial Times*, 30 January

16 Ethics and values

It is important to recognise that two major perspectives can be taken with regards to business ethics. The first of these involves thinking about the subject at the level of the individual organisation or manager, focusing on ethical decision making and professionalism, whilst the other is considerably broader and involves consideration of the impact (in ethical terms) that contemporary business activity is having on society more generally. In some respects, businesses operating in a capitalist economic system have a positive social impact, for example by creating wealth, by providing career opportunities for individuals, and in the creation of socially useful products and services. Having said this, there are also ethically questionable impacts, too, notably global warming and the concerns regarding increasing levels of social inequity. This chapter principally covers the first of the two perspectives outlined above; this decision has been made due to the nature of this text and the rising importance of the subject of ethics within the contemporary management agenda.

Previous chapters in this book have considered, in detail, the context within which organisations are operating. There are a number of important developments that influence management actions, including ethical decision making, and these include the rapid growth we have seen in competitive intensity, financialisation, ethical consumerism and the rise of ethical investment funds. In short, what must be recognised is that the business landscape has become more competitive over time, and as a result we are seeing competitive advantage endure for a considerably shorter period of time than was previously the case. There are therefore demands on organisations to reduce their cost base and maximise their short-term returns to shareholders if they are to remain attractive to investors. Consumers have simultaneously become more ethically aware over time, and studies have demonstrated that the millennial generation, in particular, is attracted to organisations offering ethically sourced and produced goods and services. Likewise, there has been a significant rise in socially responsible investment, with investors keen to associate themselves with organisations that have a positive ethical reputation. All of this has led to the subject of ethics rapidly moving up the agenda of managers and leaders within organisations.

To begin discussions within this area, this chapter first introduces the reader to the various theoretical perspectives on business ethics. By way of an introduction, the subject of 'values' is also covered before a series of consequences for organisations arising from the debates connected with business ethics are discussed. Within this part of the chapter issues such as product responsibility, pricing and relationships with employees, amongst others, are discussed in detail before attention moves to more specific debates such as the extent to which human resource development interventions can promote ethical conduct within organisations, and the links between the HRM function and corporate governance. Towards the end of this chapter the subject of professionalism will be considered, with readers introduced to debates regarding professional behaviour, and codes of conduct produced by organisations such as the CIPD. The subjects of sustainability and corporate responsibility are covered in the next chapter and so will not be discussed here in detail. This is to ensure that these subjects are given suitable prominence themselves, but readers are reminded that there are strong links between the subject material covered within these two chapters. We begin now with an exploration of the theoretical perspectives on business ethics.

Theoretical perspectives on business ethics

To set a solid stage upon which this chapter can build it is important to initially review the major theories within the field of ethics. This is, of course, a substantial field in itself and therefore this section provides a brief introduction to the key issues and debates rather than a review of all perspectives and critiques. If readers wish to follow up in more depth, the sources cited in the reference list at the end of the chapter provide a good place to begin a wider exploration of the issues introduced here. Ultimately, ethics is the branch of philosophy that concerns itself with answering questions such as 'What is good?', 'What is right?' and 'What is just?' Historically the field of ethics has been dominated by two competing schools of thought regarding how such questions should be addressed, with tensions existing at the level of the individual when there is no obvious clear-cut decision regarding a 'right' or 'wrong' answer in a given situation. What must be remembered is that individuals ultimately make decisions based on their own judgement and in full view of their upbringing, their values and cultural background, amongst other factors.

Some philosophers have taken what is known as a 'consequentialist' view of ethics, arguing that whether an act is considered ethical or not can be determined by assessing its practical impact on people. The most common form of consequentialist thinking is known as utilitarianism, whose supporters argue in favour of choosing the path of action that generates the greatest happiness for the greatest number of people. When faced with an ethical dilemma, a manager within an organisation should therefore choose the course of action that will bring the most benefit to the most people. The alternative perspective regarding ethics is known as non-consequentialism or 'deontology'. This represents a substantial departure from the consequentialist view of ethics and derives from the philosophy of figures such as Immanuel Kant. In brief, deontologists argue that some courses of action represent the 'right thing to do', even if they do not serve the interests of the greatest number. What this means is that deontologists believe that it is possible to identify enduring ethical values that are right to pursue in all circumstances. Some characterise this as being a 'human rights perspective', associating it with concepts such as justice, fairness and equality which, it is argued, are ethical in and of themselves and therefore should be supported even if in some circumstances the consequence is not to bring about the greatest benefit to the greatest number of individuals.

In addition to the major theories presented above, when attention turns to management and business ethics in particular, Walsh (2007) argues that there is a further issue relating to the extent to which commercial factors should colour decision making. Walsh (2007) suggests that actions can be divided into three groups depending on the extent to which either profitability or ethical considerations dominate. First, *lucrepathic* actions involve managers within organisations taking account only of the need to maximise profits when making decisions. This contrasts with *accumulative* actions, where the pursuit of profit is moderated in some shape or form by ethical constraints. The third group of actions are known as *stipendiary* actions, in which the objective served by a decision is not to make a profit, but where the need to do so acts as a constraint on the nature of those decisions.

Milton Friedman (1962) famously argued strongly against businesses being tempted into taking actions that are contrary to the interests of the owners of the business. His often cited expression that the only responsibility of a business is to maximise returns to shareholders, so long as the business operates within the law, is very well known indeed and, for Friedman, lucrepathic actions are not just 'ethical' but are the only course of action that can be ethical. Friedman's underlying view was that the role of businesses was not to spread wealth around within society or serve the financial interests of other stakeholders. His view was that governments and elected officials were better qualified and positioned to take on those duties. The role of a business was therefore to create as much wealth as possible, which governments could then redistribute via the tax and benefits system. The thoughts of Friedman have attracted substantial debate over time and objections to his purist position have come from supporters of corporate philanthropy and paternalistic management styles. Not only, in their view, is it right for businesses to support their communities financially, but it is often in their best interests to do so. The key learning point to take from this content so far is that making ethical decisions in the context of a commercial organisation is often not at all straightforward and clear-cut. Ethical dilemmas are common, and it is often unclear what the truly ethical course of action might be.

In illustrating these complexities, Forsythe (1980) produced a model entitled 'a taxonomy of ethical ideologies', his central argument being that individuals will fall into different categories based on their levels of relativism and idealism. Relativism in this context refers to the extent to which individuals will base personal moral philosophies on universal ethical rules, while idealism is connected with the extent to which individuals exhibit concern for others (Forsythe et al, 1988). The model therefore contains the following four categories:

- Absolutists (low relativism/high idealism).
- Exceptionists (low relativism/low idealism).
- Subjectivists (high relativism/low idealism).
- Situationists (high relativism/high idealism).

To explain these in more depth, *absolutists* tend to reject the use of the consequences of an action as a basis for moral evaluation and appeal to natural law or rationality when determining ethical judgements. These individuals would, for instance, believe that waging war is always morally wrong because it always brings more harm than good, and that abortion is always wrong because it brings a potential human life to an end. *Situationists* by contrast distrust absolute moral principles and argue that each situation needs to be examined individually. A situationist would argue that waging was is generally morally wrong, but sometimes there are situations in which it is morally necessary, and that abortion is generally immoral, but there are circumstances when it can be moral. *Exceptionists* endorse the statement that the morality of an action depends on the consequences produced by it. They would argue that waging war is not intrinsically wrong, but acts of war can be immoral, and that abortion is not immoral, but it can have immoral consequences. Finally, *subjectivists* reject the notion of universal moral principles and believe instead that moral decisions are based on individual judgements. Negative consequences do not necessarily make an action immoral. A subjectivist would argue that waging war is not intrinsically moral nor intrinsically immoral, it all depends on the point of view adopted, and similarly abortion is not immoral, but some regard it as so, whilst others do not.

As argued during the course of this section, the territory of ethics and ethics in business is complex, and very often individuals are faced with dilemmas where there may be no clear-cut 'right' or 'wrong' course of action. It is important to recognise that many factors such as an individual's upbringing, their cultural background, experiences within education, and others besides affect their judgements in respect of ethical decision making, and this is, in part, why this territory is complex and problematic to manage. The following discussions pick up on this point in considerably more depth by beginning an exploration of personal values and judgement.

Values

There has been substantial interest in the subject of values, with a literature base that has developed extensively over time. Practitioners and scholars alike have investigated the impact of personal values, and this extends to the business context where there has been considerable interest regarding how personal values influence conduct in organisations and business decision making. As far back as the mid-1960s researchers have questioned the impact of personal values on the formulation of corporate strategy (Guth and Tagiuri, 1965), arguing that whilst strategy formation on the surface may appear objective, it is in truth influenced at a fundamental level by the personal values of leaders and managers.

By way of a definition, Kluckhohn (1951: 395) argues that a value is 'a conception, explicit or implicit, distinctive of an individual or characteristic of a group... which influences the selection from available modes, means, and ends of action'. This chimes with more recent thoughts from Schwartz and Bilsky (1987: 551), who argue that values are 'concepts about desirable end states or behavior that provide guidance to individuals in evaluating and choosing between alternative courses of action across a range of situations'. In both cases, values are highlighted to be individual conceptions that ultimately influence decision making, with it being understood that the actions of individuals are influenced at a fundamental level by their underlying value set. Values in themselves arise from a combination of different influences. They tend to be acquired at an early stage of life and may be transmitted between individuals via family, culture and educational experiences (Guth and Tagiuri, 1965). For instance, in a family setting parents may transmit certain values to their children through their use of rewards, punishments and their choice of examples. What is understood is that an individual's value framework develops over time, but values are thought to be a relatively stable feature of an individual's personality. It is important to recognise that there are various conceptualisations of 'values', although Van Hoorn (2017) argues that there has been convergence in more recent times towards a standard frame of reference that encompasses ten basic values. These are:

- Power.
- Achievement.
- Hedonism.
- Stimulation.
- Self-direction.
- Universalism.
- Benevolence.
- Tradition.
- Conformity.
- Security.

Interestingly, Van Hoorn (2017) goes on to argue that these ten basic values combine to form four sub-dimensions, which are represented in Figure 16.1.

The sub-dimensions themselves are then paired up in an oppositional framework: self-enhancement versus self-transcendence, and openness to change versus conservatism. This opposition arises in the framework because of the inherent incompatibility of some values. For instance, power and achievement are argued to be incompatible with benevolence and universalism, hence the two sub-dimensions of self-enhancement and self-transcendence are argued to be in opposition to one another. What we must recognise is that individuals place differing levels of importance on different values (Weber, 2019). A value that may be of overriding importance to one individual may be considerably less important, or entirely unimportant, to another. This can occur even if an individual

Figure 16.1 The dimensions of values

Value	Sub-dimension
Power Achievement	Self-enhancement
Benevolence Universalism	Self-transcendence
Self-direction Stimulation Hedonism	Openness to change
Tradition Conformity Security	Conservatism

SOURCE: Adapted from Van Hoorn (2017: 455)

is from the same family group, same cultural background, or works in the same organisation. Values, in and of themselves, are deeply personal and we must not leap to the conclusion that because we hold a particular value strongly that this is an opinion that is automatically shared by others. In practice this is not the case. What we do know is that individuals in society, and leaders and managers within an organisational context, should take time to understand the values of others (Guth and Tagiuri, 1965).

As noted towards the start of this section, an important point to remember is that the value set held by an individual affects their decision making in a fundamental way. Ultimately, why people act the way they do is a product of their values (Weber, 2019). Interestingly, from an ethical perspective, Argandoña (2003) argues that personal values create an orientation within an individual that consequently leads to either ethical or unethical behaviour. He posits that behaviour displayed by an individual is fundamentally a manifestation of the values that the person in question holds to be most important. At a later stage of this chapter we will consider the extent to which learning events held within organisations can promote ethical behaviour. A debate we will engage with there is the extent to which it is possible for learning events to impact personal ethical conduct. To preface that, and building on the key points here, the question which must be asked is: if values are held deeply within an individual and are relatively stable over time, fundamentally affecting decision making, to what extent may an external input such as a learning event designed to improve ethical conduct actually cause behaviour change within an individual? This is certainly an interesting thought to keep in mind as readers progress deeper into this chapter.

Business ethics: consequences for organisations

As this chapter has pointed out thus far, the field of ethics is complex, and many issues arise for managers within organisations in terms of both the products and services that an organisation may provide and the relationships that it maintains with stakeholder groups. During this portion of the chapter a number of these areas are explored in depth, including product responsibility, pricing, relationships with suppliers and relationships with employees.

Product responsibility

A fascinating area of debate within the field of business ethics concerns the rights and responsibilities of customers in relation to the nature and quality of the products and services that they purchase. Is a manufacturer or service provider acting unethically if it supplies items or experiences that risk damaging the health of customers? What about goods that are unreliable or do not perform to the standards expected by the customer? Does the producer have any ethical responsibility in these situations, or should customers be expected to take responsibility for their own actions when purchasing the goods or services? An obvious example here is the production and sale of tobacco products, accepted as being responsible for fatal illness in millions of individuals each year. Alcohol is another such example in so far as it has been attributed as the direct cause of injury (and sometimes fatalities) in road accidents, and by increasing the number of incidences of violent and criminal behaviour. It can plausibly be argued that humanity would be considerably better off if these products were not manufactured and sold in the first instance.

What are your immediate thoughts in relation to the arguments above? To what extent may we term the production and sale of certain products (such as tobacco and alcohol) unethical?

The questions above should have prompted you to think quite carefully about the ethical decision making involved in the production and sale of certain products. On one level the statement at the end of the last paragraph is entirely accurate; humanity may indeed be considerably better off without some products such as tobacco and alcohol. Having said this, there are arguments the other way too. It can be argued that for *most* consumers alcohol is not damaging. Most individuals drink in moderation, they do not drive whilst under the influence of alcohol and find that it causes no health problems whatsoever. The negative effects are thus a by-product of alcohol misuse, not alcohol consumption per se. According to this argument, alcoholic drinks are therefore no more unethical as consumer products than kitchen knives, or even cars, both of which can be (and are) a cause of injury to many people. These arguments, however, cannot be made with respect to cigarettes, because their use (even in moderation) can lead to illness and reduced life expectancy. For most consumers, the product is therefore damaging. One of the key issues here is that consumers must be aware of the risks. If producers were to conceal the health impacts of cigarettes, their production and sale would most certainly be unethical.

Building on the points above, another category of goods and services consists of products that in themselves are not directly damaging to consumers, but it is often not in their interests to consume them. This is usually because other more appropriate alternatives are readily available. Such items include:

- Foods that are high in fat and sugars (cause teeth to rot and other health issues).
- Credit cards (encourage individuals to go into debt).
- Casinos and gaming machines (can be addictive and lead to loss of money).

Producers of such goods argue that these products are only damaging if overused, underused or improperly used. If used in the method intended and within some form of reasonable bounds, they can provide the consumer with a pleasurable experience, or some other tangible benefit. In addition to this, so long as proper instructions and warnings are issued where necessary, it is not unethical to offer these products for sale. It is also possible to take a middle position here by arguing that production and sale are not themselves unethical, but that producers have an ethical duty to do more than simply provide clear instructions on correct usage. They should actively monitor the use of their products and respond where they are found, in practice, to have a damaging effect.

Pricing

In many situations, the extent to which businesses and managers have substantial flexibility in terms of the pricing of their goods or services is limited. Companies are obliged to compete in an open market in which there are accepted 'going rates'. To charge substantially above this level is not possible because consumers will switch to competing options from alternative organisations. In some instances, organisations may be able to charge premium prices where goods or services have a valuable differentiation from others available in the market, but this is only possible within narrow limits. For many managers, pricing policy is not an area in which ethics can play a substantial role because pricing structures are subject to factors largely outside their control. Having said this, there are situations in which ethical judgements have to be made, and others in which consumers, rightly or wrongly, perceive that they have been 'ripped off' by unfair strategies. Examples of these include:

- Overpricing.
- Predatory pricing.
- Differential pricing.
- Overly complex pricing arrangements.
- Dynamic pricing.

Overpricing may occur where an organisation finds it has a monopoly or near monopoly within a market and takes advantage of this by pushing up prices (Hassin and Snitkovsky, 2020), creating what are termed 'super-normal' profits. Predatory pricing, by contrast, is where a larger, more powerful organisation temporarily reduces its prices drastically in an effort to damage smaller rivals in a market (Guiltinan and Gundlach, 1996). Where this occurs, smaller firms may not be in a position to respond and either lose substantial amounts of market share or go out of business altogether. Consumers may lose out in the long term because the ultimate result is higher prices over time. Differential pricing (Copeland, 1942) involves selling an identical product or service to different individuals at different prices. Unlike overpricing or predatory pricing, this is common in the public sector where it occurs, at least in part, as a result of political decision making. There is usually a solid rationale that can be used as a justification, but questions regarding the ethics of such practices persist. Examples include practices such as providing concessionary fares on public transport for older individuals irrespective of personal income, charging different fares for identical seats on planes and trains, and UK universities charging substantially different fees for home and international students. The fourth item in the bullet list, overly complex pricing arrangements, captures a variety of practices whereby organisations develop pricing structures that are deliberately opaque, taking advantage of the consumer's inability to comprehend the payment arrangements being offered for the goods or services provided. The most common examples of this occur in the financial services industry and are often connected with products such as pensions, mortgages and insurance policies. Dynamic pricing has become much more common in recent years. It is largely associated with e-commerce (see, for example, Minsker, 2014), and has a justifiable side and a very ethically questionable side as well. In its justifiable form dynamic pricing adjusts the prices of goods and services up or down depending on demand. For example, a hotel booking website may adjust prices up or down as the supply of rooms increases or decreases. This is, essentially, a sophisticated way of matching supply and demand. The ethically questionable side, however, occurs where dynamic pricing seeks to adjust prices depending on the individual consumer who is showing interest in a purchase. This involves gathering data on consumers and then charging more where individuals are perceived to have a higher disposable income.

Selling and advertising

Within the marketing space specifically, there are several examples of potentially unethical conduct. When it comes to selling products or services, Mahoney (1997) argues different techniques form part of a sliding scale. This scale starts with providing the customer with *information* about the product, progressing on to *suggestion* and then *advocacy*, where an organisation would make stronger claims regarding their offering. The fourth stage is that of *persuasion*, whilst the spectrum is ended by the final stage of *manipulation*. Mahoney notes that the first four are acceptable ethically and indeed are necessary in a competitive business landscape, but the fifth, manipulation, is not. Manipulation can take a number of forms, one of the more common being to play upon an individual's fears and emotions when attempting to sell a product or service.

Separate to direct manipulation, organisations in some instances have been found to provide misleading information about their products and services. At times it is possible to mislead without actually telling a direct lie. One method in which this is done is to use language or images that exaggerate the benefits associated with a particular product or service. Another is to omit information from a promotional campaign, leaving a one-sided impression only. A further approach involves placing all the emphasis on the benefits of a product or service, while relegating less palatable aspects to small or fine print. Another practice involves organisations advertising under false pretences regarding when a product or service will be available, or highlighting that it will have certain features when, in practice, it will not. Differently, organisations may also engage in hard selling tactics whereby substantial pressure is exerted on the consumer, for instance by having a contract ready to sign, or suggesting that there is only one opportunity for a particular deal. In these instances, consumers have little time to read sales literature or information, and to come to an informed opinion regarding their purchase decision. Such practices are therefore thought to be unethical in nature.

Relationships with suppliers

Just as we find that there are ethical issues to debate and resolve when considering how organisations interact with their customers, so too are there issues to discuss when the business itself is a customer. The first of these issues concerns payment terms, and specifically the late payment for goods or services that have been delivered to the business (see, for example, Bernstein, 2020; Šalamon et al, 2016). Very often, businesses will delay payment to their suppliers in an effort to support their cash flow position. This is particularly advantageous when the business receives money from customers on time, because monies can be invested profitably before debts are settled with suppliers. This is one reason for late payments, and such tactics are indeed ethically questionable. The other reason late payments can occur is simply where businesses are under-resourced, or have inefficient payment processing systems (such as is the case in some public sector organisations). This is more of an operational rather than an ethical issue as such, but is still something that organisations should be concerned about.

Where a supplier is heavily reliant on one or two customers for their success or survival, it is possible for the customer organisation to take advantage of the situation through over-pressurising the supplier. The key way in which this occurs is where customer organisations make demands, set standards that are very difficult for the supplier to meet or argue for substantially preferential trading terms which may include price reductions. Customer organisations can do this in highly competitive markets where there are a number of supplying organisations, and the resulting intensification of work within the supplying organisation can place staff under unreasonable pressure to perform. Common examples of this can be found in the retail sector, where major supermarkets impose terms on suppliers that can, in practice, be unreasonable (Grimmer, 2018).

Companies can also engage in phoenixing, whereby the directors of a struggling enterprise may liquidate it (ie wind it up) but then buy it back from the receivers and continue to trade under a different name. This is possible in any country that regards a company as having a distinct legal personality, separate from those of its directors. The practice is therefore quite legal in many jurisdictions, including the European Union, the USA and Canada, in part because it is thought to encourage enterprise and risk taking. Where this happens, though, creditors of the defunct business rarely receive monies that are owed to them. Some suppliers thus end up undertaking work and subsequently never get paid, even though the directors of the debtor company remain in the same business and may be more than capable of paying their debts. Similarly, in recent years there have been a number of prominent examples of asset stripping before a company has gone bust. In these circumstances company directors have taken substantial amounts of money and/or transferred the assets to a new owner just before the company in question has gone bust. Debts are thus left in the bankrupt company, while the owners of the business enrich themselves.

Relationships with employees

Whilst there are important exceptions in the case of people who are highly talented, or who have particular qualifications and experience, the employer–employee relationship is generally one that is characterised by an inequality of power. This makes it possible for employers to exploit their power unethically, and many do so. The first method in which this can be done is via the inequitable distribution of rewards. Over time, in both public and private sectors, there has been greater inequity in the compensation enjoyed by more senior individuals as compared to their more junior counterparts (see, for example, CIPD, 2019; Johnes and Virmani, 2020), with the CIPD (2019) in particular reporting that in 2018 the median FTSE 100 CEO reward package was 117 times greater than the median salary of a full-time worker in the UK (£29,574). All kinds of theoretical justifications can be put forward for inequity in reward, including the power of market rates, tournament theories, job evaluation and rewarding those who take risks, but the extent of the gap between senior and average earnings is in many cases impossible to justify on ethical grounds. It derives from power differentials within organisations and represents an abuse of that power (Newman et al, 2019).

Separate from the issue of pay in many sectors and industries, we see evidence of work intensification over time (Brown, 2012; Fitzgerald et al, 2019). Lack of job security and of alternative employment opportunities makes it possible for managers to over-work their staff, the aim being to extract greater value from them compared with the money that is paid to them. In many cases, the outcomes for employees include a less enjoyable working life and a poor work–life balance. In extreme cases, however, the result can be health-threatening levels of stress and burnout in middle age. There is often a fine line to tread between fostering a highly engaged workforce and pushing people too far for their own good. An ethical organisation will seek to protect people from themselves when it comes to work intensification, stepping in to stop them from taking on too much work.

CASE STUDY 16.1

Work intensification and employee burnout

The intensification of work over time has been a much discussed topic. The consequences of it, including employee burnout, are known to have a substantial direct financial cost, with Moss (2019) reporting that workplace stress causes $190 billion in healthcare costs in the USA per year. Similarly, in 2019 the World Health Organization reported that depression and anxiety arising from poor workplace environments globally caused $1 trillion in lost productivity (World Health Organization, 2019). In short, it is known that the intensification of work, with organisations seeking to extract increased productivity from their employees, can have a direct impact on employee health and wellbeing.

An organisation that has recognised the impact of work intensification is the professional services firm EY. In early 2014 the firm launched a new initiative focused on health in the workplace (Crawford, 2014). The initiative, titled 'Health EY', introduced mental health first aiders within the organisation, alongside the provision of monthly health education events and more targeted pathways to assist individuals with health conditions. Underlying their approach was a desire to develop conversations between managers and employees about the working environment, removing stigma from issues such as workplace burnout. EY is frequently held up as an example of best practice within the employee wellbeing arena, and its measures have helped the organisation develop a strong reputation and employer brand over time.

A third method in which managers often abuse their power when it comes to the conduct of employment relationships occurs when misinformation is communicated. In most cases, this consists of making promises that are unlikely to be fulfilled in practice (for example in connection with promotions, developmental opportunities or pay rises), or providing false assurances about future job security. The motives behind such actions can sometimes be good, and cowardice may be the reason for misinformation rather than an active wish to be dishonest, but it is unethical nonetheless. Such actions can have substantial implications for psychological contracts, with breach occurring in situations where promises are unfulfilled in practice (Rayton and Yalabik, 2014). The final major form of unethical activity that occurs in respect of employment relationships concerns undue or unjust favouritism – or indeed scapegoating. Managers inevitably have individuals within their teams for whom they have greater regard or personal affection than others. It is important, however, that personal preferences of this kind do not spill over into decision making about matters such as the distribution of work, pay rises, promotions, redundancies or development opportunities. All such decisions need to be taken fairly and should, wherever possible, be based on good evidence. Failing to deal fairly with individuals is counterproductive, and also often unethical.

Designing learning events to promote ethical conduct

So far within the confines of this chapter we have argued that the HRM function has a significant and substantial role to play in promoting ethical conduct within organisations. The question to be addressed in this section is whether, and to what extent, human resource development interventions specifically can play a role in promoting ethical conduct. This is a particularly interesting question, as the chapter has already argued that ethics and values are deeply rooted within an individual, with their upbringing, culture and many other personal factors affecting ethics and value judgements. For this reason, the extent to which external inputs, such as learning events within organisations, may shape ethical conduct is uncertain, and therefore there are important issues to discuss.

Importantly, academic research has found links between the provision of ethical learning events and improved ethical judgement and culture inside organisations (Jones et al, 2013; Warren et al, 2014). The key finding from these, and other, studies is that where organisations provide learning programmes related to the subject of ethics and values, employees are then better able to identify ethical dimensions to problems and develop moral reasoning. There are, however, a number of levels at which such programmes attempt to bring about change, with some arguably more effective than others. At the most basic level, learning events may be convened to simply provide a forum for the communication of expectations, for example ensuring that employees understand a code of conduct. Such 'awareness'-based training is useful, and has been shown to have an impact on ethical decision making, but it is in itself limited in its ability to bring about substantial change within organisations.

Treviño et al (1999) argue that ethical learning programmes may be designed with either a 'compliance' or 'values' focus. The example in the paragraph above regarding the communication of a code of conduct is an example of the former, with Treviño et al (1999) arguing that such events are unlikely to bring about substantial shifts in ethical conduct. In brief, compliance-based programmes focus on the importance of rules or laws, highlighting the consequences that ensue when these are broken, whilst values-based programmes seek to highlight the importance of values (eg honesty/integrity) in the resolution of ethical dilemmas. This, in itself, is a valuable distinction. It is important to highlight that compliance-based learning has a role to play. Employees must understand relevant processes, rules and consequences. Treviño and her colleagues do, however, highlight that it is values-based approaches to ethical learning programmes that are likely to have the most substantial impact on conduct.

PAUSE AND REFLECT

Before you move on please pause and reflect on the points above for a moment or two. Why do you think that ethical learning programmes orientated around a values-based approach are likely to have the most impact on subsequent employee conduct?

The question above should have caused you to pause and think for a few moments. The answer to the question is indeed multifaceted and does require thinking through in some depth. Essentially, the argument as to why values-based programmes are more effective is because they are firmly rooted in personal self-governance (Treviño et al, 1999). Values-based approaches are likely to orientate employees towards specific values (such as honesty and integrity) and therefore when faced with an ethical dilemma employees are more likely to choose behaviours that align with those values, rather than adopting behaviours that would be minimally required by law (as might be seen if a compliance-based approach to learning is utilised). Ultimately, values-based approaches are thought to promote ethical behaviour because they focus much more on bringing about behavioural and attitudinal change.

Just as HRM/D professionals must consider the underlying approach to ethical learning programmes, so too must they consider how the events will ultimately be facilitated. Warren et al (2014) in particular argue that HRM/D professionals must carefully reflect on the facilitation of learning events designed to bring about changes in ethical behaviours, suggesting that face-to-face events are superior to e-learning methods. This is because the events are fundamentally designed to bring about attitudinal change – something that is much better undertaken in a guided, face-to-face format, rather than an online form. Further to this, Warren et al (2014) also argue that HRM/D professionals should use a blend of experiential techniques such as role play, simulation and practical activities to build stronger engagement with the content. Irrespective of the training method and underlying approach used, however, ethical learning programmes tend to have relatively little permanence. In other words, the effects tend to wear off over time because ethical judgements are deeply rooted within individuals and their histories. As a result of this, organisations must repeat and reinforce ethical learning events at regular intervals.

HRM and corporate governance

The subject of corporate governance is a large area of academic debate in its own right. Whilst the purpose of this chapter is not to dive deeply into corporate governance theory, it is important nonetheless to consider how the HRM function interfaces with it. Human resource management functions have often been conceptualised as the 'conscience' of an organisation, but in truth the ability of the HRM function to affect corporate decision making and implement policy frameworks is constrained by prevailing corporate governance mechanisms, and the wider national business systems within which organisations operate (DunHuei and Tzu-Shian, 2016; Konzelmann et al, 2006). At its core, corporate governance is concerned with how organisations are directed and controlled (Berle and Means, 1932). It is therefore separate from day-to-day management issues, and is more concerned with the actions of board members/owners, and how these individuals and groups set the overarching values and approaches for the organisation (ICAEW, 2020).

Importantly, as briefly mentioned above, the prevailing national business system has a profound impact on corporate governance within organisations (Konzelmann et al, 2006). Where firms operate in liberal market economies, for instance, shareholder interests will be prioritised over the interests of other groups (for example employees and trade unions). Nationally distinctive approaches to corporate governance also impact organisations because they assign differing levels of importance to worker interests, strategy types and financial measures of performance, amongst other factors (Konzelmann et al, 2006). All of this means that the HRM function inside organisations must operate within a series of constraints, and therefore this may impact the nature and shape of employment policies and practices. For instance, where governance mechanisms dictate the pursuit of profit maximisation for shareholders, typical in public limited organisations operating in liberal market economies, this can have a direct impact upon HRM because there will be pressure on managers to downsize and cut costs (Ibrahim and Zulkafli, 2016). Having said this, it is also the case that HRM professionals can influence the decisions of corporate leaders in regard to employment practices, but the extent of such influence is arguably constrained, irrespective of the values that HRM practitioners might hold, or the courses of action they may prefer to implement (Martin and Gollan, 2012).

We recognise that the content presented so far in this section is rather abstract in nature. The following case study concerning the rise of RBS and its failure during the global financial crisis of 2007–09 helps to illustrate the points discussed so far in a considerably more practical manner. This case has been developed from the article produced by Martin and Gollan (2012) and illustrates how HRM, as a function, has had a greater opportunity to exert influence over the organisation's corporate governance machinery following its failure.

CASE STUDY 16.2

Corporate governance at RBS

The Royal Bank of Scotland (RBS) was a major international bank that, at its peak in 2007–08, held nearly £2.5 billion in assets and had almost 230,000 employees (Statista, 2020a, 2020b). Currently, the organisation has far lower employment and asset values, with the group rebranding itself as the NatWest Group in July 2020. Under the leadership of Fred Goodwin the organisation expanded dramatically in the early to mid-2000s, but suffered substantial losses during the global financial crisis of 2007–09, with the business ultimately rescued by the UK government in 2008. At its peak, the UK government owned 83 per cent of the organisation, although this stake has been slowly sold off over time and is considerably lower than this peak position at the time of writing.

Financial services organisations around the world, including RBS, suffered a major reputational hit as a result of the global financial crisis, with many commentators and news outlets blaming the global crash on poor corporate governance, excessive risk taking, and poor ethical decision making. The reputations of leaders within these organisations, and HRM functions as well, suffered substantial damage, which in some cases persists to this day. There is still a persistent view that financial service organisations take too many risks and exhibit characteristics associated with poor corporate governance, even though many major organisations in the financial services sector have undertaken substantial cultural change activities.

During its period of success and significant growth, RBS followed a corporate governance structure that prioritised the needs of shareholders, as did almost all organisations operating in the financial services sector. Following its substantial losses and ultimate rescue by the UK government, however, new leadership sought to improve the reputation of the organisation. Leaders began to look beyond the shareholder model in an effort to change the organisation's culture and moved towards more of a stakeholder-driven model of corporate governance. This included substantial shifts in the organisation's policy frameworks, and the emergence of a model of HRM that prioritised employee engagement, communication and involvement. The HRM function was also able to exert substantial influence over the organisation's culture, with a new model of leadership emerging that privileged soft power and encouraged leaders to reflect on their ethical behaviour. Although the organisation's culture has changed markedly, the constraints of the market create tensions between the shareholder and stockholder model. Martin and Gollan (2012), however, argue that new forms of corporate governance are emerging in the sector that better deal with these dilemmas, recognising the need to return value to shareholders, but also placing limits on some of the less positive behaviours and values displayed by organisations in previous times.

SOURCE: Adapted from Martin and Gollan (2012)

This case study demonstrates, in brief, the constraints on corporate governance structures. During its period of growth RBS pursued a governance structure that prioritised the needs and interests of shareholder groups, but this changed markedly following the financial crisis of 2007–09. One might argue that the changes then observed within the organisation's governance were only possible as the government had taken a large stake in the organisation, thus marginalising the interests of private shareholder groups. It will be interesting to follow developments within the financial services industry over the coming years, and the impacts that changing corporate governance structures may have on organisations operating within the sector.

Professionalism

Finally in this chapter it is important to consider the subject of professionalism. Professionalism is another useful approach to building and maintaining an ethical reputation, and in recent decades it has been argued that the management role has become 'professionalised', suggesting that managers increasingly see themselves as members of a profession in the same way that doctors, teachers, lawyers and architects traditionally have done. There are, however, counterarguments to this view, with some suggesting that management can never be a 'profession' in its own right, and that it is therefore inappropriate to think of the role in those terms. In common with other areas of academic debate, a great deal here depends on the definition that is used, and in the case of the term 'professional' there is no widely agreed definition.

Having said that there is no agreed definition, a number of elements typically contribute to an occupation being labelled a 'profession'. These include where individuals need to master a specialised body of knowledge or expertise that others have not mastered to the same extent. An obvious example here would be that of doctors, who must undertake substantial learning before they are certified to practise. Second, professions often require individuals to study towards qualifications, demonstrating their competence by passing examinations, and this can, in many cases, lead to admittance to a professional body. A further hallmark of a profession is that individuals tend to demonstrate substantial loyalty to their professional field, over and above the loyalty that they show towards their employer. When asked, such individuals tend to identify with their profession specifically, rather than the organisation in which they work at that particular time. Professionals also tend to be governed by a code of conduct. These codes indicate the standards of behaviour and ethics that are expected to be shown by members of the professional group. Where codes of conduct are broken, professional bodies reserve the right to suspend or eject individuals from the professional body. Finally, members of a profession often demonstrate an altruistic element to their activities, putting the best interests of their clients ahead of their own. These qualities demonstrate that professionalism is generally seen as being something of high status, and professions as bodies whose members can be trusted more than most individuals to act with integrity and detachment. Despite the points above, the term 'professional' can sometimes be used very differently, as we find, for example, in football or rugby when players commit 'professional fouls'. 'Professionals' are sometimes compared unfavourably to 'amateurs', who practise their skill simply for enjoyment, rather than for money. The term 'professional politician' is also often used in a derogatory fashion.

According to the points expressed thus far in this section, most individuals who occupy management posts are not members of a profession. Some may have professional qualifications, but in large part will not meet the other requirements discussed within the last paragraph. This does not mean, however, that managers should not conduct themselves in a professional fashion whilst at work, striving to make reasoned decisions rooted in good judgement, treating their staff politely, equitably, and with respect, whilst completing their own work to the highest standards and maintaining professional detachment. While 'acting professionally' goes beyond 'acting ethically', adherence to a code of ethics forms a very substantial part of true professionalism. Indeed, the CIPD operates a professional code of conduct that applies to all members of the body, a copy of which can be found on its website

(CIPD, 2020). This code outlines the standards of conduct expected of all CIPD members and groups under the following four principles:

- Professional competence and behaviour.
- Ethical standards and integrity.
- Representative of the profession.
- Stewardship.

The CIPD expects all members to adhere to the requirements set out within the code of conduct, with members of the public being able to submit reports where they suspect that the code may have been breached. Any breaches are investigated, and where proven the CIPD has recourse to a number of outcomes including providing advice to the member about their conduct for minor breaches through to more significant sanctions for substantial breaches of the code of conduct.

Conclusion

We now arrive at the end of our exploration of ethics and values within a business and management context. This chapter has introduced readers to the core debates concerning ethics and values in organisations and, as noted at the very start, this is a subject that has grown considerably in importance over time. Due to a variety of developments in the business environment and wider social landscape, such as ethical consumerism and ethical investment to name but two, business ethics has risen sharply up the priority list for organisations. In short, organisations cannot afford to neglect issues connected with ethics and values if they hope to succeed in meeting their ultimate goals and missions. This chapter began in earnest with a review of ethical theories, including the contributions of well-known thinkers such as Milton Friedman. One of the core learning points from this review was the understanding that leaders and managers within organisations can often be faced with dilemmas in which there is no immediate 'right' or 'wrong' answer. Judgement is therefore key, and at a personal level this is influenced by a confluence of factors including one's upbringing, cultural background and, educational experiences, amongst others. The same is true when it comes to the subject of 'values'. Values, we understand, are very personal and deeply rooted within individuals. Whilst academic theory is moving steadily towards clearer definitions of the term 'values', there is still considerable debate within this field. Values affect decision making in a fundamental way, and it is vital to remember that individuals can hold very different sets of values. A value that one person holds in high regard may be seen as considerably less important by another, and we must not leap to the conclusion that others share the same values as we do. This would be a significant oversimplification.

In the business environment, ethics and values have substantial consequences for organisations. Within the body of this chapter we reviewed a number of important areas, including product responsibility, pricing, and the sales and advertising methods used to influence purchasing decisions, as well as the relationships that we maintain with suppliers and employees. Deeper into the chapter, the role of the HRM function with respect to ethics and values was explored in more depth. Whilst reflecting on the extent to which HRD interventions may influence ethical decision making, we made the point that these learning interventions can lack permanence because our conceptions of ethics and values are very deeply held, and deeply rooted within individuals. As a result of this the impacts of learning events can wear off over time, and thus require repetition to reinforce organisational expectations regarding behaviour and conduct. The subject of corporate governance was also briefly touched upon, and readers should remember that systems of corporate governance are impacted to a very large degree by prevailing national business systems that operate to constrain organisational governance mechanisms. Finally, this chapter discussed the term 'professionalism', debating the elements that typically lead to an occupation being labelled a 'profession'. Whilst managers cannot be labelled 'professional' in the way that doctors, teachers, lawyers and architects are, this does not mean that managers cannot or should not act professionally.

We have certainly covered substantial ground within the body of this chapter and readers must recognise the growing importance of ethics and values within business and organisational contexts.

References

Argandoña, A (2003) Fostering values in organizations, *Journal of Business Ethics*, 45 (1–2), 15–28

Berle, A and Means, G (1932) *The Modern Corporation and Private Property*, Macmillan, New York

Bernstein, A (2020) Late payments: Help on how to get paid! Late payment can cause huge problems for hauliers, but there are a number of things you can do about it, *Truck and Driver*, October, 32–3

Brown, M (2012) Responses to work intensification: Does generation matter?, *International Journal of Human Resource Management*, 23 (17), 3578–95

CIPD (2019) Executive pay in the FTSE 100: Is everyone getting a fair slice of the cake?, CIPD, August

CIPD (2020) Code of professional conduct, www.cipd.co.uk/about/what-we-do/professional-standards/code (archived at https://perma.cc/T9KA-JXGP)

Copeland, M A (1942) A social appraisal of differential pricing, *Journal of Marketing*, 6 (4), 177–84

Crawford, R (2014) How to manage workplace burnout, https://employeebenefits.co.uk/issues/june-2014/how-to-manage-workplace-burnout (archived at https://perma.cc/E92Q-7NZ2)

DunHuei, H and Tzu-Shian, H (2016) The relationship between corporate governance and HRM: An empirical study of HPWS adoption, *Academy of Management Annual Meeting Proceedings*, (1), 1

Fitzgerald, S, McGrath-Champ, S, Stacey, M, Wilson, R and Gavin, M (2019) Intensification of teachers' work under devolution: A 'tsunami' of paperwork, *Journal of Industrial Relations*, 61 (5), 613–36

Forsythe, D R (1980) A taxonomy of ethical ideologies, *Journal of Personality and Social Psychology*, 39 (1), 175–84

Forsythe, D R, Nye, J L and Kelley, K (1988) Idealism, relativism, and the ethic of caring, *Journal of Psychology: Interdisciplinary and Applied*, 122 (3), 243–8

Friedman, M (1962) *Capitalism and Freedom*, University of Chicago Press, Chicago

Grimmer, L (2018) The diminished stakeholder: Examining the relationship between suppliers and supermarkets in the Australian grocery industry, *Journal of Consumer Behaviour*, 17 (1), e13–e20

Guiltinan, J P and Gundlach, G T (1996) Aggressive and predatory pricing: A framework for analysis, *Journal of Marketing*, 60 (3), 87–102

Guth, W D and Tagiuri, R (1965) Personal values and corporate strategy, *Harvard Business Review*, 43 (5), 123–32

Hassin, R and Snitkovsky, R I (2020) Social and monopoly optimization in observable queues, *Operations Research*, 68 (4), 1178–98

(Continued)

(Continued)

Ibrahim, H I and Zulkafli, A (2016) An empirical inquiry into the relationship between corporate governance and human resource management, *International Journal of Business and Economic Sciences Applied Research*, 9 (1), 7–17

ICAEW (2020) What is corporate governance? www.icaew.com/technical/corporate-governance/principles/principles-articles/does-corporate-governance-matter (archived at https://perma.cc/75DV-SPSJ)

Johnes, J and Virmani, S (2020) Chief executive pay in UK higher education: The role of university performance, *Annals of Operations Research*, 288 (2), 547–76

Jones, K P, King, E B, Nelson, J, Geller, D S and Bowes-Sperry, L (2013) Beyond the business case: An ethical perspective of diversity training, *Human Resource Management*, 52 (1), 55–74

Kluckhohn, C (1951) Values and value orientations in the theory of action, in *Toward a General Theory of Action*, ed T Parsons and E A Shils, Harper, New York, 388–433

Konzelmann, S, Conway, N, Trenberth, L and Wilkinson, F (2006) Corporate governance and human resource management, *British Journal of Industrial Relations*, 44 (3), 541–67

Mahoney, J (1997) Buyer beware: Are marketing and advertising always ethical?, in *Financial Times: Mastering management*, FT/Pitman, London

Martin, G and Gollan, P J (2012) Corporate governance and strategic human resources management in the UK financial services sector: The case of the RBS, *International Journal of Human Resource Management*, 23 (16), 3295–314

Minsker, M (2014) Dynamic pricing gains ground, *CRM Magazine*, 18 (11), 13

Moss, J (2019) Burnout is about your workplace, not your people, *Harvard Business Review* Digital Articles, November, 2–6

Newman, C M, Banning, K, Johnson, R M and Newman, J A (2019) The say on pay impact on executive pay: An analysis of pay ratios, *Journal of Business and Behavioural Sciences*, 31 (2), 192–205

Rayton, B A and Yalabik, Z Y (2014) Work engagement, psychological contract breach and job satisfaction, *International Journal of Human Resource Management*, 25 (17), 2382–400

Šalamon, T, Milfelner, B and Belak, J (2016) Late payments explained by ethical culture, *Journal of East European Management Studies*, 21 (4), 458–88

Schwartz, S H, and Bilsky, W (1987) Toward a universal psychological structure of human values, *Journal of Personality and Social Psychology*, 53 (3), 550–62

Statista (2020a) Total assets of the Royal Bank of Scotland from 2007 to 2019, www.statista.com/statistics/225077/total-assets-of-the-royal-bank-of-scotland (archived at https://perma.cc/B6GL-7EU2)

Statista (2020b) Number of employees of the Royal Bank of Scotland Group from 2005 to 2019, www.statista.com/statistics/421353/number-of-staff-of-the-royal-bank-of-scotland (archived at https://perma.cc/7EE2-T7G9)

Treviño, L K, Weaver, G R, Gibson, D and Toffler, B (1999) Managing ethics and legal compliance: What works and what hurts, *California Management Review*, 41 (2), 131–51

Van Hoorn, A (2017) Organizational culture in the financial sector: Evidence from a cross-industry analysis of employee personal values and career success, *Journal of Business Ethics*, 146 (2), 451–67

Walsh, A (2007) HRM and the ethics of commodified work in a market economy, in *Human Resource Management: Ethics and employment*, ed A Pennington, R Macklin and T Campbell, Oxford University Press, Oxford

Warren, D E, Gaspar, J P and Laufer, W S (2014) Is formal ethics training merely cosmetic? A study of ethics training and ethical organizational culture, *Business Ethics Quarterly*, 24 (1), 85–117

Weber, J (2019) Understanding the millennials' integrated ethical decision-making process: Assessing the relationship between personal values and cognitive moral reasoning, *Business and Society*, 58 (8), 1671–706

World Health Organization (2019) Mental health in the workplace, www.who.int/mental_health/in_the_workplace/en (archived at https://perma.cc/9UVV-77AZ)

17 Sustainability and corporate responsibility

Our next chapter focuses on the important areas of sustainability and corporate responsibility. As we have highlighted during previous discussions, these subjects have been rising quickly up the corporate agenda in recent times, with a growing number of organisations seeking to demonstrate their positive impacts on both society and the wider natural environment. It can be argued that the focus on sustainability and corporate responsibility has come about because the general public has become more aware of the impacts of organisational activities, particularly impacts that have damaging social and environmental consequences. For instance, the Deepwater Horizon oil spill in 2010 caused substantial financial and reputational damage to BP (Ward, 2018), and there is growing knowledge and public awareness of the environmental impact that ensues from companies extracting oil from Alberta's tar sands (Leahy, 2019). Simultaneously, consumers are becoming increasingly ethically aware (Govind et al, 2019), demanding higher ethical standards from the companies from which they purchase goods and services. In short, sustainability and corporate responsibility have rapidly grown to become critically important issues for organisations, irrespective of the sector in which they operate. It is likely that we will see continuing focus in these areas in the future, demonstrating why we, as HRM professionals, must build our knowledge and awareness within these fields.

This chapter seeks to provide the reader with an overview of sustainability and corporate responsibility. The intent is to take a broad sweep through the topics and key concepts within them, exploring case studies at various points to demonstrate the operation of theories in practice. We will begin by introducing the area of sustainability, before exploring corporate responsibility (previously known as corporate *social* responsibility) in some considerable depth. Linked to this, the chapter introduces the reader to the notion of the triple bottom line, a mechanism that sought to expand traditional methods of reporting to include both social and environmental performance alongside financial performance (Elkington, 1994). The subject of 'greenwashing' is also explored, and there are a considerable number of examples whereby organisations have been accused of exaggerating claims regarding their environmental stewardship or 'greenness' in the pursuit of increased sales and revenues. Throughout the chapter issues of direct interest to the HRM function are picked up in detail. The notion of 'green HRM' is explored in depth, together with practical methods by which HRM, as a functional area, has sought to contribute to improved social and environmental performance. The chapter closes with a brief overview of the 'circular economy', a most interesting theory that has sought to offer a radical alternative to the 'take–make–dispose' nature of our existing economic and business models. We begin now with an overview of sustainability.

Sustainability: an introduction

Organisational success can come at a price. Our business and commercial activities can have a detrimental effect on the environment, our organisations may make poor use of resources, and adverse employment conditions can generate harmful social outcomes. Organisations of all types are becoming more aware of their broader impacts, and this is increasingly being reflected in their actions. For instance, we are seeing less single-use plastic used by food retailers, and airlines are retiring older, less fuel-efficient aircraft in favour of newer models. Having said this, there are still many examples of poor organisational practice persisting, with significant environmental and social disasters being reported frequently in the media. One example of this was the collapse of a garment factory building in Bangladesh that supplied Primark (Butler, 2014), bringing to attention many substandard employment and working practices. As noted in the introductory notes to this chapter, sustainability is a subject that is rising rapidly up the corporate agenda. Increasingly, there are challenging questions being asked regarding the environmental and social outcomes accruing as a result of organisational activities across the world.

Sustainability can be defined as the desired outcome that achieves all long-term economic, environmental and social goals (Carter and Rogers, 2008). Ultimately it is about balance. There is an acknowledgement that society consumes resources in order to function, but that this resource use needs to be undertaken in such a way that maintains balance at a global level, to ensure longevity. If we consume more than we put back into the environment over time, then this, by its very nature, is unsustainable. Sustainability is indeed a broad discipline: it consists of environmental, social and economic outcomes, and recognises that these three areas are intrinsically linked to one another. The biggest question is how we protect resources for the future (both natural and human), whilst continuing to develop our way of life. These tensions are discussed frequently in the academic literature base that surrounds the subject of sustainability. Joseph et al (2020), for instance, argue vociferously that society must explicitly acknowledge the tensions in meeting economic, environmental and social priorities. Pointing out that if these areas of activity are dealt with in isolation we are unlikely to arrive at positive holistic outcomes, they suggest that acknowledging tensions between disparate goals can foster integrative thinking, and as a result stronger solutions. Indeed, the notion of corporate sustainability is, in itself, a contradictory term, as Hahn et al (2014) note. If knowledge of this fundamental tension is denied then our ability to generate holistic solutions that create positive environmental, social and economic outcomes will be hampered.

The research base that investigates sustainability is indeed large, and whilst it deals with many aspects of corporate activity, including supply chain management, business operations, design, marketing and many others, the key point to remember is that it confronts businesses with environmental and social goals, in addition to commercial outcomes (Goleman and Lueneburger, 2010). The acknowledgement of wider responsibilities arguably represents a substantial change for organisations, as well as those individuals in leadership positions. Haney et al (2020) explicitly discuss the issue of leadership, and how leaders with skills, knowledge and competencies in the area of sustainability might be developed. They argue that new leadership competencies are inevitably needed, and that leadership development programmes must explicitly recognise this. Whilst there is little current agreement on the skills, knowledge and competences that a leader needs in order to develop sustainable organisational outcomes, there is agreement that leaders must:

- Recognise and manage the tensions inherent in bringing about commercial, social and environmental outcomes.
- Be able to engage a diverse range of stakeholder groups in the pursuit of sustainable outcomes.
- Demonstrate competence in dealing with 'wicked' problems.

PAUSE AND REFLECT

To what extent do you agree with the thoughts presented above? What role might leadership development programmes play in the development of the competences required by leaders pursuing a sustainability agenda?

The questions above are intriguing, and there are many answers that might be advanced when addressing them. To an extent, the answer comes down to the nature/nurture debate. To what extent might development interventions, of any kind, impact on underlying behaviours or values related to sustainability? Arguably, values and behaviours are deeply held within an individual, solidifying during their formative years, so the extent to which HRD interventions of any kind can affect considerable change within these values and behaviours is perhaps limited. The long-standing argument is that we must hire for attitude and train for skills, and this certainly resonates here. With this in mind, leadership development programmes are likely to be best deployed in equipping leaders and managers with the practical skills needed to recognise and manage tensions, engage stakeholders and grapple with wicked problems. The role of HRD is therefore arguably about skill development, although there is certainly a role for learning programmes designed to orient all employees (including senior leaders) towards sustainable or green values.

The notion of 'wicked' problems is, itself, very interesting indeed. Whilst the purpose of this chapter is not to delve into the theory behind wicked problems as such, we know that these problems are ill-defined, complex and do not have clear solutions. It is thought that building competence within the area of systems thinking (Haney et al, 2020) – in other words, the ability to analyse situations from across a spectrum of different disciplines, with data across different scales – is increasingly important. What we do know about wicked problems, though, in an organisational context is that previous management approaches such as applying the learning from past experiences in an attempt to predict or control future activities are becoming less relevant over time (Ferdig, 2007; Rieckmann, 2012). We are beginning here to hint at areas in which the HRM function itself may contribute to the broader sustainability agenda, although we have not yet explicitly defined or otherwise explored its role in any substantial depth. One intriguing concept that has arisen relatively recently is that of 'green HRM'.

Green HRM

Just as other organisational functions have adjusted to take account of the sustainability agenda, an example being the push towards sustainable supply chain management, HRM has been no different. In many organisations, HRM frameworks and approaches have been adjusted, or are in the process of being modified, to take account of sustainability concerns, and this is being typified in many cases by the presence of what has come to be known as 'green HRM'. Green HRM, defined by Kramar (2014: 1075) as 'HRM activities which enhance positive environmental outcomes', builds from the underlying premise that our organisations must become more conscious of the environment in which they operate, understanding the impacts that their operations have. Importantly, as Dumont et al (2017) argue, we must recognise that it is employees who undertake actions within and for organisations, and therefore if these same employees are not aware of, or oriented towards, sustainable and environmentally conscious practices, organisational visions and missions regarding improvements in sustainability will not be realised. Dumont et al (2017) go on to argue that, as a result of this, organisations must develop an approach to HRM that attracts and retains individuals with compatible values, utilising learning and development programmes to build awareness of environmental and sustainability issues. In essence, the key point within the debates associated with green HRM is that we must adjust our suite of people practices so that they reflect the importance of green values.

The importance of values and the development of compatible value sets are also discussed by O'Donohue and Torugsa (2016). Reflecting the wider literature base associated with green HRM, they argue again that the HRM function has a significant and substantial role in orientating employees towards green values, and that this can only be done if all people practices (including recruitment, reward, performance management, learning and development) are aligned appropriately to broader organisational goals and objectives related to sustainability. Speaking very much to this theme, Renwick et al (2013) channel the well-known ability–motivation–opportunity (AMO) model by collapsing the broader suite of HRM policies into one of three categories: (1) developing green abilities, (2) motivating green employees, and (3) providing green opportunities. Their model argues that we must first build the necessary stocks of human capital inside our organisations through recruiting and developing employees, before inspiring commitment to green values and practices, and then enabling employees to act through involvement and empowerment mechanisms. What we must ultimately recognise is that the suite of HRM policies and practices that employees are exposed to, and the HRM framework that they operate within, has a significant impact on in-role behaviour. If an organisation seeks to pursue a sustainability agenda, then it is crucial that HRM practices are aligned with this ultimate ambition. If we see strategic drift between the organisation and its HRM framework, then it is considerably less likely that intended outcomes will be achieved.

Moving on from this introduction to sustainability, and sustainability issues within organisations, the chapter now introduces the reader to the separate but conjoined topic area of corporate responsibility.

Corporate responsibility

Whilst the term 'corporate responsibility' may have only entered the language of business relatively recently, the roots of the concept date back a considerable way indeed. There have long been discussions and debates related to the role of business in society, and there is indeed a long history of organisations demonstrating awareness of their impacts and care for the wider community. As far back as the mid- to late 1800s the Cadbury brothers focused on improving working and living conditions for their employees, providing housing and recreation areas amongst an array of other worker benefits, which were far in excess of those offered by other organisations of the time (Cadbury, 2020). History is littered with such examples, and also examples of organisations engaging in less responsible practice.

The European Commission (2020) defines corporate responsibility as 'the responsibility of enterprises for their impact on society', arguing that companies can demonstrate responsibility by 'integrating social, environmental, ethical, consumer, and human rights concerns into their business strategy and operations'. It is important to note the multifaceted nature of this concept, and that it is a much broader consideration than simply the organisation taking note of its impact on immediate stakeholder groups such as employees, suppliers and customers. Corporate responsibility is the more current term for what was previously known as 'corporate social responsibility' (CSR). Emerging in the 1960s, CSR focused largely on the responsibilities that businesses and organisations had to society, but as the definition put forward by the European Commission demonstrates, more modern interpretations of corporate responsibility substantially widen this area of focus to include environmental and ethical considerations, amongst others. This does bring a question to the forefront of our minds: what should the role of business in society be?

PAUSE AND REFLECT

Given the content covered in this section of the chapter thus far, what do you think the role of business in society should be? What examples can you find of organisations displaying responsible conduct as would be expected under the European Commission's definition of corporate responsibility?

The questions above, in tandem with the content introduced thus far during this chapter, should have given you plenty of food for thought. There are substantial debates regarding organisational conduct, and indeed a very large literature surrounding corporate governance, as we explored during part of the previous chapter. During your answer you may have thought about a variety of organisational examples as well. To provide just one example, Lush, the cosmetics and beauty firm, is often held up as a model organisation that has embedded the principles of sustainability and corporate responsibility at its core (Thomas, 2019). Its website contains a series of policy statements related to its supply chain, employment practices, tax responsibilities and charitable support, amongst many others (Lush, 2020), and it is frequently cited by a variety of media sources and commentators as an example of an organisation which displays outstanding corporate citizenship. But we have not yet discussed the expected role of business in society.

What role should business play within society? Turning back to the arguments put forward by the famed economist Milton Friedman, the only responsibility that a business has is to ensure that it utilises its resources as efficiently as possible in the pursuit of profit maximisation. So long as the organisation stays within the rules of the game (in other words, the legal framework in the country or territory in which it operates), all a business must do is increase its returns over time for its owners, normally its shareholders (Friedman, 1962). In this sense, other obligations are rather separate from the core mission of the business, although Friedman does note that businesses may seek to get involved in community, social or environmental projects, if their participation in such activities serves to meet the broader aim of the enterprise (namely profit maximisation). For instance, participating in an environmental project under Friedman's perspective could be justified if it boosted corporate reputation, and hence positively impacted upon sales and market share for the organisation in question. Anything that detracts from the pursuit of profit is therefore irresponsible conduct for the business (Nunan, 1988), as it would be an inefficient use of limited resources. However, as Morrison (2020) argues, this position is becoming increasingly precarious over time. This is because the market-based system typically does not consider environmental and social or community issues alongside financial outcomes, with a second argument being that many MNCs have grown to substantial proportions – in some cases dwarfing the gross domestic product values of some countries. In this scenario, business organisations must play a broader role, acknowledging both their social and environmental responsibilities alongside the need to generate financial returns. Having said this, there are many examples where organisations have engaged in less than responsible conduct. An example of this is Ford's handling of problems with its Pinto model in the 1970s.

CASE STUDY 17.1

The Ford Pinto

The issues with Ford's Pinto model of the 1970s are very well known and very well discussed in business ethics circles. Whilst considerable speculation and myth have developed to surround the situation, at its core there is still a very valuable lesson to learn regarding responsible organisational conduct.

Ford developed a new compact car model for its North American market in the 1970s, calling it the 'Pinto'. It was designed to be a moderately priced car, tapping into market trends that were demanding smaller, more economical family transport. Ford duly designed, built and sold its new model, with the Pinto being produced from 1970 to 1980. It generally sold well, and Ford produced just over three million vehicles during its production cycle.

Shortly after its introduction, media stories emerged regarding the car catching fire in the event of a rear end collision. Whilst this brief case study does not enable a full exploration of the safety and regulatory standards at the time, the underlying issue was related to the positioning of the car's fuel tank. In the event of an accident and rear impact, the car was at significant risk of fire, in some cases with fatal consequences for the occupants. As Leggett (1999) discusses, Ford had a new design for the fuel system which would overcome this issue, but chose not to implement it, arguing that the cost/benefit analysis was not favourable, the inference being that human life was less valuable than corporate profitability.

Ford was subject to a number of legal challenges related to this issue, with ensuing damage to its corporate reputation.

The case study above is certainly interesting, and provides a stark example of the problems that can ensue from a narrow focus on economic or financial success alone. Leggett (1999) goes on to argue that the case of the Ford Pinto shows why pure economic theory should not be used to make judgements in these sorts of instances, and it is difficult to see how organisations more broadly might balance costs and economic impact on the one hand with serious social and environmental concerns on the other. It is becoming increasingly clear that organisations must balance an intricate and diverse web of stakeholder relationships, and there are a growing number of examples of organisations that put corporate responsibility at the core of their operations. Lush, as explored earlier in this section, is one example, The Body Shop (now owned by L'Oréal) and the Co-operative Group are others. The next section of this chapter explores one method through which the broader contributions of organisations may be captured and assessed: the triple bottom line.

The triple bottom line

The 'triple bottom line' is a very well known concept within the context of sustainability and corporate responsibility. It was conceived as a solution for organisations that were seeking to measure their impacts from a broader base than allowed through traditional financial reporting mechanisms. In essence, it seeks to capture the wider effects that an organisation's activities have, specifically looking at three overlapping areas of contribution: social, environmental and financial, commonly known as 'people, planet and profits'.

Figure 17.1 The triple bottom line

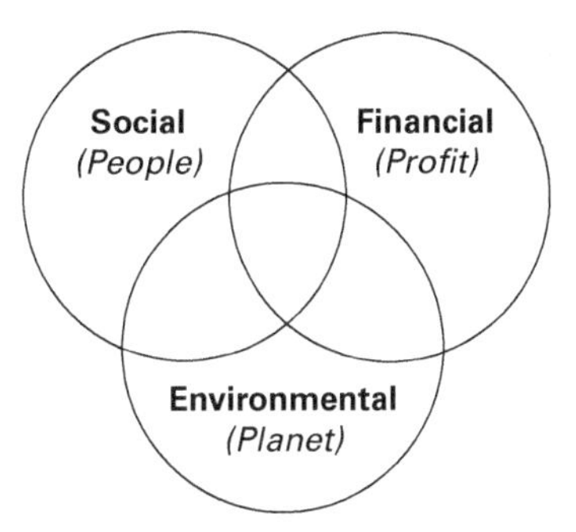

As Figure 17.1 demonstrates, the areas of contribution are thought to overlap, with sustainability residing at the very centre of the diagram where all three elements intersect. This centre point is where social performance, environmental performance and economic performance are thought to be in synch with one another, with the organisation making an optimal contribution to wider society. The concept of the triple bottom line was originally put forward by John Elkington (1994), who sought to expand the traditional reporting frameworks that existed at the time. Accountancy frameworks typically take into consideration only financial performance, with Elkington arguing that it was equally important to count human and natural capital alongside financial capital. This then gave rise to what we know as the triple bottom line, where there is equal emphasis on financial, social and environmental performance. Critics do, however, argue that the framework is rather reductive in nature, and that there is limited agreement concerning what makes up social and environmental 'performance'. Having said this, the triple bottom line concept has gathered substantial momentum over time, and has influenced the development of a range of sustainability frameworks and reporting mechanisms (Kraaijenbrink, 2019) such as the Global Reporting Initiative (www.globalreporting.org) and the Dow Jones Sustainability Indices.

More recently, Elkington (2018) has argued that the triple bottom line concept was not designed to be simply an accounting or measurement tool. As Kraaijenbrink (2019) notes, the concept was intended to provoke substantially deeper levels of thinking regarding the broad impacts of businesses and organisations. Elkington (2018) argues that the relative success of organisations in moving towards their sustainability goals cannot be measured in pure profit and loss terms, and must include a focus on both human wellbeing and the health and vitality of the natural environment. He goes on to argue that there continues to be a fixation on financial reporting and financial goals, with organisations willing to commit substantial effort to ensure they hit profitability and financial targets, but lacking the same effort in the pursuit of social or environmental goals. The question is therefore whether the triple bottom line concept has moved organisational culture forward in the pursuit of sustainable organisational development. Given evidence from organisations such as Novo Nordisk (Elkington, 2018), which reformed itself and its activities around the triple bottom line in the early 2000s, we can say that the concept has certainly had an impact, but criticism still persists.

When exploring the different areas of the triple bottom line it is important that we understand the components that come together to make up environmental, social, and financial performance. A brief summary appears in the following bullet points.

- **Environmental (planet)** What impact does the organisation have on the natural environment? Includes such items as tracking carbon footprints, use (or misuse) of natural resources, activities involving toxic materials, how waste is managed, and the activities of the organisation in restoring natural habitats affected by its operations.
- **Social (people)** What impact does the organisation have on people? How are employees, customers and suppliers, amongst others, treated? What terms and conditions of employment are utilised, and how do the activities of the organisation impact on wider communities?
- **Financial (profit)** What impact does the organisation have on economies at the local, national and global scale? We focus here on activities such as employment creation, the development of innovations, how wealth is created for shareholders, taxation, and other such issues. It is a broader category than simply assessing levels of profitability.

The key learning point to take from this overview is that there are many issues contained within the triple bottom line framework, but we must also recognise that different researchers and contributors to theory in this field do include different combinations of factors. At present this is a field of research that is continuing to develop, and we are very likely to see advances in thinking and new contributions emerging over time. However, as Longoni and Cagliano (2018) note, a growing number of organisations are utilising the triple bottom line concept to overtly integrate social and environmental responsibilities into their business strategies. Interestingly, they highlight that it is organisations with a longer-term planning focus that have typically had more success within this area. This suggests that a long-term focus is key for organisations if they are to have positive impacts on society and the planet. We now move on from the triple bottom line, turning to a related issue, the notion of 'greenwashing'.

Greenwashing

As this chapter has already shown, the subjects of sustainability and corporate responsibility have risen significantly up the agenda, for both organisations and wider society. There is increasing consumer awareness of environmental issues (Lee et al, 2018), and thus pressure is being exerted on organisations to demonstrate their environmentally friendly credentials. As the rise in popularity of fair trade products (Stratton and Werner, 2013) and other conspicuously 'green' offerings demonstrates, customers are increasing their spend on products and services that are seen as environmentally friendly, or otherwise 'good' for the planet and society. There is therefore a temptation for organisations to exaggerate claims about the environmental benefits or 'greenness' of their product and/or service offering, and this is what is known as greenwashing (Lee et al, 2018). Essentially, the argument here is that companies may be tempted to overstate the green credentials of their offerings in order to tap into an increasingly environmentally conscious consumer base. Whilst some greenwashing may be relatively minor, it can at times escalate towards organisations knowingly making false statements about their products or services in order to increase sales and ultimately revenue.

Greenwashing is an issue that has persisted for a substantial period of time. In a *Guardian* article published in August 2016, Bruce Watson discusses the actions of the oil company Chevron during the 1980s. Oil extraction and refining is a notoriously 'dirty' industry by its very nature, and at the time Chevron was very aware of the growing environmental movement. The organisation commissioned a series of advertisements, both in print and for television, emphasising its green credentials. However, as Watson, points out, the organisation was heavily criticised for its actions. In one example Watson notes that a butterfly preserve operated by Chevron was estimated to cost $5,000 a year to run, whilst the firm spent several million dollars developing advertisements and publicity material to promote its activities within the preserve. The level of spend on publicity would appear out of step with the underlying activities, and this is why commentators cite this particular issue as an example of greenwashing. It is, however, unsurprising that organisations engage in such activities as we know that changing consumer trends have forced green issues up the corporate agenda. As Berrone et al (2017) argue, though, there are substantial ethical concerns that arise from such activities. They point out that there is a fundamental issue with information asymmetry between stakeholders: in other words, one party to the ultimate transaction (the organisation) has more information than the other (the consumer). But, having said this, the researchers note that where organisations fail to 'walk the talk' (Berrone et al, 2017: 376), efforts to promote themselves as 'green' can backfire.

CASE STUDY 17.2

Greenwashing at P&G?

Procter & Gamble (P&G) is an American MNC that was founded in the 1830s. It is a very large organisation indeed with operating revenues approaching $70 billion per year, and almost 100,000 employees worldwide. The organisation operates in the fast-moving consumer goods (FMCG) sector, and encompasses familiar brands such as Head & Shoulders, Olay, Fairy, Gillette and many others. On its website the organisation conspicuously promotes corporate responsibility (P&G, 2020), arguing that it must be a force for good, from both a social and an environmental perspective. You can read more about its corporate responsibility actions at: www.pg.co.uk/doing-what-is-right.

Whilst the organisation has indeed invested significantly in improving its environmental impact, for instance via tree planting programmes and other carbon offsetting activities, aiming to become carbon neutral by 2030 (Keating, 2020), it has not found itself immune from criticism. In a recent *Financial Times* article, Evans (2020) discusses a particularly problematic situation that the organisation has found itself in. Within the article, Evans (2020) highlights an announcement by the company regarding its involvement with the NGO Conservation International programme to protect tropical mangroves in Palawan, Philippines, this involvement being part of the organisation's desire to become carbon neutral by 2030. Evans (2020) notes, however, that the US environmental organisation the Natural Resources Defense Council (NRDC) has highlighted a conflict in the organisation's operations. Whilst the efforts to attain carbon neutrality are laudable, and are to be supported, the NRDC argues that P&G is still using wood pulp for its bathroom tissue from unsustainable sources, such as Canada's boreal forest, seen as a critical carbon sink. As a result of this, the organisation arguably opens itself up to accusations of greenwashing. It is promoting its environmentally responsible actions in one part of the world, whilst still engaging with unsustainable activities in another. The argument is therefore that the organisation needs to consider the impact of its actions as a whole if it is to avoid perceptions of greenwashing in the future.

The case example above demonstrates just how difficult it can be for organisations to engage with the sustainability agenda. Whilst the efforts of P&G as an organisation when seen as a whole are indeed laudable, and the company should rightly be celebrated for its commitments to carbon neutrality, this does not mean that all of its practices are above criticism. In all cases, clear and open communication with stakeholders regarding corporate responsibility, and the steps that organisations are taking to improve their environmental (and indeed social) impacts, is key.

HRM responses to the sustainability agenda

There are several ways in which the HRM function has responded to the sustainability agenda. As discussed earlier in this chapter, the development of 'green HRM' frameworks (Dumont et al, 2017; Kramar, 2014) has represented a substantial shift in thinking, whereby HRM functions in many organisations have explicitly taken account of sustainability concerns within their policies, practices and procedures. The purpose of this present discussion is not to restate previous points made within the chapter, but to explore some of the more practical ways in which HRM functions have contributed to reducing the environmental impact of organisations. The coming discussions will first explore recent increases in home and remote working, before reviewing developments in office and workplace design.

Home and remote working

Remote working is discussed during the wellbeing chapter, where we note that substantial developments in technology have made it possible for employees to be based away from the office for part (or indeed in some cases all) of their working time. Frequently highlighted as a 'win–win' for both organisations and employees (Felstead and Henseke, 2017), we argue during the wellbeing chapter that remote and home working can have substantial benefits in terms of improving work–life balance, although equally there are times when employees who work remotely can suffer from increases in work intensification and indeed burnout. Our focus here is, however, different.

From a sustainability perspective alone, there are a number of positives that emerge from the increases in home and remote working. The first of these benefits is that individuals spend considerably less time commuting. While the stress associated with lengthy commutes is well known (see, for example, Burch and Barnes-Farrell, 2020; Legrain et al, 2015), and therefore home or remote working policies which reduce the need to commute can contribute substantially to social goals such as improving wellbeing, the other benefit is a significant reduction in carbon emissions. Where we reduce the need for individuals to travel to a place of work via cars or public transport that burn fossil fuels, the carbon footprint produced by our organisation's activities is reduced. Where individuals work remotely, organisations can also substantially reduce the energy requirements at office locations, reducing the need for lighting, heating and air conditioning, and the use of appliances or computing equipment (Dockery and Bawa, 2018). In truth, it could be argued that what is happening here is simply a shift in the location of energy consumption from the office to individuals' homes, but where organisations can scale back their requirements for office space, this can have a net positive impact overall. It has been argued that the Covid-19 pandemic has accelerated trends in home and remote working, with many individuals reporting that they would like to continue working away from the office in the future (BBC News, 2020a). Indeed, many organisations have made public statements highlighting that they have been surprised with the levels of productivity delivered by individuals working remotely, indicating that they support increased levels of remote working in the future.

Sounding a note of caution regarding increases in home working, Lopez-Leon et al (2020) indicate that remote working could lead to increases in inequality within society. They suggest that access to resources and suitable working space can be a substantial issue for some individuals, and this is supported by others, such as Rysavy and Michalak (2020), who suggest that individuals working remotely may feel excluded. They go on to note that the quality of one's internet connection can be a factor impacting productivity, and that delivering training can be rather more complex when individuals are not able to attend a central base. The key learning point to take from this discussion thus far is that there are substantial environmental benefits when individuals can work from home (alongside favourable social outcomes such as improved wellbeing), but that home and remote working requires careful thought and an appropriate policy framework to guide its implementation.

Developments in office and workspace design

Tied together with the shift towards increased levels of home and remote working have been developments in office and workspace design. Again, workplace design is discussed within the confines of the wellbeing chapter, but the content there focuses exclusively on health and wellbeing issues, rather than issues related to sustainability per se. As hinted at above, there are broad discussions ongoing at present regarding the role of the office and the need for individuals to congregate in large numbers for the entirety of the working week. Whilst these conversations have been given added impetus by the Covid-19 pandemic (BBC News, 2020b), the truth of the matter is that this is not a sudden shift in thinking. There have long been discussions regarding the size and use of office space.

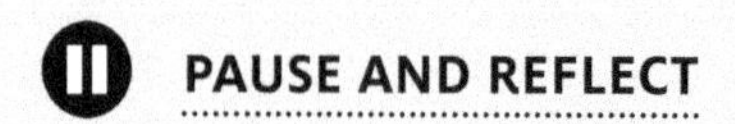

To what extent do we need to rethink the purpose and role of the 'office'?

The question above is in many ways very straightforward, but the answer to it is multifaceted and complex, with many layers of argument. Arguably, modern communication methods (email, instant messengers, video conferencing/video calls) mean that there is substantially less need for organisations to collect individuals under one roof for the purpose of undertaking work. There are obvious exceptions to this, however. Some roles cannot be performed remotely, and some job roles require individuals to access specialist machinery, or work very closely with others to accomplish their duties. In these cases, remote working is considerably less feasible, but for many knowledge workers in particular the need to attend the 'office' is arguably driven by custom and practice rather than a distinct and genuine need to congregate in the same location. Many organisations are questioning whether they need to maintain sizeable portfolios of office space, and whether it is possible to reduce their accommodation requirements. The long-time head of the major communications firm WPP, Sir Martin Sorrell, who now runs the media firm S4 Capital, has argued that he'd much rather invest money in human capital than in office space (Thomas et al, 2020), suggesting that current shifts towards home and remote working are likely to mark a turning point in how firms are organised.

Having made the point that many organisations are questioning the extent to which they need to maintain large real-estate portfolios, it is certain that the 'office' will remain in some form or another. Organisations will inevitably need a way of bringing individuals together to collaborate, so it is therefore likely that central bases will remain, even if they are used differently in the future. With this in mind, Mulville et al (2017) note that those involved in the design of office space, including HRM professionals, need to consider the architecture of the space, the materials used, and how we may nudge individuals towards positive environmental values. From a materials standpoint, window glazing can substantially impact the need for internal heating and cooling, which has knock-on effects for energy use. Mulville et al (2017) also argue that small power items such as computers and mobile devices also dramatically impact on the carbon footprint of an office. What they suggest is that organisations have a role to play in encouraging employees to display positive environmental behaviours, such as turning off lights, monitors, computers and printers when not in use, or using communal kitchen facilities such as kettles efficiently. To be most effective, these behaviours should become embedded within the organisation's culture, reinforced through induction events for new starters and continual learning events for staff, with reward mechanisms used to reinforce positive environmental behaviours.

To have a positive impact within this area, HRM professionals should seek out opportunities to become involved with the design of workspaces. HRM professionals have a substantial role to play here in terms of understanding how workspaces facilitate employee productivity and wellbeing. This is in addition to the role of the HRM professional in orientating employees towards green values. The final section of this chapter now introduces an intriguing new idea that has come to prominence within the sustainability field: the circular economy.

A brief overview of the 'circular economy'

The current model that dominates businesses and economies across the globe is based on a linear 'take–make–dispose' philosophy (Webster, 2017). In other words, resources (as an example, oil and metal ores) are extracted from the natural environment in the first instance, refined and processed into products (in this instance petrol and diesel fuel, or the steel to make the bodies of cars) and the resultant products are used until they fall into disrepair (or become outdated), at which point they are thrown away into landfill sites. This model dominates, no matter the country or industry that is observed, and arguably does not factor in the whole costs of production, such as environmental degradation, which, as we have seen so far in this chapter, is a significant consideration. As a result of dissatisfaction with the traditional linear economic system, there has been growing interest in something known as the 'circular economy' (Hopkinson et al, 2018), although there are several competing definitions of the concept and it is a framework that continues to evolve over time. As Hopkinson et al (2018) observe, however, it is a theory that is of interest to a growing number of individuals and groups because it aligns a robust business rationale with the need to decouple the creation of wealth from environmental degradation. For this reason, a substantial literature base has emerged surrounding the circular economy, and it is a concept that has piqued the interest of businesses, governments and, increasingly, broader society.

The Ellen MacArthur Foundation (2020a) has been a significant supporter and promoter of the circular economy within the UK specifically, and has produced the following interpretation of the concept:

> Looking beyond the current take–make–waste extractive industrial model, a circular economy aims to redefine growth, focusing on positive society-wide benefits. It entails gradually decoupling economic activity from the consumption of finite resources, and designing waste out of the system. Underpinned by a transition to renewable energy sources, the circular model builds economic, natural, and social capital.

The interpretation presented above demonstrates that the circular economy is a profound rethinking of current economic and business models, with Webster (2017) arguing that it is about 'roundput' (ie the notion of zero wastage) rather than 'throughput'. The Ellen MacArthur Foundation (2020a) goes on to note that the circular economy is based on three fundamental principles: namely, (1) designing waste and pollution out of systems, (2) ensuring products and materials can be kept in continuous use, and (3) the regeneration of natural systems. Essentially it argues that all 'waste' streams should be seen as inputs into a new cycle of economic activity, either through reuse, disassembly of a product to recover all of its materials, or repair. Reports from organisations such as McKinsey (2017) suggest that adopting circular economy business models could enable organisations to benefit environmentally, economically and socially.

Moves towards circular economy business models are, however, challenging for organisations (Van Loon and Van Wassenhove, 2020). It is an idea that arguably looks relatively straightforward on paper, but is considerably more difficult to put into action. This is because society is permeated with long-held beliefs, such as those built by contemporary approaches to marketing surrounding the mantra of 'new is best', and that remanufactured items are in some way inferior to brand new products (Hopkinson et al, 2018). These ingrained views have been difficult to overcome thus far. It is also important to recognise that reinventing corporate activities to align with the circular economy model is a multifaceted problem that requires leaders and managers to engage in systems thinking, and to manage significant tensions and conflicts (Hopkinson et al, 2018). Further to this, critics argue that the recovery of materials can never approach 100 per cent due to the increasing energy requirements needed (EASAC, 2015); in essence, complete reuse is prevented by the fundamental laws of physics. Finally, in terms of critique, Van Loon and Van Wassenhove (2020) argue that more work is needed to map the implementation of circular economy business models, arguing that organisations need to ensure that business models are sustainable in the short, medium and long terms, and that a supportive regulatory environment is needed to underpin the transition from a linear to a circular economy. In short, the circular economy is potentially a breakthrough idea, but one that is not without substantial challenges.

The points in the paragraph above are important, and whilst they do sound a note of caution, the notion of a circular economy is enticing. Its core principles, such as the decoupling of economic growth from the consumption of materials, the opportunity for innovation in product design, and the focus on keeping the value of raw materials high (Webster, 2017), suggest that a variety of economic, environmental and social benefits could indeed be realised over time. This is why the concept has attracted such substantial attention. From an HRM perspective, if we are to support businesses and organisations in the transition to business models based on the circular economy, we must again ensure that our suite of policies and practices align to broader business strategy and objectives. HRM professionals inevitably have a crucial role in the recruitment and retention of employees, the development of knowledge and skills associated with circular economy practices, the design of appropriate frameworks that reward employees based on the achievement of appropriate objectives, and the development of performance management systems that align with the broader ambitions of the organisation. It should be clear that without such strategic alignment it is unlikely that businesses will be successful in any move towards a circular economy model.

Whilst arguing that the operationalisation of circular economy business models is challenging, the Ellen MacArthur Foundation (2020b) provides a series of case studies indicating how practices across a range of organisations are being adapted in order to align with the principles of the circular economy. These include Biopak's (Biopak, 2019) development of compostable food packaging for the takeaway food sector, developed to overcome the use of single-use plastics, and Ahrend's (Ahrend, 2020) office furniture rental service, where organisations can rent the workspace equipment they need rather than buying new items and then disposing of them when no longer needed. These examples have varied economic, environmental and social benefits, and you may explore them in more detail if you wish by following up the weblinks provided in the chapter references. We now turn to a conclusion for this chapter.

Conclusion

Discussions within this chapter have introduced a variety of issues and debates connected with sustainability and corporate responsibility. These are items that have rapidly moved up the corporate agenda, and it is rare to see examples of organisations that do not put these issues front and centre of their public-facing company documentation. This is, in part, because consumers are increasingly aware of ethical, sustainability and environmental issues, requiring organisations to respond accordingly if they are to maintain a positive corporate reputation. In short, these issues are increasing in their prominence, and it is unlikely that we will see a reversal in this trend.

At the very start of this chapter, the notion of sustainability was defined. Drawing from Carter and Rogers (2008), sustainability was seen as being about achieving outcomes that balance long-term economic, environmental and social goals. This can indeed produce tensions for managers and leaders of organisations, and there is a clear role for leadership and management development programmes in building the necessary knowledge and skills needed to operate successfully in this context. The HRM function itself has not been immune to the rise of sustainability and corporate responsibility up the organisational agenda. Green HRM is a concept that has arisen relatively recently, with discussions noting the critical importance of alignment between HRM practices and organisational sustainability objectives. Like sustainability, corporate responsibility is an area of contemporary interest that has become increasingly important over time. Unlike sustainability, however, it is a concept that traces its roots a considerable way back into the history of management. There are many examples stretching back a very significant distance indeed where businesses have recognised that they have a wider suite of responsibilities aside from the need to return profit to shareholders, even though famed individuals such as Milton Friedman put forward a much narrower interpretation regarding the responsibilities of corporations.

The increasing importance of sustainability and corporate responsibility has led to the development of new tools, such as the triple bottom line (Elkington, 1994), designed to enable organisations to assess their broader impact on society and the environment, alongside their financial outcomes. Where these three outcome areas are in synch with one another, it is thought that organisations are making an optimal contribution to society, although there is still a degree of conceptual confusion, with differing interpretations of the triple bottom line including different combinations of factors for organisations to measure and report on. Due to the increasing importance of sustainability and corporate responsibility, we have seen examples of organisations making exaggerated claims regarding the environmental benefits of their products and services. This is known as greenwashing, and is an issue to which substantial academic interest has been devoted. From Chevron in the 1980s, through to more contemporary examples, the need to promote the 'greenness' of one's organisation has resulted in a number of corporate scandals.

Towards the end of the chapter attention turned back towards the HRM function in particular, covering the HRM responses that have been seen given the growing importance of the sustainability agenda. Taking a practical perspective, initiatives such as the rise in flexible working, and in particular home and remote working, and the design of office space were all considered. Many of these, most notably remote working, were given new impetus by the Covid-19 pandemic of 2020. Finally, readers were introduced the notion of the circular economy, a profound rethinking of current economic and business systems that are based on a linear take–make–dispose philosophy. The circular economy itself is based on the idea of zero waste, and complete reuse, in pursuit of the desire to decouple the creation of wealth from environmental degradation. It is certainly an enticing concept, and although it is not without challenges and critics, it appears to be an idea that is gaining traction relatively quickly with various organisations and government bodies alike.

References

Ahrend (2020) Ahrend, www.ahrend.com/en (archived at https://perma.cc/DD27-N9DH)

BBC News (2020a) Coronavirus: Nine in 10 would continue working from home, www.bbc.co.uk/news/uk-wales-53946487 (archived at https://perma.cc/G4HB-E6WT)

BBC News (2020b) Warnings of 'ghost towns' if staff do not return to the office, www.bbc.co.uk/news/business-53925917 (archived at https://perma.cc/643X-84LF)

Berrone, P, Fosfuri, A and Gelabert, L (2017) Does greenwashing pay off? Understanding the relationship between environmental actions and environmental legitimacy, *Journal of Business Ethics*, 144 (2), 363–79

Biopak (2019) Biopak: Products, www.biopak.com/uk/products (archived at https://perma.cc/SJP2-6FRX)

Burch, K A and Barnes-Farrell, J L (2020) When work is your passenger: Understanding the relationship between work and commuting safety behaviours, *Journal of Occupational Health Psychology*, 25 (4), 259–74

Butler, S (2014) Primark to pay £6m more to victims of Rana Plaza Factory in Bangladesh, www.theguardian.com/world/2014/mar/16/primark-payout-victims-rana-plaza-bangladesh (archived at https://perma.cc/477Y-QVY5)

Cadbury (2020) About Bournville, www.cadbury.co.uk/about-bournville (archived at https://perma.cc/NS5N-4VL5)

Carter, C R and Rogers, D S (2008) A framework of sustainable supply chain management: Moving toward new theory, *International Journal of Physical Distribution and Logistics Management*, 38 (5), 360–87

Dockery, A M and Bawa, S (2018) When two worlds collude: Working from home and family functioning in Australia, *International Labour Review*, 157 (4), 609–30

Dumont, J, Shen, J and Deng, X (2017) Effects of green HRM practices on employee workplace green behaviour: The role of psychological green climate and employee green values, *Human Resource Management*, 56 (4), 613–27

EASAC (2015) *Circular Economy: A commentary from the perspectives of the natural and social sciences*, European Academies, Science Advisory Council, Halle, Germany

Elkington, J (1994) Towards the sustainable corporation: Win-win-win business strategies for sustainable development, *California Management Review*, 36 (2), 90–100

Elkington, J (2018) 25 years ago I coined the phrase 'triple bottom line'. Here's why it's time to rethink it, *Harvard Business Review* Digital Articles, 25 June, 2–5

Ellen MacArthur Foundation (2020a) Circular economy: Concept, www.ellenmacarthurfoundation.org/circular-economy/concept (archived at https://perma.cc/T8PN-3LF7)

Ellen MacArthur Foundation (2020b) Case studies, www.ellenmacarthurfoundation.org/case-studies (archived at https://perma.cc/L8YT-V8KD)

European Commission (2020) Corporate social responsibility and responsible business conduct, https://ec.europa.eu/growth/industry/sustainability/corporate-social-responsibility_en (archived at https://perma.cc/JUM2-KX3J)

Evans, J (2020) P&G urged to match best in class to avoid 'greenwash' label, www.ft.com/content/4f807d17-2f95-4572-a7f4-50655fdda00e (archived at https://perma.cc/LP9Z-URPZ)

Felstead, A and Henseke, G (2017) Assessing the growth of remote working and its consequences for effort, well-being and work–life balance, *New Technology, Work and Employment*, 32 (3), 195–212

Ferdig, M A (2007) Sustainability leadership: Co-creating a sustainable future, *Journal of Change Management*, 7 (1), 25–35

Friedman, M (1962) *Capitalism and Freedom*, University of Chicago Press, Chicago

Goleman, D and Lueneburger, C (2010) The change leadership sustainability demands, *MIT Sloan Management Review*, 51 (4), 49–55

Govind, R, Singh, J, Garg, N and D'Silva, S (2019) Not walking the walk: How dual attitudes influence behavioural outcomes in ethical consumption, *Journal of Business Ethics*, 155 (4), 1195–214

(Continued)

(Continued)

Hahn, T, Preuss, L, Pinkse, J and Figge, F (2014) Cognitive frames in corporate sustainability: Managerial sensemaking with paradoxical and business case frames, *Academy of Management Review*, 39 (4), 463–87

Haney, A B, Pope, J and Arden, Z (2020) Making it personal: Developing sustainability leaders in business, *Organization and Environment*, 33 (2), 155–74

Hopkinson, P, Zils, M, Hawkins, P and Roper, S (2018) Managing a complex global circular economy business model: Challenges and opportunities, *California Management Review*, 60 (3), 71–94

Joseph, J, Borland, H, Orlitzky, M and Lindgreen, A (2020) Seeing versus doing: How businesses manage tensions in pursuit of sustainability, *Journal of Business Ethics*, 164 (2), 349–70

Keating, C (2020) P&G unveils 2030 'carbon neutral' goal backed by CO_2 offsetting plan, www.businessgreen.com/news/4017976/unveils-2030-carbon-neutral-goal-backed-co2-offsetting-plan (archived at https://perma.cc/U285-TS5L)

Kraaijenbrink, J (2019) What the 3Ps of the triple bottom line really mean, www.forbes.com/sites/jeroenkraaijenbrink/2019/12/10/what-the-3ps-of-the-triple-bottom-line-really-mean/#5871ddf85143 (archived at https://perma.cc/PVL9-AUQV)

Kramar, R (2014) Beyond strategic human resource management: Is sustainable human resource management the next approach?, *International Journal of Human Resource Management*, 25 (8), 1069–89

Leahy, S (2019) This is the world's most destructive oil operation – and it's growing, www.nationalgeographic.co.uk/environment/2019/04/worlds-most-destructive-oil-operation-and-its-growing (archived at https://perma.cc/7S25-4U7D)

Lee, H C B, Cruz, J M and Shankar, R (2018) Corporate social responsibility (CSR) issues in supply chain competition: Should greenwashing be regulated? *Decision Sciences*, 49 (6), 1088–115

Leggett, C (1999) The Ford Pinto case: The valuation of life as it applies to the negligence–efficiency argument, https://users.wfu.edu/palmitar/Law&Valuation/Papers/1999/Leggett-pinto.html (archived at https://perma.cc/7KQ2-4VFH)

Legrain, A, Eluru, N and El-Geneidy, A M (2015) Am stressed, must travel: The relationship between mode choice and commuting stress, *Transportation Research*, Part F (34), 141–51

Longoni, A and Cagliano, R (2018) Sustainable innovativeness and the triple bottom line: The role of organizational time perspective, *Journal of Business Ethics*, 151 (4), 1097–120

Lopez-Leon, S, Forero, D A and Ruiz-Díaz, P (2020) Recommendations for working from home during the Covid-19 pandemic (and beyond), *Work*, 66 (2), 371–5

Lush (2020) Our policies: This is what we stand for, https://uk.lush.com/tag/our-policies (archived at https://perma.cc/ZD5M-8SNE)

McKinsey (2017) Mapping the benefits of a circular economy, *McKinsey Quarterly*, 2, 12–17

Morrison, J (2020) *The Global Business Environment: Towards sustainability?*, 5th edn, Red Globe Press, London

Mulville, M, Jones, K, Huebner, G and Powell-Greig, J (2017) Energy-saving occupant behaviours in offices: Change strategies, *Building Research and Information*, 45 (8), 861–74

Nunan, R (1988) The libertarian conception of corporate property: A critique of Milton Friedman's views on the social responsibility of business, *Journal of Business Ethics*, 7 (12), 891–906

O'Donohue, W and Torugsa, N (2016) The moderating effect of 'green' HRM on the association between proactive environmental management and financial performance in small firms, *International Journal of Human Resource Management*, 27 (2), 239–61

P&G (2020) Doing what's right, www.pg.co.uk/doing-what-is-right (archived at https://perma.cc/B6MB-D7M2)

Renwick, D W, Redman, T, and Maguire, S (2013) Green human resource management: A review and research agenda, *International Journal of Management Reviews*, 15 (1), 1–14

Rieckmann, M (2012) Future-oriented higher education: Which key competencies should be fostered through university teaching and learning?, *Futures*, 44 (2), 127–35

(Continued)

(Continued)

Rysavy, M D T and Michalak, R (2020) Working from home: How we managed our team remotely with technology, *Journal of Library Administration*, 60 (5), 532–42

Stratton, J P and Werner, M J (2013) Consumer behavior analysis of fair trade coffee: Evidence from field research, *Psychological Record*, 63 (2), 363–74

Thomas, D (2019) Lush's Mark Constantine: The socially conscious capitalist, www.ft.com/content/48f70c38-201a-11ea-b8a1-584213ee7b2b (archived at https://perma.cc/J97M-UB9U)

Thomas, D, Morris, S and Edgecliffe-Johnson, A (2020) The end of the office? Coronavirus may change work forever, www.ft.com/content/1b304300-0756-4774-9263-c97958e0054d (archived at https://perma.cc/6ZEW-GDMK)

Van Loon, P and Van Wassenhove, L N (2020) Transition to the circular economy: The story of four case companies, *International Journal of Production Research*, 58 (11), 3415–22

Ward, A (2018) BP: Rebuilding trust after a disaster, www.ft.com/content/3e09d84a-489f-11e8-8ee8-cae73aab7ccb (archived at https://perma.cc/A5ZT-LGV9)

Watson, B (2016) The troubling evolution of corporate greenwashing, www.theguardian.com/sustainable-business/2016/aug/20/greenwashing-environmentalism-lies-companies (archived at https://perma.cc/L3P5-BNRU)

Webster, K (2017) *The Circular Economy: A wealth of flows*, 2nd edn, Ellen MacArthur Foundation Publishing, Cowes

18 Equality, diversity and inclusion

The subjects of equality, diversity and inclusion have moved up the corporate agenda in recent times. In 2017, the Deloitte human capital trends report noted that diversity and inclusion had become a 'CEO-level issue' (Deloitte, 2017), with high-performing organisations deriving advantage from their diversity and inclusion strategies. For organisations to leverage the benefits of such strategies, it is argued that equality, diversity and inclusion need to become mainstream issues for businesses, rather than topics that are the sole preserve of particular functions, such as human resource management.

Within this chapter we will together explore the subjects of equality, diversity and inclusion. To begin, we will explore current definitions of the concepts, considering the differences and similarities between them. We will explore the benefits that diversity can bring to organisations of all types, as well as challenges within this area. We will critically evaluate differing approaches to bringing about equality, diversity and inclusion in our organisations, whilst also examining contemporary issues such as those connected with unconscious bias, and the interesting ideas of both 'neurodiversity' and 'intersectional disadvantage'. Throughout our discussions we will link to both research and practice, and the pause and reflect activities will give you an opportunity to explore the subject in greater depth.

Context and definitions

When we reflect on the wider context, we know that our societies are becoming increasingly diverse, and along many more dimensions than before (CIPD, 2018). Alongside the more traditional categories of gender, ethnicity and disability status have arisen age diversity, sexual orientation and other such issues. At a later point in this chapter, the 2010 Equality Act will be discussed in full, so these points will not be elaborated upon at this point specifically. Importantly, scholars such as Armstrong et al (2010) argue that substantial and sustained benefits arise when organisations embrace an equality, diversity and inclusion agenda, and indeed there are many practical reasons for which our organisations are moving in this direction. We know that employees are attracted to organisations with an inclusive work environment, and therefore this can help to build a positive employer reputation (Fordham, 2016) and contribute to our organisations becoming employers of choice. Of course, it is important not to omit the legal argument here; embracing an equality, diversity and inclusion agenda can help our organisations to ensure that they meet their legal obligations, although, as we shall see later in this chapter, 'doing the minimum' within this area is unlikely to result in substantial and sustained advantage for our firms.

At this point, it is important that we arrive at clear definitions of the concepts we are speaking about. It is important to recognise that different groups and individuals define equality and diversity slightly differently, and this can cause conceptual confusion within the territory. For our purposes, though, the most relevant definition is that provided by the UK's Equality and Human Rights Commission (EHRC, 2019):

> Equality is about ensuring that every individual has an equal opportunity to make the most of their lives and talents, and believing that no one should have poorer life chances because of where, what or whom they were born, what they believe, or whether they have a disability. Equality recognises that historically, certain groups of people with particular characteristics eg race, disability, sex and sexuality, have experienced discrimination.

Important within this definition is the central idea within equality theory that no one should be treated differently on the basis of a specific protected characteristic. In the UK, the Equality Act 2010 enshrines the protections afforded to individuals, meaning that it is unlawful to discriminate on the basis of age, disability, gender reassignment, marriage/civil partnership, pregnancy and maternity, race, religion/belief and gender. Interestingly, the definition makes specific reference to the idea that 'no one should have poorer life chances' (EHRC, 2019) as a result of their circumstances, and this is an important foundational concept within the area of equality. The EHRC argues that since the UK introduced anti-discrimination legislation in the 1970s, we have made progress towards being a more equal society. They suggest that outdated attitudes and stereotypes, for example the 'stay-at-home mum', have begun to fade. But, having said that, statistical data from organisations such as the Office for National Statistics and reporting from organisations regarding gender pay gaps (CIPD, 2019) still show that females typically earn less than males, and that the lack of female representation in high-paid, board-level positions is still an issue (Terjesen and Sealy, 2016). The presence of such issues suggests that, whilst progress has been made, it would be demonstrably fallacious to argue that further steps cannot be taken to promote equality and diversity within our organisations. It has been argued that discriminatory stereotypes and attitudes persist because of a number of deep-seated psychological mechanisms, and although an exploration of these is beyond the scope of the current book, this is an area of active scholarly and practitioner research interest.

Moving on from the subject of equality, we come to diversity. Many scholars and writers within this field conflate the terms 'diversity' and 'inclusion', which again adds a level of confusion to the territory. For our purposes, the CIPD (2018) suggests a straightforward definition:

> Diversity is about recognising difference, but not actively leveraging it to drive organisational success. It's acknowledging the benefit of having a range of perspectives in decision-making and the workforce being representative of the organisation's customers.

Perhaps the key point of note within this definition is the idea that diversity is about the recognition of difference, but that this does not automatically extend to our organisations actively taking advantage of difference to drive improved performance. This presents somewhat of a problem with current approaches to diversity management, but helpfully a resolution is found when we consider the subject of 'inclusion':

> Inclusion is where difference is seen as a benefit, and where perspectives and differences are shared, leading to better decisions. An inclusive working environment is one in which everyone feels valued, that their contribution matters and they are able to perform to their full potential, no matter their background, identity or circumstances. An inclusive workplace enables a diverse range of people to work together effectively.
>
> (CIPD, 2018)

Differently to the definition of diversity, it is in the above definition of inclusion where we see the active leveraging of difference to drive organisational performance. The CIPD (2018) argues that it is through the development of inclusive working environments that organisations will ultimately see tangible benefits arising from their efforts within this area. It is also important to highlight that the definition above makes reference to differences that may be visible or non-visible (eg an individual's background), and we will explore this idea further when we examine the subject of neurodiversity. What we do know is that when organisations lack awareness of issues related to equality, diversity and inclusion, this can bring about significant costs to individuals and organisations themselves. For individuals it can involve losses in terms of pay, and potential exclusion from certain jobs, professions or occupational groups. For organisations, it may result in a narrowing of the labour pool from which employees can be selected, damage to company reputation and brand image, and potentially significant legal costs if firms are found to have breached their legal obligations. It suffices to say that this is a complex area, with nuanced and multifaceted debates – some of which we will explore as this chapter progresses. To build on our emerging understanding of equality, diversity and inclusion, we will now consider the benefits that accrue to organisations that actively engage with these topics.

The business benefits of equality, diversity and inclusion

A substantial evidence base has been built around the benefits, to both organisations and wider society, that accrue from the drive to make work more equal, diverse and inclusive. This evidence base includes contributions from academic researchers, which we will explore at a later point in this section, and professional associations such as the CIPD. In its report into diversity and inclusion in the workplace, the CIPD (2018) argues that there are a number of benefits that arise from the application of policies aimed at increasing diversity in the workplace, and a number of reasons why diverse and inclusive organisations perform better in the market. They suggest that improved corporate performance is driven, in part, because diverse and inclusive organisations:

- Tend to have more effective decision making.
- Have a more positive social reputation.
- Can be more creative and innovative.

Indeed, whilst there may be legal pressures that encourage a focus on diversity and inclusion, the positive pressure exerted by factors such as the above, and others – such as the ability to stand out as an employer of choice in the labour market – is increasingly encouraging organisations to adopt approaches that encourage equality, diversity and inclusion. One factor stands out here in amongst the reasons connected with competitive advantage and 'business' benefits, and that is the moral case or moral argument for diversity and inclusion. There are increasing calls for our organisations to be more socially responsible (Barnett, 2019), and we see this in the rise of subject areas such as corporate responsibility and ethical consumerism (see, for example, Yeow et al, 2014), amongst others. But the arguments that are often (but not exclusively) made in such work is that engagement with responsible practices such as those connected with equality, diversity and inclusion ultimately makes the organisation more financially successful. Whilst this may be the case, there is also an argument that organisations should apply equality, diversity and inclusivity-oriented work practices not for reasons of corporate advantage, but because they are morally the 'right' thing to do.

PAUSE AND REFLECT

Why might organisations of all shapes and sizes make the 'business case' for equality, diversity and inclusion more prominent in their arguments as opposed to the 'moral case'?

The question above is multifaceted and there is no straightforward answer to it. The purpose behind asking it was to stimulate thoughts and discussions connected with the reasoning that sits behind the application of workplace practices aimed to support equality, diversity and inclusion. Of course, if we subscribe to Friedman's arguments regarding the purpose of the corporation – in other words, that the only purpose of business is to improve returns to shareholders – then practices associated with encouraging equality, diversity and inclusion must be couched in those terms. Having said this, the rise of subjects such as corporate responsibility, ethical consumerism, and so on, suggest something of a movement towards the recognition that organisations of all types have important social contributions to make. Debates in these areas are contested, with multiple points of view, but it is important to raise awareness of the various reasons for which organisations engage with equality, diversity and inclusion.

Turning to the academic literature, there are significant debates and active streams of enquiry connected with understanding the business benefits that accrue from equality, diversity and inclusion. This research base is constantly developing, so this chapter can provide only a snapshot of evidence at a particular point in time. Researchers such as Armstrong and colleagues have argued that there are positive links between the application of equality and diversity practices and organisational performance (Armstrong et al, 2010). For this particular study the research team explored data from a variety of Irish service and manufacturing organisations, approaching 1,000 organisations initially and achieving full responses from 132 firms. What they found was that organisations that operated diversity and equality management systems, typically including diversity training, recruitment monitoring and robust pay and promotion systems across minority groups, perform more strongly than those organisations without such systems. They therefore concluded that there was a strong business case for investing in equality and diversity issues. It must be noted, however, that the researchers were clear to point out that their particular method of data collection and analysis did not allow for the isolation of a specific, causal relationship and so their conclusions must be seen in this light.

In very sharp contrast to the findings of Armstrong and her colleagues, research by Rebecca Riley and her team (Riley et al, 2013) argued that there was no relationship between an investment in diversity and equal opportunities and business performance. Whilst they did not collect primary data within their study, the team used the extensive data collected through the Workplace Employment Relations Study from 2004, which encompassed data from 2,295 British workplaces. Although they argue for no significant relationship between investment in equality and diversity and business performance, they also highlight that equality and diversity measures do not have significant costs for organisations either. They arrive at a conclusion that the relationship between equal opportunities and business performance is inherently complex, and dependent on organisational characteristics, but argue that further research into the area is needed in order to unpack the key variables. These contrasting studies demonstrate to us the diverse research findings that have been published within the area of equality, diversity and inclusion, and, as mentioned previously, new studies are being published at frequent intervals. Having undertaken an exploration of the business benefits that emerge from equality, diversity and inclusion, we now explore differing organisational approaches to equality and diversity.

Differing approaches to equality and diversity

Just as there are differing views on the business benefits that accrue from equality, diversity and inclusion, so there are different approaches that organisations adopt in the pursuit of more equal and diverse workforces. Broadly speaking, these approaches fall into two different categories:

- The equal opportunities approach.
- The management of diversity approach.

The first of these, the equal opportunities approach, is rooted in legislation such as the Equality Act 2010, whilst the second approach, the management of diversity, argues that legislation is insufficient and that there needs to be more concerted change in attitudes towards previously marginalized groups. A third, 'radical' approach also exists, but has substantially less support within the UK. This more extreme approach to equality and diversity argues that actions such as positive discrimination are needed in order to address problems of equality and diversity in our organisations.

The equal opportunities approach

The equal opportunities approach focuses its attention on ensuring that there is equality of opportunity for all. It is, by its very nature, a liberal approach, and focuses much more on ensuring equality of *opportunity* rather than necessarily concerning itself with equality of *outcome*, a small but crucial difference. It seeks, therefore, to prevent discrimination on the grounds of unreasonable and irrelevant criteria (eg gender, age, ethnic origin and so on), in an attempt to 'level the playing field' for all concerned. Positive action is therefore allowable in encouraging representation from minority groups in organisations, but it is important to highlight that the terms 'positive action' and 'positive discrimination' must not be conflated. In other words, organisations may put special support in place or encourage applications from disadvantaged groups, but may not discriminate in their employment decisions to favour disadvantaged groups; for example, gender could not be used as a selection criterion.

What organisations seek to do under an equal opportunities approach is to ensure that processes and procedures are set in such a way that only job-relevant criteria are used to make a decision. Differences are marginalized as much as possible, with organisations solely concerned with the ability of an individual to undertake a particular role.

While the equal opportunities approach is attractive on paper, there remain problems in its application. It has initially been criticised for taking an overly simplistic approach to the issue, with the assumption that a fair process will always lead to fair outcomes not being translated to real world results. It has therefore been suggested that an equal opportunities approach very much treats the symptoms of an underlying problem, rather than driving at the heart of the issue itself. Indeed, Taylor (2010) points out that equal opportunities approaches can give a negative perspective on difference. It is argued that the approach has an overemphasis on minorities and the disadvantaged, and therefore has little to offer the majority of individuals in work and society.

The management of diversity approach

Differently, the managing diversity approach, which emerged from the US in the 1990s, concentrates very much on improving conditions for all individuals inside an organisation. It is not aimed at any minority group or groups in particular, and is therefore argued to be more inclusive as a result. The core aim of the managing diversity approach is to support all employees based on their unique qualities and abilities, recognising the importance of individual difference in the pursuit of organisational competitive advantage. We can therefore suggest that the management of diversity approach is more attractive to both employers and employees. Whilst equal opportunities approaches tend to focus on minimising 'difference', the management of diversity approach treats 'difference' as a positive organisational asset that has significant potential organisational benefits, such as improving employee relations, building a positive reputation and brand image, and improving levels of creativity and innovation (CIPD, 2018). It is thought that the management of diversity approach brings about these advantages because all individuals inside our organisations have the ability to work to their full potential, and are given opportunities to maximise their contribution. Although there are many supporters of the management of diversity approach, there are a number of criticisms levelled at this approach to equality as well.

The first of the criticisms levelled at the management of diversity approach is whether the approach is different, in practice, to the equal opportunities model. Commentators frequently argue that it is a clear case of 'old wine in new bottles'; the ' same arguments are simply being further developed and the management of diversity approach, in reality, adds little new material to the existing debates. Linked to this, others argue that this approach to equality simply packages equal opportunities requirements with a business case that aims to make equality more 'palatable' for key organisational stakeholders. In other words, rather than emphasising legal and moral obligations, putting the business benefits front and centre within the argument is thought, by some, to appeal more to those in leadership positions within organisations (Dijk et al, 2012). The issue here is that if organisational investment into management of diversity-style approaches does not result in material business benefits, or if organisations enter into times of economic difficulty, the investment in equality may cease. Equally we may argue that if organisations operate in loose labour markets – in other words, markets where there is a surplus of employees – or if reputation has only a limited impact on increasing sales, then the business case supporting the management of diversity approach loses its power. Indeed, a review of published academic literature carried out for the Department for Business, Innovation and Skills in 2013 reinforced this viewpoint, identifying that there is no single approach that will work for all businesses, and if poorly managed, it can actually increase costs. Ultimately, the reliance on the business case is a significant criticism that can be levelled at the management of diversity approach.

In summarising the differences between the equal opportunities and management of diversity approaches to equality, Ross and Schneider (1992) developed a very useful map (Table 18.1). Essentially, they argue that the two approaches to equality differ in five distinct ways.

Table 18.1 The differences between the equal opportunities and management of diversity approaches to equality

Managing diversity	Equal opportunities
Internally driven	Externally imposed
Focused on individuals	Focused on specific groups
Celebrates difference	Tries to achieve sameness
Culture	Systems and processes
Everyone's responsibility	Driven by HR

SOURCE: Adapted from Ross and Schneider (1992)

The key learning point to take from this discussion is that there are differences in organisational approaches to equality and diversity. The extent to which the management of diversity approach differs from its predecessor, equal opportunities, is debatable, with many commentators advancing views on the levels of similarity and difference between the two schools of thought. We now turn to some of the legislation that underpins equality, diversity and inclusion.

The Equality Act 2010

The UK's anti-discrimination legislation has built up in a piecemeal fashion since the mid-1900s. Governments of the time passed numerous individual pieces of legislation, including the Equal Pay Act 1970, the Sex Discrimination Act 1975 and the Race Relations Act 1976. These individual pieces of legislation sought to protect different characteristics, but were not seen as comprehensive in themselves. In 2010 the UK's Equality Act replaced the preceding legislation with one single Act that aimed to simplify the law and remove inconsistencies, and therefore the argument was that organisations and individuals would find it easier to comply with the legislative provisions. The passing of the Equality Act also ensured that the UK was compliant with relevant European Union directives, and had caught up with recent developments in case law. At its core, the Equality Act sets out a series of different protected characteristics, which are as follows:

- Age.
- Disability.
- Race, ethnicity and national origin.
- Gender and gender reassignment.
- Marital status.
- Religion or belief.
- Pregnancy and maternity.
- Sexual orientation.

Whilst it is important to recognise that each of the different protected characteristics has different provisions within the Equality Act, the legislation sought to harmonise principles and core terminology related to discrimination. The terminology used to describe discrimination has been broken into four different areas: direct discrimination, indirect discrimination, harassment and victimisation. We now explore each in turn.

Direct discrimination

Direct discrimination is broadly the most straightforward type of detriment that an individual can suffer. It manifests itself when an individual receives unequal treatment as a direct result of one of the protected characteristics covered by the Equality Act, for example their age or national origin. When assessing such claims, courts apply the 'but for' test, which essentially seeks to understand whether the individual would have received the same treatment 'but for' their protected characteristic.

For example, if a person would have been recruited for a particular role if they had been white (or equally if they had been from an ethnic minority group) then this would be an example of direct discrimination. There is no defence for direct discrimination; the only possible exception comes into play within recruitment when for reasons of authenticity, decency or personal service, there is a requirement for either a man or a woman to perform a particular role.

Indirect discrimination

Indirect discrimination is rather more complex than direct discrimination. Indeed it can occur without organisations and individuals intending to discriminate. In legislation indirect discrimination occurs when a 'provision, criterion or practice' is applied which has the effect of disadvantaging one group more than another. For instance, if a job advertisement specifies that applicants must be able to run a certain distance under a certain time, this would potentially disadvantage older individuals who tend to have less peak physical capacity than younger individuals. This is therefore indirect discrimination. Differently to direct discrimination, there is a defence for indirect discrimination if an employer can show that they are applying a provision, criterion or practice that is a 'proportionate means of achieving a legitimate aim'. Returning to our example of running, if physical endurance is a genuine part of the job then an employer can legitimately include this within their selection criteria, perhaps incorporating some form of fitness test into their selection process. Although a greater proportion of younger individuals will likely be able to meet the required standard than older individuals, this does not make the practice unlawful, provided that the requirement is a genuinely important part of the role and fulfils a legitimate business objective.

Harassment

The Equality Act defines harassment as any form of unwanted conduct related to a protected characteristic that has the purpose of violating a person's dignity or creating an intimidating, hostile, degrading, humiliating or offensive environment for that person. In this area of the Equality Act the criteria of marriage and civil partnership and pregnancy/maternity do not apply. Harassment does not have to be intentional; it is understood to occur when a person reasonably considers conduct to have had that effect. A common example within this area is sexual harassment, which may include comments of a sexual nature, unwanted physical contact and/or sexually explicit jokes or anecdotes. The issue here is that conduct that does not offend an individual in one situation may constitute sexual harassment in another situation. Employers need to be able to show that they have taken all reasonable steps to prevent harassment. If they do not, employees may be able to claim that their employing organisations are vicariously liable for the conduct of individuals in the workplace. There are substantial debates as to what constitutes harassment, and when employers may be liable for the conduct of their employees, but organisations need to be able to show that they deal with complaints promptly and fairly, and that appropriate action is taken.

Victimisation

Finally, victimisation is defined by the Equality Act as subjecting another individual to a detriment either because they have undertaken a 'protected act', or because you believe that a person has undertaken or will undertake a protected act in the future. The Equality Act goes on to highlight that a 'protected act' includes:

- Making a claim or complaint of discrimination under the Equality Act.
- Giving evidence in support of someone else's claim under the Act.
- Making an allegation that a person has contravened the Act.
- Doing anything else in connection with the Act.

This perhaps sounds rather confusing and legalistic so let's explore a specific example. Let's assume that person A has recently bought a successful case for harassment against their organisation, with the courts finding that the employer was liable, and consequently damages were awarded. At some later point person A applies for promotion within the organisation, and their application is dismissed. If person A can demonstrate that the detriment (ie not receiving the promotion) was connected to the harassment claim, then person A could claim that they have been victimised. It is also important to point out that less favourable treatment in these circumstances does not necessarily need to be linked to a specific protected characteristic. Moving on, we now discuss the concept of inclusive work environments.

Inclusive work environments

The concept of 'inclusion' was introduced at an early stage of this chapter. At that point a definition from the CIPD was introduced, although, recognising that there are indeed many contributors to this debate, Mensi-Klarbach and Risberg (2019: 22) define inclusion as follows:

> inclusion means enabling and valuing the participation of all employees so that they can contribute fully to the organisation.

Although this definition has a slightly different emphasis to the version produced by the CIPD, the core sentiment is very much the same: enabling and encouraging contributions from all within our organisations. The subject of inclusive work environments, and what makes an inclusive work environment, has been approached from a variety of angles, with some commentators focusing on workplace design (Agarwal, 2018) and leadership (Boekhorst, 2015), whilst others focus on aspects of culture, interpersonal relations and access to opportunities. Indeed, given the expansive scope of the term 'inclusion', all such contributions have merit, and we must think broadly when considering what the term 'inclusive work environment' means.

As Agarwal (2018) argues, inclusion in the workplace begins with design. In what ways do we set our workspaces and offices up, and how does this impact upon different individuals? For example, individuals with physical disabilities may have difficulties using flights of stairs, so we therefore need to ensure that lifts are provided and that corridors are accessible to those using wheelchairs. Similarly, what about the layout of desks and other working spaces? What adjustments might we need to put into place for individuals with visual impairments? How would individuals with hearing difficulties respond to an auditory only fire alarm system? Once we begin unpacking these issues, even from simply a practical, design perspective, the plethora of considerations becomes quickly overwhelming. It is therefore very important that we take a systematic approach to this issue, as risks may arise if we overlook different subtopics within this field.

Indeed, just as organisations need to think about design of a workplace, they need to consider the culture (Fordham, 2016) as well. The policies and practices set in place by those in leadership positions can have a significant impact on the extent to which a workplace culture may be considered inclusive, just as the behaviours modelled by employees at all levels impact inclusion. Just as with the subject of design, workplace culture is a multifaceted subject and therefore we must think in broad terms to ensure that key aspects of developing an inclusive culture are not overlooked. On a different, but related, theme, Boekhorst (2015) argues that authentic leaders have an important role in fostering inclusion within the workplace, just as reward systems can reinforce certain types of behaviour. Even from this brief review it becomes apparent that the development of inclusive work environments has many different facets that we must consider.

At a later point in this chapter we will explore the important subject of neurodiversity. Within those discussions we will explore how Microsoft has adapted its recruitment techniques to better include individuals on the autistic spectrum, and this is an excellent example of an organisation adapting its HR policies and practices in order to become more inclusive. But, as Ware (2018) notes, altering recruitment and selection practices is only part of the process. Organisations also need to consider adapting their induction and onboarding processes so that they are accessible to individuals with physical and mental disabilities. If organisations do not do this, the risk is that whilst an increasingly diverse range of individuals may be recruited, retaining those diverse individuals may become an issue. Problems with onboarding and induction are frequently highlighted as reasons for substantial employee turnover during the first weeks and months of employment (see, for example, Birchall, 2013; Hartley, 2012), therefore organisations must ensure that the broad suite of HR processes and practices that guide the employee lifecycle is aligned to the need for inclusivity. As highlighted in the previous paragraph, we must remember that the construction of inclusive work environments is a multidimensional issue that encompasses a range of diverse factors in itself. If organisations achieve success within this area, however, there are substantial benefits that they may see. As we have discussed previously in this chapter, organisations may see improved corporate reputation, enhanced creativity and innovation, alongside improved employee relations climates (CIPD, 2018).

Unconscious bias

Despite the various elements of legislation that we have that seek to prevent discrimination on a variety of grounds, we know that it still persists both within our organisations and in wider society. Why might this be? Traditionally we have operated on the assumption that discrimination is a conscious choice and is therefore something that we can address through mechanisms such as legislation, codes of good practice and so on. We have assumed that 'good' people do not discriminate, they are not biased, and therefore our approaches to equality, diversity and inclusion have sought to eradicate bias by reforming 'problematic' attitudes. But, what if a greater or lesser part of our biases are rooted in the unconscious parts of our minds?

PAUSE AND REFLECT: A LOGIC PROBLEM

A father and son are involved in a car crash and are rushed to hospital. The father dies. The boy is taken to the operating room but the surgeon says, 'I can't operate on this boy, because he's my son.'

How is this possible? How do you explain this situation?

When asking this question to a wide swathe of the population, you tend to receive some very confused answers. How can the father operate on his son if he has already died in the car crash? But what some individuals fail to recognise is that the surgeon in this instance is the boy's mother. What has happened here is that individuals have made an unconscious assumption that the surgeon must be male, when in fact surgeons can equally be female. Hence the issue when addressing this question.

What we have to remember is that we are all biased, and that these biases are a product of our varying backgrounds, cultural environments, our education and the various personal experiences that we accrue throughout our lifetimes. Unconscious bias is a bias of which we are unaware; it happens when our brains make snap judgements. Indeed, the Equality Challenge Unit (2013) define unconscious bias as:

> Bias that we are unaware of, and which happens outside of our control. It is a bias that happens automatically and is triggered by our brain making quick judgements and assessments of people and situations, influenced by our background, cultural environment and personal experiences.

It is important to recognise that bias can be unconscious as well as conscious, and that it is typically unconscious bias that is more problematic for organisations and society, simply because it is hidden from view and therefore much more difficult to tackle. The first priority is therefore surfacing unconscious biases, looking for sources of potential bias in our organisations, and rooting it out within our processes and systems. Many organisations have developed unconscious bias training for their employees in a drive to deal with this problematic issue, and many HR departments have conducted audits of processes to identify where unconscious bias may operate at various stages of the employee lifecycle. For example, within recruitment processes many organisations now routinely remove names, genders and ages from CVs and application forms, and many organisations require diversity on interview panels – for example ensuring that all genders are represented in each selection decision. It is thought that such actions assist us in removing unconscious bias from our processes and therefore our workplaces become more inclusive as a result.

Neurodiversity

Neurodiversity is a term that has entered the literature connected with equality, diversity and inclusion in more recent times, and represents a substantial advance in thinking within the territory. At its core is the relatively straightforward idea that just as we differ physically in terms of our capabilities, so our brains can also be 'differently wired'. ACAS (2019) has produced a useful definition of the concept, arguing that it 'refers to the different ways the brain can work and interpret information'.

In previous times individuals with conditions that fell under the broad headings of dyspraxia, dyslexia, attention deficit hyperactivity disorder (ADHD) and autism, amongst others, were seen by society and by legislators as disabled, and thus requiring protections and adjustments to enable them to participate in typical working environments. Different from neurotypicals (ie individuals where the brain functions in the way that society typically expects), individuals with neurodiverse conditions were seen as less able. The neurodiversity movement flips this argument around, arguing that in reality such conditions are part of the normal variation of the human brain. Indeed, Temple Grandin's renowned TED talk (2010) argued that society requires different human variation, coining the thought-provoking phrase 'different, not less' to indicate the contributions made by individuals with neurodiverse conditions. Organisations such as SAP and Microsoft have indeed specifically reworked their recruitment and selection processes to increase neurodiversity within their employee base, arguing that this contributes significantly to their competitive advantage. Why might this be the case?

Both academic and practitioner studies have found that individuals with conditions falling under the banner of neurodiversity have substantial talents that can benefit organisations. Indeed, Austin and Pisano (2017) argue that these individuals often have exceptional abilities in the areas of pattern recognition, memory and mathematics – skills that are highly sought after by a substantial range of employers. Our transition to a digital economy therefore may well require greater levels of neurodiversity in the workforce. A problem though is that individuals exhibiting neurodiverse conditions may struggle to match the ideal employee 'prototype' (de Anca, 2016) because they can lack the soft or social skills that employers typically seek within their recruitment processes. For example, some neurodiverse individuals may find it difficult to work as part of a team, or they may be unable to make eye contact with recruiters. What thinkers such as Austin and Pisano (2017) and de Anca (2016), amongst many others, argue is that whilst we search out physical diversity in our organisations (eg appearance), our firms are actually intrinsically homogeneous in nature. This is an important point to reflect upon.

PAUSE AND REFLECT

Why might our organisations struggle to recruit and select neurodiverse individuals? What adjustments could we make to our workplaces in order to better incorporate individuals with neurodiverse conditions?

The above questions should have caused you to pause and think about these important questions. Accepting the argument that increased neurodiversity leads to stronger competitive advantage (Austin and Pisano, 2017; Sutherland, 2016), the struggles our organisations would seem to be having in attracting and hiring these individuals is problematic. Some of the problems that neurodiverse individuals face are typical of those faced by other individuals from minority groups. For instance, the concept of prototype matching in recruitment – which is the idea that recruiters match applicants against a standard hypothetical model, rejecting difference or individuals who do not conform (to a greater or lesser extent) to the expected prototype. It is well known that individuals tend to recruit 'in their own image', and therefore this can operate to impede neurodiversity in organisations, just as it can impede other elements of diversity and inclusion. In essence, our work environments are set up to suit neurotypical employees, and therefore individuals falling outside that frame can find it difficult to participate in work.

In attempting to overcome the problems briefly discussed above, the immediate solution is to revisit and reform HRM processes so that they better reflect the diversity of human thinking. Austin and Pisano (2017) highlight that organisations such as Microsoft and Hewlett Packard have done this with some substantial success and are seeing benefits including increased productivity, greater innovation and improved employee engagement, amongst others. The authors go on to argue that organisations best at integrating neurodiverse individuals appear to be doing a number of common things. These include adapting recruitment and selection processes, training their existing employees and managers, setting up bespoke support systems, considering career planning/management for neurodiverse individuals, and making the initiative part of the mainstream agenda within the organisation. ACAS (2019) argues that workplaces need to become more inclusive of neurodiverse individuals, highlighting that staff must feel safe to disclose neurodiverse conditions, which organisations can support by reducing the stigma associated with these disclosure decisions. In order to gain a more practical understanding of this subject, let's explore the approach that Microsoft has used to encourage neurodiversity.

CASE STUDY 18.1

Encouraging neurodiversity at Microsoft

Microsoft, as we all know, is a very successful technology company that has traditionally built its advantage in computer software, and more recently in cloud and online services. A few years ago Microsoft recognised that individuals with neurodiverse conditions could have strong skill sets that the firm needed in order to fill specific types of roles – for example, data scientists and software engineers. They learnt, though, that their hiring process presented barriers that made it difficult for neurodiverse individuals to 'get through the front door'. The company explored the reasons for this and created its Autism Hiring Program in 2015, which, it argues, helps individuals to best demonstrate their abilities to hiring managers. Subsumed now under its Inclusive Hiring Program, Microsoft has adjusted its recruitment process for neurodiverse individuals to enable them to submit a CV, complete an online skills/technical based test, and then, if successful, undertake a multi-day skills assessment at their main campus. The organisation argues that by adapting its recruitment processes and better training managers about neurodiverse conditions, it has been able to attract, engage and retain key staff needed to support its success.

You can find out more about Microsoft's Inclusive Hiring Program online via the company website (www.microsoft.com).

What we have seen in this section is that, just as diversity may manifest in physical differences, there can also be internal differences that can be, to a greater or lesser extent, hidden from view. The neurodiversity movement is relatively new, and is developing substantially in terms of the academic research base and practical examples of organisations embracing the concept. The core idea is that our brains can function and respond in differing ways, and that these differences can give individuals substantial skills in a variety of important areas. We now move on to consider one final current development in the field of diversity, equality and inclusion: intersectional disadvantage.

Intersectional disadvantage

One of the core criticisms levelled at organisational approaches to equality, diversity and inclusion, and perhaps more problematically, legislation within this area, is that it considers separate elements of disadvantage individually, rather than considering how these elements may interact. This is what is known as a 'single-axis approach' to disadvantage. Consider for a moment, however, that this may be a very simplistic approach to dealing with what is in reality a very complex issue.

For example, let's assume that we work within a very large organisation, we don't necessarily need to worry about the sector or the type of work we undertake, but we do employ individuals from all sections of society, all genders, all ethnicities, individuals with disabilities and those without, and individuals from across the generations. When looking at disadvantage from a single-axis viewpoint we see these categories as separate; so, you are either a man or a woman, disabled or non-disabled, and so on. But, when we look at identities more closely, there are in fact several overlapping layers that we need to explore. For instance, an individual could be female, but also at the same time she could be from an ethnic minority grouping, she may also be disabled, and finally she may be located within the 50–60 age bracket. When seen in these terms, this individual actually has four different elements to her identity, which may or may not interact to produce a unique set of circumstances for her. Intersectional disadvantage seeks to explore how disadvantaged identities operate together, potentially producing a much richer understanding of disadvantage in both our organisations and broader society.

PAUSE AND REFLECT

To what extent is it important for our organisations, and society more generally, to adopt an approach recognising that individuals may have many disadvantaged elements to their identity?

The activity above should have encouraged you to think carefully about disadvantage, and how individuals may experience disadvantage in our organisations, and in broader society. In addressing the question, most individuals would argue that it is important to consider intersectional disadvantage because there may well be significant impacts on work and employment outcomes for those concerned. Indeed, this is a growing area of academic interest and several researchers, and groups of researchers, are actively inquiring into this topic. For instance, Ruiz Castro and Holvino (2016) have explored how intersectional disadvantage can impact upon career advancement, and Woodhams et al (2013) have found that pay levels are substantially affected by intersectional disadvantage. Whilst this research is interesting in its own right, the question for us to unpick is why intersectional disadvantage may create these problematic outcomes.

One of the key issues to consider when starting to tackle the point above is the notion of privilege, and the idea of the 'privileged comparator'. When comparing disadvantaged individuals, we need to relate their experiences back to some form of common reference point: the 'non-disadvantaged' person. This individual is typically white, male and non-disabled. Within the territory of intersectional disadvantage, this individual is commonly referred to as the 'privileged comparator': the individual who typically suffers no detriment on account of their personal characteristics. What research evidence (see, for example, Ruiz Castro and Holvino, 2016; Woodhams et al, 2013) suggests is that as an individual moves further away from the 'privileged' group, so their employment and work outcomes degrade. So, if you are female, from an ethnic minority and also disabled you have not one but three disadvantaged elements to your identity. Researchers within this field consistently highlight that this causes substantially worse outcomes, arguing that such penalties are not 'additive' but that there is a 'snowballing' effect (Woodhams et al, 2013). In essence this means that the total penalty is not the sum total of the penalties that arise from being (1) female, (2) from an ethnic minority and (3) disabled, but there is an interaction effect that causes substantially worse outcomes. From an organisational, societal and legal perspective these findings are certainly troubling.

PERSPECTIVE FROM RESEARCH

Multiple disadvantage and wage growth: the effect of merit pay on pay gaps

One interesting study published in the field of intersectional disadvantage considered the impact of merit pay (also known as performance-related pay) on the pay gaps experienced by disadvantaged groups. The researchers who conducted this study accessed a set of longitudinal data that tracked pay over a number of years, with many thousands of individual elements of data and, following the application of controls for a number of variables, were able to isolate the impact that merit pay had on the employment outcomes (in this case pay increases) of individuals from different disadvantaged groups. The size of the data set enabled the researchers to look at pay outcomes from a number of different groups. They were able to review combinations of factors, including gender, ethnicity and disability status, uncovering a number of important findings.

Initially, the article highlighted that the pay outcomes of those groups who were furthest away from the privileged comparator were substantially worse than others. For example, if you were female and also from an ethnic minority, your pay was substantially lower than that of the comparator group (male, white). The study found that individuals from disabled groups were most disadvantaged in terms of pay, but the study did find unusual findings with relation to merit pay specifically. We might think that duc to thc nature of merit pay, it would likely operate to the benefit of the already advantaged group at work (ie the male, white, non-disabled group). The study, however, found the opposite effect. Merit pay rises were observed to close pay gaps over time, with individuals from disadvantaged groups 'catching up' over the course of the study period. Although, having said this, the rate of 'catch up' was small, and in some cases would take longer than a lifetime to equalise pay differentials. The article, and many others published on the theme of intersectional disadvantage, highlighted that new legislation is required in order to properly protect individuals with several disadvantaged characteristics at work. It is argued that legislation that operates only on the basis of individual, discrete forms of disadvantage does not offer adequate protections.

SOURCE: Woodhams et al (2015)

Discussions within this section have introduced us to the notion of intersectional disadvantage. Whilst these discussions have necessarily been brief and broad in outlook, the key learning point to take away is that diversity, equality and inclusion is a complex area of study. What we see within this field is that where individuals have multiple elements of disadvantage (eg gender, ethnicity, disability status, age and so on), their outcomes in employment and society can be significantly hampered. Academic research has found substantial career and pay penalties where individuals find themselves at the intersections of disadvantage; this is therefore an active area of both academic and practitioner thinking. We now close this chapter with a brief conclusion.

Conclusion

Having explored a number of different issues within the subject area of equality, diversity and inclusion we now reach the end of this chapter. We know that our societies and organisations are becoming increasingly diverse over time, and that diversity brings many benefits to our firms including improved creativity and innovation. Organisations that are more diverse tend to make better decisions and have more positive social reputations, and together we have explored some of the academic literature that has contributed to the business case for equality and diversity. There are differing approaches to equality and diversity in our organisations as well, the equal opportunities approach alongside the management of diversity. Whilst these approaches emphasise different elements of the debate, we must remember that there are criticisms that the management of diversity approach to equality is essentially 'old wine in new bottles'. There has nevertheless been a substantial push for our organisations to develop more inclusive work environments, and whilst exploring the subject of neurodiversity we noted how Microsoft has adapted its recruitment and selection processes to attract and employ more individuals with autism: individuals who have substantial talents that the organisation requires. Finally, we discussed the interesting issue of intersectional disadvantage, arguing here that it is rather simplistic to see disadvantage in terms of single elements of someone's identity. We must recognise that these different elements of a person's identity interact, producing substantially different employment outcomes.

References

ACAS (2019) Neurodiversity in the workplace, www.acas.org.uk/index.aspx?articleid=6676 (archived at https://perma.cc/YYR6-A9KA)

Agarwal, P (2018) How do we design workplaces for inclusivity and diversity?, www.forbes.com/sites/pragyaagarwaleurope/2018/07/19/how-inclusive-is-your-organisation-here-is-how-to-use-inclusive-design (archived at https://perma.cc/62F6-EHY3)

Armstrong, C, Flood, P C, Guthrie, J P, Liu, W, MacCurtain, S and Mkamwa, T (2010) The impact of diversity and equality management on firm performance: Beyond high performance work systems, *Human Resource Management*, 49 (6), 977–98

Austin, R D and Pisano, G P (2017) Neurodiversity as a competitive advantage, *Harvard Business Review*, 95 (3), 96–103

Barnett, M L (2019) The business case for corporate social responsibility, *Business and Society*, 58 (1), 167–90

Birchall, A (2013) Getting off on right foot can help avoid missteps, *Management Today*, August, 39

Boekhorst, J A (2015) The role of authentic leadership in fostering workplace inclusion: A social information processing perspective, *Human Resource Management*, 54 (2), 241–64

CIPD (2018) Diversity and inclusion at work: Facing up to the business case, June, CIPD, London

CIPD (2019) Guide on gender pay gap reporting, www.cipd.co.uk/knowledge/fundamentals/relations/gender-pay-gap-reporting/guide (archived at https://perma.cc/38DL-RG8S)

de Anca, C (2016) Why hiring for cultural fit can thwart your diversity efforts, https://hbr.org/2016/04/why-hiring-for-cultural-fit-can-thwart-your-diversity-efforts (archived at https://perma.cc/CK9Z-NUU4)

Deloitte (2017) *Rewriting the Rules for the Digital Age: 2017 Deloitte global human capital trends report*, Deloitte University Press

(Continued)

(Continued)

Department for Business, Innovation and Skills (2013) *The Business Case for Equality and Diversity: A survey of the academic literature*, www.gov.uk/government/publications/the-business-case-for-equality-and-diversity-a-survey-of-the-academic-literature (archived at https://perma.cc/38JL-SYHB)

Dijk, H, Engen, M and Paauwe, J (2012) Reframing the business case for diversity: A values and virtues perspective, *Journal of Business Ethics*, 111 (1), 73–84

EHRC (Equality and Human Rights Commission) (2019) Understanding equality, www.equalityhumanrights.com/en/secondary-education-resources/useful-information/understanding-equality (archived at https://perma.cc/NBW5-Z9FW)

Equality Challenge Unit (2013) *Unconscious Bias and Higher Education*, ECU

Fordham, L (2016) Fostering an inclusive culture can have a positive impact in the workplace, *Employee Benefits*, July/August, 15–18

Grandin, T (2010) The world needs all kinds of minds, www.ted.com/talks/temple_grandin_the_world_needs_all_kinds_of_minds (archived at https://perma.cc/R3K8-XB3B)

Hartley, D (2012) Driving efficiency through people, *Chief Learning Officer*, 11 (6), 40–1

Mensi-Klarbach, H and Risberg, A (2019) Diversity in Organizations: Concepts and practices, 2nd edn, Red Globe Press, London

Riley, R, Metcalf, H and Forth, J (2013) The business case for equal opportunities, *Industrial Relations Journal*, 44 (3), 216–39

Ross, R and Schneider, R (1992) From equality to diversity: A business case for equal opportunities, Pitman, London

Ruiz Castro, M and Holvino, E (2016) Applying intersectionality in organizations: Inequality markers, cultural scripts and advancement practices in a professional services firm, *Gender, Work and Organization*, 23 (3), 328–47

Sutherland, A (2016) Time to celebrate neurodiversity in the workplace, *Occupational Health*, 68 (11), 11

Taylor, S (2010) *Resourcing and Talent Management*, CIPD, London

Terjesen, S and Sealy, R (2016) Board gender quotas: Exploring ethical tensions from a multi-theoretical perspective, *Business Ethics Quarterly*, 26 (1), 23–65

Ware, G (2018) Enhance your onboarding: Three building blocks to create a more inclusive work environment, *Small Business Cleveland*, 29 (6), 12

Woodhams, C, Lupton, B and Cowling, M (2013) The snowballing penalty effect: Multiple disadvantage and pay, *British Journal of Management*, 26 (1), 63–77

Woodhams, C, Lupton, B, Perkins, G and Cowling, M (2015) Multiple disadvantage and wage growth: The effect of merit pay on pay gaps, *Human Resource Management*, 54 (2), 283–301

Yeow, P, Dean, A and Tucker, D (2014) Bags for life: The embedding of ethical consumerism, *Journal of Business Ethics*, 125 (1), 87–99

19 Wellbeing

Wellbeing is a subject that is of increasing importance in our workplaces. Commentators from both practitioner and academic backgrounds are contributing to the increasing debates regarding wellbeing, and we are seeing frequent reports on the subject through news outlets and professional associations such as the CIPD (CIPD, 2019). All of this commentary argues that organisations could, in many cases, do more to encourage wellbeing at work, and that being 'well' at work contributes significantly to an individual's broader quality of life. Whilst different individuals and groups focus their discussions on differing aspects of the wellbeing territory, the common thread is that organisations can derive substantial benefits when they invest in wellbeing initiatives.

During the course of this chapter we will explore a number of facets of the wellbeing territory. We will initially examine the broad subject area and settle upon a definition of what we mean through the term 'wellbeing', before we consider the benefits that organisations may derive by investing in such initiatives. We will then move on to consider how a variety of elements of workplace design might impact wellbeing, as well as exploring the subjects of stress, anxiety and resilience. We will spend some time exploring the concept of 'stress' specifically, and debate the extent to which either too much, or conversely too little, stress can be problematic. To draw us towards a conclusion to this chapter, we will cover the important subjects of presenteeism at work and burnout. Burnout in particular is a topic that is moving increasing into the spotlight, in both the press and academic outlets, and it is from there that we will stand back from our broader discussions and tease out a conclusion. To begin our exploration let's set the stage by reviewing common definitions of wellbeing and the various facets that make up the subject.

Context: the increasing importance of workplace wellbeing

Wellbeing is a term that is used, and sometimes misused, in a variety of contexts. In the academic sphere individuals have explored wellbeing from a variety of perspectives, including the wellbeing of whole nations (Anderson et al, 2019), cities (Lee et al, 2018), and, more importantly for our purposes, organisations themselves (Moore and Piwek, 2017). The agreement amongst scholars is that wellbeing, howsoever defined, is a positive quality that individuals, organisational leaders and indeed governments should strive to develop. Having said this, there is also substantial commentary that suggests that our working lives are becoming increasingly stress-filled (Newton and Teo, 2014; Stich et al, 2019); indeed, the article from Stich et al (2019) explores the role of email as a stressor in the workplace, a subject that is frequently covered by a variety of news outlets. Of course, the diverse nature of the commentary that surrounds wellbeing as a concept makes arriving at a definition somewhat problematic, but given that our focus is the workplace, it is logical for us to focus our attention there. In this context, the International Labour Organization (2019) provides a useful and practical definition of workplace wellbeing:

> Workplace wellbeing relates to all aspects of working life, from the quality and safety of the physical environment, to how workers feel about their work, their working environment, the climate at work and work organization.

As we can see from this definition, there are a number of facets of workplace wellbeing, including physical safety – in other words, that we are kept safe from physical harm at work – alongside more ethereal issues, such as how individuals 'feel' about their work and the environment that they work within. Other organisations such as the CIPD (2016) support this multifaceted approach to workplace wellbeing by arguing that the subject of wellbeing has a number of 'domains':

- Health.
- Work.
- Personal growth.
- Values/principles.
- Collective/social.

In common with the ILO definition above, the CIPD highlights the importance of keeping individuals physically safe at work, but goes into substantially more depth in relation to how work contributes to (or detracts from) personal wellbeing. In particular it highlights the importance of effective line management and the provision of autonomy at work, as well as fair and transparent reward. It also argues that employee voice positively contributes to a sense of collective wellbeing and that providing access to lifelong learning opportunities and career progression can fulfil our desires for personal growth, another important aspect of wider wellbeing. In a similar vein to the work of the CIPD, Hesketh and Cooper (2019) argue that there are four key tenets of wellbeing: psychological, physiological, societal and fiscal. In line with the CIPD they suggest that individuals need to be physically well, and free from injury and disease, but that psychological wellbeing goes much further than simply being 'happy' at work and in wider life. There is indeed a significant debate as to what constitutes wellbeing from a psychological perspective, but a common view is that whilst having a positive outlook is important, there is more to it than this. Ryff and colleagues (2004) speak about eudaimonic wellbeing, which suggests that six different areas make up broader notions of psychological wellbeing:

- Self-acceptance.
- Environmental mastery.
- Positive relationships.
- Personal growth.
- Purpose in life.
- Autonomy.

Whilst most of these are self-explanatory it is important to highlight that the term 'self-acceptance' means that an individual has a positive attitude towards themself, and that environmental mastery revolves around individuals making the best use of opportunities that present themselves, and having a sense of control over factors and activities outside of oneself. Moving back to Hesketh and Cooper's (2019) tenets of wellbeing we encounter the term 'societal' wellbeing. This area is ultimately concerned with the broader quality of life experienced by individuals in a particular area and can be measured in a variety of ways. Finally, financial wellbeing is concerned with access to resources, predominantly money, and recognition that there are times in an individual's life when there can be significant financial stress, for instance managing on low pay at the start of one's career. What we can see from this brief review is that wellbeing is indeed a multifaceted concept. There are various conceptions of wellbeing, what it is, what facets make up wellbeing, and similarly a variety of views in terms of how we should measure it. With such a diverse literature exploring wellbeing, it is important for us to put some boundaries in place to guide our exploration. In terms of discussions in this chapter, we will be looking exclusively at wellbeing within the workplace. This is not to disregard the importance of concepts such as societal wellbeing, but is in recognition of the fact that our focus here is on work, employment and the workplace.

Whilst we have so far explored some of the conceptualisations of wellbeing in the workplace, we haven't yet explored the broader context within which our work sits. This is important to cover at this point so that we can better situate our upcoming discussions. We know that the world is becoming increasingly complex, with the world of work transformed by technology, and being much more volatile than it was previously. Casey (2014) coined the acronym VUCA to highlight that the modern workplace environment is increasingly characterised by volatility, uncertainty, complexity and ambiguity. From a wellbeing perspective, this presents some challenges. For instance, if our world of work is increasingly uncertain, how might we plan for career development and lifelong learning, key aspects of the 'personal growth' domain expressed by the CIPD? It is arguably increasingly difficult to generate and sustain excellent levels of wellbeing in a rapidly shifting external context. Alongside this, we know that there are some substantial societal issues that impact wellbeing in the workplace. Whilst speaking from a predominantly American perspective, Stringer (2016) notes a number of trends including rising levels of obesity, occupational stress, musculoskeletal problems, and the impacts of smoking.

She also highlights that employees in many countries typically work a substantial number of hours over and above their employment contracts, which can include checking emails out of hours, taking telephone calls or being unable to prevent our minds racing through work-related problems. Furthermore, whilst many individuals have some level of autonomy over how and where they work – for instance, there has been a substantial rise in home working on either a full- or part-time basis – Stringer suggests that our jobs have become increasingly sedentary, negatively impacting our physical health. The upshot of this is that we have a cocktail of issues that may positively or negatively impact upon our wellbeing at work, and these issues may manifest themselves differently in different contexts.

The key point to take away from this initial discussion is that we have a variety of ways of defining wellbeing, and differing views in terms of the components that make up wellbeing both in the workplace and in broader society. The academic literature connected with wellbeing is developing rapidly, but it is yet to settle on clear and inclusive definitions and models of the concept at this particular point in time. What is important to recognise is that the changing context of work and employment, and indeed changes in the contents of jobs themselves, can have significant impacts upon individual wellbeing. We will now move on to consider specific aspects of the wellbeing territory, beginning with the benefits that accrue to organisations that take a proactive approach to wellbeing.

Organisational benefits of wellbeing

If we initially accept the view that wellbeing is a positive concept and something that our organisations (and indeed individual employees themselves) should strive to improve, it is important to question the tangible benefits that arise from organisational investment in this area. Whilst there are certainly moral arguments that can be made here, for instance the view that investing in employee wellbeing is the correct and proper thing for our organisations to do, isolating specific business benefits means that the case becomes that much stronger. Indeed, Whitehouse (2019) argues that there is substantial evidence suggesting that investing in employee health and wellbeing brings significant benefits to organisations, realised through improved productivity and lower rates of sickness absence. This is not an isolated perspective, either; in the US context Stringer (2016) highlights very similar benefits indeed. She argues that there is a direct link between investment in wellbeing and business performance because where employees are healthier, organisations experience lower healthcare and associated insurance costs, employees demonstrate stronger levels of engagement and they are also more productive, thus producing stronger corporate performance. When we review the academic literature we see a very similar argument as well. For instance, Baptiste (2007) highlights that HRM practices have a substantial impact on employee wellbeing, and where organisations conspicuously invest in the wellbeing of their staff, they generate stronger ultimate outcomes. Interestingly, however, she argues that we mustn't maintain a sole focus on organisational outcomes at the expense of employees. It is ultimately important to recognise that organisations derive results through their people; hence we should have a dual focus on both employee and business benefits.

The thoughts and themes above are not isolated to those particular authors; others such as Hubbard (2014) and Colley et al (2016) highlight similar arguments that positively relate wellbeing initiatives to stronger business performance. The latter article is particularly fascinating as it links access to outdoor, green space during the working day to improved employee outcomes. Critics, however, argue that where wellbeing initiatives have ill-defined goals it can be very difficult to measure business impact. Whitehouse (2019) therefore argues that our first step must be to define the ultimate ambition of our wellbeing programme. For instance, does our organisation want to reduce sickness absence, improve productivity, encourage stronger employee resilience, or perhaps a combination of these issues? Clarity concerning the ultimate goal is thought to be of paramount importance. Whitehouse (2019) does go on to argue that there has, so far, been relatively little academic research into the impacts of wellbeing initiatives. Reviewing journal databases, this would appear to be true. In particular, it would appear that more quantitative, statistical testing of the link between wellbeing and organisation performance is needed. Whilst there is an enticing logic that investing in employee wellbeing produces stronger organisational results, and an emerging literature base that generally supports this view, we could argue that the academic research base connected with this subject could be rather stronger. Let's now explore a specific organisational example to solidify some of the points that we have covered thus far.

CASE STUDY 19.1

Wellbeing in the NHS

The National Health Service (NHS) is an extremely large organisation employing some 1.5 million people in the UK. As we might imagine, the organisation employs a very diverse mix of staff, both on the front line of delivering healthcare and in a multitude of support and professional services divisions and roles.

Back in 2009 the Boorman Review isolated and discussed the importance of prioritising health and wellbeing in the NHS. The review produced a number of recommendations and argued quite clearly that a healthier workforce would mean fewer avoidable periods of sickness absence, reductions in levels of presenteeism, and a more efficient staffing system, which would ultimately benefit the key organisational outcome: patient care. Boorman's recommendations led to the development of a number of initiatives in health and wellbeing, including developments within sickness absence procedures and a wellbeing framework.

We know, however, that despite the recommendations, staff within the NHS are under significant pressure. Reports from The Health Foundation (2019) note evidence of low morale and high levels of occupational stress and burnout. We must recognise that healthcare is a demanding area of work, and that staff working in the NHS are subject to a multitude of pressures that do not impact individuals working in other organisations. The Health Foundation goes on to note that staffing pressures have exacerbated health and wellbeing problems and highlights that staff survey results suggest a decline in recent years in overall perceptions of health and wellbeing at work.

The NHS is, however, attempting to tackle the issues it faces. A joint report from The Health Foundation, The King's Fund and the Nuffield Trust (Beech et al, 2019) suggested further recommendations connected to improving staff retention and improving the work environment for employees. There have separately been other initiatives to look at improving staff rest periods, and providing staff with access to green outdoor space where possible (The Health Foundation, 2019).

What we can learn from the above example is that health and wellbeing at work are multifaceted and complex. Within organisations such as the NHS they can be a particular challenge given operational pressures and indeed the nature of the work itself. We can see that there is, and has been, a desire to improve wellbeing within the organisation for a substantial period of time (the Boorman report was published in 2009), but that whilst there is evidence of separate initiatives gaining traction, the staff survey results point to a more complex picture. As we mentioned earlier, there is an enticing logic to organisational investment in wellbeing. Our base assumption is that where employers invest in the wellbeing of their staff, they will see stronger organisational outcomes, and indeed there is emerging research evidence to show that this is the case. We must recognise, however, that more academic enquiry is needed into this area, and the example from the NHS demonstrates the complexities in operationalising a wellbeing agenda.

Workplace design

Building from our discussions above we will now review some of the intricacies of wellbeing at work, starting with the subject of workplace design. The layout of our offices, even the desks and work equipment that we use, all have the potential to impact our health and wellbeing. Stringer (2016), for instance, highlights how important it is that people can physically move during the working day, rather than spending long periods of time seated, either at one's own desk or in meeting spaces. In order to encourage movement, she suggests that organisations can invest in standing desks, where individuals stand rather than sit, or consider making stairwells more attractive than lifts, thus encouraging individuals to take the stairs rather than the lift between floors. Whitehouse (2019) suggests similar points, arguing that organisations could make more use of standing desks as well as activity trackers, alongside considering the choices available within any on-site food outlets. For instance, it is well known that Google offers staff free food at work, but the organisation attempts to promote healthy choices by displaying healthy options more prominently, and making sweet treats harder to find. Google hasn't removed unhealthy snacks, as this action could, in itself, promote a backlash over perceived meddling in individual choice; it instead engages with nudge theory to promote healthier options.

Whilst the interior of the office is important to consider, so is the location of the office itself. For instance, employers seeking to promote health and wellbeing at work may attempt to locate their company bases near transport infrastructure such as cycle routes or train stations, promoting alternatives to commuting by car. Company owners/managers may also seek to locate their premises near recreation and leisure facilities, such as gyms or fitness centres, thus providing employees with the opportunity to easily access these facilities. Separately, the media frequently report the benefits offered on site by many large American corporations such as Google, Facebook and Microsoft. Many such organisations have developed extensive campuses with a diverse range of facilities on site. Adopting a critical perspective, however, we could develop an argument that bundling facilities into an office/campus location might encourage individuals to spend excessive time on site and therefore provide employees with less time to 'switch off', which we know is a substantial factor in overall health and wellbeing. We must also recognise that increasingly work takes place in many locations other than the traditional office.

Remote working, facilitated by substantial improvements in various technologies such as virtual private networks, video conferencing/meetings and online collaboration tools such as Slack, has grown substantially in recent times (Felstead and Henseke, 2017). Whilst, of course, there are exceptions to this, and roles where individuals have to be physically present in order to undertake their duties – for instance an airline pilot – many individuals undertake work away from a traditional office base for at least some part of their working week (De Menezes and Kelliher, 2017). Remote working is often said to be a 'win–win' for employers and employees (Felstead and Henseke, 2017), with it thought to enable improvements in work–life balance and wellbeing for employees, whilst encouraging greater levels of organisational commitment and engagement, benefiting employers. It is argued, however, that these benefits do not come without costs, with Felstead and Henseke (2017) suggesting that remote working can create work intensification and promote an 'always on' culture where employees struggle to switch off from their working tasks and duties. This is particularly the case where individuals work from home, as the boundaries between 'work' and 'home' can blur to a very great extent.

PAUSE AND REFLECT

What are your own views regarding remote working? To what extent might it bring about a positive or negative impact on employee health and wellbeing?

Hesketh and Cooper (2019) argue, again, that there can be substantial advantages and drawbacks to remote working. From a positive stance it can provide very substantial levels of autonomy and flexibility for workers, assisting those with caring responsibilities to a very great extent. It also reduces the amount of time that individuals spend commuting, thus saving money, benefiting the environment, and allowing greater levels of productivity for remote workers. For the organisation, remote working also enables a reduction in office costs, potentially enabling moves to smaller office locations. From a negative stance, remote working, when undertaken over a considerable period of time, can lead to social isolation, feelings of loneliness and potentially abandonment, as employees do not have frequent contact with co-workers. We also must remember that employers have substantially less control over the environment of remote workers. For instance, in an office environment an individual may use a desk and chair specifically designed to improve posture, whilst when working remotely the same employee may decide to work whilst lounging on a sofa. Employers operating more formalised processes regarding remote working may require employees to detail their working equipment and location before a manager will agree to a remote working arrangement, but not all organisations undertake such assessments.

Within our discussions here we must also recognise the rise of so-called co-working or shared working spaces. Discussed from both a practitioner and academic perspective, co-working spaces are work locations where individuals, irrespective of their employer, can work either independently or collaboratively in some form of shared office space. Co-working spaces are thought to bridge the gap between the traditional 'office' and working remotely, with Burt (2018) arguing that this is one of the most significant new trends in working practices since the establishment of the modern office. Whilst these are arguably more suited to micro and small businesses, start-ups or entrepreneurs, Burt indicates that some larger organisations are also enabling employees to make use of co-working spaces. Exploring the academic research base, there would appear to be good reasons for this. For instance, a study conducted by Weijs-Perrée et al (2019) highlights that whilst remote working is increasingly popular, individuals are still seeking networking opportunities and the ability to collaborate with others, hence the need for a space in which to do this. The authors also note that not all individuals want to work from their home environment and prefer the atmosphere created whilst working with others, but away from a traditional office. The drive towards increased use of co-working spaces shows little sign of diminishing at this time, and this will be an interesting trend to follow in the coming years.

Stress, anxiety and resilience

The subjects of stress, anxiety and resilience are very important issues within the wellbeing territory. There is evidence, both from academic research and anecdotally, that levels of stress and anxiety are rising, both within the world of work and within society more broadly. This section of the chapter will introduce you to the key debates, beginning with stress.

Stress

The word 'stress' is used very frequently in our language when describing a great many aspects of our lives. Within the workplace context, the Health and Safety Executive (2019) offer a useful formal definition for us. They argue that workplace stress is:

> The adverse reaction people have to excessive pressures or other types of demand placed on them at work.

The stress response inside the body has been discussed frequently, and we know that a series of chemicals mix to produce what we feel as 'stress', creating symptoms such as increases in our heart rate and alterations in our nervous systems. What we must recognise, though, is that stress is not always a negative; there are instances where a degree of stress is indeed helpful, and can have a positive impact on our performance. Hesketh and Cooper (2019) argue that moderate and short-lived episodes of stress can bring about challenge and stimulation, which can help to increase our focus and improve our productivity. It is when stress becomes overwhelming that we experience negative side effects, which the Health and Safety Executive (2019) indicates can include greater instances of sickness absence, poorer workplace performance, increased arguments and more emotional reactions. Stress at work can be caused by a number of issues, for instance where individuals are unable to cope with the demands of their work, where managers do not allow individuals sufficient control over their work, or where we don't provide individuals with enough information to undertake their work tasks. Similarly, episodes of change can be very stressful times within organisations, particularly where individuals lack clarity about their own role and responsibilities. What we must recognise though is that different individuals have different capacities to adsorb and deal with stress: what for one individual is not stressful may be overwhelming to another.

The issue in our organisations is how we can seek to deal with stress in the workplace. Stringer (2016) argues that one highly effective tactic is to provide individuals with autonomy at work, so employees have control over how, when and where they undertake work activities. This can overcome some of the issues mentioned in the paragraph above, as can the provision of workplace wellness programmes, ensuring that employees utilise all of their allocated annual leave, and that they are taking appropriate breaks from the workplace. As indicated towards the start of this chapter, significant attention has been devoted to understanding the role of technology, and in particular email, as a source of workplace stress (Stich et al, 2019). Whilst there are benefits to enabling individuals to work remotely, and technology can positively facilitate this, stress can build where work and life blur in an uncontrollable or unmanageable way. The Health and Safety Executive (2019) indicates that problems are best tackled early on, and it is important for line managers to check in with employees on a frequent basis, to understand the range of tasks that individuals are undertaking and to spot any signs of excessive stress.

Anxiety

Anxiety is understood to be a reaction to stress. Just as we have seen an increase in the commentary connected with stress, we have also seen increased focus on anxiety and its impact in both the workplace and broader social life. Whilst anxiety shares many of the same physical symptoms as stress, it is typically longer term in nature and the symptoms don't fade once the stressor/triggering factor has subsided. The NHS (2018) highlights that anxiety is a feeling of unease, such as worry or fear, noting that it can range quite substantially in intensity. The NHS goes on to point out that generalised anxiety disorder has many causes, including substantial stressful or traumatic experiences, long-term health conditions, drug and alcohol misuse and imbalances in our bodily chemistry.

Anxiety in the workplace has been the subject of significant attention, with McCarthy et al (2016) defining it in terms of feelings of nervousness and uneasiness regarding one's role or job performance. We know that levels of anxiety connected to work are increasing, and Cheng and McCarthy (2018) argue that this is problematic because it can give rise to a number of adverse outcomes including lower levels of performance, unacceptable risk taking and unethical workplace behaviours. Interestingly, the authors argue that whilst most attention has been focused on the problematic and negative outcomes that arise as a result of anxiety, the condition may also have some positive consequences. This is rather counterintuitive as commentary connected with anxiety in the general media has an almost universally negative outlook. Nonetheless, Cheng and McCarthy (2018) argue that individuals experiencing anxiety may be more sensitive to feedback and thus better at monitoring both themselves and their surroundings, and that from an evolutionary perspective anxiety helps us to take actions to avoid harm. These thoughts are indeed intriguing, and cast new light on the debates connected with anxiety, but we must not forget that for individuals experiencing substantial levels of anxiety, particularly in connection with their work and employment, the effects can be profound and debilitating.

In terms of how individuals and organisations may tackle anxiety-related issues, similar advice to tackling workplace stress applies. It is important that employees take care of their physical health; this includes their diet, exercise and sleep. Organisations must ensure that workloads are reasonable and that employees have as much control as possible over how, when and where they undertake their work. Encouraging open conversations between employees and line managers is similarly very important in developing a supportive culture, as is ensuring a sensible balance between in-work and out-of-work activities.

Resilience

Just as there has been significant and substantial commentary surrounding both stress and anxiety, so too has there been a growing focus on the subject of resilience. Hesketh and Cooper (2019) argue that there are again several definitions and interpretations of the term, but that all individuals will, at some point in their lives, require reserves of resilience to deal with challenging situations and events. Perhaps one of the clearest and most unambiguous definitions of resilience at the level of the individual comes from the work of Haglund et al (2007: 889), who argue that resilience is:

> The ability to successfully adapt to stressors, maintaining psychological wellbeing in the face of adversity.

The core idea here is that individuals face a multitude of stressors, whilst at work and in their broader lives, and that resilience is about rebounding during times of struggle. Commentators have similarly discussed resilience at the level of the organisation, suggesting that definitions are much the same as for resilience at the level of the individual (CIPD, 2011), focusing very much on the capacity of an organisation to 'weather the storm', or adapt to new challenges.

At the individual level it can be very difficult to keep a positive mindset during times of trouble, but Hesketh and Cooper (2019) argue that individuals can develop their capacity to be resilient, although there is little agreement on any one set of techniques or approaches that work across a wide swathe of the population. It is commonly argued that physical condition can impact upon someone's resilience; in other words, getting enough sleep and exercise, and looking after one's diet, can be important in promoting one's level of resilience. Others argue that mindfulness techniques, cognitive behavioural therapy and self-efficacy training can help individuals to become more resilient in the face of stressors (CIPD, 2011). In short, we know that there is a variety of techniques that are thought to foster resilience in individuals, but the effectiveness of any one of these will depend very much on the context and situation of the individual.

Practitioner and industry commentary argues that individuals from the millennial generation are less resilient than previous generations (Andriote, 2018; Lipman, 2016). It is argued that performance expectations, increased parental expectations and pressures emanating from social media and other related technology have contributed to a decline in resilience amongst this generation. Some argue that this generation is less able to cope with 'bumps in the road', despite resilience being an important trait that recruiters and organisations specifically search for, and it being known that resilience is important for success in education and life more broadly (Yeager and Dweck, 2012). This view may, however, be rather simplistic since we may equally argue that the quality of 'resilience' may be developed over the course of one's life and through exposure to a broader range of experiences. Naturally, therefore, we would expect younger members of society to be less resilient, because they do not have the same level of accumulated life experience. Like so many discussions in this chapter, and the broader textbook as a whole, the subject of resilience is multifaceted and it is thus simplistic to settle wholly on one side of this argument.

Presenteeism

Presenteeism is a growing phenomenon in the modern workplace, with ever more frequent news stories and comment pieces highlighting both the scale, and problematic nature of the issue. It is defined, quite simply, as going to work when ill or otherwise unfit to undertake work (Lohaus and Habermann, 2019). Ultimately it occurs when an individual feels compelled, either through internal or external forces, to undertake their normal work tasks when they should, in reality, be recovering and recuperating. We might well question at this particular juncture how significant this issue really is. Is it something that has been the subject of a media frenzy and has been unnecessarily hyped up, or is it something that is a real issue? Fortunately the CIPD's (2019) health and wellbeing at work survey helps to shed light on the scale of the issue. In this particular survey 86 per cent of respondents (based on 1,078 responding organisations) noted that they had observed presenteeism in their organisation over the last 12 months, up from 72 per cent in 2016, and a relatively low 26 per cent in 2010. This data would therefore suggest that presenteeism is on the rise, but when thinking critically we have to question whether we are becoming better at spotting and recognising presenteeism, or if the rise demonstrates a genuine increase in the issue. The CIPD goes on to argue that the more worrying issue is not necessarily wholly related to the rise in presenteeism, but the fact that only a minority of organisations were, in the survey responses, actively seeking to tackle presenteeism at work. In the survey, only one in ten organisations said that presenteeism was viewed as a priority for the company's management/ownership board.

In attempting to answer one of the questions posed above concerning whether presenteeism is on the rise, or if we are just getting better at spotting it in the workplace, it is useful to turn towards the academic research base that is developing around the subject. Lohaus and Habermann (2019) have reviewed a wide swathe of the research related to presenteeism and have noted that it is indeed a worldwide phenomenon, citing studies from the US, the UK, Europe, Asia and many other countries around the world. They suggest that absenteeism and presenteeism are often linked subjects and suggest that the latter can be defined as 'the loss in work productivity due to a person's health problems' (Lohaus and Habermann, 2019: 44). It is argued that the phenomenon has grown as levels of job insecurity have risen; in essence, people fear that if they are seen to be absent from work, they are more at risk of losing their jobs, particularly in loose labour markets. It is also thought that sick employees may derive self-esteem benefits from continuing to work, because working distracts them from their illness, so in this sense presenteeism may have, in some instances, arisen from something of a positive decision-making process. Building on this, Johns (2010) suggests that the type of illness or personal issue will affect an individual's decision to attend work or stay at home and that there is a range of personal factors and work-specific factors that influence an individual's decision as to whether they wish to work or not.

One of the key issues arising for organisations as a result of presenteeism is the impact that it has on productivity (Hemp, 2004). Hemp argues that whereas with absenteeism you can more easily quantify the lost productivity when an employee does not turn up, you cannot always tell how productivity is affected when an employee is present, but has an illness or other condition impacting upon their performance. This is arguably particularly the case in service and knowledge sectors, where it is less easy to measure the tangible output of an individual's efforts. Whilst the costs of presenteeism are less easy to measure than those associated with absenteeism, they are substantial; for instance, a People Management article (Burt, 2017) estimated that presenteeism costs UK employers £4,000 per employee per year. Interestingly, however, Johns (2010) argues that there may be financial benefits that arise from presenteeism. This is a little counterintuitive, but it does stand to reason. The argument from Johns is that absenteeism produces zero productivity, a 100 per cent loss of the employee for the period of their absence, but that presenteeism ensures that the organisation derives at least some benefit from having the employee turn up, even if they are not working at full capacity. There are, of course, a myriad of factors and reasons that impact on an individual's ability to attend work when suffering from either an acute or chronic condition, and the advice from bodies such as the CIPD and ACAS is that organisations should provide employees with appropriate time and support to recover. Where time away from the workplace is needed, it should be given. Consider the following example.

PERSPECTIVE FROM PRACTICE

Presenteeism at work

Note: The below example is fictional; any resemblance to an actual situation or person is entirely coincidental and not an intentional act on the part of the author.

Suzanne has always been a very efficient and highly effective employee. Working in sales for a technology company, she frequently leads the monthly sales charts, and her customers value her highly diligent approach and extensive product knowledge. Suzanne has worked for her company for over ten years, and takes pride in the fact that she is almost always present in the office. Her only period of absence was the result of an injury sustained on a skiing holiday some years ago. Last week, however, Suzanne contracted an illness that, whilst not especially contagious to others, impacted her quite substantially, resulting in a period of low energy, mental fuzziness and some very severe headaches. Whilst her doctor suggested that she take a few days' leave from work, Suzanne didn't want to be seen to be absent, so continued to go to the office even though she was not in the best physical condition. Whilst she was normally full of energy, she noticed that she because excessively tired by around lunchtime each day and consequently spent most of her afternoons browsing holiday sites on the internet, before she felt she had reached a time at which she could leave the office. On one occasion she did not respond to an urgent request from an important client, which may lead to the customer establishing a relationship with a competing organisation in the near future.

PAUSE AND REFLECT

What are your initial thoughts on the example above? Could the company have done anything to support Suzanne, and was she right in her decision to attend work?

Whilst the example presented above is entirely fictional, it has a basis in reality for a great many individuals. There are often times when we try to muddle through or 'grin and bear it', and the decision to attend work or not is, as suggested earlier, impacted by a wide range of personal and external factors. In this instance the argument from Johns (2010) might well apply; Suzanne did give the organisation some productivity (during the mornings) and so one view is that her decision was correct. A competing view would be that Suzanne should have heeded the advice given from the medical practitioner, particularly as we can see that her lack of full engagement may have damaged an important client relationship. In terms of whether the company could have done anything to support Suzanne, we first need to consider whether Suzanne's condition was sufficiently visible, or whether she declared it to her line manager. In some instances illness or other conditions may not be immediately apparent to an untrained eye, and therefore the company would rely on Suzanne disclosing her illness to her manager. In supporting Suzanne the company would then need to make a decision in accordance with any policy framework that they have connected with sickness or absence. This might have included sending her home for a specified period, or adjusting her work – perhaps allowing her to work from home for a few hours each day, recognising the positive benefits that individuals derive from work (Lohaus and Habermann, 2019), but always with the employee's health the first and foremost consideration. Inevitably this requires a judgement call on the part of the employee and organisation.

A separate issue that merits discussion within this section is a more recent development called 'leavism'. The CIPD (2019) notes that leavism involves employees using their allocation of annual leave (holiday) to catch up on work tasks. Indeed, the CIPD's health and wellbeing survey highlights that this issue had, in some form, been spotted in almost two-thirds of the responding organisations. As a result of these statistics, Hesketh and Cooper (2019) argue that organisations cannot rely on traditional measures of sickness absence alone in an effort to understand the full picture related to the overall wellbeing of their workforce. In attempting to tackle the issue, the CIPD (2019) argues that organisations should focus more on outputs rather than inputs at work. In other words, organisations should not log time spent in the office, or time sitting at one's desk; instead we should focus on outcomes such as project delivery, task quality and so on. Before moving on, let's take a few moments to think about other ways in which organisations may tackle the issue of presenteeism.

Before putting concrete steps in place, Webber (2018) highlights the need for organisational managers and leaders to develop an evidence-based understanding of the causes and factors impacting presenteeism at work. It is only with a solid evidence base that organisations will be able to implement robust solutions to tackle the issue of presenteeism, and we must remember that different workplaces will likely have different blends of factors and issues that need to be unpacked. A clear first step is then to encourage open and honest conversations between employees and their managers regarding any issues that might be impacting an employee during the course of their work. Following these conversations, organisations can then put in place reasonable adjustments to assist their employees. For instance, allowing staff who perhaps have contracted an infectious virus, but are otherwise able to work, to work from home might be appropriate in some instances. For staff who are perhaps caring for a sick relative, it might be appropriate to again allow them to work from home, or to vary their start or finish times so that they can arrange other support for their dependant. The overwhelming message from the body of advice suggests that open communication, followed up with reasonable, pragmatic adjustments, can help to tackle the issue of presenteeism. Of course, where employees are unfit to work, policy frameworks operated by employers should enable a period of complete absence from the workplace, and for this to be managed fairly, consistently and with a strong sense of mutual trust.

Burnout

Finally in this chapter, we come to the subject of burnout, a term that is becoming more commonplace in modern society. In the work context, burnout refers to a state of mental, physical and emotional exhaustion that is typically caused by overexertion and can have damaging consequences for employee motivation, engagement and productivity. Academic studies have been conducted into the issue of burnout, for instance Jyoti and Rani (2019), and a variety of links have been made between HRM practices and the tendency of employees to encounter the symptoms of burnout. Interestingly, burnout is not something that only affects individuals in traditional workplaces; studies by Taylor et al (2019) find significant instances of burnout in sporting contexts, linking burnout to the presence of workaholism. Indeed, workaholism is thought to be a significant issue and one contributing factor to burnout, with Oates (1971: 11) defining this simply as 'the compulsion or the uncontrollable need to work incessantly'. Some academics argue that workaholism and engagement exist on the same scale, with engagement in one's work seen as a positive attribute, but unhealthy attachment to work, as characterised by workaholism, seen as a negative issue that both employers and employees need to understand and tackle.

Whilst there is substantial debate surrounding the issue of burnout, which Maslach et al (1996: 20) usefully define as 'a state of exhaustion in which one is cynical about the value of one's occupation and doubtful of one's capacity to perform', we do know that the causes of burnout (Taylor et al, 2019) include factors such as:

- Extended working hours.
- Excessive pressure.
- The emotional demands of one's job.
- An overload of work demands and lack of control over work tasks.
- Lack of feedback from supervisors.
- Lack of social support.

Ultimately, what we see here is that burnout is triggered where jobs become overwhelming for individuals, and there is a consequential negative feeling regarding an individual's capacity to cope. Burnout in this sense is often characterised as chronic stress that leads to cynicism on the part of the employee in relation to both their work and their organisation, and substantial feelings that goals cannot be met, or that work cannot be accomplished. It is obvious therefore that burnout is a problematic issue that has substantial impacts on both employees and employers. Stringer (2016) highlights that a particular problem is that individuals frequently spend too much time attached to work and work tasks, noting that on average workers in the USA and Canada log 1,705 hours annually, whilst workers in the UK log 1,650. These numbers appear small in comparison to some Asian countries, for instance Taiwan (2,144) or Singapore (2,287), but remembering that these are average figures puts them into context. Indeed, Stringer (2016) goes on to point out that surveys have highlighted that an overwhelming percentage of workers in the USA put in an average of an extra day per week over and above the requirements of their employment contracts. Similar surveys across other contexts have found remarkably similar results.

Aside from the volume of working hours, and other factors highlighted in this chapter previously such as increases in stress levels, we must also consider some of the other factors that cause burnout. It is well known that jobs in organisations of all shapes and sizes are placing increasing demands on individuals. This happens at the higher end of the labour market, and also towards the lower end of the spectrum. For instance, there are frequent news stories regarding the pressures placed on individuals working in warehouses for organisations such as Amazon and Sports Direct (BBC News, 2016, 2019). Pressure can occur from organisations requiring individuals to undertake work at a specific pace; indeed, this has been specifically cited by workers in the case of Amazon. Similarly, we know that employees are at greater risk of burnout where they have a lack of autonomy to decide how their work tasks are undertaken. Academic research has been conducted into the role of social support and how this can mitigate burnout (for instance, see Thomas and Lankau, 2009) and has found that increased socialisation at work can minimise emotional exhaustion and thus reduce the symptoms of burnout. In essence, where employees feel supported by their immediate managers and direct colleagues, there is less risk of burnout. We must, however, recognise that this is a complex picture with a diverse range of factors that can impact our employees.

PAUSE AND REFLECT

To what extent do you agree, or alternatively disagree, with the causes of burnout discussed so far in this section?

There is, of course, a variety of answers that you could develop in response to the activity above based upon your own personal experiences and views. The common thread to our discussions though is that pressures in work, and working life, are intensifying and this is arguably making burnout rather more likely in the modern workplace.

Schulte (2019) has discussed the pressures in modern working life, and argues that employees frequently have to navigate periods of work–life conflict. She found in her research that whilst almost every organisation spoken to highlighted the importance of work–life balance, work nonetheless had a tendency to spill over into family time, weekends and even periods of holiday. Schulte argues that some individuals wear 'busyness' as a badge of honour, in that it is a sign of being an important employee, and that managers and organisational leaders frequently fail to model good work–life behaviours. In order to tackle what she terms the 'busyness paradox', she argues that first our organisations need to recognise the power of social signals. For instance, if we see an email from our manager at 9 pm we may assume that they have been working solidly all day and that we therefore need to do the same. In fact, the reality may be that the individual in question took time off during the day and came back to answer a message later in the evening. Schulte then argues that we need to build slack into our work schedules. Humans, we know, are notoriously poor at estimating the time needed to complete a task. We tend to significantly underestimate the time that we need in order to reach goals. The argument therefore is that we need to build additional time into our work schedules so that we do not become overloaded with demands. Finally, there is an argument that organisations need to increase transparency over workloads. Whilst there is debate regarding the accuracy of workload planning tools, such systems and processes do provide useful data regarding the allocation of work across a group of individuals. They can therefore provide useful insights to managers and organisational leaders.

Conclusion

Our discussions in this chapter have highlighted the growing importance of wellbeing within work and employment. We have explored a variety of elements of the wellbeing territory and you will have an emerging understanding of some of the key debates within this field. Whilst the academic research base surrounding wellbeing and its impact in the workplace is developing, we do know that organisations investing in wellbeing typically see improvements in productivity, lower rates of sickness absence, and in contexts such as the USA specifically a reduction in the costs of employee healthcare and health insurance. Health and wellbeing at work can be substantially impacted by workplace design. Whilst we know that our workplaces are changing – indeed, there have been notable rises in remote working and the use of co-working spaces – office location and layout, and the environment directly outside the office, must be taken into consideration from a wellbeing perspective. Building on from this, there is substantial debate about the issue of work–life balance and how technology has impacted our ability to 'switch off' from the workplace, with some arguing that technology has directly impacted upon rising levels of stress and anxiety at work. Issues such as presenteeism, leavism and burnout are frequently discussed in both practitioner and academic literature, with commentators highlighting a variety of factors contributing to these issues, including increasing pressure and demands at work, as well as reductions in levels of job security. The debates concerning wellbeing at work, and how we ensure the wellness of our employees, are certainly fascinating and it is important to monitor and stay in touch with them into the future.

References

Anderson, G, Farcomeni, A, Pittau, M G and Zelli, R (2019) Multidimensional nation wellbeing, more equal yet more polarized: An analysis of the progress of human development since 1990, *Journal of Economic Development*, 44 (1), 1–22

Andriote, J M (2018) Young are more aware of mental health, yet less resilient, www.psychologytoday.com/gb/blog/stonewall-strong/201808/young-are-more-aware-mental-health-yet-less-resilient (archived at https://perma.cc/LTK8-3C4Q)

Baptiste, N R (2007) Tightening the link between employee wellbeing at work and performance: A new dimension for HRM, *Management Decision*, 46 (2), 284–309

BBC News (2016) Sports Direct: Former employees speak out, www.bbc.co.uk/news/uk-36864345 (archived at https://perma.cc/DZS5-NFHG)

BBC News (2019) Amazon workers launch protests on Prime Day, www.bbc.co.uk/news/business-48990482 (archived at https://perma.cc/9Q6H-EE97)

Beech, J, Bottery, S, Charlesworth, A, Evans, H, Gershlick, B, Hemmings, N, Imison, C, Kahtan, P, McKenna, H, Murray, R and Palmer, B (2019) Closing the gap: Key areas for action on the health and care workforce, The Health Foundation, March

Burt, E (2017) Presenteeism costs more than £4,000 per employee, www.peoplemanagement.co.uk/experts/research/presenteeism-costs-4000-pounds-per-employees (archived at https://perma.cc/7X2R-CULT)

Burt, E (2018) You call this work? Co-working spaces were once the strict preserve of freelancers and 'creative' types. But now some of the country's

(Continued)

(Continued)

biggest employers are getting in on the act, *People Management*, July/August, 36–41

Casey, G W (2014) Leading in a VUCA world, *Fortune*, 169 (5), 75

Cheng, B H and McCarthy, J M (2018) Understanding the dark and bright sides of anxiety: A theory of workplace anxiety, *Journal of Applied Psychology*, 103 (5), 537–60

CIPD (2011) Research insight: Developing resilience, CIPD, London

CIPD (2016) *Growing the Health and Well-being Agenda: From first steps to full potential – CIPD Policy Report: January 2016*, CIPD, London

CIPD (2019) *Health and Well-being at Work Survey Report: April 2019*, CIPD, London

Colley, K, Brown, C and Montarzino, A (2016) Restorative wildscapes at work: An investigation of the wellbeing benefits of greenspace at urban fringe business sites using 'go-along' interviews, *Landscape Research*, 41(6), 598–615

De Menezes, L M and Kelliher, C (2017) Flexible working, individual performance, and employee attitudes: Comparing formal and informal arrangements, *Human Resource Management*, 56 (6), 1051–70

Felstead, A and Henseke, G (2017) Assessing the growth of remote working and its consequences for effort, well-being and work–life balance, *New Technology, Work and Employment*, 32 (3), 195–212

Haglund, M E M, Nestadt, P S, Cooper, N S, Southwick, S M and Charney, D S (2007) Psychobiological mechanisms of resilience: Relevance to prevention and treatment of stress-related psychopathology, *Development and Psychopathology*, 19 (3), 889–920

Health and Safety Executive (2019) Work-related stress, www.hse.gov.uk/stress (archived at https://perma.cc/N75V-PR7D)

Hemp, P (2004) Presenteeism: At work – but out of it, *Harvard Business Review*, 82 (10), 49–58

Hesketh, I and Cooper, C (2019) *Wellbeing at Work*, Kogan Page, London

Hubbard, J (2014) Well-being: The new priority for business, *Finweek*, September, 10–17

International Labour Organization (2019) Workplace well-being, www.ilo.org/safework/areasofwork/workplace-health-promotion-and-well-being/lang–en/index.htm (archived at https://perma.cc/E7VC-GF59)

Johns, G (2010) Presenteeism in the workplace: A review and research agenda, *Journal of Organizational Behavior*, 31, 519–42

Jyoti, J and Rani, A (2019) Role of burnout and mentoring between high performance work system and intention to leave: Moderated mediation model, *Journal of Business Research*, 98, 166–76

Lee, W H, Ambrey, C and Pojani, D (2018) How do sprawl and inequality affect well-being in American cities?, *Cities*, 79, 70–7

Lipman, V (2016) 3 reasons resilience is an especially valuable quality for millennials, www.forbes.com/sites/victorlipman/2016/01/30/3-reasons-resilience-is-an-especially-valuable-quality-for-millennials/#790e3dbc39bc (archived at https://perma.cc/ZKE4-2FCE)

Lohaus, D and Habermann, W (2019) Presenteeism: A review and research directions, *Human Resource Management Review*, 29, 43–58

Maslach, C, Jackson, S E and Leiter, M P (1996) *MBI: The Maslach burnout inventory manual*, Consulting Psychologists Press, Palo Alto, CA

McCarthy, J M, Trougakos, J P and Cheng, B H (2016) Are anxious workers less productive workers? It depends on the quality of social exchange, *Journal of Applied Psychology*, 101, 279–91

Moore, P and Piwek, L (2017) Regulating wellbeing in the brave new quantified workplace, *Employee Relations*, 39 (3), 308–16

Newton, C and Teo, S (2014) Identification and occupational stress: A stress-buffering perspective, *Human Resource Management*, 53 (1), 89–113

NHS (2018) Overview: Generalised anxiety disorder in adults, www.nhs.uk/conditions/generalised-anxiety-disorder/ (archived at https://perma.cc/NR2G-3BJE)

Oates, W (1971) Confessions of a workaholic: The facts about work addiction, World Publishing Co., New York

(Continued)

(Continued)

Ryff, C D, Singer, B H and Love, G D (2004) Positive health: Connecting wellbeing with biology, *Philosophical Transactions of the Royal Society*, 359, 1383–94

Schulte, B (2019) Preventing busyness from becoming burnout, *Harvard Business Review* Digital Articles, 15 April, 2–7

Stich, J-F, Tarafdar, M, Stacey, P and Cooper, C (2019) Appraisal of email use as a source of workplace stress: A person-environment fit approach, *Journal of the Association for Information Systems*, 20 (2), 132–60

Stringer, L (2016) *The Healthy Workplace*, McGraw-Hill Education, New York

Taylor, E A, Huml, M R and Dixon, M A (2019) Workaholism in sport: A mediated model of work–family conflict and burnout, *Journal of Sport Management*, 33 (4), 249–60

The Health Foundation (2019) Improving the health and wellbeing of 15m NHS employees, www.health.org.uk/news-and-comment/newsletter-features/improving-the-health-and-wellbeing-of-1.5m-nhs-employees (archived at https://perma.cc/WFX8-KKTD)

Thomas, C H and Lankau, M J (2009) Preventing burnout: The effects of LMX and mentoring on socialization, role stress, and burnout, *Human Resource Management*, 48 (3), 417–32

Webber, A (2018) Only minority of employers tackling presenteeism, warns CIPD, despite tripling of cases, www.personneltoday.com/hr/employers-tackling-presenteeism-tripling-cases (archived at https://perma.cc/LV2Q-6WU8)

Weijs-Perrée, M, van de Koevering, J, Appel-Meulenbroek, R and Arentze, T (2019) Analysing user preferences for co-working space characteristics, *Building Research and Information*, 47 (5), 534–48

Whitehouse, E (2019) Let them eat fruit: Or maybe not? Too much employee wellbeing takes place on a superficial level. We asked the experts how HR teams can move beyond gimmicks and help staff genuinely change, *People Management*, March, 42–6

Yeager, D S and Dweck, C S (2012) Mindsets that promote resilience: When students believe that personal characteristics can be developed, *Educational Psychologist*, 47 (4), 302–14

20 Working internationally

This chapter explores the important subject of working internationally. As many of the chapters in this book have demonstrated, the world is becoming increasingly interconnected, and our communities and organisations are becoming more diverse over time. The ability to work successfully in an international context is therefore a foundation of personal and professional success. We must initially recognise that the term 'working internationally' covers a variety of different issues in and of itself. Taking it at face value, we know that a growing number of individuals migrate for improved work, education and economic prospects. We also know that individuals may work as expatriate managers for a more limited time within multinational organisations. Working internationally can, however, encompass a range of other issues. For instance, many organisations, even very small organisations, can operate on a global basis (see, for example, Copeland, 2006; Dimitratos et al, 2014), meaning that individuals may work in global teams without ever leaving home themselves. As organisations become increasingly diverse, the ability to work with others different from oneself is a fundamental contributor to success in the contemporary business landscape.

Initially, this chapter sets the context by revising and recapping some of the key features of international business. The first main section of content explores the role and importance of international business, briefly reminding the reader of the importance of multinational enterprise to the global economy, as well as the various ways in which business systems vary around the world. Cultural diversity will also be discussed during this initial section of content, as it is important to remember that cultural differences have a substantial impact on work and society. Attention then moves to HRM in particular, with the chapter outlining and exploring differences in HRM in four different contexts: the USA, the UK, Japan and China. This 'tour' through HRM systems will demonstrate how HRM practices vary very considerably around the globe as a result of institutional differences and cultural diversity. The expatriate process is then explored in some depth. Expatriation occupies a pivotal position in the international business and international human resource management literatures, and this chapter introduces the reader to the concept, and the intricacies of the expatriate process, whilst also examining a number of reasons why expatriate assignments can fail. To close discussions, the final section of content explores how individuals may successfully operate in the international context, focusing on both cross-cultural communication skills and the notion of intercultural or cross-cultural competence.

Context: the nature of international business

The internationalisation of business presents significant and substantial complexity for individuals seeking to operate in an increasingly global context. When operating internationally, individuals and organisations must contend with differing economic, regulatory and legal systems, which can have a substantial impact on the operations of organisations. But, are differences being overemphasised here? May we argue that, in reality, the essentials of management are rather more universal than we may initially anticipate? Indeed, the overarching capitalist system that dominates much of global business is a powerful constraint on the diversity of individual national business systems, with very similar demands being placed on managers no matter where they operate. Managers must provide returns to shareholders, they must create a successful organisational structure and develop enduring relationships with stakeholders. These aspects do not vary. Having said this, there is indeed substantially more complexity when operating on an international basis as a result of a blend of various political, economic, technological, environmental, socio-cultural and legal issues. Whilst we may argue that the rules of the management 'game' are similar irrespective of context, working and operating internationally are inherently more complex than operating solely on a domestic basis.

Companies that operate across international borders are an important part of the international business arena. Whilst the literature base makes reference to a variety of terms to describe them – 'global companies', 'international companies', 'trans-national businesses' – for the purposes of this chapter the term 'multinational company' (MNC) is used because it is arguably the most popular and inclusive term. In terms of a definition, a multinational company describes a single organisation that operates internationally, employing people in more than one country. Whilst we may think that multinational companies are a product of recent technological developments and increases in global business activity, they are not a new phenomenon. Exploring the historical record, the earliest MNCs were born from the trading and economic activities of the major European powers at the time, with an example being the British East India Trading Company (Blakemore, 2019). These organisations were in many cases very successful indeed at providing returns to shareholders, and in many cases had very long histories. What we would characterise as modern MNCs, however, have been made possible by developments in technology such as factory and shipping systems, advances in computing and the internet, and deregulatory activity that has taken place across the globe.

Adding complexity to the operations of MNCs are the distinct business systems that exist around the world (Whitley, 1999). As Whitley, and other authors such as Hall and Soskice (2001) note, there are divergent models of capitalism. Whereas, for instance, the UK and USA are characterised as liberal market economies, where companies are permitted substantial freedom and central regulation is limited, the coordinated market economy model that dominates much of Europe operates a more stringent stakeholder-based model. Trade unions therefore, as an example, have substantially more power in these contexts, and there tends to be more central regulation of economic activity. The management practices that MNCs use in one context, for example the USA, may therefore be entirely inappropriate in another context, such as Germany. The following case study demonstrates these differences in the context of the airline industry.

CASE STUDY 20.1

Divergent national business systems, the airline industry and 9/11

The terrorist attacks in New York on 11 September 2001 (9/11) had substantial impacts and implications for a variety of areas of our lives, and the operations of our organisations. Perhaps nowhere experienced the effects more profoundly, at least in the short term, as the airline industry. Immediately following the events of 9/11, airlines around the world grounded their fleets. Flights were stopped, initially for security reasons, but even upon the resumption of air travel shortly thereafter, passenger numbers were substantially below previous levels, with a significant reduction in business for the airlines. This meant that financial turnover was reduced, and many organisations began to experience losses. How did the various airlines around the world react?

The reaction differed, and these differences can be traced back to the different national business systems in which they operated. Carriers such as British Airways in the UK and American Airlines in the USA focused on immediate cost reduction. The airlines cut back services and made large numbers of roles redundant in an effort to shore up short-term share prices and demonstrate to the market that they would return to profitability as swiftly as possible. This was a consequence of the prevailing liberal market economy (LME) philosophy present in the UK and the USA, where there was a shorter-term focus and organisations were more sensitive to falls in share prices. By contrast, Air France and Lufthansa, major airlines in France and Germany respectively, felt lower shareholder pressure due to the prevailing coordinated market economy (CME) system. This enabled a more coordinated institutional response where working hours were reduced and routes were restructured in an effort to protect employment.

What we see here is that the presence of different national business systems led to very different organisational reactions to the same event and pressures. What we are not arguing here is that one approach is superior to the other; we are simply noting that they are different. Whilst this example dates from 2001, we may see similar events transpire in response to the Covid-19 pandemic of 2020. It will be interesting to observe the reactions of organisations located in different national business systems and track the actions that they take in response to the impact of the virus.

Just as there are different varieties of capitalism, legal systems vary substantially across the globe – the Romanistic system is very different from the Germanic, which itself is very different from the Chinese system. Again, management practices that work very well in one context cannot be uprooted and transported overseas without having the same underlying legal principles in place. Finally, cultures vary very substantially across the world as well. Considerable academic interest has been invested in understanding both national and organisational culture, and the work of researchers such as Geert Hofstede (Hofstede et al, 2010) and Fons Trompenaars (Trompenaars and Hampden-Turner, 1997) is very well known and widely cited. Whilst it is not the aim of this particular book to provide a detailed overview of national and organisational culture, it is worth briefly recapping the work of Geert Hofstede in particular.

Geert Hofstede was very much a pioneer of research into culture (Hofstede et al, 2010). He conducted extensive studies, the first of which was within the American MNC IBM, where he captured data from a large number of managers and employees around the world. This led to the development of the first four scales within his framework, with the fifth and sixth scales emerging from more recent research. Whilst Hofstede's work is not without criticism, it has been replicated by others and is very widely cited and used in both business and academic fields. Table 20.1 provides an overview of his dimensions of culture.

Table 20.1 Hofstede's cultural dimensions

Dimension	Description
Power distance	Attitudes towards authority, the distance between individuals in a hierarchy and the acceptance (or otherwise) of the uneven distribution of power.
Uncertainty avoidance	Tolerance of uncertainty and ambiguity and attitude towards risk taking.
Individualism/collectivism	The extent of independence or interdependence in a society, and whether identity is grounded within the individual or the group.
Masculinity/femininity	Preferences either for achievement, material rewards and distinct gender roles, or cooperation, quality of life and less defined gender roles.
Short-term/long-term orientation	The relative concern between the present and the future, and a focus on either short-term or long-term goals and objectives.
Indulgence/restraint	The extent to which a society permits the relatively free gratification of needs versus the control of impulses and desires.

Hofstede argues that cultures vary across these dimensions, and tools on the Hofstede website (Hofstede Insights, 2020) enable comparisons between countries. In short, the key learning point to take from this section of text is that the international business landscape is complex, with companies and individuals operating across international borders needing to navigate through a number of different issues if they are to be successful. The next section of content very much builds on this introductory material and discusses how different national varieties of HRM have arisen in response to different blends of institutional and cultural factors.

National varieties of HRM

Having refreshed understandings of international business, and some of the key underlying points concerning cultural diversity and institutional difference, attention now turns to how these differences manifest themselves in relation to HRM. This portion of the chapter will undertake something of a whistle-stop tour through various HRM systems, explaining the differences that can be seen in various contexts. We begin by exploring HRM as it operates in the United States of America, commonly seen as the birthplace of the field, before exploring in turn the UK, Japan, and finally China.

United States of America

The USA is often argued to be the birthplace of what we understand as human resource management (Harzing and Ruysseveldt, 2003). It is argued that HRM as a concept emerged as a response to a number of factors, including an increasingly competitive business environment, both from indigenous firms and overseas competitors, allied to other issues such as declines in overall levels of trade union membership and changes in the broader labour market brought about by the rise of so-called 'white collar' occupations. These changes stimulated the need to manage individuals in a different way, and hence what we see as HRM was born. The USA tends to have a very weak system of employment protection. It is an environment characterised by hard models of HRM whereby labour is obtained as cheaply as possible, exploited as fully as possible, and then let go if no longer needed. This is evidenced by the presence of 'at will' employment in many parts of the country.

Discussed by a variety of scholars and commentators, including Gely et al (2016), Radin and Werhane (2003), and Sallmen (2018), the 'at will' doctrine essentially entitles individuals to a day's pay for a day's work, with no requirement for an organisation to continue an employment relationship if they deem it unnecessary. Individuals who work under employment at will arrangements can find their contracts terminated with no notice, with organisations not required to provide a reason to justify their termination decision. As Gely et al (2016) note, employment at will arrangements are heavily debated and there is considerable interest in how such employment arrangements will persist, or change, in the future, with many arguing that such contractual relationships are inherently unfair (Radin and Werhane, 2003). These arrangements have persisted because of the underlying hard model of HRM present in the USA, and the strong non-interventionist position adopted by successive governments. Having said this, it is important to note that the USA has strong anti-discrimination legislation, with substantial penalties for individuals and/or organisations found to have breached these requirements.

Whilst a hard model of HRM has persisted in the USA for some time, we do see changes occurring in the competitive landscape. These changes are causing a move towards softer models of HRM based more on commitment, engagement and employers setting out clear value propositions to attract the most talented and capable employees. Examples of this include Facebook, Google and other major technology companies that are expending significant effort to demonstrate that they are attractive places to work. Whilst these organisations are having notable success, others are finding it substantially more difficult to transition from the hard model of HRM that has persisted for some considerable time. Indeed, there are questions as to whether it is possible for a soft model of HRM to dominate in a liberal market economy.

United Kingdom

Turning back home, the UK has evolved a distinctive model of HRM, which has been shaped, in part, by both national legislation and legislation arising from the European Union. As a result of this, it will be interesting to observe what, if any, changes in employment legislation and practice appear over time as a result of the UK's departure from the European Union at the end of January 2020. At the time of writing, the UK government has indicated that it will protect employment rights, but it is arguably inevitable that we will observe some degree of divergence between UK and EU employment legislation, and perhaps the broader approach to HRM, in the coming years (for more information see CIPD, 2020a).

Historically within the UK, the state has been reluctant to intervene in the employment relationship, meaning that the UK has generally utilised a 'voluntarist' arrangement with regard to employment practices. The economy of the UK is dominated by the model of private enterprise and the country has a large service and knowledge-based sector, with employers having substantial latitude when it comes to developing employment practices. Legislation, from both UK and EU sources, has, however, had a substantial impact in shaping employment practices and approaches to HRM.

PAUSE AND REFLECT

Please take a moment to reflect upon employment legislation within the UK. Can you find any specific areas where UK-based employment law has been influenced by legislation arising from the European Union?

How did you get on with the question above? The intent here was to offer you an opportunity to reflect upon the sources of influence upon employment legislation, and broader models of HRM utilised within the UK. There are several examples where the UK's prior ties with the European Union have directly impacted employment legislation. One good example of this is the Working Time Directive (European Commission, 2020a), which sought to manage working hours across the EU. EU member states were able to implement this directive in a locally sensitive manner, but you can see clear linkages between the contents of the EU directive and the UK's rules regarding working hours (Gov.uk, 2020a). This is just one example of the influence of EU rules upon employment legislation and practice within the UK; there are many others.

Moving on, the UK model of HRM is further characterised by a substantial focus on the individualisation of the employment relationship, with substantial use of practices such as individual performance-related pay. UK organisations have, on average, underinvested compared to international competitors in training and learning programmes, and this has contributed to something of a productivity puzzle in recent times. Media outlets frequently run stories and headlines arguing that the UK continues to demonstrate poor growth in productivity, which, in turn, negatively impacts economic growth. The causes of this are diverse, but include an underinvestment in the workforce via learning and development, poor levels of investment in new technology, resistance to change and low levels of funding allocated to research and development. Having said this, since 2018 there have been signals that this trend may be reversing (see Partington, 2018), although data from the Office for National Statistics demonstrates that this trend is not consistent across the economy (ONS, 2019a).

Human resource management as a function area does, however, tend to punch above its weight in the UK. The CIPD (CIPD, 2020b) is the largest global body of HRM professionals and it is influential, frequently being cited by a variety of media outlets. The CIPD has produced a substantial amount of research into work and employment, and its influence assists in raising the profile of the HRM profession. It is often said, however, that within the UK we tend to see islands of best HRM practice (which are often foreign owned) in a broader sea of pragmatism.

Japan

Turning east, we come to Japan, and a very different model of HRM. Japan established many of its employment and working practices in the years following the Second World War, in the late 1940s and 1950s. The original foundation of its employment system was built on a number of key principles, including the promise from organisations of lifetime employment, wages that were determined by seniority, and the presence of trade unions within each enterprise. As a result of the desire to maintain lifetime employment for individuals, organisations historically paid very careful attention to the selection of staff, with individuals screened against numerous criteria before any offer of employment was made. To this day, Japanese organisations do not tend to dismiss employees unless absolutely necessary. There are, however, criticisms of the lifetime employment philosophy, with individuals such as Tanaka (1981) arguing that the presence of surplus workers can be seen as unemployment within the organisation itself. Whilst commentators in the late 1990s mused over the decline of lifetime employment, it has been argued to be a central pillar around which many other successful Japanese HRM practices have evolved, such as slow evaluation and promotion, collective responsibility and multi-skilling (Hirakubo, 1999). Pointing to a decline in lifetime employment more recently, however, Ono (2010) has highlighted that a minority (20 per cent) of individuals are engaged under lifetime employment practices, with this number continuing to fall as organisations strive to become more flexible.

Japanese organisations have embodied, and continue to embody, the principles of Atkinson's (1984) flexible firm model, typically utilising a permanent 'core' staff, supplemented by a more flexible 'periphery' where individuals are engaged through fixed-term arrangements, part-time contracts, consultancy agreements and so on. In some cases this has enabled lifetime employment practices to remain for the 'core' workers, with organisations able to be flexible and responsive to changing market conditions via their changing use of 'peripheral' staff. We have also seen a number of management practices emerge in Japan and spread very rapidly around the world. Quality circles, a form of bottom-up management, where employees regularly meet to discuss and solve problems related to their jobs, and just-in-time production, which focuses on strict inventory and supply chain management, are examples of practices attributed to the success of Japanese organisations that have been globally influential.

Despite the historical success and substantial growth of the Japanese economy, the Asian slowdown of the late 1990s significantly affected the country, with many years of poor (or negative) growth since (Ono, 2019). As a result of this, we have seen many Japanese firms begin to experiment with what we would typically see as 'western' HRM practices, such as performance-based pay and competitive promotion. To what extent this experimentation will continue, and whether we will see the emergence of a more distinct hybrid 'Japan–west' model of HRM, is an open question at this time. What we have certainly seen through the case of Japan, however, is that the country has exported many successful HRM and broader management practices, and that the country itself has been influenced by HRM and management systems that have developed in other parts of the world.

China

China is a very interesting case indeed. There is presently substantial debate regarding HRM practices in China, and specifically whether a distinct Chinese model of HRM has emerged, or will emerge at some point in the future. What is certain is that the country again has a very different model of employment and working practices that, similar to the Japanese system, has previously offered lifetime employment and a significant system of welfare support, known as the 'iron rice bowl' (Xie et al, 2020). Whilst reforms have sought to modernise employment and working practices, the core tenets of the iron rice bowl are still seen to persist in Chinese society (Xie et al, 2020).

Historically the Chinese government has exercised strict controls over industries. Many enterprises were state owned and in some cases jobs were inherited (Berkowitz et al, 2017), with roles passing from father to son, or from mother to daughter. As Berkowitz et al (2017) explain, in this context it can be very difficult for an employee to leave a job, and in many cases impossible for an organisation to fire an unwanted worker. Alongside this, HRM policies, such as those connected with pay, have tended to be set centrally and applied across entire industries. Wage scales in China tend to be complex but with low differentials between the lowest and highest paid within an organisation – an outcome of the prevailing collectivist (Hofstede et al, 2010) culture. Organisations also typically provide substantial employee benefits, including healthcare, paid company holidays and company housing. Due to the centrally determined nature of work and employment practices, managers have traditionally had limited discretion over how to manage their operations, although this has changed somewhat in recent years.

What we currently see in China, driven by economic reforms, are increasing levels of competition (both from locally owned firms and overseas enterprise). Managers are being given substantially more discretion to manage their operations in a way that they see fit, and there is now substantially more open competition amongst individuals for jobs (*The Economist*, 2020), promotions and pay increases. Western practices such as performance appraisal and performance-related pay are also being adopted in some contexts, although adapted to suit the cultural environment. Trade unions in China do, however, continue to play a substantially different role to that which is commonly observed in western countries. The primary purpose of trade unions in China is to mediate in disputes and provide access to welfare support for staff. Additionally, trade unions provide a forum in which workers can voice their ideas and concerns, but there is no ability for trade unions to call a strike, or apply other such pressure to influence managerial decision making. As mentioned at the start of this section, there continues to be debate as to whether a distinct Chinese model of HRM exists, or will emerge in the future. Commentators are speculating about various future directions for Chinese HRM, and it will be interesting to observe these developments as they unfold over time.

We now build on the content introduced during this portion of the chapter by considering some of the global variations within HRM practice.

Global variation of HRM practice

As the previous section has demonstrated, HRM practices vary very substantially around the globe. Whilst there are debates concerning the extent to which we may see practice in any single area of HRM converging towards a dominant model, differences are likely to persist. This is because HRM as a functional area is very sensitive to changes in the institutional and cultural context (Wilkinson and Wood, 2017). To illustrate this in further detail, variations in practice associated with employee reward and employee voice are now discussed in greater depth.

Employee reward

Employee reward is a particularly pertinent area to discuss, as wage levels, expected employee benefits and the tolerance of differentials between the rewards accrued by those at the top of an organisation's hierarchy versus those towards the bottom vary substantially across contexts. Taking first of all average wage levels, the Office for National Statistics reports that the UK's median weekly earnings for full-time employees reached £585 per week in 2019 (ONS, 2019b), whereas the US Bureau of Labor Statistics (2020) highlights that the median weekly earnings for full-time employees in the USA were $1,002 (approximately £760) in the second quarter of 2020. Contrast this with the average weekly wage in China of 1,585 yuan (approximately £180) (Statista, 2018), or the average weekly earnings in India of 5,000 rupees (approximately £50) (Statista, 2017). Whilst these are clearly not direct equivalences as the data were collected at different times, and along differing scales, the numbers do show rather substantial divergence around the world. Further to this, in some contexts, such as the USA and the UK, it is very normal for pay at the very top of an organisation to be orders of magnitude greater than average employee earnings.

For example, the CIPD reported in 2019 (CIPD, 2019) that the average FTSE 100 chief executive earned 117 times UK median earnings in 2018. The same is, however, not true around the world. In China, a substantially more collectivist society, wages may be set along complex scales, but differentials between highest and lowest earners are substantially lower. In this context we would simply not see a senior leader earning 117 times the wages of an average employee.

Employee benefits vary considerably around the world as well. In the UK, full-time employees are entitled, by law, to 5.6 weeks holiday per year; in practice this works out to be four working weeks plus bank holidays (Gov.uk, 2020b). In the United States, however, there is no requirement under US law for employers to offer paid leave of any kind, although in practice many organisations offer two weeks (and in some cases more) paid time off (Edwards and Rees, 2017). In China, by contrast, employees who have worked for more than one year but less than ten are entitled to five days' paid leave; for those who have worked for between ten and twenty years this rises to ten working days, and to fifteen days for those who have worked for more than twenty years (Thompson Reuters, 2020). In addition to this, Chinese organisations frequently organise company paid holiday trips and may also provide paid housing for their staff too (Edwards and Rees, 2017). Again, legal entitlements and cultural norms vary very substantially and what is deemed appropriate and credible in one location may be inappropriate (or indeed illegal) in another. For instance, an American MNC setting up a base in the UK could not apply a holiday policy developed in the USA as this would typically not meet the UK's legal minimums. The organisation could, however, bring across a model of company-paid private health insurance for UK staff if they wished. Typically, American organisations have very much more developed health insurance as the USA lacks a national health service as found in the UK. In the UK context such an offering may be seen positively by staff and could act as an employee attraction and retention tool.

Employee voice

Employee voice is another area where HRM practices vary substantially around the world. There are significant differences in the extent to which employees are (and indeed expect to be) involved in decisions that affect their employment. In terms of a definition, Edwards and Rees (2017: 253) provide a clear overview of voice, arguing that it is 'a general, over-arching concept that refers to the various ways in which employees take part in decision-making in organisations and also the degree of influence they have over the same'. In some contexts, such as the UK and the USA, where there is lower power distance (Hofstede et al, 2010), employees typically expect to be involved and are more willing to vocalise their suggestions or concerns directly to management. In other contexts, notably where power distance is higher, such as many Asian contexts, employees are considerably less willing to speak up (Hsiung and Tsai, 2017). In these contexts employees expect to be directed by managers who have stronger perceived authority. This is captured to an extent in the typical role of trade union bodies in China, which, as we saw earlier in this chapter, play a role in mediating disputes and provide a forum for the collection of ideas, but do not have the ability to call for strike action or apply other such pressure to organisations. This role is considerably different from the one that trade unions typically play in western societies.

Legislation can also impact substantially on employee voice (Edwards and Rees, 2017). As an example of this, the EU has issued a number of directives over time, such as those connected with collective redundancies in the late 1990s, the transfer of undertakings (TUPE) consultation measures in the early 2000s, and the European Works Council Directive of 1994 (revised in 2009), which have all directly targeted the expansion of employee voice mechanisms, with the latter being particularly interesting. The European Works Council Directive enables the formation of bodies within organisations representing the European employees of a company and can be triggered by employers themselves, or by a request from at least 100 employees operating in two countries (European Commission, 2020b). The intent here from the EU was to provide stronger avenues of employee voice as companies became increasingly transnational, and there are many examples in the media of the role played by employee works councils in shaping key business decisions. An example of this is provided in the case study below.

CASE STUDY 20.2

BMW's Works Council

BMW's Works Council has a long history. It was formed in 1996 (Whittall, 2010) as a result of EU-wide legislation dating from 1994 and has been involved in the operation and management of the company since that time, having input into major strategic decisions. BMW's Works Council draws its membership from the wider workforce, with employees elected onto the board at regular intervals. The Works Council currently has 23 members (BMW Group, 2020), and has a substantial role in developing and monitoring agreements between the company and its employees, together with shaping the wider direction of the business alongside senior managers.

One of the biggest challenges facing all automotive companies at the present time is the electrification of vehicles, with substantial debates occurring around the specific architectures that cars will use in the future. Some manufacturers are developing electric-only platforms, whilst others are developing models that can be powered either by electric, hybrid or conventional internal combustion engines. These decisions are complex, and will inevitably impact the future success of car makers.

Until recently BMW has pursued a course of action that has involved hybrid platforms – ie car models that can have electric-only propulsion, a combination of electric and conventional power, or conventional diesel or petrol engines. Its Works Council has, however, pushed the organisation to pursue electric-only platforms, arguing that developing their own proprietary technology will be an important strategic advantage in the future (Taylor, 2020). Employee representatives have argued that the electric-only platform provides the organisation with greater future opportunities, and it appears that the organisation will formally pivot its future strategy to align with the views emerging from the Works Council.

As the case study demonstrates, the intent behind the European Works Council Directive is not solely the upholding of employer–employee agreements related to collective bargaining arrangements, health and safety issues, and so on. When operating effectively, these groups should collaboratively work with company managers and other stakeholders to positively shape the future direction of the organisation. It will be interesting to follow the above example over time to see whether, and to what extent, BMW realises improved results as a consequence of the change in strategic direction prompted by its Works Council.

Expatriation

Expatriation, as a concept, occupies a pivotal position within the literature base surrounding both international management and international human resource management. Whilst there are differing interpretations of the concept, Hollinshead (2009: 72), drawing from Harzing (2004), defines it as 'a parent country national (PCN) working in [the] foreign subsidiaries of the MNC for a pre-defined period, usually two to five years'. International organisations may decide to use expatriate managers for a number of different reasons (Reiche and Harzing, 2011). These reasons may include where organisations cannot locate knowledge or skills within the local workforce, where there is a desire to develop managerial competence and experience within the expatriate manager themselves, or to strengthen the lines of communication within the various disparate parts of the international organisation. In truth, expatriate assignments may achieve multiple objectives at the same time, with Reiche and Harzing (2011) indicating that these assignments may have various 'control', 'knowledge transfer' and 'network building' elements to them. The expatriation process, irrespective of its broader aims and objectives, consists of three linked steps:

- Pre-assignment preparation.
- Assignment.
- Post-assignment (repatriation).

Initially, there is a phase of selecting the appropriate individual for the international assignment and preparing them effectively for the experience. This is then followed by the start and continuation of the international assignment itself, before the international manager is brought back home at the end of the process. This final stage (repatriation) is, however, often overlooked and substantial issues can occur at this stage caused by difficulties in reintegrating the international manager back into the home country environment. We will discuss these difficulties in greater depth at a later point in this section.

The first element of the pre-assignment preparation stage is the recruitment and selection of the individual who will undertake the expatriate assignment. This process follows essentially the same process as domestic recruitment and selection in that we prepare a role description and person specification, before advertising and then selecting an individual from the pool of applicants (who could be internal to our firm or from the broader domestic labour market). Having said this, we must keep in mind issues of cultural adaptability as well as technical or skill-based requirements during the recruitment and selection process. It is vital to recognise that while technical skills and knowledge are important, expatriate assignments often fail because individuals lack cultural sensitivity and the ability to operate effectively in a different context (Haile and White, 2019). Following the recruitment and selection of an appropriate individual it is then important that the organisation provides appropriate learning and development interventions to ensure that the expatriate manager is properly equipped to undertake their overseas role. Frequently organisations (and expatriate managers themselves) underestimate the level of adjustment to a different culture and working environment, and where effective learning and development provision is not on hand it becomes increasingly likely that the expatriate assignment will not achieve its aims and objectives. When determining the learning and development provision needed, specialist trainers must consider issues such as the novelty of the job and culture, and the degree to which the expatriate manager must interact with others in their role (Mendenhall et al, 1995). A 'one size fits all' approach to learning and development is highly unlikely to be successful because each expatriate assignment will have a unique combination of factors.

PAUSE AND REFLECT

Please pause here and reflect for a moment or two. Is it sufficient for an organisation to provide training only to the expatriate manager? Are there other individuals here for whom the organisation should consider providing learning and development interventions?

How did you get on with the questions above? The key underlying point here is to think very carefully about the impacts of the expatriation process. There is indeed an argument that could be made for training individuals in the host country in cross-cultural awareness and other such important issues, and there is also a larger issue here related to the expatriate's family. Consider, for instance, that the expatriate manager selected for an assignment by your organisation has a spouse or partner, and perhaps one or more children. Frequently organisations will send whole family units overseas rather than separate an expatriate manager from their family bubble for the period of the international assignment. If this is the situation, and irrespective of whether the spouse or partner of the expatriate manager is employed by the same organisation or not, learning and development must still be provided. Where we might provide the expatriate manager with language training, development events designed to improve cultural awareness, or other provision aimed at improving their ability to operate in diverse environments, we must also provide those learning and development opportunities to the family unit, if we are sending the unit as a whole overseas with the expatriate manager. As an aside, should we seek to send a spouse/partner and children with the expatriate manager, we must also then give consideration to issues such as housing, schooling, a work visa and job opportunities for the spouse/partner.

Whilst recruitment, selection and training are crucial issues at the pre-assignment stage, we must also consider reward management. Rewarding international managers can be highly complex and organisations must give consideration here to issues surrounding equity, both in relation to the internal/domestic labour market and the pay structures/systems in the host country. At the most basic level reward packages must be legally compliant, but beyond that the organisation must decide whether to utilise a 'balance sheet' or 'going rate' approach (Özbilgin et al, 2014). In brief, the 'balance sheet' approach seeks to base market value in the *home* country, maintaining the international manager's existing living standards and reward package, with the addition of any supplements or incentives deemed necessary, or agreed through contractual negotiations. By contrast, the 'going rate' approach bases market value in the *host* country, with the reward package for the international manager based on local pay scales and the reward arrangements of other expatriates. The advantage of this arrangement is that it is arguably a more straightforward solution, but it lacks parity with the pay arrangements at home. MNCs can, and do, successfully utilise both approaches to design reward packages for international managers (Watson and Singh, 2005). Decisions regarding the appropriateness of either approach will be influenced by a combination of firm-specific and labour market factors (both home-based and host country-based).

As mentioned at an earlier point in this section, repatriation can be considerably more complex than we might initially anticipate. Organisations tend to assume that the complexities surrounding expatriation arise from the initial selection and preparation, followed by the assignment itself, with repatriation, ie 'coming home', the most straightforward part of the process. In reality this is often not the case. What organisations often fail to realise is that while the expatriate is away their home country undergoes changes, whilst, at the same time, the mental 'maps' that the expatriate holds also change as they become accustomed to working in what might be a very different context. What this can therefore mean is that it can be as difficult to return home as it is to begin an overseas assignment. If the expatriate manager is not effectively reintegrated then they may be less effective at their job role, or may leave the organisation altogether, negating other positive outcomes from the assignment. Having said this, the expatriate process, when properly managed, can hold significant benefits for managers and organisations alike (Reiche and Harzing, 2011). It can bring significant career development for managers, organisations can benefit from new knowledge flows, and there can also be better coordination between teams as a result of strengthened relationships and stronger personal bonds within the organisation.

Expatriate failure

The subject of failure within expatriation is a much-discussed topic (Reiche et al, 2019). There has been a variety of academic studies exploring the rates of expatriate failure, and both the causes and impacts of it. Studies conducted by different academics arrive at different results, and this is largely because they differ in terms of their definition of 'failure' (Reiche et al, 2019). Without a common understanding of failure, it is therefore difficult to draw comparisons between the various pieces of research exploring expatriate failure. 'Failure' in expatriation is a term that can mean a number of things. We might argue that failure is simply where an expatriate manager returns early from the overseas assignment, or where the individual concerned leaves the organisation entirely, either during or at the end of the assignment (Hollinshead, 2009). There is, however, a more subtle issue that can easily be overlooked: underperformance in one's role. If the expatriate manager is not able to meet agreed objectives, does that mean that the assignment should be termed a failure? Of course, there may be factors outside the expatriate manager's direct control (for instance, market conditions) that influence relative success during the overseas assignment, but underperformance should be scrutinised as this can indeed point to failure within the expatriate process.

As far back as 1995 Anne-Wil Harzing discussed variations in the failure rates of expatriate assignments. Indeed, she noted that underperformance, or simply performing 'adequately' during the overseas assignment, was potentially more damaging than expatriate managers who return early. She argued that failure rates quoted in various research studies ranged from 5 to 10 per cent up to 50 per cent of assignments, although the lack of reliable data suggests that we should refrain from attempting to put an exact percentage figure on the numbers of overseas assignments that fail in practice. Much more recently, Haile and White (2019) have explored expatriate failure within multinational corporations, and whilst they again do not seek to ascribe a percentage figure to the phenomenon, noting that there can be several competing definitions of failure, they do cite research from Peng (2018) highlighting that the costs of each episode of expatriate failure can be very significant, in the order of $250,000 to $1,000,000 per individual. These costs are substantial and mean that the underlying causes of expatriate failure must be understood.

Most reasons for expatriate failure are connected, in some way, with cultural issues (Harzing, 1995; Haile and White, 2019). These may include a lack of adjustment or sensitivity to the new culture, a lack of acceptance from locals and/or language difficulties. In essence, an expatriate manager may find they are unable to operate successfully in the overseas context. Often, expatriate failure is driven, at least in part, by poor selection and preparation. As noted earlier in this section, organisations must select individuals who display high levels of cultural sensitivity and adaptability, not just excellent technical skills. Allied to this, where organisations do not provide appropriate training, we know that failure is more likely. Having said this, no matter how well prepared an individual may be, they will almost always encounter 'culture shock' as they transition to the overseas working environment. Culture shock, as explained by Fitzpatrick (2017), is caused by exposure to a new environment and the consequential need for us to adjust our behaviour. Organisations can reduce the extent of culture shock through proper preparation (eg cross-cultural training, language training, pre-visits), but it is argued that it is like the common cold: there is no defence, and individuals can catch it again and again as they move between contexts. Indeed, Howard (1974) indicates that returning expatriates can experience culture shock when they return to their home base. This is the result of the expatriate having become acclimatised to working in the overseas environment and consequently needing to go through a period of adjustment as they return home.

The key learning point to draw is that the risks and costs arising from expatriate failure are significant, and organisations must therefore carefully consider how they recruit, select and prepare individuals for overseas assignments. Organisations must consider adaptability and cultural sensitivity alongside technical ability during the recruitment and selection phase, properly prepare individuals following the selection decision and recognise the broader complexities of the employment relationship with individuals who are posted abroad.

We now move on from expatriation to explore the last main concept to be introduced in this chapter: how we might work effectively in an international context.

Working effectively in an international context

A key theme that has run through discussions within this chapter is that our increasingly connected world means that more and more of us will be required to work in an international context. For some, as discussed in the last section, this may involve formal periods of overseas working via expatriate assignments; for others, 'working internationally' may take on different forms, for example working in increasingly diverse teams with individuals from around the world. Irrespective of the form of international working, the increasing need to operate successfully in an international context means that growing emphasis is being placed on cross-cultural communication skills and what is known as 'intercultural competence', two subjects which will now be explored in some depth.

Cross-cultural communication

The ability to communicate effectively across contexts is perhaps the biggest challenge when working internationally (Browaeys and Price, 2019). The well-known cross-cultural practitioner Pellegrino Riccardi (2020) describes the challenges involved in cross-cultural communication as often being a matter of perception; what is 'accepted' and 'familiar' in one context is very different in another. Individuals may perceive things in very different ways, and therefore to operate successfully in an international context individuals must develop an understanding of other cultures, and how these influence communication in particular. Drawing from Hofstede's cultural dimensions (Hofstede et al, 2010), and in particular power distance, to illustrate this issue, in a context characterised by low power distance such as the UK or USA, it would be normal for employees to speak up and potentially challenge the decisions of their managers. In a context typified by high levels of power distance, such as China or many other Asian cultures, however, such actions would be considered very unusual, and very much against the cultural norm. What we are not doing here is ascribing a value judgement to these situations; there is no 'right' or 'wrong'. The key learning point is to recognise that cultures vary substantially around the world, and these variations impact communication norms and expectations.

Whilst the aim of this chapter is not to provide an overview of communication studies, it must be recognised that communication is not just the words that are said, or the message that is conveyed through written text. Communication has many elements, including the message itself, but also body language, and the tone and pace of voice. Non-verbal communication varies across cultures, just as expectations concerning verbal communication change, and can be a substantial source of irritation where individuals misunderstand what is 'accepted' and 'familiar' (Pellegrino Riccardi, 2020) in a given context. As an example of this, within the French culture it is very normal to use expansive gestures whilst speaking, whereas in the Netherlands the Dutch are generally much more reserved, using significant gestures only when emotions are strong (Browaeys and Price, 2019). Without an understanding of the variations in non-verbal communication, an individual from the Dutch culture may perceive the French as overly emotional, and an individual from a French cultural background may perceive the Dutch as distant and emotionally cold. The key word here is 'perceive'. Again, a value judgement is not attached to these differences in non-verbal communication; there is no 'right' or 'wrong', it is simply that different cultures have evolved to have substantial differences in non-verbal communication.

Just as there are differences in non-verbal communication across cultures, perceptions of personal space also vary (Browaeys and Price, 2019). Personal space often has different definitions, but can typically be seen as the invisible zone that surrounds an individual. When another individual enters this zone uninvited, we tend to feel very uncomfortable, with the size of what we deem personal space influenced by a large range of factors including gender, age and cultural background. Some cultures such as the Arab peoples, Indians and individuals from Pakistani descent typically have small personal space. Individuals from these cultures expect close contact between people. By contrast, individuals from Asian or North European cultural backgrounds typically have large areas of personal space, and expect greater distance between people during interactions. If one's personal space is invaded then communication between the parties can be negatively impacted as this can cause substantial levels of stress and unease. Again, it is important for international managers to understand the differences expected in various cultural contexts, taking time to understand the behavioural norms and expectations.

PAUSE AND REFLECT

As an aspiring international manager, what might you do in order to increase your awareness of communication differences between cultures?

The question above was intended to provide an opportunity for you to consider how you may begin developing your cross-cultural communication skills. One of the very best ways to develop cross-cultural communication skills is immersion in other cultures, through travelling, reading and developing a diverse friendship network where you can get to know more about cultures other than your own. For instance, it may be possible to join local groups or networks connected with a particular common interest, or a professional network such as the CIPD's series of local branches around the UK. The key learning point to take from this section of the chapter is that communication (both verbal and non-verbal) varies substantially across cultures. Individuals cannot assume that their behavioural norms are the same as those of others.

Intercultural or cross-cultural competence

The final subject of investigation in this chapter is the notion of intercultural or cross-cultural competence. Research has highlighted that intercultural competence is an important factor in the success of international managers (Dodd, 2007), and the increase in cultural heterogeneity around the world has meant that the ability to work in a culturally diverse environment is increasingly prized by many organisations (Dias et al, 2020). The term intercultural or cross-cultural competence is, however, rather ambiguous, as Dias et al (2020) explain, with some individuals seeing it as being 'sensitive' to different cultures, whereas others define it as working effectively in a culturally diverse environment. Charleston et al (2018: 3069), drawing from Adler and Bartholomew (1992), provide an inclusive explanation, arguing that 'a cross-culturally competent person is someone who can learn about foreign cultures; perspectives and approaches; is skilful in working with people from other cultures; can adapt to living in other cultures and knows how to interact with foreign colleagues'. Within this definition there are three key aspects: the ability to learn about overseas cultures, being skilful in working with a range of diverse individuals and having the ability to adapt to living in different environments.

Charleston et al (2018) explain that working in a cross-cultural context is complex, and that individuals vary in their ability to develop intercultural competence. They go on to argue that in order to develop cross-cultural competence individuals must be curious about themselves and others, they must be passionate about their work and learning about different cultures, show adaptability and a tolerance of uncertainty as well as open-mindedness, and they must be able to display empathy for others. In addition, the researchers also argue that individuals must be excellent communicators who can adjust their style of communication to suit a given context. With these points in mind, it is apparent that individuals predisposed to working effectively in an international context will be very aware of themselves, and also sensitive to changing contexts around them.

A first step in developing cross-cultural competence is expanding one's knowledge of other cultures, again through a variety of methods including reading and assimilating information about other cultures, and immersing oneself in a variety of different cultural environments. Again, it is important to understand that culture is a relative term. There is no 'better' or 'worse' culture; as mentioned previously in this section of the chapter, we are not seeking to ascribe a value judgement to differing cultures – through our process of learning we are simply seeking to understand more about other cultures rather than focusing on arriving at a judgement about them. In order to further develop our cross-cultural competence, we may also intentionally seek out opportunities to work in diverse groups, with others from backgrounds different from our own. From an HRM perspective we may wish to think about how we may encourage diversity within our own organisations, providing opportunities for individuals to work within diverse groups, or specific learning opportunities designed to build cultural knowledge. These processes, however, need continuous support and investment; the development of intercultural competence is, and should be, an ongoing effort within both ourselves and our organisations.

Conclusion

The ability to work effectively in an increasingly international context is arguably fundamental to personal and organisational success. Due to changes in technology, increases in the extent of globalisation and favourable regulatory changes, businesses are increasingly operating across international borders – and this is not just the case for large MNCs; we are seeing small and often micro-organisations trade and employ individuals around the globe. Business systems and national cultures, however, differ around the world. A method of management that works well in one context is unlikely to work effectively in another where you may have substantial differences in business regulation, legal frameworks, or indeed cultural expectations regarding the interactions between people. All of this demonstrates the importance of sensitivity to local conditions, and emphasises the importance of being flexible and adaptive in our approach to working with others who may be very different from ourselves. Organisations are increasingly seeking (and prizing) individuals who can work collaboratively in a cross-cultural setting, emphasising the importance of developing your ability to communicate across cultures and to demonstrate intercultural or cross-cultural competence.

Just as business systems differ around the world, so too do systems of HRM. It is arguably the case that HRM, as a functional area within our organisations, is influenced to a greater extent by changes in culture and underlying business systems than other areas of our firms, such as finance or operations. What we see around the world are different national varieties of HRM and systems of HRM that are adapted to operate successfully in a particular set of institutional and cultural circumstances. Again, when operating in an international context, no matter the size or scale of our organisations, we must demonstrate awareness of these differences, and understand the underlying conditions that have brought about these different models of HRM. This chapter has introduced you to some of the key debates and areas of interest, and has formed a foundation from which you can continue to build your knowledge and skills.

References

Adler, N J and Bartholomew, S (1992) Academic and professional communities of discourse: Generating knowledge on transnational human resource management, *Journal of International Business Studies*, 23, 551–69

Atkinson, J (1984) Manpower strategies for flexible organisations, *Personnel Management*, 16, 28–31

Berkowitz, D, Hong, M and Shuichiro, N (2017) Recasting the iron rice bowl: The reform of China's state-owned enterprises, *Review of Economics and Statistics*, 99 (4), 735–47

Blakemore, E (2019) How the East India Company became the world's most powerful business, www.nationalgeographic.com/culture/topics/reference/british-east-india-trading-company-most-powerful-business (archived at https://perma.cc/9U8Z-QHRQ)

BMW Group (2020) Works council, www.bmwgroup-werke.com/landshut/en/our-plant/works-council.html (archived at https://perma.cc/8LLV-HHVU)

Browaeys, M-J and Price, R (2019) *Understanding Cross-cultural Management*, 4th edn, Pearson, Harlow

Charleston, B, Gajewska-De Mattos, H and Chapman, M (2018) Cross-cultural competence in the context of NGOs: Bridging the gap between 'knowing' and 'doing', *International Journal of Human Resource Management*, 29 (21), 3068–92

CIPD (2019) *Executive Pay in the FTSE 100 – Research Report: August 2019*, CIPD, London

CIPD (2020a) Employment law: UK, EU and Brexit, www.cipd.co.uk/knowledge/fundamentals/emp-law/about/eu-impact (archived at https://perma.cc/Z6AQ-JE3V)

(Continued)

(Continued)

CIPD (2020b) About us, www.cipd.co.uk/about (archived at https://perma.cc/K3T3-3D9F)

Copeland, M V (2006) The mighty micro-multinational, *Business 2.0*, 7 (6), 106–14

Dias, D, Zhu, C J and Samaratunge, R (2020) Examining the role of cultural exposure in improving intercultural competence: Implications for HRM practices in multicultural organizations, *International Journal of Human Resource Management*, 31 (11), 1359–78

Dimitratos, P, Amarós, J E, Etchebarne, M S and Felzensztein, C (2014) Micro-multinational or not? International entrepreneurship, networking and learning effects, *Journal of Business Research*, 67 (5), 908–15

Dodd, C H (2007) Intercultural readiness assessment for pre-departure candidates, *Intercultural Communication Studies*, 16 (2), 1–17

Edwards, T and Rees, C (2017) *International Human Resource Management*, 3rd edn, Pearson, Harlow

European Commission (2020a) Working conditions: Working Time Directive, https://ec.europa.eu/social/main.jsp?catId=706&langId=en&intPageId=205 (archived at https://perma.cc/9BRE-9B8M)

European Commission (2020b) Employee involvement: EU directives in action, https://ec.europa.eu/social/main.jsp?catId=707 (archived at https://perma.cc/S58X-VGAE)

Fitzpatrick, F (2017) Taking the 'culture' out of 'culture shock': A critical review of literature on cross-cultural adjustment in international relocation, *Critical Perspectives on International Business*, 13 (4), 278–96

Gely, R, Cheramie, R and Chandler, T (2016) An empirical assessment of the contract based exception to the employment-at-will rule, *Employee Responsibilities and Rights Journal*, 28 (1), 63–78

Gov.uk (2020a) Maximum weekly working hours, www.gov.uk/maximum-weekly-working-hours (archived at https://perma.cc/5DX3-RA2M)

Gov.uk (2020b) Holiday entitlement, www.gov.uk/holiday-entitlement-rights (archived at https://perma.cc/ZCE2-W78J)

Haile, S and White, D (2019) Expatriate failure is a common challenge for multinational corporations: Turn expatriate failure to expatriate success, *International Journal of Business and Public Administration*, 16 (1), 27–40

Hall, P and Soskice, D (2001) An introduction to the varieties of capitalism, in *Varieties of Capitalism: The institutional basis of competitive advantage*, ed P Hall and D Soskice, Oxford University Press, Oxford

Harzing, A-W K (1995) The persistent myth of high expatriate failure rates, *International Journal of Human Resource Management*, 6 (2), 457–74

Harzing, A-W K (2004) Composing an international staff, in *International Human Resource Management*, 2nd edn, ed A-W Harzing and J Van Ruysseveldt, SAGE, London

Harzing, A-W and Ruysseveldt, J (2003) *International Human Resource Management: Managing people across borders*, SAGE, London

Hirakubo, N (1999) The end of lifetime employment in Japan, *Business Horizons*, 42 (6), 41–6

Hofstede, G H, Hofstede, G J and Minkov, M (2010) *Cultures and Organizations: Software of the mind*, 3rd edn, McGraw-Hill, Maidenhead

Hofstede Insights (2020) Compare countries, www.hofstede-insights.com/product/compare-countries (archived at https://perma.cc/V5YT-G2NK)

Hollinshead, G (2009) *International and Comparative Human Resource Management*, McGraw-Hill, Maidenhead

Howard, C G (1974) The returning overseas executive: Cultural shock in reverse, *Human Resource Management*, 13 (2), 22–6

Hsiung, H-H and Tsai, W-C (2017) The joint moderating effects of activated negative moods and group voice climate on the relationship between power distance orientation and employee voice behaviour, *Applied Psychology*, 66 (3), 487–514

Mendenhall, M, Punnett, B J and Ricks, D (1995) *Global Management*, Blackwell, Cambridge, MA

Ono, H (2010) Lifetime employment in Japan: Concepts and measurements, *Journal of the Japanese and International Economies*, 24 (1), 1–27

Ono, Y (2019) Japanese economy: Two lost decades and how many more?, *Intereconomics*, 54 (5), 291–6

ONS (2019a) Labour productivity, UK: October to December 2019, www.ons.gov.uk/employmentandlabourmarket/peopleinwork/labourproductivity/bulletins/labourproductivity/octobertodecember2019 (archived at https://perma.cc/V4Y6-E3EJ)

(Continued)

(Continued)

ONS (2019b) Employee earnings in the UK: 2019, www.ons.gov.uk/employmentandlabourmarket/peopleinwork/earningsandworkinghours/bulletins/annualsurveyofhoursandearnings/2019 (archived at https://perma.cc/7D2D-Y8K4)

Özbilgin, M F, Groutsis, D and Harvey, W S (2014) *International Human Resource Management*, Cambridge University Press, New York

Partington, R (2018) UK productivity jumps at fastest rate for six years, www.theguardian.com/business/2018/jan/05/uk-productivity-jumps-at-fastest-rate-for-six-years (archived at https://perma.cc/X9C5-QNVX)

Pellegrino Riccardi (2020) www.pellegrino-riccardi.com (archived at https://perma.cc/RH2N-8ME5)

Peng, MW (2018) *Global Business*, Cengage, Boston, MA

Radin, T J and Werhane, P H (2003) Employment-at-will, employment rights, and future directions for employment, *Business Ethics Quarterly*, 13 (2), 113–30

Reiche, B S and Harzing, A W (2011) International assignments, in *International Human Resource Management*, 3rd edn, ed A-W Harzing and A H Pinnington, SAGE, London

Reiche, B S, Harzing, A-W and Tenzer, H (2019) *International Human Resource Management*, 5th edn, SAGE, London

Sallmen, J (2018) Non-competes, consideration, and common sense: A temporarily revocable arrangement to preserve 'afterthought' agreements in at-will employment, *University of Pittsburgh Law Review*, 79 (3), 543–61

Statista (2017) Average wage per month in select countries in 2017 and 2040, www.statista.com/statistics/974101/average-wage-per-month-select-countries (archived at https://perma.cc/4QNT-ARD2)

Statista (2018) Average annual wages in China from 2008 to 2018, www.statista.com/statistics/743522/china-average-yearly-wages (archived at https://perma.cc/WU3P-R8EM)

Tanaka, F J (1981) Lifetime employment in Japan, *Challenge*, 24 (3), 23–9

Taylor, M (2020) Works Council is pushing BMW towards a stand-alone EV platform, www.forbes.com/sites/michaeltaylor/2020/06/29/works-council-is-pushing-bmw-towards-a-stand-alone-ev-platform (archived at https://perma.cc/E7YC-6LA8)

The Economist (2020) Millions of Chinese students brace themselves for joblessness, www.economist.com/china/2020/05/02/millions-of-chinese-students-brace-themselves-for-joblessness (archived at https://perma.cc/2KKX-6TS8)

Thompson Reuters (2020) Employment and employee benefits in China: Overview, https://uk.practicallaw.thomsonreuters.com/1-503-3245?transitionType=Default&contextData=(sc.Default) (archived at https://perma.cc/5BXY-W4XV)

Trompenaars, F and Hampden-Turner, C (1997) *Riding the Waves of Culture: Understanding cultural diversity in business*, 2nd edn, Nicholas Brealey Publishing, Boston, MA

US Bureau of Labor Statistics (2020) Usual weekly earnings summary, www.bls.gov/news.release/wkyeng.nr0.htm (archived at https://perma.cc/6XWK-72L8)

Watson Jr, B W and Singh, G (2005) Global pay systems: Compensation in support of a multinational strategy, *Compensation and Benefits Review*, 37 (1), 33–6

Whitley, R (1999) *Divergent Capitalisms: The social structuring and change of business systems*, Oxford University Press, Oxford

Whittall, M (2010) The problem of national industrial relations traditions in European works councils: The example of BMW, *Economic and Industrial Democracy*, 31 (4), 70–85

Wilkinson, A and Wood, G (2017) Global trends and crises, comparative capitalism and HRM, *International Journal of Human Resource Management*, 28 (18), 2503–18

Xie, J, Cherrie Zhu, J, Fan, D and Zhang, M M (2020) The 'iron rice-bowl' regime revisited: Whither human resource management in Chinese universities?, *Asia Pacific Journal of Human Resources*, 58 (2), 289–310

INDEX

Lightning Source UK Ltd.
Milton Keynes UK
UKHW051101011222
413137UK00005B/40